EMC 2008
14th European Microscopy Congress
1–5 September 2008, Aachen, Germany

Martina Luysberg · Karsten Tillmann
Thomas Weirich

Editors

EMC 2008

14th European Microscopy Congress
1–5 September 2008, Aachen, Germany

Volume 1:
Instrumentation and Methods

 Springer

Martina Luysberg
Karsten Tillmann
Ernst Ruska-Centre for Microscopy and
Spectroscopy with Electrons
Institute of Solid State Research
Research Centre Jülich
52425 Jülich
Germany
m.luysberg@fz-juelich.de
k.tillmann@fz-juelich.de

Thomas Weirich
RWTH Aachen
Central Facility for Electron Microscopy
Ahornstr. 55
52074 Aachen
Germany
weirich@gfe.rwth-aachen.de

ISBN 978-3-540-85154-7   e-ISBN 978-3-540-85156-1

DOI 10.1007/978-3-540-85156-1

Library of Congress Control Number: 2008934066

Typesetting: digital data supplied by the authors
Production: le-tex publishing services oHG, Leipzig, Germany
Cover design: eStudioCalamar S.L., F. Steinen-Broo, Girona, Spain

Printed on acid-free paper

9 8 7 6 5 4 3 2 1

springer.com

Preface Volume 1: Instrumentation and Methods

This volume contains all instrumentation and methods related contributions presented at the *European Microscopy Congress 2008* held in Aachen on 1 – 5 September 2008. Fourteenth in the series of quadrennial conferences, the meeting focused on latest developments in the field of advanced electron microscopy techniques together with the application to materials science and life sciences. The international character of this series of congresses was once again emphasised – over 1000 contributions from 46 countries representing all areas of Europe and the globe have been submitted for presentation to the programme committee.

Main methodical and instrumental topics of *EMC 2008* covered fundamentals and prospects of aberration correctors, detectors, monochromators, phase plates and spectrometers as well as demonstrations of quantitative microscopic and spectroscopic investigations in all types of solid state and soft matter materials. The multifariousness of contributions thus represented a state-of-the-art overview of the microscopic techniques now being applied to a wide range of problems. The contributions brought together in this volume cover fundamental treatments of electron optical instrumentation and methods, advanced numerical image analysis and processing techniques, as well as latest developments of sample preparation methods which are only beginning to find application.

The *European Microscopy Congress 2008* was conjointly organised by the European Microscopy Society (EMS), the German Society for Electron Microscopy (DGE), and local microscopists from RWTH Aachen University and the Research Centre Jülich. Hosted by the Eurogress Conference Centre, the congress organisation was underpinned by the meticulous work carried out by Tobias Caumanns, Achim Herwartz, Helga Maintz, Evi Münstermann, Daesung Park, Thomas Queck, Stefanie Stadler, and Sarah Wentz as well as the staff of Aachen Congress who deserve very special thanks.

Martina Luysberg, Karsten Tillmann, and Thomas Weirich

Editors, Volume 1 of the EMC 2008 Proceedings

Content

I   Instrumentation and Methods

I1   TEM and STEM instrumentation and Electron Optics

I1.1   Aberration Correctors

I1.2 Filters, Spectrometers, Monochromators and Sources

I1.3 Phase Plates and Detectors

I2   TEM and STEM methods

I2.1   *Quantitative HRTEM and STEM*

I2.2 *Quantitative diffraction and crystallography*

I2.3   Holography

I2.4   Tomography

I2.5   EELS/EFTEM

I2.6 In-situ TEM and dynamic TEM

I3 SEM/FIB Instrumentation and Methods

I3.1 SEM instrumentation and methods

I3.2   FIB and Dual Beam - instrumentation and methods

I4   Other Microscopies

I4.1   Light and X-rays

I4.2   Scanning Probe Microscopy

I4.3   Field Ion Microscopy and Atom Probe

I4.4   Surface Analytical Techniques

I5   Image analysis and Processing

I5.1   New image processing developments in Materials Science and beyond

Aberration corrected STEM and EELS:
Atomic scale chemical mapping

A.L. Bleloch[1,2], M. Gass[1,2], L. Jiang[1,2], B. Mendis[1,3], K. Sader[1,4] and P. Wang[1,2]

1. SuperSTEM Laboratory, STFC Daresbury, Warrington, Cheshire, WA4 4AD, UK
2. Department of Engineering, University of Liverpool, Liverpool L69 3GH, UK
3. Department of Physics and Astronomy, University of Glasgow, Glasgow, G12 8QQ
4. Institute for Materials Research, University of Leeds, Leeds, LS2 9JT, UK

a.l.bleloch@liverpool.ac.uk
Keywords: STEM, EELS, Chemical mapping

Scanning transmission electron microscopy (STEM) has enjoyed a recent surge in activity for two reasons. A long awaited improvement in the STEM performance of conventional TEM/STEM instruments coincided with the advent of aberration correction. This improvement was not for any fundamental reason – from the principle of reciprocity, the STEM resolution should be at least as good as the TEM resolution for the same objective lens. Lattice resolution high angle annular dark field (HAADF) STEM images are now a much more routine part of the analytical arsenal applied to materials characterisation.

For aberration corrected instruments, the popularity of the STEM geometry has two motivations. The first is the monotonic dependence of the HAADF signal on the sample thickness. This allows the quality of the data to be more easily assessed during acquisition – if it looks better it almost always is better. It does not, however, mean that data analysis, modelling and simulation should be any less rigorous. The second is that the gain in spatial resolution in STEM imaging brings almost the same gain in the spatial resolution of inelastic signals, in particular electron energy loss spectroscopy (EELS). A recent flurry of papers in high profile journals have reported atomic scale chemical mapping in each case of perovskite structures [1-3].

The ultimate aim of an analytical technique is to identify and map where individual atoms are in three dimensions. Hitherto, this was almost always achieved by diffraction methods averaging over many identical structures and, where it can be used, this remains the most powerful approach. However, the need to characterise the atomic positions in non-periodic structures is being driven by our ability and need to engineer and understand complex non-periodic structures at this scale. The application of HAADF STEM and EELS to characterise a range of material systems will be used to exemplify the capabilities of this technology. This will include imaging of single gold atoms in silicon nanowires, characterisation of nano-particles for heterogeneous catalysis and nanotubes in macrophage cells. Atomic column by atomic column mapping will also be discussed with particular emphasis on the experimental conditions needed. Figure 1 shows a line profile across a Ruddleston-Popper defect in a perovskite material showing contrast in the EELS signal at the individual atomic columns.

M. Luysberg, K. Tillmann, T. Weirich (Eds.): EMC 2008, Vol. 1: Instrumentation and Methods, pp. 1–2, DOI: 10.1007/978-3-540-85156-1_1, © Springer-Verlag Berlin Heidelberg 2008

2

Progressing from two dimensional projected chemical maps to three-dimensional information is particularly difficult at the atomic scale with, as yet, no clear experimental solutions. Figure 2 shows a three dimensional reconstruction of a gold nano-particle using a novel reconstruction algorithm as a step toward atomic mapping in three-dimensions.

1. M. Bosman, et al., Phys. Rev. Lett. **99**, 086102 (2007).
2. K. Kimoto, et al., Nature **450**, 702 (2007).
3. D. A. Muller, et al., Science **319**, 1073 (2008).
4. L. J. Allen, Nature Nanotechnology **3**, 255 - 256 (2008)
5. P. Wang, et al., presented at EMAG(2007), Glasgow
6. We would like to thank the EPSRC for financial support of the SuperSTEM facility.

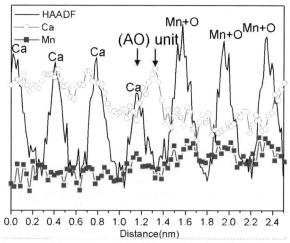

Figure 1. The signal integrals of Ca-$L_{2,3}$ (-o-) and Mn-$L_{2,3}$ (-■-) edges extracted from an EELS line scan of the perovskite structure $CaMnO_3$ together with the HAADF signal line scan (black line) across a Ruddleston-Popper defect. The beam is in the [100] direction and the line scan is along [001]. The switch from the Ca to the Mn columns can be seen together with an anomalous Ca signal at the defect.[5]

Figure 2. A three-dimensional reconstruction of a 15nm gold particle using a novel algorithm that uses multiplicative back-projection.

An update on the TEAM project - first results from the TEAM 0.5 microscope, and its future development

U. Dahmen[1], R. Erni[1], C. Kisielowki[1], V. Radmilovic[1], Q. Ramasse[1], A. Schmid[1], T. Duden[1], M. Watanabe[1], A. Minor[1], and P. Denes[2]

1. National Center for Electron Microscopy, LBNL, Berkeley CA 94720
2. Advanced Light Source/Engineering Division, LBNL, Berkeley CA 94720

UDahmen@LBL.gov

Keywords: Aberration correction, sub-Angstrom TEM/STEM, buried defects

Recent advances in aberration-correcting electron optics have led to increased resolution, sensitivity and signal to noise in atomic resolution microscopy. Building on these developments, the TEAM project was designed to optimize the electron microscope around aberration-corrected electron optics and to further advance the limits of the instrument and the technique [1]. The vision for the TEAM project is the idea of providing a sample space for electron scattering experiments in a tunable electron optical environment by removing some of the constraints that have limited electron microscopy until now. The resulting improvements in resolution, the increased space around the sample, and the possibility of exotic electron-optical settings will enable new types of experiments. The TEAM microscope will feature unique corrector elements for spherical and chromatic aberrations, a novel AFM-inspired specimen stage, a high-brightness gun and numerous other innovations that will extend resolution down to the half-Angstrom level.

The project is a collaboration of several Department of Energy-funded efforts [1] and two commercial partners (FEI and CEOS). Led by the National Center for Electron Microscopy, the project pursues several key developments in parallel, with each partner responsible for a specific set of tasks. The machine is being implemented in two stages – TEAM 0.5 in 2008 and TEAM I in 2009.

The TEAM 0.5 instrument is a double Cs-corrected microscope with a hexapole aberration corrector on the imaging side that fully corrects aberrations up to third order and partially up to fifth order, and an improved hexapole aberration corrector on the probe side that fully corrects aberrations up to fifth order with an information transfer to 0.05 nm. The machine is equipped with a specially developed high brightness gun and a Wien-type monochromator. Following installation at NCEM in January 2008 and subsequent site acceptance, the instrument is currently undergoing a series of tuning and alignment procedures in preparation for the start of user operations in October 2008.

This talk will illustrate initial applications of the TEAM 0.5 instrument to the analysis of different materials. The performance of the instrument [2] in both TEM and STEM operating modes is illustrated in Figure 1. An application of the enhanced depth resolution at large convergence angles in the STEM mode is shown in Figure 2. Other applications will be demonstrated, including defects in Au, ZnO, GaN, Al-Li based alloys, graphene sheets and grain boundaries in Au bicrystals [3].

M. Luysberg, K. Tillmann, T. Weirich (Eds.): EMC 2008, Vol. 1: Instrumentation and Methods, pp. 3–4, DOI: 10.1007/978-3-540-85156-1_2, © Springer-Verlag Berlin Heidelberg 2008

4

1. Lawrence Berkeley National Laboratory, Argonne National Laboratory, Oak Ridge National Laboratory and the FSMRL at the University of Illinois. For more details, see http://ncem.lbl.gov/TEAM-project/index.html
2. C. Kisielowski, B. Freitag, M. Bischoff, H. van Lin, S. Lazar, G. Knippels, P. Tiemeijer, M. van der Stam, S. von Harrach, M. Stekelenburg, M. Haider, S. Uhlemann, H. Müller, P. Hartel, B. Kabius, D. Miller, I. Petrov, E. A. Olson, T. Donchev, E.A. Kenik, A. Lupini, J. Bentley, S. Pennycook, I.M. Anderson, A.M. Minor, A.K. Schmid, T. Duden, V. Radmilovic, Q. Ramasse, M. Watanabe, R. Erni, E.A. Stach, P. Denes, U. Dahmen, submitted for publication. See also other presentations at this meeting.
3. The TEAM project is supported by the Department of Energy, Office of Science, Basic Energy Sciences. NCEM is supported under Contract # DE-AC02-05CH11231.

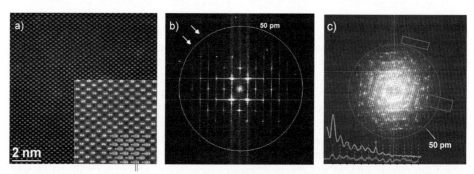

Figure 1. Performance of TEAM 0.5 microscope in STEM and TEM. Aberration-corrected high-resolution STEM image of GaN in [211] orientation (a), showing the 0.63Å distance between Ga dumbbells clearly resolved (see inset model). The corresponding diffractogram in (b) shows Fourier components beyond the 50 pm marker indicated by the circle. The Fourier diffractogram from high resolution TEM images in (c) shows Young's fringes extending beyond the 50 pm mark indicated by the circle.

Figure 2. HAADF STEM images of a faceted grain boundary in a Au bicrystal viewed along the [111] direction (a). A buried boundary segment becomes visible only when the probe is focused 7 nm into the sample (b). A schematic of this geometry is shown in (c). 300 kV, probe semi-convergence angle: 35.6 mrad, inner detector angle ~50 mrad.

Synchrotron based X-ray Microscopy:
state of the art and applications

J. Susini

European Synchrotron Radiation Facility
X-ray imaging Group, Grenoble, France

susini@esrf.fr

Keywords: X-ray microprobe, synchrotron, X-ray fluorescence, X-ray absorption spectroscopy

The dramatic growth of nanoscience and nanotechnologies is currently fostering the development of high spatial resolution analytical techniques [1]. In this context, synchrotron based techniques have the potential to become mainstream tools for analysis, imaging and spectroscopy in a number of applications, as exemplified by the current evolution of hard X-ray microscopy. Hence, considering the concomitant developments of laboratory instruments and dedicated synchrotron beamlines worldwide [2], a very competitive context can be anticipated for the coming years. Within this perspective, synchrotron based analytical techniques (diffraction, imaging and micro-spectroscopies) will play an important role by offering unique capabilities in the study of complex systems. Ultimately, this complexity can be envisioned in three dimensions: composition, time and space.

The main fields of applications are driven by the unique attributes of X-ray microscopy in this spectral range [3]: i) access to K-absorption edges and fluorescence emission lines of medium-light elements and L,M - edges of heavy materials for micro-spectroscopy, chemical or trace element mapping; ii) higher penetration depths compared to soft X-rays allowing imaging of thicker samples; iii) favourable wavelengths for diffraction studies and iv) generally large focal lengths and depths of focus which are advantageous for the use of specific sample environments (in-situ, high pressure, controlled temperatures....). Typical experiments can be broadly divided into two categories: i) morphological studies which require high spatial resolution and are therefore well adapted to 2D or 3D transmission full-field microscopy. ii) studies dealing with co-localization and/or speciation of trace elements in heterogeneous matrices at the sub-micron scale. Scanning X-ray microscopy, in transmission and/or X-ray fluorescence modes, tends to be better suited for the latter cases, which often require both low detection limits and spectroscopic analysis capabilities. Compared to other techniques, synchrotron X-ray fluorescence microprobes display a unique combination of features [4]. Associated with high detection efficiency, the radiation damage is minimized and accurate quantification is possible.

Over the past three decades, interests for X-ray microscopy has been revived, fostered by several major advances in X-ray sources and X-ray optics. X-ray imaging techniques largely benefit from the high brilliance of X-ray beams produced by third generation synchrotron sources which offer a control of the brightness, spectrum, geometry, polarization and coherence of the beam. Driven by these unprecedented

M. Luysberg, K. Tillmann, T. Weirich (Eds.): EMC 2008, Vol. 1: Instrumentation and Methods, pp. 5–6, DOI: 10.1007/978-3-540-85156-1_3, © Springer-Verlag Berlin Heidelberg 2008

properties of X-ray beams, concomitant progresses have been made in X-ray optics. Above all, recent advances in manufacturing techniques have enlarged the accessible energy range of micro-focusing optics and offer new applications in a broad range of disciplines. Hard and soft X-ray focusing optics are now reaching almost similar level of performances with focused beam sizes below 20nm [5]. However, most of these remarkable achievements remain at the demonstration stage and are still far from routine. Besides they have to be fully integrated into stable and reliable X-ray microscopes.

The possibility of in-situ experiments remains a unique attribute of synchrotron based analytical methods. Photon penetration depth of hard X-rays enables specific sample environments to be developed to study realistic systems in their near-native environment rather than model systems. The ability to perform in-situ analysis with environmental chambers offering high or low temperature conditions, high pressure, or preserving sample hydration explains the increasing interest from communities such as Planetary and Earth sciences, environmental science and microbiology.

A natural evolution of the elemental (X-ray fluorescence) or structural (X-ray diffraction) mappings are the extension towards in-depth third dimension. The knowledge of the elemental variation along this third dimension intrinsically improves the data quality and therefore its interpretation. Compared to the standard absorption X-ray Computed Tomography, Fluo-tomography is more challenging since it is limited by self-absorption and matrix effect. A multimodal strategy has been developed to overcome these physical limitations and is based on an algorithm solution which relies on the combination of several signals (transmission, fluorescence and scattering) to derive the volumic distribution of elements [6]. Similar approach has been used for X-ray diffraction computed tomography as local structural probe for heterogeneous diluted materials [7].

This lecture aims at giving a short overview of the main development trends of synchrotron based X-ray microscopy and spectro-microscopy. Following a brief description of their characteristics, the strengths and weaknesses of spectro-microscopy techniques will be discussed and illustrated by examples of applications in materials sciences, biology and environmental science.

1. F. Adams, L. Van Vaeck, R. Barrett, Spectrochim. Acta **B60** (2005), p.13.
2. M. Howells, C. Jacobsen and T. Warwick in "Science of Microscopy", Vol. II, ed. P.W. Hawkes and J.C.H. Spence, (Springer) (2007), p. 835.
3. J. Susini, M. Salomé, B. Fayard, R. Ortega & B. Kaulich, Surf. Rev. Lett. **9** (2002), p. 203.
4. P.M. Bertsch, D.B. Hunter, Chem. Rev. **101** (2001), p. 1809.
5. W. Chao, B.D. Harteneck, J. A. Liddle, A. H. Anderson & D. T.Attwood, Nature **435** (2005), p.1210.
6. B. Golosio, A. Somogyi, A. Simionovici, P. Bleuet, J. Susini, L. Lemelle, Applied Physics Letters **84**(12) (2002), p. 2199.
7. P. Bleuet, E. Welcomme, E. Dooryhee, J. Susini, J.-L. Hodeau, P. Walter, Nature Materials, in press.
8. The author wishes to thank M. Salomé, R. Tucoulou, G. Criado-Martinez, P. Bleuet, M. Cotte, J. Cauzid, S. Bohic, P. Cleotens, A. Solé, C. Larabell, J. Maser, S. Vogt for their contributions.

High-resolution spectro-microscopy with low-voltage electrons and double aberration correction

Thomas Schmidt[1,2], Helder Marchetto[2], Rainer Fink[3], Eberhard Umbach[1,2,4],
and the SMART collaboration

1. Experimentelle Physik II, Universität Würzburg, D-97074 Würzburg, Germany
2. Fritz-Haber-Institut der Max-Planck Gesellschaft, D-14195 Berlin, Germany
3. Physikalische Chemie II, Universität Erlangen, D-91058 Erlangen, Germany
4. Forschungszentrum Karlsruhe and Karlsruhe Institute of Technology, D-76344 Eggenstein-Leopoldshafen, Germany

eberhard.umbach@vorstand.fzk.de
Keywords: spectro-microscopy, aberration correction, dynamic surface studies

The greatest success of surface science, best symbolized by the Nobel Prize in Chemistry awarded to Gerhard Ertl in 2007, is the achievement of a nearly complete understanding of complicated two-dimensional surface systems. This success is primarily based on sophisticated surface methods, ranging from various spectroscopic techniques over diffraction methods to microscopic techniques. Methods such as x-ray photoemission, angle-resolved UV photoemission, x-ray absorption, Auger spectroscopy, low energy electron diffraction, x-ray photoelectron diffraction, surface x-ray diffraction, and scanning electron microscopy have revolutionized surface science during the past three decades and have lead to a detailed understanding of all kinds of surfaces and adsorbates on an atomic level. However, it always remained a dream of surface science to simultaneously combine the microscopic view, including very high spatial resolution, with as much spectroscopic information as possible and with diffraction data. Moreover, all this information should be obtainable in ultra-high vacuum, and it should be received at the same time, from the same sample spot, and in a dynamic fashion, i.e. as movie, especially in the case of reactions. These requirements are met by an instrument named SMART, which stands for "Spectro-Microscope with Aberration correction for many Relevant Techniques". The concept, realization, and achieved performance of this novel instrument will be outlined.

Spatially resolved spectroscopic information can be obtained by so called micro-spectrocopes (scanning instruments) or spectro-microscopes (imaging instruments). Numerous successful approaches have been developed, and some are commercially available today, which are called Low Energy Electron Microscopes (LEEM) or (X-ray) Photoelectron Emission Microscopes (X-PEEM). The heart of these instruments is usually a very simple low-energy electron microscope consisting of an objective and a projection lens system and a 2-dim detector. However, due to chromatic and spherical aberrations these instruments are rather limited in spatial and energy resolution.

A collaboration of four university groups, BESSY, and Zeiss has designed and built a new type of spectro-microscope that combines a "normal" LEEM/PEEM instrument with aberration corrections (both, chromatic AND spherical aberration correction) and

M. Luysberg, K. Tillmann, T. Weirich (Eds.): EMC 2008, Vol. 1: Instrumentation and Methods, pp. 7–8, DOI: 10.1007/978-3-540-85156-1_4, © Springer-Verlag Berlin Heidelberg 2008

with high energy resolution using an imaging energy filter (Omega filter as UHV version). The instrument is installed at the high-brilliance synchrotron source BESSY II and uses electrons, UV, or tuneable x-ray photons as source. Theoretically, a spatial resolution of 0.5 nm at an energy resolution of less than 100 meV is achievable at (simultaneously!) more than a hundred times higher transmission compared to conventional LEEM/PEEM instruments due to a fairly large acceptance angle. This instrument called SMART has been developed in three stages with novel experiments after completion of each stage. It is now, after the move to a new, high-flux density synchrotron beam-line, in the final stage of improving and commissioning.

The SMART allows to combine nearly simultaneously various electron microscopy techniques (e.g. LEEM, PEEM, X-PEEM, MEM) with various high-resolution electron spectroscopy and electron diffraction techniques from small sample spots (e.g. nano-XPS, -ARUPS, -XAS, -LEED). A spatial resolution of 3 nm, an electron energy resolution of 150 meV at very low background, and an x-ray absorption energy resolution of 30 meV at the carbon 1s edge have been achieved. Electron diffraction experiments with unprecedented transfer widths (lateral coherence) could be performed with sub-μm spatial resolution. The present performance of the SMART instrument is close to the design parameters, the promised goals could essentially be reached.

Several experiments on the growth properties of organic thin films, their dependence on substrate, surface topography, and temperature were performed and new properties of organic layers revealed. The discovery of the internal structure of organic microcrystallites shows the potential of the instrument and will be briefly presented. Single domains of molecular adsorbates in the monolayer range could be investigated with respect to their size, density, molecular orientation, domain orientation, and growth or desorption dynamics. Similar molecules resulted in very different growth behaviour, which also depends on surface orientation, substrate step density, and preparation conditions. Movies showing the growth dynamics as well as statistical evaluations are possible using a single preparation cycle. Prominent examples of unexpected results on organic layers like PTCDA and NTCDA on Ag and Au (111) surfaces will be presented to elucidate the potential of the new instrument.

1. Extensive financial support by the Bundesministerium für Bildung und Forschung and continuing support by the Fritz Haber Institute are gratefully acknowledged.

Developments of aberration correction systems for current and future requirements

M. Haider, H. Müller, S. Uhlemann, P. Hartel and J. Zach

CEOS GmbH, Englerstr. 28 D-69126 Heidelberg, Germany

haider@ceos-gmbh.de

Keywords: aberration correction, high resolution, TEM, STEM, instrumentation

The hardware correction of aberrations of the objective lens of an electron microscope was a long ongoing process from the first ideas till the first applicable system. The advantages of aberration correction could the first time successfully demonstrated more than 10 years ago. The first corrector with which an improvement of resolving power could be demonstrated was a Cc/Cs corrector system for a SEM [1]. However, the most remarkable correction was the compensation of the spherical aberration of a modern commercially available high resolution TEM by means of a Hexapole-corrector [2]. This first Cs-corrected TEM really changed the complete field of high resolution transmission electron microscopy with unprecedented images. A Cs-correction system for a scanning transmission electron microscope by means of a quadrupole-octupole corrector could first be demonstrated by Krivanek [3]. Today, hexapole-correctors to compensate the spherical aberration of the objective lens are available for almost all commercial TEM and STEM.

When achieving the compensation of the spherical aberration and improving the point resolution of TEM and STEM the overall stability of the basic instrument turned out to be the new barrier on the path to highest resolution. Hence, the basic instruments had to be improved and this was successfully done by the emergence of new models. These new instruments are the reason to go now for the next generation of correctors and there are several possibilities to improve the overall performance of TEM and STEM. These new types of correctors are:

1. The partly or full compensation of higher order axial aberrations for STEM by means of advanced Hexapole correctors in order to tackle the dominating six-fold astigmatism A_5 which mainly limits the illumination angle in STEM [4].

2. In order to use the full capability of new large area CCD-cameras (4k x 4k or even 8k x 8k) when aiming for sub-Ångstrom resolution the compensation of off-axial aberrations is becoming mandatory. Hence, a aplanatic system with a large field of view and a resolution below 1 Å will be developed.

3. The compensation of the chromatic aberration in TEM in combination with the correction of the spherical aberration and the off-axial coma in order to achieve aberration free imaging at 50 pm resolution and a large field of view is requested for the TEAM project [5]. This correction system, which is currently the most complex and most challenging project, is under development and should be finished within one year. It will set new milestones in the field of high resolution electron microscopy when successfully finished. This correction system (Fig. 1) is already incorporated into a FEI

M. Luysberg, K. Tillmann, T. Weirich (Eds.): EMC 2008, Vol. 1: Instrumentation and Methods,
pp. 9–10, DOI: 10.1007/978-3-540-85156-1_5, © Springer-Verlag Berlin Heidelberg 2008

Titan TEM and is currently under alignment. The compensation of Cc could already been shown at 200 kV accelerating voltage and other voltages - of 80 kV and 300 kV - will follow.

4. Besides the compensation of the chromatic aberration in TEM the compensation of Cc is also important for a probe forming system if one considers a Schottky emitter. The chromatic aberration does not change the FWHM of an electron probe if the spherical aberration is compensated but it changes dramatically the diameter which contains a certain fraction of the total electron current (e.g. 50%). This amount of beam current is confined in a certain diameter and this figure is important for high density probes as used for EELS, for example.

1. J. Zach and M.- Haider Optik **98** (1995) 112
2. M. Haider et al. Nature **392** (1998) 768.
3. O. Krivanek, N. Delby and A.R. Lupini, Ultramicroscopy **78** (1999) 1
4. H. Mueller et al Micr. Microanal. **12** (2006) 442
5. http://ncem.lbl.gov/TEAM-project/index.html
6. *"This work was conducted as part of the Transmission Electron Aberration-Corrected Microscopy (TEAM) project funded by the Department of Energy, Office of Science, USA"*

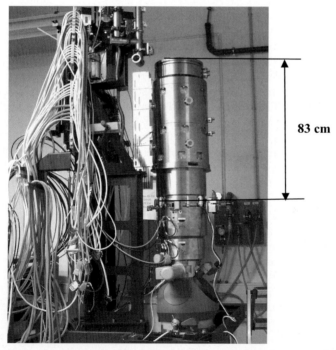

83 cm

Figure 1. A picture of the C_C-corrector taken when incorporating it into a FEI Titan. This corrector has an overall length of 83 cm which shifts all parts of the objective lens and above by this amount. The outer diameter of the C-COR had to be increased from 30 cm for the central part to 35 cm.

STEM Aberration Correction: an Integrated Approach

Ondrej Krivanek, Niklas Dellby, Matt Murfitt, Christopher Own, and Zoltan Szilagyi

Nion Co., 1102 8th St., Kirkland, WA 98033, USA

krivanek.ondrej@nion.com

Keywords: aberration correction, STEM, sample stages

Progress in aberration correction of electron microscopes has been rapid in the last decade. CEOS and Nion, the two companies chiefly responsible for the advent of practical aberration correctors, were started 12 and 11 years ago, respectively, at a time when aberration correction still seemed an impractical dream to many electron microscopists. The first sub-Å resolution, directly interpretable aberration-corrected images were published in 2002 [1]. Today, there are more than 100 aberration-corrected electron microscopes in the world, and several more are installed each month. The resolution attained has reached 50 pm (0.5 Å) in both STEM and TEM, and the electron current in an atom-sized electron probe can now be such that atomic-resolution STEM/EELS elemental maps can be acquired in less than one minute [2].

To make such abilities widely available and to advance the field further, Nion has developed a new dedicated STEM. The aim of the project was to produce an electron microscope that could be configured for many different uses, and whose principal elements (electron source, probe-forming column, sample stage, image/diffraction /EELS collecting column, various detectors) were all up to the high standards set by successful aberration correction. Accordingly, the microscope uses a cold field emission gun (CFEG) as its source, with a brightness $B \sim 2 \times 10^9$ A/cm^{2}str at 100 kV (i.e, about 5-10x brighter than a Schottky gun) and employs an advanced C_3/C_5 probe corrector. It also has an ultra-stable sample stage based on new construction principles, and post-sample optics that can collect >75% of the available EELS signal while giving better than 0.5 eV energy resolution. It employs UHV construction methods, and it can be baked at 140 C. The column vacuum is in the 10^{-10} torr region, the gun vacuum is in the low 10^{-11} torr region, and viewing initially clean samples is completely free of contamination. The microscope has been described in detail elsewhere [3]. Here we give just one example of its construction, and show recent results.

Fig. 1 shows the double gimbal of a new type of the microscope's double-tilt sample cartridge. The cartridge is inserted into the OL polepiece from the side, under computer control, and is decoupled from the outside world during operation. The stage, which moves the cartridge in X, Y and Z, has achieved drift rates <1 Å per 100s, and minimum mechanical steps < 10 Å. The double gimbal uses ball bearings to achieve exceptionally friction-free and therefore smooth, responsive and backlash-free tilting, with minimum tilt steps < 0.05° [4]. The ball bearing races have an O.D. of just 1 mm, making the bearings good candidates for the smallest ball bearings ever employed.

The excellent precision of the sample movement means that it can be modelled accurately by a computer, allowing operation modes such as computer-driven doubly-eucentric tilting, and point-and-click shifting and tilting of the sample, whereby one

M. Luysberg, K. Tillmann, T. Weirich (Eds.): EMC 2008, Vol. 1: Instrumentation and Methods, pp. 11–12, DOI: 10.1007/978-3-540-85156-1_6, © Springer-Verlag Berlin Heidelberg 2008

clicks on a desired new centre of an image or a desired pole in a diffraction pattern, and the corresponding mechanical adjustments are carried out automatically.

Figures 2 to 4 show examples of recent work carried out with the new microscope, such as high resolution imaging at 60 kV (a primary voltage known not to cause displacement damage in C), and EELS spectroscopy.

1. P.E. Batson, N. Dellby and O.L. Krivanek, Nature **418** (2002) 617-620.
2. D.A. Muller et al., Science **22** (2008) 1073-1076.
3. O.L. Krivanek et al., Ultramicroscopy **108** (2008) 179-195.
4. O.L. Krivanek, G.C. Corbin and Z. Szilagyi, US patent application (2008).

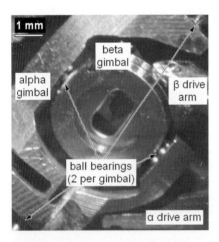

Figure 1. The double gimbal mechanism. 0.4 mm Ø balls are used. $\alpha = 0$, $\beta = 25°$.

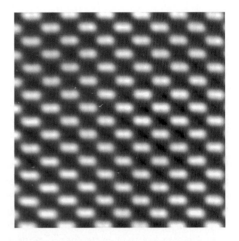

Figure 2. HAADF image of [110] Si recorded at 60 kV primary voltage.

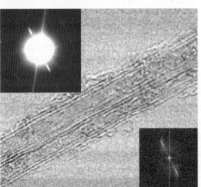

Figure 3. 60 kV STEM BF image of a C nanotube. Inserts show a parallel-beam diffraction pattern and an FFT, both from the above area. Courtesy Dr. M. Kociak.

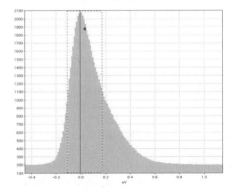

Figure 4. EELS zero loss peak recorded at 100 kV. Full width at half-maximum is 0.27 eV; the peak shows the asymmetry typical of CFEG energy distribution.

Applications of aberration-corrected TEM-STEM and high-resolution EELS for the study of functional materials

G.A. Botton[1,2], C. Maunders[1,2], L. Gunawan[1,2], K. Cui[1], L.Y. Chang[2] and S. Lazar[2,3]

1. Department of Materials Science and Engineering, McMaster University, 1280 Main St. West, Hamilton L8S 4M1, Canada
2. Canadian Centre for Electron Microscopy, Brockhouse Institute for Materials Research, McMaster University, 1280 Main St. West, Hamilton L8S 4M1, Canada
3. FEI Company, Acthseweg Noord 5, 5600 KA Eindhoven, The Netherlands.

gbotton@mcmaster.ca

Keywords: Aberration corrected (S)TEM, Ultrahigh resolution microscopy, EELS, Quantum dots, CdTe, Multiferroic materials, LaBaCuO, $Ba(RuTi)O_3$, nanowires

The first applications of high-resolution aberration-corrected microscopy were demonstrated 10 years ago in key publications [1-3] and earlier conferences showing the potential impact of the technique in the study of materials. Aberration-corrected STEM imaging results demonstrating sub-Å resolution were presented soon after [5,6].

In our laboratory, aberration-corrected microscopy has been used in a variety of projects related to the study of defects of oxides, thin films, nanowires, nanoparticles and a variety of quantum structures. With a FEI Titan 80-300 Cubed equipped with an image corrector, located in an ultrastable environment offering low thermal fluctuations, low vibrations and electromagnetic fields, this instrumentation has demonstrated low drift rates (below 0.3Å over 3 minutes) and, as a consequence, high-resolution images with significantly increased (figure 1). The stability of the microscope is demonstrated in examples where the geometric phase analysis method can be applied in high-angle annular dark field (HAADF) STEM images of quantum wires (figure 2). Similarly, HAADF images (even without probe correction) have been used to determine the polarity in CdTe nanowires grown on sapphire and the growth defects in multiferroic materials (figure 3). With such stability, single scan STEM images with high signal-to-noise ratio and low instabilities can be readily acquired. Early results on a double-corrected and monochromated FEI Titan 80-300HB equipped with a SuperTwin lens (at factory acceptance tests) have demonstrated better than 1.4Å STEM resolution with a monochromated beam of 0.14eV energy resolution. A second instrument equipped with a CryoTwin, an image corrector and a monochromator has demonstrated better than 1Å information limit (at factory acceptance tests). Results of these instruments installed in our laboratory will be presented. Particular emphasis will be given to the enhanced contrast and benefits of aberration-correction for solving materials science problems.

1. Haider, M., Rose, H., Uhlemann, S., Schwan, E., Kabius, B. and Urban, K. (1998) Ultramicroscopy, Vol. 75, pp. 53-60.
2. Haider, M., Rose, H., Uhlemann, S., Kabius, B. and Urban, K. (1998) J. of Electron Microscopy, Vol. 47, pp. 395-405.
3. Haider, M., Uhlemann, S., Schwan, E., Rose, H., Kabius, B. and Urban, K. (1998) Nature, Vol. 392, pp. 768-769.

M. Luysberg, K. Tillmann, T. Weirich (Eds.): EMC 2008, Vol. 1: Instrumentation and Methods, pp. 13–14, DOI: 10.1007/978-3-540-85156-1_7, © Springer-Verlag Berlin Heidelberg 2008

14

4. Krivanek, O. L., Dellby, N. and Lupini, A. R., Ultramicroscopy, **78**, (1999) 1-11.
5. Krivanek, O. L., Nellist, P. D., Dellby, N., Murfitt, M. F. and Szilagyi, Z. Ultramicroscopy, **96**, (2003) 229-237.
6. We are grateful to J. Preston, R. Hughes, Y. Zhao, B. Gaulin, H. Dabkowska (LBCO sample), M. Robertson, A. Pignolet, O. Gautreau (BLTO sample), J. Etheridge and J.L. Rouviere for providing several valuable samples and for valuable discussions.

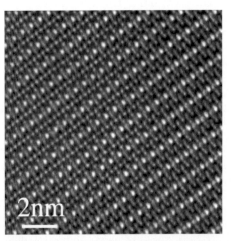

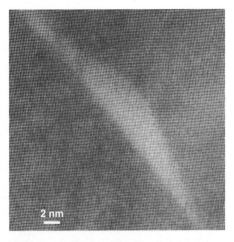

Figure 1. Aberration-corrected HRTEM micrograph of a $La_{2-x}Ba_xCuO_4$ crystal viewed down the [011] zone axis.

Figure 2. High-angle annular dark-field image of an InAs quantum wire deposited on InGaAlAs.

Figure 3. HAADF image of $Ba_{4-x}La_xTi_3O_{12}$ (x=0.75). The bright columns contain Bi atoms. Clearly visible are the double Bi-O layers and the perovskite slabs separating these layers. Highlighted in the box are stacking defects.

Aberration corrected TEM and STEM for dynamic *in situ* experiments

Pratibha L. Gai[*] and Edward D. Boyes[#]

Departments of Chemistry and Physics (*)
and Departments of Physics and Electronics (#)
The University of York, The York-JEOL NanoCentre, Science Park, Heslington, York
YO10 5BR, UK

pgb500@york.ac.uk and eb520@york.ac.uk

Keywords: aberration-correction, in situ, dynamic

Aberration correction is particularly beneficial in dynamic in-situ experiments because there is rarely if ever the opportunity to record from a scene which may be continuously changing a systematic through focal series of images for subsequent data reconstruction. It is therefore necessary to extract the maximum possible information from single images which may be in a continuously changing sequence. In addition we need to minimise the electron dose exposure to ensure minimally invasive conditions and to avoid secondary effects such as contamination.

As well as an instrument configuration able to accommodate a hot stage, the priorities for aberration correction also change. As well as benefiting from improved resolution at the 1Å end of the range and below, it becomes important to be able to set the conditions to avoid the previous intrusive contrast transfer function (CTF) and defocus sensitive oscillations in image contrast also in the more customary spatial resolution range from 1Å to 3Å or so; which of course is where the atomic neighbourhoods lie. This is especially important in studying the surfaces of nanoparticles which is a particular topic of interest, e.g. in considering the possible origins (and process control knobs to turn for control) in heterogeneous catalyst design and processing (intentional and inadvertent, such as deactivation) related to sensitivity and selectivity outcomes. This is an aspect of nanotechnology to which electron microscopy is perhaps uniquely well qualified to contribute.

We show in both theory [1] and in practice [2] it is possible to combine the larger gap (HRP) lens polepiece needed to take a standard hot stage [3] with the requisite high levels of imaging performance, including demonstrated spatial resolution at or below 1Å in both TEM and HAADF STEM modes. This is an example of using aberration correction to combine the limited added space required for in-situ experiments with a high level of imaging performance with which such facilities were previously incompatible, and thereby to extend considerably application specific and relevant TEM and STEM experimental capabilities. The system is in practice stability limited and some of the practical steps necessary to deliver this powerful combination of capabilities will be covered in the presentation, as will specific examples of the new tool in action. These considerations quickly set a limit to how far the lens gap can be stretched without beginning to compromise performance too seriously, taking into

M. Luysberg, K. Tillmann, T. Weirich (Eds.): EMC 2008, Vol. 1: Instrumentation and Methods,
pp. 15–16, DOI: 10.1007/978-3-540-85156-1_8, © Springer-Verlag Berlin Heidelberg 2008

16

consideration realistically attainable stabilities in internal electronics and mechanics, and in key external environmental factors. All things in moderation but with exciting results! The 'stretched' instrument configuration is considered to be a pre-requisite for further in-situ developments adding additional facilities and capabilities [4, 5] to the exciting and still relatively new generation of aberration corrected TEM and STEM instruments, with in our case both capabilities combined on the JEOL 2200 FS platform.

1. www.maxsidorov.com/ctfexplorer/index.htm
2. P L Gai and E D Boyes, these proceedings
3. Gatan model 628
4. E D Boyes and P L Gai, Ultramicroscopy, **67** (1997) 219
5. P.L. Gai, E.D. Boyes, S. Helveg, P.Hansen, S.Giorgio, C.Henry, MRS Bull. **32** (2007) 1044.

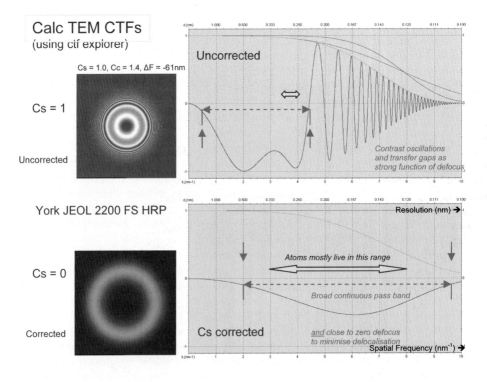

HREM study of the SrTiO₃ Σ3 (112) grain boundary

K.J. Dudeck[1], N. Benedek[2] and D.J.H. Cockayne[1]

1. Department of Materials, Oxford University, Parks Road Oxford. OX1 3PH
2. Department of Materials, Imperial College London, Exhibition Road, London SW7 2AZ, UK

Karleen.Dudeck@materials.ox.ac.uk

Keywords: SrTiO₃, quantitative HRTEM, Σ3 (112) grain boundary

Grain boundaries play a dominant role in determining electronic and structural properties of polycrystalline ceramic materials such as SrTiO₃. In order to develop an understanding of how grain boundary structure affects bulk properties, quantitative atomistic structural information is required. Due to their highly controlled nature, low angle coincident site lattice grain boundaries in the form of bicrystals provide ideal candidates for quantitative studies, as experimental results can be compared to modelled structures [1].

HRTEM and HAADF STEM are ideal tools for structural characterisation, however, due to the complex relation between specimen structure and recorded images, image interpretation is not always straightforward. In this work aberration corrected transmission electron microscopy and image simulation have been combined to determine quantitative information about the structure of the symmetric tilt SrTiO₃ Σ3 (112) boundary. Initially the [111] projection has been studied. Aberration correction allows for spherical aberration to be tuned to optimal values for imaging of light elements [2], while image simulation via multislice calculations allows for comparison between experimental images and modelled structures [3].

Figure 1 presents a typical HRTEM image of the grain boundary taken under optimal aberration corrected conditions, with a small negative spherical aberration and small positive (over) focus defocus condition [4]. In the case of the Σ3 (112), the results of Density Functional Theory (DFT) simulations suggest that there are two distinct and nearly equally energetically favourable geometric configurations – one that contains Sr-Ti-O at the grain boundary plane (a mirror symmetric state) and one that contains only O at the grain boundary plane (mirror symmetric with a translation along [111], i.e. with mirror glide symmetry). The two model structures are shown in Figure 2. Simulated images of the model structures, as well as an enlarged image of the experiment grain boundary image, are shown in Figure 3.

Atomic column positions are being determined with high precision and accuracy with the aim of comparing experimentally determined structures with model structures. Attention has been paid to careful data collection and image processing to improve signal-to-noise ratio of the images in order to obtain high precision quantitative data.

1. V. P. Dravid and V. Ravikumar, Interface Science **8** (2000) 177.
2. C. L. Jia, M. Lentzen and K. Urban, Microscopy and Microanalysis **10** (2004) 174.
3. J. M. Cowley and A. F. Moodie, Acta Crystallographica **10** (1957) 609.

M. Luysberg, K. Tillmann, T. Weirich (Eds.): EMC 2008, Vol. 1: Instrumentation and Methods, pp. 17–18, DOI: 10.1007/978-3-540-85156-1_9, © Springer-Verlag Berlin Heidelberg 2008

18

4. L. Y. Chang, A. I. Kirkland and J. M. Titchmarsh, Ultramicroscopy **106** (2006) 301.

5. Financial support by the European Commission under contract Nr. NMP3-CT-2005-013862 (INCEMS) is acknowledged.

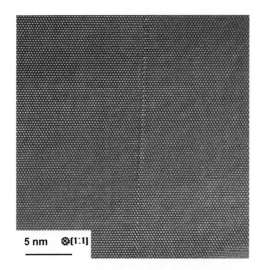

Figure 2. SrTiO$_2$ Σ3 (112) mirror symmetric (top) and translation (bottom) state model structures viewed in [111] projection.

Figure 1. Typical HRTEM image of SrTiO$_2$ Σ3 (112) grain boundary viewed in [111] projection.

Figure 3. The experimental grain boundary structure (top) shows periodic bright spots at the grain boundary. Simulated images of the two model structures (mirror symmetric on bottom left and translation state on bottom right) used as inputs for refining atomic positions.

A Method to Measure Source Size in Aberration Corrected Electron Microscopes

C. Dwyer[1,2], J. Etheridge[1,2] and R. Erni[3]

1. Monash Centre for Electron Microscopy, Monash University, Victoria 3800, Australia
2. Department of Materials Engineering, Monash University, Victoria 3800, Australia
3. National Centre for Electron Microscopy, Lawrence Berkeley National Laboratory, CA 94720, USA

joanne.etheridge@mcem.monash.edu.au
Keywords: coherence, STEM, Ronchigram

The ability to correct spherical aberrations in electron optics [1,2], along with improvements in instrumental stability [3], have enabled sub-Ångström imaging to be achieved in both TEM and STEM. This has introduced the possibility of obtaining new and local information about a specimen via an exciting range of new experiments. However, extracting the maximum amount of information from such experiments requires experimental data to be interpreted quantitatively by means of comparisons with simulations. This, in turn, requires that the microscope's performance is well-characterized. In particular, experimental data is strongly influenced by the degree of coherence of the electron beam. Currently, however, there is no established method for measuring this. In the present work, we will describe a method for measuring the spatial coherence of the electron beam arising from finite source size and instrumental instabilities. This method involves the analysis of a Ronchigram obtained from a crystalline material.

The effect of finite source size on Ronchigrams was first considered by Cowley and Moodie [4], and a method for measuring the source distribution assuming the weak phase object approximation was given more recently by James and Rodenburg [5]. It can be shown that, under quite general circumstances, the intensity of a set of interference fringes in a Ronchigram is damped according to the Fourier transform of the effective source intensity distribution [6]. In the present work, this effect was used to measure the degree of spatial coherence by incorporating it into Ronchigram simulations which were then matched to experiment. An example is given in Figure 1 which shows a comparison of experimental and calculated Ronchigrams of diamond <210>. The effective source size in the specimen plane deduced from this analysis is about 0.7 Å (FWHM). The shape of the source was found to be Gaussian within experimental error.

As an example of the effect of finite effective source size on ADF-STEM images, Figure 2 shows simulated ADF-STEM image profiles of diamond <210> for different effective source sizes. The incorporation of finite source size is seen to lead to a significant reduction in the image contrast. This effect will be described in more detail elsewhere at this conference.

1.   O.L. Krivanek et al., Ultramicroscopy **78** (1999), p. 1.
2.   H. Rose, Ultramicroscopy **78** (1999), p. 13.

M. Luysberg, K. Tillmann, T. Weirich (Eds.): EMC 2008, Vol. 1: Instrumentation and Methods,
pp. 19–20, DOI: 10.1007/978-3-540-85156-1_10, © Springer-Verlag Berlin Heidelberg 2008

3. B. Freitag et al., Ultramicroscopy (2005) 102, p. 209.
4. J. M. Cowley and A. F. Moodie, Proc. Phys. Soc. B, **LXX** (1957), p. 497.
5. E.M. James and J. M. Rodenburg, Appl. Surf. Sci. **111** (1997), p. 174.
6. J. Etheridge, C. Dwyer and R. Erni; C. Dwyer and J. Etheridge; 2 papers in preparation.

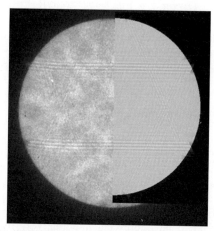

Figure 1. Experimental (left) and computed (right) Ronchigram of diamond <210> exhibiting interference fringes corresponding to 0.89 Å (horizontal fringes) and 0.73 Å (oblique fringes) taken at 300 keV with a convergence semi-angle ~25 mrad. The degree of partial spatial coherence corresponds to an effective source size of about 0.7Å (FWHM), corresponding to a final probe intensity distribution in the specimen plane of 0.81 Å, in the absence of spherical aberration.

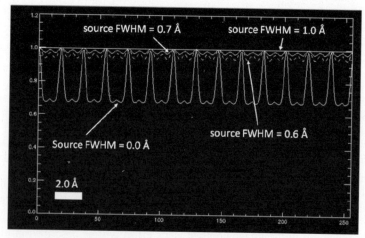

Figure 2. Simulated ADF-STEM image profiles of diamond <210> using different effective source sizes. The beam energy is 300 keV and the convergence semi-angle is 25 mrad. The probe direction is <400>.

Determining resolution in the transmission electron microscope: object-defined resolution below 0.5Å

B. Freitag[1] and C. Kisielowski[2]

1. FEI Company, Building AAE, Achtseweg Noord 5, Eindhoven, The Netherlands
2. National Center for Electron Microscopy, Lawrence Berkeley National Laboratory, One Cyclotron Rd., Berkeley CA 94720, USA

Bert.Freitag@fei.com
Keywords: Aberration-corrected, sub-Angstrom resolution, S/TEM

The Transmission Electron Aberration-corrected Microscope (TEAM) project was initiated by the US Department of Energy as a collaborative effort to redesign the electron microscope around aberration-corrected optics [1], and is aimed at achieving 50 pm resolution. But the ability to resolve deep sub–Ångstrom spacing entails a number of unresolved questions that can now be addressed. Among them is an ongoing debate about the physical meaning of resolution. Traditional strategies include the recording of Young's fringes, the detection of image Fourier components from STEM images, the demonstration of a suitable peak separation in periodic lattices or signal width measurements from images of single atoms, to name a few. The drawback is that seemingly conflicting results are produced [e.g. 2]. Further, these methods define resolution through a selectable object, unlike light microscopy where resolution is instrument-defined. Two limitations of this approach are electron channeling [3, 4] and elastic scattering at single crystals [5]. The TEAM Project adopted a pragmatic view of information transfer below 50 pm: detecting Young's fringes in TEM and (660) image Fourier components from gold (111) STEM images at 48 pm. Recently the TEAM 0.5 prototype microscope achieved this goal [1].

Figure 1 shows two amplitude images of channelling waves that were reconstructed from 30 experimentally recorded lattice images from gold crystals imaged in [110] and [111] direction with the TEAM 0.5 microscope. Gold atoms are spaced by 0.29 nm in [110] direction but by 7 nm in [111] direction. Their different spacing alters electron channelling significantly. It is seen from the line profiles in Figure 2 that the full width at half maximum of the signals is 67± 4 pm and 46 ± 3pm for Au [110] and Au [111], respectively. The results agree with predictions from multi-slice calculations that also predict similar effects for HAADF STEM images from gold [110] and gold [111]. Therefore, we conclude that the TEAM 0.5 microscope can resolve column spacings below 0.5 Å, close to the Rayleigh resolution limit, if the objects are carefully chosen and prepared. However, in general, electron channelling can limit the object-defined resolution to values above 50 pm in both STEM and TEM images.

M. Luysberg, K. Tillmann, T. Weirich (Eds.): EMC 2008, Vol. 1: Instrumentation and Methods, pp. 21–22, DOI: 10.1007/978-3-540-85156-1_11, © Springer-Verlag Berlin Heidelberg 2008

22

1. http://ncem.lbl.gov/TEAM-project/index.html
2. P.E. Batson, Ultramicroscopy 96, 239 (2003)
3. A.J. den Dekker, S.Van Aert, D. Van Dyck, A. Van der Bos, P. Geuens, Ultramicroscopy 89, 275 (2001)
4. S. Van Aert, A.J. den Dekker, D. Van Dyck, A. van den Bos, Ultramicroscopy 90, 273 (2002)
5. L. Reimer, Transmission Electron Microscopy, Springer, Optical Sciences, Berlin 1993, p. 152
6. The authors acknowledge support of the National Center for Electron Microscopy, Lawrence Berkeley Lab, which is supported by the U.S. Department of Energy under Contract # DE-AC02-05CH11231.

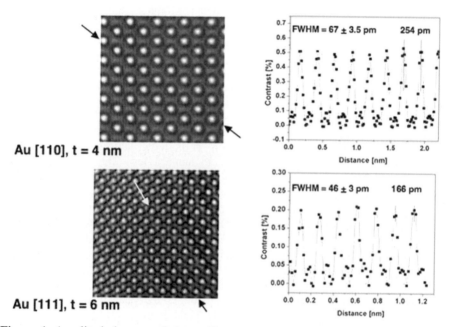

Figure 1. Amplitude images of channelling waves that were reconstructed from lattice images of gold [110] and gold [111] crystals recorded with the TEAM 0.5 microscope. The crystal thickness t is indicated. The extracted signal width from the associated line profiles proves that an object-defined resolution below 50 pm is obtained in Au [111] that increases to 67 pm in case of Au [110].

Direct measurement of aberrations by convergent-beam electron holography (CHEF)

C. Gatel[1], F. Houdellier[1] and M.J. Hÿtch[1]

1. CEMES-CNRS, 29 rue J. Marvig, 31055 Toulouse, France

gatel@cemes.fr
Keywords: aberrations, holography

Herring *et al.* introduced a new form of holography where convergent-beam diffraction disks are interfered with the aid of an electron biprism [1]. The specimen is shifted in height with respect to a convergent beam focused on the object plane, similar to large-angle convergent-beam diffraction (LACBED). An electron biprism deviates the diffracted disks so that they overlap in the focal plane allowing interference fringes to form (Figure 1). Optimum conditions require a field emission gun (FEG), rotatable biprism and imaging filter. We have recently shown how the technique can be used to measure dynamic phase changes across the diffracted disk [2]. Here, we will concentrate on the use for measuring aberrations.

For thin specimens, the local phase $\phi^{chef}(\mathbf{k})$ of the interference fringes as a function of the position in the disk, $\mathbf{k}$, is directly related to the aberrations of the objective lens:

$$\phi^{chef}(\mathbf{k}) = \gamma(\mathbf{k}+\mathbf{g}) - \gamma(\mathbf{k})$$

where $\gamma(\mathbf{k})$ is the complex transfer function of the lens as a function of scattering vector, and $\mathbf{g}$ the reciprocal lattice vector of the diffracted beam [2]. The specimen needs to be sufficiently thin to avoid dynamic phase changes across the diffracted disk.

To extract the phase plate from the measured phase, two methods can be used. The first is to simulate the CHEF phase for a given set of aberration coefficients and to fit with the experimental data. This method is possible when the number of significant aberrations is small – i.e. for conventional microscopes having large spherical aberration coefficients (C_s). The second method is more general and involves fitting a polynomial function to the experimental phase. The polynomial coefficients can then be converted into the aberration coefficients by suitable matrix transformations. Theoretically, just two CHEF patterns are necessary to give a unique solution for the aberrations. To measure coefficients up to order n, the polynomial needs to be of order *n-1*.

To test our theory, experiments were performed using the SACTEM-Toulouse, a Tecnai F20 ST (FEI), fitted with objective lens aberration corrector (CEOS), rotatable electron biprism (FEI) and imaging filter (Gatan Tridiem). CHEF patterns were acquired on a 2k camera (Gatan) at 200kV and for various values of aberrations. The local phase of the interference fringes was measured using GPA software [3]. The polynomial fitting procedure and aberrations calculations are obtained using in-house

M. Luysberg, K. Tillmann, T. Weirich (Eds.): EMC 2008, Vol. 1: Instrumentation and Methods, pp. 23–24, DOI: 10.1007/978-3-540-85156-1_12, © Springer-Verlag Berlin Heidelberg 2008

24

routines developed for Gatan DigitalMicrograph. Direct measurements of aberrations to 6th order are shown in Fig. 2.

1. R.A. Herring, G. Pozzi, T. Tanji, and A. Tonomura, Ultramicroscopy **60** (1995), p. 153.
2. F. Houdellier and M.J. Hÿtch, Ultramicroscopy **108** (2008), p. 285.
3. GPA Phase 2.0 (HREM Research Inc.) a plug-in for DigitalMicrograph (Gatan).
4. We thank the EU Integrated Infrastructure Initiative ESTEEM for support (Reference 026019 ESTEEM) and P. Hartel (CEOS) for fruitful discussions.

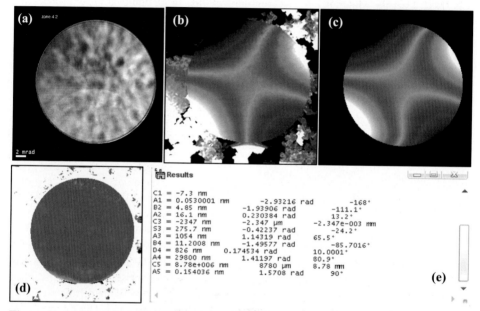

Figure 1. (a) Experimental CHEF pattern from the interference of the (000) and (220) beams in Si, disk radius 7.8 mrad; (b) experimental CHEF phase; (c) polynomial-fitted phase; (d) difference between experimental and fitted phases; (e) results of aberrations measurements.

Atomic Structure of BiFeO₃-BiCrO₃ film on (111) SrTiO₃ Grown by Dual Cross Beam Pulsed Laser Deposition

L. Gunawan[1], R. Nechache[2], C. Harnagea[2], A. Pignolet[2] and G.A. Botton[1]

1. Dept. of Materials Science and Engineering & Brockhouse Institute for Materials Research, McMaster University, 1280 Main St. West, Hamilton, ON L8S 4M1, Canada
2. INRS Énergie, Matériaux et Télécommunications, 1650, boulevard Lionel-Boulet, Varennes, QC J3X 1S2, Canada

gunawal@mcmaster.ca
Keywords: double perovskite oxides, multiferroics, aberration corrected TEM, HR-EELS

The existence of unexpected magnetic ordering at room temperature in the multiferroic double perovskite Bi_2FeCrO_6 (BFCO) thin film synthesized by conventional pulsed laser deposition (PLD) has been attributed to the exchange interaction between the alternating planes of Fe and Cr cations along [111] direction.[1] The properties of this material have partially confirmed the theoretical predictions by Baettig and Spaldin in terms of magnetic moment and polarization at 0 K.[2] In order to observe the exchange interaction as the function of the distance of separation between Fe and Cr cations, the composition of the thin film was modified by controlling the $BiFeO_3$ / $BiCrO_3$ ratio during deposition using dual crossed beam PLD.

The epitaxial $Bi_4FeCr_3O_{12}$ (BFO/BCO 1/3) film was grown on a (111)-oriented single crystal $SrTiO_3$ (STO) substrate using dual crossed beam PLD. The structural characterization of the films was carried out with an aberration-corrected scanning transmission electron microscope FEI Titan 80-300 Cubed. The BFO/BCO 1/3 film is ~80 nm thick and oriented with its (111) planes parallel to the substrate's surface.

The cross-sectional TEM image taken at zone axis $[121]_{STO}$ in Fig. 1(a) shows the layered structure along the growth direction. However, a disordered zone of few monolayers (~ 2 nm thick) was uniformly present along the substrate-film interface. Despite the interfacial disorder, the layer-by-layer structure of the BFO/BCO 1/3 film is still maintained during deposition. The high quality of the epitaxial growth is demonstrated by the smooth and flat top surface of the film, with only occasional single unit cell surface steps, as shown in Fig. 1(b).

The atomic resolution annular dark field (ADF) image (Fig. 2) taken at $[121]_{STO}$ zone axis clearly shows the layered Bi atoms along the [111] direction. Based on these structural observations and electron energy loss spectra acquired with a monochromated aberration-corrected TEM, the differences in the unit cell structure of BFO/BCO 1/3 with BFO/BCO 1/1 and ordering of the Fe and Cr atoms will be discussed.

1. R. Nechache, L.-P. Carignan, L. Gunawan, C. Harnagea, G. A. Botton, D. Ménard, and A. Pignolet, J. Mater. Res. **22** (2007) p. 2102.
2. P. Baettig and N. A. Spaldin, Appl. Phys. Lett. **86** (2006) p. 012505.
3. R. Nechache, C. Harnagea, L. Gunawan, L.-P. Carignan, C. Maunders, D. Ménard, G. A. Botton, and A. Pignolet, IEEE TUFFC **54** (2007) p. 2645

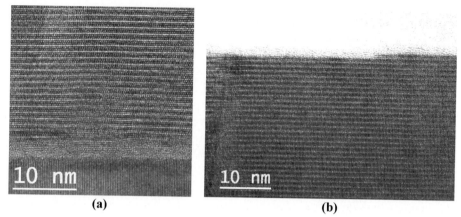

(a) (b)

Figure 1. High-resolution TEM images of (a) interface of BFO/BCO 1/3 film and STO substrate at ZA [121]$_{STO}$ and (b) top surface of BFO/BCO 1/3 film at ZA [101]$_{STO}$

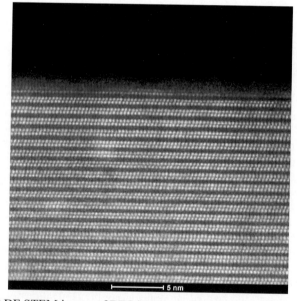

Figure 2. The ADF STEM image of BFO/BCO 1/3 film taken at ZA [121]$_{STO}$.

Demonstration of C_C/C_S-correction in HRTEM

P. Hartel[1], H. Müller[1], S. Uhlemann[1], J. Zach[1], U. Löbau[1], R. Höschen[2] and M. Haider[1]

1. CEOS GmbH, Englerstr. 28, D-69126 Heidelberg, Germany
2. MPI für Metallforschung, Heisenbergstr. 3, D-70569 Stuttgart, Germany

haider@ceos-gmbh.de
Keywords: chromatic correction, aberration correction, quadrupole-octupole corrector

The next major step in transmission electron microscopy (TEM) towards improved resolution and usability is the correction of the axial chromatic aberration C_C. Our quadrupole-octupole-type C_C/C_S-corrector consists of ten quadrupole stages. Two of them produce crossed electric/magnetic quadrupole fields for C_C-correction and octupole fields for C_S-correction. The design allows to compensate for all axial aberration coefficients up to fourth order. The fifth-order axial aberrations are sufficiently small to allow for a spatial resolution of 50pm at 200 and 300kV [1,2]. Moreover, all off-axial aberrations up to third order which increase linearly with the field of view are adjustable. Particularly, the azimuthal off-axial coma of the magnetic objective lens can be compensated. The corrector was built for the TEAM project [3] and is incorporated in a FEI TITAN 80-300 electron microscope below the objective.

The correction of the axial aberrations up to third order is demonstrated in Figure 1 by means of a Zemlin tableau. The aberration coefficients have been calculated from the tilt induced defocus and astigmatism [4].

Successful C_C-correction is proven in Figure 2. The solid line is a fourth-order polynomial fit to the measured defocus values (circles) introduced by a change dE of the primary electron energy $E_0 = 200$kV. The linear term of the fit represents the coefficient $C_C = C_{\alpha K}$ of first degree. The total C_C of objective and corrector is reduced from 2.7mm (dashed line) to below 0.01mm. The results are in good agreement with the theoretical prediction (chain dotted line) for the chromatic aberration coefficients of second and third degree (experiment: $C_{\alpha K K} = 3.91$mm, $C_{\alpha K K K} = 884$mm; theory: 3.68mm and 754mm, respectively). The characteristic of the defocus-over-dE function is that of an apochromat [2] which provides the same focus for three different electron energies.

The practical difference with and without C_C-correction becomes obvious from Figure 3 at images of gold clusters on carbon: Atomic resolution and focus are kept although the electron energy has been changed by dE = 50eV from (a) to (b) without refocusing. Without C_C-correction the same image would look like (c) due to the equivalent defocus $C_1 = -dE/E_0 \cdot C_C$.

1. M. Haider, H. Müller, S. Uhlemann, J. Zach, U. Löbau and R. Höschen, Ultramicroscopy **103** (2008), p. 167.
2. H. Rose, Ultramicroscopy **93** (2002), p. 293.
3. Transmission Electron Aberration-Corrected Microscope Project, http://www.lbl.gov/LBL-Programs/TEAM/.
4. S. Uhlemann and M. Haider, Ultramicroscopy **72** (1998), p. 109.

M. Luysberg, K. Tillmann, T. Weirich (Eds.): EMC 2008, Vol. 1: Instrumentation and Methods, pp. 27–28, DOI: 10.1007/978-3-540-85156-1_14, © Springer-Verlag Berlin Heidelberg 2008

28

	Value	Angle	Confidence
C1	-470.7 nm	–	7.203 nm
A1	4.156 nm	-38.5 °	8.161 nm
A2	30.25 nm	-75.9 °	45.13 nm
B2	19.48 nm	139.6 °	31.57 nm
C3	3.622 µm	–	1.573 µm
A3	1.092 µm	134.7 °	1.322 µm
S3	775.4 nm	124.2 °	585.6 nm

Figure 1. Zemlin tableau taken with 24mrad outer tilt at 200kV with calculated axial aberration coefficients up to third order.

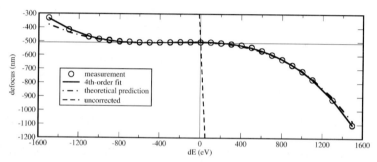

Figure 2. Verification of C_C-correction at 200kV. The defocus as a function of a change of the primary energy is shown. The slope at dE = 0 represents the chromatic aberration coefficient C_C.

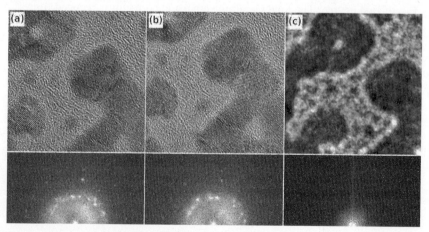

Figure 3. Demonstration of C_C-correction. Below each image its diffractogram is displayed. (a) nominal energy of 200keV, (b) energy offset of +50eV, (c) underfocus of 675nm (equivalent defocus for an energy offset of +50eV if C_C is not corrected).

New electron diffraction technique using Cs-corrected annular LACDIF: comparison with electron precession

Florent Houdellier[1], and Sara Bals[2]

1. CEMES-CNRS, 29 rue Jeanne Marvig, 31055 Toulouse, France
2. EMAT-universiteit Antwerpen, Groenenborgerlaan 171, B-2020 Antwerpen, Belgium

florent@cemes.fr

Keywords: Cs corrector, electron diffraction, dynamical effects

Electron precession, a technique which was proposed in 1994 by Vincent and Midgley [1], has become a powerful tool due to the following advantages:
- The intensities of the diffracted beams are integrated over a hollow cone. Zone axis patterns (ZAP) are therefore perfectly symmetric and can be simulated using simple kinematical approach. Indeed, the two-beam conditions are mostly verified for every diffracted beam. Provided the precession angle is very large (about 3°), the kinematical forbidden reflections can be identified.
- Precession patterns display more reflections in the Zero-Order Laue Zone (ZOLZ) but also in the High-Order Laue Zones (HOLZ) in comparison to conventional diffraction patterns.

A new technique combining the Large-Angle Convergent Beam Electron Diffraction (LACBED) configuration with a microscope equipped with an imaging C_s corrector allows us to obtain a very good quality and symmetry of the spot patterns in imaging mode. This new technique, called LACDIF [2] displays, similar to precession patterns, a large number of reflections where the intensities are now integrated over the filled cone. The ZAP symmetry is well equilibrated with respect to the transmitted spot but the intensity is integrated over all the orientations within the incident cone including the orientations which are along and close to the zone axis where a strong multi-beam behaviour prevails. This explains why the forbidden reflections remain visible for all convergence angles. Figure 1.b shows the LACDIF [130] Silicon ZAP with A=2° convergence angle where the (002) reflection is clearly present.

In order to suppress the kinematical forbidden reflections, the integration must be realized using a high angle hollow cone which can be provided by an annular condensor aperture. This new annular LACDIF configuration, presented in Fig. 1.a, has been studied using the SACTEM-Toulouse, which is a TECNAI F20 fitted with an imaging Cs-corrector and a GIF Tridiem. Two platinium annular apertures with a 100 μm and 200 μm diameter (A=1,3° and 2,6° convergence angle respectively) have been prepared using a Focused Ion Beam (FIB) in a similar way as explained in [3].

For example, we studied the evolution of the (002) and (006) kinematical forbidden reflections in the [130] Si ZAP with the convergence angle. As can be seen on the Fig 1.c and Fig 1.d these reflections disappear when the biggest aperture is inserted, leading to a kinematical-like pattern equivalent to precession (illustrated by the arrows

M. Luysberg, K. Tillmann, T. Weirich (Eds.): EMC 2008, Vol. 1: Instrumentation and Methods, pp. 29–30, DOI: 10.1007/978-3-540-85156-1_15, © Springer-Verlag Berlin Heidelberg 2008

in Figure 1). An application of space group determination in a TiAlNb alloys will also be presented.

1. R. Vincent and P. A. Midgley, Ultramicroscopy **53** (1994), p. 271.
2. J.P. Morniroli, F. Houdellier, C. Roucau, J. Puiggalí, S. Gestí and A. Redjaïmia, Ultramicroscopy **108** (2008), p. 100.
3. S. Bals, B. Kabius, M. Haider, V. Radmilovic and C. Kisielowski, Solid State Communications **130** (2004), p. 675
4. The authors thank the European Union for support under the IP3 project ESTEEM (Enabling Science and Technology through European Electron Microscopy, IP3: 0260019). The authors are grateful to the Fund for Scientific Research-Flanders (Contract No. G.0180.08N)

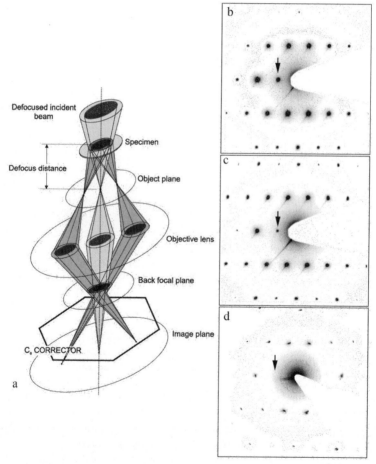

Figure 1. Description of the annular LACDIF configuration: (a) electron ray-path with the imaging Cs-corrector. (b) Silicon [130] LACDIF ZAP with 2° convergence angle. Annular LACDIF in the same conditions with A=1,3° and 2,6° are reported in (c) and (d).

The newly installed aberration corrected dedicated STEM (Hitachi HD2700C) at Brookhaven National Laboratory

H. Inada[1], Y. Zhu[2], J. Wall[2], V. Volkov[2], K. Nakamura[3], M. Konno[3], K. Kaji[3], K. Jarausch[1] and R.D. Twesten[4]

1. Hitachi High Technologies America, Inc., Pleasanton CA USA
2. Center for Functional Nanomaterials, Brookhaven National Laboratory, NY USA
3. Nanotech Products Business Gr., Hitachi High Technologies Corp., Ibaraki Japan
4. Gatan Inc., Pleasanton CA USA

inada-hiromi@naka.hitachi-hitec.com
Keywords: Aberration corrector, STEM, Cold field emission, EELS

The Hitachi HD2700C was recently successfully installed at the newly established Center for Functional Nanomaterials, Brookhaven National Lab (BNL). It was the first commercial aberration corrected electron microscope manufactured by Hitachi. The instrument is based on HD2300, a dedicated STEM developed a few years ago to complete with the VG STEMs [1]. The BNL HD2700C has a cold-field-emission electron source with high brightness and small energy spread, ideal for atomically resolved STEM imaging and EELS. The microscope has two condenser lenses and an objective lens with a gap that is slightly smaller than that of the HD2300, but with the same ±30° sample tilts capability. The projector system consists of two lenses that provide more flexibility in choosing various camera lengths and collection angles for imaging and spectroscopy. There are seven fixed and retractable detectors in the microscope. Above the objective lens is the secondary electron detector to image surface morphology of the sample. Below are the Hitachi HAADF and BF detector for STEM, and a Hitachi TV rate (30frame/sec) CCD camera for fast observations and alignment. The Gatan 14bit 2.6k×2.6k CCD camera located further down is for diffraction (both convergent and parallel illumination) and Ronchigram analysis. The Gatan ADF detector and EELS spectrometer (a specially modified high energy resolution Enfina spectrometer incorporating full 2nd and dominant 3rd order corrected optics and low drift electronics, a 16bit 100×1340 pixel CCD) are located at the bottom of the instrument. The CEOS probe corrector has been modified and optimized for this instrument. Other features of the microscope include remote operation, double shielding of the high tension tank and an anti-vibration system for the field emission tank. The entire instrument is covered with a telephone-booth-like metal box (Figure.1) to reduce acoustic noise and thermal drift. The instrument was delivered in July, 2007. Within two weeks of the start of the installation, 0.1nm resolution of HAADF-STEM image was achieved. Results from the commissioning are shown.

Our ultimate goal is quantitative electron microscopy. We note that in Z-contrast imaging when the convergent angle of the illumination is large (most of the cases with the aberration corrector on) insufficient large collection angle can result in contrast reversal in ADF images. Furthermore, since most software automatically adjust contrast and brightness for optimum viewing (qualitative images), rather treating the

M. Luysberg, K. Tillmann, T. Weirich (Eds.): EMC 2008, Vol. 1: Instrumentation and Methods, pp. 31–32, DOI: 10.1007/978-3-540-85156-1_16, © Springer-Verlag Berlin Heidelberg 2008

signals as quantitative data (numbers), we will discuss how to retrieve the absolute values of all detector signals to eliminate any ambiguity in interpretation [2]

1.   1 K. Nakamura, et al proc. 16th International Microscopy Congress, 2006, Sapporo, Japan, p.633.
2.   2 Supported by US Department of Energy BES, under contract No. DE-AC02-98CH10886.

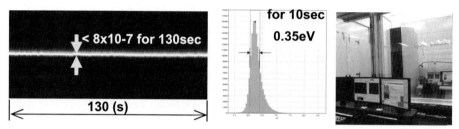

Figure 1. EELS energy time trace for zero loss peak for 130s duration (left), indicates stability of less than 8×10^{-7} for 130sec acquisition and corresponding zero loss peak spectrum FWHM of 0.35eV for acquisition time of 10 sec with cold field emission source acceleration voltage of 200kV, emission current of 1µA and EELS collection angle of 10mrad for 10s duration time. Microscope is installed in a dedicated microscope room (right).

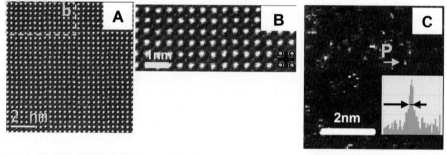

Figure 2. (A) HAADF image of $BaTiO_3$ (raw data), (B) Enlarged "b" area in A, ADF collection angle $\beta = 53$mrad, (C) HAADF image of Uranium atoms, intensity profile of single Uranium atom "P" in C, FWHM of the peak represents 0.08nm.

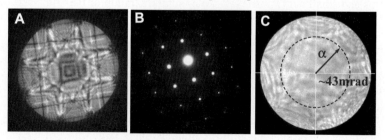

Figure 3. (A) CBED pattern of Si (100), (B) Nano beam diffraction of Si (100). Convergence angle $\alpha = 28$mrad, $\beta = 1.7$mrad, (C) Ronchigram of imaging condition semiangle of flat area is 43mrad

Uranium single atom imaging and EELS mapping using aberration corrected STEM and LN2 cold stage

H. Inada[1], J. Wall[2], Y. Zhu[2], V. Volkov[2], K. Nakamura[3], M. Konno[3], K. Kaji[3], K. Jarausch[1]

1. Hitachi High Technologies America, Inc., Pleasanton CA USA
2. Center for Functional Nanomaterials, Brookhaven National Laboratory, NY USA
3. Nanotech Products Business Gr., Hitachi High Technologies Corp., Ibaraki Japan

inada-hiromi@naka.hitachi-hitec.com
Keywords: Aberration corrector, STEM, EELS, Resolution, Single atom, Cold stage

Single heavy atoms on a thin carbon substrate represent a nearly ideal test specimen to evaluate STEM performance [1,2]. The single atoms approximate point scatters when imaged with the STEM large angle annular detector. (This is not necessarily true when using small angle scattering in TEM to make a phase contrast image.) The high scattering power relative to the substrate gives a high signal-to-noise ratio, even with relatively low beam current. The thinness of the sample eliminates any issues regarding depth of focus or channelling effects. The specimen was prepared in a manner similar to negative staining, except with a much lower concentration of Uranyl Acetate. The sample shown consisted of tobacco mosaic virus (TMV) on a 2nm thick carbon film substrate supported by holey film. The sample was rinsed several times with 0.01% Uranyl Acetate (compare to 2% normally used for negative staining) and air dried.

Imaging at high magnification (high dose) causes the TMV to disintegrate and the Uranyl Acetate to decompose to Uranyl Oxide, forming crystals in areas of high concentration and single atoms and short polymers in areas of low concentration. The atom-atom spacing is typically 0.34nm, so even atoms in aggregates give close to substrate signal between atoms, permitting measurement of full width at half maximum on most atoms in a given field. Observation at room temperature resulted in contamination which was particularly problematic due to the thinness of the substrate. Cooling to -160°C eliminated the contamination. The images presented show clear single atoms and small aggregates with atom-atom separation of 0.34nm. The stability from one scan line to the next and from scan to scan is less than desirable, but not so bad as to limit resolution measurement. Furthermore, uranium atoms have a strong EELS signal at ~100eV, permitting identification of even single spots. The line profiles obtained by simultaneous spectrum imaging and HAADF (Fig. 1 G & H) show very similar appearance with features less than 0.2nm FWHM. Based on measurements with the Hitachi HD2700C, CEOS aberration corrector and Gatan high resolution Enfina spectrometer, we conclude that the instrument is capable of producing a probe size is <0.1nm and suitable for high-resolution mapping. Since other specimens give very stable images, we conclude that the image instability observed is a property of the uranium atoms on the carbon substrate. We plan to study this further as a function of

M. Luysberg, K. Tillmann, T. Weirich (Eds.): EMC 2008, Vol. 1: Instrumentation and Methods, pp. 33–34, DOI: 10.1007/978-3-540-85156-1_17, © Springer-Verlag Berlin Heidelberg 2008

dose, dose-rate, specimen temperature and accelerating voltage. We will also compare the absolute value of observed spots (% scattering) with simulations [3].

1. A.V. Crewe, J.S. Wall, & J. Langmore, Visibility of single atoms. Science 168, 1338, (1970).
2. P. E. Batson, N. Dellby & O. L. Krivanek. Sub-ångstrom resolution using aberration corrected electron optics, Nature 418, 617 (2002).
3. Supported by US Department of Energy BES, under contract No. DE-AC02-98CH10886.

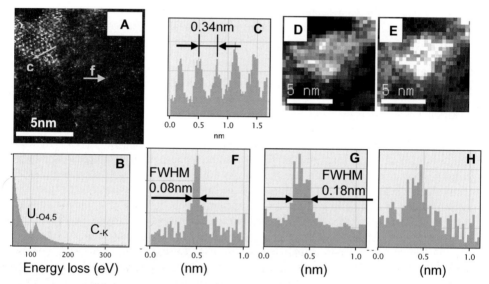

Figure 1. A: HAADF-STEM image of Uranium atoms and clumps (probe convergence 28mrad, collection angle of 53-280mrad). B: EELS spectrum showing U-O edge and C-K edge. C: Intensity profile along line "c" in A showing 0.34nm spacing. D: HAADF intensity map of Uranium clump (25x25 pixel, 0.4nm pixel spacing). E: Simultaneous spectrum imaging of area D using Gatan Enfina spectrometer set for 96eV-150eV loss with a spectrometer acceptance angle of 30mrad. F: Intensity profile of single Uranium atom "f" in A, FWHM of the peak represents 0.08nm. G: HAADF intensity profile of a small object (image not shown) 0.03nm/pixel. H: Simultaneous spectrum imaging of uranium O edge signal same area in G.

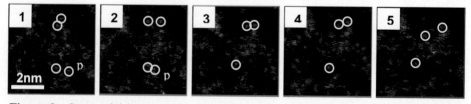

Figure 2. Sequential image acquisition of Uranium single atom motions. Single atom moves by irradiating electron beam. During this study specimen is cooled temperature of -160 degrees C with liquid nitrogen cooling stage. (256x256 pixel, 0.04nm/pixel, 64μs/pixel)

Sub-Ångstrøm Low-Voltage Electron Microscopy – future reality for deciphering the structure of beam-sensitive nanoobjects?

U. Kaiser[1], A. Chuvilin[1], R.R. Schröder[2], M. Haider, and H. Rose

1. Electron Microscopy Group of Material Science, Ulm University, Ulm, Germany
2. Cluster of Excellency "CellNetworks", Bioquant, University of Heidelberg, Germany
3. CEOS Company Heidelberg, Germany

ute.kaiser@uni-ulm.de

Keywords: HRTEM, aberration correction, low voltage

In recent years, the scientific discoveries and technological developments of greatest economic impact have largely been those situated at the interface between the classical disciplines of physics, chemistry, biology and engineering. In this overlap region, nanoscale objects constitute the primary focus of numerous interdisciplinary research efforts. Of course, the various functions of nano-objects can be understood completely only by investigating them at the atomic level.

Nowadays new developments in aberration-corrected transmission electron microscopy (TEM) allow for direct imaging at the atomic scale (see the difference between Fig. 1a and b). When operating at a voltage of 300 kV, the resolution goes down to 80pm (for the case that the spherical aberration of the objective is corrected [1,2]) or even 50pm (design goal for the case that the spherical and chromatic aberration are corrected [3,4]). In the ideal case of an infinitely radiation-resistant specimen, the resolution of the image is solely determined by the instrumental resolution and therefore the increase of accelerating voltage will give highest resolution provided that one can tolerate the loss in contrast.

However, real samples are always damaged by the electron beam and thus by the tolerable duration for observing the sample (that is the tolerable electron dose D exposed on the sample). Maximum D depends on physical processes occurring due to interaction of electrons with the sample, where the contribution of the processes depends on the energy of the electron beam used for observation and on the materials characteristic (compare Fig. 1c and d). At lower accelerating voltage, atom displacement processes (knock-on damage) and surface etching are strongly reduced, however, other processes as heating (compare Fig. 1e and f) and ionisation are increased. The achievable specimen resolution becomes a function of the instrumental resolution limit d_i, the tolerable dose D, the contrast C and the signal to noise ratio S/N [4].

Highest phase contrast and maximum S/N at the Gaussian image plane can be achieved for weak phase objects by eliminating the axial aberrations (correcting for spherical and chromatic aberration) and shifting the phase of the non-scattered wave by $\pi/2$ (inserting a phase plate at the diffraction plane or at an image of this plane [5]) and by lowering the accelerating voltage of the electrons.

M. Luysberg, K. Tillmann, T. Weirich (Eds.): EMC 2008, Vol. 1: Instrumentation and Methods, pp. 35–36, DOI: 10.1007/978-3-540-85156-1_18, © Springer-Verlag Berlin Heidelberg 2008

36

These theoretical findings will be explained in detail by calculations and experiments. The experimental results will include phase contrast imaging of classical semiconductors, nanocarbons, polymers and organic crystals at 300kV and 80kV accelerating voltage using TITAN 80-300 instrument equipped with a corrector for the spherical aberration of the objective. We conclude that reducing the acceleration voltage much lower than 80kV may allow atomic resolution for a variety of beam sensitive materials in future if new components such as a dedicated C_C/C_S corrector and a phase plate can be implemented into a highly mechanical and electrical stable microscope platform.

1. Rose H, Optik **85** (1990) 19
2. Haider M, Rose H, Uhlemann S, Schwan E, Kabius B, Urban K, Ultramicrosc. 75 (1998) 53.
3. Rose H, Nuc. Insrum. Meth. Phys Res. A 519 (2004) 12
4. M. Haider TITAN userclub meeting, Eindhoven. January 2008
5. Rose H, Microscopy and Microanalysis **13** (2007) 134-135
6. Schultheiss K, Perez-Fillard F, Barton B, Gerthsen D, Schroeder R, Rev. Scient. Instr. **77** (2006) 033701
7. We acknowledge Dr. S. Roth, D. Obergfell, A. N. Khlobystov, M. Haluska, S. Yang for sample preparation.

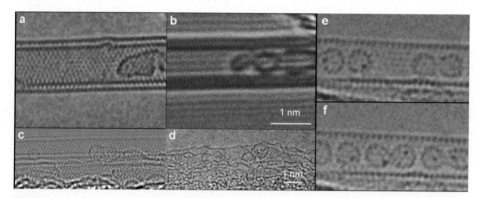

Figure 1. (a,b) show the effect of the C_S corrector on the example of a DWCNT at 80kV operating voltage (a) with C_S corrector (b) C_S corrector turned off, (c,d) show the effect of acceleration voltage on the example of a single wall CNT (c) at 80kV after 20min of observation (d) at 300kV totally damaged after 10sec. (e,f) show the effect of heating to molecules on the example of $(Dy@C_{82})@$ SWCNT, see the movement of the peapods and starting coalescence within 20 seconds of observation.

Detecting and resolving individual adatoms, vacancies, and their dynamics on graphene membranes

J.C. Meyer[1], C. Kisielowski[2], R. Erni[2], and A. Zettl[1]

1. Materials Sciences Division, Lawrence Berkeley National Laboratory and Department of Physics, University of California at Berkeley
2. National Center for Electron Microscopy, Lawrence Berkeley National Laboratory, One Cyclotron Rd., Berkeley CA 94720, USA

email@jannikmeyer.de

Keywords: Graphene, defects, aberration-corrected electron microscopy, single atom detection

Graphene is a single atomic layer of graphite that has only recently become experimentally accessible in an isolated form [1]. We prepare graphene into free-standing membranes [2,3], i.e., a crystalline foil with a thickness of only one carbon atom. These membranes are highly promising for TEM studies, since (a) this membrane itself is of tremendous scientific interest, (b) adsorbates can be studied against a highly transparent and crystalline background, and (c) the precisely known structure is an ideal tuning and calibration tool for electron microscopy developments. Graphene membranes are very stable at low acceleration voltages, we and present results obtained on a conventional TEM (Jeol 2010 at 100kV) as well as initial tests on the new aberration-corrected TEAM 0.5 microscope operated at its lower limit of 80kV. The TEAM 0.5 microscope achieves sub-Angstrom resolution even at 80kV, thus resolving every single carbon atom in the graphene lattice.

Individual low-atomic number adatoms on a clean graphene membrane are detected, as well as individual vacancies or molecular scale adsorbates. The dynamics of these objects is shown, such as the replacement of vacancies from mobile adsorbates. By using the aberration-corrected microscope, even subtle effects such as the relaxation of the graphene lattice near an adatom or vacancy can be detected. Further, the binding configuration of an individual adatom with the graphene membrane can be discerned. Damage occors predominantly at the edges of the sheet, where individual carbon atoms can be seen to move around inbetween individual exposures. Fig. 1 shows graphene membranes prepared across 1 μm holes in a carbon film. Fig. 2 shows atomic resolution images of graphene samples. Fig. 2a shows a single- vs. a two-layer graphene membrane in one image for comparison. Since single-layer graphene is only half a unit cell in the c-direction of graphite, it has a unique signature both in the direct image as well as in a diffraction pattern. Fig. 2b shows an adatom that is centered between the positions of two carbon atoms of the graphene lattice, as is expected for a carbon adatom [4]. We expect that a controlled placement of atoms and molecules on graphene, in combination with aberration-corrected electron microscopy, will open new avenues to observe chemical interactions or structural modifications of low contrast molecules or nanoobjects.

M. Luysberg, K. Tillmann, T. Weirich (Eds.): EMC 2008, Vol. 1: Instrumentation and Methods, pp. 37–38, DOI: 10.1007/978-3-540-85156-1_19, © Springer-Verlag Berlin Heidelberg 2008

38

1. K. S. Novoselov, Science **306** (2004) p. 666.
2. J. C. Meyer et al., Nature **446** (2007) p. 60.
3. J. C. Meyer et al., Appl. Phys. Lett. in press (to appear March or April 2008).
4. K. Nordlund et al., Phys. Rev. Lett. **77** (1996) p. 699.
5. Acknowledgments: The TEAM project is supported by the Department of Energy, Office of Science, Basic Energy Sciences. NCEM is supported under Contract # DE-AC02-05CH11231. This work was supported by the Director, Office of Energy Research, Office of Basic Energy Sciences, Materials Sciences and Engineering Division, of the U.S. Department of Energy under contract No. DE-AC02-05CH11231. AZ gratefully acknowledges support from the Miller Institute of Basic Research in Science.

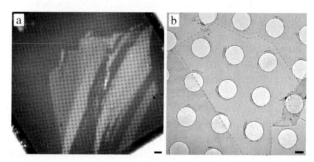

Figure 1. (a) Optical micrograph of graphene sheets on a perforated carbon film (c-flat, 1 μm hole size), and (b) low magnification TEM view of a graphene sample. The single-layer regions are indicated by red dashed lines. Scale bars: (a) 2 μm, (b) 500 nm.

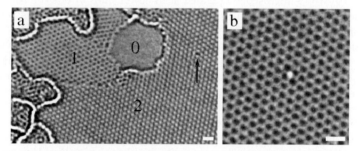

Figure 2. Atomic resolution images of graphene samples, obtained with the TEAM 0.5 microscope at 80 kV, Cs=-15μm. Atoms appear white. (a) Sample showing a zero- (vacuum), one- and two-layer graphene area. Arrow points to a vacancy. (b) Adatom on a single-layer graphene membrane. The hexagonal arrangement of the carbon atoms is indicated by the red hexagon. The adatom is centered between two carbon atoms of the graphene lattice, and induces a small distortion into it. Scale bars are 0.5 nm.

Scanning confocal electron microscopy in a double aberration corrected transmission electron microscope

P.D. Nellist[1], E.C. Cosgriff[1], G. Behan[1], A.I. Kirkland[1], A.J. D'Alfonso[2], S.D. Findlay[2*] and L.J. Allen[2]

1. Department of Materials, University of Oxford, Parks Road, Oxford, OX1 3PH, UK
2. School of Physics, University of Melbourne, Victoria 3010, Australia
* Now at School of Engineering, University of Tokyo, Yayoi 2-11-16, Bunkyo-ku, Tokyo, 113-8656, Japan

peter.nellist@materials.ox.ac.uk

Keywords: three-dimensional imaging, confocal microscopy, aberration correction

The development of spherical aberration correctors has allowed the numerical aperture of electron lenses to be increased by typically 2-4 times, allowing significant improvements in resolution. The depth of field of an image is inversely proportional to the square of the numerical aperture, and therefore the reduction in depth of field in aberration corrected microscopes has been dramatic. The depth of field of a spherical aberration corrected scanning transmission electron microscope (STEM) may only be a few nm, creating an opportunity to perform nanoscale optical sectioning, as has previously been demonstrated [1, 2]. The optical transfer function (OTF) for incoherent STEM optical sectioning (Figure 1a), however, shows a large missing cone region which severely restricts the depth resolution for extended objects.

By using a TEM fitted with both pre- and post-specimen aberration correctors it is possible to generate confocal trajectories [3], similar to those used in the ubiquitous confocal scanning optical microscope, to create an aberration-corrected scanning confocal electron microscope (SCEM). The addition of a post-specimen lens further reduces the contribution of scattering away from the focal plane (Figure 2). If we assume the scattering to be completely incoherent then the SCEM OTF (Figure 1b) does not contain a missing cone. Therefore SCEM maintains good depth resolution even for extended objects. In practice, SCEM imaging using elastically scattered electrons is predominantly coherent and imaging using inelastically scattered electrons shows partial coherence effects. Theory and calculations of SCEM image contrast will be presented.

The accurate mutual alignment of both the pre- and post-sample aberration correctors is crucial [3], and methods to achieve this will be described. In the absence of a sample, a beam cross-over is observed in the detector plane that is the image of a probe at the sample (Fig. 3). Experiments on SCEM optical sectioning are currently under way and latest experimental results will be presented.

1. A.Y. Borisevich, A.R. Lupini, and S.J. Pennycook, *Proc. Natl. Acad. Sci.* **103** (2006) 3044.
2. K. Van Benthem, et al., *Appl. Phys. Lett.* **87** (2005) 034104.
3. P.D. Nellist, et al., *Appl. Phys. Lett.* **89** (2006) 124105.
4. The authors would like to acknowledge the support of JEOL UK, the EPSRC, The Department of Materials at the University of Oxford and the Australian Research Council.

M. Luysberg, K. Tillmann, T. Weirich (Eds.): EMC 2008, Vol. 1: Instrumentation and Methods, pp. 39–40, DOI: 10.1007/978-3-540-85156-1_20, © Springer-Verlag Berlin Heidelberg 2008

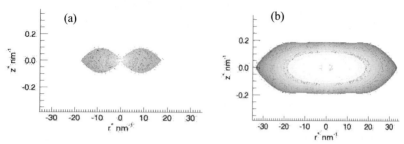

Figure 1. The log of the OTFs for incoherent (a) STEM and (b) SCEM imaging for the Oxford-JEOL 2200MCO instrument (200 kV; beam semiangle of convergence 22 mrad). The OTFs are plotted as a function of the reciprocal radial, r^*, and reciprocal longitudinal, z^*, coordinates. Note the expanded z^* scale.

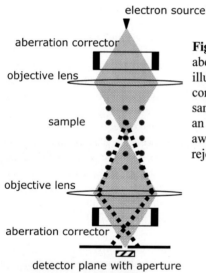

Figure 2. A schematic diagram of an aberration-corrected SCEM. The sample is illuminated in a similar way to an aberration-corrected STEM. Aberration-corrected post-sample optics then image the focused spot to an aperture in the detector plane. Scattering away from the confocal plane is further rejected by the detector plane aperture.

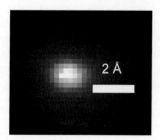

Figure 3. The image formed in the detector plane with the Oxford-JEOL 2200MCO instrument in SCEM mode. The scale marker corresponds to the magnification setting of the lower column. The effective FWHM of the probe image formed is 1.1 Å.

Design of apochromatic TEM composed of usual round lenses

S. Nomura

Nomura Electron Optics Laboratory, 2196-414 Hirai, Hinode-machi Nishitama-gun, Tokyo-to 190-0182 Japan

nomurasetsuo @yahoo.co.jp
Keywords: TEM, aberration correction, aplanatic, achromatic

The article proposes an apochromatic TEM composed of usual round lenses. The TEM has the following features. (1) The number of elements to be controlled is quite a few: additional installing of three round lenses changes the conventional TEM to the apochromatic one. (2) All the lenses are magnetic, providing aplanatic ($Cs=0$) and also achromatic ($\Delta f=0$) high voltage electron microscopes without concerning electrical break down difficulties.

Premising applications to "weak lenses", Scherzer derived the equation describing that round lenses can never be aplanatic. He states the premise by describing the calculation processes in detail [1]. He made the partial integration to calculate the spherical aberration at the image. Use of the partial integration convenience requires that the lens is so weak that zi, the image plane coordinate, is outside the lens. The partial integration is carried out, for example, by equating $\left[B^2(z)r'(z)r^3(z)\right]_{zo}^{zi} = 0$ for a magnetic

lens, where $r(z)$ is the electrons trajectory [2]. To satisfy the eqution, zo and zi must be outside the lens, where $B^2(zi) = B^2(zo) = 0$. The TEM is a strong lens whose object is in the magnetic field. For a "strong" lens, accurate aberrations are figured out only by solving the trajectory equation without making the partial integration. The method has been made in designing the light-optical instruments; the computer ray tracing.

When a lens is installed in front of the objective lens and the lens (Cs-correction lens) is excited so strong to produce a beam crossover inside, the lens produces the spherical aberration used for compensating the aberration produced by the objective lens. Figure 1 illustrates the situation. The figure draws the schematics of the compound lens and the computed trajectory $r(z)$ of electrons entering from the left. The excitation of 22 A/√V makes a beam crossover at z_{crs} ,which is near the lens center. The electrons go further changing their inclinations by the strong lens field. They leave the Cs-correction lens, travel straightforward, and enter the objective lens. The lens focuses the electrons onto zi. The objective lens produces the image of $r(zo')$. The size is $r(zo')\times M_{obj}$, where M_{obj} is the magnification of the objective lens focusing action. The objective lens, on the other hand, produces the spherical aberration $Cs_{obj}\beta^3$. Then the distance of the ray from the axis at zi is expressed as $r(zi)=M_{obj} \cdot r(zo')+Cs_{obj}\beta^3$. The amounts, $M_{obj} \cdot r(zo')$ and $Cs_{obj}\beta^3$ are those having different signs as shown in the figure. Then we have a chance of reducing $r(zi)$ to zero by changing the Cs-correction lens excitation $Excs$, introducing changes in zo' and M_{obj} and finding suitable M_{obj}.

M. Luysberg, K. Tillmann, T. Weirich (Eds.): EMC 2008, Vol. 1: Instrumentation and Methods, pp. 41–42, DOI: 10.1007/978-3-540-85156-1_21, © Springer-Verlag Berlin Heidelberg 2008

42

When the optical system is made aplanatic by using a compound lens, the chromatic aberration increases inevitably. The inreased aberration coefficient Cc, however, is reduced to the value close to Cc_{obj} by making the compound lens achromatic, where Cc_{obj} is the coefficient intrinsic to the objective lens. The achromatizing can be made by installing a Cc-correction lens in front of the Cs-correction lens and exciting the lens to harmonize the focal lengths of two rays different in energy.

Figure 2 depicts the schematics of the apochromatic 100 kV-TEM designed by the above considerations . The specimen is placed at zo=0.0 mm. Figure 3 plots Cs andΔf vs.$Excs$. When the Cs-correction lens is excited at $Excs$=6.56

A/√V, the TEM becomes apochromatic (Cs=0 and Δf=0). In Figure 4 the point spread functios, PSF, of the usual (uncorrected), aplanatic and apochromatic TEMs designed present are depicted. Table I describes the optical parameters of the TEMs.

1. Scherzer, Z. Physik, 101(1936) 593.
2. P. Grivet, "Electron Optics", Pergamon press(1965), Oxford, London, New York.

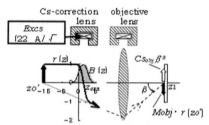

Figure 1. Preparation of the spherical aberration for use to compensate the objective lens aberration.

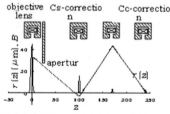

Figure 2. Schematic drawings of the apochromatic 100 kV TEM and the ray path emerged from the object at zo=0.0 mm with the inclination of 32 mrad.

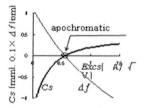

Figure 3. Focal length difference Δf (100 keV ±0.3 eV) and Cs vs. Cs-correction lens excitation $Excs$.

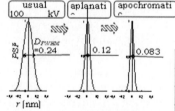

Figure 4. Comparison of PSF among the 100 kV-TEMs designed present.

Table I . Comparisons of Cs, Cc and D_{FWHM} among the TEMs.

	uncorrected	aplanatic	apochromatic
Cs	0.36 mm	0.00	0.00
Cc	-0.72 mm	-1.4	-0.78
D_{FWHM}	0.24 nm	0.12	0.083

Back-Scattered Electron microscopy in Aberration corrected Electron microscope

E. Okunishi, Y. Kondo, H. Sawada, N. Endo, A. Yasuhara, H. Endo, M. Terao, and T. Shinpo

Electron Optics Division, JEOL Ltd.

okunishi@jeol.co.jp

Keywords: Back-Scattered Electron, Aberration correction, STEM, HAADF, Z-contrast

An atomic sized probe with large current, provided by spherical aberration corrector, allows us to determine a site of atom column and specify its atomic species using electron energy loss spectrometer in recent analytical microscope [1-3]. Z contrast in scanning transmission electron microscopy (STEM) is commonly obtained by high angle annular dark field electron microscopy (HAADF). On the other hand, back-scattered electron (BSE) microscopy also provides Z contrast, which is commonly used in scanning electron microscopy for bulk specimen. The investigations of BSE imaging, with aberration corrected microscope, have been remained because of low yield of BSE. We report BSE imaging with a with large probe current provided by aberration corrected microscope.

The experiments were carried out using a JEM-2100F (JEOL), equipped with a STEM Cs corrector (CEOS GmbH) and a parallel EELS (Gatan). Back-scattered electrons were detected by Micro Channel Plate (MCP), which is located between an objective lens and a condenser mini lens. A specimen used in this experiment was $SrTiO_3$ single crystal thinned by argon ion milling. BSE and HAADF images were acquired simultaneously to compare under the same condition. The probe size and current were estimated to be 0.2 nm and 500 pA, respectively.

BSE and HAADF images from thin and thick samples were observed. BSE images shown in Figs. 1 (a) and 2 (a) were improved using the radial difference filter (developed by HREM Research Inc.) to reduce the background due to detection noise. The atomic column was clearly observed in BSE image. In Figs. 1 (a) and 2 (a), bright and dark dots corresponds to strontium and titanium atom columns respectively, judging from the contrast in HAADF images shown in Figs. 1 (b) and 2 (b).

The inset values (t/λ) in Figs 1 (c) and 2 (c) are the ratio of sample thickness to mean free path length measured from with EEL spectra shown in Figs. 1(c) and 2(c). Both sample thicknesses are roughly estimated to be 27 and 600 nm. It indicates that a BSE image reveals Z contrast in fairly thick specimen as well as a HAADF image. The lower panels located under BSE and HAADF images in Figs. 1(a), 1(b), 2(a) and 2(b) show the averaged profiles of image signals along the direction of the arrow shown in Fig. 1(a). The higher peak corresponds to sum of signal from strontium and oxygen, and the lower peak corresponds to that of titanium and oxygen signals as shown in Fig. 1 (b). The Z contrast was clearly detected in the BSE images similar to one in the HAADF images. The contrast degreased for thicker sample, which is the same tendency as

M. Luysberg, K. Tillmann, T. Weirich (Eds.): EMC 2008, Vol. 1: Instrumentation and Methods, pp. 43–44, DOI: 10.1007/978-3-540-85156-1_22, © Springer-Verlag Berlin Heidelberg 2008

HAADF images. This work showed the possibility of atomic resolution back-scattered electron microscopy.

1. E. Okunishi et al.: Microsc. & Microanl. **12** 1150 (2006)
2. D. Muller et al. : Science **319** 1073 (2008)
3. M. Bossman et al.: PRL **99** 086102 (2007)

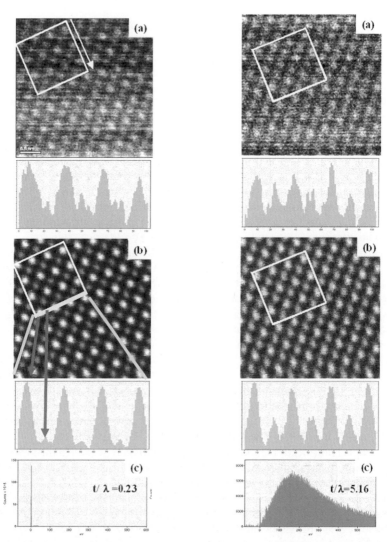

Figure 1. (a) BSE image, (b) HAADF image and (c) EELS spectrum of of thin SrTiO₃

Figure 2. (a) BSE image, (b) HAADF image and (c) EELS spectrum of of thick SrTiO₃

Structure determination of H-encapsulating clathrate compounds in aberration-corrected STEM

Q.M. Ramasse[1], N.L. Okamoto[2], D. Morgan[2], D. Neiner[3], C.L. Condron[3], J. Wang[3], P. Yu[3], N.D. Browning[2, 4] and S. Kauzlarich[3]

1. Lawrence Berkeley National Laboratory, National Centre for Electron Microscopy, 1 Cyclotron Road, Berkeley CA 94720, U.S.A.
2. Dept. of Chemical Engineering and Materials Science
3. Dept. of Chemistry, University of California-Davis, Davis CA 95616, U.S.A.
4. Livermore National Laboratory, 7000 East Avenue, Livermore CA 94550, U.S.A.

QMRamasse@lbl.gov

Keywords: STEM, aberration correction Z-contrast, image analysis, hydrogen storage, crystallography, TEAM

In the quest for alternative energy sources, hydrogen has long been touted as one of the most likely candidates to replace fossil fuels, provided practical solutions for its storage and transport can be found. Clathrate hydrates, amongst other nano-porous materials, have shown remarkable potential for hydrogen storage, adsorbing up to 7.5%wt H, albeit in extreme pressure and temperature conditions [1]. Upon crystallising, these compounds form a three-dimensional host matrix where guest atoms can be accommodated within distinct polyhedral "cages". Recent work exhibited a sodium silicide clathrate Na_xSi_{46}, consisting of a silicon matrix with two distinct Na guest sites, labelled *2a* and *6d*, stable at room temperature and with promising hydrogen encapsulation properties [2]. Magic angle spinning nuclear magnetic resonance was initially used to show that in the growth conditions specific to this work some of the *6d* sites were Na-deficient and H-rich.

When observed in (001) projection, the 6d guest site, at the centre of the larger of two polyhedral cages, is nicely isolated from neighbouring atoms. Thanks to the angstrom-sized probes obtained on an aberration-corrected VG HB501 dedicated STEM, it was possible to determine unequivocally its occupancy through quantitative comparison of Z-contrast images with simulations: fig. 1. In the case of Na_xSi_{46}, it emerged that a third of the *6d* sites were Na-deficient consistent with the $Na_4(H_2)_2Si_{46}$ stoichiometry derived independently from EDX analyses of the compound.

The same methodology was applied to determine the precise structure and stoichiometry of more complex cases, where different sites exhibit different levels of guest deficiencies. In the case of the K_xSi_{46} type-I clathrate for instance, rotating the structure into several other zone axes such as (111), see figure 2, and (110), to project the *2a* and *6d* sites into different positions revealed that both sites were indeed K-deficient. A careful crystallographic averaging of the experimental images was performed to insure the most quantitative match with the simulations [3]. The world-class capabilities of the new TEAM microscope [4, 5] were also put to the test for higher order zone axes where a half-angstrom probe and very short depth of focus became necessary to obtain a

M. Luysberg, K. Tillmann, T. Weirich (Eds.): EMC 2008, Vol. 1: Instrumentation and Methods, pp. 45–46, DOI: 10.1007/978-3-540-85156-1_23, © Springer-Verlag Berlin Heidelberg 2008

46

good separation between the various sites: see figs 1 and 2. This general methodology of pseudo three-dimensional structure determination is currently being extended to other structures: type III clathrates, and Ge/Ga-based compounds.

1. G. Antek *et Al.*, J. Am. Chem. Soc., **128** (2006), pp. 3494–3495.
2. D. Neiner *et Al.*, J. Am. Chem. Soc., **129** (45) (2007), pp. 13857–13862.
3. 2dx Software Package. B. Gibson et Al., J. Struct. Biology, **157** (2007) pp. 67–72
4. Dr. B. Freitag's expertise must be gratefully acknowledged for the images obtained on the TEAM 0.5 microscope during testing at the FEI factory, Eindhoven.
5. C. Kisielowski et Al., submitted (2008).
6. This work was carried out in part at the National Centre for Electron Microscopy, under the auspices of the U.S. Department of Energy under contract number DE-AC02-05CH11231 and DE-FG02-03ER46057 and was supported by the Centre of Excellence for Chemical Hydrides under contract number DE-FC36-05GO15055

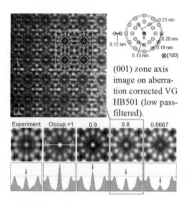 

Figure 1. Determining the *6d* site occupancy (coloured in red) through comparisons with image simulations. Left: image taken on an aberration-corrected VG HB501 (low pass filtered for clarity); right: unprocessed image of a neighbouring region, obtained at 300kV on the TEAM 0.5 microscope, showing the stability of the structure.

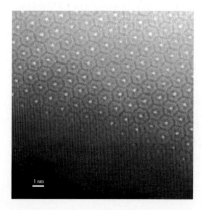

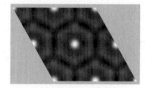

Figure 2. Left: experimental image of the (111) projection of K_xSi_{46} obtained on the TEAM 0.5 microscope [4]; right: crystallographic average performed with the 2dx software package [3].

Performance of R005 Microscope and Aberration Correction System

H. Sawada[1,3], F. Hosokawa[1,2], T. Kaneyama[1,2], T. Tomita[1,2], Y. Kondo[1,2], T. Tanaka[1,3], Y. Oshima[1,3], Y. Tanishiro[1,3], N. Yamamoto and K. Takayanagi[1,3]

1 CREST, Japan Science and Technology Agency, 4-1-8 Honcho, Kawaguchi, Saitama 332-0012, Japan
2 JEOL Ltd., 3-1-2 Musashino, Akishima, Tokyo, 196-8558, Japan
3 Tokyo Institute of Technology, 2-12-1-H-51 Oh-okayama, Meguro-ku, Tokyo 152-8551, Japan

hsawada@jeol.co.jp

Keywords: Half angstrom (50 pm), Aberration correction, Electron microscope, Cold field emission gun, High angle annular dark field image

To achieve 50 pm resolution (R005: Resolution 0.05 nm), we have been developing a 300 kV high-resolution scanning transmission electron microscope (STEM)/transmission electron microscope (TEM) equipped with a cold-field emission gun (FEG) (Fig.1(a))[1]. Energy spread (FWHM) of the cold-FEG was measured to be 0.32 eV (Fig.1(b)). Spherical aberration correction systems have been incorporated into STEM and TEM systems of the microscope. The aberration corrector is based on a hexapole-type correction system, which has asymmetrical dodeca-poles [2]. We developed a method to measure aberrations for a STEM corrector. In the method, the aberrations are obtained from sets of auto-correlation functions on Ronchigram pattern of an amorphous specimen. For the TEM corrector, a diffractogram tableau is used to measure the aberration of image forming system. Aberrations up to the fifth order were measured and the aberrations are automatically adjusted using the newly developed software in both TEM and STEM systems.

Using the cold-FEG, we observed a high angle annular dark field (HAADF) image of a GaN crystal [3]. Each pair of bright dots like dumbbells in Figs.2 (a) and (b) correspond to two Ga atomic columns projected along the [211] direction. 63 pm distance between Ga atomic columns was resolved at a HAADF image of a [211] GaN crystalline specimen (Fig. 2(c)) and its Fourier transform confirmed not only the ,306 spot corresponding to 63 pm spacing but also the ,417 spot corresponding to 53 pm spacing (Fig.2(d-f)).

This work was supported by the Japan Science and Technology Agency (JST) under the CREST project.

1. CREST project (development of 0.05nm resolution EM for materials analysis),
2. http://www.busshitu.jst.go.jp/kadai/year01/team05.html
3. F. Hosokawa, K. Takayanagi, et. al, Proc. IMC16 (2006) 582.
4. H.Sawada, et al., Jpn. J.Appl. Phys. 46(2007) L568.

M. Luysberg, K. Tillmann, T. Weirich (Eds.): EMC 2008, Vol. 1: Instrumentation and Methods, pp. 47–48, DOI: 10.1007/978-3-540-85156-1_24, © Springer-Verlag Berlin Heidelberg 2008

48

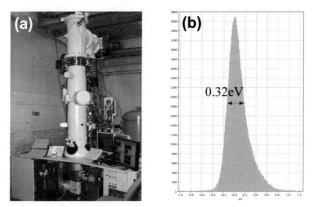

Figure 1. (a) Appearance of Spherical Aberration Corrected 300kV FETEM (R005). (b) Energy spread (FWHM) of the cold-FEG (0.05s acquisition).

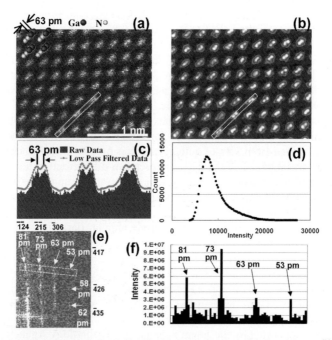

Figure 2. HAADF image of GaN [211]. (a) Raw image. (b) Low pass filtered image. Scheme of the atomic structure of GaN viewed from the [211] direction are superimposed on the upper left part of Fig.2 (a). (c) Intensity profiles from the white frames in Figs.2 (a) and (b). (d) Intensity histogram from Fig.2 (a). (e) Fourier transform pattern of Fig.1 (a). (f) Intensity profile from the frame in Fig.2 (e).

Optimum operation of Schottky electron sources: brightness, energy spread and stability

P. Kruit[1], M.S. Bronsgeest[1] and G.A. Schwind[2]

1 Delft University of Technology, Lorentzweg 1, 2628CJ, Delft, The Netherlands.
2 FEI company, Hillsboro, Oregon 97124.

p.kruit@tudelft.nl

Keywords: electron sources, brightness, energy spread, Coulomb interactions, ring collapses

Many modern electron microscopes have a Schottky electron source, named after the Schottky effect, that is the effective lowering of the workfunction by the strong electric field at the tip. Practical Schottky emitters consist of a sharp W tip of typical radius 500nm, covered by ZrO_2 and heated to 1800K [1]. These emitters are popular because they combine a high brightness and low energy spread with good current stability and a long life time.

There is a subtle interplay between these parameters: one obtains a higher brightness at higher temperature and extraction potential. However, this is at the cost of a larger energy spread and possibly a lower lifetime because the tip may sharpen and crystallize at very high fields. At the other side of the scale, at the minimum energy spread, the risk is that the tip dulls through a "ring collapse" process.

The intrinsic brightness and energy spread as a function of temperature and field can be predicted from emission theory with a reasonable accuracy.

However, the practical brightness and energy spread is influenced by stochastic Coulomb interactions [2,3], which are notoriously difficult to calculate in the accelerating field of sharp emitter tips.

We have compared experimental values with approximate theories, both for brightness [4] and for energy spread [5], see figures 1-3. The results can be used for the optimum choice of emitter form and operation parameters.

It is well known that the form of the tip can change under the influence of temperature and field [1,6], thus in turn affecting the beam properties. Figure 4 gives a few encountered end forms. We have studied these form changes both theoretically and experimentally with the aim of distinguishing between reversible and irreversible form changes and hopefully finding parameter regimes within which the tips can be operated safely, yet at high brightness and low energy spread.

1. L.W. Swanson and G.A.Schwind, in *Handbook of Charged Particle Optics*, edited by J.Orloff (CRC New York, 1997), Chapter 2.
2. H.Boersch, Z. Phys. **139**, 1954, p115
3. P.Kruit and G.H. Jansen, in *Handbook of Charged Particle Optics*, Chapter 7.
4. A.H.V.van Veen etal, J.Vac.Sci.Technol. B **19**, 2001, p2038
5. M.S.Bronsgeest, J.E.Barth, G.A.Schwind, L.W.Swanson and P.Kruit, J.Vac.Sci.Technol.B **25**, 2007, p2049
6. S.Fujita and H.Shimoyama, Phys.Rev.B **75**, 2007, 235431

M. Luysberg, K. Tillmann, T. Weirich (Eds.): EMC 2008, Vol. 1: Instrumentation and Methods, pp. 49–50, DOI: 10.1007/978-3-540-85156-1_25, © Springer-Verlag Berlin Heidelberg 2008

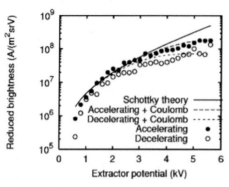

Figure 1. Brightness of the Schottky electron emitter, showing the influence of Coulomb interactions at high currents (from [4]).

Figure 2. Energy broadening caused by Boersch effect for different tip radii (from [5]).

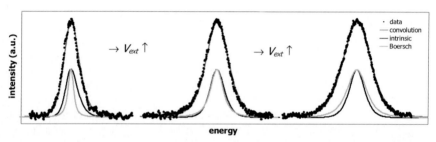

Figure 3. Energy spectra from 550nm tip radius operated at 1800K at (from left to right) 0.2, 0.6, 1.0 mA/sr, showing the relative contributions from intrinsic energyspread and Boersch effect. Note how the form of the distribution also changes (from [5]).

Figure 4. The shape of Schottky emitter tips varies widely after operation. These examples are the result of operation outside of the manufacturers recommended operating conditions.

The MANDOLINE filter and its performance

E. Essers, D. Mittmann, T. Mandler and G. Benner

Carl Zeiss NTS GmbH, Carl-Zeiss-Str. 56, 73447 Oberkochen, Germany

essers@smt.zeiss.com

Keywords: EELS, MANDOLINE filter, SESAM

The MANDOLINE filter, proposed by Uhlemann and Rose [1], was developed within the SESAM (Sub-Electronvolt Sub-Ångström Microscope) project [2]. This highly sophisticated in-column imaging energy filter recently passed its final acceptance tests, surpassing all its specifications considerably.

The analytical performance of the MANDOLINE filter is seen in the Zeiss SESAM microscope, which furthermore includes a highly stable 200 kV high voltage supply, the CEOS electrostatic Omega-type monochromator [3,4] and the highly stable current supply [5] of the filter.

Figure 1 shows the main components of the MANDOLINE filter. Nine 12-pole elements enable the correction of second- and third-order aberrations.

The EEL spectra in Figures 2 and 3 are taken with a 45 meV monochromator energy-selection slit width. For an exposure time of 30 ms, Figure 2 shows only 50.8 meV energy resolution (FWHM) of the zero-loss peak. This demonstrates the excellent stability of both the high voltage and the MANDOLINE filter current. In Figure 3, the energy resolution (FWHM) for an exposure time of 1 s is measured to be only 62 meV. Furthermore the drop of the zero-loss peak is 10^{-1} after 67 meV, 10^{-2} after 175 meV and 10^{-3} after 410 meV. Concerning the energy resolution (and drop of the zero-loss peak) further strong improvements are expected for the scheduled next type of high voltage supply. It will also further improve the reproducibility and thereby make accessible an energy resolution even below the width of the monochromator energy-selection slit by the use of deconvolution techniques.

In the Si [111] CBED pattern of Figure 4, diffraction angles > 83 mrad are transferred at only $\Delta E = 410$ meV energy slit width. These large angles and the associated spherical aberration disc (introduced by $C_{S,Objective} = 1.2$ mm) can be transferred at the same time due to the exceptionally high transmissivity of the MANDOLINE filter even at small energy slit widths down to 200 meV. This allows for Electron-Spectroscopic Imaging of large fields of view with small energy windows and Spectrum Imaging investigations in the TEM-Mode with high energy resolution for applications in the low-loss range.

The measurement of the transmissivity and the isochromaticity will be explained.

1. S. Uhlemann, H. Rose, Optik **96** (1994), p. 163.
2. M. Rühle et al., Microsc. Microanal. **6**, Suppl. 2: Proceedings (2000), p. 188.
3. F. Kahl, H. Rose, Proc.11th Europ.Cong.Electron Microsc. Vol.1 (1996), p. 478.
4. S. Uhlemann, M. Haider, Proc.15th Int.Cong.Electron Microsc. Vol.3 (2002), p. 32.
5. The MANDOLINE filters current supply was developed by R. Höschen, MPI Stuttgart.

M. Luysberg, K. Tillmann, T. Weirich (Eds.): EMC 2008, Vol. 1: Instrumentation and Methods, pp. 51–52, DOI: 10.1007/978-3-540-85156-1_26, © Springer-Verlag Berlin Heidelberg 2008

Figure 1. Schematic drawing of the MANDOLINE filter. The path of the optical axis intersects itself twice within the first magnet, which has a homogeneous magnetic field. The other two magnets have inhomogeneous magnetic fields, focussing the ray continuously in both principal sections. While electrons with nominal energy pass the slit aperture along the optical axis, electrons with an energy loss are kept back by the slit aperture.

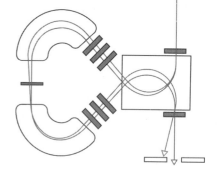

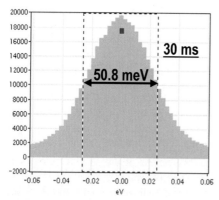

Figure 2. EEL spectrum of the zero-loss peak showing 50.8 meV energy resolution for 30 ms exposure time.

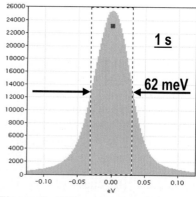

Figure 3. EEL spectrum of the zero-loss peak showing 62 meV energy resolution for 1 s exposure time.

Figure 4. CBED pattern of Si [111] taken at $\Delta E = 410$ meV energy slit width. Diffraction angles > 83 mrad are transferred in all directions with the FOLZ ring at 73 mrad well included even at this narrow energy slit width. The Distortion can be seen to be well below 1%.

Using a monochromator to improve the resolution in focal-series reconstructed TEM down to 0.5Å

P.C. Tiemeijer[1], M. Bischoff[1], B. Freitag[1], C. Kisielowski[2]

1. FEI Company, PO Box 80066, 5600 KA Eindhoven, The Netherlands
2. National Center for Electron Microscopy, Lawrence Berkeley National Laboratory, One Cyclotron Road, Berkeley, CA 94270, USA

p.tiemeijer@fei.com

Keywords: monochromator, brightness, focal-series reconstruction

The resolution of present-day spherical-aberration corrected TEMs is limited by the chromatic aberration of the objective lens to about 0.7Å at 300kV. The resolution can be improved by Cc correction or by reducing the energy spread in the illumination with a monochromator. We used TEAM 0.5, a special Titan column built within the TEAM project [1], and on this microscope we succeeded to improve the resolution to 0.5Å by monochromation, not only in single images but also in images reconstructed from focal series. Such reconstructed images are free of possible contrast reversals due to incorrect focusing and possible artifacts due to non-linear interferences. However, focal series are more demanding to acquire than single images because of the higher demand on the stability of the column, and because they must be taken over some focus interval and this significantly increases the demands on the parallelness or coherence of the beam.

The coherence of the beam is closely related to brightness of the source. In STEM, brightness determines how much beam current can be squeezed in a small probe for a given convergence angle. Similarly, in TEM, brightness determines how much beam current can be squeezed in a parallel beam for a given coherence angle.

We estimate that, in order to show 0.5Å in a focal series of a typical length of 40nm, the brightness *after monochromation* must be at least ~$3 \cdot 10^8$ A/cm^2/sr. This is not easy to obtain because this brightness is comparable to what a standard Schottky FEG delivers at 300kV and 4kV extraction voltage *before monochromation* (the brightness after monochromation is at least a factor 3 to 10 lower than the brightness before monochromation because the monochromator removes part of the energy spectrum, and the source can be blurred by aberrations or Coulomb interactions in the monochromator). We used a prototype 'high brightness' gun (currently being developed within the TEAM project) and a monochromator designed for minimum brightness loss [2], and we obtained an exceptionally high brightness >$5 \cdot 10^8$ A/cm^2/sr at $\Delta E = 0.13$eV.

In our set-up, the source was focused on the specimen to a probe of about 100nm diameter and 0.1mrad convergence angle, and this probe was dispersed by the monochromator to a line of about 1µm length, as shown in Figure 1. In this way, the non-roundness of the dispersed probe is only present in the specimen plane, and not in the reciprocal plane. This ensures that the power spectra of the images are round, not only close to focus but also at the end of the focal series, as demonstrated in Figure 2. Figure 3 shows an example of an exit wave reconstructed from a focal-series of 30 steps at 0.9nm focus intervals. The Fourier Transform of its phase shows information down to 0.5Å. [3].

M. Luysberg, K. Tillmann, T. Weirich (Eds.): EMC 2008, Vol. 1: Instrumentation and Methods, pp. 53–54, DOI: 10.1007/978-3-540-85156-1_27, © Springer-Verlag Berlin Heidelberg 2008

54

1. http://ncem.lbl.gov/TEAM-project/
2. P.C. Tiemeijer, Ultramicroscopy **78** (1999), p. 53.
3. The TEAM project is supported by the Department of Energy, Office of Science, Basic Energy Sciences.

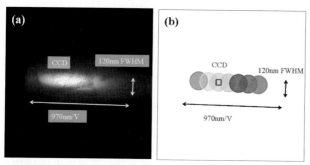

Figure 1. The dispersed probe is focused on the specimen. In this example, the convergence was α=0.09mrad RMS and J=20A/cm^2 on the area sampled by the CCD.

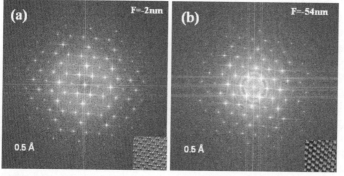

Figure 2. Fourier transform of images of Ge<110>. Settings: ΔE=0.13eV, J=40A/cm2, α=0.13mrad RMS, 1s exposure. (a) 2nm underfocus, (b) 54nm underfocus.

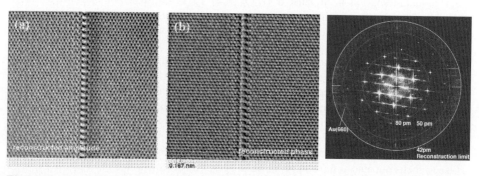

Figure 3. Focal-series reconstruction of a grain boundary in Au<111> (ΔdE=0.13eV, J=90A/cm2, α=0.18mrad RMS, 1s exposure). (a) Amplitude, (b) phase, (c) FT of phase.

First performance measurements and application results of a new high brightness Schottky field emitter for HR-S/TEM at 80-300kV acceleration voltage

B. Freitag[1], G Knippels[1], S. Kujawa[1], P.C. Tiemeijer[1], M. Van der Stam[1], D. Hubert[1], C. Kisielowski[2], P. Denes[2], A. Minor[2] and U. Dahmen[2]

1. FEI Company, Building AAE, Achtseweg Noord 5, Eindhoven, The Netherlands
2. National Center for Electron Microscopy, Lawrence Berkeley National Laboratory, California

Bert.Freitag@Fei.com

Keywords: HR-S/TEM, field emitter, brightness

The performance of a transmission electron microscope is determined by the optical performance of the lenses, the stability of the column and last but not least by the performance of the electron source.

Over the last years great improvements in electron optics due to spherical aberration correction [1] and monochromators [2] and in the stability of the Titan80-300 column have led to Sub-Ångström imaging performance in HR-S/TEM [3]. For a desired SNR in the HR-TEM image the brightness is limiting the spatial coherence of an ideal illumination system. Consequently the improvement of the electron source is an evolutionary next step to explore the frontier in electron microscopy. For the TEAM project [4], which is based on the Titan platform, a significant increase in brightness over the conventional Schottky field emitter was needed in order to achieve the 0.5Å resolution target.

The Schottky field emitter is designed to deliver high brightness and a relatively high total current with a good energy resolution for TEM and STEM application. Both the brightness and the total current are of importance for a flexible TEM/STEM column to cover the entire range of applications a modern electron microscope can provide. Especially in TEM the brightness is not the only performance parameter for mid-range magnification applications (EFTEM, Lorentz microscopy, etc.) but the total current of the source becomes one of the dominant performance parameters. Moreover the lifetime and long-term stability of electron emission without the need for `flashing` the tip are important and a prerequisite for complex optical or time consuming experiments, high throughput applications and ease of use. Nevertheless higher brightness is required to explore the limits in atomic resolution imaging or to gain speed and resolution in spectroscopy or dynamic experiment applications. This complex mix of requirements has led to the development of a new ultra stable Schottky emitter design with higher performance for S/TEM applications.

The new Schottky field emitter delivers a significantly higher brightness in combination with still a large total current. No compromise in the long time and short time emission stability of better than 1% has been made. This performance translates to ultra high current values in Ångström probe size on a probe Cs-corrected Titan at

M. Luysberg, K. Tillmann, T. Weirich (Eds.): EMC 2008, Vol. 1: Instrumentation and Methods, pp. 55–56, DOI: 10.1007/978-3-540-85156-1_28, © Springer-Verlag Berlin Heidelberg 2008

300kV acceleration voltage. As a result, the exposure times in monochromized HR-TEM applications with the new design are better than in non monochromized HR-TEM imaging using the standard Schottky field emitter. Using the ultra stable TEAM project columns as a platform [4], the present contribution demonstrates the performance of this new emitter with a number of applications to different materials (see figs.1,2).

1. M. Haider, et. al, Ultramicroscopy 81 (2000) 163
2. P.C. Tiemeijer, Inst. Phys. Conf. Ser. 161 (1999) 191.
3. S. Kujawa, B. Freitag, D. Hubert, Microsc. Today 13 (2005) 16
4. http://ncem.lbl.gov/TEAM-project/index.html

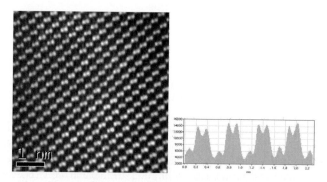

Figure 1. Monochromized Cs-corrected HR-TEM image of Ge<110> at 80kV. The dumbbell structure can be clearly resolved. The exposure time is only 1s. The intensity profile shows the germanium dumbbell distance clearly resolved with a count rate of 12000 counts/pixel in the image acquired with the monochromator switched on ($\Delta E=0.2eV$).

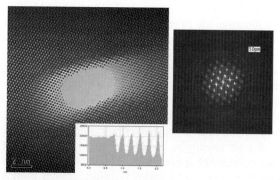

Figure 2. Monochromized Cs-corrected HR-TEM image of Au<110> at 300kV. Fourier components up to 50pm are transmitted. The exposure time is only 1s. The intensity profile shows a count rate of 15000 counts/pixel and the high contrast level in the image acquired with the monochromator switched on ($\Delta E=0.2eV$).

Image Information transfer through a post-column energy filter detected by a lens-coupled CCD camera

U. Luecken[1], P. Tiemeijer[1], M. Barfels[2], P. Mooney[2], B. Bailey[2] and D. Agard[3]

1. FEI Electron Optics, Achtseweg Noord 5, 5651 GG Eindhoven, The Netherlands
2. Gatan, Inc., 5794 W. Las Positas Blvd., Pleasanton, CA 94588, USA
3. University of California, San Francisco, San Francisco, CA 94143-2440, USA

uwe.luecken@fei.com
Keywords: TEM, Filtering, CCD Camera

The current work addresses the suitability of post-column energy filtering for high-resolution low-dose work for which information transfer to high-resolution is critical. The primary issue affecting information transfer in a post-column energy filter is the challenge of maintaining alignment of the axes of the two electron-optical systems. Another issue is the performance of the detector. Experiments were performed using a 300kV TF30 with liquid-He stage (FEI Polara) fitted with a 1Kx1K fiber-optically coupled CCD camera (Gatan 794 MultiScan) and a transmission-scintillator lens-coupled back-illuminated 4Kx4K CCD camera-equipped post-column energy filter (Gatan UltraCam with Tridiem optics)[1,2].

Information transfer has traditionally been demonstrated using Thon ring visibility. The present comparison makes use of image spectral signal to noise ratio (SSNR) ([3], [4]), division of the SSNR by detector DQE [2] and dose compensation [2] to provide a more quantitative comparison of electron optical factors above and below the post-column filter alongside a comparison of the detectors and overall system performance. A platinum nano-crystal / amorphous iridium specimen was used to provide high-contrast Thon rings to high resolution with a local magnification reference (FFT of images shown in Figure 1a, b).

It was demonstrated that it was possible to produce a consistent co-alignment through proper management of cross-over position. Pt/Ir Thon rings recorded with the above-GIF and below-GIF CCDs (with a 1K sub area selected from the UltraCam 4Kx4K for comparison) show comparable appearance though with slightly different character due to the different shapes of the DQE curves (Figure 2a). Dividing SSNR curves derived from the Thon ring images by the DQE [2, p. 687] for the above-GIF and below-GIF CCDs removes detector effects and shows a very good match of electron optical envelope across spatial frequencies up to 4/5 of the Nyquist frequency (Figure 2b).

The results shown in Figures 1 to 3 demonstrate transfer of over 3000 independent pixels through the electron optics of the Polara-GIF system into the detector. This is the first time such a transfer has been shown. Figures 3a and 3b show maintenance of strong envelope in a highly defocused state for the Polara-GIF system as well confirming suitability for high-resolution low-dose work.

M. Luysberg, K. Tillmann, T. Weirich (Eds.): EMC 2008, Vol. 1: Instrumentation and Methods, pp. 57–58, DOI: 10.1007/978-3-540-85156-1_29, © Springer-Verlag Berlin Heidelberg 2008

58

1. Bailey, B. et al, Microsc. Microanal. 10(Suppl 2) (2004) 1204.
2. Mooney, P., Meth. in Cell Biol., Vol. 79, Academic Press, San Diego, CA, 2007, pp 661-719.
3. Zhang, P. et al., J. Struct. Biol.143 (2003) 135.
4. Booth, C. R. et al, J. Struct. Biol. 147 (2004) 116.

Figure 1a. 794 Thon rings; 1Å pixels on both detectors at 300kV.

Figure 1b. UltraCam Thon rings with comparable specimen, dose and pixel size.

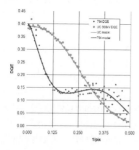

Figure 2a. DQEs of above- and below GIF detectors referenced to respective pixel sizes (794 Nyquist is 20lp/mm, UltraCam Nyquist is 33lp/mm).

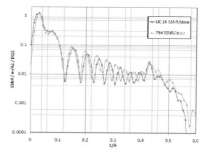

Figure 2b. Compensated SSNRs above and below GIF showing no significant differences in envelope.

Figure 3a. High-defocus SSNR image (marks indicate increments of 10 Thon rings).

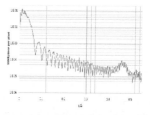

Figure 3b. Profile taken at 230kx and 1μm defocus shows preservation of coherence.

Third-rank computation of electron and ion optical systems with several and rotated Wien filters

Karin Marianowski and Erich Plies

Institute of Applied Physics, University of Tübingen,
Auf der Morgenstelle 10, D-72076 Tübingen, Germany

karin.marianowski@uni-tuebingen.de

Keywords: charged particle optics, Wien filter, third-rank aberrations

The Wien filter (WF) is a well-known element of charged particle optics which is used as beam splitter, velocity analyser, monochromator, imaging energy filter as well as mass separator. It is also proposed as a corrector for spherical and axial chromatic aberrations of round lenses. If the WF consists of homogeneous ExB-fields, i.e. pure dipole fields, it acts as a direct-vision prism with a superimposed cylinder lens. Using an inhomogeneous WF with a certain electric or magnetic quadrupole component it acts like a round lens. While the dispersion of the WF can be cancelled for example by an intermediate image in its mid-plane, the WF, based on dipole fields which normally curve the optic axis, still has second-order aberrations. To correct some of these second-order aberrations or to make a system achromatic, i.e. dispersion-free, two WF [1-3] or four WF [4,5] are used. However, if one corrects the aberrations of second order and the dispersion of first rank, the system is limited by the aberrations of the next order or rank. Therefore it is very much desirable to have a program to calculate the total contribution of these remaining aberrations.

We started from the existing commercial program ABERW (ABERrations of systems containing Wien filters) [6] of Mebs Ltd., London, which is a second-rank code. First this program was extended by Haoning Liu and one of the authors (E.P.) for systems consisting of "normal" Wien filters (E_x, B_y) and Wien filters which are rotated by 90° (E_y, -B_x) using the general theory published in [7]. This additional option carried out in ABERW2 is necessary for a novel highly symmetric monochromator with four WF [8]. In order to check one method against another for higher confidence in computing, in the next step the code was extended in such a way that the total contribution of the second-rank aberrations is computed by another method proposed in [9]. Finally the algorithm of [9] was used to compute the total contribution of the third-rank aberrations. The new extended program does not only enable us to compute this contribution in the energy selection plane or in an image plane but also results in the complete knowledge of the contribution of all second- and all third-rank aberrations along the optic axis. This is extremely helpful to see where aberrations arise and where they are compensated for or corrected. The program can handle not only several homogeneous and inhomogeneous WF ("normal" and rotated) which may be strong (long) or weak (short), but also round lenses, separate multipole elements and rotated multipole elements. Furthermore, there is no fringing field restriction.

To test the overall program, a simple and therefore analytically treatable system consisting of a single infinitely long WF was studied. Thereby, the total contributions of

M. Luysberg, K. Tillmann, T. Weirich (Eds.): EMC 2008, Vol. 1: Instrumentation and Methods,
pp. 59–60, DOI: 10.1007/978-3-540-85156-1_30, © Springer-Verlag Berlin Heidelberg 2008

the second- and third-rank aberrations in the first Gaussian image plane as well as the second-rank fundamental rays along the optic axis were examined.

Additionally, we will present current simulation results on a 4-WF-monochromator for electrons, see Figure 1, and on a novel stigmatic and achromatic mass separator [10] for focussed ion beam applications.

1. J. Teichert and M. A. Tiunov, Meas. Sci. Technol. **4** (1993), 754.
2. M. Tanaka, M. Terauchi, K. Tsuda et al., Microsc. Microanal. **8**, Suppl. 2 (2002), 68.
3. G. Martínez and K. Tsuno, Ultramicroscopy **100** (2004) 105.
4. Y. Kawanami et al., Nucl. Instrum. Meth. Phys. Res. **B 37/38** (1989), 240.
5. X. D. Liu and T. T. Tang, Nucl. Instrum. Meth. Phys. Res. **A 363** (1995), 254.
6. H. Liu, SPIE Vol. **3777** (1999), 92.
7. E. Plies and D. Typke, Z. Naturforsch. **33A** (1978), 1361 and **35A** (1980), 566.
8. E. Plies and J. Bärtle, Microsc. Microanal. **9**, Suppl. 3 (2003), 28.
9. E. Plies, Ultramicroscopy **93** (2002), 305.
10. E. Plies, Proc. 11th International Seminar on Recent Trends in Charged Particle Optics and Surface Physics Instruments, Brno, Czech Republic (2008).
11. We kindly acknowledge the help of Mebs Ltd. in general and Haoning Liu in special.

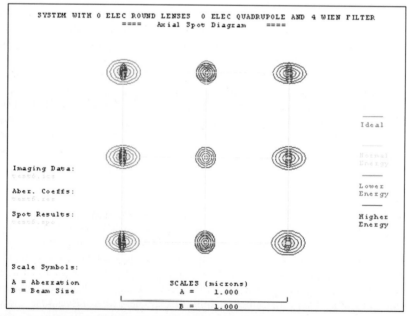

Figure 1. Spot diagrams in the dispersion-free image plane of a 4-WF-monochromator generated by the modified programs ABERW2 and PLOTSW2. Beam energies are 999.95 eV, 1000.00 eV and 1000.05 eV. The aperture is 1.5 mrad.

Wavelength dispersive soft X-ray emission spectroscopy attached to TEM using multi-capirary X-ray lens

S. Muto[1], K. Tatsumi[2] and H. Takahashi[2]

Department of Materials, Physics and Energy Engineering, Nagoya University
Furo-cho, Chikusa-ku, Nagoya 464-8603, Japan

s-mutoh@nucl.nagoya-u.ac.jp

Keywords: wavelength dispersive x-ray spectrometer, soft x-ray emission, multi-capirary

There is now growing interest in localized chemical bonding states in complex metals and insulators such as those containing light elements, rare earths or transition metals because their key spectral features appear in the soft X-ray or vacuum UV regions, which is not easily accessible to by SOR light sources and often includes crucial information to understand their electronic, optical, magnetic or ionic effects in the materials. For this purpose electron energy-loss spectroscopy (EELS) and soft X-ray emission spectroscopy (SXES) attached to TEM are now unique and indispensable tools for nanotechnology fields.

SXES provides the partial DOS (pDOS) of the occupied states, complimentary to EELS, which provides pDOS of the unoccupied states. For TEM-SXES to be realized, the characteristic X-ray emitted from the sample must be dispersed with the energy resolution of ~1 eV or less. A wavelength dispersive X-ray spectrometer (WDXS) using diffracting gratings and CCD was thus recently developed [1]. This WDX spectrometer, however, exhibits worse energy resolution for higher spectral energy regions and effective solid angle collecting the X-ray is very small. On the other hand, the WDX system attached to SEM has been long developed, but its large scale geometrical configuration hampers its implementation to TEM.

We developed a compact WDX system for TEM; as shown in Figures 1(a) and (b), the emitted characteristic X-ray is collected and parallelized by multi-capirary X-ray lens (MCX), recorded by a set of a diffracting crystal and a gas flow-type detector. By exchanging 3-4 diffracting crystals with different lattice spacings the system covers approximately from 150 to 2500 eV with the energy resolution of ~1 eV.

In Figure 2 is shown the experimental oxygen K-line (O 2p → 1s) SXES spectrum from an Al2O3 thin film, together with the O 2p pDOS calculated with the DV-X☐ molecular orbital (MO) method. The experimental spectrum is qualitatively well reproduced by the theoretical prediction. As another example, Ni L-line and O K-line from a NiO polycrystalline thin film is shown in Figures 3(a) and (b). The relative intensity ratio of the L_α to $L_{\beta1}$ peaks provides the valency of the transition metal. The peak split in the O K-line corresponds to the crystal field splitting of the Ni 4d into the e_g and t_{2g} bands. From this splitting the energy resolution of this spectrometer is estimated to be ~1 eV. The MO calculation within the one-electron approximation did not provide correct peak intensity ratio, even by taking the anti-ferromagnetic ordering of Ni into account. We then try a multiplet calculation including the many-body effects.

M. Luysberg, K. Tillmann, T. Weirich (Eds.): EMC 2008, Vol. 1: Instrumentation and Methods, pp. 61–62, DOI: 10.1007/978-3-540-85156-1_31, © Springer-Verlag Berlin Heidelberg 2008

62

We also attempt site-specific electronic structure measurements in a strongly correlated system under the electron channelling condition. Those results will be presented.

1. M. Terauchi and M. Kawana, Ultramicrosc. 106 (2006) 1069.
2. The MCX-WDXS for TEM was developed in collaboration with Dr. H. Soejima and Mr. T. Kitamura of Shimadzu Co.

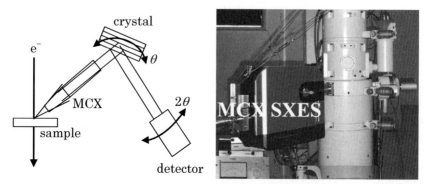

Figure 1. (a) Block diagram of MCX-WDX system for TEM. (b) Photo of MCX-WDX system attached to 200CX TEM.

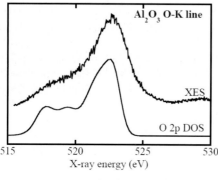

Figure 2. SXES spectrum of O K-line from Al_2O_3 thin film (upper) and O 2p pDOS calculated by molecular orbital method (lower).

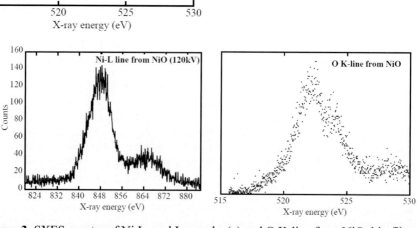

Figure 3. SXES spectra of Ni L_α and $L_{\beta1}$ peaks (a) and O K-line from NiO thin film (b).

Miniature electrostatic-magnetostatic column for electrons

C. Rochow, T. Ohnweiler and E. Plies

Institute of Applied Physics, University of Tübingen,
Auf der Morgenstelle 10, D-72076 Tübingen

christoph.rochow@uni-tuebingen.de
Keywords: miniaturised column, electrostatic, permanent magnetic

Simulations are presented for a miniaturised low-energy electron optical column consisting of an electron emitter, two electrostatic lenses, a magnetostatic snorkel lens, an electrostatic deflection and scanning unit, and a secondary electron detector. The two electrostatic lenses are manufactured from conventional components [1] such as platinum apertures as used in conventional electron microscopy and a common Schottky emitter module is used as an electron source. The double deflector is based on two electrostatic eight-poles allowing high scanning frequencies. The secondary electrons are detected by means of a conventional scintillator-photomultiplier combination.

Figure 1 shows the schematic configuration of the column. The system is approximately 85 mm in height and is currently being built. The detection system and the double deflector unit have been placed in the intermediate region between condenser (figure 2) and objective lens (figure 3). The linertube is set to a potential of 10 kV relative to the sample resulting in the advantage of increasing the immersion ratio of the two electrostatic lenses. Thus the aberrations are reduced and the detection efficiency is increased by accelerating the secondary electrons to an energy suitable for a scintillator-photomultiplier combination. A permanent magnetic snorkel lens similar to the add-on lens described in [2] is placed below the sample. The system has been optimised for primary energies of less than 1 keV and a physical working distance of 1 mm.

The calculations show that, assuming a virtual source size of about 20 nm and an energy width (FWHM) of 1.2 eV, a probe size of about 30 nm is achievable even when both electrostatic lenses are operated in internal deceleration mode and no electric field at the sample is applied (figure 4).

The secondary electrons leaving the sample are accelerated through the objective lens towards the detection system. Because of the relatively strong magnetic field at the sample, high care has been taken during simulation of the trajectories of the secondary electrons. Placing the scintillator with a 1 mm bore 30 mm above the sample and assuming a cosine distribution for the starting angles as well as a Maxwellian energy distribution with a maximum at 2 eV, simulations show a collection efficiency of 30% up to 70% for starting energies up to 10 eV depending on the electric field strength at the sample.

The presented simulations for a miniaturised low-energy electron optical column show that a resolution of 30 nm is achievable. The theoretically interesting internal acceleration mode of the lenses has also been studied and shows better performance of the system. The possibility of in-column detection of the secondary electrons has been

M. Luysberg, K. Tillmann, T. Weirich (Eds.): EMC 2008, Vol. 1: Instrumentation and Methods,
pp. 63–64, DOI: 10.1007/978-3-540-85156-1_32, © Springer-Verlag Berlin Heidelberg 2008

64

confirmed by simulation of their trajectories as well as by calculating the collection efficiency.

1. C.-D. Bubeck et al., Nucl. Instrum. Meth. Phys. Res. **A 427** (1999), 104.
2. A. Khursheed, N. Karuppiah and S. H. Koh, Scanning **23** (2001), 204.
3. Simulations of fields and trajectories were carried out with the program EOS of E. Kasper.

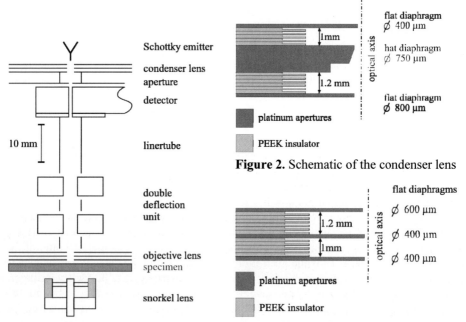

Figure 1. Schematic system overview

Figure 2. Schematic of the condenser lens

Figure 3. Schematic of the objective lens

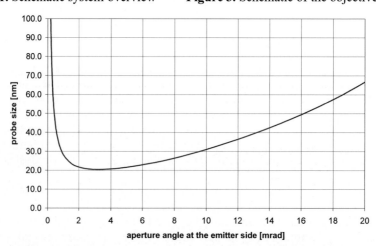

Figure 4. Achievable probe size for a final beam energy of 1 keV

Comparison of monochromated electron energy-loss with X-ray absorption near-edge spectra: ELNES vs. XANES

T. Walther[1,2] and H. Stegmann[3]

1. Dept. Electronic and Electrical Engineering, Mappin Street, University of Sheffield, Sheffield S1 3JD, UK
2. formerly at: Center of Advanced European Studies and Research (caesar), Ludwig-Erhard-Allee 2, D-53175 Bonn, Germany
3. Carl Zeiss Nano Technology Systems GmbH, Carl-Zeiss-Str. 56, D-73447 Oberkochen, Germany

t.walther@sheffield.ac.uk

Keywords: monochromator, electron energy-loss near-edge structure (ELNES), x-ray absorption near-edge structure (XANES)

We have recorded electron energy-loss near-edge spectra (ELNES) in a mono-chromated 200kV transmission electron microscope and compare the results with X-ray absorption near-edge spectra (XANES) acquired previously by another group on a high-resolution synchrotron radiation beamline.

The Zeiss Libra 200FE[CRISP] instrument has a Schottky-type field-emission gun (FEG) and an in-column corrected Omega type imaging filter (90° filter). The monochromator in the microscope is of the electrostatic Omega type [1] and an aberration corrector based on the hexapole design is incorporated for the probe-forming lenses [2]. The potential benefit of aberration correction for chemical analysis of nano-structures lies in the increase of the beam current density in nano-probe or scan mode, by selecting larger condenser apertures and thus allowing more off-axis electrons to contribute to the beam, while the monochromator improves both the energy resolution and the chromatic probe spread [3]. In our experiments we operated the instrument in image mode where the aberration correction of the illumination has no effect.

The combination of monochromator (MC) and 90° filter yields a spectroscopic system with total energy resolution expressed as the full width at half maximum (FWHM) of the zero loss peak (ZLP) ranging from 0.1eV (0.5 μm MC slit aperture and exposure times up to ~0.1s) to 0.6eV (60μm MC slit and exposure times up to some minutes). Figure 1a displays a ZLP spectrum recorded at 0.1s exposure with 11.8 meV/channel dispersion (FWHM: 100±6meV) [4]. Figure 1b depicts a plot of the FWHM as a function of the MC slit aperture size, showing linearity from 1-5μm slit width that makes it easy to judge system performance. The spectra shown in Figure 2 were recorded with the 1.5μm MC slit, giving ~0.2eV energy resolution (0.17eV FWHM measured for the ZLP, but the exposures for the Mn L-edge have been ~1min, rather than ~0.1s as for the ZLP, and degrade the effective resolution by 0.03eV [5]).

Electron transparent samples were produced by crushing powder material obtained from Sigma-Aldrich (St. Louis, USA) in ethanol. ELNES data were recorded in image mode at 200kV primary energy with 16.4 meV/channel dispersion, 2mrad convergence

M. Luysberg, K. Tillmann, T. Weirich (Eds.): EMC 2008, Vol. 1: Instrumentation and Methods, pp. 65–66, DOI: 10.1007/978-3-540-85156-1_33, © Springer-Verlag Berlin Heidelberg 2008

semi-angle, $10^6 Am^{-2}$ current density, 11mrad collection angle and an aperture of 147nm effective diameter in the spectrometer entrance plane.

In Figure 2a we show the as-recorded Mn L-edges of the four manganese oxides $Mn^{II}O$, $Mn^{III}_2O_3$, $Mn^{II/III}_3O_4$ and $Mn^{IV}O_2$ with differing oxidation states and crystallography. The agreement of the ELNES with the XANES data from [6] is very good: the same numbers of peaks and shoulders are resolved and all spectral features occur at the same energy (within experimental uncertainty of ±0.3eV of the calibration in ELNES energy). Minor changes in relative peak heights may be due to different ways of data collection (in XANES: direct X-ray yield without aperture, in ELNES: small collection angle). The reason for the excellent agreement of ELNES and XANES despite the differences in energy resolution (0.2eV for ELNES vs. 0.1eV for XANES) is that both resolutions lie below the natural line width of the Mn L-edge of ~0.3eV [7].

1. S. Uhlemann and M. Haider, Proc. 15th Int. Cong. Electr. Microsc., Durban, **3** (2002) 327
2. M. Haider, G. Braunshausen and E. Schwan, Optik **99** (1995) 167
3. T. Walther and H. Stegmann, Microsc. Microanal. **12** (2006) 498
4. T. Walther, H. Schmid and H. Stegmann, Proc. 16th Int. Microsc. Cong., Sapporo, **2** (2006) 833
5. T. Walther, E. Quandt, H. Stegmann, A. Thesen and G. Benner, Ultramicroscopy **106** (2006) 963
6. B.Gilbert, B.H. Frazer, A. Belz *et al.*, J. Phys. Chem. A **107** (2003) 2839
7. M.O. Krause and J.H. Oliver, J. Phys. Chem. Ref. Data **8** (1979) 329
8. We thank H. Schmid, University of Bonn, for the provision of the specimen material.

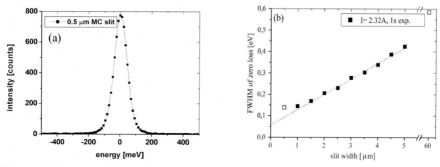

Figure 1. (a) ZLP with smallest MC slit aperture and (b) influence of MC slit width

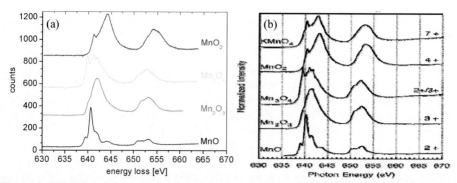

Figure 2. ELNES at 0.2eV resolution (a) compared to XANES with 0.1eV resolution (b)

A hybrid electron energy loss spectrometer with simultaneous serial and parallel detection

Jun Yuan[1,2], Zhiway Wang[2], Shu Hu[2], and Ling Xie[2]

1. Department of Physics, University of York, Heslington, York, YO10, 5DD, U.K.
2. Beijing National Electron Microscopy Centre, Tsinghua University, Beijing, 100084, China

jy518@york.ac.uk
Keywords: electron energy loss spectrometer, serial detection, parallel detection

Electron energy loss is a physical process whose probability decreases sharply with the amount of energy loss involved in any single inelastic scattering event [1]. This physical fact means that the electron energy-loss spectrum (EELS), after correcting for multiple scattering effects, has a very strong intensity at low energy loss but the intensity decays rapidly with the increase in energy loss. When the electrons not losing any significant energy (the so-called zeroloss beam) is also considered, the dynamical range of EELS is huge and often exceeds the capability of any physical detectors.

Experimental techniques have been developed to extend the dynamical range of EELS collected. They can be divided into time-division and energy division solutions. In the time-division solutions, we collect the intense low loss part of EELS at a given time with very short exposure and collect the high energy loss part of EELS at another time with an extended exposure. This is probably the most widely used method compatible with most existing EELS spectrometer. The process can be automated to produce a composite spectrum of a high dynamical range not possible in a single exposure time [2]. The disadvantage is that the specimen or the energy calibration of the spectrometer can be changing with time. In the energy division solutions, multiple detectors are employed to detect different energy parts of EELS simultaneously. They are more complex but more versatile. Here we present a hybrid two detector electron energy loss spectrometer. The system is hybrid because we have employed a CCD for the efficient detection of high energy loss signals and a serial detector for the high fidelity acquisition of intense low energy loss signals including the zeroloss beam. We have developed a novel serial detection system by extending the wire detector method pioneered by Dennis McMullan [3]. In this method, instead of using a slit to select the energy of the electrons beams and blocking all other electrons, we use a micrometer wire to select electrons of interest for serial acquisition and allowing all other electrons to pass through for parallel detection by a CCD detector. By scanning electrons of different energies across the sensing wire, we achieve the serial acquisition. Although serial detection is a form of time division solution which is inherently inefficient, it is ideal for low loss signal detection where the signal is usually too strong for the sensitive CCD detector geared to detection of much weaker signals. In fact, the sampling frequency of the low loss signal by the serial detector can be in kilohertz, much shorted than the typical readout time of a few seconds of the slow scan CCD detector, making

M. Luysberg, K. Tillmann, T. Weirich (Eds.): EMC 2008, Vol. 1: Instrumentation and Methods, pp. 67–68, DOI: 10.1007/978-3-540-85156-1_34, © Springer-Verlag Berlin Heidelberg 2008

the acquisition of the low loss signal by the serial detector and the high loss signal by CCD detector virtually simultaneous in practice. The fast refreshing rate of the low loss signal also introduced the possibility of real-time energy drift correction [4]. Progress in the development of this hybrid dual detector EELS spectrometer will be reported.

1. R. F. Egerton Electron Energy-loss Spectroscopy in the Electron Microscope, Plenum Press, New York (1996).
2. P. Thomas, J. Scott, M. MacKenzie, S. McFadzean, J. Wilbrink and A. Craven, 'Near-simultaneous Core- and Low-loss EELS Spectrum-Imaging in the STEM using a Fast Beam Switch', Microscopy and Microanalysis, 12, (2006), p1362-1363
3. D. McMullan, Beam stabilizer for an electron spectrometer, *Electron Microsc. Anal.* **138** (1993), pp. 511–514.
4. Z. Wang, S. Hu, C. Xu, D. McMullan and J. Yuan, An energy stabilized post-column electron energy-loss spectrometer for transmission electron microscopy, *Inst. Phys. Conf. Ser.* (2007).

In-focus phase contrast: Present state and future developments

R.R. Schröder[1,*], B. Barton[1], K. Schultheiß[2], B. Gamm[2], D. Gerthsen[2]

1. Max-Planck-Institute of Biophysics, Max-von-Laue Str. 3, 60438 Frankfurt /Main, Germany
2. Laboratory for Electron Microscopy, University Karlsruhe, 76128 Karlsruhe, Germany
* present address: Bioquant, CellNetworks Cluster of Excellency, University Heidelberg, Im Neuenheimer Feld 267, 69120 Heidelberg, Germany

rasmus.schroeder@bioquant.uni-heidelberg.de
Keywords: in-focus phase contrast, phase plates, anamorphotic Hilbert phase plate

The advantages of in-focus phase contrast over the conventional bright-field defocus contrast of weak phase objects has recently been the topic of several studies. Different implementations of carbon-film based phase plates[1] and electro-static devices such as einzel lenses[2] or a drift tube[3] have been proposed and tested.

Even though carbon-film phase plates have traditionally been connected with the occurrence of image aberrations e.g. by charging effects, film phase plates have proven to be experimentally easier to operate than their matter-free electrostatic relatives[4].

The main problem of electrostatic phase plates is the size of the electrodes generating the field. Usually such phase plates are placed in a diffraction plane of the microscope, where their supporting structures and electrodes themselves obstruct the electron beam[2]. This leads to an unacceptable loss of information, which has so far prevented the use of electrostatic phase plates for imaging. It is therefore necessary to make the devices smaller and/or to magnify the diffraction pattern. Both avenues are currently pursued, with the PACEM project[5] as an example for an advanced electron optical design of a dedicated phase contrast TEM.

A radically different design of an obstruction-free electrostatic phase plate and dedicated optics was proposed recently[6]. Fig. 1 shows an anamorphotic phase plate with three electrodes (A, Zernike-type) and a related two electrode design (B, Hilbert-type). Simulations of CTF and object contrast (Figs. 2, 3) demonstrate that the proposed Hilbert electrode design reduces the constraints on the aspect ratio of the anamorphotic diffraction plane and will be better suited for its integration into aberration corrected microscopes.

1.   Zernike- and Hilbert-type carbon phase plates as tested and published by Danev and Nagayama
2.   Multilayer electrostatic einzel lenses as published by e.g. Majorovits et al., Ultramicroscopy, 107 (2007) 213, or Hsieh et al., Proc. Micros. Microanalysis, 2007.
3.   Cambie et al., Ultramicroscopy, 107 (2007) 329.
4.   e.g. single particle reconstruction of ice-embedded GroEL: Danev et al., J. Struct. Biol., 2008, and tomography of thick plastic sections: Barton et al., Proc. Micros. Microanalysis, 2007.
5.   Matijevic et al., abstract at this conference.
6.   Schröder et al., Proc. Micros. Microanalysis, 2007.

M. Luysberg, K. Tillmann, T. Weirich (Eds.): EMC 2008, Vol. 1: Instrumentation and Methods, pp. 69–70, DOI: 10.1007/978-3-540-85156-1_35, © Springer-Verlag Berlin Heidelberg 2008

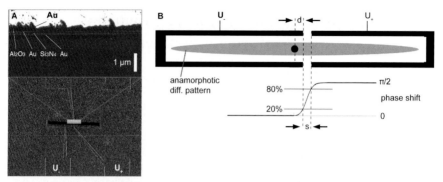

Figure 1. A - SEM picture of an anamorphotic phase plate with Zernike-type electrodes as seen along the electron optical axis and side-on (red box). B - Electrode design and diffraction pattern location in an anamorphotic Hilbert-type phase plate. Black dot = unscattered central beam.

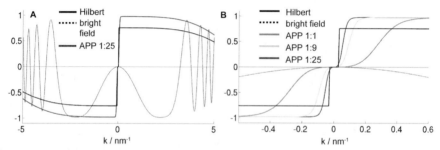

Figure 2. Contrast Transfer Functions (CTFs) for bright-field (Scherzer defocus), carbon-film Hilbert phase plate (52% transmission) and anamorphotic Hilbert phase plate (APP) as electrostatic device. Due to their asymmetry both Hilbert- and electrostatic APP phase plates generate an anti-symmetric CTF. The red, green and blue CTFs correspond to anamorphotic phase plates of different aspect ratio. A - CTFs at 0.2nm resolution. B - CTFs at 1.6nm resolution.

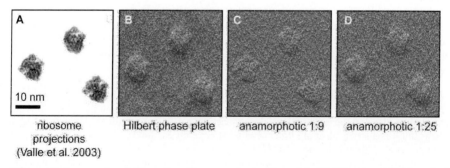

Figure 3. Image simulations for ribosome projections (A, EMD entry 1055) embedded in an ice layer (30% noise) for a carbon-film Hilbert phase plate (B) and electrostatic Hilbert phase plate (C, D) at varying aspect ratios.

The Detective Quantum Efficiency of Electron Area Detectors

R. Henderson, G. McMullan, S. Chen and A.R. Faruqi

MRC Laboratory of Molecular Biology, Hills Road, Cambridge CB2 0QH, U.K.

rh15@mrc-lmb.cam.ac.uk
Keywords: detector, film, CCD, CMOS, MTF, DQE

Recent progress in detector design has created the need for a careful side-by-side comparison of the modulation transfer function (MTF) and resolution-dependent detective quantum efficiency (DQE) of existing electron detectors, including film, with detectors based on new technology. I will present the results of measurements of the MTF and DQE of several detectors at 120 and 300keV. We have used the method of Meyer & Kirkland [1] modified to use analytically fitted curves, as shown in Figure 1. Figure 2 shows some of the MTF and DQE measurements for 300 keV electrons. Computer simulations have also been carried out, showing good agreement with the experimental results. We will conclude that the DQE to be expected from direct detection by back-thinned CMOS designs is likely to be equal to or better than film at 300 keV.

1. R. R. Meyer and A. I. Kirkland, Ultramicroscopy **75** (1998), p. 23.

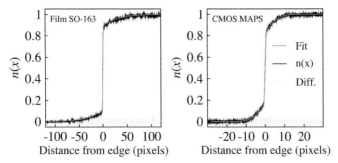

Figure 1. Comparison of experimental and Gaussian fitted edge spread functions for 300keV electrons for (a) SO-163 film (6 μm pixels) and (b) CMOS/Maps chip (25 μm pixels).

M. Luysberg, K. Tillmann, T. Weirich (Eds.): EMC 2008, Vol. 1: Instrumentation and Methods, pp. 71–72, DOI: 10.1007/978-3-540-85156-1_36, © Springer-Verlag Berlin Heidelberg 2008

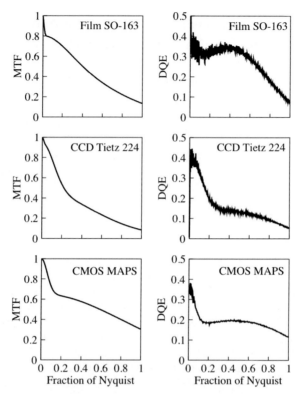

Figure 2. MTF (left) and DQE (right) for 300 keV electrons. Top panel : SO-163 film with an open-backed filmholder (6 μm pixels). Middle panel : Tietz 224 phosphor/fibre optics/CCD detector (24 μm pixels). Bottom panel : CMOS/MAPS detector (25 μm pixels).

Experimental observation of the improvement in MTF from backthinning a CMOS direct electron detector

G. McMullan[1], A.R. Faruqi[1], R. Henderson[1], N. Guerrini[2], R. Turchetta[2], A. Jacobs[3] and G. van Hoften[4]

1. MRC Laboratory of Molecular Biology, Hills Road, Cambridge CB2 0QH, U.K.
2. STFC Rutherford Appleton Laboratory, Chilton, Didcot OX11 0QX, U.K.
3. Technical University of Eindhoven, NL-5600 MB, Eindhoven, Netherlands.
4. FEI Electron Optics, NL-5600 MD Eindhoven, Netherlands.

rh15@mrc-lmb.cam.ac.uk
Keywords: detector, CMOS, MTF, backthinning

The modulation transfer functions (MTFs) of prototype backthinned CMOS direct electron detectors have been measured at 300keV. The images have contributions from two components, which arise from the initial passage of the electron through the epilayer and from the subsequent backscattering. At zero spatial frequency, in nonbackthinned 700 μm thick detectors, the backscattered component makes up over 40% of the total signal but, by backthinning to 100, 50 or 35 μm, this can be reduced to 25, 15 and 10% respectively. For the 35 μm backthinned detector, this reduction in backscatter increases the MTF by 40% for spatial frequencies between 0.1 and 1.0 Nyquist. Backthinning makes the sensor more flexible and slightly curled (Figure 1), without changing the behaviour of the transistors and all the other electronic components.

Figure 1. Single detector chip, backthinned to 35μm

M. Luysberg, K. Tillmann, T. Weirich (Eds.): EMC 2008, Vol. 1: Instrumentation and Methods, pp. 73–74, DOI: 10.1007/978-3-540-85156-1_37, © Springer-Verlag Berlin Heidelberg 2008

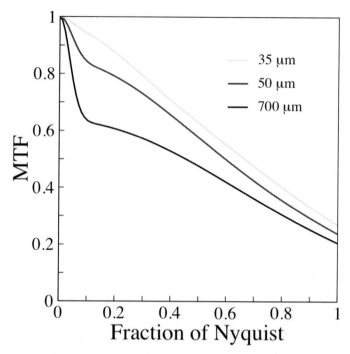

Figure 2. The MTF curves for 35 μm (top line), 50 μm (middle line) and 700 μm (bottom line) thick detectors.

High speed simultaneous X-ray and electron imaging and spectroscopy at synchrotrons and TEMs

Lothar Strüder[1,2,3]

1. MPI für extraterrestrische Physik, Giessenbachstr., 85741 Garching Germany
2. MPI-Halbleiterlabor, Otto-Hahn-Ring 6, 81739 Munich, Germany
3. Universität Siegen, Fakultät für Physik, ENC, 57234 Siegen, Germany

lts@hll.mpg.de
Keywords: imaging spectroscopy, CCDs, active pixel sensors, radiation hardness, EDX

Silicon is a good detector material for ionizing radiation as it has sufficient stopping power for e.g. particles and X-rays. In a standard configuration with 500 µm thick Silicon X-rays up to 25 keV can be efficiently converted into signal charges as well as e.g. electrons up to 300 keV. We have developed a variety of position, energy and time resolving detectors with a sensitive thickness ranging from 50 µm to 500 µm and sensitive areas from 1 cm² to 30 cm². They are being used in basic research (e.g. astrophysics, high energy physics, material sciences) as well as in industrial applications (e.g. in SEM, TEM, handheld XRF spectrometers). All the devices are based on the principle of "sideward depletion": Silicon Drift Detectors (SDDs), fully depleted pn-junction Charge Coupled Devices (pnCCDs) and the more recent Active Pixel Sensors based on the Depleted P-channel Field Effect Transistor (DePFET). They all exhibit remarkable properties for the detection of ionizing radiation, especially for photons and electrons: (a) low noise operation (2 – 5 electrons rms) close to room temperature, (b) simultaneous fast signal processing leading to high count rates, (c) sub-pixel position resolution, (d) high radiation tolerance due to the lack of active MOS structures and low energy thresholds for X-ray photons (50 eV) and electrons (500 eV) due to the back-illuminated device structures. All the above properties are simultaneously fulfilled in one device. The choice on whether a SDD, a pnCCD or a DePFET is adequate, is given by the main focus of a measurement: e.g. collecting, sensitive area, position resolution or count rate capability.

	pnCCD	DePFET APS
detector thickness	50 µm to 450 µm	50 µm to 450 µm
readout noise (rms) at – 40°C, room temperature	1.8 electrons 15 electrons	3.2 electrons 8 electrons
FWHM for X-rays	120 eV @ 5.9 keV	125 eV @ 5.9 keV
frame rate (frames per sec.)	1.000	1.000
energy range (X-rays)	50 eV up to 25 keV (QE=20%)	50 eV up to 25 keV (QE=20%)
energy range (electrons)	500 eV up to 250 keV, if fully stopped in the silicon	
existing formats	up to 1024 x 512	up to 256 x 256
pixel sizes	36 µm up to 150 µm	24 µm up to 625 µm

M. Luysberg, K. Tillmann, T. Weirich (Eds.): EMC 2008, Vol. 1: Instrumentation and Methods, pp. 75–76, DOI: 10.1007/978-3-540-85156-1_38, © Springer-Verlag Berlin Heidelberg 2008

Experiments will be reported from electron emission spectroscopy, electron autoradiography, transmission electron microscopy and dedicated experiments at the world´s first X-ray free electron laser facility FLASH, showing the full bandwidth and capabilities of the new detectors for electrons and X-rays.

Figure 1. Fully depleted (sensitive) pnCCD with a format of 1024 x 512 pixel and a pixel size of 75 x 75 μm^2. The detec-tors are fabricated on 150 mm high purity Silicon wafers with a thickness of 500 μm. Every pnCCD unit is equipped with 1024 on-chip amplifiers located at the end of the CCD co-lumns, leading to a frame rate of approxi-mately 250 fps.

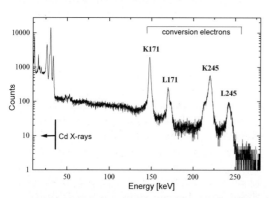

Figure 2. Position resolved electron and X-ray spectrum from the electron emission channeling measurements. The position was resolved with 60 μm (rms) and the energy of the X-rays around 30 keV with 200 eV (FWHM) and the electrons at 150 keV with 3 keV (FWHM).

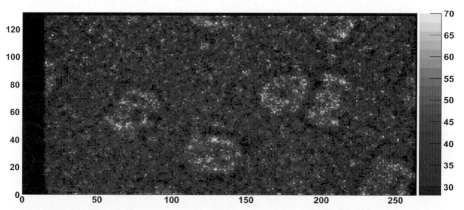

Figure 3. Transmission electron microscope image of a biological sample recorded with a very low electron current (down to 100 electrons per mm^2 and sec) on a FEI CM12 with 120 keV electron energy.

The image intensity in Zernike mode with electrons

M. Beleggia[1]

1. Institute for Materials Research, University of Leeds, Leeds LS2 9JT, UK

m.beleggia@leeds.ac.uk

Keywords: phase contrast, phase plates, Zernike contrast, electron microscopy

By using Zernike-type phase plates, suitably located within the microscope column, the object phase is transferred into the image amplitude and, hence, image intensity. Information transfer relies on controlled phase shifting of all diffracted beams while leaving the transmitted beam unaltered [1]. The advent of viable electron-optical phase shifting elements [2] paved the way to experimental practice. To interpret correctly the recorded intensity profiles and reveal phase information, it is essential to rely on a simple expression for computing the image intensity associated to phase objects.

The simplest phase plate for electrons is an extremely thin film of electron-transparent materials such as Carbon, with a tiny hole at the center, positioned in the diffraction plane of the microscope objective lens. While in general Zernike phase contrast is associated to quarter-wave shifts ($\pi/2$), this is not necessarily the case. Enhanced image contrast may be obtained by increasing the Zernike phase angle up to π when the phase object cannot be considered "weak". Tuning properly the phase plate according to the particular object under scrutiny requires the possibility of varying continuously the Zernike phase angle. This might be accomplished by utilizing electrostatic phase plates [2], where the phase shift is induced by an external voltage.

While the standard expression describing image intensity in Zernike mode given by Born and Wolf [3] (hereafter referred to as "BW expression") is a precious source of information, its validity is limited to weak phase objects. This fact may not be widely recognized because Born and Wolf appear to have neglected emphasizing the assumptions implicit in their treatment. In a forthcoming article [4] I illustrate in detail the shortcomings of the BW expression when employed to simulate intensity profiles associated to strong periodic phase objects. Troubles of the BW expression include violation of charge conservation (the intensity does not average to 1) and the unphysical possibility of observing constant phases. In the same article [4] I propose an alternative formula with more general validity:

$$I(x, \phi_Z) = 1 + 2\gamma^2 - 2\gamma^2 \cos\phi_Z - 2\gamma \cos[\phi_\gamma + \varphi(x)] + 2\gamma \cos[\phi_\gamma + \phi_Z + \varphi(x)], \quad (1)$$

where the complex number $\gamma_0 = \gamma \exp(-i\phi_\gamma)$ is the zero-order coefficient of the Fourier series representing the phase object, e.g.

$$\gamma_0 = \frac{1}{2L} \int_{-L}^{L} e^{i\varphi(x)} dx. \quad (2)$$

Here, I analyze an ideal sinusoidal phase grating as visualized by an electron microscope equipped with a Zernike phase plate according to equation (1). The image intensity for phase gratings with $\Phi = \pi/10$, $\pi/4$, $\pi/2$, where ϕ_Z varies from 0 to π, is

M. Luysberg, K. Tillmann, T. Weirich (Eds.): EMC 2008, Vol. 1: Instrumentation and Methods, pp. 77–78, DOI: 10.1007/978-3-540-85156-1_39, © Springer-Verlag Berlin Heidelberg 2008

78

displayed in Figure 2(a-c) as density plots. Intensity profiles for the optimal Zernike angle yielding maximum contrast are extracted along the dashed lines in Figure 2 and superimposed to the density plots. For phase gratings, the optimization criterion

$$\phi_Z^{opt} = \frac{\pi}{2} + \operatorname{arc\,cot}\left(\frac{\sin \Phi}{\gamma - \cos \Phi}\right) \tag{3}$$

may be obtained analytically by searching for the peak intensity as a function of ϕ_Z. The resulting optimal angles for the phase gratings here analysed are indicated in Figure 2.

The effect of more realistic phase plates where partial or total masking of the object spectrum occurs needs to be accounted for, and work is in progress in this direction [5].

1. F. Zernike, Physica **9** (1942), p. 686; Physica **9** (1942), p. 974.
2. S.-H. Huang et al., J. Electron Microsc. **55** (2006), p. 273.
3. M. Born and E. Wolf, Principles of Optics, 6th ed., Pergamon Press (1980), p. 427.
4. M. Beleggia, Ultramicroscopy (2008) submitted.
5. I gratefully acknowledge the Royal Society for financial support.

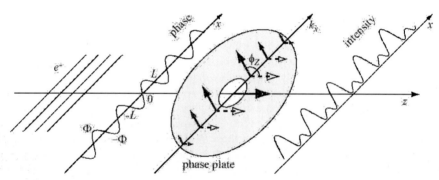

Figure 1. Scheme of image formation in Zernike phase contrast mode. A plane wave illuminates a phase grating; in the focal plane, all diffracted beams are phase shifted at an angle ϕ_Z; the resulting image intensity is recorded on the screen.

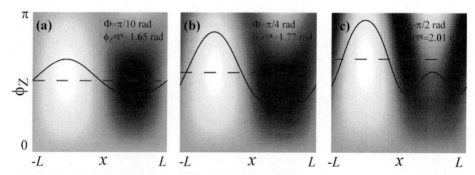

Figure 2. Image intensity associated to phase gratings as visualized with a range of Zernike phase angles from p/2 to p. Superimposed are the intensity profiles with the optimal Zernike phase angle according to equation (3).

Application of a Hilbert phase plate in transmission electron microscopy of materials science samples

M. Dries[1], K. Schultheiß[1], B. Gamm[1], H. Störmer[1] and D. Gerthsen[1],
B. Barton[2], R.R. Schröder[2]

1. Laboratorium für Elektronenmikroskopie, Universität Karlsruhe (TH),
D-76128 Karlsruhe,Germany
2. Max-Planck-Institut für Biophysik, Max-von-Laue-Str. 3,
D-60438 Frankfurt am Main, Germany

dries@lem.uni-karlsruhe.de
Keywords: transmission electron microscopy, phase plate, phase contrast

Several suggestions involving physical phase plates in a transmission electron microscope have emerged recently to enhance contrast of (weak) phase objects. One option is a (half-plane) Hilbert phase plate which is positioned the back focal plane of the objective lens [1].

The phase plate consists of a thin amorphous carbon (a-C) film which deposited by electron-beam evaporation on a cleaved mica surface. The deposited film is desolved from the substrate, floated on the surface of distilled water and placed on a standard 150 mesh copper grid. Rectangular holes are cut into the film by a focused-ion-beam system. The film thickness is adjusted to produce a phase shift of π using the equation $\Delta\phi = C_E V_0 t$ (V_0: mean inner Coulomb potential of carbon, t: film thickness). At an electron energy of 120 keV, a film thickness of 36 nm is required.

Both, the actual phase shift and thickness of the a-C film were determined and V_0 was derived for our a-C material. The phase shift close to the edge of the film is measured by electron holography in a transmission electron microscope (Fig. 1(a)). The film thickness was determined either by electron energy-loss spectroscopy using the equation $t = \lambda_i \ln(I_{tot}/I_0)$ (λ_i: inelastic mean-free path, I_{tot}: total electron intensity, I_0: zero-loss electron intensity) [2] (Fig. 1(b)) or alternatively by the preparation and analysis of a cross-section TEM sample of the a-C film that was deposited in the same evaporation run on a Si substrate. With the measured phase shift and film thickness, V_0 of the a-C film was determined to be 8.9 V.

The phase plate is placed in the back-focal plane of the objective lens such that the transmitted beam propagates near the phase-plate edge. Fig. 2(a) shows a Fourier-transformed image of an a-C test sample which shows a shift of the Thon rings with a nearly cosine-type phase contrast transfer with respect to Thon rings in the narrow stripe marked by arrows.

To test imaging with a Hilbert phase plate for a material science specimen, amorphous Nb_2O_5 was used which serves a as dielectric in niobium electrolytic capacitors. Fig. 2(b) shows a conventional bright-field TEM image of Nb_2O_5. The bright stripe at the sample edge is associated with contamination. Fig. 2(c) shows an image of

M. Luysberg, K. Tillmann, T. Weirich (Eds.): EMC 2008, Vol. 1: Instrumentation and Methods,
pp. 79–80, DOI: 10.1007/978-3-540-85156-1_40, © Springer-Verlag Berlin Heidelberg 2008

the same specimen area obtained with a Hilbert phase plate at a defocus $\Delta f \approx 0$. The latter image shows structures near the specimen edge which are not visible by a conventional TEM imaging. The observed contrast fluctuations in the thinnest part of the TEM sample close the edge may be related to artefacts induced by the ion milling during the final stage of the sample preparation.

1. R. Danev and K. Nagayama: J. Phys. Soc. Jpn. **73** (2004) 2718.
2. M. Wanner, D. Bach, D. Gerthsen, R. Werner and B. Tesche: Ultramicroscopy **106** (2006) 341.

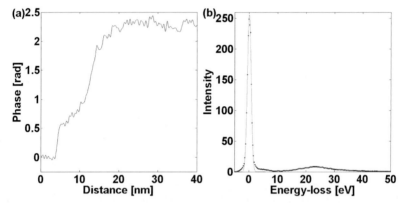

Figure 1. (a) Phase shift determination by electron holography. (b) Film thickness determination by Energy Electron-loss Spectroscopy.

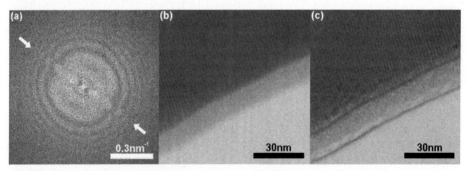

Figure 2. (a) Fourier-transformed image of an amorphous carbon test sample obtained with a Hilbert phase plate. A nearly cosine-type phase contrast transfer is visible with Thon rings which are shifted with respect to the Thon rings in the narrow stripe marked by the arrows. (b) Conventional bright-field TEM image of amorphous Nb_2O_5 close to the sample edge. (c) Image of the same specimen area taken with the phase plate inserted. Contrast fluctuations near the specimen edge are visible which cannot be recognized in (b).

Optimal Imaging Parameters in Cs-Corrected Transmission Electron Microscopy with a Physical Phase Plate

B. Gamm[1], K. Schultheiss[1], D. Gerthsen[1], R.R. Schröder[2]

1. Laboratorium für Elektronenmikroskopie (LEM), Universität Karlsruhe (TH), D-76128 Karlsruhe
2. Max-Planck-Institut für Biophysik, Max-von-Laue Str. 3, D-60438 Frankfurt/Main

gamm@lem.uni-karlsruhe.de

Keywords: phase contrast, phase plate, double-hexapole corrector

The double-hexapole Cs-corrector [1] has substantially improved resolution and interpretability of images in transmission electron microscopy. The device allows the adjustment of arbitrary, even negative Cs-values with a precision of ~ 1 μm. Lentzen et al. [2] derived optimum imaging parameters $C_{S,Len} = \frac{64}{27}\left(\lambda^{-3}u_{\text{inf}}^{-4}\right)$ and $Z_{Len} = -\frac{16}{9}\left(\lambda^{-1}u_{\text{inf}}^{-2}\right)$ (u_{inf}: spatial frequency corresponding the information limit) for imaging weak-phase objects with a Cs-corrected microscope without a physical phase plate. These settings result in high contrast transfer at intermediate and large spatial frequencies and optimized delocalisation but transfer at low spatial frequencies is limited due to the sine shape of the phase contrast transfer function (PCTF). Combining Cs-correction with a physical phase plate in the back focal plane of the objective lens could overcome this problem. Several different concepts for physical phase plates were proposed and realized successfully in the recent past [3-5]. In analogy to Zernike's λ/4 phase plate in light microscopy, such phase plates generate a phase shift χ_{PP} between the scattered and unscattered electrons. The total phase shift of the electrons is then given by $\chi_{tot}(u) = \pi\left(Z\lambda u^2 + 1/2 C_s\lambda^3 u^4\right) + \chi_{pp}$ resulting in a cosine shape of the PCTF for $\chi_{PP} = \pi/2$.

Optimal imaging conditions are achieved theoretically if the phase shift due to the objective lens defocus and lens aberrations are zero for all spatial frequencies, i.e. for Z=0, C_S=0. Residual aberrations and varying, non-zero local defocus values due to the surface topography of the sample will prevent adjustment of these settings over a large field of view. This motivates our theoretical study in which we test the sensitivity of Cs and Z with respect to maintaining optimum imaging conditions.

As a measure for phase contrast the PCTF is integrated up to the information limit

$$PC = \int_0^{u_{\text{inf}}} E_t E_S \left|\sin(\chi_{PP})\right| du$$ (E_s, E_t: envelope functions due to the partial temporal and spatial

coherence of the electron wave). PC is compared with the optimal value $PC_{id} = \int_0^{u_{\text{inf}}} E_t E_S du$

obtained with a phase plate and $\chi_{tot} = \pi/2$. Fig. 1 shows PC(Cs,Z) for a 200 keV microscope with an information limit of 0.12 nm with (Fig. 1a) and without phase plate

M. Luysberg, K. Tillmann, T. Weirich (Eds.): EMC 2008, Vol. 1: Instrumentation and Methods, pp. 81–82, DOI: 10.1007/978-3-540-85156-1_41, © Springer-Verlag Berlin Heidelberg 2008

(Fig. 1b). The black lines in Fig. 1a denote parameters at which PC is reduced by 5% (10%) with respect to PC_{id}. In Fig. 1b ($\chi_{PP}=0$) the black lines show a 5% (10%) deviation of the PC value with respect to PC for $C_{S,Len}$ and Z_{Len}. Figs. 1a,b demonstrate that the values of PC is significantly higher for a microscope with a phase plate. However, apart from phase contrast, delocalisation is important for high-resolution imaging. Delocalisation is determined by $R = \left|\max\left(\frac{d\chi}{du}\right)\right|$. The white lines in Fig. 1 denote parameters where delocalisation is $\frac{16}{27} u_{inf}$ for $C_{S,Len}$, and Z_{Len}. Regions inside the white lines correspond to imaging conditions with small delocalization. With respect to the ranges of Z and Cs, where optimum conditions are maintained, we note: the Z range is mainly limited by the PC reduction and the C_S range mainly by delocalisation. Fig. 1a shows that an extended range of Z and Cs values exists around the optimal values at Cs=0 and Z = 0 with high PC and small delocalization. Without a phase plate the situation is more complex because optimum conditions are not at Cs=0 and Z=0. If we consider the Cs and Z ranges with the same reduction in PC with respect to the optimal parameters and same maximum delocalisation, we find a significantly lower sensitivity of the these parameters in a microscope with a physical phase plate – even if obstructing structures, e.g. due to an electrode [4], are taken into account.

1. M. Haider, H. Rose, S. Uhlemann, E. Schwan, B. Kabius, K. Urban, Nature 392 (1998) 768.
2. M. Lentzen, B. Jahnen, C.L. Jia, A. Thust, K. Tillmann, K. Urban, Ultramicroscopy 92 (2002) 233.
3. R. Danev, K. Nagayama, Ultramicroscopy 88 (2001) 243.
4. K. Schultheiss, F. Perez-Willard, B. Barton, D. Gerthsen, R.R. Schröder, Rev. Sci. Instrum. 77 (2006) 033701.
5. R. Cambie, K.H. Downing, D. Typke, R.M. Glaeser, Jian Jin, Ultramicroscopy 107 (2007) 329.
6. This work is funded by the Deutsche Forschungsgemeinschaft (DFG) under project Ge 841/16 and Sch 824/11

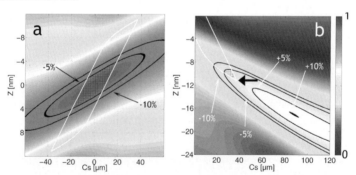

Figure 1. Colour-coded phase contrast integral as a function of Z and Cs for a 200 keV microscope with an information limit of 0.12 nm a) with phase plate and b) without phase plate. 5% and 10% deviation of PC from the respective optimal conditions are indicated by black lines. Delocalisation of R=16/27 u_{inf} is indicated by white lines. Lentzen parameters are indicated by the black arrow in b)

Electron optical design of the
Phase Aberration Corrected Electron Microscope

M. Matijevic[1], S. Lengweiler[1], D. Preikszas[1], H. Müller[2], R.R. Schröder[3], G. Benner[1]

1. Carl Zeiss NTS GmbH, 73447 Oberkochen, Germany
2. CEOS GmbH, Englerstr. 28, 69126 Heidelberg, Germany
3. MPI für Biophysik, Max-von-Laue-Str. 60438 Frankfurt am Main

matijevic@smt.zeiss.com
Keywords: Phase contrast, Boersch phase plate, Aberration corrector

Thin biomolecular specimens are weak phase objects. Therefore image contrast is described by the so called Phase Contrast Transfer Function (PCTF) which depends on the phase introduced by the spherical aberration and defocus. The PCTF describes the signal transfer as a function of the spatial frequency. Due to the sine-type transfer function artefacts arise from contrast reversal and cancellation of low spatial frequencies. Substantial improvement of the quality of the images can be achieved by using a Phase Plate (PP) integrated in a C_S-corrected microscope. The PP normally placed in the back focal plane of the objective helps to transfer lower frequencies while the C_S corrector is essential for the transfer of higher frequencies. This offers the possibility of artefact free imaging over a wide range of spatial frequencies. In this paper we describe the electron optical design of the Phase Aberration Corrected Electron Microscope (PACEM), which is developed in collaboration with the Max-Plank-Institute for Biophysics in Frankfurt. First results will be presented.

In the Boersch PP, the unscattered beam is shifted by an electrostatic potential produced by a self supporting, ring shaped mini lens [1]. Compared to the thin film phase plate this design offers the advantages of a freely adjustable phase shift and a charge-free and contamination-free operation. Unfortunately the unavoidable ring and the support structure cut of some spatial frequencies. To minimize this effect the Boersch PP is placed in a conjugate plane of the objective focal plane magnified by a Diffraction Magnification Unit (DMU).

The PACEM is equipped with a FEG with Schottky Emitter and an electrostatic omega type monochromator designed by CEOS. The objective lens has a large pole piece gap according to the needs for cryo- and tomography applications. A corrected Omega filter allows zero loss filtering and spectroscopic investigations (EELS, ESI).

The (DMU) [2] is arranged below the C_S-corrector whereas the final adapter lens of the C_S-corrector is used as first lens (DL1) of the DMU. The DL1 magnifies the diffraction plane located at the centre of the second hexapole (HP2) into the centre of the second diffraction lens (DL2), This arrangement provides an effective focal length f_{eff} of 15mm in DL2, where the PP is arranged. The DL2 transfers the object image into the SA plane. Special care has to be taken in the design of this lens to achieve the target lateral resolution of 0.2nm. Furthermore a monochromator is essential to reduce the energy spread to ≤ 0.3 eV in order to reduce the damping of the chromatic envelope.

M. Luysberg, K. Tillmann, T. Weirich (Eds.): EMC 2008, Vol. 1: Instrumentation and Methods,
pp. 83–84, DOI: 10.1007/978-3-540-85156-1_42, © Springer-Verlag Berlin Heidelberg 2008

Due to the increase in length by the DMU of more than 300mm the column is placed in the ultra stable frame [4] with suspension in the objective area (Fig. 2)

1. E. Majorovits et al, Ultramicrospy 107 (2007) 213-226.
2. G. Benner and M. Matijevic, EU Patent, EP 1 835 532 A2.
3. E. Essers et al., proceedings 15th Int. Cong. on Electron Microscopy Vol. 3 (2002).
4. We kindly acknowledge the financial support of the DFG

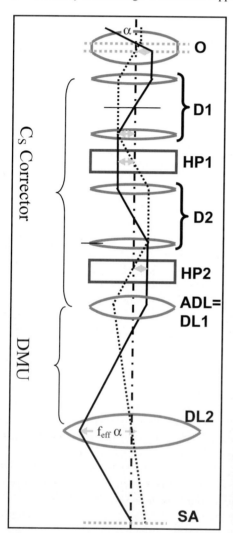

Figure 1. Electron optical set-up of the Diffraction Magnification Unit (DMU) below the C_S Corrector which provides an effective focal length of 15mm in the DL2.

Figure 2. PACEM Column in ultra stable frame.

Direct electron detectors for TEM

G. Moldovan, X. Li, P. Wilshaw and A.I. Kirkland

Department of Materials, University of Oxford, Parks Road, Oxford, OX1 3PH, UK

grigore.moldovan@materials.ox.ac.uk
Keywords: direct electron detection, transmission electron microscopy, resolution, efficiency

Recent advanced in electron sources and optics for Transmission Electron Microscopy (TEM) have revealed severe shortcomings in electron detection. Current detection systems are based on a phosphor-coupled charge-coupled-device approach, which suffers from a number of severe limitations that include weak Modulation Transfer Function (MTF) and poor Detective Quantum Efficiency (DQE). As these limitations are intrinsic to indirect sensor technology attention is now turning towards the use of novel sensor technologies based on direct electron detection [1,2]. The present work reports on Monte Carlo simulations of electron-sensor interaction to explore the improvements brought to MTF and DQE by this new generation of detectors. Attention is centered on a model detector structure and 200keVelectrons.

The model geometry considered here is a thin Si electron sensitive layer divided into square pixels supported by a thick Si wafer (Figure 1a). This simplified structure is similar to that of integrating Hybrid Active Pixel Sensors (HAPS), where the read-out is embedded in the support [1]. A secondary model structure without the supporting wafer is also considered, representing integrating Monolithic Active Pixel Sensors (MAPS), where the readout is fabricated in the sensitive layer [2]. This second geometry is also illustrative of counting Double Sided Strip Detectors (DSSD), where the readout is placed laterally. Simulations of electron trajectories and the resultant distribution of charge generated in the sensor reveal that electrons suffer significant displacement that tends to produce charge in more than one pixel (Figures 1b-2a). Typical charge packages are well above the usual noise levels in HAPS, MAPS and DSSD (Figure 2b). The support is found to increase the yield of electrons backscattered into the sensitive layer and enlarge the fraction of electrons that produce charge in several pixels.

Resultant MTFs and DQEs show remarkable improvements in performance of direct electron sensors in comparison with that of existing detectors (Figure 3). For integrating sensors, overall performance is significantly improved if the support is removed, as wider point spread and larger variance is induced by backscattering. Optimum performance is expected from counting detectors, as on-the-fly discrimination can be used to reject events in which more than one pixel are triggered by the same electron.

Experimental results obtained with free-standing DSSD that confirm these simulations will also be discussed.

1. A.R. Faruqi et al, Nuclear Instr. and Methods in Physics Research A **546** (2005), p. 170.
2. G. Deptuch et al., Ultramicroscopy **107** (2007), p. 674.
3. This work was supported under EPSRC grant EP/C009509/1 and EU Framework 6 project 026019 ESTEEM.

M. Luysberg, K. Tillmann, T. Weirich (Eds.): EMC 2008, Vol. 1: Instrumentation and Methods, pp. 85–86, DOI: 10.1007/978-3-540-85156-1_43, © Springer-Verlag Berlin Heidelberg 2008

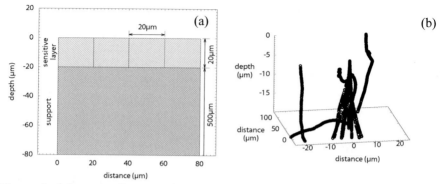

Figure 1. Schematic diagram of model structure considered (a) and trajectory of 10 electrons in the sensitive layer (b). Note the two electrons backscattered by the support.

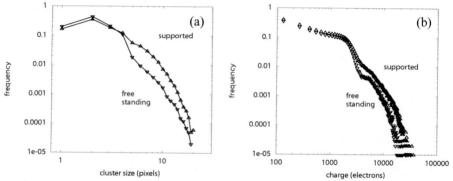

Figure 2. Log-log distributions of number of pixels with integrated well-electrons above noise for each high-energy electrons (a) and well-electrons collected in each pixel of the detector (b).

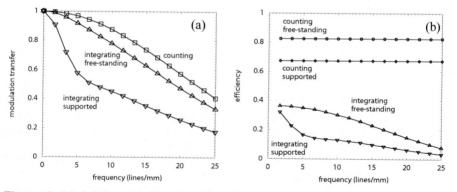

Figure 3. Modulation transfer function (a) and detective quantum efficiency (b) of integrating and counting direct electron detectors. Overall expected performance is above that of existing indirect-detectors, with the counting DSSD optimum.

A Newly Developed 64 MegaPixel camera for Transmission Electron Microscopy

H.R. Tietz[1]

1. TVIPS GmbH, 82131 Gauting, Germany

Hans.Tietz@tvips.com

Keywords: digital camera, transmission electron microscopy, image processing

A new technology has been developed to produce a 64 MPixel monolithic image sensor with fiber optic coupling to the scintillator for applications in Transmission Electron Microscopy. For fiber optic coupled cameras a large pixel size is mandatory. Especially for use in transmission electron microscopy, the size of the pixel in combination with the scintillator thickness defines the contrast transfer function of the detector.

It was obvious that with today's CCD (charge coupled device) manufacturing capabilities, it is not possible to produce such a monolithic sensor. The possibility of combining several smaller CCD chips to produce a mosaic sensor was not a choice, since this always results in missing gaps between the individual sensors. We decided to develop a CMOS (complementary metal oxide semiconductor) based sensor where wafer sizes of 8-12" are standard technology. The decreased feature size in CMOS technology makes it possible to build sensors with large pixel size and high fill factor. The fast prototyping possibilities of CMOS and the flexible means of designing various pixel geometries or dedicated image sensors are two more advantages of CMOS technology. CMOS image sensors have mainly penetrated the low-end applications because the image quality of most sensors is considerably lower than that of their CCD counterparts. The reason for the low image quality of CMOS sensors is the pixel architecture but a newly developed type, the so called "active pixel" can achieve comparable noise and sensitivity values to CCD's. This type has an amplifier stage in the pixel which reduces the capacity on the photodiode node, resulting in low noise and high light-to- voltage conversion gain.

The pixel architecture chosen is a classical 3-transistor pixel on a pitch of 15.6 µm. This results in a very high fill factor of 72% of the pixel area. The sensor area contains 8200 x 8200 square pixels which is subdivided into 8 x 8 areas, each 1024 x 1024 pixel large. The read out is done on 8 channels synchronously. The horizontal reset and select lines can be driven from left or right or both. This is important to implement the 2 different readout modes. The sensor can be readout in the so called "CDS" mode (correlated double sampling) or "rolling shutter mode" for fast operation.

For very low noise and high sensitivity single image acquisition, CDS mode is used. In Figure 1 the sequence of a single exposure is shown. The first operation is the reset of all pixels followed by a readout into the internal memory (reset image) and then the sensor is exposed and read out a second time. Now the reset image is subtracted from the exposed frame and subsequently transferred to the PC. This procedure ensures very

M. Luysberg, K. Tillmann, T. Weirich (Eds.): EMC 2008, Vol. 1: Instrumentation and Methods, pp. 87–88, DOI: 10.1007/978-3-540-85156-1_44, © Springer-Verlag Berlin Heidelberg 2008

low noise similar to the typical read out noise of slow scan CCD cameras. The typical read out time for the full sensor is 5 sec. For fast readout the sensor can be operated in the rolling shutter mode where line by line is first readout and then reset. The minimum exposure time is then defined by the total read out time of the sensor or sensor sub area. Frame rates of up to 7 frames/sec at 1024 x 1024 pixel can be reached.

All camera electronics are integrated in the camera head. It is built around a FPGA (field programmable gate array) and a 512 Byte memory. The firmware allows individual sub area read out in steps of 1024 x 1024 pixel, software binning by factor 1, 2, 3 and 4. The camera is controlled by a FireWire interface and the data transfer is done by a 16 bit fast parallel interface.

Fiber optic coupling, thermoelectric cooling and mechanical design:
Due to the very large image area (128 mm x 128 mm) a new technology has been developed for the fiber optic coupling. With this new technology, a homogenous resolution over the whole imaging area has been achieved. A thermoelectric cooler stabilizes the chip temperature at -10 to -15°C. The camera head with the scintillator on top is shown in Figure 2.

The camera has been integrated into the TVIPS EM-MENU 4.0 image processing package for transmission electron microscopy. A special Fourier module has been developed to overcome memory restrictions of 32bit Windows operating systems. The camera is also a new device in the CAMC4 COM module. This allows the TVIPS cameras to be controlled by external software.

The camera has been installed on a JEM-2010, on a JEM-Z2100FC and on a Titan KRIOS.

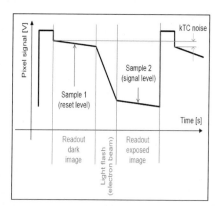

Figure 1. CDS sequence.

Figure 2. Camera head with phosphor scintillator.

Characterization of a fiber-optically coupled 8k CCD/CMOS device

D. Tietz[1], H. Tietz[2], S. Nickell[1], W. Baumeister[1] and J.M. Plitzko[1]

1. Max-Planck-Institute of Biochemistry, Department of Molecular Structural Biology, Am Klopferspitz 18, D-82152 Martinsried, Germany
2. TVIPS, Tietz Video and Imaging Processing Systems, Eremitenweg 1, D-82131 Gauting, Germany

tietz@biochem.mpg.de

Keywords: Distortion correction, CCD, CMOS, fibre optics, cryo-electron microscopy

High-end CCD cameras are the most popular recording devices in electron microscopy, especially for applications where automation is mandatory, e.g. cryo-electron microscopy and tomography. However, photographic film is still a highly attractive medium for recording data when large numbers of particle-images need to be collected, like in single particle applications, mainly due to its larger array (10000 13000 pixel). The array size in commercially available CCD cameras varies typically between 1024 (1k) and 4096 (4k) square pixel, with a pixel size between 15 and 30 μm, thus they are much smaller than photographic film. Moreover, at higher acceleration energies (>100 kV) they are characterized with a large point-spread function relative to their pixel size, thus the usable number of pixels is significantly smaller than the actual array size.

In our study we have tested the performance of a recently developed 8192x8192 pixel (8k) CCD/CMOS camera (TVIPS, Gauting, Germany) attached to the Titan Krios TEM (FEI Company, Eindhoven, The Netherlands) installed at the MPI for Biochemistry, Martinsried, Germany (Figure 1). The active area is as large as 128x128 mm, with a physical pixel size of 15.6 μm. Based on the CMOS design and eight readout ports a full frame image can be digitized and transferred within only 5 seconds, making it ideal for high-throughput applications [1].

However, due to its very large dimensions, geometric distortions (mainly shear distortions generated during the production process) induced by the fibre optical part of the camera have to be taken into consideration to a greater extend than, for example, in smaller 4k arrays. Therefore, we have focused our work on the analysis and correction of the geometric distortions of this 8k CMOS camera. For this purpose we have used a line grating (400 lines/inch) which was directly placed on the whole CCD stack for an initial light-optical measurement. The subsequent analysis and correction method is based on the procedures typically used in electron holography [2-4]. We will report about our recent findings regarding the performance characterization and the distortion analysis and correction procedure for this 8k camera.

M. Luysberg, K. Tillmann, T. Weirich (Eds.): EMC 2008, Vol. 1: Instrumentation and Methods, pp. 89–90, DOI: 10.1007/978-3-540-85156-1_45, © Springer-Verlag Berlin Heidelberg 2008

1. A. Korinek, F. Beck, S. Nickell, W. Baumeister and J.M. Plitzko, EMC2008 abstract volume.
2. W. D. Rau, Ein on-line Bildverarbeitungssystem für die Bildebenen-Off-Axis Holographie mit Elektronen (1994)
3. I. Daberkow, K.H. Herrmann, L.Liu, W.D. Rau and H. Tietz, Ultramicroscopy 64 (1996) 35-48.
4. D. C. Ghiglia and M. D. Pritt, Two-Dimensional Phase Unwrapping: Theory, Algorithms and Software (Wiley, 1998).
5. This work was supported by the 3DEM Network of Excellence and the High-throughput-3DEM Grant within the Research Framework Program 6 (FP6) of the European Commission.

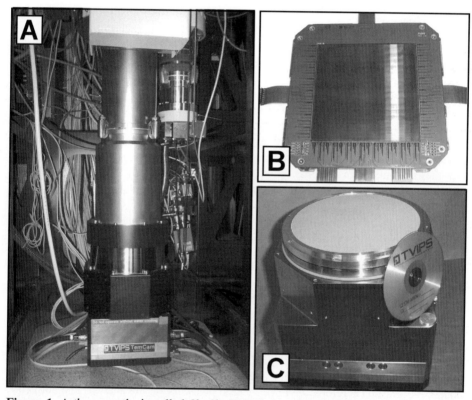

Figure 1. A the recently installed 8kx8k CMOS Camera (TVIPS, Gauting, Germany) mounted at the Titan Krios (FEI Company, Eindhoven, The Netherlands) installed at the MPI of Biochemistry in Martinsried, Germany. **B** the 8192x8192 CMOS Chip covering an area of 128x128 mm. **C** the complete assembly of the CMOS camera including the scintillator, the fibre optical stack (190x100 mm) and the integrated CMOS chip as seen in B.

Direct Single-Electron Imaging using a pnCCD Detector

Alexander Ziegler[1], Robert Hartmann[2], Robert Andritschke[3], Florian Schopper[3], Lothar Strüder[3], Heike Soltau[2] and Jürgen M. Plitzko[1]

1. Max-Planck Institute of Biochemistry, Department of Molecular Structural Biology, am Klopferspitz 18, D-82152 Martinsried, Germany
2. PNSensor GmbH, Römerstrasse 28, 80803 München, Germany
3. Max-Planck Institute Halbleiterlabor, Otto-Hahn Ring 6, D-81739 München, Germany

ziegler@biochem.mpg.de

Keywords: Single-Electron Imaging, cryo-electron tomography, CCD camera

Cryo-electron microscopy and especially cryo-electron tomography are widely applicable for structural studies in biology, but the major limitation is the extreme sensitivity of frozen hydrated biological samples to electron radiation [1-3]. Thus low-dose techniques have to be used, distributing the total electron dose (typically in the range of 50 electrons/Angstrom) to many single images, demanding highly sensitive recording devices. Phosphor-coated fiber coupled CCD cameras are typically used where the primary electrons are converted to visible light. This indirect detection strategy is characterized by significant levels of readout and dark current noise and the number of optical interfaces within the complete camera assembly results in multiple scattering and a subsequent loss in resolution. Since the total applied dose can't be increased in cryo-EM studies, the improvement in camera performance with noiseless recording, a higher sensitivity and at the same time higher spatial resolution is therefore be definitely beneficial for low-dose cryo-applications [4,5].

In this investigation we characterize a pnCCD detector that is highly suitable for direct single-electron detection in the TEM [6]. The detector is a 132x264 50µm-pixel array, cooled to –30°C and it was tested on a Philips CM12 at 120kV. A sharp knife-edge was placed directly on top of the pixel array, such that the LSF, MTF, NPS and DQE could be evaluated properly. Additional imaging and sensitivity measurements were conducted and preliminary results show that the novel detector is highly sensitive such that ultra-low dose experiments are feasible.

The experimental results (LSF, MTF and DQE) are shown in Fig 1, where it can be seen that in particular the MTF and the DQE exhibit promising features; remarkably better than the transmission of conventional detectors.

The sensitivity of this pnCCD detector is such that every single electron is detected. Individual imaging frames of 5ms exposure time each are collected and integrated over to obtain the final image and every single one of these frames registers 100-1000 electrons. Although the total time to obtain a full image is much longer than with conventional cameras (up to 100 sec), the extremely low electron impact on the sample provides perfect conditions for ultra-low dose imaging, as the average electron dose is 50-100 electrons/pixel ($2-5 \times 10^{-10}$ electrons/Angstroem^2) for an entire 100sec exposure.

M. Luysberg, K. Tillmann, T. Weirich (Eds.): EMC 2008, Vol. 1: Instrumentation and Methods, pp. 91–92, DOI: 10.1007/978-3-540-85156-1_46, © Springer-Verlag Berlin Heidelberg 2008

92

1. R. M. Glaeser et.al., J. of Microscopy 112 (1978) 127-138
2. H. G. Heide et al., Ultramicroscopy 16 (1985) 151-160
3. R. Henderson, Quart. Rev. Biophys. 37 (2004) 3-13
4. R. Faruqi et.al., Quart. Rev. Biophys. 33 (2000) 1-27
5. G. McMullan et.al., Ultramicroscopy 107 (2007) 401-413
6. R. Hartmann et.al., Nucl. Instr.& Meth. Phy. Res A568 (2006) 118-123
7. We thank the European Union for financial support within the 3DEM network of excellence (NoE).

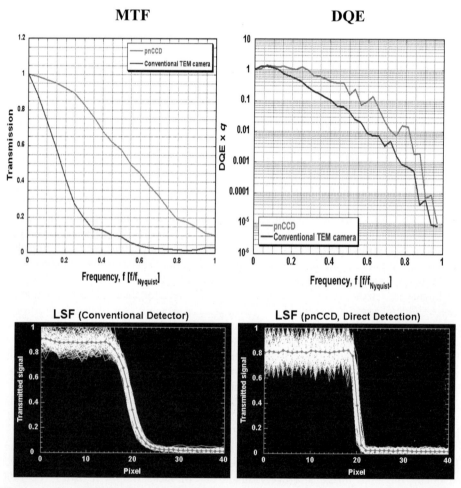

Figure 1. The MTF, DQE and LSF for the pnCCD detector are shown. The data for the LSF and the MTF were not fitted except for finding and re-aligning of the knife-edge to a vertical line on the recorded images.

Quantitative TEM and STEM Simulations

C.T. Koch

Stuttgart Center for Electron Microscopy
Max Planck Institute for Metals Research, Heisenberstr. 3, 70569 Stuttgart, Germany

koch@mf.mpg.de
Keywords: image simulation, HAADF-STEM, thermal diffuse scattering, Fresnel contrast

Recent advances in the development of electron optics have pushed the available information limit in high resolution electron microscopy (HRTEM) and scanning transmission electron microscopy (STEM) well beyond the 1Å boundary. With increasing resolution multiple elastic scattering of electrons within the sample and thermal diffuse scattering make a direct image interpretation increasingly difficult, calling for quantitative methods to simulate images to compare with the experiment.

I will report on software for simulating TEM and STEM images that was designed under the following considerations:

- Potential slicing should work for arbitrary samples (e.g. interfaces, defects, etc., not just perfect crystals in zone axis) and arbitrarily thin slices.
- It should be possible to compute images for arbitrary orientations, not just low-index zone axes of single crystals.
- Atomic scattering factors should be accurate up to the large angles needed for STEM simulations, but also at low scattering angles, reflecting charge distribution.
- TEM and STEM simulations should be quantitative.
- Accuracy over speed, but still try to be as fast as possible.
- Allow very large simulations to be run on any computer.

In addition to employing a special algorithm design these goals have been realized by:

- Computing the 3-dimensional potential around every atom and properly assigning slices of the atomic potential to slices of the multislice simulation. This 'slicing of atoms' allows correct simulations for arbitrarily tilted crystalline specimen.
- Use the frozen phonon approximation for TEM and STEM simulations (optional).
- Use scattering factors for neutral atoms and ions tabulated up to $\sin(\theta)/\lambda=6$ [2], allowing for different (arbitrary) charges on every atom by interpolation.

I will present TEM, STEM, and electron diffraction simulations of dislocations, grain boundaries, and (charged) point defects. Using selected examples I will demonstrate the effect of charge distribution on (Fresnel) TEM contrast of interfaces and of the ADF detector configuration on the amount of diffraction contrast in ADF STEM images.

1. This software can be downloaded at www.mf.mpg.de/en/organisation/hsm/koch/stem/
2. D. Rez, P. Rez, and I. Grant, Acta Cryst.A**50** (1994) 481
3. Financial support from the European Commission under contract nr. NMP3-CT-2005-013862 (INCEMS) is acknowledged. I want to thank Prof. Peter Rez very much for providing the tabulated scattering factors [2] in electronic form.

M. Luysberg, K. Tillmann, T. Weirich (Eds.): EMC 2008, Vol. 1: Instrumentation and Methods, pp. 93–94, DOI: 10.1007/978-3-540-85156-1_47, © Springer-Verlag Berlin Heidelberg 2008

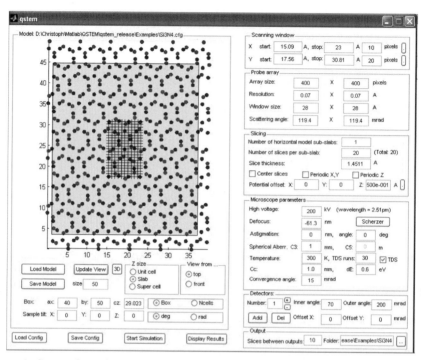

Figure 1. Screenshot of the user interface in STEM mode demonstrating the setup of the scan window and the size of the super cell needed for the computation.

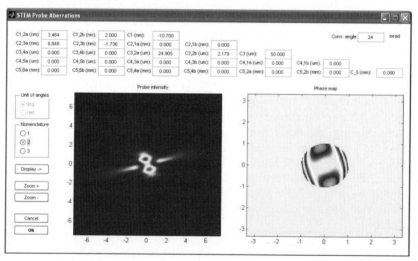

Figure 2. The software allows the user to define all aberrations up to C_5. The user can switch between different nomenclatures for aberration coefficients.

Quantitative determination of the chemical composition of an alloy by High Angle Annular Dark Field imaging

V. Grillo[1], F. Glas[2] and E. Carlino[1]

1. Laboratorio Nazionale TASC-INFM-CNR, Area Science Park, S. S. 14, Km 163.5, 34012 Trieste, Italy
2. CNRS, Laboratoire de Photonique et de Nanostructures, Route de Nozay, 91460 Marcoussis, France

grillo@tasc.infm.it

Keywords: Quantitative HAADF, Z-contrast, Static Displacement

Scanning transmission electron microscopy (STEM) high angle annular dark field (HAADF) imaging is now a powerful methodology for studying the structural and chemical properties of solid-state matter at atomic resolution. As the HAADF image intensity depends strongly on the atomic number of the elements in the atomic columns of the specimen, important qualitative elemental and structural information can be achieved by direct inspection of the atomic resolution HAADF image contrast. Nevertheless, if quantitative information on the chemistry of the specimen is sought for, it becomes necessary to couple the experimental results with HAADF image simulations.

In this context we propose a method to measure the chemistry of a specimen from the quantitative evaluation of the relevant HAADF image intensity [1]. The approach can be applied to many material systems. As the method is based on the comparison between the intensity of the unknown alloy with that of an alloy with known composition, it is of particular interest for a large class of specimens such as, for example, heterostructures.

The method requires the accurate calculation of the variation of the HAADF image intensity as a function of specimen composition and thickness. Other parameters have to be taken into account in the calculations, depending on the features of the material to be investigated [1]. For example, to properly measure the composition of InGaAs quantum wells (QWs) grown on (001) GaAs substrate (see below), it is necessary to take into account, among others parameters, the influence of the atomic static displacements (SDs) on the HAADF image contrast [2]. To this aim the realistic atomic structure for different In compositions in the presence of SDs was calculated by applying the valence force field model [2]. The resulting cells have been then used as input for the HAADF image calculation.

The need of accurate calculation of complex structure requires the use of the multislice method within the frozen phonon framework. Unfortunately, for realistic complex structures, this approach requires an extremely large calculation time. Hence, we developed a parallel version of the multislice algorithm reducing the computation time by orders of magnitude while retaining the accuracy of the approach [3].

M. Luysberg, K. Tillmann, T. Weirich (Eds.): EMC 2008, Vol. 1: Instrumentation and Methods, pp. 95–96, DOI: 10.1007/978-3-540-85156-1_48, © Springer-Verlag Berlin Heidelberg 2008

Fig 1a shows the calculated calibration curves relating the In concentration to the InGaAs HAADF image intensity, after normalization to GaAs, for different specimen thicknesses. The results have been applied to experiments on a sample containing three QWs grown by molecular beam epitaxy with calibrated compositions of 5%, 12% and 24% and prepared in the <110> cross section geometry following a procedure reported elsewhere [2]. In particular, fig 1b is an atomic resolution HAADF image of the QW with 24% of In content. Together with the calibration curve in fig 1a, the intensity in fig 1b can be used to produce the composition map in fig 1c. Fig 1d shows the same composition data averaged in the direction parallel to the well together with the composition profile calculated without taking into account the influence of the SDs on the image contrast. As demonstrated in fig. 1d the role of SDs on the HAADF image contrast cannot be neglected whenever accurate chemistry quantification has to be performed by HAADF imaging. Other applications of this methodology, along with the details of the approach, will be presented during the talk.

1. E. Carlino, V.Grillo, Phys. Rev. B. **71** (2005) p. 235303
2. V. Grillo, E. Carlino and F. Glas, Phys. Rev. B **77** (2008) p. 054103
3. E. Carlino, V. Grillo, and P. Palazzari, in Microscopy of Semiconducting Materials 2007 edited by A. G. Cullis and P. A. Midgley, IOP Conf. Proc. Institute of Physics, Bristol, in press

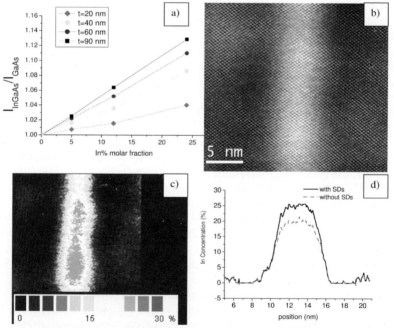

Figure 1. a) Calibration curves relating HAADF normalized intensity to In concentration at various thicknesses; b) Atomic resolution HAADF images of the $In_{0.24}Ga_{0.76}As$ QW; c) Composition map; d) Laterally averaged profile obtained from b), the dashed line was obtained not accounting for SDs.

The benefits of statistical parameter estimation theory for quantitative interpretation of electron microscopy data

S. Van Aert[1], S. Bals[1], L.Y. Chang[2], A.J. den Dekker[3],
A.I. Kirkland[2], D. Van Dyck[1] and G. Van Tendeloo[1]

1. Electron Microscopy for Materials Science (EMAT), University of Antwerp, Groenenborgerlaan 171, 2020 Antwerp, Belgium
2. Department of Materials, University of Oxford, Parks Road, Oxford, OX1 3PH, United Kingdom
3. Delft Center for Systems and Control, Delft University of Technology, Mekelweg 2, 2628 CD Delft, The Netherlands

Sandra.Vanaert@ua.ac.be

Keywords: quantitative structure determination, statistical parameter estimation theory

In principle, electron microscopy can provide accurate and precise numbers for the unknown structure parameters of materials under study. As an example, atomic column positions can be determined with accuracy and precision that is orders of magnitude better than the resolution of the microscope. This requires, however, a quantitative, model-based method. In our opinion, the maximum likelihood (ML) method is the most appropriate one since it has optimal statistical properties. This contribution aims to explain the basic principles and limitations of the method. Moreover its practical applicability and usefulness will be demonstrated using recent experimental examples.

Successful application of the ML method requires the availability of the joint probability density function of the observations (the recorded pixel values) and its dependence on the structure parameters to be estimated. This dependence is established by the availability of an expectation model, which is a model describing the expectation values of the observations. In fact, the expectation model includes all the ingredients needed to perform a computer simulation of an image or restored exit wave and it is parametric in the quantities of interest, such as the atomic column positions. A thus parameterized joint probability density function can be used to obtain ML estimates of the parameters. Details of this model-based method are given in [1-2].

In order to evaluate the limitations of the proposed method when applied to restored exit waves, a simulation study has been carried out. More specifically, the effects of model misspecification, noise, the restoration method, and amorphous layers have been quantified. The presence of amorphous layers, which is hard to avoid in realistic experimental conditions, has been found to be the most important factor limiting the reliability of atomic column position estimates. Figures 1(a) and 1(b) show the phase of a simulated exit wave of $SrTiO_3$ $\langle 001 \rangle$ without amorphous layers and with an amorphous layer both on the top and bottom of the sample, respectively. The symbols indicate the estimated atomic column positions. Table I shows the corresponding root mean-square errors, measuring the spread of the estimated positions about the true positions, which have been used as an input for the simulations. These numbers clearly

M. Luysberg, K. Tillmann, T. Weirich (Eds.): EMC 2008, Vol. 1: Instrumentation and Methods, pp. 97–98, DOI: 10.1007/978-3-540-85156-1_49, © Springer-Verlag Berlin Heidelberg 2008

98

show the strong influence of amorphous layers on the estimated atomic column positions. Moreover, amorphous layers seem to have a greater impact on the reliability of the position measurement of light element atomic columns (O) than on that of heavier element atomic columns (Sr and TiO). Furthermore, it can be shown that the root mean-square errors increase with increasing thickness of the amorphous layers and that they are independent of the restoration method. However, despite of the strong influence of amorphous layers, errors in the picometer range are still feasible.

Application of the proposed method to experimentally restored exit waves shows that atomic column positions and interatomic distances can be determined with a precision in the picometer range [3]. These results not only apply to well-resolved structures but also to projected distances beyond the information limit of the microscope which can be determined with picometer range precision. Moreover, the results are confirmed by means of a quantitative X-ray powder diffraction analysis. Other examples demonstrate the present possibilities for quantitative interface characterization using high angular annular dark field scanning transmission electron microscopy images [4] where the method allows us to distinguish between atomic columns with neighbouring atomic numbers.

1. A.J. den Dekker, S. Van Aert, A. van den Bos and D. Van Dyck, Ultramicroscopy **104** (2005), p. 83.
2. S. Van Aert, A.J. den Dekker, A. van den Bos, D. Van Dyck and J.H. Chen, Ultramicroscopy **104** (2005), p. 107.
3. S. Bals, S. Van Aert, G. Van Tendeloo and D. Ávila-Brande, Physical Review Letters **96** (2006), p. 096106-1
4. M. Huijben, G. Rijnders, D.H.A. Blank, S. Bals, S. Van Aert, J. Verbeeck, G. Van Tendeloo, A. Brinkman and H. Hilgenkamp, Nature Materials **5** (2006), p. 556.
5. The authors acknowledge financial support from the European Union under the Framework 6 program under a contract for an Integrated Infrastructure Initiative. Reference 026019 ESTEEM. S. Van Aert and S. Bals are grateful to the Fund for Scientific Research (FWO).

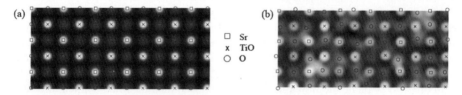

Figure 1. Phase of a simulated exit wave of SrTiO$_3$ $\langle 001 \rangle$ (a) without amorphous layer and (b) with an amorphous layer both on the top and bottom of the sample. The symbols indicate the estimated atomic column positions for the different atom types.

Table I. Estimated root mean-square errors for the Sr, TiO and O atomic columns obtained from the phase of a simulated exit wave (a) without amorphous layers and (b) with an amorphous layer both on the top and bottom of the sample.

RMSE (pm)	Sr	TiO	O
a	0.05	0.05	0.05
b	5.3	5.7	21.8

First time quantification of the HRTEM information-limit reveals insufficiency of the Young's-fringe test

J. Barthel, A. Thust

Institute of Solid State Research and Ernst Ruska-Centre for Microscopy and Spectroscopy with Electrons, Research Centre Jülich, Germany

ju.barthel@fz-juelich.de

Keywords: HRTEM, resolution, information limit, Young's fringes, partial temporal coherence

The information limit of a transmission electron microscope refers to the smallest object detail which is transferred in a linear way by a high-resolution transmission electron microscope. By definition, the information limit corresponds to the spatial frequency where the temporal coherence envelope drops to a value of $1/e^2 = 13.5\%$ [1].

Since a long time the Young's fringe test is used as a standard method to determine the information limit [2]. However, due to several stringent methodological considerations, the Young's fringe test is totally unsuitable for a serious quantification of the information limit [3]. Apart from the unjustified neglect of non-linear contrast contributions, this method reveals only a resolution restriction due to the cumulation of very different effects, such as the microscope's information limit, the object scattering-function, mechanical vibrations, and the modulation-transfer-function of the detector.

We present a new quantitative method, which allows one to measure directly and seperately the information limit of transmission electron microscopes from diffractograms of high-resolution micrographs. The micrographs are recorded from thin amorphous objects under tilted illumination. Our measurement principle is based on the fact that large beam tilts cause an anisotropic deformation of the diffractogram damping envelope and the appearance of an additional "holographic" background contribution (Fig. 1). These two effects are primarily caused by the partial temporal coherence of the electron beam. The information limit is determined by fitting a model function based on a Gaussian focal distribution to the envelope and the background extracted from experimental diffractograms (Fig. 2).

The new method was applied to measure the information limit of two instruments installed at the Ernst Ruska-Centre in Jülich. Results of these measurements are listed in Table I. In general, the information limit measured by our new method differs significantly from the result of the traditional Young's fringe method. The most prominent discrepancy occurs at 80 kV accelerating voltage: Whereas the traditional (and wrong) interpretation of the Young's fringe test leads to an unexpectedly good information limit near 1.1 Å, the true information limit determined by our new method is only 1.8 Å at 80 kV.

1. K.J. Hanßen, L. Trepte, Optik **32** (1971), p. 519.
2. J. Frank, Optik **44** (1976), p. 379.
3. J.C.H. Spence, "Experimental High Resolution Electron Microscopy", Oxford University Press (1988).

M. Luysberg, K. Tillmann, T. Weirich (Eds.): EMC 2008, Vol. 1: Instrumentation and Methods, pp. 99–100, DOI: 10.1007/978-3-540-85156-1_50, © Springer-Verlag Berlin Heidelberg 2008

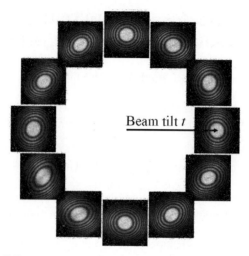

Figure 1. Series of diffractograms arranged according to the azimuth of the beam-tilt. The long-scale intensity anisotropy of the diffractograms due to the partial temporal coherence, rotating with the beam-tilt azimuth, is clearly visible.

(a)   (b)

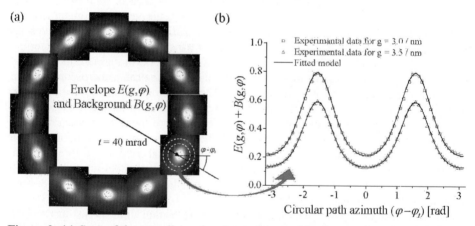

Figure 2. (a) Sum of the two-dimensional envelope and background extracted from the diffractogram series of Fig. 1. (b) Two intensity profiles along the selected circular paths shown in (a).

Table I. Measured information limit for two C_S-corrected high-resolution transmission electron microscopes installed at the Ernst Ruska-Centre in Jülich.

Instrument	Information limit	Young's fringe test
FEI Titan 80-300 (80 kV)	1.86 (±0.09) Å	~1.1 Å
FEI Titan 80-300 (200 kV)	1.11 (±0.01) Å	~1.0 Å
FEI Titan 80-300 (300 kV)	0.83 (±0.02) Å	~0.8 Å
Philips CM200 (200 kV)	1.10 (±0.02) Å	~1.3 Å

Quantitative Investigations of the Depth of Field in a Corrected High Resolution Transmission Electron Microscope

J. Biskupek[1], A. Chuvilin[1], J.R. Jinschek[2], and U. Kaiser[1]

1. Electron Microscopy Group of Material Science, Ulm University, Ulm, Germany
2. FEI Company, Eindhoven, Netherlands

johannes.biskupek@uni-ulm.de
Keywords: HRTEM, aberration correction, image simulation, depth sectioning

Aberration correction in modern TEMs moves the interpretable resolution level to the ultimate information limit of the microscope in the sub-Ångström region. It was expected for HRTEM lattice imaging that the use of "Scherzer focus" settings will lead once again to directly interpretable images. Reduced aberrations directly result in larger opening apertures and thus in smaller depth of field. It is extremely difficult to determine experimentally the correct focus; even the smallest default focus step in a TEM may be too coarse. The acquisition of focus series followed by exit-wave restoration have been recommended for choosing the focus and, in this way, helping to find the perfect imaging condition by defocus post-propagation via the reconstruction software.

Here, we present investigations of the depth of field of aberration corrected TEM at different high tension settings (FEI Titan 300kV and 80kV) by analysing focus series of images of ultra small (~2nm) Si particles within a SiO_2 matrix (details about the preparation of the particles see [1]). Fig.1. depicts the drastic influence of different defocus values. A small focus change (Δf) of only 8nm leads to an almost completely disappearance of the contrast of the Si nanoparticle marked as NC#1 in Fig.1. However, the Si nanoparticles marked NC#2 is clearly in focus (Fig.1b). The experiments are accompanied by calculations using molecular dynamics for model preparation and multi-slice algorithm for image simulation. This systematic study includes the investigation of the influence of high tension, coherence (beam convergence), and signal-to-noise ratio (electron dose) on the contrast and the depth of focus. Fig.2 shows calculated HRTEM images (HT=300kV, C_S=-5µm). The model contains 2.5nm sized Si nanocrystals of different orientations in different depth positions within the matrix of amorphous SiO_2. Small focus changes of few nm already lead to an appearance or disappearance of the contrast of nanocrystals, respectively. The image calculations show that Scherzer focus conditions result only in a focussed image of the sample surface and are not useful if ultra small structures within amorphous volumes have to be imaged. We revealed by experiments and simulations that aberration correction allows z-resolution of less then 1nm. Increasing the semi-opening angle (reducing aberrations) while lowering HT (e.g. 20mrad at 300kV, 42mrad at 80kV for 1Å resolution) enhances the z-resolution even more and allows to obtain atomic 3D information about the sample. We will present first results on atomic depth sectioning as additional approach to TEM-tomography.

1. A. Romanyuk, Oelhafen, V. Melnik, Y. Olikh, J. Biskupk, U. Kaiser, M. Feneberg, K. Thonke, App. Phys. Lett. submitted
2. We are grateful to A. Romanyk of Uni Basel for supplying Si nanocrystals within SiO_2 on Si substrates. We acknowledge the support of DFG

M. Luysberg, K. Tillmann, T. Weirich (Eds.): EMC 2008, Vol. 1: Instrumentation and Methods, pp. 101–102, DOI: 10.1007/978-3-540-85156-1_51, © Springer-Verlag Berlin Heidelberg 2008

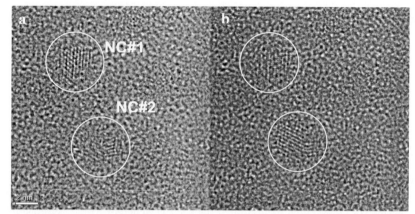

Figure 1. Exit-wave reconstruction of experimental HRTEM images (300kV) of Si nanocrystals in SiO$_2$. (a) and (b) differ by a focus change of 8.0 nm. (a) NC#1 is in focus, (b) NC#2 is in focus

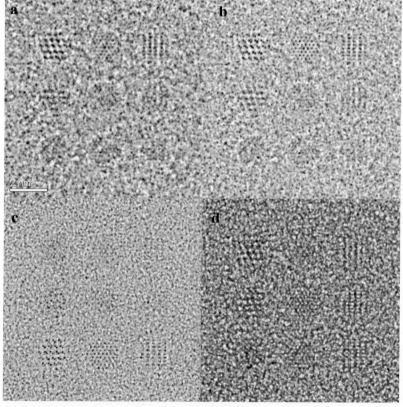

Figure 2. Calculated focus series (300kV, C$_S$=-5μm) of Si nanocrystals (2.5nm size, [110], [111], [112] orientations) at different sample depths (depths of 2.8nm, 5.6nm, 8.4nm) within SiO$_2$ (11nm thick). (a) defocus Δf =-15.0nm, (b) Δf =-13nm (c) Δf =-4.0nm, (d) Δf = +2.0nm.

Quantitative characterisation of surfaces on bi-metallic Pt nanoparticles using combined exit wave restoration and aberration-corrected TEM

L.Y. Chang[1,2], C. Maunders[1], E.A. Baranova[3], C. Bock[3], and G. Botton[1]

1. Canadian Centre for Electron Microscopy, Brockhouse Institute for Materials Research and Dept of Materials Science and Engineering, McMaster University, 1280 Main St. West, Hamilton, ON, L8S 4L7, Canada.
2. Dept. of Materials, University of Oxford, Parks Road, Oxford, UK. OX1 3PH
3. Institute for Chemical Processes and Environmental Technology, National Research Council, Montreal Road, Ottawa, ON, K1A 0R6, Canada

lanyun.chang@materials.ox.ac.uk

Keywords: nanoparticles, surfaces, exit wave restoration

The active components of heterogeneous catalysts are often small metallic particles, whose reactivity and selectivity depend on the structure of their surfaces. In particular, the surface defect structures such as steps and kinks play a key role in their catalytic properties [1]. However, little is known about the detailed surface structures on nanoparticles (<10nm) at atomic resolution, as commonly used surface characterisation techniques are not suitable for nanoparticles of such size range.

High-Resolution Transmission Electron Microscopy (HRTEM) is a firmly established technique to study the atomic structures of nanoparticles. However, HRTEM images suffer from objective lens aberrations and generally are very difficult to interpret. The resolution, interpretability, and signal to noise ratio of HRTEM can be improved by combining (objective lens) aberration correctors and exit wave restoration [2]. The restored phase of the exit wave function gives a directly interpretable representation of the atomic structures, and such information can then be used to quantitatively analyse the detailed surface structures for nanoparticles.

In the present work, we have examined the PtRu nanoparticles, which have been developed for tuning the efficiency for fuel cells [3]. Figure 1 shows the HRTEM image of one complex PtRu nanoparticle acquired with a Titan 80-300 Cubed located at the Canadian Centre for Electron Microscopy. The surface terraces, steps and also multiple twinned structures in the nanoparticle can be seen immediately in the image. In order to quantitatively understand local topologies on the surfaces of nanoparticles at atomic resolution, an iterative surface atomic position refinement scheme, as illustrated in Figure 2, has been developed to determine the surface atomic positions at sub-pixel resolution. Such information can, in turn, provide input into synthesis and structure-activity correlation of these systems.

M. Luysberg, K. Tillmann, T. Weirich (Eds.): EMC 2008, Vol. 1: Instrumentation and Methods, pp. 103–104, DOI: 10.1007/978-3-540-85156-1_52, © Springer-Verlag Berlin Heidelberg 2008

1. J. M. Thomas et al. in "Principles and practice of heterogeneous catalysis", (VCH, Cambridge), 1997.
2. A. I. Kirkland, L. Y. Chang, S. Haigh and C. J. D. Hetherington, Current Applied Physics, **8** (2008), p. 425.
3. E. Baranova, C. Bock, D. Ilin, D. Wang and B. MacDougall, Surface Science, **600** (2006), p. 3502.

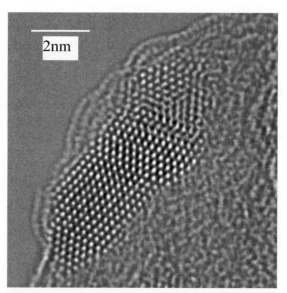

Figure 1. High-resolution image of a PtRu nanoparticle, taken from aberration-correced Titan 80-300 Cubed.

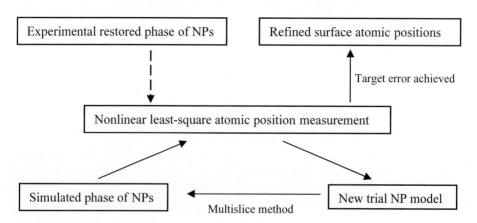

Figure 2. Schematic diagram of the nonlinear least-square surface atomic position refinement for nanoparticle (NP) systems.

An HAADF investigation of AlAs-GaAs interfaces using SuperSTEM

A.J. Craven[1,2], P. Robb[1,2] and M. Finnie[1]

1. Dept Physics & Astronomy, University of Glasgow, Glasgow, G12 8QQ, UK
2. SuperSTEM Laboratory, Daresbury Laboratory, Daresbury, WA4 4AD, UK

a.craven@physics.gla.ac.uk
Keywords: High angle annular dark field, aberration corrected STEM, interfaces

The quality of the interfaces in III-V heterostructures deposited by molecular beam epitaxy (MBE) plays a key role in the performance of semiconductor devices. For example, for AlAs grown on a (001) GaAs substrate, the surface prior to deposition of the AlAs would be As terminated and flat and the next layer formed would be Al. In reality, the growth can result in "islands" where an extra monolayer (ML) forms in some areas and not others. Here, one ML is defined as one unit of the structure i.e. GaAs or AlAs. These islands are bounded by terraces where AlAs and GaAs meet in the same plane. The width of the interface is then controlled by the number of levels of island present, as indicated schematically in Figure 1a. Alternatively, the interface could be ideally planar but there could be some diffusion of species across it. Thus, when looking down [110], the atomic columns formed from the Group III atoms would contain random distributions of Ga and Al atoms and the average content would vary across the interface. Its width is then controlled by the diffusion, as illustrated schematically in Figure 1b.

Since high angle annular dark field (HAADF) imaging gives true atomic resolution when used in an aberration corrected scanning transmission electron microscope such as SuperSTEM1 [1], its use to image the dumbbells along the [110] direction should give key information about the interface quality. However, since the signal is the sum of the scattering along the column, it is difficult to differentiate between random stepping and diffusion. One way to do this is to investigate how the interface width varies with the specimen thickness. In the case of diffusion, the step width should be independent of the thickness whereas, in the case of terracing, the width should tend to zero once the specimen is thinner than the dimensions of a typical island.

A heterostructure was grown on (001) GaAs giving GaAs-on-AlAs and AlAs-on-GaAs interfaces with a range of separations. A specimen was prepared by standard cross-sectioning using dimpling and ion milling and finished with 400eV ions. HAADF images were recorded using SuperSTEM1 with a probe half-angle of 24mrad and a HAADF detector with 70mrad inner and 210mrad outer half-angles. An example is given in Figure 2. The Fourier transform (FT) of the images showed reflections out to a spacing of 1Å. The background under the dumbbells was removed by applying a mask that removed the central spot of the FT and then performing an inverse FT. A constant value was then added to make the intensity between the dumbbells zero. Line profiles were taken along [001] using a width of 10 pixels normal to the line, as shown in Figure 2. Profiles were taken along 20 adjacent lines of dumbbells and the odd and

M. Luysberg, K. Tillmann, T. Weirich (Eds.): EMC 2008, Vol. 1: Instrumentation and Methods, pp. 105–106, DOI: 10.1007/978-3-540-85156-1_53, © Springer-Verlag Berlin Heidelberg 2008

even numbered traces were averaged separately to give the shapes of the two interleaving sets of dumbbells. The intensity ratio of the group III and group V columns was calculated for each dumbbell and plotted against position, as in Figure 3. The width of the interface was defined as the distance over which the ratio rose from 5% to 95% of the overall change. The width of the interface for AlAs-on-GaAs increased from ~3ML to ~6ML as the thickness increased from 40nm to 100nm whereas the width was constant at ~3ML for GaAs-on-AlAs.

Frozen phonon multislice calculations using the Kirkland code [2] showed that ideal interfaces would have a width of zero ML. When the probe is incident on a Ga column, the calculations also showed that the electrons channelled on the column are scattered away after ~20nm. Any further increase in the HAADF signal is from scattering off other columns and forms the background under the dumbbells. However, for an Al column, channelled electrons are still present after >100nm. Thus the fact that the experiment showed that the width of one interface increased steadily with thickness while the other one was independent of it could also be explained if the surface of the wafer were not exactly parallel to (001), as shown schematically in Figure 1c. On the side the beam meets GaAs first, the width is determined by the first 20nm of material and after this the interface will appear independent of thickness. Where the beam meets AlAs first, the width will increase with increasing thickness. Hence careful consideration must be given to the channelling depth when interpreting interface widths.

1. M. Falke, U. Falke, A Bleloch, S. Teichert, G. Beddies and H.J. Hinneberg, App. Phys. Letts. **86** (2005) 203103.
2. E.J. Kirkland "Advanced Computing in Electron Microscopy" (Plenum Press, New York) (1998).
3. We would like to thank Dr M Holland for growing the wafer, the SuperSTEM team for help and advice and the EPSRC for funding under grant GR/S41036.

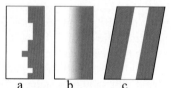

Figure 1. Schematic interfaces between GaAs (grey) and AlAs (white).

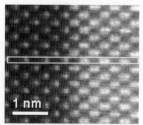

Figure 2. HAADF image of AlAs-GaAs interface.

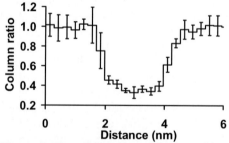

Figure 3. Plot of the ratios of the intensities of the III column to the V column in a dumbbell across GaAs to AlAs to GaAs.

Spatial Coherence and the Quantitative Interpretation of Atomic Resolution Images

C. Dwyer[1,2], J. Etheridge[1,2]

1. Monash Centre for Electron Microscopy, Monash University,Victoria 3800, Australia
2. Department of Materials Engineering, Monash University, Victoria 3800, Australia

christian.dwyer@mcem.monash.edu.au
Keywords: coherence, contrast, HREM, STEM, Ronchigram

The inability to obtain a quantitative computational match to the absolute intensity in experimental images and diffraction patterns obtained using the transmission electron microscope [1-3] has placed fundamental limits on the information that can be obtained about the specimen. It has also frustrated attempts to provide rigorous and meaningful definitions of image resolution by which instrument performance can be measured. Whilst it has been shown that a number of factors will mitigate contrast in images [4-10], such as amorphous surface layers and inelastic scattering, none of these factors can account for the magnitude of the mismatch between observation and calculation.

In the present paper, we examine the effect of partial spatial coherence on the intensity distribution in various imaging and diffraction experiments, including point projection phase contrast images and annular dark field images. For a sufficiently small, quasimonochromatic electron source, there is an angular range within the aperture of the probe-forming lens which will be illuminated with electron waves that are transversely coherent. Using field emission guns and under routine imaging conditions, this angular range is only a small percentage of a typical aperture size, so that the illumination is only partially coherent across the aperture.

The effect of partial coherence can be incorporated accurately by treating each point of the source as generating a coherent wave field and then summing incoherently over all source points with an appropriate weighting. Using this approach, the effect of partial coherence on various image types can be calculated. Figures 1a and b, respectively, show indicative calculations of ADF STEM images and point projection phase contrast images (Ronchigrams) of <210> zone axis diamond taken under identical illumination conditions. The degree of source coherence ranges from perfect coherence (i) to significant incoherence (iii). The image contrast is observed to decrease dramatically with reduction in coherence.

The degree of partial coherence for a Cs-corrected Titan 80-300 TEM with a Schottky field emission gun has been measured (reported elsewhere at this conference). The images in Figure 1 (ii), a and b, correspond approximately to this condition and illustrate the significant damping of contrast in experimental images using Schottky field emission sources.

1. M.J. Hytch, W.M. Stobbs, Ultramicroscopy **53** (1994) 191.
2. C.B. Boothroyd, R.E. Dunin-Borkowski, W.M. Stobbs, C.J. Humphreys, in: D.C. Jacobson, D.E. Luzzi, T.F. Heinz, M. Iwaki (Eds.), Proceedings of the MRS Symposium, vol. **354** (Pittsburgh) (1995), p. 495.

M. Luysberg, K. Tillmann, T. Weirich (Eds.): EMC 2008, Vol. 1: Instrumentation and Methods, pp. 107–108, DOI: 10.1007/978-3-540-85156-1_54, © Springer-Verlag Berlin Heidelberg 2008

3. K.V. Hochmeister, F. Phillipp, in: C.O.E.S.O. Microscopy (Ed.), Electron Microscopy 96, vol. **1**, Brussels, (1998) p. 418.
4. C.B. Boothroyd, J. Microscopy **190** (1998) 99.
5. D.M. Bird, M. Saunders, Ultramicroscopy **45** (1992) 241.
6. J.M. Zuo, J.C.H. Spence, M. O'Keeffe, Phys. Rev. Lett. **61** (1988) 353.
7. R.E. Dunin-Borkowski, R.E. Schaublin, T.Walther, C.B. Boothroyd, A.R. Preston, W.M. Stobbs, in: D. Cherns (Ed.), Proceedings of the Electron Microscopy and Analysis Group Conference on EMAG, vol. **147**, (IOPP Bristol) (1995) p. 179.
8. J.C.H. Spence, Mater. Sci. Eng. R **26** (1999) 1.
9. C.B. Boothroyd, Ultramicroscopy **83** (2000) 159.
10. W.Qu, C.B.Boothroyd and A. Huan Applied Surface Science **252** (2006) 3984

(a)

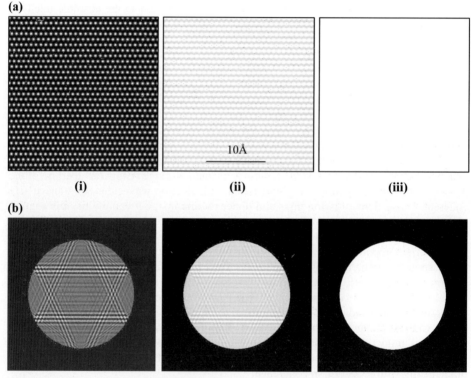

(i)   **(ii)**   **(iii)**

(b)

Figure 1. (a) Computed ADF-STEM images (top row) and (b) Ronchigrams (bottom row) of diamond <210> for (i) perfect coherence (effective source size is zero), (ii) partial coherence (effective source size of 0.7Å, corresponding approximately to a Schottky FEG on a Titan 80-300) and (iii) weak partial coherence (effective source size is 1.06 Å). In each case the beam energy is 300keV, the convergence semi-angle is 25mrad and the specimen thickness is 191 Å. The absolute intensity has been scaled equivalently in all images while maintaining a maximum intensity corresponding to white in each image.

Analysis of HRTEM diffractograms from amorphous materials: a simple and minor (but not explained so far?) question revisited

T. Epicier

Université de Lyon; INSA-Lyon, MATEIS, umr CNRS 5510, bât. B. Pascal, F-69621 Villeurbanne Cedex

Thierry.epicier@insa-lyon.fr
Keywords: HREM, diffractograms, amorphous materials

High Resolution Transmission Electron Microscopy (HRTEM) of non-crystalline or amorphous materials generally serves to the structural characterization of the material itself [1-4], or more practically to the measurement of optical parameters such as the coefficient of spherical aberration C_s and the defocus value δf, according to more or less sophisticated approaches based on the analysis of diffractograms as with the Krivanek's method [5]. To be accurate, these techniques require however the validity of the Weak Phase Object (WPO) approximation: it was demonstrated out by Gibson [6], and further confirmed in details [7] that breakdown of the 'projection' approximation (i.e. the sample is considered to have a zero-thickness), implicitly included in the WPO approximation, occurs in relatively thin amorphous materials, leading to false C_s and δf determinations.

However, these investigations left unexplained the fact that experimental diffractograms frequently present a strong background intensity, easily evidenced on rotationally-averaged profiles, as depicted in figure 1 a-b). According to the theory of HRTEM image formation in the case of a phase object, minima of this profile correspond to the 'cut-offs' of the contrast transfer function of the microscope $CTF(q)$ (where q is the spatial frequency), and in principle should thus be exact zeros.

It is emphasized here that this feature is simply due to a non-uniform thickness t of the examined area, combined to the breakdown of the 'projection' assumption [6]. In the frame of the kinematical theory (dynamical interactions between scattered electrons are weak enough to be ignored even at relatively large thickness in the case of amorphous materials [9]), and neglecting any astigmatism, it will be shown that the intensity of the rotationally-averaged profile of a diffractogram, or the modulus of the Fourier transform of the HRTEM image $D(q)$, can be written as:

$$D(q) \propto \frac{1}{(t_{max}-t_{min})} \int_{t_{min}}^{t_{max}} |CTF(q, \delta f+\tau)| \, d\tau \qquad /1/$$

where the proportionality includes the envelope function describing partial coherence of the electron beam and the structure factor of the object. The important feature in relation /1/ is the thickness τ, which induces a propagation phase shift for

M. Luysberg, K. Tillmann, T. Weirich (Eds.): EMC 2008, Vol. 1: Instrumentation and Methods, pp. 109–110, DOI: 10.1007/978-3-540-85156-1_55, © Springer-Verlag Berlin Heidelberg 2008

electrons scattered in the direction **q**, the effect of which is similar to applying a 'effective' defocus $\delta f + \tau$.

Figure 1c) illustrates the present simple model. It is worth noting that the rotationally-average experimental profile exhibits minima which are clearly not zeros, which indicates that it cannot be matched by a unique thickness as it will be explicitly demonstrated. Taking the thickness variation into account (fig. 1 c) firstly modifies the minima of the CTF (which raise to non-zero values), and secondly shifts the positions of the CTF extrema, which can then fit correctly that of the experimental profile. As a consequence it is also evident that the correct evaluation of C_s and δf from such data requires to account for the non-uniformity of the thickness, which is then evidenced by the non-zero background of the rotationally-average profile from the diffractogram [10].

1. G.Y. Fan and J.M. Cowley, Ultramicrosc., 21, (1987), 125.
2. A. Saeed, P.H. Gaskell and D.A. Jefferson, Philos. Mag. B., 66, 2, (1992), 171.
3. G. Mountjoy, J. Non-Cryst. Solids, 293-295, (2001), 458.
4. A. Hirata, Y. Hirotsua, T.G. Niehb, T. Ohkuboc, N. Tanaka, Ultramicrosc., 107, (2007), 116.
5. O.L. Krivanek, Optik, 45, (1976), 97.
6. J.M. Gibson, Ultramicrosc., 56, (1994), 26.
7. H.S. Baik, T. Epicier, E. Van Cappellen, Eur. Phys. J. AP., 4, (1998), 11.
8. B. Van de Moortele, T. Epicier, J.L. Soubeyroux, J.M. Pelletier, Phil. Mag. Letters, 84, 4, (2004), 245.
9. L.D. Marks, Ultramicrosc., 25, (1988), 25.
10. The CLYM (Centre Lyonnais de Microscopie) is gratefully acknowledged for the access to the 2010F microscope.

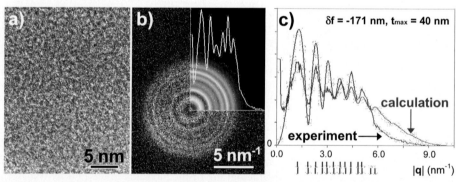

Figure 1. HRTEM imaging of a bulk metallic glass $Zr_{46.25}Ti_{8.25}Cu_{7.5}Ni_{10}Be_{27.5}$ (see [8] for details). a): HRTEM micrograph. b): corresponding numerical diffractogram, corrected from the background at high spatial frequencies. The upper-right quarter shows the reconstructed image from the rotationally-averaged profile shown in inset. c): 'match' of the extrema of the experimental profile (vertical bars below the diagram, upper black row) with that of expression 1 (lower blue row) with a thickness varying linearly from 0 to 40 nm (microscope JEOL 2010F, with $C_s = 0.51$ mm, spread-of-focus Δ and convergence half-angle θ_c of the incident beam taken respectively equal to 7 nm and 0.3 mRad).

HAADF-STEM image simulation
of large scale nanostructures

P. Galindo[1], J. Pizarro[1], A. Rosenauer[2], A. Yáñez[1], E. Guerrero[1], S.I. Molina[3]

1. Departamento de Lenguajes y Sistemas Informáticos, CASEM, Universidad de Cádiz, Campus Río San Pedro, s/n, 11510 Puerto Real, Cádiz, Spain.
2. Univ. Bremen. Inst.Solid State Physics. P.O. Box 330440, Bremen, 28334 Germany
3. Departamento de Ciencia de los Materiales e I.M. y Q.I., Facultad de Ciencias, Universidad de Cádiz, Campus Río San Pedro, s/n, 11510 Puerto Real, Cádiz, Spain.

pedro.galindo@uca.es
Keywords: HAADF-STEM simulation images, Z-contrast, Finite Element Analysis

The combination of strain measurements[1] and the analysis of High Angle Annular Dark Field (HAADF) Scanning Transmission Electron Microscopy (STEM) images constitutes a powerful approach to investigate strained heterostructures on the nanometer scale and has proved to be extremely powerful for characterizing nanomaterials[2,3].

STEM simulation tools can be used to obtain quantitative information at near atomic resolution[4]. However, image simulation requires an enormous amount of computing power compared with conventional TEM image simulation, since the dynamical scattering has to be evaluated at each scanning point. STEM image simulation of a few unit cells can take hours and the simulation of nanostructures, where millions of atoms are involved, is unfeasible in the state of the art personal computers. To overcome this problem, a new parallel HAADF-STEM simulation software has been developed, following the theory as described by Kirkland[5] and implementing thermal diffuse scattering using Ishizuka's FFT multislice approach[6]. The software is able to simulate images from nanostructures represented by about 1 million atoms in a couple of days. The software runs in the University of Cadiz cluster, having 320 nodes and 3.8 Tflops.

In order to run the software we need a detailed description of the nanostructure at the atomic level. In this work, a finite element analysis (FEA) commercial package has been used to describe the geometry of the nanostructure (subdomains, boundaries, constraints, symmetries, etc.) and to define the local parameters (composition, initial strains, elastic constants,...). The displacement field is obtained by solving the equations of the anisotropic elastic theory. The supercell is created by replicating the substrate unit cell coordinates across three dimensions of space and modifying atom properties using imported values from the FEA model at equilibrium, i.e. atom coordinates according to the local displacement field, occupancy or Debye-Waller factors according to the local composition, and so on. Anyway, other approaches such as molecular dynamics can be applied also, in order to generate the input supercell to the HAADF-STEM software.

In this paper we summarize the results obtained by applying this approach to generate HAADF-STEM simulated image from InAs self-assembled nanowires grown by solid-source molecular beam epitaxy at 515 °C on InP (001) substrates. Model size was defined to be 15x15x40 nm. The compositional distribution across the nanowire has been obtained using spatially resolved electron energy loss spectroscopy (EELS) and aberration-corrected high resolution HAADF-STEM imaging and introduced in the FEA model. Once the equations of the anisotropic elastic theory are solved, the supercell is generated and sliced into layers (5.8687 Å / layer). The parallel HAADF-STEM software performs in two stages: first, the phase grating is calculated for each layer, and afterwards, each pixel of the final image is obtained by running a complete multislice simulation followed by the integration of intensities across the annular HAADF detector. The overall simulation process was 48 hours for a 512x512 pixels resolution image and 135 hours for a 1024x1204 pixels resolution image.

1   P Galindo, S. Kret, A. Sanchez, J Laval, A. Yañez, J. Pizarro, E. Guerrero, T. Ben, S. Molina, Ultramicroscopy **107** 1186-1193 (2007)
2   S. I. Molina, T. Ben, D. L. Sales, J. Pizarro, P. L. Galindo, M. Varela, S. L. Pennycook, D. Fuster, Y. Gonzalez, and L. Gonzalez. Appl. Phys. Lett **91**,132112 (2007)
3   S. I. Molina, T. Ben, D. L. Sales, J. Pizarro, P. L. Galindo, M. Varela, S. L. Pennycook, D. Fuster, Y. Gonzalez, and L. Gonzalez. Nanotech. **17**, 5652-5658 (2006)
4   G.R. Antis, S. C. Anderson, C. R. Birkeland and D. J. H. Cockayne, Scanning Microscopy vol 11, 287-300 (1997)
5   E. J. Kirkland, Plenum Press, New York, ISBN 0-306-45936-1 (1998)
6   K. Ishizuka, Ultramicroscopy **90**, 71-83 (2002)

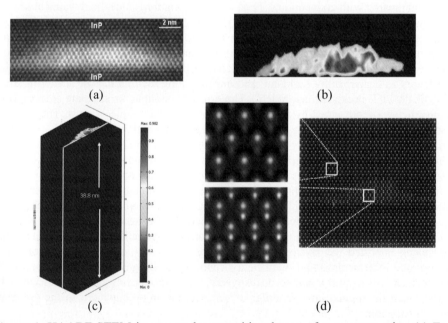

Figure 1. HAADF-STEM images and compositional map of a quantum wire: (a) Experimental image. (b) Compositional map. (c) 3D FEA model. (d) 1024x1024 resolution HAADF-STEM simulated image, showing details inside and outside the nanowire.

Aberration-corrected HRTEM study of incommensurate misfit layer compound interfaces

M. Garbrecht[1], E. Spiecker[1], W. Jäger[1], K. Tillmann[2]

1. Mikrostrukturanalytik, Institut für Materialwissenschaft, Christian-Albrechts-Universität zu Kiel, D-24143 Kiel, Germany
2. Ernst Ruska-Centrum und Institut für Festkörperforschung, Forschungszentrum Jülich GmbH, D-52425 Jülich, Germany

mag@tf.uni-kiel.de
Keywords: misfit layer compounds, incommensurate interfaces, aberration-corrected HRTEM

The development of spherical aberration (C_s) correctors for transmission electron microscopes offers new exciting opportunities for atomic scale investigations of solids. A main advantage of C_s-corrected HRTEM over conventional HRTEM is, that by tuning C_s to small negative values (NCSI imaging), image delocalisation can be reduced below the typical separation of atomic columns simultaneously with an optimisation of image contrast. This is an important feature when investigating interfaces in crystals, where the translation symmetry of the perfect crystal is broken and a wide range of different spatial frequencies contribute to the image contrast.

The transition metal dichalcogenide (TMDC) misfit layer compounds $(MX)_nTX_2$ (M = La, Sn, Pb or rare earths; X = S, Se, Te; T = Ta, Ti, Nb, V; n = 1.08....1.19) are ideal model systems for studying incommensurate interfaces because they exhibit a high density of such due to their particular microstructure [1]. They consist of sandwich-like stacks of alternating double atomic MX layers of distorted rocksalt structure and layers of TX_2. For the system $(PbS)_{1.14}NbS_2$ studied here, a nominal structure is schematically shown in Figure 1.

To investigate this class of compounds by HRTEM is challenging because of special structural features. The layer interfaces are incommensurate in one in-plane direction because of the different periodicities of the adjacent lattices. Also, light (S) and heavy atom columns (alternating Pb and S) occupy opposite positions at each interface.

We could already demonstrate that direct imaging of the complete projected crystal structure, including atomic column positions at the interfaces, becomes possible by applying C_s-corrected HRTEM performed with a Titan 80-300 instrument operated under NCSI-conditions [2][3]. Further investigations (Figure 2) reveal that interface phenomena, such as inhomogeneities, layer distortions and stacking disorder, that so far could only been investigated by spatially averaging X-ray diffraction methods, can be quantitatively analysed on a local scale.

Another characteristic of the atomic structure can be studied by imaging the misfit layer compound in a zone axis tilted 45° about the c-axis (using correspondingly cut samples). In this projection the contrast of Pb and S atom columns in the PbS sublattice becomes separated making it possible to quantitatively evaluate the protrusion of Pb atoms from the PbS layer due to interaction with the S atoms of NbS_2.

M. Luysberg, K. Tillmann, T. Weirich (Eds.): EMC 2008, Vol. 1: Instrumentation and Methods, pp. 113–114, DOI: 10.1007/978-3-540-85156-1_57, © Springer-Verlag Berlin Heidelberg 2008

1. G.A. Wiegers and A. Meerschaut, Mat. Sci. For. **100-101**, 101-172 (1992).
2. E. Spiecker, M. Garbrecht, C. Dieker, U. Dahmen, and W. Jäger, Microsc. and Microanal. **13 (Suppl. 3)**, 424 (2007).
3. M. Garbrecht, E. Spiecker, W. Jäger, K. Tillmann, "Quantitative Electron Microscopy for Materials Science", edited by E. Snoeck, R. Dunin-Borkowski, J. Verbeeck, and U. Dahmen (Mater. Res. Soc. Symp. Proc. **Volume 1026E**, Warrendale, PA, 2007), 1026-C10-01.

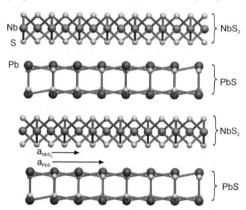

Figure 1. (colour) Schematic drawing of the atomic structure of $(PbS)_{1.14}NbS_2$. The crystal consists of an alternating stacking of two atomic sheets of cubic PbS and a sandwich layer of hexagonal NbS_2. The two substructures are coherent along the first interface direction, but incommensurate along the perpendicular one.

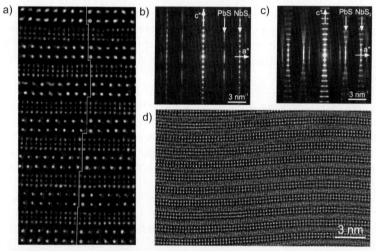

Figure 2. Stacking disorder and long-period undulations in $(PbS)_{1.14}NbS_2$. a) HRTEM image revealing the stacking disorder in the PbS substructure. b) Electron diffraction pattern along the commensurate crystal direction showing a streaking of the PbS reflections along c*. c) Electron diffraction pattern along the coherent direction showing an angular spreading of reflections due to a rotation of lattice planes in undulated layers. d) HRTEM-image of the undulated layers taken under NCSI-conditions.

Influence of atomic displacements due to elastic strain in HAADF-STEM simulated images

E. Guerrero[1], A. Yáñez[1], P. Galindo[1], J. Pizarro[1], S.I. Molina[2]

1. Departamento de Lenguajes y Sistemas Informáticos, CASEM, Universidad de Cádiz, Campus Río San Pedro, s/n, 11510 Puerto Real, Cádiz, Spain
2. Departamento de Ciencia de los Materiales e I.M. y Q.I., Facultad de Ciencias, Universidad de Cádiz, Campus Río San Pedro, s/n, 11510 Puerto Real, Cádiz, Spain

elisa.guerrero@uca.es
Keywords: Strain Analysis, HAADF-STEM simulation images, Finite Element Analysis

In this work, we analyze the influence of atomic displacements due to elastic strain in HAADF-STEM simulated images. This methodology is demonstrated on an InP/InAs$_x$P$_{(1-x)}$ nanowire heterostructure theoretical model under strained/unstrained conditions. Since the HRTEM image simulation requires an enormous amount of computing power, all simulations have been obtained using a parallel version of HAADF-STEM simulation software[1] running in the University of Cadiz supercomputer (3.8 Tflops). This program is able to simulate HAADF-STEM images from nanostructures containing more than one million of atoms in a few days.

The model consists of a 15x15x20 nm InP/InAs$_x$P$_{(1-x)}$ nanowire heterostructure. The compositional distribution across the nanowire has been set to 100% of As content in order to be able to compare theoretical and observed strain quantitavely.

Two different simulations were run. The first one considers atoms at their original crystalline theorctical positions, without taking into consideration any strain. A more precise simulation is made calculating the displacement field in the nanostructure using Finite Element Analysis by solving the equations of the anisotropic elastic theory, and modifying accordingly atomic positions as a previous step to the HAADF-STEM simulation.

The resulting images from both simulations have been compared in terms of peaks intensities, showing some differences that depend directly on the strain considered in the original crystalline heterostructure.

An interesting analysis has been the evaluation of the differences between the theoretical strain calculated by the FEM model, and the simulated strain calculated on final images using different strain mapping techniques [2, 3].

1. P Galindo, etc. To be presented at EMC2008, Aachen, Germany.
2. P Galindo, S. Kret, A. Sanchez, J Laval, A. Yañez, J. Pizarro, E. Guerrero, T. Ben, S. Molina, Ultramicroscopy **107** 1186-1193 (2007)
3. M. J. Hÿtch, E. Snoeck and R. Kilaas, Ultramicroscopy 74 (1998) 131–146

M. Luysberg, K. Tillmann, T. Weirich (Eds.): EMC 2008, Vol. 1: Instrumentation and Methods, pp. 115–116, DOI: 10.1007/978-3-540-85156-1_58, © Springer-Verlag Berlin Heidelberg 2008

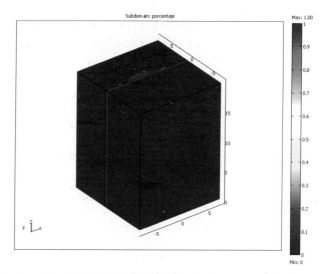

Figure 1. 3D FEA model of a 15x15x20 nm InP/InAs$_x$P$_{(1-x)}$ nanowire heterostructure

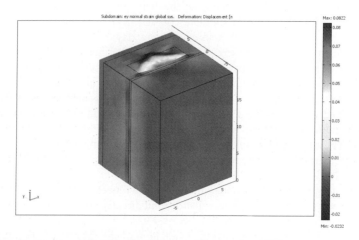

Figure 2. 3D displacement field in the heterostructure using FEA

Effects of electron channeling in HAADF intensity

M. Haruta, H. Komatsu, H. Kurata, M. Azuma, Y. Shimakawa and S. Isoda

Institute for Chemical Research, Kyoto University, Uji, Kyoto 611-0011, Japan

haruta@eels.kucir.kyoto-u.ac.jp
Keywords: HAADF-STEM, electron channeling, double perovskites

Atomic resolution imaging by HAADF-STEM can be applied to analyze atomic structures of materials directly, since this technique provides incoherent contrast correlating with the atomic number of constituent elements. In the present research, however, we report the strange contrasts making difficult to interpret intuitively the HAADF-STEM image of double perovskites oxide La_2CuSnO_6. Double perovskites oxide ($A_2BB'O_6$) is that two kinds of B site are ordered array material. In the majority of double perovskites oxides ($A_2BB'O_6$), B and B' cations are arranged in the rock-salt configuration, whereas Cu^{2+} and Sn^{4+} are arranged in a layered configuration only in La_2CuSnO_6 as an ambient pressure phase [1-3]. The unit cell dimensions of La_2CuSnO_6 are $a = 0.8510$ nm, $b = 0.7815$ nm, $c = 0.7817$ nm, $\alpha = 90.00°$, $\beta = 91.15°$ and $\gamma = 90.00°$. The layered crystal structure model determined by x-ray analysis is shown schematically in Figure 1. The La-La distance around Cu is shorter than that around Sn because the CuO_6 octahedrons are slightly distorted by the Jahn-Teller effect.

The layered double perovskites La_2CuSnO_6 films were grown heteroepitaxially on $SrTiO_3$ substrates by a pulsed laser deposition technique [4]. The growth temperature was 670°C with an oxygen partial pressure at 0.1 Torr, which are the conditions for the ordered structure. Cross-sectioned samples were thinned down to electron transparency by ion milling. High-resolution HAADF-STEM images were observed with a JEM-2200FS and a JEM-9980TKP1 which is equipped with Cs aberration corrector.

An observed HAADF-STEM image projected along b-axis of layered structure is shown in Figure 2(a). The contrast of Sn column appears brighter than La columns, although the atomic number of La is larger than that of Sn. Additionally, each contrast of La columns is different alternatively as shown in Figure 2(b, c). Figure 3(a) shows the intensity of HAADF image at each atomic site as a function of thickness calculated by multislice method [5], in which the electron probe is fixed on each atomic column. The intensity of Sn column almost consistently appears brighter than those of La columns and the intensity of La2 column is weaker than that of La1 column, reproducing the experimental results. These can be attributed to the different channeling process of electrons, because the La atoms are not arrayed straight along b-axis but have dumbbell structure in the projected plane, separated by 0.057 and 0.045 nm in La1 and La2 columns, respectively. Figure 3(b) shows the simulated channeling peak intensity at each column. The intensities on La sites are significantly weak compared to that on Sn site and the channeling intensity of La2 column is weaker than that of La1column. Therefore, the electron channelling breaks the Z-contrast in HAADF images.

M. Luysberg, K. Tillmann, T. Weirich (Eds.): EMC 2008, Vol. 1: Instrumentation and Methods, pp. 117–118, DOI: 10.1007/978-3-540-85156-1_59, © Springer-Verlag Berlin Heidelberg 2008

1. M. T. Anderson and K. T. Poeppelmeier, Chem. Mater. **3** (1991), 476.
2. M. T. Anderson, K. B. Greenwood, G. A. Taylor and K. R. Poeppelmeier, Prog. Solid State Chem. **22** (1993), 197.
3. M. Azuma, S. Kaimori and M. Takano, Chem. Mater. **10** (1998), 3124.
4. A. Masuno, M. Haruta, M. Azuma, H. Kurata, S. Isoda, M. Takano and Y. Shimakawa, Appl. Phys. Lett. **89** (2006), 211913.
5. K. Ishizuka. Ultramicroscopy **90** (2002), 71.
6. We acknowledge to Mrs. Kawasaki, Otsuka and Miss. Nishimura of TRC for preparing the cross-section sample.

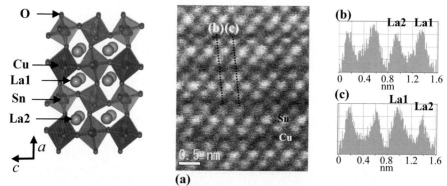

Figure 1. A layered structure model of La_2CuSnO_6.

Figure 2. (a) HAADF-STEM image of the regularly layered structure of La_2CuSnO_6. (b, c) intensity profile of La columns along the dotted lines in (a).

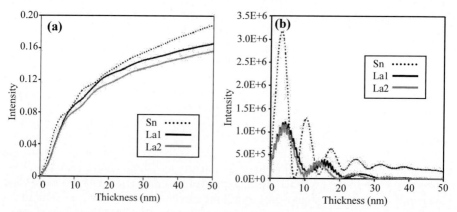

Figure 3. (a) Intensity at each atomic column in the simulated HAADF image. (b) Simulated channelling peak intensity at each atomic column.

Analysis of the mechanism of N incorporation in N-doped GaAs quantum wells

M. Herrera[1], Q.M. Ramasse[2], N.D. Browning[1,3], J. Pizarro[4], P. Galindo[4], D. Gonzalez[5], R. Garcia[5], M.W. Du[6], S.B. Zhang[7], and M. Hopkinson[8]

1. University of California-Davis, Dept. of Chemical Engineering and Materials Science, Davis CA 95616, U.S.A.
2. Lawrence Berkeley National Laboratory, 1 Cyclotron Road, Berkeley CA 94720, U.S.A.
3. Lawrence Livermore National Laboratory, 7000 East Avenue, Livermore CA 94550, U.S.A
4. University of Cadiz, Sección Departamental del Departamento de Lenguajes y Sistemas Informáticos, 11510 Puerto Real, Cadiz, Spain
5. University of Cadiz, Dep. De Ciencia de los Materiales, 11510 Puerto Real, Cadiz, Spain
6. Materials Science & Technology Division, Oak Ridge National Laboratory, Oak Ridge, TN 37831, U.S.A.
7. Department of Physics, Applied Physics, and Astronomy, Rensselaer Polytechnic Institute, Troy, NY 12180, U.S.A.
8. University of Sheffield, Dept. of Electronic and Electrical Engineering, Sheffield S1 3JD, U.K

mherrera@ucdavis.edu
Keywords: GaAsN, HAADF-STEM

The quaternary compound GaInNAs has attracted recent interest due to the possibility of obtaining laser diodes in the range 1.3-1.55 μm. The main characteristic of the GaInNAs/GaAs system is a strong negative bowing parameter that causes a rapid decrease of the bandgap by the addition of relatively small amounts of N (<5%) to GaInAs. Although the solubility of N in Ga(In)As is extremely low, GaInNAs layers with N content as high as 5% have been reported. Because of the small size of N, the incorporation of this element in As-sites produces a strong local strain that highly destabilizes the structure and could favour the formation of alternative N-containing complexes, instead of simple substitution on the arsenic site. As such, the optimization of the optoelectronic properties of this alloy is still a challenge, as witnessed by the observation of a strong reduction in the photoluminescence intensity associated with non-radiative recombination centres. In order to overcome these limitations, a full understanding of the incorporation mechanism associated with N alloying and its resulting atomic distribution in the alloy is essential.

In this work, the structural effect of the doping of GaAs with N has been studied at atomic scale by High Angle Annular Dark Field Scanning Transmission Electron Microscopy (HAADF-STEM). Interestingly, our analysis has shown a brighter contrast coming from the doped area in comparison to the GaAs substrate, as it can be observed

M. Luysberg, K. Tillmann, T. Weirich (Eds.): EMC 2008, Vol. 1: Instrumentation and Methods, pp. 119–120, DOI: 10.1007/978-3-540-85156-1_60, © Springer-Verlag Berlin Heidelberg 2008

in Fig. 1. The HAADF-STEM technique is sensitive to the Z number of the atoms in a perfect crystal, therefore the substitution of As atoms by N in GaAsN quantum wells (QWs) is expected to show darker contrast with regard to GaAs, contrary to what our results show. We have also analyzed the evolution of contrast with N content, in the range 0.1%-2.5%N, finding that the contrast increases with the composition in N. This means that there should be a N-related physical phenomenon behind the contrast reversal.

In order to find the origin of this contrast at atomic scale, high resolution HAADF-STEM images of the GaAsN QWs (as shown in Fig. 2 for 2.5%N) have been taken on an aberration-corrected VG HB501 dedicated STEM, and this experimental images have been simulated using the approximation from Ishizuka [1] to calculate the thermal diffuse scattering (TDS). For this, we have considered relaxed GaAsN complexes where N is located either in substitutional or in interstitial positions, according to the theoretical calculations by Zhang et al. [2]. Our simulations show that actually the addition of both substitutional and interstitial N to GaAs QWs produces brighter contrast in comparison to GaAs. This result highlights the importance of local distortions of the lattice in HAADF-STEM imaging, explaining our results as the effect of the static atomic displacements in the TDS in the sample. Further simulations are in progress in order to find out the evolution of contrast with composition for the different complexes with the aim to compare them with our experimental images and eventually find out the real location of N in GaAsN quantum wells.

1.   K. Ishizuka, Ultramicroscopy **90**, (2002) p. 71.
2.   S. B. Zhang and S-H. Wei, Phys. Rev. Lett. **86(9)**, (2001) p. 1789.
3.   This work was supported by a grant from the European Union under contract MOIF-CF-2006-21423 and in part by DOE grant number DE-FG02-03ER46057.

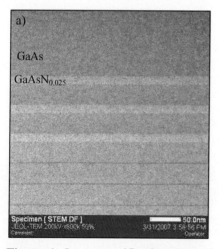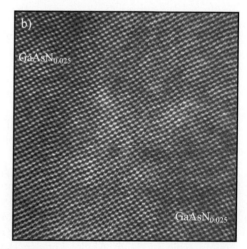

Figure 1. Low magnification (a) and high resolution (b) Z contrast images of the GaAsN quantum wells with N content of 2.5% N.

Coherence of high-angle scattered phonon loss electrons and their relevance to TEM and STEM ADF Stobbs Factors

R.A. Herring

Center for Advanced Materials & Related Technologies, Mechanical Engineering, University of Victoria, Victoria Canada V8W 3P6

rherring@uvic.ca

phonon loss electrons, thermal diffuse scattered electrons, electron holography, coherence

The degree of coherence (γ) and lateral coherence (δ) of high-angle phonon loss electrons scattered to angles (> 16 mrad) from a Germanium crystal were measured [1] using energy-filtered Diffracted Beam Interferometry/Holography (DBI/H) [2] that requires a TEM with an electron source with good coherence, an electron biprism for holography, an imaging energy filter (GIF) for separating the zero loss electrons and phonon loss electrons from the plasmon loss electrons, ionization loss electrons, etc., as well as, the condenser aperture focused on the diffraction plane for separating the zero loss electrons from the phonon loss electrons. The phonon loss electrons dominate the scattered electron beam intensity at high scattering angles >16 mrad [3], which depends on the material. A γ of ~0.1 was measured in Figure 1 at ~ 7 mrad. The fringes disappeared when the interfering beams were separated from the exact overlay position producing a measurement of the lateral coherence of ~0.075 nm on the specimen plane.

Recently, a Stobbs factor has been identified for STEM ADF imaging that required scaling of the contrast of simulated images to match the experimental images [4]. This STEM ADF Stobbs factor likely resulted from the coherence of the phonon loss electrons not being properly accounted for, which from the beginning have been considered incoherent [5].

The results presented here are pertinent to both STEM ADL imaging and TEM lattice imaging where the former uses 100% phonon loss electrons and the latter uses ~12% phonon loss electrons. STEM ADL images often comprise of ordered alloys having multiple different sources of phonon loss electrons that will each contribute delocalized information. The coherence will cause these sources to enhance the atomic column contrast when they are in phase and reduce the atomic column contrast when they are out of phase. Likewise with TEM lattice images, the coherence of the phonon loss electrons will enhance the lattice fringe contrast when they are in phase and reduce the lattice fringe contrast when they are out of phase with the zero-loss electrons and plasmon loss electrons. To this end, knowing the intensity and coherence of all sources of electrons is necessary to understand the contrast of fringes in high resolution atomic images produced in electron microscopes.

1. R.A. Herring, J. Appl. Phys. Submitted.
2. R.A. Herring, Ultramicroscopy 104 (2005) 261.

M. Luysberg, K. Tillmann, T. Weirich (Eds.): EMC 2008, Vol. 1: Instrumentation and Methods, pp. 121–122, DOI: 10.1007/978-3-540-85156-1_61, © Springer-Verlag Berlin Heidelberg 2008

122

3. Z.L. Wang Micron 34 (2003) 141.
4. M.P. Oxley, M. Varela, T.J. Pennycook, K. van Benthem, S.D. Findlay, A.J. D'Alfonzo, L.J. Allen and S.J. Pennycook, Phys. Rev. B 76 (2007) 064303.
5. S. J. Pennycook and D.E. Jesson, Phys. Rev. Lett 64 (1990) 938.
6. FEI and Bert Freitag are gratefully acknowledged for the use and operation, respectively, of their Titan microscope.   Grants from UVic, NSERC and the Canadian Foundation for Innovation are also gratefully appreciated.

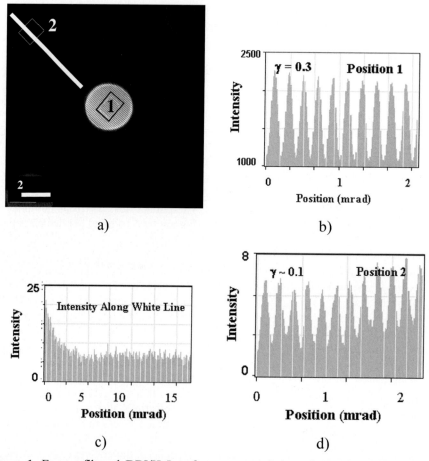

a)   b)

c)   d)

Figure 1. Energy-filtered DBI/H Interferogram consisting of zero-loss electrons and phonon loss electrons using the main beam (000) and the 111 diffracted beam of Ge in a), having a fringe intensity profile inside the high intensity area limited by the condenser aperture of $\gamma \sim 0.3$ at position 1 in b), an intensity distribution along the white line, which initially falls off exponentially then remains relatively constant at the high scattering angles in c), and the fringe intensity profile at position 2 having at a scattering angle of ~ 7 mrad and $\gamma \sim 0.1$.

Strain measurements in electronic devices by aberration-corrected HRTEM and dark-field holography

F. Hüe[1,2], F. Houdellier[1], E. Snoeck[1], V. Destefanis[3], J.M. Hartmann[2], H. Bender[4],
A. Claverie[1] and M.J. Hÿtch[1]

1. CEMES-CNRS, 29 rue Jeanne Marvig, 31055 Toulouse, France
2. CEA-LETI, 17 rue des Martyrs, 38054 Grenoble, France
3 STMicroelectronics, 850, Rue Jean Monnet 38926 Crolles, France
4. IMEC, Kapeldreef 75, 3001 Leuven, Belgium

florian.hue@cemes.fr
Keywords: strain, high-resolution, holography, semiconductors

The recent introduction of strained silicon in electronic devices has led to tremendous performance increases [1]. Strain can be engineered in active areas by different processes such as biaxial and uniaxial technologies. Measuring strain at the nanometre scale is therefore essential and has lead to the development of techniques like convergent-beam electron diffraction (CBED) and nanobeam diffraction [2]. Now, high-resolution transmission electron microscopy (HRTEM) can map strains in devices [3] and the new technique of dark-field holography appears extremely promising [4].

We show how strains can be mapped by these two methods. TEM specimens are prepared by focussed ion beam (FIB) and more traditional techniques (tripod and PIPS). Specimen thicknesses are 80 nm for HRTEM and 200 nm for dark-field holography. Observations are carried out on a the SACTEM-Toulouse, a Tecnai (FEI) 200kV TEM equipped with a Cs corrector (CEOS), rotatable biprism and 2k CCD camera (Gatan). High-resolution imaging is performed at nominal magnifications such as 97000 with exposure times of 2 seconds and strain fields are extracted by geometric phase analysis [5] using GPA Phase 2.0 (HREM Research Inc.) for DigitalMicrograph (Gatan).

Different devices have been investigated: p-MOSFET with recessed $Si_{80}Ge_{20}$ source and drain [3], (Figure 1), and strain-balanced {$c-Si_{0.6}Ge_{0.4}$ / t-Si} x 10 superlattices on $Si_{0.8}Ge_{0.2}$ virtual substrates for multi-channel MOSFET architectures [6] (Figure 2). The experimental results are compared with finite-element modelling to estimate the impact of thin foil relaxation on strain measurements. The result with the HRTEM method is a measurement precision of 0.3% and a spatial resolution of 4 nm. The field of view is typically 200 nm x 200 nm, but could be increased with a larger camera providing the contrast remains uniform. Dark-field holography is the more powerful technique but requires a biprism and a Lorentz-type lens. Measurements can be obtained with minimal thin-film relaxation, a precision of 0.05% and large fields of view: 1 μm x 300 nm.

1. S.E. Thompson, G.Y. Sun, Y.S. Choi, and T. Nishida, Trans. Elec. Dev. **53** (2006), p. 1010.
2. B. Foran., M.H. Clark, & G. Lian, Future Fab International **20** (2006), p. 127.
3. F. Hüe, M.J. Hÿtch, H. Bender, F. Houdellier, and A. Claverie, PRL (2008) accepted.
4. M.J. Hÿtch, F. Houdellier, F. Hüe and E. Snoeck, Patent Application FR N° 07 06711.
5. M. J. Hÿtch, E. Snoeck, and R. Kilaas, Ultramicroscopy **74** (1998), p. 131.

M. Luysberg, K. Tillmann, T. Weirich (Eds.): EMC 2008, Vol. 1: Instrumentation and Methods, pp. 123–124, DOI: 10.1007/978-3-540-85156-1_62, © Springer-Verlag Berlin Heidelberg 2008

124

6. C. Dupré et al., Sol. State Electron. (2008), in press.
7. The authors thank the European Union for support through the projects PullNano (Pulling the limits of nanoCMOS electronics, IST: 026828) and ESTEEM (Enabling Science and Technology for European Electron Microscopy, IP3: 0260019).

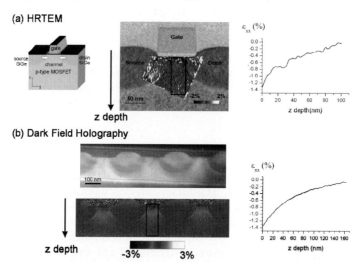

Figure 1. Strain mapping of ε_{xx}: (a) in a 45nm p-MOSFET channel, measured by HRTEM; (b) in a 80nm p-MOSFET channel, measured by dark-field holography.

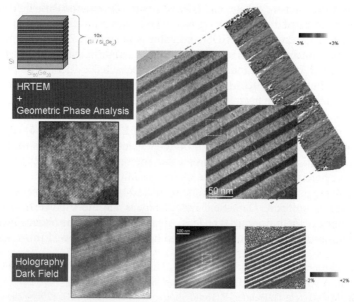

Figure 2. Comparison of HRTEM and dark-field holography for measuring ε_{zz} in a $Si_{0.8}Ge_{0.2}$ virtual substrate / t-Si 40 nm / (c-$Si_{0.6}Ge_{0.4}$ 13.5 nm / t-Si 16.5 nm)×10 stack.

PPA: An Improved Implementation of Peak Pairs procedure as a DM plug-in for Strain Mapping

K. Ishizuka[1], P. Galindo[2], J. Pizarro[2] and S.I. Molina[3]

1. HREM Research, Inc. 14-48 Matsukazedai, Higashimastuyama, 355-0055 JAPAN
2. Departamento de Lenguajes y Sistemas Informáticos, Universidad de Cádiz, Campus Río San Pedro, s/n, 11510 Puerto Real, Cádiz, Spain.
3. Departamento de Ciencia de los Materiales e I.M. y Q.I., Facultad de Ciencias, Universidad de Cádiz, Campus Río San Pedro, s/n, 11510 Puerto Real, Cádiz, Spain.

ishizuka@hremresearch.com
Keywords: Peak Pairs, Strain Mapping, Digital Micrograph

In any branch of science dealing with materials and their behaviour, most people are interested in measuring deformation and strain. Recent advances in digital imaging, together with improved resolution of microscopes have offered the possibility of locally determining the elastic strain of materials at subnanometric scale using HRTEM images. In order to determine strain field mapping, several methodologies have been described in the literature[1], which are based mainly on two different approaches, real space [2] and Fourier space algorithms [3].

Peak Pairs is a recently introduced real space algorithm for strain mapping [2]. It works on a filtered image, locating pairs of peaks along a predefined direction and calculating the local discrete displacement field at each pair. In order to reduce errors in the location of pairs, an affine transformation defined by a pair of basis vectors is applied. This transformation greatly reduces potential errors in the determination of pairs. Subsequently, by derivating the calculated displacement field, the strain field is obtained.

Peak Pairs offers some advantages when compared to Fourier space approach, such as reduced memory and CPU requirements, given that 2-D complex Fourier transforms are not needed at all. This allows the calculation of strain maps of greater images, that is to say, wider areas and/or higher resolution. Another advantage is that its behaviour is unaltered in the presence of defects as well as in the strain mapping of nonhomogeneous surfaces. Nevertheless, we consider that both approaches are useful for strain determination, each having different properties, advantages and pitfalls, and should be considered as complementary tools in strain mapping.

PPA is a plug-in for DigitalMicrograph [TM] (DM, Gatan Inc.) for performing Peak Pairs Analysis developed by HREM Research under collaboration with original developer, Dr. Pedro L. Galindo from the Univ. of Cádiz (Spain). This means that results are fully compatible with the other functions present in DM. For example, the displacement images produced, or strain maps, can be analysed as desired. PPA allows the visualisation of lattice distortion. This distortion can be measured using the profile tools and statistical tools in specified areas, as has been seen. Each pixel in the image is also a measure of the local deformation and lattice orientation. PPA offers some improvements with respect to older implementations of Peak Pairs, such as an improved functionality

M. Luysberg, K. Tillmann, T. Weirich (Eds.): EMC 2008, Vol. 1: Instrumentation and Methods, pp. 125–126, DOI: 10.1007/978-3-540-85156-1_63, © Springer-Verlag Berlin Heidelberg 2008

and an easier-to-use interface. It also includes some image processing options useful for strain determination, such as Adaptive Wiener filter, Bragg filter and Low Pass filter. PPA also includes a function for geometric distortion correction. All optical systems, CCD cameras and scanners (for digitising negatives for example) distort the images they form. Fortunately, these geometric distortions are usually fixed for a given system, and it is therefore possible to eliminate these distortions [4,5] by measuring them (usually only once) using the image of a perfect crystal and then correcting subsequent images.

1   S. Kret, P. Ruterana, A. Rosenauer, D. Gerthsen, Phys. Status Solidi (b) 227 (2001) 247
2   P Galindo, S Kret, A M Sánchez, J-Y Laval, A Yáñez, J Pizarro, E Guerrero, T Ben and S I Molina, Ultramicroscopy 107 (2007) 1186-1193
3   M. J. Hÿtch, E. Snoeck and R. Kilaas, Ultramicroscopy 74 (1998) 131–146.
4   A M Sanchez, P L Galindo, S Kret, M Falke, R Beanland, P J Goodhew. Journal of Microscopy, 221 (2006) 1-7
5   A M Sanchez, P L Galindo, S Kret, M Falke, R Beanland and P J Goodhew. Microscopy and Microanalysis, 12 (2006) 285-294

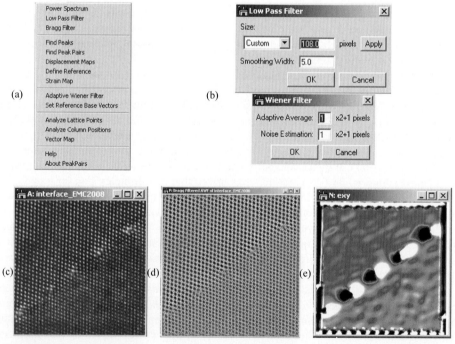

Figure 1. (a) PPA main menu (b) Two Filter Dialog Boxes (c) [110] HRTEM image of a CdTe/GaAs interface (d) Peak Pairs detection results (e) Calculated e_{xy} strain map

Domain structure in Delithiated LiFePO$_4$, a cathode material for Li ion Battery Applications

M. Kinyanjui[1,2], A. Chuvilin[1], U. Kaiser[1], P. Axmann[2], M. Wohlfahrt-Mehrens[2]

1. Electron Microscopy Group of Materials Science, University of Ulm, Albert Einstein Allee 11, 89081 Ulm, Germany
2. Centre for Solar Energy and Hydrogen Research, Helmholtzstr. 8, 89081 Ulm, Germany

michael.kinyanjui@uni-ulm.de

Keywords: HRTEM, GPA, phase transitions, LiFePO$_4$, Li ion batteries

Delithiation is a basic step in the operation of the Li ion battery and it involves the extraction of Li ions from the lattice of the cathode material and insertion into the lattice of the anode. LiFePO$_4$ is a cathode material whose wide application in high energy Li ion batteries, is limited by low ion and electron diffusion [1]. Partially delithiated LiFePO$_4$ grains are known to exist in a two phase state characterized by the delithiated, Li$_{1-x}$FePO$_4$ and lithiated LiFePO$_4$ phases. It has been proposed that poor mobility of Li ions across the interface between these two phases plays an important role in limiting Li diffusion within the LiFePO$_4$ grain [3]. Therefore, investigating the two-phase domains in LiFePO$_4$ is a crucial step towards a better understanding of the Li diffusion in LiFePO$_4$.

We have studied the phase structure in partially delithiated LiFePO$_4$ grains using Geometrical Phase Analysis (GPA) method [4, 5]. In a distorted lattice, quantitative information regarding structural deformations can be obtained from the phase changes of the Fourier coefficients g_i in the corresponding HRTEM image. In the absence of structural distortions, the phase of g-component is constant, while in a distorted lattice the phase is modified by the displacement field $u(x, y)$; the phase is then $-2\pi g.u(x, y)$. In the resulting phase map, the intensity at a point is then proportional to the displacement. All HRTEM images were obtained using the Cs corrected Titan 80-300kV microscope. Figure 1 shows the bright field image of a partially delithiated LiFePO$_4$ grain. Figure 1(b) shows the HRTEM image used to determine the structural variations along the Li diffusion paths in the [010] direction. The amplitude image in Figure 1(c) shows the changes in intensity. Figure (d) shows the changes in the phase along the [010] direction. The contrast changes in the phase image represent a phase shift of π between the regions with dark and bright contrast. This is a result of atomic plane rearrangement and lattice parameter variation along the Li ion diffusion path. We will discuss these results in the context of the methods applied, LiFePO$_4$ phase transitions and Li ion diffusion during Li battery operations.

1. A.K. Padhi, K.S Nanjundaswamy and J.B.Goodenough, Journal Electrochem. Soc. 144 (1997) 1188
2. D.Morgan, A. Van der Ven and G. Ceder, Electrochem. Solid-State Lett. 7 (2004)

M. Luysberg, K. Tillmann, T. Weirich (Eds.): EMC 2008, Vol. 1: Instrumentation and Methods, pp. 127–128, DOI: 10.1007/978-3-540-85156-1_64, © Springer-Verlag Berlin Heidelberg 2008

128

3. C. Delacourt, L. Laffont, R. Bouchet, C. Wurm, J.B. Leriche, M. Mocrette, J.M Tarascon, C. Masquelier, J. Electrochem. Soc., 152, 5, (2005) A913
4. M. Hytch, Microsc. Microanal. Microstruct. 8 (1997) 41
5. A. K. Gutakovskii, A. Chuvilin, S.A. Song, Bulletin of the Russian Academy of Sciences. Physics 71, 10, (2007) 1426
6. The authors acknowledge the BMBF for the funding in the project REALIBATT.

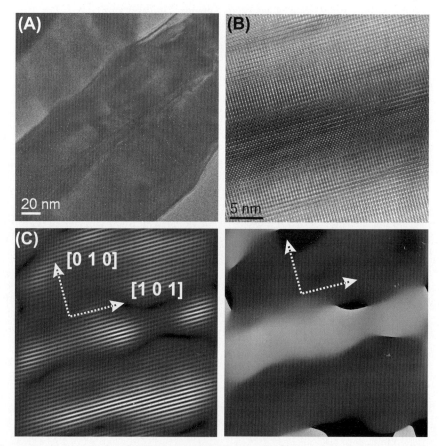

Figure 1: GPA analysis of delithiated LiFePO4 grains. (a) A bright field image showing the domain morphology (b) the corresponding HRTEM image (c) Amplitude image showing intensity changes along [010] (d) phase image showing the phase changes along [010] direction.

New Approach to Quantitative ADF STEM

J.M. LeBeau[1], S.D. Findlay[2], L.J. Allen[3] and S. Stemmer[1]

1. Materials Department, University of California, Santa Barbara, CA 93106
2. Institute of Engineering Innovation, School of Engineering, The University of Tokyo, Tokyo, 113-8656, Japan
3. School of Physics, University of Melbourne, Victoria 3010, Australia

lebeau@mrl.ucsb.edu

Keywords: Quantitative STEM, annular dark-field

High-angle annular dark-field scanning transmission electron microscopy (HAADF STEM or Z-contrast) has been shown to be remarkably sensitive to atomic number (Z). However, HAADF images are currently formed on an arbitrary intensity scale, thereby limiting the possibility of truly quantitative imaging. Recently, it was reported that a mismatch exists between experimental and simulated image contrast in HAADF STEM [1]. Without an absolute scale, it is impossible to determine the cause of the discrepancy [2]. Additionally, an absolute scale would facilitate composition mapping at atomic resolution. Here we demonstrate that the HAADF detector can measure the incident beam intensity to normalize Z-contrast images onto an absolute intensity scale. We report on a practical approach that ensures that the detector does not saturate and is sufficiently linear over the intensity range of interest. An FEI Titan 80-300 STEM/TEM equipped with a super-twin lens ($C_s \sim 1.2$ mm) operating at 300 kV was used for this study.

We discuss the practical aspects of achieving quantified HAADF images. Characterization of the detector and acquisition/quantification of the image intensities will be described. In addition, the detection efficiency across the ADF detector has been investigated. As shown in Figure 1, the efficiency is not uniform and can vary drastically near the hole in the detector. By normalizing the atomically resolved signal, we demonstrate quantified HAADF imaging of a $SrTiO_3$ single crystal. Figure 2 displays an experimental image from a region ~ 200 Å thick and a corresponding Bloch wave image simulation that takes into account spatial incoherence. As can be seen, a quantitative match exists between simulations and experiments.

Local sample thickness was determined by the electron energy loss spectroscopy (EELS) log-ratio method [3]. We will show that the combination of information from the HAADF background signal and the thickness determination by EELS can be used to provide improved estimates of thickness.

1. D. O. Klenov and S. Stemmer, Ultramicroscopy **106**, 889 (2006).
2. D. O. Klenov, S. D. Findlay, L. J. Allen, and S. Stemmer, Phys. Rev. B **76**, 014111 (2007).
3. R. F. Egerton, Electron Energy-Loss Spectroscopy in the Electron Microscope, 2nd ed. (Plenum Press, New York, 1996).
4. S. S. and J. M. L. acknowledge the U.S. DOE for support of this research (DE-FG02-06ER45994). L. J. A. acknowledges support by the Australian Research Council. S. D. F. is supported as a Japan Society for Promotion of Science (JSPS) fellow.

M. Luysberg, K. Tillmann, T. Weirich (Eds.): EMC 2008, Vol. 1: Instrumentation and Methods, pp. 129–130, DOI: 10.1007/978-3-540-85156-1_65, © Springer-Verlag Berlin Heidelberg 2008

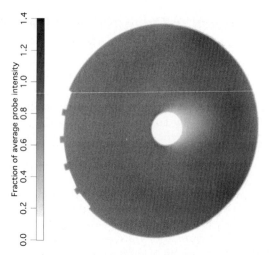

Figure 1. The efficiency across the surface of the HAADF detector is not uniform. Near the hole in the detector, a drastic decrease in efficiency is seen on the right side. The output is plotted in terms of the fraction of an average output measured at four perpendicular points across the detector. The elliptical shape results from the detector being placed at an angle relative to the optic axis.

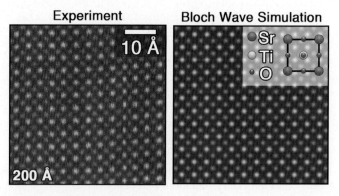

Figure 2. Experimental HAADF image of $SrTiO_3$ along <100> for a region approximately 200 Å thick (left) compared with the corresponding Bloch wave simulation (right). Spatial incoherence is included in the simulation by convolution using a Gaussian envelope with a FWHM of 0.8 Å. Note that the image intensities in both cases are over the same range and normalized to the incident beam intensity.

Reconstruction of the projected crystal potential in high-resolution transmission electron microscopy

M. Lentzen[1,2] and K. Urban[1,2]

1. Institute of Solid State Research, Research Centre Jülich, 52425 Jülich, Germany
2. Ernst Ruska Centre, Research Centre Jülich, 52425 Jülich, Germany

markus.lentzen@fz-juelich.de
Keywords: potential reconstruction, high-resolution electron microscopy, sub-angstrom resolution

In high-resolution transmission electron microscopy information on the object under investigation can be derived from recorded image intensities or, by means of wave function reconstruction [1–3], from exit wave functions. Two processes, however, severely hamper the direct interpretation of image intensities or exit wave functions with respect to the object structure: 1. Dynamical electron scattering along atomic columns of a crystalline material leads in general to a non-linear relation between the atomic scattering power and the local modulation of the exit wave function. 2. Lens aberrations of the electron microscope impose phase changes on the wave function passing from the object plane to image plane and induce an unwanted contrast delocalisation [4].

The interpretation of reconstructed exit wave functions with respect to the object structure can be strongly improved by simulating the effects of dynamical electron diffraction using test object structures [5]. The simulation of microscopic imaging [5] can help in a similar way in the interpretation of experimentally recorded image intensities by injecting knowledge of imaging parameters, in particular lens aberration parameters. It is, however, highly desirable to find ways for the structure interpretation in high-resolution electron microscopy which do not rely on simulations of dynamical electron scattering and microscopic imaging with a possibly excessive number of test object structures.

The reconstruction of the projected crystal potential from an exit wave function is one of the ways to avoid tedious simulations, and it is the logical extension of already successfully implemented exit wave function reconstruction methods. The two reconstruction steps combined would lead from the recorded image intensities over the reconstructed exit wave function to the reconstructed projected crystal potential, and this measurement of the potential would allow a direct structure interpretation.

In the past decades few attempts have been made to solve the potential reconstruction problem for the case of thick objects including the effects of dynamical diffraction. In a first attempt by Gribelyuk [6] an estimate at the projected potential $U(r)$ was iteratively refined using forward multislice iterations [7] and evaluations of the difference between the simulated and the reconstructed exit plane wave functions, $\psi_{sim}(r)$ and $\psi_{exp}(r)$. In the second attempt by Beeching and Spargo [8] the same scheme was applied using reverse multislice iterations and comparing the respective entrance plane wave functions. Both attempts fail already at small specimen thickness, mostly

M. Luysberg, K. Tillmann, T. Weirich (Eds.): EMC 2008, Vol. 1: Instrumentation and Methods, pp. 131–132, DOI: 10.1007/978-3-540-85156-1_66, © Springer-Verlag Berlin Heidelberg 2008

due to improper use of the weak phase object approximation for the first estimate and for the correction steps.

The attempt by Lentzen and Urban [9] successfully solved the potential refinement problem within the least-squares formalism by implementing a search for the projected potential along the gradient of the figure of merit, $S^2 = \int | \psi_{sim}(r) - \psi_{exp}(r) |^2 \, dr$, with respect to the potential. Together with use of the channelling model of electron diffraction [10,11] a successful reconstruction of the crystal potential was achieved for non-periodic objects over a wide thickness range.

In this work the former treatment of Lentzen and Urban [9] has been extended to include phenomenological absorption, which turned out to be important already for thin objects of few nanometres in thickness. Accuracy and stability of the refinement algorithm are greatly improved by a new formulation for the gradient of S^2 with respect to the potential. If $\psi_e(r, t')$ denotes the difference of $\psi_{exp}(r)$ and $\psi_{sim}(r)$ being back-propagated through the object from the exit plane t to a plane t' inside the crystal, and if $\psi_s(r, t')$ denotes the entrance plane wave being propagated through the object from the entrance plane 0 to the same plane t' inside the crystal, then

$$\delta U(r) = \frac{1}{\pi \lambda t^2} \mathrm{Im}\left\{ (1 - i\kappa) \int_0^t \psi_s^*(r,t') \, \psi_e(r,t') \, dt' \right\}$$

denotes the correction of the projected potential for each iteration step, with t the object thickness, λ the electron wavelength, and κ the phenomenological absorption parameter ranging from 0 to around 0.2.

Simulation studies assuming an $YBa_2Cu_3O_7$ crystal of several nanometres in thickness show that the use of the new gradient formulation improves the convergence of the refinement algorithm considerably in comparison to the use of the gradient presented in [9] for the case of absorption being present. The reconstruction algorithm is successful even at larger object thickness, where the exit wave function exhibits strongly differing modulations in amplitude and phase at the Y and Ba atom columns of strong scattering power, the Cu-O atom columns of medium scattering power, and the O atom columns of weak scattering power.

1. H. Lichte, Ultramicroscopy **20** (1986), p. 293.
2. W. Coene, G. Janssen, M. Op de Beeck and D. Van Dyck, Phys. Rev. Lett. **69** (1992), p. 3743.
3. A. Thust, W.M.J. Coene, M. Op de Beeck and D. Van Dyck, Ultramicroscopy **64** (1996), p. 211.
4. W. Coene and A.J.E.M. Jansen, Scan. Microsc. Suppl. **6** (1992), p. 379.
5. P.A. Stadelmann, Ultramicroscopy **21** (1987), p. 131.
6. M.A. Gribelyuk, Acta Cryst. A**47** (1991), p. 715.
7. J.M. Cowley and A.F. Moodie, Acta Cryst. **10** (1957), p. 609.
8. M.J. Beeching and A.E.C. Spargo, Ultramicroscopy **52** (1993), p. 243.
9. M. Lentzen and K. Urban, Acta Cryst. A**56** (2000), p. 235.
10. K. Kambe, G. Lehmpfuhl and F. Fujimoto, Z. Naturforsch. A**29** (1974), p. 1034.
11. F. Fujimoto, Phys. Status Solidi A**45** (1978), p. 99.

Three-dimensional atomic-scale structure
of size-selected nanoclusters on surfaces

Z.Y. Li[1], N.P. Young[1], M. Di Vece[1], S. Palomba[1], R.E. Palmer[1], A.L. Bleloch[2], B.C. Curley[3], R.L. Johnston[3], J. Jiang[4] and J. Yuan[4,5]

1. Nanoscale Physics Research Laboratory, School of Physics and Astronomy, University of Birmingham, Birmingham B15 2TT, U.K.
2. SuperSTEM Laboratory, STFC Daresbury, Daresbury WA4 4AD, U.K.
3. School of Chemistry, University of Birmingham, B15 2TT, U.K.
4. Beijing Electron Microscopy Centre, Tsinghua University, Beijing 100084, China
5. Department of Physics, University of York, Heslingt, York YO10 5DD, U.K.

ziyouli@nprl.ph.bham.ac.uk

Keywords: STEM, simulation, size-selected clusters

An unambiguous determination of the three-dimensional structure of nanoparticles is challenging. Electron tomography requires a series of images taken for many different specimen orientations. This approach is ideal for stable and stationary structures. But ultrasmall nanoparticles are intrinsically structurally unstable and may interact with the incident electron beam, constraining the electron beam density that can be used and the duration of the observation. Here we demonstrate that high-angle annular dark field (HAADF) in a spherical-aberration-corrected scanning transmission electron microscopy (STEM), coupled with imaging simulation, can be used to obtain a snap-shot of the size, shape, orientation and atomic arrangement of size-selected metallic nanoclusters.

A detailed example is shown for size-selected Au_N (N=309±6) clusters [1], where Au_{309} is known to be a possible 'magic number' nanocluster. The Au_{309} clusters are formed by gas-phase condensation of sputtered atoms in a rare-gas atmosphere, size-selected by a lateral time-of-flight mass spectrometer and soft-landed on an amorphous carbon support [2]. Fig. 1b displays a high-resolution HAADF image taken from a Au_{309} cluster, showing a clear five-fold symmetry in the atomic arrangement. An illustrative line intensity profile from the centre of this cluster to one of the corners is also shown, where five peaks and a shoulder (marked by the arrow) are apparent, with the peak intensity decreasing towards the corner. Using a simple kinematical approach, the simulated HAADF image of the Ino-decahedral Au_{309} cluster is obtained and shown in Fig. 1c together with the intensity profile (the solid red curve). The correspondence between the simulated profile and the experimental profile, with respect to both the peak positions and the relative peak intensities, is remarkable, indicating the correct identification of the atomic column structure. We have also conducted a full dynamical calculation using the multislice method. The corresponding line profile is shown by the dashed line in Fig. 1c. The similarity between the two simulated line profiles confirms the validity of the simple kinematical approximation for HAADF image simulation of the Au_{309} clusters and that the quantization of the HAADF intensity correlates directly

M. Luysberg, K. Tillmann, T. Weirich (Eds.): EMC 2008, Vol. 1: Instrumentation and Methods, pp. 133–134, DOI: 10.1007/978-3-540-85156-1_67, © Springer-Verlag Berlin Heidelberg 2008

with the quantization of the number of the atoms. This later point is verified by our experimental confirmation of linear dependence of integrated HAADF intensity as a function of size of clusters (for N up to ~1500) [3].

Statistical analysis of Au_{309} clusters reveals that not only they can be identified with Ino-deahedral, but also cuboctahedral or icosahedral geometries. Comparison with theoretical modelling of the system suggests the multiplicity of cluster geometries being consistent with energetic considerations. It is found that the difference in total energy between different geometries was less than 1.2 eV from the most stable to the least stable. More over, there are many local energy minima and the energy barriers between these structures are small. Given the narrow size distribution of the deposited clusters, our results may shed light on the relative structural stabilities of the various cluster isomers in the gas phase, information that has solicited many theoretical investigations but little hard experimental evidence.

Close comparison between Fig.1b and c highlights a discrepancy in the atom columns at the edge of the cluster. We interpret this as evidence for increased fluctuations and motion of cluster surface atoms relative to the core atoms, on a time-scale shorter than the period for the data acquisition.

We will also demonstrate the wide applicability of this approach to other supported ultrasmall metal cluster systems, by showing examples on size-selected Ag and Pd nanoclusters.

1.   Z.Y. Li, N.P. Young, M. Di Vece, S. Palomba, R.E. Palmer, A.L. Bleloch, B.C. Curley, R.L. Johnston, J. Jiang, J. Yuan, *Nature* **415** (2008) 46-48.
2.   S. Pratontep, S.J. Carrol, C. Xirouchaki, M. Streun, R.E. Palmer, *Rev. Sci. Instrum.* **76** (2005) 045103-045109
3.   N.P. Young, Z.Y. Li, Y. Chen, S. Palomba, M. Di Vece, R.E. Palmer, *submitted.*

Figure 1 (a) 3D atom density profile of Au_{309}. A hard-sphere model for an Ino-decahedral structure is shown with the electron beam parallel to the five-fold axis. (b) Experimental intensity line profile taken from the central atom column of the cluster to one of the corners (indicated in inset with red line. (c) Simulated HAADF image obtained with a simple kinematical approach, of an Au_{309} cluster with Ino-decahedral geometry. An intensity profile (solid curve) across one ridge (indicated in inset) is compared with the result from a full dynamical multislice calculation.

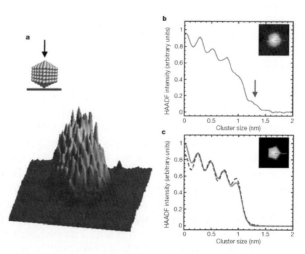

Direct retrieval of a complex wave from its diffraction pattern

A.V. Martin and L.J. Allen

School of Physics, University of Melbourne, Victoria 3010, Australia

andrewvm@ph.unimelb.edu.au

Keywords: coherent diffractive imaging, phase retrieval

Since the resolution obtainable by electron imaging is currently limited by the aberrations in the objective lens, lensless imaging offers the prospect of resolution limited only by the wavelength of the electrons, which is on the order of a picometer. Lensless imaging involves recovering the exit surface wave function from a diffraction pattern. This is currently very topical in electron microscopy [1]. Knowledge of the wave function in the exit surface plane is essential to obtain structural information about the sample.

Recovering the wave function from a single diffraction pattern typically uses a constraint in the real space plane to establish a system of nonlinear equations. The standard approaches to solving these equations are iterative methods such as the hybrid-input-output method [2] and difference map methods [3,4]. With iterative methods there is no guarantee that they will converge and, if so, to a unique solution. Alternatively if recovery of the exit surface wave function from a single diffraction pattern can be posed as a linear problem, then we can apply a direct method to solve the equations and it is guaranteed that there is a unique solution to the problem.

Under certain illumination conditions, which we will present, an unknown scattered wave function can be uniquely recovered via the solution of a set of linear equations. These equations are obtained by constructing the autocorrelation function of the wave function at the exit surface from a diffraction pattern. These conditions can be met if, for example, the unknown sample is of a finite size and is contained within half of a well-characterized illumination beam which contains spatial variation outside a prescribed region. An illustrative example with a circular beam is shown in Figure 1. The method can potentially be applied in optical and X-ray imaging, as well as in electron imaging. We will discuss how the method can be applied to the geometry of a scanning transmission electron microscope. We will also discuss the sensitivity of the method to experimental realities like spatial incoherence, noise, and uncertainties in the characterization of the incident beam. We envisage that this method will be applicable to the imaging of nanoparticles and structures at atomic resolution.

1. J.M. Zuo, I.A.Vartanyants, M. Gao, R. Zhang, and L.A. Nagahara, Science **300** (2003), p. 1419.
2. J.R. Fienup, Appl. Opt. **21** (1982), p. 2758.
3. V. Elser, J. Opt. Soc. Am. A **20** (2003), p. 40.
4. W. McBride, N.L. O'Leary, and L.J. Allen, Phys. Rev. Lett. **93** (2004), p. 233902.

M. Luysberg, K. Tillmann, T. Weirich (Eds.): EMC 2008, Vol. 1: Instrumentation and Methods, pp. 135–136, DOI: 10.1007/978-3-540-85156-1_68, © Springer-Verlag Berlin Heidelberg 2008

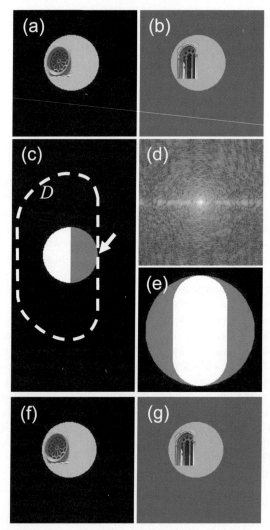

Figure 1. The image and phase of a test object are shown in (a) and (b). A schematic of the illumination conditions is shown in (c), with the area known to contain the object shown in white. Outside the region labeled D, bounded by the dashed line, the illumination must not be constant. This requirement is satisfied because the edge of the illumination lies on the boundary of D at the point indicated by the white arrow. The diffraction pattern generated by the test object is shown in (d). A schematic of the domain of the autocorrelation function is shown in (e) and the grey region shows where the unknown scattered wave function is linearly related to the autocorrelation function. The recovered wave function is shown in (f) and (g).

HRTEM evaluation of iron in acid treated ground vermiculite from Santa Olalla (Huelva, Spain)

N. Murafa[3], C. Maqueda[1], J.L. Perez-Rodriguez[2], J. Šubrt[3]

1. Instituto de Recursos Naturales y Agrobiología (CSIC) Apdo 1052, 41080 Sevilla. Spain
2. Instituto de Ciencia de Materiales de Sevilla (UNSE-CSIC) Americo Vespucio s/n, 41098 Sevilla, Spain
3. Institute of Inorganic Chemistry AS CR, 250 68 Řež, Czech Republic

murafa@iic.cas.cz

Keywords: vermiculite, amorphous silica, structural iron, acid leaching, electron microscopy, akaganeite (β-FeOOH)

The decrease in the particle size of clay minerals is important for many industrial applications. The porosity studies showed the highest specific surface area of the residue obtained with clay minerals by leaching. Selective leaching with acid has been used for preparing porous silica from various clay minerals including vermiculite [1-4]. Vermiculite is a 2: 1 phyllosilicate mineral with substitutions in tetrahedral (Si^{4+} by Al^{3+} and Fe^{3+}) and octahedral (Mg^{2+} by Al^{3+}, Fe^{3+} and Fe^{2+}) sheets. The presence of iron in this mineral play an important role in the preparation of an amorphous silica with very high specific surface area, which is higher than other clay minerals including vermiculites without iron in their structure [1, 2]. This highest specific surface area is attributed to the presence of iron in the residue coming from structural iron. The X-ray diffraction analysis shows presence of akaganeite (β-FeOOH).

Taking the about considerations into account, the aim of this work was to use the High Resolution Transmission Electron Microscopy (HRTEM) to study how the iron oxide is included in the amorphous silica formed during acid leaching of Santa Olalla vermiculite. It was found that the sample consist of amorphous silica and akaganeite microcrystals, the structure of akaganeite was confirmed by electron diffraction analysis. The particle morphology of the akaganeite microcrystals correspond well to β-FeOOH precipitated from the solution, is also common an increased content of Cl^- in the akaganeite particles. The leached vermiculite residue contained also small amount of Ti^{4+}, which is accumulated into the akaganeite microcrystals.

To evaluation of iron in acid treated ground vermiculite from Santa Olalla (Huelva, Spain) the HRTEM measurements (Figure 1) were carried out using the 300 kV JEOL JEM 300 UHR electron microscope with LaB_6 electron source and equipped with Semi STEM and EDS. The measurements were performed at 300 kV, the sample was prepared from suspension in ethanol treated in ultrasonic bath for 4 min, the Cu grid was immersed into the suspension and dried at laboratory temperature in air. The EDS measurements were performed with Oxfords Instruments EDS detector.

M. Luysberg, K. Tillmann, T. Weirich (Eds.): EMC 2008, Vol. 1: Instrumentation and Methods, pp. 137–138, DOI: 10.1007/978-3-540-85156-1_69, © Springer-Verlag Berlin Heidelberg 2008

138

1. C. Maqueda, A.S. Romero, E. Morillo and J.L. Pérez-Rodríguez, Effect of grinding on the preparation of porous materials by acid-leached vermiculite, Journal of Physics and Chemistry of Solids, Volume 68, Issues 5-6, May-June 2007, Pages 1220-1224
2. K. Okada, N. Arimitsu, Y. Kameshima, A. Nakajima and Kenneth J.D. MacKenzie, Solid acidity of 2:1 type clay minerals activated by selective leaching, Applied Clay Science, Volume 31, Issues 3-4, March 2006, Pages 185-193
3. H. Suquet, R. Franck, J.F. Lambert, F. Elsass, C. Marcilly and S. Chevalier, Catalytic properties of two pre-cracking matrices: a leached vermiculite and a Al-pillared saponite, Applied Clay Science, Volume 8, Issue 5, January 1994, Pages 349-364
4. J. Temuujin, K. Okada and Kenneth J. D. MacKenzie, Preparation of porous silica from vermiculite by selective leaching, Applied Clay Science, Volume 22, Issue 4, February 2003, Pages 187-195
5. Acknowledgement: This work was supported by the Ministry of Education, Youth and Sports of the Czech Republic, project № LC 523.

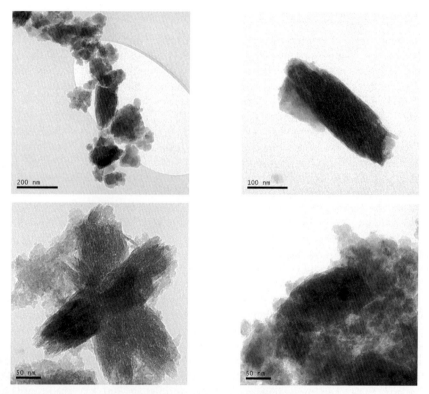

Figure 1. TEM micrograph of acid leaching vermiculite

Bloch wave analysis of depth dependent strain effects in high resolution electron microscopy

P.D. Nellist[1], E.C. Cosgriff[1], P.B. Hirsch[1] and D.J.H. Cockayne[1]

1. Department of Materials, University of Oxford, Parks Road, Oxford, OX1 3PH, UK

peter.nellist@materials.ox.ac.uk

Keywords: bloch waves, image simulations, strain imaging

High resolution electron microscopy (HREM) of crystals is usually carried out with the crystal aligned along a high symmetry direction. For perfect crystals the columns are straight, but crystal defects can result in strain fields that bend the atomic columns. A question then arises of how one can interpret the peaks and troughs that are observed in high-resolution electron microscope data. Of course it is possible to proceed by simulating HREM images of the sample containing a defect using standard multislice codes, and then matching the simulated images to the experimental data. Such calculations do not elucidate the basic physical processes that result in image changes due to the crystal imperfections.

An approach that can reveal more about the process of electron scattering from crystals is through the use of Bloch waves. Finding Bloch-wave solutions to the time-independent Schrödinger equation for the fast electron allows n-beam dynamical scattering to be computed [1]. The excitations of the Bloch wave solutions can be found by wavefunction matching at the entrance surface of the crystal, and for a perfect crystal the excitations do not change through the thickness of the crystal. In the presence of a displacement, the Bloch wave solutions referenced to the new lattice position will be excited by the wavefunction arising from the Bloch waves in the lattice prior to the displacement (figure 1). The resulting Bloch wave excitations will be different from those in the undisplaced crystal, and these changes we describe as interband scattering. In practice, continuous displacement through the crystal thickness requires wavefunction matching at each atomic layer, thereby resulting in effectively continuous interband scattering.

To illustrate the types of transition that are possible, consider the imaging of a body-centred cubic metal along the [111] direction. We consider the direct beam and the 6 scattered {110}-type beams to form a 7-beam model. In the absence of absorption we can readily calculate the Fourier components of the 7 Bloch wave solutions using the approach described in [1]. The amplitudes and intensities of the Bloch waves plotted in real-space are shown in figure 2. By analogy with atomic orbitals, they have been labeled s, p, d and f depending on their rotational symmetry. Only the s-type waves have a non-zero zeroth order Fourier component, and therefore these are the only ones that can be excited by plane-wave illumination along the column direction. If the atomic columns are bent by crystal imperfections then it can be shown that the first-order interband scattering is only into the p-type waves [2]. A selection rule applies that is analogous to the dipole selection for atomic excitations. Briefly, the selection rule can be explained by noting that a column displacement will result in the Bloch waves in

M. Luysberg, K. Tillmann, T. Weirich (Eds.): EMC 2008, Vol. 1: Instrumentation and Methods, pp. 139–140, DOI: 10.1007/978-3-540-85156-1_70, © Springer-Verlag Berlin Heidelberg 2008

140

the displaced crystal being illuminated by a wave that now contains antisymmetric components relative to the displaced lattice. These antisymmetric components excite the antisymmetric p-type waves, and the amount of interband scattering depends on the nature of the defect.

The result of the interband transitions will be to displace the peaks (or troughs) of an image away from the atom locations in the exit-surface of the sample, but the degree of displacement requires a careful calculation of the amount of interband scattering. The results of such calculations will be presented.

1. P. Hirsch, A. Howie, R. Nicholson, D.W. Pashley, and M.J. Whelan, Electron Microscopy of Thin Crystals. 2nd (revised) ed. 1977, Malabar: Krieger.
2. P.D. Nellist, E.C. Cosgriff, P.B. Hirsch, and D.J.H. Cockayne, Philosophical Magazine 88 (2008) 135.
3. The authors would like to thank the Head of the Department of Materials, University of Oxford, for provision of facilities and funding.

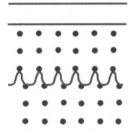

illuminating plane wave

wave at defect depth has asymmetry with respect to displaced crystal

Figure 1. A schematic diagram to show the origin of interband transitions. In the undistorted crystal the wave will reflect the symmetry of the lattice. In the displaced crystal the Bloch waves are referenced with respect to the displaced lattice, and are therefore illuminated by a wave that is no longer symmetric with respect to the displaced Bloch waves. Asymmetric Bloch waves are then excited.

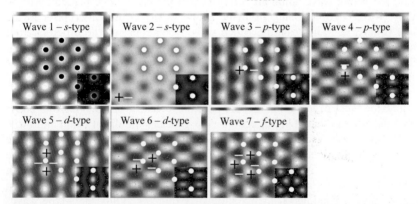

Figure 2. Spatial distribution of the amplitude and intensity (inset) of the Bloch waves for illumination along the symmetry axis [111] and superimposed on atom columns of a bcc lattice passing through the lattice points. Positive and negative amplitudes are indicated by + and -. The amplitude greyscales are scaled symmetrically on either side of zero so that black and white correspond to equal magnitudes of opposite sign, with grey showing zero amplitude.

Quantitative local strain analysis of Si/SiGe heterostructures using HRTEM

V. Burak Özdöl[1], F. Phillipp[1], E. Kasper[2] and P.A. van Aken[1]

1. Stuttgart Center for Electron Microscopy, Max Planck Institute for Metals Research, Heisenbergstr. 3, 70569 Stuttgart, Germany
2. Institute for Semiconductor Engineering, University Stuttgart, Pfaffenwaldring 47, 70569 Stuttgart, Germany

ozdol@mf.mpg.de

Keywords: strain analysis, Si/SiGe, HRTEM

Semiconductor device technology beyond 65 nm exploits strain engineering to boost device performance in the face of limits to device scaling. Via altering the electronic band structure, strain is used to increase the carrier mobility in the channel region of MOSFET devices. To support the nano-device development, local strain measurement techniques with high accuracy at nanometer spatial resolution are required. High-resolution transmission electron microscopy (HRTEM) together with advanced image processing techniques is the powerful candidate to fulfil these demands.

In this present work, we employed HRTEM and two different image processing softwares, LAttice DIstortion Analysis (LADIA) [1,2] and Geometric Phase Analysis (GPA) [3,4], to map the strain distribution in Si/SiGe heterostructures due to lattice mismatch.

A SiGe film with 30% Ge and a nominal thickness of 30 nm was grown by molecular-beam epitaxy on a (001)-oriented Si substrate at 500 °C and subsequently capped by a Si layer of 200 nm. Specimens for HRTEM were prepared by standard Ar-ion milling and tripod polishing. Special care was taken to reduce amorphous surface layers and specimen bending. HRTEM experiments were performed on a Jeol ARM 1250 and a Jeol 4000EX with point resolutions of 1.2 Å and 1.7 Å, respectively. It was found that the material had grown completely coherently without the formation of extended defects, i.e., a fully strained state was preserved. Figure 1a shows a high-resolution image across the entire SiGe layer from the Si substrate to the Si capping layer with an enlarged view across the coherent interface (Figure 1b). Figures 1c and 1d display the strain maps obtained from LADIA and GPA, respectively. Here, we defined the strain as deformation in the [001] direction perpendicular to the interface plane, which is the only important component, with respect to the non-distorted Si lattice. From both methods 1.8 ± 0.1% strain within the layer is obtained, which is slightly lower than theoretically expected (2.0%). At both interfaces a pronounced strain gradient is observed. As shown in Figure 1e for the lower Si/SiGe interface this gradient extends over ca. 6 nm. Similar observations are reported by other authors [5,6]. Such gradients may arise from transients in the Ge flux and interdiffusion or segregation of Ge during the layer growth.

1. K. Du, Y. Rau, N. Y. Jin-Phillipp, F. Phillipp, J. Mater. Sci. Tech., **18**/2 (2002), p. 135-138.
2. K. Du, F. Phillipp, Journal of Microscopy, **221**/1 (2006), p. 63-71.

M. Luysberg, K. Tillmann, T. Weirich (Eds.): EMC 2008, Vol. 1: Instrumentation and Methods, pp. 141–142, DOI: 10.1007/978-3-540-85156-1_71, © Springer-Verlag Berlin Heidelberg 2008

142

3. M.J. Hÿtch, E. Snoeck, R. Kilaas, Ultramicroscopy, **74**/3 (1998), p. 131-146.
4. S. Kret, P. Ruterna, A. Rosenauer, D. Gerthsen, phys. stat. sol., **227**/1 (2001), p. 247-295.
5. M.J. Hÿtch, F. Houdellier, Microelectronic Engineering, **84**/3 (2006), p. 460-463.
6. W. Neumann, H. Kirmse, I. Häusler, R. Otto, Journal of Microscopy, **223**/3 (2006), p. 202.

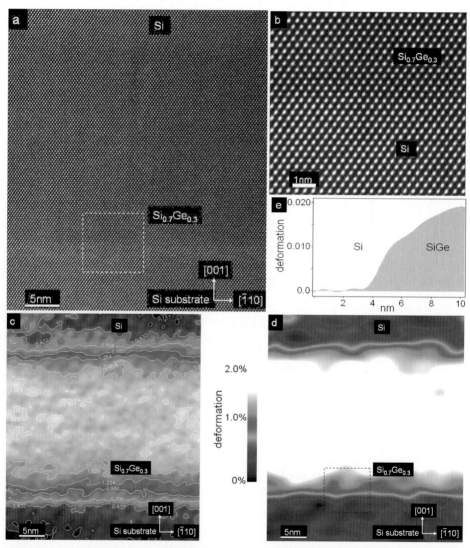

Figure 1. a) HRTEM image of Si/Si$_{0.7}$Ge$_{0.3}$ heterostructure and b) enlargement of the interface region indicated as a dashed square in a). 2D strain maps calculated with c) LADIA and d) GPA. Strain evolution across the lower interface is given as integrated line profile in e).

Displacement field analysis around hydrogen implantation induced platelets (HIPs) in semi-conductors

F. Pailloux, M.-L. David, L. Pizzagalli and J.-F. Barbot

Laboratoire de Physique des Matériaux, UMR 6630 CNRS-Université de Poitiers, SP2MI, 86962 FUTUROSCOPE-CHASSENEUIL

frederic.pailloux@univ-poitiers.fr

Keywords: HRTEM, ion implantation, image processing

Ion implantation is widely used in micro-electronics engineering for many purposes: doping of semi-conductors, smart-cut® process, local strain relaxation and even lattice defects engineering. Under light elements (H, He) implantation high doses (greater than $1 \times 10^{16} cm^{-2}$), many semi-conductors (Si, Ge, GaAs, SiC, GaN, …) experience the formation of cavities (vacancies clusters and/or gas bubbles). The formation of extended planar defects (such as hydrogen induced platelets, HIPs) is among the most encountered phenomenon when hydrogen is implanted into silicon wafers at moderate temperature [1]. Many transmission electron microscopy (TEM) works have been devoted to characterize the type, morphology and distribution of these defects as a function of experimental conditions. Atomistic calculations have also been performed to propose an atomic model for these defects [2]. A precise knowledge of the structure of HIPs lying in the {111} planes has been achieved [1]. However, very few quantitative studies have been performed regarding HIPs lying in {001} planes. The structure of these particular defects is thus still unknown.

In this study, we investigate how the coupling of atomistic calculations, image simulation and geometrical phase shift analysis (GPA) of HRTEM contrast, could provide a way to determine the structure of {001} HIPs. Fig.1 shows the typical strain field around such a defect. The GPA method has been proven to be sensitive to displacement field as low as 3 pm [3]. One can thus expect this kind of analysis to be sensitive enough to discriminate different core configurations by analysing the amplitude and/or extent of the strain field around the defect. The strain fields determined by ab initio calculations for different configurations of the defect will be compared to those measured by the GPA method on images simulated by the multislice method. This procedure will lead us to estimate the accuracy of the method in order to further analyze experimental HRTEM pictures of {001} HIPs.

1. S. Muto, S. Takeda and M. Hirata, Phil. Mag. A 72 (1995), 1057.
2. N. Martsinovitch, M.I. Heggie and C. Ewels, J. Phys.: Condens. Matter 15 (2003), 2815.
3. M. Hÿtch, J.-L. Puteaux, J.-M. Penisson, Nature 423(2003), 270

M. Luysberg, K. Tillmann, T. Weirich (Eds.): EMC 2008, Vol. 1: Instrumentation and Methods, pp. 143–144, DOI: 10.1007/978-3-540-85156-1_72, © Springer-Verlag Berlin Heidelberg 2008

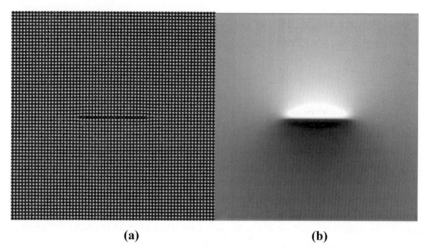

(a)   **(b)**

Figure 1. (a) Model image of a gaz inclusion inserted in periodic lattice. (b) Strain field measured by the GPA method around the defect of fig. 1(a); this phase image was obtained by considering the planes parallel to defect.

Thickness effects in Tilted Sample Annular Dark Field Scanning Transmission Electron Microscopy

A. Parisini[1], V. Morandi[1] and S.A. Mezzotero[2]

1. CNR-IMM Sezione di Bologna, via P. Gobetti 101, 40129 Bologna, Italy
2. Dipartimento di Fisica, Università di Bologna, v.le B. Pichat 6/2, 40127 Bologna, Italy

parisini@bo.imm.cnr.it

Keywords: Z-contrast, STEM, dopant profiles

In previous reports [1-4], we have recently shown how a modification, based on a well defined sample tilt procedure, of the Z-contrast Annular Dark Field Scanning Transmission Electron Microscopy (ADF-STEM) technique pioneered by Pennycook and co-workers [5] may lead to the quantitative determination of the dopant depth distribution in Ultra Shallow Junctions (USJ) in Si.

This technique, referred to as Tilted Sample ADF-STEM (TSADF-STEM), has been introduced as a routine characterization technique able to define in cross-sectional samples of standard thickness by an appropriate signal discrimination, an in-depth dopant distribution at a sub-nm level. Tests performed on low energy (3-5 keV) As implanted Si samples indicated a sensitivity of the TSADF-STEM technique slightly better than 1 at. % and an attainable precision of the dopant quantitative determination procedure of about 10 % [4].

TSADF-STEM was also found to accurately describe the As distribution after annealing treatments at intermediate temperatures. In particular, this technique allowed, for the first time, to quantitatively account for the As pile-up observed in the very first nm from the sample surface as a result of the dopant segregation occurring during the solid phase epitaxial regrowth [4]. Previous experiments were performed at a fixed sample thickness of about ~ 100 nm. An example of an experimental TSADF-STEM image obtained at this sample thickness in an annealed sample presenting a surface As pile-up is reported in Fig. 1a. Although an extensive study of the thickness dependence of the TSADF-STEM contrast is still in progress we report in Fig. 1b and c, preliminary results obtained on the same annealed sample in regions of increasing thickness. The known delta-like surface As distribution [1-4] makes this sample rather ideal to test the effects of the beam broadening on the attainable depth resolution in Z-contrast profiles. In figure 1, the thickness, measured by Parallel Electron Energy Loss Spectroscopy (PEELS), is expressed in relative units, t/λ, λ being the total inelastic electron mean free path. In the corresponding contrast profiles reported in Fig. 1d, it is observed that on increasing the sample thickness a progressive reduction of the surface peak height and a displacement in depth of the overall profile is observed. A detailed analysis of these results will be reported and discussed. Here we bound ourselves to note that from this behaviour, the effect of the beam broadening appears to prevent accurate analyses of the dopant profile only for thickness values $t/\lambda \geq 1.5$.

M. Luysberg, K. Tillmann, T. Weirich (Eds.): EMC 2008, Vol. 1: Instrumentation and Methods, pp. 145–146, DOI: 10.1007/978-3-540-85156-1_73, © Springer-Verlag Berlin Heidelberg 2008

146

1. M. Ferri, S. Solmi, A. Parisini, M. Bersani, D. Giubertoni and M. Barozzi, J. Appl. Phys.,**99** (2006) , p. 113508-1.

2. Parisini, D. Giubertoni, M. Bersani, M. Ferri, V. Morandi and P. G. Merli, Proceedings of the 8th Multinational Congress on Microscopy, Prague, 18-21 june 2007, p. 43.

3. Parisini, D. Giubertoni, M. Bersani, V. Morandi, P. G. Merli and J. A. van den Berg, in Quantitative Electron Microscopy for Materials Science, edited by E. Snoeck, R. Dunin-Borkowski, J. Verbeeck, and U. Dahmen, Mater. Res. Soc. Symp. Proc., **1026E** (2007), 1026-C09-04.

4. A.Parisini, V. Morandi, P. G. Merli, D. Giubertoni, M. Bersani and J. A. van den Berg, submitted to Appl. Phys. Lett. (2008).

5. S. J. Pennycook, S. D. Berger, and R. J. Culbertson, J. Microsc., **144** (1986), p. 229.

6. This work was supported by the EU project "European Integrated Activity of Excellence and Networking for Nano and Micro- Electronics Analysis", ANNA, contract n. 026134(RII3).

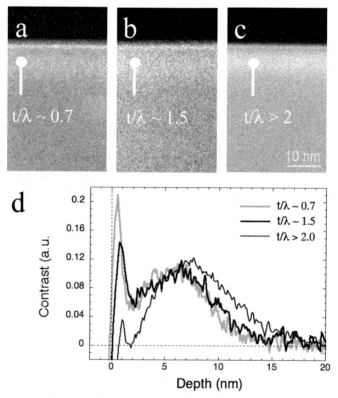

Figure 1. (a-c): TSADF-STEM images obtained in regions of increasing thickness of a sample implanted at 5 keV with 2×10^{15} As^+/cm^2 and annealed at 800 °C for 3 min. (d): TSADF-STEM contrast profiles obtained on (a-c).

A novel emission potential multislice method to calculate intensity contributions for thermal diffuse scattering in plane wave illumination TEM

A. Rosenauer[1], M. Schowalter[1], J.T. Titantah[2] and D. Lamoen[2]

1. Institut für Festkörperphysik, Universität Bremen, Otto-Hahn-Allee 1, 28359 Bremen, Germany
2. Departement Fysica, Universiteit Antwerpen, Groenenborgerlaan 171, 2020 Antwerpen, Belgium

rosenauer@ifp.uni-bremen.de

Keywords: Thermal diffuse scattering, Multislice simulation, Frozen phonon simulation

Thermal diffuse scattered electrons significantly contribute to high-resolution transmission electron microscopy images. Their intensity adds to the background and is peaked at positions of atomic columns. In this contribution we suggest an approximation to simulate intensity of thermal diffuse scattered electrons in the object exit plane using an emission potential multislice algorithm which is computationally less intensive than the frozen lattice approximation or the mutual intensity approach.

Denoting the incident wave in the center plane of slice n by $\Psi(n,\mathbf{r})$, each position $\mathbf{r}$ in this plane is considered as a TDS point source emitting a wave with intensity

$$J_{\mathrm{TDS}}(n,\mathbf{r}) = 2\sigma\Delta z V_{TDS}(\mathbf{r})\Psi(n,\mathbf{r})\Psi^*(n,\mathbf{r}),$$

where σ is the interaction constant, Δz the slice thickness and $V_{TDS}(\mathbf{r})$ the imaginary part of the projected crystal potential of the slice. $V_{TDS}(\mathbf{r})$ is caused by thermal diffuse scattering (TDS) and is obtained from the imaginary part of the atomic scattering amplitudes computed according to Ref. [1]. The Bragg scattered wave $\Psi(n,\mathbf{r})$ is calculated with a conventional multislice algorithm. The wave from a TDS point source is assumed incoherent with respect to waves from other TDS point sources and the Bragg scattered wave. Propagation of a wave from a TDS point source in the center plane of the slice n to the object exit plane is also computed with a conventional multislice approach. We formally describe the result of this computation by a response function $G_n(\mathbf{r},\mathbf{r}')$ which gives the intensity at position $\mathbf{r}'$ in the object exit plane that is generated from a point source at position $\mathbf{r}$. Due to spreading of the electron wave emitted from a TDS point source in the crystal, this method requires use of supercells with sizes larger than the extensions of response functions used. The response function $G_n(\mathbf{r},\mathbf{r}')$ is normalized according to $\frac{1}{A}\int_A G_n(\mathbf{r},\mathbf{r}')\mathrm{d}^2\mathbf{r}'=1$, where A is the area of the slice. The TDS intensity in the object exit plane at position $\mathbf{r}'$ is computed by

$$I_{TDS}(\mathbf{r}') = \sum_n \frac{1}{A}\int_A J_{TDS}(n,\mathbf{r})G_n(\mathbf{r},\mathbf{r}')\mathrm{d}^2\mathbf{r}.$$

M. Luysberg, K. Tillmann, T. Weirich (Eds.): EMC 2008, Vol. 1: Instrumentation and Methods,
pp. 147–148, DOI: 10.1007/978-3-540-85156-1_74, © Springer-Verlag Berlin Heidelberg 2008

Intensity patterns are computed for Au (s. Figure 1) and InSb for different crystal orientations. These results are compared with intensities from the frozen lattice approximation based on uncorrelated vibration of atoms as well as with the frozen phonon approximation for Au. The frozen phonon method (s. Figure 2) uses a detailed phonon model based on force constants we computed by a density functional theory approach. The comparison shows that our suggested absorptive potential method is in close agreement with both the frozen lattice and the frozen phonon approximations.

1.   A. Weickenmeier, H. Kohl, Acta Cryst. A **47** (1991), p. 590.

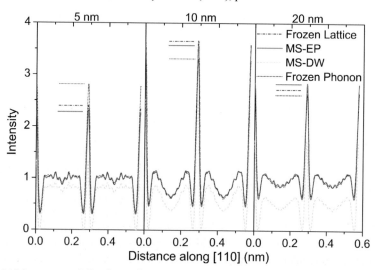

Figure 1. Linescans of the intensity in the object exit plane across a crystal unit cell computed for Au[100] with different approaches: Emission-potential multislice (MS-EP), frozen lattice, conventional multislice (MS-DW) and frozen phonon. The three columns of graphs correspond to specimen thicknesses of 5, 10 and 20 nm. The lines mark the intensity maxima.

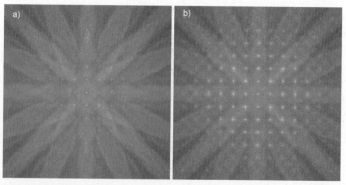

Figure 2. Simulated diffraction patterns for Au[100], where a) was computed with the frozen lattice approximation and b) with the frozen phonon approach.

Three-dimensional HREM Structure Retrieval

Z. Saghi, X. Xu and G. Möbus

Dept. of Eng. Materials, University of Sheffield, Mappin Street, Sheffield, S1 3JD, UK.

z.saghi@sheffield.ac.uk
Keywords: tomography, HREM, simulations

The structural analysis of nanoobjects is usually achieved by 2D HREM imaging interpreted by image simulations. Electron tomography [1] is a non-destructive technique capable of the 3D chemical and structural characterization of sub-micron sized materials with nanometer resolution. The combination of electron tomography with HREM [2] is explored here for the 3D structure retrieval of a perfect 2.5nm thick CeO_2 supercell, simulating an octahedral nanoparticle corner (Figure 1). HREM multislice calculations were obtained with the EMS software package [3]. Three different microscope technologies in Scherzer focus were simulated: (1) JEM-ARM 1250-Stuttgart LaB_6 (V=1250kV, Cs=1.6mm, focus spread Δ=10nm, beam convergence α=0.3mrad); (2) Aberration corrected JEM-2200FS-AC FEG (V=200kV, Cs=0.05mm, Δ=2nm, α=0.1mrad); (3) JEM-3010 LaB_6 (V=300kV, Cs=1.1mm, Δ=12nm, α=0.5mrad). The CeO_2 supercell was tilted over a 45° range with 5° increment and expanded according to its cubic symmetry to a tilt range of 180°. At each tilt, the complex exit wave function and the non-linear image intensity were stored.

Figure 2(a,a1,a2) are the phase images of the 0° exit wave functions for the three microscopes respectively. These images differ by the microscope voltage only. In all configurations, the projected atoms are well resolved. Figure 2(b, b1, b2) are the 0° HREM images: the JEM-ARM 1250 and JEM-2200FS-AC give similar Scherzer images, with the advantage of the flatter Ewald sphere at 1250kV hardly visible. The JEM-3010 images are severely blurred and the projected oxygen columns not resolved.

Tomographic reconstruction by filtered backprojection was applied to both sets of tilt series. Figure 2(c,c1,c2) are the top views of the 3D voxel projected reconstructions of the supercell from the HREM tilt series, and the positions of atoms can be compared to the model in Figure 1(b). Figure 2(d,d1,d2) are slices through the volume at the thickest position, and can be compared to the oxygen atom distribution in Figure 1(c). The JEM-ARM 1250 reconstruction retrieves all atoms (Figure 2(c,d)), while the oxygen atoms in the JEM-2200FS reconstruction are less well resolved at high thicknesses (Figure 2(c1,d1)). In both configurations, an artificial double peak appears at cerium positions, at thicknesses above 1nm (Figure 2(c,c1)). This artefact is not apparent in the 3010 reconstruction (Figure 2(c2)), where all Ce atoms appear as single, though blurred peaks, but none of the oxygen atoms is present (Figure 2(d2)).

In summary, HREM tomography of a cone shaped object allows to explore the thickness limit of tomography by (i) weak-phase object (WPO)-HREM (all atom positions plus scattering potentials correct), (ii) intermediate range HREM (atom positions only or one sublattice only), and (iii) complete failure due to dynamical scattering and non-linear interferences. Comparison of the input images with re-projected voxel-data allows quantifying the amount of non-linearity in the tomographic

M. Luysberg, K. Tillmann, T. Weirich (Eds.): EMC 2008, Vol. 1: Instrumentation and Methods, pp. 149–150, DOI: 10.1007/978-3-540-85156-1_75, © Springer-Verlag Berlin Heidelberg 2008

150

projection. The basic reason for HREM tomography to partially operate beyond the WPO limit relates to tomography of discrete data spaces [4] and porous structures [5].

1. J. Frank, "Electron tomography", 2nd ed. Springer, (Plenum, New York) (2007).
2. G. Möbus and B.J. Inkson, Microscopy & Microanalysis **9** 02 (2003), p.176.
3. P.A. Stadelmann, Ultramicroscopy **21** (1987), p. 131.
4. G.T. Herman and A. Kuba, "Discrete Tomography", (Birkhäuser, Boston) (1999).
5. This work was supported by a grant from EPSRC, GR/S85689/01, UK.

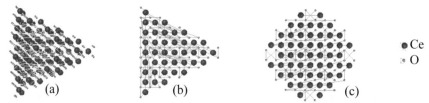

Figure 1. 3D distribution of O and Ce atoms in the CeO_2 Supercell model (a), top (b) and side (c) views of (a).

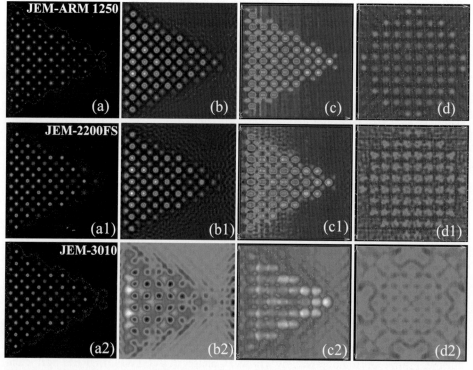

Figure 2. Imaginary part of the 0° exit wave functions with the JEM-ARM 1250 (a), JEM-2200FS (a1) and JEM-3010 (a2). 0° inverted Scherzer images (b,b1,b2). Voxel projection (c,c1,c2) and slice through (d,d1,d2) the tomographic reconstructions from the Scherzer tilt series.

Description of electron microscope image details based on structure relaxations with enhanced interaction potentials

K. Scheerschmidt

Max Planck Institute of Microstructure Physics, D-06120 Halle, Germany

schee@mpi-halle.de

Keywords: Bond Order Potential, Molecular Dynamics, Image Simulation

The imaging of crystal defects by transmission electron microscopy (high resolution HRTEM or conventional bright- or dark-field TEM) is well known and routinely used. Although the image calculations always tend to establish standard rules of interpretation, a direct phenomenological analysis of electron micrographs is mostly not possible, thus requiring the application of image matching techniques by calculating both the Fourier imaging including the microscope aberrations and the interaction process. Especially for the latter good structure models are necessary.

Modern computing allows predicting materials properties by using quantum-theoretical ab initio calculations with a minimum of free parameters, which, however, are very restricted to small structures only. Thus the classical molecular dynamics (MD) method using suitably fitted many-body empirical potentials have to be applied to simulate atomic processes with nanoscopic relevance. The quality of the relaxations and its physical relevance, however, depend strongly on the interaction potential used in the MD simulation. Fig. 1 shows resulting structural differences and Fig. 2 the different image contrast details for a SiGe/Si island expected for different empirical potentials.

The potential of Tersoff [1] with different parametrizations has the shape of a bond order where the bonds are weighted by the many body interactions over neighbours, but it is a pure empirical potential only. Thus, to enhance MD, we use the analytic bond order potential (BOP) based on the tight-binding model [2], as it preserves or mimics the essential quantum mechanical nature of atomic bonding, which allows better scattering models for the TEM-interaction process [3], too. Besides it analyticity, the main advantages of BOP are to achieve O[N] scaling with the particle number, which enables to simulate larger structures nearly with the same effort as Tersoff, and to have a good transferability to other materials systems. In addition, further enhancements are based on higher level approximations as BOP4, which is given up to the fourth-level continued fraction of the Greens function and necessary to describe correctly the π-bonds, and the extension of the analytic angular terms according to its physical relevance, which gives the higher stiffness of the BOP4+ [5] as demonstrated in Fig. 1.

1. J. Tersoff, Phys. Rev. B **38** (1989) 9902 & Phys. Rev B **39** (1989) 5566.
2. D. G. Pettifor, I.I. Oleinik, Phys. Rev. B **59** (1999) 8487.
3. K. Scheerschmidt, V. Kuhlmann, Proc. Int. Conf. DFTEM2006, Vienna (2006), 167-170.
4. K. Scheerschmidt, V. Kuhlmann, Int. J. Mat. Res. **98** (2007) 11.
5. V. Kuhlmann, K. Scheerschmidt, Phys. Rev. B **75** (2007) 014306.

M. Luysberg, K. Tillmann, T. Weirich (Eds.): EMC 2008, Vol. 1: Instrumentation and Methods, pp. 151–152, DOI: 10.1007/978-3-540-85156-1_76, © Springer-Verlag Berlin Heidelberg 2008

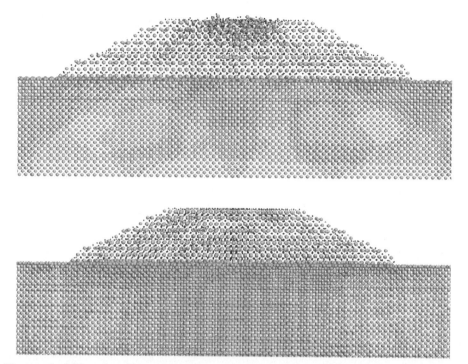

Figure 1. MD relaxation of an free-standing SiGe/Si island ([110] views) after annealing 0K>900K>0K with a Tersoff potential (top) and an analytical BOP4+ (bottom).

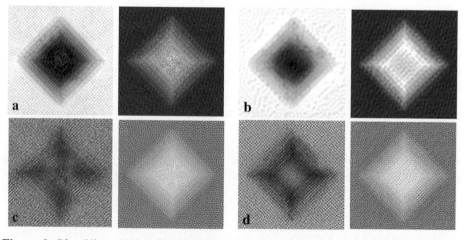

Figure 2. Plan-View 200kV bright-field TEM images (amplitude and phase, left and right, resp.) of the quantum dots of Fig. 1 relaxed with Tersoff (a,c) and BOP4+ (b,d) potentials for different sample thickness: 15nm (a,b) and 43nm (c,d).

Computation and parametrization of Debye-Waller temperature factors for sphalerite type II-VI, III-V and group IV semiconductors

M. Schowalter[1], A. Rosenauer[1], J.T. Titantah[2] and D. Lamoen[2]

1. Institut für Festkörperphysik, Universität Bremen, Otto-Hahn Allee 1, 28359 Bremen, Germany
2. EMAT, Universiteit Antwerpen, Groenenborgerlaan 171, 2020 Antwerpen, Belgium

schowalter@ifp.uni-bremen.de

Keywords: ab initio calculations, Debye-Waller factors, semiconductors

In (S)TEM comparison between experimental images and simulated images are widely used to retrieve quantitative information from TEM or STEM images. Such simulations are based on the Fourier components of the Coulomb potential, which are typically taken from publications as e.g. Ref. [1]. Another factor entering simulations of the TEM or STEM images is the Debye-Waller factor. Though being of less importance in the simulation of TEM images, the Debye Waller factor plays an significant role in the simulation of STEM images, not only since it strongly influences the amplitude of beams that are scattered to large angles, but also because the Debye-Waller factor is proportional to the root mean square (RMS) displacement of the atoms due to the finite temperature.

In this work we compute force constant matrices between the atoms of supercells within the density functional theory (DFT) formalism using the so called direct method of Parlinski et al. [2]. The DFT computations were carried out using the WIEN2k code [3]. The Fourier transform of the force constant matrices provides the dynamical matrices for any arbitrary phonon k-vector. Diagonalizing the dynamical matrix yields phonon frequencies and phonon polarization vectors belonging to the respective k-vector. By calculating phonon frequencies along certain high symmetry directions or on a grid in the Brillouin zone phonon dispersion relations or phonon density of states can be derived. The temperature dependence of the Debye-Waller factors is found from

$$B_\upsilon(T) = \frac{4\pi^2\hbar}{m_\upsilon} \int_0^\infty \coth(\frac{\hbar\omega}{2k_B T})\frac{g_\upsilon(\omega)}{\omega}d\omega$$

where g_υ is the generalized phonon density of states, T is the temperature, k_B and $\hbar$ are Boltzmann's constant and Planck's constant divided by 2π, respectively and m_υ is the mass of atom υ. In general the Debye-Waller factor is a 3x3 matrix for each atom type, but in the case of the cubic materials in this work the matrices are diagonal and the entries on the diagonal are the same. Therefore the matrix indices were suppressed in the last equation.

For non-homonuclear materials polarization effects give rise to a splitting of the optical phonon modes at the Γ point. In order to take into account polarization effects Born effective charges were computed using the ABINIT code [4]. From these charges

M. Luysberg, K. Tillmann, T. Weirich (Eds.): EMC 2008, Vol. 1: Instrumentation and Methods, pp. 153–154, DOI: 10.1007/978-3-540-85156-1_77, © Springer-Verlag Berlin Heidelberg 2008

the non-analytical part of the dynamical matrix was obtained which was added to the analytical part before diagonalization. Fig. 1a compares measured phonon frequencies (red dots) with calculations, in which the polarization effects were neglected (green dashed lines) and taken into account (black lines). At the Γ point the splitting of the optical modes becomes visible and the comparison with the experimental phonon frequencies shows that the polarization effects have to be taken into account for an accurate description of the phonon frequencies.

The temperature dependence of the Debye-Waller factors was calculated in the temperature range from 1 K to 1000K in steps of 1 K and then was fitted using the following procedure: For each temperature data point we derived a characteristic frequency $\omega_c(v;T)$ from the relation

$$B_v(T) = \frac{4\pi^2\hbar}{\omega_c(v;T)\cdot m_v}\coth(\frac{\hbar\omega_c(v;T)}{2k_BT}).$$

This equation can be found by applying the mean value theorem to the first equation. The characteristic frequency $\omega_c(v;T)$ then is fitted using the gaussian $\omega_c(v;T)=Aexp(-T^2/\sigma^2)+B$ with fit parameters A, B and σ.

In Fig. 1b the Debye-Waller factors for Ga and As are shown as circles and squares. The fit curves obtained using the above described approach are shown as lines as well.

1. A. Rosenauer, M. Schowalter, F. Glas and D. Lamoen, Phys. Rev. B **72** (2005), p. 085326.
2. K. Parlinski, Z.Q. Li, Y. Kawazoe, Phys. Rev. Lett. **78** (1997), p. 4063.
3. P. Blaha et al. WIEN2k, An Augmented Plane Wave+Local Orbitals Program for calculating crystal properties (2001), ISBN 3-9501031-1-2.
4. X. Gonze et al., Comp. Mat. Sci., **25** (2002), 478

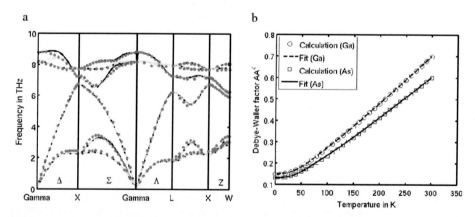

Figure 1. a) Comparison of a measured phonon dispersion relation with a computed dispersion for GaAs. b) Debye-Waller factors of Ga and As with the respective fits curves.

Structural Investigation of Amorphous/Crystalline Interfaces by Iterative Digital Image Series Matching

K. Thiel[1], N.I. Borgardt[2], T. Niermann[3] and M. Seibt[1]

1. IV. Physikalisches Institut der Universität Göttingen and Sonderforschungsbereich 602, Friedrich-Hund-Platz 1, 37077 Göttingen, Germany
2. perm. address: Moscow Institute of Electronic Technology, TU MIET, Passage 4806, bldg. 5, Zelenograd, 124498 Moscow, Russia
3. now at: Institut für Optik und Atomare Physik, TU Berlin, Straße des 17. Juni 135, 10623 Berlin, Germany

seibt@ph4.physik.uni-goettingen.de
Keywords: quantitative hrem, interfaces, distribution function (?)

In our work interfaces between covalently bonded crystalline and amorphous materials were studied with regard to the induced ordering in the amorphous material in the interfacial region by means of high-resolution transmission electron microscopy (HREM).

For quantification of the influence of the long-range ordered crystal onto the structure of the amorphous material we used the iterative digital image matching procedure [1] as a basis, which compares experimental and simulated images. The structural description for the material within the simulation is usually done on the basis of a supercell, for which the actual atom positions must be known. Since amorphous materials can only be described meaningful by statistical distribution functions, that approach cannot be used directly. Instead, the approach of Borgardt *et al.* [2] was used, which relies on the quantitative comparison of *averaged* experimental and simulated images. This means that the experimental images are segmented into N stripes perpendicular to the interface with width d, which is the period of the lattice parallel to the interface, and subsequently averaged over equivalent points of these stripes (see Fig. 1a & b). The intensity components, which are correlated with the period of the lattice image, could thus be separated from the statistical intensity fluctuations, which are characteristic for images of amorphous materials.

For the description of the structure of the amorphous material in the vicinity of the interface a three-dimensional distribution function $\rho_{3D}(\mathbf{r})$ was taken as a basis, which specifies the probability to find an atom in the amorphous material, if $\mathbf{r}=0$ is the position of an atom in the crystal. Its two-dimensional projection, $\rho(x,y)$, can be determined using iterative digital image matching techniques on averaged experimental and simulated interface images. Borgardt *et al.* have already shown, that within this so-called APP ("averaged projected potential") – approximation averaged images of the interface can be *directly* simulated with sufficient accuracy by means of conventional multislice calculations [2].

M. Luysberg, K. Tillmann, T. Weirich (Eds.): EMC 2008, Vol. 1: Instrumentation and Methods, pp. 155–156, DOI: 10.1007/978-3-540-85156-1_78, © Springer-Verlag Berlin Heidelberg 2008

156

In order to separate the influence of the delocalization, which is inherent to the imaging process, from the ordering, which is induced by the crystalline atoms at the interface, onto the structure determination, we analyzed 20 images of a through-focus series *simultaneously*.

The construction of the two-dimensional model for the amorphous material in the vicinity of the interface itself is based on a recursive scheme, which was proposed from Borgardt *et al.* [3]. Here, the crystal structure is continued into the amorphous material, but Gaussian distributions are used for the bond angles and the bond length. Thus it is achieved, that the crystalline order decreases with increasing distance from the interface until it finally ends up into a homogeneous distribution (see Fig. 1c).

In order to make this analysis a quantitative one and to be able to mark differences between two distribution functions emanating from two different regions of the interface as significant, it is necessary to determine the errors of $\rho(x,y)$. To do so, we used the so-called Bootstrap-method. Technically, this method uses the actual measurement, which is now called *dataset* and consists here of the N stripes of the averaging region, to generate any number R of *synthetical* datasets, by drawing N data points/stripes at a time with replacement from the original dataset. As a consequence of the replacement, these synthetical datasets differ from the original one. Here, due to the following averaging of the stripes the actual order of the stripes is of no importance. Performing the iterative digital image series matching procedure on each of these new datasets results in R two-dimensional distribution functions $\rho^{(r)}(x,y)$, which then can be statistically evaluated. Fig. 1d displays the resulting error distribution exemplary.

In order to decide whether the difference in some pixel is significant we make use of the statistical discrepancy. It is defined as the difference between two measured values and classified significant, if its value is bigger or equal the sum of their errors. This means that values greater or equal 1 express significance.

1. G. Möbus, R. Schweinfest, T. Gemming, T. Wagner, and M. Rühle, J. Microsc. **190**, Pts ½, 109 (1998)
2. N.I. Borgardt, B. Plikat, M. Seibt, and W. Schröter, Ultramicroscopy **90**, 241 (2002)
3. N.I. Borgardt, B. Plikat, W. Schröter, M. Seibt, and T. Wagner, Phys. Rev. B **70**, 195307 (2004)

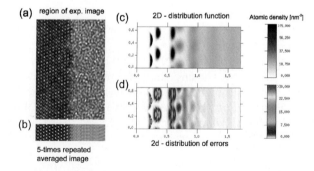

Figure 1. Results of the iterative digital image series matching procedure with error analysis by the bootstrap method.

Novel carbon nanosheets as support foils for ultrahigh resolution TEM studies of nanoobjects

A.S. Sologubenko[1], A. Beyer[2], C. Nottbohm[2], J. Mayer[1], and A. Gölzhäuser[2]

1. Central Facility for Electron Microscopy RWTH Aachen, Ahornstrasse 55, 51074 Aachen and Ernst Ruska Centre for Microscopy and Spectroscopy with Electrons, Research Centre Jülich 52425 Jülich, Germany
2. Fakultät für Physik, Universität Bielefeld, Postfach 100131, 33501 Bielefeld, Germany

sologubenko@gfe.rwth-aachen.de

Keywords: self-assembled monolayer, ultrathin support foil, nanosheet, transmission electron microscopy, scanning transmission electron microscopy, nanoclusters

The resolution in the newest generation of transmission electron microscopes reaches about 0.08 nm [1], which is sufficient to resolve single atomic columns in any crystalline material. In the case of the imaging of nanoscale clusters, however, the experimental resolution frequently does not reach such values [2]. The main difficulty arises from the necessity to put the clusters onto a sheet of a mechanical support for imaging by the electron beam. For nanoclusters with a size comparable to the thickness of a support, the scattering of electrons from the latter diminishes the image contrast and impedes the direct interpretation of transmission electron (TEM) or scanning transmission electron (STEM) micrographs. To increase the object contrast over the support contrast, thin foils of an amorphous material of light elements are usually used as substrates. The foil thickness, thickness modulations, microstructure, chemical inertness and stability under exposure to the high-energy electron beam are considered when choosing support for a particular task. Most commonly used foils are amorphous carbon, silicon oxide or silicon nitride films spanned on a copper, nickel or gold grids. The thickness of these foils is usually 10-20 nm that results in drastic deterioration of the phase contrast and sets a limit for the imaging of nanoclusters smaller than 2 nm. But a further reduction of the foil thickness strongly affects the mechanical stability and largely increases the sensitivity of the foil against beam damage.

In present work we propose and demonstrate the novel support foil material, the free-standing nanosheet of self-assembled monolayers (SAMs). The nanosheets are produced by the electron beam induced cross-linking of the biphenyls molecules self-assembled on the solid substrate, with subsequent dissolution of the substrate. The thickness of the nanosheet is defined by the length of the biphenyl molecules and is about 1 nm. Afterwards, the free-standing nanosheets are transferred onto a Quantifoil TEM grid, copper or gold TEM grid spanned over with an amorphous carbon film that contains regular sized opennings. As a result, within the openings, the TEM grid is covered only by about 1 nm thick nanosheet.

The composed system of the "Quantifoil plus the nanosheet" was studied by HRTEM and HRSTEM. The comparison of the image contrast obtained from two areas

M. Luysberg, K. Tillmann, T. Weirich (Eds.): EMC 2008, Vol. 1: Instrumentation and Methods, pp. 157–158, DOI: 10.1007/978-3-540-85156-1_79, © Springer-Verlag Berlin Heidelberg 2008

of the same TEM specimen, the Quantifoil plus nanosheet (Fig. 1a, b) and the free-standing nanosheet (Fig.1c, d), demonstrates remarkable enhancement of the phase contrast of the gold nanoclusters on the free-standing nanosheet area. The TEM and STEM studies prove that the nanosheets exhibit exceptional structural homogeneity and are fully stable under the electron beam exposure.

1. M.A. O'Keefe, C.J.D. Hetherington, Y.C. Wang, E.C. Nelson, J.H. Turner, C. Kisielowski, J.O. Malm, R. Mueller, J. Ringnalda, M. Pan, A. Thust, Ultramicroscopy **89** (2001) 215.
2. Konkar, S. Lu, A. Madhukar, S.M. Hughes, A.P. Alivisatos, Nanoletters **5** (2005) 969.

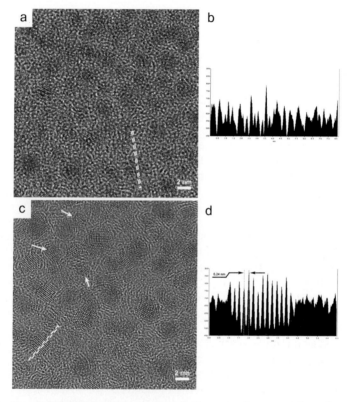

Figure 1. (a, c) HRTEM micrographs of Au-nanoclusters, taken from the same specimen in the aberration corrected FEI Titan 80-300 TEM under identical imaging conditions (Cs= - 14 μm, Z = 40 nm); (a) on the Quantifoil plus nanosheet (~ 15 nm thickness), (c) on the freestanding nanosheet (~ 1 nm thickness). (b, d) Integrated line profiles of the two gold clusters of about the same size and orientation with respect the beam direction. The clusters are marked in a and in c.

Quantitative HRTEM studies of reconstructed exit-plane waves retrieved from C_S-corrected electron microscopes

M. Svete[1], W. Mader[1]

1. Institute of Inorganic Chemistry of the University of Bonn, Römerstraße 164, 53117 Bonn, Germany

msvete@uni-bonn.de

Keywords: aberration-corrected TEM, EWACS, silicon nitride

Quantitative high-resolution TEM requires the analysis of lattice images or reconstructed exit-plane waves (EPW) with respect to imaging parameters as well as with respect to orientation and thickness of the specimen. The analysis is always based on the comparison of recorded images with simulated electron intensities, or of retrieved waves with electron waves calculated from structure models [1].

The retrieval of unknown objects is significantly simplified if the so-called geometrical parameters of the object, i.e. crystal thickness t and crystal tilt $\tau = (\tau_x, \tau_y)$, are known. Finding the correct values in a parameter space with at least three dimensions is a time consuming task. To facilitate this procedure the EWACS software package has been developed which partially automates this process. In most cases the image data contains parts of perfect crystalline regions of which crystallographic data is available. Hence these regions can be used to determine the geometrical parameters in order to reduce the parameter space.

In this contribution β-Si_3N_4-crystals are imaged along the c axis in the image-side C_S-corrected FEI Titan 80-300 electron microscope, installed at the Ernst-Ruska-Centre in Jülich. This microscope is equipped with a double-hexapole type C_S-corrector [2]. The motivation is to image the projected nitrogen sub-lattice structure where the smallest projected distance is ca. 0.95 Å. Therefore, the microscope was set to optimum defocus conditions, i.e. negative C_S-value of -13μm und positive defocus of 6 nm (Figure 1).

In a following step the method of exit-plane wave reconstruction from a focal-series has been used to benefit from a better signal to noise ratio and from further correction of residual aberrations by software [3]. Also the determination of geometrical parameters is facilitated by using the EPW compared to the analysis of single images. The reconstructed wave contains the desired information in a more direct way as it is obvious from the Fourier transformed EPW (Figure 1).

The single image as well as the exit-plane phase shows a remarkable deviation from the inversion symmetry of the projected potential. This is consistent with simulations where a small specimen tilt of $\tau = (0.6°, 0.5°)$ is included (Figure 2).

1.   G. Möbus et al., Journal of Microscopy **90** (1997), p. 109.
2.   M. Haider, H. Rose et al., Ultramicroscopy **75** (1998), p. 53.
3.   A. Thust, W.Coene et al., Ultramicroscopy **64** (1996), p. 109.
4.   We would like to thank M. Luysberg, L. Houben and C. Tillmann for their help. We kindly acknowledge the financial support by the DFG (project number MA 1020/13-1).

M. Luysberg, K. Tillmann, T. Weirich (Eds.): EMC 2008, Vol. 1: Instrumentation and Methods, pp. 159–160, DOI: 10.1007/978-3-540-85156-1_80, © Springer-Verlag Berlin Heidelberg 2008

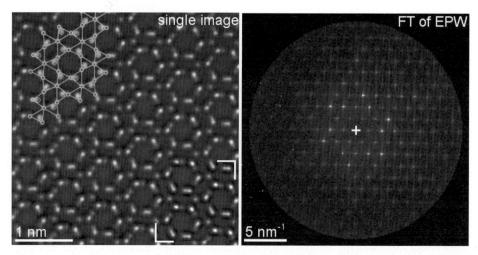

Figure 1. Single image of β-Si$_3$N$_4$ in <0001> and modulus of the Fourier transformed exit-plane wave. The single image shows an asymmetry of the resolved dumbbells. This asymmetry is also present in the simulated image inset if a specimen tilt is applied. Similar to real diffraction patterns a small specimen tilt off the zone-axis causes an asymmetric intensity distribution of the reflections in the Fourier transformed EPW.

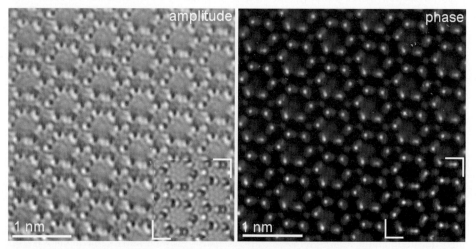

Figure 2. Reconstructed exit-plane wave as amplitude and phase. The asymmetry mentioned in Figure 1 is also apparent in the reconstructed EPW. The geometrical parameters for all simulation insets are $t = 4.7$ nm and $\tau = (0.6°, 0.4°)$.

Geometrical phase analysis of the 1:1 cation ordered domains in complex perovskite ferroelectrics

C.W. Tai and Y. Lereah

Department of Physical Electronics, School of Electrical Engineering,
Faculty of Engineering, Tel Aviv University, Ramat Aviv 69978, Israel

cwtai@eng.tau.ac.il
Keywords: geometrical phase analysis, HRTEM, ferroelectric relaxor

The crystal structures and microstructures of Pb-based complex perovskite ferroelectrics, $Pb(B'_{1/2}B''_{1/2})O_3$, have been studied extensively due to the technological importance as well as scientific interest. It is known that the degree of 1:1 B-site cation order strongly influences on the physical properties of most of these oxides [1]. The normal ferroelectric and relaxor behaviour are found in a material when it is in the ordered and disordered states, respectively.

High-resolution transmission electron microscopy (HRTEM) imaging is one of the techniques to examine the ordered/disordered domains, especially for the nano-sized ordered domains in the disordered matrix. The superstructure ½ ½ ½ reflections are generated by the cations occupying the crystallographic equivalent B-sites alternately along <111>. But the Bragg-filtered image by selecting the superstructure reflections is limited to distinguishing the region is whether ordered or disordered [2]. Other reflections contributed by the fundamental structure, such as 100 and 110 reflections, is unable to disclose any detail of the cation ordering. The combination of HRTEM and geometrical phase analysis (GPA) can reveal the strain distribution quantitatively [3]. In this study, the strain induced by small deformation of crystal lattice near an anti-phase boundary (APB) in a well-known order-disorder ferroelectric/relaxor, $Pb(Sc_{1/2}Ta_{1/2})O_3$ (PST) [1], is revealed by the above two techniques for the first time.

Figure 1a is a HRTEM image of a region in PST ceramic. The corresponding FFT (power spectrum) of the HRTEM image is shown in Figure 1b. In Figure 1c, the Bragg-filtered image of Figure 1a, which is obtained by selecting a superstructure reflection ½ ½ ½, shows the lattice fringes of two ordered regions separated by an APB. The π-shift of the lattice fringes across the two ordered domains is illustrated visibly. GPA of the same region is shown in Figure 2. The displacement field was calculated by selecting g_1=001 and g_2=110 reflections, which have no relation to the cation ordering. The variation of the calculated strains in the ordered regions is ~1%. We consider this as noise, which was verified by a [111] HRTEM image of a detect-free region in PST. Various strains (ε_{xx}, ε_{yy} and ε_{xy} ~±10 % or more) and rotation (ω_{xy} ~±10°) are detected in the vicinity of the APB. This demonstrates that localize distortion exists at the APB. We believe that such distortion can also be found in the disordered regions in other complex perovskite ferroelectrics/relaxors. This observation is particularly important when considering the random field model of the order-disorder ferroelectrics or relaxors [4].

M. Luysberg, K. Tillmann, T. Weirich (Eds.): EMC 2008, Vol. 1: Instrumentation and Methods, pp. 161–162, DOI: 10.1007/978-3-540-85156-1_81, © Springer-Verlag Berlin Heidelberg 2008

162

1. N. Setter and L. E. Cross, J. Appl. Phys. **51** (1980) p. 4356.
2. X. Pan, W. D. Kaplan and M. Rühle and R. E. Newnham, J. Am. Ceram. Soc. **81** (1998) p. 597.
3. M. J. Hÿtch, E. Snoeck and R. Kilaas, Ultramicroscopy **12** (1998) p. 131.
4. V. Westphal, W. Kleemann and M. D. Glinchuk, Phys. Rev. Lett. 68 (1992) p. 847.
5. We are grateful to Dr. K. Z. Baba-Kishi of The Hong Kong Polytechnic University for providing samples. This study was supported by European Commission under Grant No. 29637.

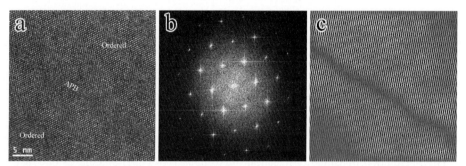

Figure 1. (a) HRTEM image of a region in PST ceramic. Superstructure ½ ½ ½ reflections can be seen in the power spectrum of (a) found in (b). (c) Lattice fringes of (½ ½ ½) showing an anti-phase boundary between two ordered domains.

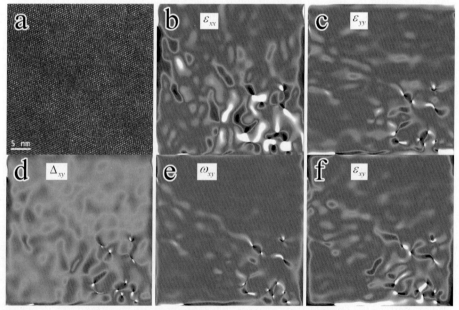

Figure 2. Geometrical phase analysis of the ordered and disordered regions in PST ceramic. Various strains and rotations are detected in the regions near the anti-phase boundary. Other defects in the ordered domains are also observed.

The Stobbs factor in HRTEM: Hunt for a phantom?

A. Thust

Institute of Solid State Research, Research Centre Jülich, D-52425 Jülich, Germany

A.Thust@fz-juelich.de

Keywords: HRTEM, Stobbs factor, factor-of-three problem, contrast mismatch, image simulation

A long standing problem in high-resolution transmission electron microscopy (HRTEM) was first publicly formulated by Hÿtch and Stobbs in 1994, who found that experimental images lack in contrast approximately by a factor of three compared to simulated images of equal mean intensity [1]. Since then, numerous not too successful attempts were made to explain this huge contrast discrepancy, which is often called Stobbs-factor problem, contrast-mismatch problem or factor-of-three problem.

In 2001 Rosenfeld and Thust found that the contrast discrepancy occurs already in experiment as a seeming contradiction, even without involving any image simulations [2]. Performing simple two-beam interference experiments, where a GaAs crystal is (mis)used as a beam splitter to generate two beams with amplitudes A and B, they showed that the recorded modulation amplitude 2AB of the mutual interference term is by far too weak when compared to the separately measured beam intensities A^2 and B^2. Since the beam intensities agreed almost perfectly with simulations, strong evidence for locating the problem in the formation and recording of the 2AB interference term arose.

While the interference mechanism itself, including coherent and partially coherent effects, is well examined, little attention has been paid to the acquisition of the finally resulting intensity modulation by photographic film plates or CCD cameras, involving the modulation transfer function (MTF) of the recording device. It is symptomatic that the MTF was not considered in Ref. [1], and that even today image simulation packages do not provide the possibility to include differently shaped MTFs at least phenomenologically, not to mention MTFs that have been experimentally determined.

Whereas the initial evaluation of the Rosenfeld-Thust experiment was based on the optimistic assumption of a Gaussian-shaped MTF dropping to 10% at Nyquist frequency, a recent reassessment including the meanwhile measured MTF does no longer justify to state a contrast mismatch problem of the discussed order.

Likewise, investigations with a focal series of 30 images of [110]-oriented $SrTiO_3$ acquired recently with a C_S-corrected FEI Titan 80-300 electron microscope do not leave room for a noteworthy mismatch factor when incorporating the measured MTF of a Gatan UltraScan 1000 CCD camera into image simulations. The dramatic contrast-flattening effect of the MTF is demonstrated by Figures 1 and 2. Already for the generous and unnecessarily high experimental sampling rate of 97 pixels/nm, a Stobbs factor of roughly 2 would seemingly arise when the MTF is erroneously neglected.

1. M.J. Hÿtch and W.M. Stobbs, Ultramicroscopy **53** (1994) 191.
2. R. Rosenfeld, PhD thesis RWTH Aachen (2001), published in: Berichte des Forschungs-zentrums Jülich 3921, ISSN 0944-2952.

M. Luysberg, K. Tillmann, T. Weirich (Eds.): EMC 2008, Vol. 1: Instrumentation and Methods, pp. 163–164, DOI: 10.1007/978-3-540-85156-1_82, © Springer-Verlag Berlin Heidelberg 2008

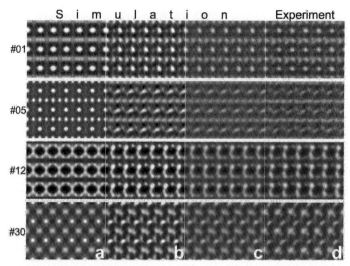

Figure 1. HRTEM image patches comprising 3×3 SrTiO₃ unit cells projected along the [110] zone axis. The patches #1, #5, #12, and #30 belong to different focal values and are shown exemplarily for a series of 30 images, which was recorded with a FEI Titan 80-300 microscope operated at 300 kV using a spherical aberration of $C_S = -25$ µm. All patches are displayed on the same common grey scale extending between intensity values 0.35 (black) and 2.2 (white). (a) Simulation for an object thickness of 2.8 nm, (b) simulation including additionally residual lens aberrations, (c) simulation including residual lens aberrations and the measured CCD-camera MTF, (d) experimental images.

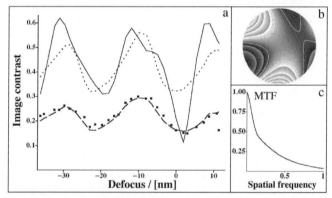

Figure 2. (a) Solid line: Simulated image contrast for an object thickness of 2.8 nm calculated for the focal values of the experimental SrTiO₃ series. Dotted line: Simulated image contrast including additionally residual aberrations. Dashed line: Simulated image contrast including residual aberrations and MTF. Full squares: Experimental image contrast. (b) Phase plate describing residual aberrations up to 12.5 nm⁻¹, sawtooth jumps every ±π/2. (c) MTF of a Gatan UltraScan 1000 CCD camera measured at 300 kV by the knife-edge method, frequency axis in units of the Nyquist frequency.

Measuring coherence in an electron beam for imaging

T. Walther, K. Atkinson, F. Sweeney and J.M. Rodenburg

Department of Electronic & Electrical Engineering, University of Sheffield, Sir
Frederick Mappin Building, Mappin Street, Sheffield S1 3JD, UK

t.walther@sheffield.ac.uk
Keywords: coherence, partial spatial coherence, interference, electron microscope

Transmission electron microscopy applies either parallel or focused electron beams
for illuminating the object, and all high resolution methods such as lattice imaging,
electron holography or diffractive imaging rely on interference effects which necessitate
coherent electrons. In lattice imaging the effects of partial spatial and partial temporal
coherence are usually dealt with by applying Gaussian envelopes to the contrast transfer
function to describe the dampening of higher spatial frequencies. This means the
coherence is only modelled indirectly, for partial spatial coherence by a convergence
angle and for temporal coherence by a defocus spread due to chromatic broadening. To
measure the coherence function would enable a full quantification of the contrast
observed in these interference experiments and can be done by the following methods:

1. Young's slits experiment: interference of electrons that pass fine holes nano-
   machined into a thick specimen using e.g. a focused ion beam instrument [1]
2. electron diffraction: interference between Bragg diffracted discs in convergent
   beam electron diffraction of thin crystalline specimens [2]
3. electron holography: interference of two phase-related electron beams created
   by splitting the electron beam by an electrostatic bi-prism [3]
4. comparison of the defocused far-field ring pattern formed behind an
   illuminated aperture with simulations

The last method is the probably only one that does not necessitate a specimen to act
as a beam splitter which, if not perfectly stable and well characterised, could falsify the
results. It also has the advantage of working in real space. In the following, first
experiments using this method are described, with some preliminary results for a JEOL
2010F field-emission transmission electron microscope. The microscope has been
operated in either imaging or convergent beam diffraction mode with a condenser
aperture of 20µm fixed size. The convergence semi-angles of illumination, α, have been
determined using gold particles on carbon and silicon as references (TEM: α=2.0mrad,
CBD: α=4.6mrad). We recorded series of images of the defocused probe onto a charge-
coupled device (CCD) camera mounted behind a Gatan imaging energy filter at nominal
magnification of 40k×. The GIF induces an additional magnification so that the
effective magnification was ~10^6×, yielding a sampling of 0.0234nm/pixel. Parameters
that influence the coherence and were varied included the settings of anode A1
(extraction voltage), anode A2 (focusing anode) and the condenser mini lens CM (on in
TEM, off in CBD mode). The result is that the beam coherence is best for low values of
A1, A2 and the CM lens switched off. For the lowest settings of A1 and A2 investigated
the coherence length extends to ~10nm, defined as the maximum distance from the spot

M. Luysberg, K. Tillmann, T. Weirich (Eds.): EMC 2008, Vol. 1: Instrumentation and Methods,
pp. 165–166, DOI: 10.1007/978-3-540-85156-1_83, © Springer-Verlag Berlin Heidelberg 2008

centre where fine fringe details are still observable in the defocused image. The intensity then drops to ¼ of its value under standard illumination conditions.

1.   C. Jönsson, Z. Phys. **161** (1961) 454-474
2.   B.C. McCallum and J.M. Rodenburg, Ultramicroscopy **52** (1993) 85-99
3.   F.F. Medina and G. Pozzi, J. Opt. Soc. Am. A **7** (1990) 1027-1033

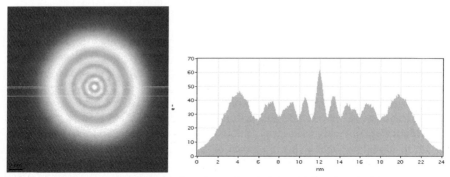

Figure 1. Experimental image of electron probe (left) and intensity profile (right).

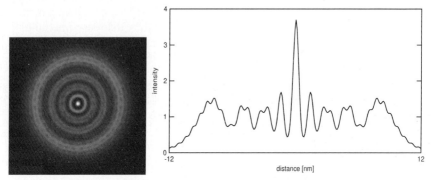

Figure 2. Simulated fully coherent beam for U=197kV, α=2.0mrad, f=5.6μm (left) and profile (right); same scale as in Figure 1. This suggests $r_c \approx$5nm in the above experiment.

Figure 3. Experimental images of most coherent electron probe obtained (α=3.4mrad), recorded at various defocus values of the condenser lens. While the number of fringes various, their contrast stays constant and is a measure of the coherence; here $r_c \approx$10nm.

Argand plot: a sensitive fingerprint for electron channelling

A. Wang, S. Van Aert and D. Van Dyck

EMAT, Department of Physics, University of Antwerp, Groenenborgerlaan 171, 2020 Antwerp, Belgium

amy.wang@ua.ac.be
Keywords: channelling theory, Argand plot, elastic and inelastic scattering

In this abstract it will be shown that exit waves are very useful in order to obtain quantitative information on the projected atom column positions and on the composition of these columns. According to the channelling theory [1], an atom column acts as a channel for the incoming electrons in which the electrons scatter dynamically without leaving the column. This theory gives a simple mathematical expression for the complex exit wave including only the lowest bound state of the column potential, which is called the 1s state. Its amplitude is peaked at the atom column positions whereas the phase, which is a constant over the column, is proportional to the average mass density of the column [2]. A convenient way to visualize these effects is by plotting the complex value of the exit wave for each pixel located at a projected atom column position in a so-called Argand plot with the x- and y-coordinate corresponding to the real and imaginary value, respectively [2, 3]. It turns out to be more convenient to shift the exit wave over a vector corresponding with the entrance wave so as to yield a fingerprint of the electron interactions only. The column position is chosen since its value is most sensitive to absorption, the Debye-Waller factor and the scattering factors. For a wedge-shape structure containing different thicknesses, the Argand plot is expected to be a circular locus which starts at the origin (for zero thickness) and which has its center on the x-axis. The radius of the circle is then the amplitude of the scattering wave, which should be identical for different atom column types if the columns are well-separated. This can also be proved using an empirical scaling behaviour which is valid for all the 1s states [4]. The constant angular increment corresponds to an extra atom in a column, which should be different for different atom column types.

In order to test the above mentioned predictions of the channelling theory, we simulated exit waves based on the multislice (MS) theory [5]. Figure 1 shows the Argand plots of wedge-shape structures of Si, Cu, Se and Au in [100] orientation with the spatial frequency range chosen large enough to avoid calculation artefacts. It is clearly shown that the radii of the circles corresponding to different column types are approximately the same and that the angular increment is a constant for which the value depends on the atom column type as predicted by the channelling theory. This plot can be used as a test for the number of beams (spatial frequencies) required for accurate MS calculations. Figure 2(a) shows Argand plots for Au[100] obtained from assuming different spatial frequency ranges in the MS simulation. It is also a fingerprint in the case of absorption which is also shown in the figure. The absorption is assumed to be

M. Luysberg, K. Tillmann, T. Weirich (Eds.): EMC 2008, Vol. 1: Instrumentation and Methods, pp. 167–168, DOI: 10.1007/978-3-540-85156-1_84, © Springer-Verlag Berlin Heidelberg 2008

168

10% of the atomic potential (see [6] for more details on this approach). From figure 2(a) it is clear that both an insufficient spatial frequency range as well as absorption lead to a change of a circle into a spiral. Then we shift the circle (without absorption) and spiral (with absorption) to their centers and calculate the phase and the logarithm of the amplitude for each point. This results in a "Channelling plot" as shown in figure 2(b). It is shown that there is a constant increment of the phase since it is proportional to the thickness. Without absorption, the data give a nearly horizontal line meaning that the radius is constant. In the case of absorption, we observe an inclined line, since the radius is exponentially decreasing with the thickness. From the slope of this inclined line, we can determine the ratio between the inelastic and elastic potential.

Experimental data will be collected from off-axis holography exit waves. From this technique, the inelastic scattering signals are reduced leaving only the elastic scattering signals. Then by using the Channelling plot, it will be possible to measure the degree of absorption coming from the elastic scattering only. If a significant absorption value is measured, this may provide a proof for the Stobbs factor.

1. D. Van Dyck and M. Op de Beeck, Ultramicrscopy **64** (1996), p. 99.
2. S. Van Aert, P. Geuens, D. Van Dyck, C. Kisielowski and J.R. Jinschek, Ultramicroscopy **107** (2007), p. 551.
3. W. Sinkler and L.D. Marks, Ultramicroscopy **75** (1999), p. 251.
4. D. Van Dyck and J.H. Chen, Solid State Communications **109** (1999), p. 501.
5. J.M. Cowley and A.F. Moody, Acta Cryst. **10** (1957), p. 609.
6. C.J. Humphries and P.B. Hirsch, Philos. Mag. **18** (1968), p. 115.
7. A. Wang and S. Van Aert are grateful to the Fund for Scientific Research – Flanders.

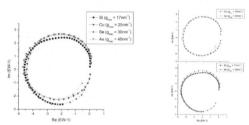

Figure 1. Argand plots for Si, Cu, Se and Au in [100] orientation obtained from a simulated exit wave using the multislice theory.

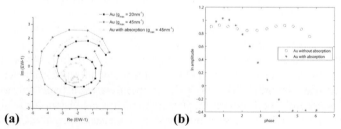

Figure 2. (a) Argand plots for Au[100] showing the effect of different spatial frequency ranges in the multislice simulation and the effect of absorption. (b) A Channelling plot shows that a horizontal line is obtained from the circle (without absorption) and that the line is inclined in the case of absorption.

Atomic-resolution studies of In₂O₃–ZnO compounds on *aberration*-corrected electron microscopes

Wentao Yu[1], Lothar Houben[2], Karsten Tillmann[2], and Werner Mader[1]

1. Institute of Inorganic Chemistry, University of Bonn, Römerstr. 164, D-53117 Bonn, Germany
2. Ernst Ruska-Centre and Institute of Solid State Research, Forschungszentrum Jülich GmbH, D-52425 Jülich, Germany

wentao.yu@uni-bonn.de
Keywords: Aberration correction, HAADF, In₂O₃–ZnO Compounds

ZnO based materials have many interesting electrical, optical and piezoelectric properties and are widely used for device applications. In order to tailor its properties, ZnO is commonly doped with other oxides. In this contribution, we present structural studies on In_2O_3–ZnO compounds by using image and probe side aberration-corrected FEI Titan80-300 electron microscopes installed at the Ernst Ruska-Centre in Jülich.

It is known that ZnO and In_2O_3 form homologous compounds of type $In_2O_3(ZnO)_m$ where the In ions are ordered on the close-packed basal planes and on pyramidal planes of ZnO for $m \geq 7$ [1], thereby forming an inversion domain structure in the ZnO lattice (see Fig. 1a and Fig. 2a). However, the co-ordination of In ions and the structure of the inversion domain boundaries (IDBs) is poorly known since information on these boundaries relies on imaging of ZnO along the $[2\bar{1}\bar{1}0]$ zone axis [2,3], where the Zn atoms are separated by more than 2.8 Å. The new instrumentation now available enables high-resolution phase contrast and HAADF imaging in the $[1\bar{1}00]$ orientation of ZnO, where cation columns with 1.62 Å spacing can be resolved (Fig. 1a). The phase contrast image (Fig. 1a) demonstrates clear evidence for displacements of cation columns which lead to a bending of type $\{2\bar{1}\bar{1}0\}$ and $\{0002\}$ lattice planes across the pyramidal IDBs resulting in a rigid body shift of the domains of ca. 0.2 Å and 0.5 Å, respectively. From the image in Fig. 1a an atomic model with relaxations of cations at the pyramidal IDBs is deduced where the In ions are in a bi-pyramidal co-ordination (Fig.1b).

The HAADF image (Fig. 2a) clearly exhibits the distribution of the In ions. The In density per area and, hence, the peak signal associated with a pyramidal IDB is lower compared to the basal IDB. The contrast of the basal IDBs observed in Fig. 2a proves the presence of a close-packed In monolayer whereas at the pyramidal IDBs the In atoms appear to be distributed over ca. 3 atom column distances (Fig. 2b/1). A simulated HAADF image (code ref. [4]) at the pyramidal IDB based on the bi-pyramidal model with the 3 column distribution of In ions is shown in Fig.2b/3. The intensity line profiles across the pyramidal IDB of the experimental HAADF image and the simulated image are displayed in Fig.2b/2 and Fig.2b/4, respectively. [5]

1. C. Li, Y. Bando et al., J. Solid State Chem. **139** (1998) 347
2. C. Li, Y. Bando, M. Nakamura, N. Kimizuka, Micron **31** (2000) 543

M. Luysberg, K. Tillmann, T. Weirich (Eds.): EMC 2008, Vol. 1: Instrumentation and Methods, pp. 169–170, DOI: 10.1007/978-3-540-85156-1_85, © Springer-Verlag Berlin Heidelberg 2008

170

3. Y. Yan, S. J. Pennycook et al., Appl. Phys. Lett. **73** (1998) 2585
4. C. Koch, Ph.D. Thesis, Arizona State University, 2002
5. Special thanks to Dr. C. Koch for assisting with the HAADF image simulation program. Financial support by the DFG (project number MA 1020/13-1) is kindly acknowledged.

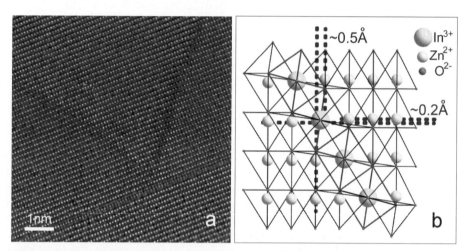

Figure 1. Atom column resolved phase contrast image of $In_2O_3(ZnO)_{30}$ in [1$\bar{1}$00] of ZnO showing strain and lattice shifts at the pyramidal IDBs (a). Deduced bi-pyramidal model for In ions at the pyramidal IDBs with relaxations of the cations (b).

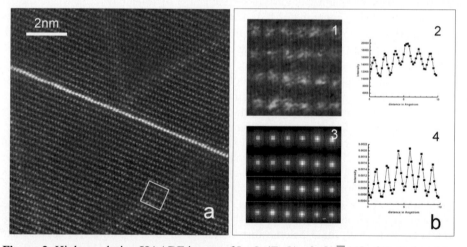

Figure 2. High-resolution HAADF image of $In_2O_3(ZnO)_{30}$ in [1$\bar{1}$00] of ZnO showing a close-packed In monolayer at basal IDB and In distribution of ca. 3 columns at the pyramidal IDB (a). Experimental HAADF image /1/, simulated image /3/, intensity line profiles of experimental /2/ and simulated /4/ HAADF images (b).

Advances in automated diffraction tomography

U. Kolb[1], T. Gorelik[1], E. Mugnaioli[1], G. Matveeva[1] and M. Otten[2]

1. Institute of Physical Chemistry, Johannes Gutenberg-University, Welderweg 11, 55099 Mainz, Germany
2. FEI, Building AAE, P.O.Box 80066, 5600 KA Eindhoven, The Netherlands

kolb@uni-mainz.de

Keywords: electron diffraction, STEM, nanodiffraction, automation, tomography

Crystal structure solution by means of electron diffraction or investigation of special structural features needs high quality data acquisition followed by data processing which delivers cell parameters, space group and in the end a 3D data set. The final step is the structure analysis itself including structure solution and subsequent refinement.

Electron diffraction under parallel illumination has been traditionally performed in a TEM using selected-area electron diffraction [1]. 3D diffraction data were collected by manual tilts of a crystal around a selected low indexed crystallographic axis. In order to increase the spatial resolution by decreasing the beam diameter to a few nanometers (nano electron diffraction: NED) we use a small C2 condenser aperture (10 μm)). Working in TEM mode with such a small, low-intensity beam makes it difficult to image the crystal and there is the potential for beam shifts caused by switching between TEM imaging and diffraction. To overcome the problem, scanning transmission electron microscopy (STEM), can be used to image the crystal. One major advantage of this approach is that the microscope remains in diffraction mode. In addition the place where the diffraction pattern should be taken from can be selected with high accuracy – in a stationary image without any electrons hitting the area of interest.

In order to switch between imaging and diffraction with as small changes as possible imaging is performed in less convergent microprobe STEM using small C2 aperture of 10 μm. This provides a "quasi-parallel" beam for coherent electron diffraction where only a slight defocus is necessary to reduce the convergence angle further resulting in a diffraction pattern with well separated spots. Because the collection of electron diffraction data is a time-consuming task and therefore difficult to adapt to many samples we developed the first automated electron diffraction tool which combines NED and STEM imaging using a high angular annular dark field detector (HAADF). The automated diffraction pattern collection allows us to sample more of the reciprocal space than before because beam damage is reduced significantly in comparison with more conventional approaches to diffraction.

The automated diffraction tomography module (ADT) for data acquisition is based partly on existing tomography software [2]. The typical approach during experimental work is to find a suitable crystal, select it with a marker (spot or allowed area) and tilt it, according to a given sequence, around the goniometer axis. ADT does not need a crystallographic axis oriented along the goniometer axis. Automated data processing routines, programmed in Matlab, provide automated cell parameter determination after a peak search. Additionally, the use of the complete patterns including the data between

M. Luysberg, K. Tillmann, T. Weirich (Eds.): EMC 2008, Vol. 1: Instrumentation and Methods, pp. 171–172, DOI: 10.1007/978-3-540-85156-1_86, © Springer-Verlag Berlin Heidelberg 2008

the reflections allows the detection of special structure effects such as partial disorder, twinning, and superstructures (see Figure 1) [3]. The module has been already successful used on nanoparticles with a size down to 5 nm (e.g. Au cluster using 15nm spot) as well as on a number of beam sensitive samples or crystals embedded in organic matrix.

1. U. Kolb and G.N. Matveeva, Electron crystallography on polymorphic organics, Z. Krist. spezial issue: Electron crystallography, **218 (4)** (2003) 259.
2. U. Kolb, T. Gorelik, C. Kübel, M.T. Otten and D. Hubert, Towards automated diffraction tomography: Part I -Data acquisition, Ultramicroscopy, **107(6-7)** (2007) 507.
3. U. Kolb, T. Gorelik, M.T. Otten, Towards automated diffraction tomography. Part II - Cell parameter determination, Ultramicroscopy (2008).

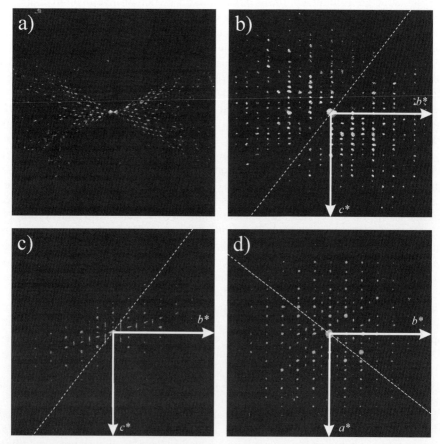

Figure 1. Projections of full-data integrated reciprocal space of two modifications of NaAl(WO$_4$)$_2$,; a) HT along the tilting axis; the data were collected over a 60° interval – from -30° till 30°; b) HT along the a* crystallographic axis; c) LT along a* crystallographic axis; d) LT along c*. Tilt axis = white dashed line.

Identification/ fingerprinting of nanocrystals by precession electron diffraction

S. Nicolopoulos[1], P. Moeck[2], Y. Maniette[1,3], P. Oleynikov[4]

1. NanoMEGAS SPRL Blvd Edmond Machtens 79, B-1080 Brussels Belgium
2. Portland State University, Dept of Physics P.O.Box 751,OR 97207-0751,USA
3. TVIPS GmbH, Eremitenweg 1, D-82131 Gauting Germany
4. Stockholm University, Svante Arrhenius v., 12, 10691 Stockholm Sweden

info@nanomegas.com

Keywords: precession electron diffraction, structural fingerprinting, nanocrystals

Structural identification of unknown materials are usually carried out by meand of standard X-Ray(powder and single crystal)techniques.Nowadays,standard identification techniques by means of powerful PDF algorithms may identify ab-initio several phases present in X-Ray powder diffraction patterns. However, identification of novel nanocrystal phases in powder pattern can be very difficult (if not impossible) due to important peak broadening for crystal size < 5 nm. On the other hand, although transmission electron microscopy and diffraction (TEM) is ideal technique for nanocrystal structure analysis, ED intensities usually deviate a lot from ideal kinematical values and space group extinctions are violated for crystals thicker than few nm.

Precession electron diffraction (PED) developed recently [1] is extremely promising technique that reduces a lot dynamical effects in ED patterns, rendering possible ab-initio nano-structure analysis for several compounds: complex oxides, catalysts, minerals and metals [2]. Precession device can be adapted to any TEM (100-400 KV, LaB6 or FEG)[3]. PED as tool for direct space/point group identification for nanocrystals [2] by comparing FOLZ/ZOLZ shifts and periodicity differences is unique for nanocrystal identification/fingerprinting where traditional X-Ray or conventional SAED techniques fail.

As an example case, X-Ray powder diffraction for both cubic maghemite (γ-Fe2O3 S.G P4132, a = 0.833 nm) and magnetite (Fe3O4, S.G Fd-3m, a = 0.839 nm) is very similar as both structures possess almost same lattice constant and are spinel ferrites with quite similar atomic arrangements (differences only in Fe+2 occupancies). SAED patterns are not reliable for structure identification as they are affected by dynamical diffraction and space group extinctions are usually not observed.

Experimental (PED) patterns have been taken in a Tecnai 12(120 KV TEM) equipped with "spinning star" precession device and precession angle 2° and sample a mixture of maghemite-magnetite nanocrystals. Figure 1,2 show experimental (PED) patterns and (kinematical precession) simulated ones (eMap software [3]). (PED) patterns show "wide" thickness FOLZ reflections (when they tilted away from [001] zone axis) allowing direct comparison (regarding shift/periodicities) with ZOLZ reflections. Both (PED) experimental patterns are only compatible with [001] Fd - - partial extinction symbol compared with theoretical patterns in [4,5], therefore revealing Fd-3m as the

M. Luysberg, K. Tillmann, T. Weirich (Eds.): EMC 2008, Vol. 1: Instrumentation and Methods, pp. 173–174, DOI: 10.1007/978-3-540-85156-1_87, © Springer-Verlag Berlin Heidelberg 2008

174

only S.G possibility, fingerprinting the studied nanocrystal as magnetite.Patterns corresponding to S.G P4₁32 (extinction symbol P4₁- - maghemite structure) are completely incompatible with experimental results, as also shown in our simulations. We have obtained similar results with asbestos (crocidolite structure) crystals, where (PED) patterns allow to identify inambiguously crystal structures and orientations. We conclude that (PED) is very useful tool for identifying unknown nanocrystals by direct space/point group identification.

1. R.Vincent, P.Midgley Ultramicroscopy **53** (1994), p. 271-282.
2. New Fronteers in Electron Crystallography Ultramicroscopy vol. **107**, June/July (2007)
3. http:// www.nanomegas.com application notes, www.analitex.com (eMap simulation)
4. J.P.Morniroli and J.W.Steeds Ultramicroscopy **45** (1992) p.219-239
5. J.P.Morniroli Atlas of Zone Axis Patterns Juin 1992

Figure 1. PED [001] of magnetite nanocrystal (left), [001] magnetite simulated kinematical pattern (center) and maghemite [001] simulated kinematical pattern (right).

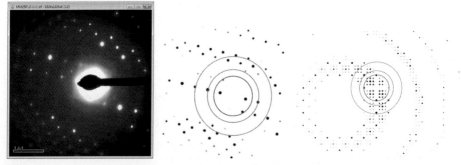

Figure 2. Magnetite PED taken with small tilt away from [001] of magnetite (left), kinematical PED simulation for magnetite (center) and maghemite structure (right).

On the Origin and Asymmetry of High Order Laue Zone Lines Splitting in Convergent Beam Electron Diffraction

A. Béché[1], L. Clément[2], J.L. Rouvière[1]

1. CEA/INAC/SP2M/LEMMA, 19 rue des Martyrs, 38 054 Grenoble, France
2. ST-Microelectronics, 850 rue Jean Monnet, 38 926 Crolles Cedex, France

jean-luc.rouviere@cea.fr

Keywords: HOLZ Lines, TEM, CBED, Splitting, Lattice Rotation

In the last five years, a lot of work has been done on the use of CBED HOLZ lines to quantitatively measure strain at nanometre scale [1-4]. HOLZ lines splitting have been observed in some situations and were related to stress relaxation in thin TEM lamella [1]. In this study we work further on the origin of HOLZ lines splitting and propose an explanation of lines asymmetry which appears in some experiments [5]. A model system, composed of a 30 nm $Si_{0.7}Ge_{0.3}$ epitaxial layer embedded in silicon, was experimentally observed (Figure 1a). Part of TEM samples were realized by tripode polishing with such a weak angle ($0.5°$) that faces can be considered as parallel. Others, presenting asymmetric HOLZ line splitting, were obtained using the SACT method [6]. CBED patterns were recorded on a FEI Titan operating at 300kV and energy filtered with a slit of 10eV in different zone axis ([8,11,0] and [2,3,0]).

During the thinning process to electron transparency, stress stored in lamella relaxes. This leads to a curvature of the atomic plane along the electron beam [1] resulting in the splitting of HOLZ line. Simulated multiple beams CBED patterns of the (6 -4 10) lines were obtained by two different algorithms: one using the usual displacement formulation and another one using the distortion field (Figure 1b). Both formulations give the same results. The distortion model allows separating the rotation contribution (Figure 1c) from the strain one (Figure 1d). In the present sample, two different regions can be defined. (Region *i*) Far from the SiGe interfaces, rotations dominate and allow reproducing correctly the splitting obtained with complete distortion field (Figure 1f). (Regions *ii*) Close to the interfaces, rotations are no more preponderant and splitting disappears because only a small fraction of the crystal is bent (Figure 1e).

Using these results, asymmetry of splitting can be investigated. For tripode polished samples, rotation occurs around the [1,1,0] axis which is parallel to the SiGe layer (direction y in Figure 1a and doted line in Figure 1g). As this axis is a symmetric axis for CBED pattern recorded along [8,11,0], splitting of HOLZ line appears symmetric for each couple of lines, like lines (6,-4,10) and (6,-4,-10) in Figure 1g. However, with cross-section samples with a less controlled geometry, like those realised by SACT, relaxation occurs in a more complicated way and the above symmetric lines are no more symmetric (Figure 1g). This asymmetry can be interpreted by supposing that the bending of all the planes can be approximated by a unique rotation. We note that atomic planes oriented parallel to the rotation axis are fully split whereas those perpendicular to the rotation axis present fine lines. Such analysis allowed us to position the rotation axis

M. Luysberg, K. Tillmann, T. Weirich (Eds.): EMC 2008, Vol. 1: Instrumentation and Methods, pp. 175–176, DOI: 10.1007/978-3-540-85156-1_88, © Springer-Verlag Berlin Heidelberg 2008

176

of asymmetric CBED patterns, like the one of Figure 1g. In conclusion, the asymmetry in the splitting of symmetric lines can be due to either the sample geometry (as in Figure 1g) or to a complex 3D stress field of the structure (as in [5]). The control of the sample geometry is then compulsory to determine the complete 3D stress field. With a controlled sample, the analysis of the asymmetric CBED pattern in terms of a unique local rotation should help in characterising quickly stress distribution in complex microelectronic devices.

1. L. Clement et al., Applied Physics Letter (2004), p.651-653.
2. A. Morawiec, Ultramicroscopy (2007), p. 390-395.
3. P. Zhang et al., Applied Physics Letter (2006).
4. F. Houdellier, C. Roucau, L. Clement, J.L. Rouviere, Ultramicroscopy (2006), p. 951-959.
5. A. Benedetti and H. Bender, MSM 2007, **In press**.
6. S.D. Walck and J.P. McCaffrey, Thin Solid Film (1997), p.399-405.

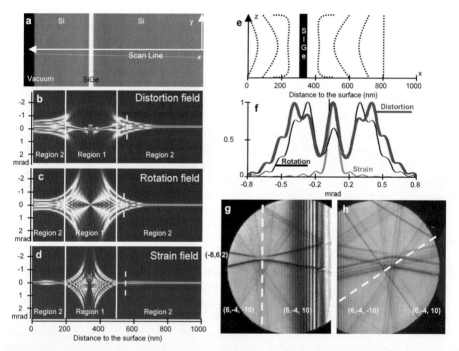

Figure 1a. General view of the sample. **1b.1c.1d.** Calculated splitting map of the line (6 -4 10) depending on the position of the probe under the surface (using respectively distortion, rotation and strain fields). **1e.** Profile of lattice distortion when approaching the SiGe stressed layer. **1f.** (6 -4 10) Line profile corresponding to the dot line in Figure b, c and d. **1g.** CBED pattern in [8,11,0] zone axis realized on tripode polished sample. Rotation axis is represented by the dashed line. **1h.** CBED pattern in [8,11,0] zone axis realized on SACT sample. Rotation axis is represented by the dashed line.

Precession electron diffraction: application to organic crystals and hybrid inorganic-organic materials

E.G. Bithell[1], M.D. Eddleston[2], C.A. Merrill[1,3], W. Jones[2] and P.A. Midgley[1]

1. Department of Materials Science and Metallurgy, University of Cambridge, Pembroke Street, Cambridge, CB2 3QZ, UK
2. Department of Chemistry, University of Cambridge, Lensfield Road, Cambridge, CB2 1EW, UK
3. Materials Department, University of California, Santa Barbara, 93106-5050, USA

egb10@cam.ac.uk

Keywords: precession electron diffraction, structure determination, MOF

It is now accepted that precession electron diffraction (PED) [1, 2] has a significant contribution to make to the nanoscale characterisation of materials. In this presentation we explore the extent to which data may be derived from acutely beam sensitive materials, specifically organic materials and hybrid inorganic-organic frameworks.

Figure 1a shows a PED pattern from an organic film of crystalline para-terphenyl [3]. Given our interest in structure determination from electron diffraction data, it is significant that the focussed (~10nm) probe PED pattern contains weak but clear reflections associated with the low temperature ordered form [4] (pseudo-monoclinic, a=1.601nm, b=1.109nm, c=1.353nm, β=92.0°). In a selected area diffraction pattern sampling a wider area (~1μm^2) these reflections were visible only as streaks (Figure 1b).

The choice of recording time is critical to the quality of the data: too short a time, and the diffracted intensities will be lost in low intensity noise; too long a time and the intensity data will be dominated by scattering from beam damaged material. We have achieved a very significant improvement in the quality of the data collected using an electron imaging plate system capable of recording intensity data to 32 bit accuracy [5]. The PED pattern recorded on an image plate and shown in Figures 1c and 1d was obtained from a coordination polymer [6], chromium ethylenediphosphonate (P6cc, a=1.311nm, c=0.926nm). Both low order and very high order reflections have been recorded in the same pattern within the usefully quantifiable intensity range.

Even under extremely low electron flux, such samples may retain their integrity for only a few seconds. It is often impossible to position a probe accurately, to orient the sample precisely, or to use long acquisition times. We have approached this problem by developing a procedure for the collection and analysis of large numbers of diffraction patterns, rather than by attempting to collect a smaller quantity of well oriented patterns. This is facilitated by the use of precession diffraction, for which small tilts away from zone axes are less critical than is the case for conventional diffraction conditions. Additional results will be presented to demonstrate the extent to which crystallographic information is retained and can be analysed in PED patterns of electron beam sensitive materials such as these. We will also discuss experimental approaches which maximise the likelihood of successful structure solution.

M. Luysberg, K. Tillmann, T. Weirich (Eds.): EMC 2008, Vol. 1: Instrumentation and Methods, pp. 177–178, DOI: 10.1007/978-3-540-85156-1_89, © Springer-Verlag Berlin Heidelberg 2008

178

1. R. Vincent and P.A. Midgley, Ultramicroscopy **53** (1994), p. 271-282.
2. D.L. Dorset, C.J. Gilmore, J.L. Jorda and S. Nicolopoulos, Ultramicroscopy **107** (2007), p. 462-473.
3. W. Jones, J.M. Thomas, J.O. Williams and L.W. Hobbs, J. Chem. Soc. Faraday Trans. **71** (1975), p. 138-145.
4. J.L. Baudour, Y. Delugeard and H. Cailleau, Acta Crystallographica **B32** (1976), p. 150-154.
5. We are grateful to ISS Group Services Ltd. for the loan of a Ditabis Micron Vario image plate system.
6. P.M. Forster and A.K. Cheetham, Topics in Catalysis **24** (2003), p. 79-86.
7. Dr. E.G. Bithell is grateful for support in the form of a Daphne Jackson Fellowship funded by Lucy Cavendish College, Cambridge, the Thriplow Charitable Trust and the Isaac Newton Trust. Prof. P.A. Midgley thanks the EPSRC for financial assistance.

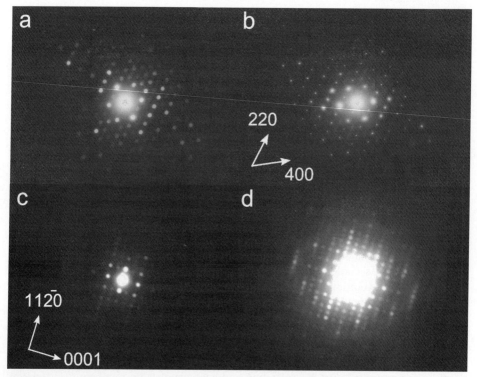

Figure 1. (a) para-terphenyl PED pattern obtained at a temperature of 92K using a Philips CM30 TEM; (b) selected area electron diffraction pattern obtained under comparable conditions to Figure 1a; (c) and (d) Chromium ethylenediphosphonate PED patterns obtained at a temperature of 92K, with the intensity scaled by a factor of four between the two patterns. Precession conditions were achieved using a Nanomegas Spinning Star system: (a) and (b) were recorded on Kodak SO-163 film, (c) and (d) on Ditabis imaging plates.

Structural studies of amorphous materials using RDF, RMC and DFT refinement

K. Borisensko[1], Y. Chen[1], G. Li[1], D.J.H. Cockayne[1] and S.A. Song[2]

1. Department of Materials, University of Oxford, Parks Road, Oxford OX1 3PH, UK
2. AE Center, Samsung Advanced Institute of Technology, Yongin 446-712, Korea

david.cockayne@materials.ox.ac.uk
Keywords: amorphous, RDF

The technique of electron diffraction reduced density function (RDF) analysis [1,2] is now sufficiently mature to provide a reliable tool for refinement of model structures of amorphous nanovolumes using reverse Monte Carlo (RMC) and density functional theory (DFT) methods commonly used for bulk materials using neutron and X-ray diffraction.

The first requirement is to obtain diffraction data to high scattering angles (q) which conforms to single scattering. The maximum sample thickness can be determined from multislice calculations, and can be verified by observing how well the experimental intensity I(q) fits single scattering. Figure 1 shows the diffraction pattern from a thin GeSbTe film, and the single scattering curve obtained using the atomic scattering factors, weighted for the known film composition. The fit at high q is excellent, with I(q) oscillating about the curve at low q. The reduced intensity S(q) can be obtained directly from the difference between these two curves.

The RDF obtained by Fourier transformation of S(q) (Figure 2) shows peaks which approximate to nearest neighbour distances (NND) (they would be NND in single element materials). With sufficient care taken in calibrating the scale of q, and in determining the centre of the diffraction pattern for azimuthal averaging, 0.02A accuracy in the peak positions in G(r) can be obtained.

RMC refinement of model structures can be carried out against either S(q) or the RDF (G(r)). Fourier transformation of G'(r), in which data below some lower cutoff has been removed from G(r), removes the oscillations caused by termination of data in S(q) at high angles. Constraints can be placed upon bonding distances, and DFT tests can reject configurations which are physically unreasonable. Figure 3 shows a flow diagram which we have implemented. It has been applied to several systems including glassy metals and the rapid phase change material GeSbTe, using both CASTEP and VASP in the DFT step. As an example, the refined structural model of GeSbTe shows motifs of fourfold rings which are closely related to the {100} fourfold rings found in the crystalline structure of GeSbTe. The belief is that these rings, by organization and disorganization, act as the link between the crystalline and amorphous phases, allowing the observed rapid transition between the two structures.

1. D.J.H. Cockayne and D.R.McKenzie, Acta Cryst. A 44 (1988) p.870.
2. D.J.H. Cockayne, D.R.McKenzie and D. Muller, Microanalysis, Microscopy, Microstructure 2 (1991) 359.

M. Luysberg, K. Tillmann, T. Weirich (Eds.): EMC 2008, Vol. 1: Instrumentation and Methods, pp. 179–180, DOI: 10.1007/978-3-540-85156-1_90, © Springer-Verlag Berlin Heidelberg 2008

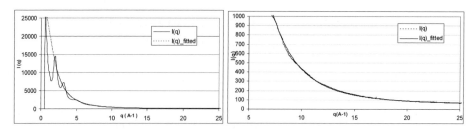

Figure 1. (Left) I(q) and single scattering curve (I(q) fitted) for a GeSbTe thin film; (Right) enlargement showing fit at high q.

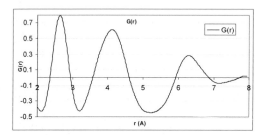

Figure 2. RDF from Figure 1

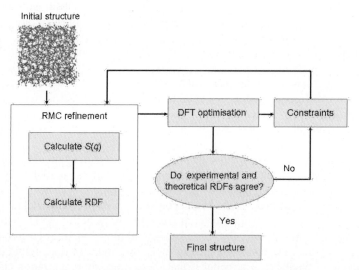

Figure 3. Flow diagram showing implementation of refinement procedures

A Nanoprobe Electron Diffraction Study of Surface Phases in LiCoO$_2$

F. Cosandey[1], J.F. Al-Sharab[1], N. Pereira[2], F. Badway[2] and G.G. Amatucci[1,2]

1 Materials Science and Engineering Dept., Rutgers University, Piscataway NJ 088541
2 Energy Storage Research Group, Rutgers University, Piscataway NJ 08854

cosandey@rci.rutgers.edu

Keywords: Diffraction, simulation, LiCoO$_2$

The LiCoO$_2$ compound with trigonal structure (R-3m), is the most common positive electrode material used in Li-ion batteries. In this paper, we report nanoprobe electron diffraction results of two surface phases in LiCoO$_2$ formed by two different heat treatments conditions. The nanoprobe diffraction patterns were taken at 200 kV with a Topcon TEM and using a 5nm probe size of convergence semi-angle α=2.4 mrad. Kinematical simulations of the nanoprobe diffraction patterns were done using JEMS program [1]. The diffraction patterns were analyzed with respect to published crystallographic data of stoichiometric LiCoO$_2$ and Li deficient Li$_x$CoO$_2$ phases [2, 3].

Nanoprobe diffraction patterns of a LiCoO$_2$ sample annealed at 1010°C for 4hrs followed by quenching (LCOQHT) are shown in Figure.1. The patterns were taken at the edge of the particle (Figure 1a) and 40 nm from the edge (Figure 1c). The pattern of this surface phase with two-fold rotation axis can be indexed as the cubic spinel Li$_x$CoO$_2$ (Fd-3m) phase in a [011] zone axis orientation as shown in the simulated pattern shown of figure 1b. The diffraction pattern of figure 1c is a superposition of matrix trigonal LiCoO$_2$ (R-3m) phase in a [241] orientation (Figure 3a) and the cubic spinel surface phase.

Nanoprobe diffraction patterns of a LiCoO$_2$ sample annealed at 1010°C for 4hrs followed by a second anneal at 850°C for 4hrs and slow cooling (LCOSCHT) are shown in figure 2. The patterns were taken at the edge of the particle (Figure 2a) and 60 nm from the edge (Figure 2c). This surface phase (Figure 2a) can be indexed as the monoclinic Li$_x$CoO$_2$ (P2/m) phase in a [011] zone axis orientation as shown in the simulated pattern of figure 2a. The strong reflections of the nanoprobe diffraction pattern of figure 2c belong to the matrix trigonal LiCoO$_2$ (R-3m) phase in a [241] orientation as shown in the simulated pattern of figure 3a. The weaker reflections in figure 2b are formed by the superposition of two orientation variants of the monoclinic surface phase. The variant I (Figure 2b) and variant II (Figure 3c) are obtained by a 180° rotation along the zone axis with the final superposed diffraction pattern shown in figure 3c in agreement with the experimental diffraction pattern of figure 2c.

1   http://cimewww.epfl.ch/people/stadelmann/jemsWebSite/jems.html.
2   Y. Shao-Horn et al., J. Electrochem. Soc., 150(3) (2003) p. A366
3   H. Gabrisch et al. J. Electrochem. Soc., 151(6) (2004) p. A891

M. Luysberg, K. Tillmann, T. Weirich (Eds.): EMC 2008, Vol. 1: Instrumentation and Methods,
pp. 181–182, DOI: 10.1007/978-3-540-85156-1_91, © Springer-Verlag Berlin Heidelberg 2008

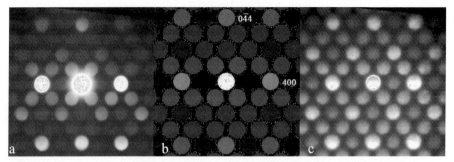

Figure 1. Nanoprobe diffraction patterns of LCOQHT sample taken (a) at the edge and (c) 40 nm from the edge with (b) simulated pattern for cubic spinel Li_xCoO_2 (Fd-3m) in a [011] zone axis.

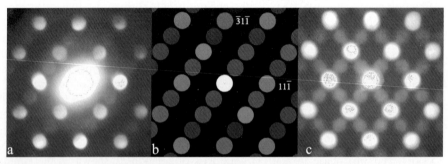

Figure 2. Nanoprobe diffraction patterns of LCOSCHT taken (a) at the edge and (c) 60 nm from the edge with (b) simulated pattern for monoclinic Li_xCoO_2 (P2/m) in a [011] zone axis (Variant I).

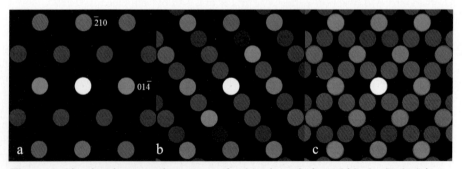

Figure 3. Simulated nanoprobe patterns for (a) trigonal phase $LiCoO_2$ (R-3m) in a [241] zone axis orientation, (b) monoclinic Li_xCoO_2 (P2/m) in a [011] zone axis orientation (Variant II) and (c) monoclinic $LiCoO_2$ (P2/m) obtained from the superposition of the two variants I and II.

Structural features of RF magnetron sputter deposited Al-Fe and Al-Cu thin films

S. Lallouche, M.Y. Debili

Laboratory of Magnetism and Spectroscopy of Solids LM2S, Department of physics, Faculty of Science, Badji-Mokhtar Annaba University BP12 23200 Annaba.

mydebili@yahoo.fr

Keywords: Thin films, grain size, stress, microstrain, sputtering, Al-Fe, Al-Cu

Microcrystalline and nanocrystalline materials can be currently produced by several elaboration methods and the resulting metal has a polycrystalline structure without any preferential crystallographic grain orientation. Gas-phase deposition processes using physical or chemical methods (PVD or CVD) are currently used to produce thin films coatings for mechanical engineering industries. Aluminium and its alloys with their low density and easy working occupy a significant place in the car industry, aeronautics and food conditioning. The on-glass slides sputter-deposited aluminium-based alloys thin films, such Al-Mg [1], Al-Ti [2,3], Al-Cr [4] and Al-Fe [5,3], exhibit a notable solid solution of aluminium in the films and microhardnesses higher than that of corresponding traditional alloys.

This work deals with Al-Fe and Al-Cu thin films, deposited on glass substrates by RF (13.56 MHZ) magnetron sputtering and annealing up to 500°C. Film thickness of coatings was approximately the same 3-4 µm. The microstrain determination has been performed thanks to Williamson-Hall method.

The residual stress of the films changes from tensile to compressive after crystallisation of amorphous Al-Fe thin films, and with aluminium content in Al-Cu thin films at 500°C.

Structural features as grain size, dislocation density and lattice parameter were also investigated using X-Ray diffraction and transmission electron microscopy.

The chemical analysis of atomic Fe or Cu composition in Al-Fe and Al-Cu films was made by X-ray dispersion spectroscopy. The microstructure of the films was studied by X-ray diffraction (XRD) and transmission electronic microscopy (TEM).

Two methods were used for the quantitative approach of the size of the grains:

The first is the application of Scherer formula [7] .This one is based on the measure of the width of the field of diffraction of x-ray via the measurement of the angular width Δ (2 θ). The average dimension of crystallites being given by $< D> = 0{,}9\ \lambda\ /\ \Delta\ (2\theta)\cos\theta$, where λ is the wavelength of the radiation used, θ is the angular position of a line of diffraction and Δ (2θ) its width with half intensity expressed in radian. This method assumes the exploitation of diagrams obtained in mode focusing $\theta/2\theta$ with a low divergence of the incidental beam. In order to limit the errors, of the diagrams on aluminium and iron with coarse grains (several micrometers) allowed to free itself from the instrumental widths of the lines (111) α Al and (110) c.c. which was used. The

M. Luysberg, K. Tillmann, T. Weirich (Eds.): EMC 2008, Vol. 1: Instrumentation and Methods, pp. 183–184, DOI: 10.1007/978-3-540-85156-1_92, © Springer-Verlag Berlin Heidelberg 2008

184

results, which rise from this method, provide a good estimate of the size of grains when the latter have a dimension lower than the micrometer.

The second method consists in evaluating the size of the grains starting from the images in dark field obtained by transmission electron microscopy, this method applies some is the smoothness of the microstructure. Figure (1).

1   R.D. Arnell, R.I. Bates, Vacuum, **43** (1992), p. 105.
2   T.Uesugi , Y. Takigawa , K.Higashi Materials Science Forum Vols.561-565 (2007) p 997
3   F.Sanchette, Tran Huu Loï, C. Frantz, Surf. Coat. Technol., **74-75** (1995), p. 903.
4   F.Sanchette, Tran Huu Loï, A. Billard, C. Frantz, Surf. Coat. Technol., **57** (1993), p. 179.
5   H. Yoshioka, Q. Yan, K. Asami, K. Hashimoto.Materials Science and Engineering, A 134 (1991) p. 1054-1057.
6   Guinier, Théorie et Technique de la radiocristallographie, 3ème éd. Dunod Paris (1964) p. 461.

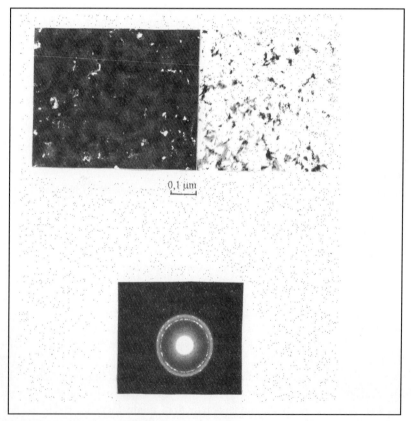

Figure 1. Dark field and bright field transmission electron micrographs from Fe-92.5%atAl thin film and its associated SAED ring pattern.

Structural Investigation of a Layered Carbon Nitride Polymer by Electron Diffraction

M. Döblinger[1], B.V. Lotsch[2], L. Seyfarth[3], J. Senker[3] and W. Schnick[1]

1. Department Chemie und Biochemie, LMU München, Butenandtstr. 5-13, 81377 München, Germany
2. Lash Miller Chemical Laboratories, Department of Chemistry, University of Toronto, 80, St George Street, Toronto, Ontario M5S 3H6
3. Lehrstuhl für Anorganische Chemie I, Universität Bayreuth, Universitätsstr. 30, 95440 Bayreuth, Germany

markus.doeblinger@cup.uni-muenchen.de

Keywords: electron crystallography, carbon nitride

Graphitic carbon nitride has attracted continuous interest because of its potential use as a precursor for ultrahard materials [1]. Down to present days, the synthesis of truly binary carbon nitride C_3N_4 has always been spoiled by the presence of additional salts or hydrogen. Inclusion of the latter 'defects' likely results in incomplete condensation of the network forming molecules triazine (C_3N_3) and heptazine (C_6N_7), which is typically accompanied by amorphisation and denitrification. For the characterisation of the nanocrystalline and disordered character of the resulting light-element materials, electron microscopy is particularly well suited, be it with respect to synthesis optimisation or structure analysis.

We used a synthesis route based on melamine, $C_3H_6N_6$, at 630°C under the autogenous pressure of ammonia. Powder X-ray diffraction reveals a mainly crystalline layered material exhibiting 2D planar defects, the layer distance being 3.2 Å (graphite: 3.3 Å). Electron diffraction (ED) was performed using a FEI Titan 80-300 equipped with a Gatan USC 1000 CCD camera. Phase analysis showed the coexistence of several structurally related layered phases. The predominant phase in this material was already described previously [2]. Electron diffraction revealed a layered structure made up from chains of heptazine linked via N(H)-bridges, thereby forming a closely packed 2D array.

The structure described in the present study appears as a minority phase in the same material. A diffraction pattern of the *hk0* plane is shown in Figure 1 a, after subtraction of the diffuse background. Assuming hexagonal or trigonal symmetry, indexing the *hk0* patterns yields the cell parameters a = b= 12.4 Å, γ = 120°. Reflection intensities were extracted with ELD [3]. Using the kinematical approximation, the structure was solved by SIR–97 [4] in the plane group *p*31m and refined with Shelxl-97 [5] to residuals of R1 = 0.09 and wR2 = 0.26. Fixing hydrogen atoms at their corresponding sites, but without further constraints, the refinement delivers realistic distances and geometries involving all carbon and nitrogen atoms. A kinematical simulation of the *hk0* plane (Figure 1 b) shows very good agreement with the experimental diffraction patterns. The structure can be described as a 2D network based on heptazine units connected by N(H)-bridges. However, instead of maximum dense condensation of the heptazine

M. Luysberg, K. Tillmann, T. Weirich (Eds.): EMC 2008, Vol. 1: Instrumentation and Methods, pp. 185–186, DOI: 10.1007/978-3-540-85156-1_93, © Springer-Verlag Berlin Heidelberg 2008

186

building blocks via trigonally coordinated tertiary nitrogen atoms, the heptazine units are connected via imide bonds, thereby generating triangular voids with dimensions extending into the nanometer range. The resulting cavities are filled with a single melamine molecule, which is anchored to the framework by hydrogen bridges. Further work on this structure will include structure calculations and a refinement on possible stacking variants.

1   D. M. Teter, R. J. Hemley, Science 271 (1996) p. 53.
2   B.V. Lotsch, M. Döblinger, J. Sehnert, L. Seyfarth, J. Senker, O. Oeckler, W. Schnick, Chem. Eur. J. 13 (2007), p. 4969.
3   X. D. Zou, Y. Sukharev, S. Hovmöller, Ultramicroscopy 49 (1993) p. 147.
4   A. Altomare, M.C. Burla, M. Camalli, G.L. Cascarano, C. Giacovazzo, A. Guagliardi, A.G.G. Moliterni, G. Polidori, R. Spagna, J. Appl. Cryst. 32 (1999) p. 115.
5   G.M. Sheldrick, Acta Cryst. A64 (2008) p. 112.
6   We kindly acknowledge financial support that was granted from the Deutsche Forschungsgemeinschaft DFG (projects SCHN 377/12–1 and SE 1417/2–1).

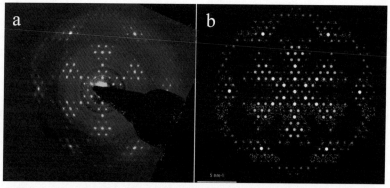

Figure 1. Experimental (a) and simulated (b) *hk0* diffraction pattern.

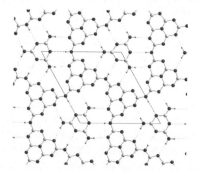

Figure 2. Projection of the new structure; black: N, gray: C. Hydrogen atoms (dark grey) and bonds are included to illustrate the relation between melamine and the framework structure.

Measuring the particle density of a nanocrystal deposit using DF images and a reciprocal space analysis

P. Donnadieu[1]

1. SIMAP, INPGrenoble-CNRS-UJF, BP 75, 38402 Saint Martin d'Hères

patricia.donnadieu@ltpcm.inpg.fr
Keywords: nanocrystals, DF imaging, reciprocal space analysis

Nanocrystals deposits are attracting interest because of their potential applications. For instance, for small size and high density, Si nanocrystals can have application as floating gate memory and CoPt nanoparticles as high density magnetic recording media. For controlled elaboration, quantitative characterization is required. In particular, the density is a key parameter. Therefore, it is worth trying to measure the particle density using TEM, namely the Dark-Field (DF) imaging mode. DF images result from the selection by the objective aperture of a part of the diffracted intensity. They give then only a partial view of a set of particles which has to be corrected to restore the global information. The aim of this work is to establish a reliable correction factor to overcome a major limitation of DF images. For an isotropic assembly, i.e. the nanocrystals have all possible orientations, the correction factor is given by the ratio of the selected intensity to the total diffracted intensity. In reciprocal space, the intensity diffracted by a nanocrystal assembly is described by a (hkl) shell with surface $4\pi/d_{hkl}^2$, The shell surface is divided by m the multiplicity of the (hkl) line to avoid overcounting. So the reciprocal surface describing the intensity diffracted by the nanocrystal assembly is given by $4\pi/md_{hkl}^2$.

In reciprocal space, the intensity selected by the objective aperture is determined by ϕ the aperture diameter and $1/L$ the deviation to the Bragg angle due to the particle size L. If only one (hkl) ring is selected, the correction factor is given by the ratio of the objective aperture selection to the while sphere: i.e. $4\pi L/(\phi m d_{hkl}^2)$. If the aperture selects several close rings, the correction must take in account the total multiplicity. The correction factor to apply to the DF image formed with an objective aperture centred on a ring with mean radius 1/d is then:

$$C = \frac{4\pi . L}{\phi . d^2} \frac{1}{\sum m_i}$$

Fig.1a illustrates the reciprocal analysis developed above. Fig.1b shows the diffraction pattern of Mo_5Si_3 crystallite deposit very appropriate to test the reliability of the above correction factor. Fig. 2a displays the Bright Field (BF) image of the nanocrystal deposit and its related DF image (Fig.2b). According to the BF images, the particles are rather monodispersed (L=7 nm ± 1 nm) and are forming a rather hexagonal packing. Assuming a perfect packing, the density can be easily calculated from the distance between particles (D ~ 9 nm): $2/(D^2\sqrt{3}) = 1.4\ 10^{12}/cm^2$. On the DF image (Fig. 2b), the apparent particle density is $2.2\ 10^{10}/cm^2$. The DF image was obtained using an objective aperture of diameter $\phi = 1\ nm^{-1}$, centred on a mean distance d = 0.23 nm and

M. Luysberg, K. Tillmann, T. Weirich (Eds.): EMC 2008, Vol. 1: Instrumentation and Methods, pp. 187–188, DOI: 10.1007/978-3-540-85156-1_94, © Springer-Verlag Berlin Heidelberg 2008

selecting two rings both with a multiplicity m=12. Hence, for 7 nm large particle, the correction factor is C = 69. The corrected density is then: 1.5 10^{12}/cm^2. This value is in good agreement with the one estimated assuming an hexagonal packing and within the incertitude due to the particle size, the main source of errors for this method.

The correction factor is established with 1/L for the deviation to the Bragg angle, consequently only the brightest particles have to be taken in account which simplifies the image analysis. It is worth noting that the method relies on basic knowledge of DF image formation and electron diffraction. Of course, the method can be safely applied only to isotropic assembly of nanocrystals showing a narrow size distribution.

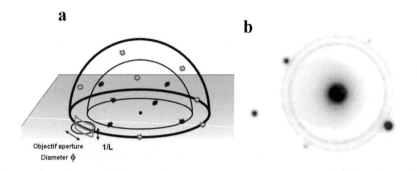

Figure 1. a) Schematic representation of DF imaging in the reciprocal space. b) Diffraction rings characteristic of isotropically dispersed Mo$_5$Si$_3$ crystallites.

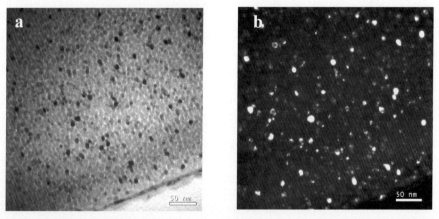

Figure 2. a) BF image of the Mo$_5$Si$_3$ nanocrystals forming an almost hexagonal packing on the substrate. b) DF image showing 29 bright particles on a surface S = 0.13 μm^2, the apparent density is then 2.2 10^{12}/cm^2 .

Towards a quantitative understanding of precession electron diffraction

A.S. Eggeman, T.A. White and P.A. Midgley

Department of Materials Science and Metallurgy, University of Cambridge, Pembroke Street, Cambridge, CB2 3QZ, UK

ase25@cam.ac.uk

Keywords: precession electron diffraction, charge flipping, electron crystallography

The development of direct methods [1] or charge-flipping algorithms [2] has allowed the 'phase-problem' of X-ray diffraction to be solved and enabled rapid, accurate solutions of unknown crystal structures. For electron diffraction to use such techniques the data should be as kinematical as possible, i.e. that the observed intensity of a diffracted beam is proportional to the square of the structure factor for that reflection. The strength of interaction between electrons and matter means that, in general, this condition applies only for specimens that are very thin or comprised of low atomic number elements. In an attempt to reduce dynamical effects the electron precession technique was introduced [3]. It has been found [4] that large precession angles yield 'more kinematical' data but very large angles can lead to the overlap of high-order and zero-order Laue zones and, through the spherical aberration of the probe forming lens, sometimes unacceptably large probe sizes. Here we investigate the transition to kinematical data in more detail using a complex oxide as a test structure.

Electron diffraction patterns were recorded with precession angles ranging from 0 to 55 mrad at the [001] zone axis of $Er_2Ge_2O_7$; a selection of these are shown in Figure 1. Qualitatively, increasing the precession angle appears to reduce the effects of dynamical scattering, and this is confirmed in a quantitative manner in Figure 2 where an un-weighted intensity residual, calculated from intensities of sequential diffraction patterns, is plotted. There is a marked decrease in the residual as the precession angle is increased and appears to plateau at a precession angle of ~30-35 mrad, suggesting that a further increase in precession angle will not improve significantly the kinematic nature of the pattern. To confirm this, a charge flipping algorithm was used to achieve a structure solution from the precession data. A plot of the phase-residual is also shown in Figure 2; this is calculated from the difference between the structure factor phases retrieved from these structure solutions and those from an ideal structure solution [5]. There is a significant increase in agreement between the true phases of the structure factors and the phases assigned to diffracted intensities at precession angles above 35 mrad. This is in reasonable agreement with the experimentally derived intensity residual.

1. D. Harker and J. S. Kasper. Acta Cryst **1**, (1948), 70.
2. G Oszlanyi and A. Suto, Acta Cryst **A60**, (2004), 134.
3. R. Vincent and P. A. Midgley, Ultramicroscopy. **53**, (1994) 271.
4. J. Ciston, et al, Ultramicroscopy (2007), doi:10.1016/j.ultramic.2007.08.004.

M. Luysberg, K. Tillmann, T. Weirich (Eds.): EMC 2008, Vol. 1: Instrumentation and Methods, pp. 189–190, DOI: 10.1007/978-3-540-85156-1_95, © Springer-Verlag Berlin Heidelberg 2008

190

5. Y. I. Smolin, Sov. Phys Cryst. **15**, (1970), 36
6. The authors thank the EPSRC and FEI Company for financial assistance.

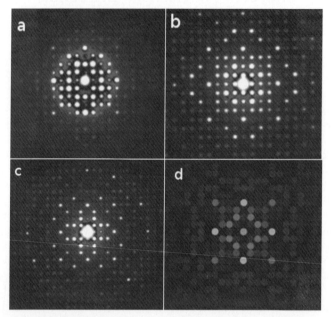

Figure 1. Electron diffraction patterns of $Er_2Ge_2O_7$ recorded parallel to the [001] zone axis with precession angles of a) 0 mrad, b) 20 mrad and c) 47 mrad; d) shows the calculated kinematical diffraction pattern for comparison.

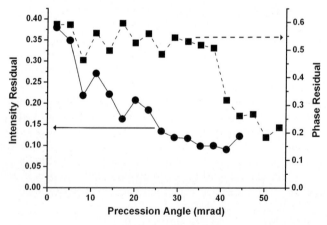

Figure 2. Intensity residual calculated from the difference in intensities recorded from sequential precessed diffraction patterns and the phase residual calculated from the difference between the phases of the correct structure factors and those derived from structures solved by a charge flipping algorithm.

Electron crystallography by quantitative CHEF

F. Houdellier and M.J. Hÿtch

CEMES-CNRS, 29 rue J. Marvig, 31055 Toulouse, France

florent@cemes.fr

Keywords: CBED, electron holography, electron cristallography

Quantitative convergent-beam electron diffraction (CBED) is a well established technique for determining the amplitude and phase of structure factors, in particular the so-called three-phase structure invariant of non-centrosymmetric crystals [1,2]. However, the technique requires careful fitting of the two-dimensional experimental data with results from N-Beam dynamical theory. We will show how similar information can be retrieved more directly by diffracted-beam holography [3]. An investigation of the Stobbs factor [4] will ensue by comparing dynamical simulations with experimental amplitudes and phases.

In energy-filtered diffracted-beam holograph (CHEF) convergent-beam diffraction disks are interfered with the aid of an electron biprism [5]. The specimen is shifted in height with respect to a convergent beam focused on the object plane, similar to large-angle convergent-beam diffraction (LACBED), and an electron biprism deviates the diffracted disks so that they overlap in the focal plane. We use a modified configuration where the beam is focussed on the sample like CBED (Fig. 1) so that the interference fringes are free of the shadow image from the defocused image of the sample [5]. Optimum conditions require a field-emission gun (FEG), rotatable biprism and imaging filter.

Experiments were performed using the SACTEM-Toulouse, a Tecnai F20 ST (FEI), fitted with an objective-lens aberration corrector (CEOS), rotatable electron biprism (FEI) and imaging filter (Gatan Tridiem with 2k CCD camera). The local phase of the interference fringes was measured using GPA software [6]. Two test specimens have been studied. In the first case, a silicon sample is used to understand the influence of the Stobbs factor (Figure 1). CHEF experiments were performed using different diffracted beams for different thicknesses.

To determine the three-phase structure invariant, a GaAs sample was used as a non-centrosymetric test crystal. The sample was then oriented in a non-systematic 3-beam condition for which the second Bethe approximation remains valid (Figure 2). CHEF patterns were obtained by interfering the transmitted and diffracted CBED disks. We will show how the three-phase structure invariant can be obtained by measuring phase changes at particular positions in the pattern. For the example shown in Figure 2, the value obtained of 10° is consistent with that measured by CBED analysis. A systematic comparison will be presented between CHEF and Q-CBED measurements.

1   J.M. Zuo, R. Hoier and J.C.H. Spence, Acta. Cryst. A**45** (1989), p. 839.
2   R. Hoier, L.N. Bakken, K. Marthinsen and R. Holmestad, Ultramicroscopy **49** (1993), p. 159.
3   R.A. Herring, G. Pozzi, T. Tanji and A. Tonomura, Ultramicroscopy **60** (1995), p. 153.

M. Luysberg, K. Tillmann, T. Weirich (Eds.): EMC 2008, Vol. 1: Instrumentation and Methods, pp. 191–192, DOI: 10.1007/978-3-540-85156-1_96, © Springer-Verlag Berlin Heidelberg 2008

4   M.J. Hÿtch and W.M. Stobbs, Ultramicroscopy **53** (1994), p. 191.
5   F. Houdellier and M.J. Hÿtch, Ultramicroscopy **108** (2008), p. 285.
6   GPA Phase 2.0 (HREM Research Inc.) a plug-in for DigitalMicrograph (Gatan).
7   The authors thank the European Union for support under the IP3 project ESTEEM (Enabling Science and Technology through European Electron Microscopy, IP3: 0260019)

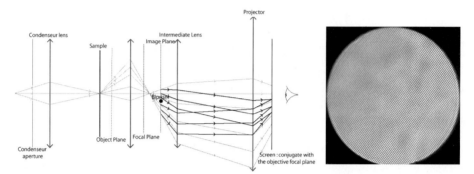

Figure 1. Optical configuration for CHEF experiments in CBED mode (black line = ray paths after the biprism; green and red lines = transmitted and diffracted ray paths without biprism). CHEF pattern from the interference between (000) and (220) disks in a thin crystal of Si.

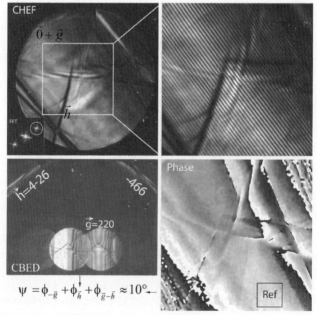

Figure 2. CBED and CHEF patterns obtained in a GaAs crystal oriented near a non-systematic three-beam condition. The CHEF phase, from which the three-phase structure invariant is determined, is also reported and compares well with the value of 10° obtained from CBED analysis.

Precession Electron Diffraction for the characterization of twinning in pseudo-symmetrical crystals: case of coesite

D. Jacob[1], P. Cordier[1], J.P. Morniroli[2], H.P. Schertl[3]

1. Laboratoire de Structure et Propriétés de l'Etat Solide - UMR CNRS 8008
Université des Sciences et Technologies de Lille – Bât. C6
59655 Villeneuve d'Ascq Cedex, France
2. Laboratoire de Métallurgie Physique et Génie des Matériaux, UMR CNRS 8517
USTL and ENSCL
Bât. C6, 59655 Villeneuve d'Ascq Cedex, France
3. Institut für Geologie, Mineralogie und Geophysik, Ruhr-Universität Bochum,
D-44780 Bochum, Germany

damien.jacob@univ-lille1.fr
Keywords: Electron precession, TEM, coesite

Recent developments in electron diffraction call for a reappraisal of the possibilities offered by spot patterns. Among them is the Vincent-Midgley Precession Electron Diffraction (PED) technique [1], which recently became available on many TEM thanks to hardware implementations. Giving the possibility to measure integrated electron diffraction intensities, this technique was originally developed for electron crystallography applications. In this work, we take advantage of the intensity data provided by PED to propose the first characterization of a structural crystal defect, namely a twin, using this technique.

Coesite is a high-pressure polymorph of silica that exhibits a monoclinic symmetry (space group $C12/c1$) with $a \sim c$ and $\beta = 120.34°$ [2]. While monoclinic in symmetry, the crystal is pseudo-hexagonal, and conventional selected-area electron diffraction (SAED) can easily lead to incorrect indexing. Indeed, when the twin is observed edge on, no differences are observed on conventional patterns taken from both parts of the twins, indicating that all the twin reflections nearly coincide (figures 1a and b). Due to the pseudo-hexagonal symmetry, the indexation of the patterns is very ambiguous. As a matter of fact, examination of the spots position gives rise to four possible zone axes: $\pm[\bar{2}1\bar{2}]$, $\pm[\bar{2}\bar{1}2]$, $\pm[0\bar{1}2]$ and $\pm[012]$. With electron precession, spots intensity modulation makes the patterns distinguishable (figures 1c and d) and comparison with Jems dynamical simulations [3] leads to index the zone axis patterns as $[\bar{2}1\bar{2}]$ and $[\bar{2}\bar{1}2]$ (figures 1e and f). Another example of univocal indexation based on the examination of spot intensity (and not on their positions only) is given in figure 2 for the parallel $[\bar{1}10]$ and $[\bar{1}0\bar{1}]$ directions. The patterns are only distinguishable using precession. In particular, the kinematically forbidden 001 reflection (00h with h odd) is not visible on the PED pattern (arrowed in figure 2c). Distinction between the precession patterns is also easily made using their symmetry (twofold rotation axis versus 2mm symmetry).

M. Luysberg, K. Tillmann, T. Weirich (Eds.): EMC 2008, Vol. 1: Instrumentation and Methods, pp. 193–194, DOI: 10.1007/978-3-540-85156-1_97, © Springer-Verlag Berlin Heidelberg 2008

194

Finally using various orientations of the samples, sets of quasi-parallel directions in both variants are determined and, using classical analysis based on stereographic projections, the twin law is fully characterized. It is described as a mirror along the (021) plane. This result is fully consistent with our previous work based on large-angle convergent-beam electron diffraction [4]. Nevertheless, the PED offers the great advantage of simplicity.

1. R. Vincent and P. Midgley, Ultramicroscopy **53** (1994), p. 271.
2. L. Levien and C.T. Prewit, American Mineralogist **66** (1981), p. 324.
3. P.A. Stadelmannn, http://cimewww.epfl.ch/people/stadelmann/jemswebsite/jems.html
4. D. Jacob, P. Cordier, J. P. Morniroli, H. P. Schertl, European Journal of Mineralogy **20** (2008), p.119.

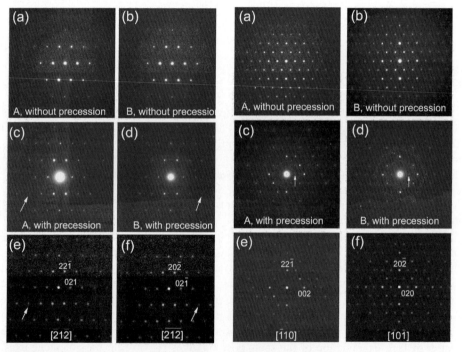

Figure 1. Experimental patterns taken on each side of the twin (variants A and B) observed edge on. Without precession: (a) and (b). With precession: (c) and (d). Jems simulated patterns: (e) and (f).

Figure 2. Experimental patterns taken on each part of the twin for [$\bar{1}$10] and [$\bar{1}0\bar{1}$] orientations. Without precession: (a) and (b). With precession: (c) and (d). Jems simulated patterns: (e) and (f).

Electron precession characterization of pseudo-merohedral twins in the LaGaO$_3$ perovskite

G. Ji[1], J.P. Morniroli[1], G.J. Auchterlonie[2] and D. Jacob[3]

1. Laboratoire de Métallurgie Physique et Génie des Matériaux, UMR CNRS 8517, USTL and ENSCL, Bât C6, Cité Scientifique, 59655 Villeneuve d'Ascq, France
2. Centre for Microscopy and Microanalysis, the University of Queensland, Brisbane, Australia
3. Laboratoire de Structure et Propriétés de l'Etat Solide, UMR CNRS 8008, USTL and ENSCL, Bât C6, Cité Scientifique, 59655 Villeneuve D'Ascq, France

Jean-Paul.Morniroli@univ-lille1.fr
Keywords: electron precession, twins, perovskite

Pseudo-merohedric twins are frequently observed in the orthorhombic LaGaO$_3$ perovskite [1] because this perovskite displays some very small symmetry departures from an ideal cubic ABO$_3$ perovskite. The characterization of the twin law (i.e. the identification of a rotation axis [uvw]$_R$ and a rotation angle α_R around it) requires the coherent identification of two couples of parallel [uvw] directions, for example, two couples of zone axis electron diffraction patterns performed on each side of the twin boundary. Due to the small symmetry departures, these patterns look very similar and cannot be surely identified from conventional electron diffraction patterns (Selected-Area Electron Diffraction or Microdiffraction)[2].

To solve this problem, we used the electron precession technique proposed by Vincent and Midgley [3]. With this technique, the integrated intensities of the diffracted beams can be taken into account as well as very weak differences of intensities.

Figures 1a gives two sets of electron precession patterns obtained on each sides of the studied twin boundary. These patterns contain some weak extra reflections typical of the symmetry departures (circled reflections). In addition, some of them display a clearly visible difference of intensity (arrowed reflections) which can be used to identify without ambiguity the actual zone axes by comparison with the theoretical simulated patterns (Figure 1b). In the present case, the [012] and [210] zone axes were identified and the twin law was determined to be a 180° rotation around the [-1-1-1] direction following the mathematic method described in Ref. [4]. Comparatively, the sets of the conventional microdiffraction patterns taken with the same experimental conditions (not shown here) do not display any typical intensity differences for the extra reflections.

In conclusion, electron precession allows an easy and sure identification of the zone axis patterns patterns located on each side of the LaGaO$_3$ pseudo-merohedral twin and thus provide a way to deduce the twin law. This technique can be applied to any other pseudo-merohedral or merohedral twins.

1. W.L. Wang and H.Y. Lu, Journal of the American Ceramic Society **89** (2006), p. 281.
2. J.P. Morniroli, G.J. Auchterlonie, J. Drennan and J. Zou, Journal of Microscopy, in press.

M. Luysberg, K. Tillmann, T. Weirich (Eds.): EMC 2008, Vol. 1: Instrumentation and Methods, pp. 195–196, DOI: 10.1007/978-3-540-85156-1_98, © Springer-Verlag Berlin Heidelberg 2008

196

3. R. Vincent and P.A. Midgley, Ultramicroscopy **53** (1994), p. 271.
4. J.P. Morniroli and F. Gaillot, Interface Science **4** (1996), p. 273.

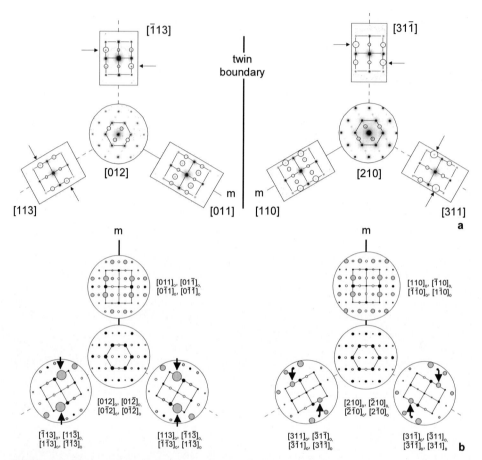

Figure 1. Characterization of a twin in the LaGaO$_3$ perovskite from electron precession microdiffraction patterns.

a - Sets of experimental electron precession patterns performed on each side of the twin boundary. The weak extra reflections are circled. Note the difference of intensity of the arrowed extra reflections and the presence of a mirror m.

b - Simulated sets of electron precession patterns located around the <012> and <210> zone axis forms. For the sake of visibility, the extra reflections are magnified ten times.

Kikuchi electron double diffraction

R.K. Karakhanyan, K.R. Karakhanyan

Armenia, Yerevan, 0025, Al.Manoogyan 1, Yerevan State University,
Department of solid state physics

rkarakhanyan@yandex.ru
Keywords: electron, diffraction, Kikuchi pattern

Although electron double diffraction phenomenon is studied very well, Kikuchi electron double diffraction is little known [1]. The purpose of the present work is to show the role of this phenomenon in the Kikuchi patterns formation.

Kikuchi electron double diffraction means that the diffracted beams as well as the primary electron beam can become the sources of Kikuchi pattern [2-4]. In this case the diffracted beams are exposed to inelastic scattering which take place for the primary electron beam. In consequence secondary Kikuchi pattern with sources in diffracted beams are formed. The resulting Kikuchi pattern will be a superposition of the primary pattern with the source in the incident beam and of the secondary patterns with the sources in the diffracted beams. On diffraction pattern the zero and diffraction reflections are traces of these beams. Around each of these reflections one and the same Kikuchi pattern is produced. But each of the secondary patterns is deviated from the primary one parallel to itself, either to one or the other side, depending on the diffracted beam taken as source of the secondary Kikuchi pattern.

As specimens for investigations were used single crystalline silicon and germanium films prepared by chemical etching of bulky crystals. The transmission Kikuchi patterns were obtained in an EG - 100M electron diffraction camera with the incident electron beam almost parallel to the different crystallographic directions.

The forbidden 222, 666, 002, 006 Kikuchi lines (which are analogs of the forbidden Bragg reflections) are present on the obtained Kikuchi patterns. It is showed, for example, that the forbidden excess and deficient 222 lines are due to the excess and deficient 444 lines of the secondary Kikuchi pattern with source in the diffracted 111 beam. The forbidden 002 and 006 Kikuchi lines appear due to 004 and 008 lines of the secondary Kikuchi pattern with the source in the diffracted 131 beam.

On the basis of Kikuchi electron double diffraction the formation of unindexed Kikuchi lines too can be easily explained. When electron beam is incident along the [111] axis three unindexed lines run along the middle line of the 220 Kikuchi bands. These unindexed lines are formed when the 220 diffracted beams are taken as source of secondary Kikuchi patterns and are conditioned by the 440 lines of these secondary Kikuchi patterns. In is founded that the unindexed Kikuchi line which runs along the middle line of the forbidden 222 lines appears only when double-diffracted beams are taken as source of the secondary Kikuchi patterns. This shows that the double-diffracted beams can participate in the Kikuchi patterns formation.

It is obtained the Kikuchi patterns with the enhancement of the excess lines in the vicinity of reflection with indices different those of the Kikuchi lines [5]. The Kikuchi

M. Luysberg, K. Tillmann, T. Weirich (Eds.): EMC 2008, Vol. 1: Instrumentation and Methods, pp. 197–198, DOI: 10.1007/978-3-540-85156-1_99, © Springer-Verlag Berlin Heidelberg 2008

patterns formation elementary mechanism [1] does not predicts such enhancement, but this enhancement can be explained by means of the Kikuchi electron double diffraction. One must assume that some intense diffracted beam with the Kossel cone of excess intensity leads to the above-mentioned enhancement.

It is observed the mutual contrast reversal of the excess and deficient Kikuchi lines [6]. The contrast is reversed not only along certain segments (the partial contrast reversal), but also over the entire length of these lines (the complete contrast reversal). The mutual contrast reversal of the Kikuchi lines is explained with due regard for the Kikuchi electron double diffraction. This phenomenon occurs when the excess Kikuchi line passes through or in the vicinity of an intense spot reflection with indices different from the indices of the Kikuchi excess line. This means that the intense diffracted beam coincides with the Kossel cone of deficient intensity and brings about the mutual contrast reversal of the Kikuchi lines. It is noted that the larger the contrast of a given pair of the Kikuchi lines, i.e., the larger the difference between the intensities of electron waves along the cones of excess and deficient intensities, the higher should be the intensity of the diffracted beam in order to change this difference in favour of the cone of the deficient intensity and to achieve the mutual contrast reversal of the Kikuchi lines.

The complete contrast reversal take place when the excess line pass through or in vicinity of a number of the intense spot reflections. This implies that numerous diffracted beams, whose inelastic scattering leads to an increase of electron beam intensity along the Kossel cone of the deficient intensity, pass in vicinity of this cone. As the diffracted beams are adjacent to each other, the intensity also increases in adjacent segment of the Kossel cone of the deficient intensity, which results in the mutual contrast reversal of the excess and deficient Kikuchi lines over their entire length.

Thus the Kikuchi electron double diffraction leads to the forbidden and unindexed Kikuchi lines formation as well as to the enhancement of the excess lines in the vicinity of the spot reflection with indices different from those of the Kikuchi lines and to the mutual contrast reversal of the Kikuchi excess and deficient lines.

1. D.B. Williams and C.B. Carter, Transmission electron microscopy, II, Diffraction, Plenum Press, New York, (1996).
2. R.K. Karakhanyan, P.A. Grigoryan and P.A. Bezirganyan, Kristallografiya, **24**, (1979), p. 817.
3. R.K. Karakhanyan and P.L. Aleksanyan, Kristallografiya, **32**, (1987), p. 1256.
4. R.K. Karakhanyan, P.L. Aleksanyan and A.O. Aboyan, Kristallografiya, **41**, (1996), p. 567.
5. R.K. Karakhanyan, P.L. Aleksanyan and J.K. Manucharova. Phys. Stat. Sol. (a), **121**, (1990), p. K1.
6. R.K. Karakhanyan and P.L. Aleksanyan, Crystallography Reports, **44**, (1999), p. 398.

The structure of the complex oxide PbMnO$_{2.75}$ solved by precession electron diffraction

H. Klein

Institut Néel, Université Joseph Fourier and CNRS, 25 av. des Martyrs, BP 166, 38042 Grenoble Cedex 9

holger.klein@grenoble.cnrs.fr

Keywords: electron crystallography, TEM, precession electron diffraction

X-ray diffraction has been used for nearly a century to solve crystal structures and the methods that have been developed make it the prominent method of structure resolution today. Structures have been solved from single crystals of a few tens of μm in diameter or even less using synchrotron radiation techniques. Powders, even if they contain more than one phase, have also been used for X-ray diffraction structure determination. However, with the trend in fundamental research and applications towards materials on the nanometre scale, X-ray diffraction reaches its limits for structure resolution more and more often. Single crystals of less than 1 μm^3 are not suitable for X-ray diffraction experiments and powders always present the difficulty of peak overlap that in the case of complex phases often prevents the determination of the cell parameters, without even speaking of structure determination. In these cases electron crystallography can be a powerful tool for the determination of the atomic structures of crystals. Its advantages are the fact that an individual nanometre-sized particle can be used as a single crystal for electron diffraction and in the possibility to obtain real space images of atomic resolution. The "classical" electron diffraction suffers from multiple scattering which prevents a simple relationship between the measured intensities and the structure factor amplitude. Therefore precession electron diffraction, which largely reduces multiple diffraction, has to be used to obtain reliable diffraction data.

In this contribution we present the structure solution of a powder sample of the complex oxide PbMnO$_{2.75}$ synthesized at high temperature and under high pressure. Due to its large cell parameters this monoclinic structure (space group $A2/m$, $a = 3.2232$ nm, $b = 0.3831$ nm, $c = 3.5671$ nm, $\beta = 130°$) suffers from severe peak overlap in the X-ray powder diffraction pattern and the cell parameters could not be determined by this technique. SAED yielded the cell parameters and the space group of this phase [1]. Due to the small b parameter and the presence of the mirror plane, atomic positions have either $y = 0$ or $y = 1/2$. The problem can therefore be treated as a 2 dimensional one in the **ac**-plane.

Figure 1 shows the precession electron diffraction pattern of the zone axis [0 1 0]. The data contained in this single pattern is sufficient for the structure determination. The intensities of 991 independent reflections corresponding to lattice spacings of 0.08 nm $< d <$ 1 nm were determined from this pattern. A structure solution was found using the direct methods program SIR97 [2]. The 30 most intense peaks in the resulting

M. Luysberg, K. Tillmann, T. Weirich (Eds.): EMC 2008, Vol. 1: Instrumentation and Methods, pp. 199–200, DOI: 10.1007/978-3-540-85156-1_100, © Springer-Verlag Berlin Heidelberg 2008

electron density map contained all of the 29 independent cation positions (the remaining peak corresponding to an oxygen position). The value for the y coordinate was subsequently chosen in accordance with the crystal chemistry of the related perovskite phase.

The atomic positions found in this study are well in registry with those proposed by Bougerol *et al.* from a refinement of powder X-ray diffraction data [1]. The mean distances between the positions in the two models are 0.026 nm for Pb atoms and 0.036 nm for Mn atoms. These values show a good agreement between the two models, especially when taking into account the standard deviation of the result of the X-ray refinement of 0.01 nm for Pb and 0.03 nm for Mn.

1. C. Bougerol, M.F. Gorius and I. Grey, J. Solid State Chem. 169 (2002), p. 131
2. http://www.ic.cnr.it/registration_form.php

Figure 1. Precession electron diffraction pattern of the [0 1 0] zone axis of $PbMnO_{2.75}$. The directions of the reciprocal lattice vectors **a*** and **c*** are shown by arrows. Due to the complex structure most of the reflections are too weak to be seen on this pattern.

Software Precession Electron Diffraction

C.T. Koch[1], P. Bellina[1], P.A. van Aken[1]

1. Stuttgart Center for Electron Microscopy, Max Planck Institute for Metals Research,
Heisenberstr. 3, 70569 Stuttgart, Germany

koch@mf.mpg.de
Keywords: electron diffraction, precession electron diffraction, microscope control

Precession electron diffraction (PED) [1] has recently become a very useful technique for the determination of crystal symmetry and structure. The advantage of this technique is commonly seen in the fact that electron diffraction patterns recorded with a precessing illumination can more frequently be interpreted using kinematic scattering theory.

We have developed a software tool (a script running within Gatan Digital-Micrograph) that automatically controls the illumination tilt and image shift to produce PED patterns on the viewing screen as well as on the CCD camera. The calibration of the illumination tilt and diffraction shift used for compensation is fully automated.

While being much slower (about 25 different tilt angles per second on a RS232-controlled Zeiss TEM), the advantage of controlling the electron beam by software rather than hardware lies in the fact, that very versatile experiments can be devised. In addition to recording the PED pattern (Fig. 1) it is also possible to extract 3-dimensional diffraction information by recording a whole stack of diffraction patterns, one for each illumination tilt angle. From such a 3-dimensional diffraction data set one can easily extract individual rocking curves for each reflection (Fig. 2a). The range over which these rocking curves can be extracted is comparable to that of LACBED experiments, but with the advantage that all reflections are recorded simultaneously. A more memory preserving variation of this experiment is to compensate the illumination tilt only partially (this can also be achieved on a hardware PED device by de-aligning the de-scan adjustments). The PED patterns now consist of diffraction rings instead of spots. These rings can be chosen to have a diameter smaller than the Bragg spacing and will therefore not overlap. They nevertheless correspond to tilt angles much larger than the distance between Bragg spots and would produce strongly overlapping discs in a comparative CBED experiment. Figure 2b shows a zero-loss filtered under-compensated pattern that has been recorded in a Zeiss EM912. The scan compensation has been set to 90%, implying that the actual tilt amplitude around the ring is 10 times that of the ring radius.

The applications of this technique that will be discussed include fitting of structure factor amplitudes and specimen thickness by interpreting precession rocking curves (Fig. 2a) with 2-beam dynamical theory and N-beam dynamical theory [2], as well as obtaining information about surface reconstruction in $SrTiO_3$.

1. R. Vincent and P.A. Midgley, Ultramicroscopy **53** (1994) 271
2. C.T. Koch and J.C.H. Spence, J. of Physics A **36** (2003) 803

M. Luysberg, K. Tillmann, T. Weirich (Eds.): EMC 2008, Vol. 1: Instrumentation and Methods,
pp. 201–202, DOI: 10.1007/978-3-540-85156-1_101, © Springer-Verlag Berlin Heidelberg 2008

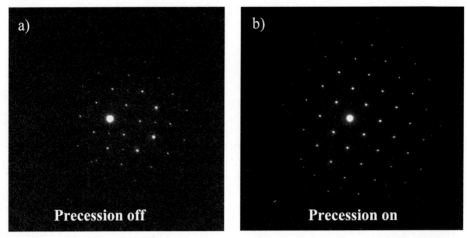

Figure 1. a) Electron Diffraction pattern of SrTiO3 (100), off zone axis. b) Precession electron diffraction pattern produced by turning on software precession. The tilt angle during precession was set to about 42 mrad.

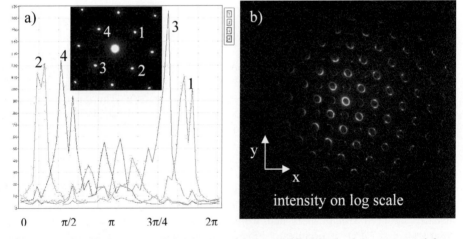

Figure 2. a) Rocking curves of the 4 <011> spots specified in the inset extracted from the set of 50 diffraction patterns that produced the PED pattern shown in Fig. 1a. The abscissa defines the angle around each of the diffraction rings (starting on the positive x-axis defined in the inset of Fig 2b and running counter clock-wise), and the ordinate is in arbitrary units. b) The information given in a) can also be extracted from a PED pattern recorded with under-compensated de-scan. The diffraction intensity as a function of illumination tilt can be extracted from this single image. Here only 90% of the illumination tilts have been compensated for by diffraction shift. The intensity is shown on a logarithmic scale in order to enhance the contrast.

A new method for electron diffraction based analysis of phase fractions and texture in thin films of metallic nano-crystals

J.L. Lábár, P.B. Barna, O. Geszti, R. Grasin, G. Lestyán, F. Misják, G. Radnóczi, G. Sáfrán, L. Székely

Research Institute for Technical Physics and Material Science, H-1121, Budapest, Konkoly-Thege M. u. 29-33, HUNGARY

labar@mfa.kfki.hu

Keywords: nano-crystalline, quantitative analysis, phase fractions, texture, SAED, TEM

X-ray diffraction (XRD) based determination of phase fractions from powder samples has been common since the introduction of the Rietveld-method. Similar method did not exist for electron diffraction (ED), mainly due to the dynamic nature of ED. The last decade saw a boom in nano-technology and most of the transmission electron microscopy (TEM) laboratories frequently deal with nanocrystalline thin films. For these samples the kinematical approximation is acceptable.

A method, based on selected area electron diffraction (SAED) was developed to extract quantitative phase and texture information from SAED ring-patterns, recorded from nano-crystalline thin films in the TEM. An XRD-like one-dimensional (1D) distribution is deduced first from the ring pattern, by averaging the intensity circularly. As a result, the original SAED rings are converted to peaks in this 1D distribution [1]. Intensities of the peaks are extracted by fitting pre-defined functional shapes to both the background and the peaks (see Fig. 1). The optimum parameters of the curves are determined from best fitting of the model curve to the experimental distribution. The parameters are optimized by the downhill-SIMPLEX method [2]. The experimental intensities, obtained by the fitting procedure, are compared to the theoretical intensities, calculated from the kinematical approximation for each of the previously identified crystalline phases present. Volume fractions of the individual nano-crystalline phases are given as the result. Analysis of the results obtained on numerous samples proved that introduction of two corrections was necessary. On the one hand, the Blackman-correction was introduced to modify the calculated intensities due to dynamic effects [3]. On the other hand, truly random orientation distribution proved to be rare in thin films, so analysis of texture was also incorporated into the method. At the moment, fibre-texture, observed from the direction of the texture axis can only be quantified with the present implementation of our method. However, this seems to be the most frequent texture form in thin films. The fraction of a phase with truly random orientation distribution and the fraction of the same phase with perfectly textured orientation distribution are computed separately and their volume fracions are determined separately during the least-square fitting.

Figure 1 shows an example of a thin film of sequentially evaporated two components: 10 nm Ag and 10 nm Cu on a self-supporting substrate of 5 nm

M. Luysberg, K. Tillmann, T. Weirich (Eds.): EMC 2008, Vol. 1: Instrumentation and Methods, pp. 203–204, DOI: 10.1007/978-3-540-85156-1_102, © Springer-Verlag Berlin Heidelberg 2008

amorphous-Carbon. Analysis result is 51 vol% Ag / 49 vol% Cu. Nominal and measured volume fractions are in a very good agreement.

1. Lábár JL, Microscopy and Analysis, Issue 75 (2002) 9-11
2. Nelder J.A., Mead R.. Comp. J. 7 (1965) 308-313
3. Blackman M., Proc. Roy. Soc. London A173 (1939) 68-82

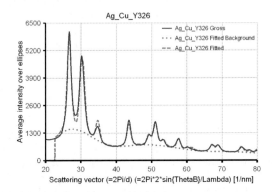

Figure 1. Measured (-) and fitted (--) intensity distributions of diffracted intensities and background (..) in an Ag/Cu 50/50 vol% bilayer.

Local structures of metallic glasses studied by experimental RDF and model refinement

G. Li[1], K.B. Borisenko[1],Y. Chen[1], E. Ma[2] and D.J.H. Cockayne[1]

1. Department of Materials, University of Oxford, Parks Road, Oxford OX1 3PH
United Kingdom
2. Department of Materials Science and Engineering, Johns Hopkins University,
Baltimore, MD 21218 USA

guoqiang.li@materials.ox.ac.uk

Keywords: metallic glass, electron diffraction, radial distribution function, ductility

Metallic glasses (MGs) have interesting mechanical, magnetic and acoustic properties and have the potential to be used in applications such as consumer electronics, sporting goods and healthcare [1,2]. However, most MGs have poor ductility, which threatens their reliability. Some MGs, on the contrary, can sustain considerable plastic strains. The structural origin of such behavior is not understood at present. Recently there has been considerable interest in what local structural features control the plasticity and how they respond to pressure [3] and deformation [2]. Of particular interest is the structure inside the shear bands of nanometer size which is the dominant mode of plastic deformation for MGs [4,5].

It is very difficult to determine the atomic structure in local regions from X-ray or neutron diffraction data. Electrons have a relatively large scattering cross section and can be focused with lenses to illuminate chosen nanoscale volumes in the transmission electron microscope (TEM). In this study, such an electron RDF analysis [6,7] has been used to determine the local structure of chosen small volumes in several MGs, and atomistic models have been refined against the data using the reverse Monte Carlo (RMC) method and density functional theory (DFT) calculations.

TEM samples were prepared from Al-La-Ni, Cu-Hf-Al, and Zr-Ti-Cu-Al MGs. Using a parallel 300 keV electron beam, selected area diffraction (SAD) and nano-beam diffraction (NBD) patterns were collected up to a scattering vector of q $(=4\pi\sin\theta/\lambda)$ around 20 Å^{-1}. Considerable effort was taken to keep the microscope in identical electron optical conditions during data collection so that the RDFs were comparable. The RDFs were then extracted from the diffraction patterns following procedures described in [6,7]. As an example, Figure 1a shows a diffraction pattern from the Zr-Ti-Cu-Al MG. Figure 1b is the corresponding reduced intensity function and Figure 1c is the RDF curve extracted. The RDFs were used to refine atomic models using RMC accompanied by DFT calculations. EDX and EELS attached to the TEM were also used to monitor the local composition of the volumes examined.

Three types of RDF with different first and second peak positions were observed in the Al-La-Ni MG, indicating three distinct amorphous structures (see Figure 2). RMC refinements suggested that the Al-Al, Al-La, and Al-Ni RDFs are slightly different for each of the three amorphous structures. This finding suggests polyamorphism in MGs, which is attracting attention recently [3,8,9]. The structural heterogeneity in the compositionally uniform, fully glassy alloy is likely to have effects on the shear band

M. Luysberg, K. Tillmann, T. Weirich (Eds.): EMC 2008, Vol. 1: Instrumentation and Methods, pp. 205–206, DOI: 10.1007/978-3-540-85156-1_103, © Springer-Verlag Berlin Heidelberg 2008

formation and propagation [10,11], and thus have implications for the improvement of plasticity of MGs. The amorphous structures inside the shear bands in Cu-Hf-Al, and Zr-Ti-Cu-Al glasses were also studied by RDF analysis and model refinement.

1. A. Inoue and N. Nishiyama, MRS Bulletin **32** (2007), p. 651.
2. A. L. Greer and E. Ma, MRS Bulletin **32** (2007), p. 611.
3. H. W. Sheng, H. Z. Liu, Y. Q. Cheng, J. Wen, P. L. Lee, W. K. Luo, S. D. Shastri and E. Ma, Nature Mater. **6** (2007), p. 192.
4. Y. Zhang and A. L. Greer, Appl. Phys. Lett. **89** (2006), p. 071907.
5. Qi-Kai Li and M. Li, Appl. Phys. Lett. **88** (2006), p. 241903.
6. D. J. H. Cockayne and D. R. Mckenzie, Acta. Cryst. A **44** (1988), p. 870.
7. D. J. H. Cockayne, Annu. Rev. Mater. Res. **37** (2007), p. 159.
8. C. Way, P. Wadhwa, R. Busch, Acta Mater. **55** (2007), p. 2977.
9. W.K. Luo and E. Ma, J. Non-cryst. Solids **354** (2008), p. 945.
10. Y.H. Liu, G. Wang, R.J. Wang, D.Q. Zhao, M.X. Pan and W.H. Wang, Science **315** (2007), p. 1385.
11. J. Das J, M.B. Tang, K.B. Kim, R.Theissmann, F. Baier, W.H. Wang, J. Eckert, Phys. Rev. Lett. **94** (2005), p. 205501.

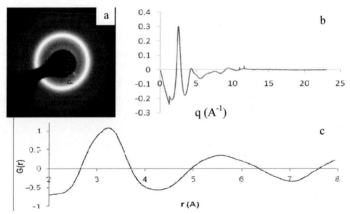

Figure 1. (a) Diffraction pattern from Zr-Ti-Cu-Al MG, (b) reduced intensity function from (a), and (c) RDF curve.

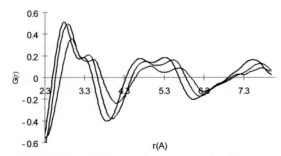

Figure 2. Three types of RDF from Al-La-Ni with different first and second peak positions.

Three groups of hexagonal phases and their relation to the i-phase in Zn-Mg-RE alloy

M.R. Li[*], S. Hovmöller, X.D. Zou

Structural Chemistry, Stockholm University, SE-106 91 Stockholm, Sweden

mingrunl@struc.su.se

Keywords: icosahedral quasicrystal, Zn-Mg-RE alloy

In the Zn-rich Zn-Mg-RE alloys three groups of ternary hexagonal phases $(Zn,Mg)_{58}RE_{13}$, $(Zn,Mg)_5RE$, Zn_6Mg_3RE (also named Z-phase) are found to coexist with the icosahedral quasicrystal. In a study of the ternary Zn-Mg-Sm system, Drits et al. [1] first found the hexagonal Zn_6Mg_3Sm phase with $a = 14.62$ Å and $c = 8.71$ Å. Later, this hexagonal phase has also been found in the Zn-Mg-Y [2] and Zn-Mg-Gd [3] alloys and their crystal structures were determined by the single crystal X-ray diffraction analysis [2,3]. Recently, we found two new hexagonal phases, Zn_3MgY phase with $a = 9.082$ Å and $c = 9.415$ Å and $(Zn,Mg)_4Ho$ phase with $a \cong 14.3$ Å and $c \cong 14.1$ Å [4-7] by substituting Mg for Zn in Zn_5Y and $Zn_{58}Ho_{13}$, respectively. Their crystal structures were studied by single crystal X-ray and electron diffraction. Now these three phases have been extended to RE =Y, Sm, Gd, Dy, Ho, Er, Yb (see Table 1). Since these three phases have a high percentage of icosahedral coordination and therefore are crystalline approximants of the icosahedral i-phase. The connections of ICO (icosahedra) in these hexagonal phases are given in Table 1. The strong spots in the [100] electron diffraction patterns (EDPs) show pseudo ten-fold symmetry (see Fig. 1), due to the presence of two or four penetrated icosahedra (I2(P) or I4(P), respectively) in the structures, along the <100> directions.

On the other hand, there is pairs of face-sharing ICO (I2(F)) clusters along the [001] direction. The phase reactions in order of decreasing temperature are (see Fig. 2):

(a) L + $(Zn,Mg)_{58}RE_{13}$ → $(Zn,Mg)_5RE$;
(b) L + $(Zn,Mg)_5RE$ → i;
(c) L + $Zn_{17}Y_2$ → i;
(d) i → Z.

Reactions b and c have been observed earlier by Tsai et al [8,9].

1. M.E. Drits, L.L. Rokhlin, V.V. Kinzhibalo, Izv. Akad. Nauk SSSR, Metally., No. 6 (1985), 194.
2. H. Takakura, A. Sato, A. Yamamoto, A.P. Tsai, Philos. Mag. Lett., 78 (1998) 263.
3. K. Sugiyama, K. Yasuda, T. Ohsuna, K. Hiraga, Z. Kristallogr., 213 (1998) 537.
4. K.H. Kuo, D.W. Deng, J. Alloys Compd., 376 (2004) L5.
5. D.W. Deng, K.H. Kuo, Z.P. Luo, D.J. Miller, M.J. Kramer, D.W. Dennis, J. Alloys Compd., 373 (2004) 156.
6. M.R. Li, D.W. Deng, K.H. Kuo, J. Alloys Compd., 414 (2006) 66.
7. M.R. Li, K.H. Kuo, J. Alloys Compd. J. Alloys Compd., 432 (2007) 81.

8. A.P. Tsai, A. Niikura, A. Inour, T. Masumoto, J. Mater. Res., 12 (1997) 1468.
9. E. Abe, A.P. Tsai, Phys. Rev. Lett., 83 (1999) 753.

Table 1. Some structurally related hexagonal Zn-Mg-RE phases and the icosahedral chain (ICO) presented in the structures (F stands for Face-sharing and P for Penetrating icosahedra).

Phase	Zn_6Mg_3RE			$(Zn,Mg)_{5-6}RE$			$(Zn,Mg)_4RE$		
RE	Composition, exp	a, Å	c, Å	Composition, exp	a, Å	c, Å	Composition, exp	a, Å	c, Å
Y	$Zn_{65.72}Mg_{27.92}Y_{6.86}$ [2]	14.579	8.687	$Zn_{60.68}Mg_{18.28}Y_{21.04}$ [7]	9.0822	9.416	$Zn_{71}Mg_{10}Y_{19}$	14.3[a]	14.1[a]
Sm	$Zn_{64.9}Mg_{28.4}Sm_{6.7}$ [3]	14.619	8.708	–	–	–	$Zn_{70}Mg_{11}Sm_{19}$	14.3[a]	14.1[a]
Gd	$Zn_{63.8}Mg_{28.6}Gd_{8.6}$ [3]	14.635	8.761	$Zn_{65.6}Mg_{17.8}Gd_{16.6}$	9.0962	9.411	$Zn_{72}Mg_9Gd_{19}$	14.3[a]	14.1[a]
Dy	–	–	–	–	–	–	$Zn_{70}Mg_{12}Dy_{18}$	14.3[a]	14.0[a]
Ho	–	–	–	$Zn_{66}Mg_{17}Ho_{17}$	9.1[a]	9.4[a]	$Zn_{68.4}Mg_{12.7}Ho_{18.9}$[b]	14.259	14.007
Er	–	–	–	$Zn_{61.9}Mg_{24.1}Er_{13.9}$	8.9079	9.269	$Zn_{70.8}Mg_{10.6}Er_{18.6}$[b]	14.185	13.957
Yb	–	–	–	$Zn_{67.7}Mg_{16.3}Yb_{16.0}$	9.0616	9.449	$Zn_{71.9}Mg_{9.1}Yb_{19.0}$[b]	14.429	14.252
ICO		I4(P)	I2(F)		I2(P)	I2(F)		I4(P)	I2(F) + CN15

[a] Parameters determined by SAED (Selected-area electron diffraction)

[b] Single crystal X-ray diffraction structural analysis made in the present investigation

Figure 1. SAED patterns (a) of what along two-fold axis; arrowed spots belong to the face-centered icosahedral structure, and of the hexagonal (b) $(Zn,Mg)_{58}Y_{13}$, (c) $(Zn,Mg)_5Y$, and (d) Z phases along the [100] direction. The reciprocal $(bc)^*$ cell is outlined. Arrowed spots arise from the interconnection of icosahedra along the fivefold axis in the <100> directions of these phases.

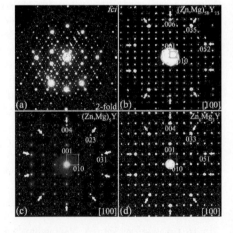

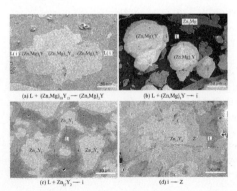

(a) L + $(Zn,Mg)_{58}Y_{13}$ → $(Zn,Mg)_5Y$ (b) L + $(Zn,Mg)_5Y$ → i

(c) L + $Zn_{17}Y_2$ → i   (d) i – z

Figure 2. Scanning electron microscopy (SEM) images showing phase transitions: (a) L + $(Zn,Mg)_{58}RE_{13}$ → $(Zn,Mg)_5RE$; (b) L + $(Zn,Mg)_5RE$ → i; (c) L + $Zn_{17}Y_2$ → i; (d) i → Z.

Diffraction analysis of incommensurate modulation in "chain-ladder" composite crystal $(Sr/Ca)_{14}Cu_{24}O_{41}$

O. Milat, K. Salamon, S. Tomić, T. Vuletić, and T. Ivek

Institute of Physics, Zagreb, Bijenička 46, Croatia

milat@ifs.hr

Keywords: modulated structure, composite crystal, electron diffraction

The "chain-ladder" compounds with formula $(Sr/La/Ca)_{14}Cu_{24}O_{41}$ exhibit a rather complex crystallographic structure due to their composite character. These materials consist of an alternating stacking of two distinct types of layers forming interpenetrated subsystems [1]. One subsystem is formed of $(Sr/La/Ca)_2Cu_2O_3$ layers with a ladder-like structure and orthorhombic unit cell (*Fmmm*), while the other contains layers of CuO_2 chains with orthorhombic unit cell (*Amma*) [2]. All lattice parameters slightly vary with La/Ca for Sr substitution, but the one dimensional incommensurate modulation results from the misfit between the *c* lattice parameters of these layer elements. In pure Sr compound: $c_{Ch} = 0.273$nm, $c_{Ld} = 0.393$ nm $\approx \sqrt{2} \cdot c_{Ch}$; the *a* parameter ($\approx 1.147$ nm) is the same in both lattices, as well as *b* (≈ 1.341 nm) pointing along stacking direction. For high Ca for Sr substitution in $Sr_3Ca_{11}Cu_{24}O_{41}$: $c_{Ch} = 0.276$ nm, $c_{Ld} = 0.391$ nm, so that the ratio of these parameters is always close to $\sqrt{2}$. The c_{Ld}/c_{Ch} ratio also varies from 1.416 for pure undoped $Sr_{14}Cu_{24}O_{41}$, to 1.445 for highly doped $Sr_xCa_{14-x}Cu_{24}O_{41}$[3].

As La for Sr substitution reduces intrinsic hole-doping (six hole per formula unit in pure $Sr_{14}Cu_{24}O_{41}$), a lot of doubt has emerged recently about charge carriers ordering into a Wigner "hole crystal" on a $5c_{Ld}$ or $3c_{Ld}$ superlattice, found by Resonant X-ray Scattering [4][5]. The "hole crystal" carrying charge and spin, was allocated onto "spin-ladder" subsystem based on a superficial argument of being commensurate to "ladder" lattice, while not commensurate to "chain" lattice.

This modulated structure has been analysed by X-ray diffraction using 4-dim approach, in case of $M_{14}Cu_{24}O_{41}$ ($M= Bi_{0.04}Sr_{0.96}$) [6].

Here, we present the results of detailed electron diffraction study of complete reciprocal space. As shown in Figure 1, all spots of all relevant zones indicated in Figure 2, can be indexed as structural reflections according to: $H = ha^* + kb^* + lc_{Ld}^* + mc_{Ch}^*$, and they need not to be assigned to some additional "exotic" modulation on top of the complex composite structure. All spots with non-zero *m*-index are more or less streaked with diffuse feature perpendicular to c^* revealing disorder of the phase of modulation in the "chain" lattice [1]. Extinctions in the [1000]* and [0100]* zones of Figure 1 (a)&(e), reveal the superspace symmetry. Each spot along c^* reciprocal axis shifts in position depending on the c_{Ld}/c_{Ch} ratio [3], and in particular those assigned as 004̲6 and 005̲7, at $q_{004\underline{6}} = 0.333c_{Ld}^*$, and at $q_{005\underline{7}} = 0.08c_{Ld}^*$ in Figure 3, respectively. The RXS intensities observed at these positions [4] were probably misinterpreted as an evidence of "non-structural" satellite reflections revealing "hole crystal" superlattice: $\Lambda_3 = 3c_{Ld} = 1/q_{004\underline{6}}$ for $Sr_3Ca_{11}Cu_{24}O_{41}$; $\Lambda_5 = 5c_{Ld} = 1/q_{005\underline{7}}$ for $Sr_{14}Cu_{24}O_{41}$ ($c_{Ld}/c_{Ch} = 1.441$).

M. Luysberg, K. Tillmann, T. Weirich (Eds.): EMC 2008, Vol. 1: Instrumentation and Methods, pp. 209–210, DOI: 10.1007/978-3-540-85156-1_105, © Springer-Verlag Berlin Heidelberg 2008

1. O. Milat, G. Van Tendeloo, S. Amelinckx, M. Mebhod, R. Deltour, Acta. Cryst. **A48** (1992), 618
2. T. Siegrist, L.F. Schneemeyer, S.A. Shunnsine, J.V. Waszeck, R.S. Roth, Mat. Ress. Bull. **23** (1988) 1429
3. Z. Hiroi, S. Amelinckx, G. Van Tendeloo, Phys. Rev. **B54** (1996), 15849
4. A. Rusydi, P. Abbamonte, H. Eisaki, Y. Fujimaki, G. Blumberg, S. Uchida, G.A. Sawatzky, PRL **97** (2006), 016403.
5. P. Abbamonte, G. Blumberg, A. Rusydi, A. Gozar, P.G. Evans, T. Siegrist, L. Venema, H. Eisaki, E.D. Isaacs, G.A. Sawatzky, Nature **431** (2004) 1078.
6. A. Frost Jensen, V. Petricek, F. Krebs Larsen, E.M. McCarron III, Acta Cryst., **B53** (1997), 125

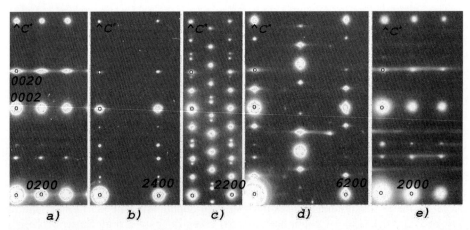

Figure 1. Tilting series of diffraction zones around c-axis for $Sr_3Ca_{11}Cu_{24}O_{41}$, assigned by 4 index notation: (a) - $(0, 2k, 2l, 2m)$; (b) – $(2h, 4h, 2l, 2m)$; (c) – (h, h, l, m); (d) – $(3k, k, l, m)$; (e) – $(2h, 0, 2l, 2m)$. Traces of corresponding sections of reciprocal space are indicated in diffraction pattern along the common c-axis of Figure 2.

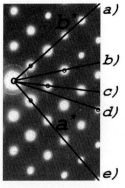

Figure 2. (a* b*) reciprocal plane; marked lines indicate zones in Fig. 1., correspondingly.

Figure 3. reciprocal lattice raw of $(0, 0, l, m)$ reflections from EDP of Fig. 1(c) indicating spots: $00\underline{4}6$ at $q* = 0.333c*_{Ld}$, and $005\underline{7}$ at $q* = 0.08c*_{Ld}$, which could be incorrectly assigned to $3c_{Ld}$, and $12.5c_{Ld}$ superlattice cell.

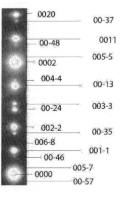

Contribution of electron precession to the study of crystals displaying small symmetry departures

J.P. Morniroli[1], G. Ji[1], D. Jacob[2] and G.J. Auchterlonie[3]

1. Laboratoire de Métallurgie Physique et Génie des Matériaux, UMR CNRS 8517, USTL, ENSCL, Bât C6, Cité Scientifique, 59655 Villeneuve d'Ascq, France
2. Laboratoire de Structure et Propriétés de l'Etat Solide, UMR CNRS 8008, USTL, Bât C6 Bât C6, Cité Scientifique, 59655 Villeneuve d'Ascq, France
3. Centre for Microscopy and Microanalysis, The University of Queensland, Brisbane, Australia

Jean-Paul.Morniroli@univ-lille1.fr
Keywords: Electron Diffraction, Electron precession, Crystal Symmetry

Many crystals display some small symmetry departures from a high symmetry. For example, a crystal has a tetragonal or an orthorhombic structure very close to a cubic structure. In this case, many zone axis electron diffraction patterns look similar and cannot surely be identified with conventional Selected-Area Electron Diffraction (SAED) or microdiffraction.

To quantify this aspect, let us consider the case of the $LaGaO_3$ perovskite. Its crystal structure, actually orthorhombic, is very close to an ideal cubic perovskite [1]. In the general case, observed when u≠v≠w and non-zero, a <uvw> zone axis form of the ideal cubic perovskite contains 48 equivalent [uvw] directions (directions having the same parameter $P_{[uvw]}$ but different orientations). It gives two types of non-superimposable and mirror related diffraction patterns as shown on figure 1a for the example of the <123> zone axis form.

With the real orthorhombic perovskite, the situation becomes more complex since the 48 <123> equivalent cubic directions are transformed into six orthorhombic zone axis forms : <133>, <331>, <551>, <115>, <422> and <224> each of them containing 8 equivalent directions. As a result, two sets of 6 slightly different and mirror related (labelled A and B) diffraction patterns are obtained as shown on figures 1b and c. They differ from the <123> cubic patterns by the presence of weak extra reflections (to be more visible, these reflections are magnified 10 times on figure 1) which are located at two different positions:
- in the middle of the small edge of the parallelograms drawn with respect to the cubic reflections on figure 1b for the <133>, <331> <511> and <115> zone axis forms. Note that the intensity of the extra reflections is typical in each of these four patterns.
- in the middle of both sides of the parallelograms as well as in the middle of the parallelograms for the two <422> and <224> zone axis forms (Figure 1c). The intensity of the extra reflections is also typical of the zone axis.

With conventional electron diffraction (SAED or microdiffraction), the diffracted intensities are too strongly modified by dynamical effects (multiple diffraction) and/or by thickness variations and crystal misorientations in the diffracted area, so that only the

M. Luysberg, K. Tillmann, T. Weirich (Eds.): EMC 2008, Vol. 1: Instrumentation and Methods, pp. 211–212, DOI: 10.1007/978-3-540-85156-1_106, © Springer-Verlag Berlin Heidelberg 2008

positions of the reflections (the "net" symmetry) can be trusted. This means that a zone axis cannot be surely identified among the four <133>, <331> <511> and <115> zone axis forms or among the two <422> and <224> zone axis forms.

This is no longer the case with electron precession [2] since both the position and the intensity of the reflections (the "ideal" symmetry) can be taken into account. Thus, the observation of the intensity of these typical additional reflections are the basis of the zone axis identification described in the present paper.

Experimental precession diffraction patterns from both $LaGaO_3$ and coesite (a mineral with a monoclinic structure very close to a hexagonal structure [3]) crystals will be given to prove the validity of this approach. They also prove the interest to describe a zone axis pattern of a crystal displaying small symmetry departures with respect to its corresponding pattern in the ideal symmetrical crystal.

1. P.R. Slater, J.T.S. Irvine, T. Ishihara and Y Takita, Journal of Solid State Chemistry **139** (1998), p. 143
2. R. Vincent and P. Midgley, Ultramicroscopy **53** (1994), p. 271.
3. L. Levien and C.T. Prewit, American Mineralogist **66** (1981), p. 324.

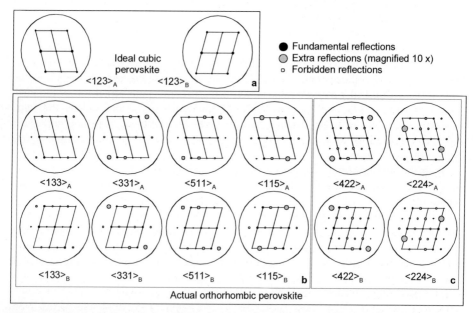

Figure 1. Description of the diffraction patterns produced by: a - the 48 equivalent <123> zone axes from the ideal cubic perovskite, b, c - the corresponding <133>, <331>, <511>, <115>, <422> and <224> zone axes from the orthorhombic perovskite. For the sake of simplicity, only the ZOLZ reflections are shown. To increase their visibility, the weak extra reflections are magnified 10 times.

The symmetry of microdiffraction electron precession patterns

J.P. Morniroli[1], P. Stadelmann[2]

1. Laboratoire de Métallurgie Physique et Génie des Matériaux, UMR CNRS 8517, USTL, ENSCL, Bât C6, Cité Scientifique, 59500 Villeneuve d'Ascq, France
2. CIME-EPFL, Station 12, CH1015 Lausanne, Switzerland

Jean-Paul.Morniroli@univ-lille1.fr

Keywords: Electron diffraction, Electron precession, Point Groups

Conventional microdiffraction patterns are special CBED patterns having a very small beam convergence. Therefore, the CBED symmetries tabulated by Buxton et. al. [1] could also be applied to them. Four symmetries are available from CBED patterns:
- the Bright-Field (BF) symmetry (the symmetry within the transmitted disk),
- the Whole-Pattern (WP) symmetry (the symmetry of a patterns displaying at least one High-Order Laue Zone (HOLZ)),
- the Dark-Field (DF) symmetry (the symmetry within a disk in Bragg condition),
- the +/- g symmetry (the symmetry between two opposite hkl and -h-k-l disks)
These four symmetries lead to 31 three-dimensional (3D) diffraction groups if the CBED patterns display at least one HOLZ or to 10 two-dimensional (2D) projection diffraction groups if the patterns only display the ZOLZ. These diffraction groups are connected with the 32 point groups through a table given in ref 1.

The diameter of the disks on microdiffraction patterns is too small to allow the observation of any internal symmetries within the disks. As a result, the WP symmetry can be obtained but the BF, DF and +/-g are no longer available. Note that the ZOLZ symmetry of a microdiffraction pattern is connected with the 2D projection diffraction groups. Taken into account these limitations, a table was given in ref 2 which connects, the ZOLZ and WP "ideal" symmetries (the "ideal" symmetry takes into account both the position and the intensity of the reflections on a pattern) with the 32 point groups.

Nevertheless, for this table to be valid, the experimental microdiffraction patterns should be perfectly aligned since the diffracted intensities are very strongly connected with the crystal orientation. The diffracted intensities also depend on the specimen thickness so that a very small probe is required in order to avoid any thickness changes in the diffracted area. These experimental conditions are very difficult to fulfil.

The situation changes with microdiffraction electron precession [3] because the integrated intensities of the diffracted beams are available on these patterns and can be used to identify with confidence the "ideal" symmetry. It is justified to consider that the microdiffraction "ideal" symmetries are also valid for electron precession. This assessment was checked with simulations performed with the Jems software [4].

As a matter of fact, the main issue is connected with the non-centrosymmetrical point groups. Is it possible to detect the non-centrosymmetrical point groups from electron precession patterns? On CBED patterns, this detection is based on the

M. Luysberg, K. Tillmann, T. Weirich (Eds.): EMC 2008, Vol. 1: Instrumentation and Methods, pp. 213–214, DOI: 10.1007/978-3-540-85156-1_107, © Springer-Verlag Berlin Heidelberg 2008

observation of subtle differences of intensity which are visible on some couples of opposite hkl and -h-k-l reflections. For example, let us consider the case of the GaAs with non-centrosymmetrical point group F-43m. The 002 and 00-2 reflections on the CBED pattern on figure 1a display a clear difference of aspect.

On the corresponding precession pattern (Figure 1b), the two 002 and 00-2 disks displays the same aspect and there is no clear and visible difference of intensities between between. To clarify this point, Jems dynamical simulations were performed for different precession angles and for different specimen thicknesses. They indicate that the difference of intensity between the two opposite reflections 002 and 00-2 is usually very weak, especially for large precession angles so that it appears difficult to base an experimental identification of a non-centrosymmetrical point group on such a tiny effect. Additional simulations for other non centro-symmetrical crystals lead to the same conclusion. Therefore, it is more realistic to connect the "ideal" symmetry of the microdiffraction precession patterns rather with the 11 Laue classes than with the 32 point groups as given in Table I.

1. B.F. Buxton, J.A. Eades, J.W. Steeds and G.M. Rackham, Phil. Trans. Royal Soc. of London **A281** (1976), p. 181.
2. J.P. Morniroli and J.W. Steeds, Ultramicroscopy **45** (1992), p. 219.
3. R. Vincent and P. Midgley, Ultramicroscopy **53** (1994), p. 271.
4. P.A. Stadelmann, http://cimewww.epfl.ch/people/stadelmann/jemswebsite/jems.html

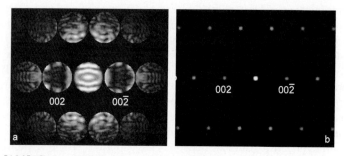

Figure 1. [130] GaAs CBED (a) and microdiffraction precession patterns (b). Jems simulated patterns.

Table 1. Connection between the "ideal" symmetry of microdiffraction precession patterns and the Laue class. ZOLZ symmetries are given between parentheses.

Laue class	1̄	2/m	mmm	4/m	4/mmm	3̄	3̄m	6/m	6/mmm	m3̄	m3̄m
					and absence of 3m (6mm)		and absence of 4mm (4mm)			and presence of 2mm (2mm)	and presence of 4mm (4mm)
Highest "ideal" symmetry	1 (2)	2 (2)	2mm (2mm)	4 (4)	4mm (4mm)	3 (6)	3m (6mm)	6 (6)	6mm (6mm)	3 (6)	3m (6mm)

Measurement of GaAs structure factors from the diffraction of parallel and convergent electron nanoprobes

Knut Müller[1], Marco Schowalter[1], Andreas Rosenauer[1], Dirk Lamoen[2], John Titantah[2], Jacob Jansen[3], Kenji Tsuda[4]

1. Institut für Festkörperphysik – Bereich Elektronenmikroskopie, Universität Bremen, Otto-Hahn-Allee 1, D-28359 Bremen
2. EMAT, Universiteit Antwerpen, B-2020 Antwerpen
3. National Centre for HREM, Kavli Institute of Nanoscience, Delft University of Technology, NL-2628 AL Delft
4. Institute of Multidisciplinary Research for Advanced Materials, Tohoku University, Sendai 980-8577, Japan

mueller@ifp.uni-bremen.de
Keywords: structure factor, bonding, GaAs, refinement, chemical sensitivity, diffraction

Present techniques in quantitative Transmission Electron Microscopy (TEM) frequently rely on the comparison of experimental and simulated data. For example, the evaluation of the chemical composition in ternary semiconductor nanostructures (CELFA, [1]) is based on both the computation and the measurement of the contrast in (200) lattice fringe images. This technique takes advantage of the chemical sensitivity of the 200 reflections in zincblende crystals. High accuracy of the CELFA method requires a precise knowledge of structure factors (SF) to simulate the correct contrast.

Conventionally, theoretical structure factors are calculated from atomic scattering amplitudes (ASA) obtained by the isolated atom approximation [2]. However, the redistribution of electrons due to the bonding in crystals may significantly affect ASA and SF of low order, such as 002. We recently published modified ASA obtained by density functional theory (DFT) for the 002 reflection in $In_xGa_{1-x}As$ alloys [3]. Here, we present a new approach to measure SF from diffraction patterns formed by parallel electron nanoprobes. To judge the plausibility of this method, the results will be compared with those we obtained from convergent beam electron diffraction (CBED) and from DFT.

The high efficiency in the refinement of e.g. atom positions, Debye parameters and specimen thickness using parallel beam electron diffraction (PBED) has motivated the development of the software package ELSTRU by J. Jansen during the last decade. We use the subroutine GREED to extract the background-corrected Bragg intensities from experimental PBED patterns. The subroutine MSLS [4] is then used to refine orientation, specimen thickness and Debye-Waller factors using up to 15 data sets simultaneously. The refinement of low-order structure factors is finally performed by the Bloch-wave refinement program Bloch4TEM with the MSLS results as input. The applicability and the precision of this procedure was initially checked with simulated diffraction patterns containing thermal diffuse background. We deduced conditions to recognize physically plausible refinement results for the thicknesses and the Debye-parameters, because the simultaneous refinement of both quantities may be ambigous

M. Luysberg, K. Tillmann, T. Weirich (Eds.): EMC 2008, Vol. 1: Instrumentation and Methods, pp. 215–216, DOI: 10.1007/978-3-540-85156-1_108, © Springer-Verlag Berlin Heidelberg 2008

for bad initial guesses. Since varying specimen thickness and –orientation may be neglected only if the probe diameter is in the range of 10nm, we fabricated C2 apertures between five and 10μm in diameter by focused ion beam etching. The refinement result for the 200 SF corresponds to the minimum of the so called R-value, which sums all squared differences of experimental and simulated 200 intensities. Figure 1 (left) shows the R-value as a function of the 200 SF for one experimental GaAs PBED pattern near the [010] zone axis. It can clearly be seen that the measured value (Bloch4TEM result) is about 20% larger compared with the 200 SF expected for isolated atoms.

Additionally, we recorded GaAs CBED patterns near the [035] zone axis with the 200 disc center in Bragg condition. The program package MBFIT [5] was used to correct for the background, elliptical distortions of the discs and for the refinement. Specimen thickness and –orientation, as well as the Debye parameters and the 200 SF were optimised during the MBFIT refinement. Figure 1 (right) shows observed and simulated discs and their difference for two cases. Row A corresponds to a simulation which uses ASA for isolated atoms. Row B shows the final refinement result, clearly revealing a minimisation in the difference between simulation and experiment. The resulting 200 SF was increased about 27% during the refinement.

Note that both measurements may not be compared directly since the Debye parameters are different, but in both cases, the resulting 200 SF deviate less than 7% from what we expect from DFT calculations. The results shift about 20% or more to more positive values, compared with 200 SF obtained from ASA for isolated atoms.

1. A. Rosenauer, W.Oberst, D. Litvinov, D. Gerthsen, A. Förster and R. Schmidt, Phys. Rev. B **61** (2000), p. 8276
2. A. Weickenmeier, H. Kohl, Act. Cryst. A **47** (1991), p. 590
3. A. Rosenauer, M. Schowalter, F. Glas and D. Lamoen, Phys. Rev. B **72** (2005), 085326
4. J. Jansen, D. Tang, H. W. Zandbergen and H. Schenk, Act. Cryst. A **54** (1998), p. 91
5. K. Tsuda and M. Tanaka, Act. Cryst. A **55** (1999), p. 939
6. We kindly thank the Deutsche Forschungsgemeinschaft for supporting this work under contract number RO2057/4-1 and the FWO-Vlaanderen under contract number G.0425.05.

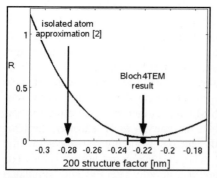
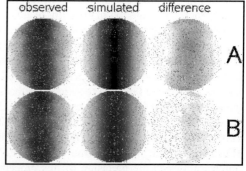

Figure 1. *Left:* R-value curve for the PBED method. Arrows indicate the 200 SF obtained from isolated atoms and from the refinement. *Right:* 200 discs obtained from the CBED method. Experiment, simulation and difference are shown for isolated atoms (row A) and the refinement result (row B).

Differential Electron Diffraction

P.N.H. Nakashima[1]

1. Monash Centre for Electron Microscopy and Department of Materials Engineering, Monash University, Victoria 3800, Australia

Philip.Nakashima@eng.monash.edu.au
Keywords: quantitative CBED, inelastic/elastic scattering, crystal thickness, structure factors

A new approach to electron diffraction is presented that eliminates almost all of the inelastic scattering signal without any energy filtering hardware. It makes use of changes in specimen thickness alone to cancel the thickness insensitive signal in electron diffraction patterns and is readily demonstrated in convergent beam electron diffraction (CBED). Most notably, the method removes the signal due to thermal diffuse scattering (TDS) as well as Borrmann effects [1] and can be used in conjuction with hardware energy filtering. The present work explains and explores the new method in application to the measurement of structure amplitudes (structure factors) by pattern-matching refinement of CBED patterns.

The proportion of electrons scattered inelastically from a material can be described as follows:

$$I_{inel} \propto 1 - e^{-H/\lambda}. \tag{1}$$

The specimen thickness is given by H, and λ represents the mean free path for inelastic scattering of electrons within a particular material. In the regime of quantitative CBED, H > 1000Å and $\lambda \sim$ 1000Å. For a change in thickness of 200Å, the accompanying change in the inelastic scattering signal is less than 3%. In practice, changes in thickness of much less than 200Å are possible, even in crushed specimens, via simple translation of the specimen. This means that a 3% change in inelastic scattering signal is an upper limit in thickness difference CBED experiments. For this reason, subtracting two electron diffraction patterns that differ only in the thickness of the specimen from which they were collected, results in a difference pattern where only thickness-sensitive and largely elastic information remains.

The thickness difference approach to CBED (illustrated in figure 1) was recently demonstrated in a number of applications [2 – 4]. In application to quantitative CBED, it was shown that the technique improves the precision of the measured structure factors over conventional measurements [4]. This was even the case when energy-filtered data were used in conventional quantitative CBED, whilst the new method used unfiltered CBED patterns. The present work also explores the effectiveness of this differential technique when applied to energy-filtered CBED data (as in figure 1).

1. P. Goodman, *Z. Naturforsch.* **28A** (1973), 580.
2. P.N.H. Nakashima, A.F. Moodie and J. Etheridge, *Acta Cryst.* A**63** (2007), 387.
3. P.N.H. Nakashima, A.F. Moodie and J. Etheridge, *Ultramicroscopy* (2008), *in press*.
4. P.N.H. Nakashima, *Phys. Rev. Lett.* **99** (2007), 125506.

M. Luysberg, K. Tillmann, T. Weirich (Eds.): EMC 2008, Vol. 1: Instrumentation and Methods, pp. 217–218, DOI: 10.1007/978-3-540-85156-1_109, © Springer-Verlag Berlin Heidelberg 2008

218

5. The author thanks the Royal Melbourne Institute of Technology (RMIT) Microscopy and Microanalysis Facility, Prof. D. McCulloch, Assoc. Prof. J. Etheridge and Prof. A. Moodie and the Australian Research Council for supporting this work (ARC DP0346828).

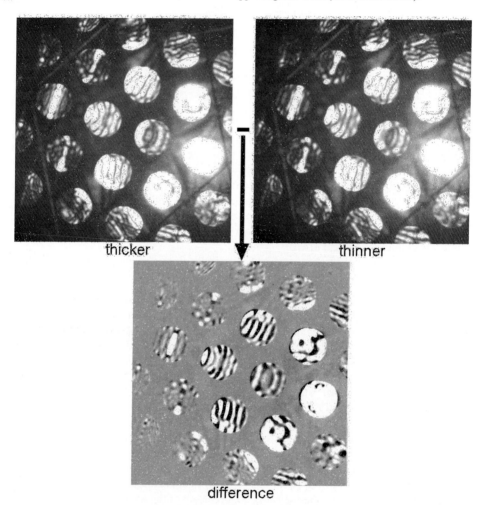

Figure 1. An example of the thickness difference CBED technique, applied to energy-filtered data. The patterns at the top were collected from α-Al_2O_3, about [0 0 1], at 203kV with a 6eV wide energy-selecting slit centred on the zero-loss peak in the electron energy loss spectrum. Even with energy filtering, a small but significant structured diffuse background remains. However, the difference between the two patterns collected from regions of slightly different specimen thickness, shows very little remaining background and a strong difference signal within the discs. Much of the signal due to TDS has been removed.

Atomic Structure Determination by "Observing" Structural Phase in 3-Beam CBED Patterns.

P.N.H. Nakashima[1,2], A.F. Moodie[1,2], J. Etheridge[1,2]

1. Monash Centre for Electron Microscopy, Monash University, Victoria 3800, Australia
2. Department of Materials Engineering, Monash University, Victoria 3800, Australia

Philip.Nakashima@mcem.monash.edu.au
Keywords: "Phase Problem", CBED, 3-phase invariant, atomic structure determination

Almost 80 years ago, Lonsdale demonstrated a method for structure determination [1] that relies entirely on knowledge of the signs of a number of 3-phase invariants of a structure. However, the phases of structure amplitudes cannot be measured by standard structure analysis techniques and can only be deduced from the distribution of intensities in a diffraction pattern. This is the infamous "Phase Problem". This difficulty in measuring structural phase has, up until now, prevented broad application of the powerful yet simple technique of Lonsdale for structure determination [1]. This situation has recently changed for the case of centrosymmetric crystals, with the development of an easy method for the direct measurement of structural phase.

Recent work [2 & 3] demonstrated the practicability of the theory of Moodie [4] and Moodie et al. [5] for "observations" of structural phase in centrosymmetric crystals from the intensity distribution within one disc in a 3-beam CBED pattern. Another consequence of this theory is the ability to measure the magnitudes of the associated structure amplitudes. We present our application of the derived method to α-Al_2O_3 and present a set of three-phase invariants and the magnitudes of the associated ten structure amplitudes.

This is a new approach to structure determination that has as its starting point, directly measured phases. This is in contrast to conventional structure analysis, which starts with measurements of magnitudes. We demonstrate the ease with which 3-phase invariants can be determined by inspection and discuss how this can be used to determine an outline structure of the crystal according to the unambiguous and direct approach of Lonsdale [1].

1. K. Lonsdale, *Proc. R. Soc.* **123A** (1929), 494.
2. P.N.H. Nakashima, A.F. Moodie and J. Etheridge, *Acta Cryst.* A63 (2007), 387.
3. P.N.H. Nakashima, A.F. Moodie and J. Etheridge, *Ultramicroscopy* (2008), *in press*.
4. A.F. Moodie, Chem. Scripta **14** (1978-79), 21.
5. A.F. Moodie, J. Etheridge and C.J. Humphreys, Acta Cryst. A52 (1996), 596.
6. We acknowledge the Australian Research Council for supporting this work (ARC DP0346828).

M. Luysberg, K. Tillmann, T. Weirich (Eds.): EMC 2008, Vol. 1: Instrumentation and Methods, pp. 219–220, DOI: 10.1007/978-3-540-85156-1_110, © Springer-Verlag Berlin Heidelberg 2008

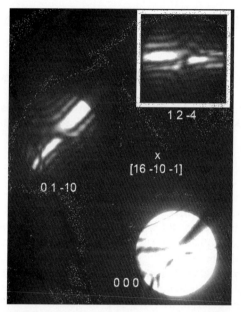

Figure 1. A 3-beam CBED pattern from the [16 -10 -1] zone with the reflections 1 2 -4 and 0 1 -10 at the Bragg condition. Deflections of the "bright bands" in the Bragg satisfied reflections determine the sign of the 3-phase invariant for this triplet ($V_{0\,1\,-10}$, $V_{1\,2\,-4}$ and the coupler, $V_{1\,1\,6}$) as shown in more detail in figure 2 for beam 1 2 -4.

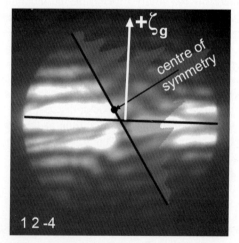

Figure 2. The centre of symmetry of the "two-beam-like" intensity oscillations "deflects" in the positive ζ_g direction (where the solid black line perpendicular to the ζ_g axis is the locus of the Bragg condition for this reflection). This indicates that the sign of the three phase invariant, $V_{1\,1\,6} \cdot V_{0\,1\,-10} / V_{1\,2\,-4}$, is positive.

Automatic space group determination using precession electron diffraction patterns

P. Oleynikov[1], S. Hovmöller[1] and X.D. Zou[2]

1. Structural chemistry, Stockholm University, SE–10691 Stockholm, Sweden
2. Berzelii Centre EXSELENT on Porous Materials, Stockholm University, SE–10691 Stockholm, Sweden

oleyniko@struc.su.se
Keywords: space groups, electron diffraction, precession

A set of algorithms for automatic High Order Laue Zone (HOLZ) indexing and possible set of space groups extraction were developed and implemented in the "Space Group Determinator" program. The symmetry analysis is performed using Morniroli-Steeds tables [1 and 2]. The developed program becomes extremely useful for the analysis of precession electron diffraction patterns (PEDs, see Fig. 1) [3]. "Space Group Determinator" was successfully tested on both simulated and experimental diffraction patterns with different zone axis orientations [4].

There are several advantages using PEDs:

- The intensities extracted from PEDs are less dynamical, especially for main zone axes;
- There are more reflections with higher resolution visible (depending on the precession angle);
- The width of a HOLZ band (if visible) can be significantly larger than on a corresponding SAED pattern.

The last statement is especially important for the correct space group or set of space groups determination. The possibility to observe several reflection lines within HOLZ makes the plane lattice shifts extraction easier. The knowledge of FOLZ shift with respect to the ZOLZ and the possible differences in periodicities provides very important information (see Fig. 2). The corresponding plane shifts in **a*** and **b*** directions between ZOLZ and HOLZ can be used together with tables from [1] for finding lattice centering, glide planes and partial symmetry symbol. This information can be treated systematically and implemented in the automatic procedure.

1   J.P. Morniroli, J.W. Steeds, Ultramicroscopy **45** (1992), p. 219–239.
2   J.P. Morniroli, A. Redjaïmia, S. Nicolopoulos, Ultramicroscopy **107** (2007), p. 514–522.
3   R.J. Vincent, P.A. Midgley, Ultramicroscopy **53**, 3 (1994), p. 271–282.
4   P. Oleynikov, PhD thesis, Stockholm University, (2006), p. 73–78.

M. Luysberg, K. Tillmann, T. Weirich (Eds.): EMC 2008, Vol. 1: Instrumentation and Methods, pp. 221–222, DOI: 10.1007/978-3-540-85156-1_111, © Springer-Verlag Berlin Heidelberg 2008

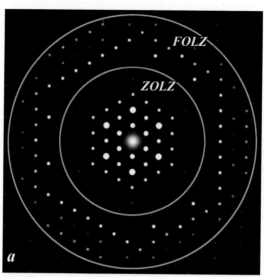

Figure 1. The simulated precession electron diffraction pattern of Ca₃SiO₅ (hatrurite mineral). Space group *R3m*, [001] zone axis. Notice that the spacing between diffraction spots is equal in the ZOLZ and FOLZ, but the lattices are shifted relative to each other by (½, ⅔) of the basic vectors in the ZOLZ. This is only possible for *I*-centered cubic and for rhombohedral crystals.

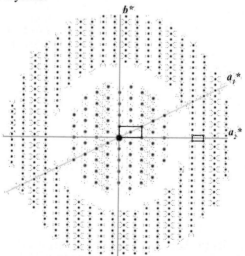

Figure 2. An example of a cubic system, [110] zone axis, partial extinction symbol is *I..d* or orthorhombic system, any of [100], [010] or [001] zone axes, partial extinction symbols are *Fd..*, *F.d.* and *F..d* respectively. Notice that the lattice spacing along both **a*** and **b*** is twice as long for FOLZ as for ZOLZ. This must be caused by a *d*-glide plane with its normal parallel to the electron beam. The relative shifts of the two lattices along the **a*** direction are due to *I* or *F*-centering

Compositional dependence of the (200) electron diffraction in dilute III-V semiconductor solid solutions

O. Rubel, I. Nemeth, W. Stolz and K. Volz

Department of Physics and Material Sciences Center, Philipps-University Marburg, Renthof 5, 35032 Marburg, Germany

oleg.rubel@physik.uni-marburg.de
Keywords: dark-field transmission electron microscopy, structure factor, scattering, density functional theory, valence force field

The quality of semiconductor mixed crystals is strongly dependent on the homogeneity of the distribution of the chemical constituencies. Cross-sectional dark-field transmission electron microscopy (DF TEM) provides an opportunity to access the chemical composition of the structures on the nanoscale.

The change in the intensity of the (200) electron beam reflection induced by incorporation of various isovalent impurities in GaAs and GaP is studied theoretically. Calculations are performed in the framework of the kinematical scattering theory, i.e. assuming the intensity of a diffracted beam to be proportional to the square of the corresponding structure factor. The effect of redistribution of the electron density on the electron scattering amplitudes is included to the model along with the effect of the local lattice distortions associated with the impurity sites.

We propose a way to calculate the local distortions analytically and to introduce them in a simple form to the expression for the structure factor. In the case of a *dilute* ternary alloy $AB_{1-x}C_x$, the structure factor corresponding to the diffraction indexes $\mathbf{g} = (h,k,l)$ can be approximated as

$$F(\mathbf{g}) \approx x f(C) + (1-x) f(B) + [1 - 8x(\pi \varepsilon \mathbf{g} \cdot \boldsymbol{\rho}_0)^2] f(A) \cos(2\pi \mathbf{g} \cdot \boldsymbol{\rho}_0), \qquad (1)$$

where f's are the electron scattering factors for individual chemical species, x is the impurity content, $\boldsymbol{\rho}_0 = (1/4, 1/4, 1/4)$ is the position vector of atomic spices in the ideal lattice, and ε is the local lattice distortion associated with the impurity. The latter is defined as

$$\varepsilon = \frac{r_{AC} - r_{AB}^0}{r_{AB}^0}, \qquad (2)$$

where r_{AB}^0 is the equilibrium interatomic spacing in the binary compound AB and r_{AC} is the bond length of atom C substituted on the place of atom B in the compound AB. The strain of anion-cation bond lengths around some isolated isovalent impurities in GaAs and in GaP are presented in Table 1.

Equation (1) is an alternative to the simulations, which invoke demanding computations for atomic relaxation, and enables quantitative prediction of the

M. Luysberg, K. Tillmann, T. Weirich (Eds.): EMC 2008, Vol. 1: Instrumentation and Methods, pp. 223–224, DOI: 10.1007/978-3-540-85156-1_112, © Springer-Verlag Berlin Heidelberg 2008

compositional variation of the structure factor for *dilute* alloys taking into account static atomic displacements (see Figure 1).

The implications of the results for quantification of composition fluctuations in heterostructures using DF TEM will be discussed. The effect of nitrogen and boron incorporation on the intensity of the (200) reflection is found to be partly compensated by the static atomic displacements they cause. Neglecting this effects would lead to a significant underestimate of the impurity content. The redistribution of the electron density is found to be less crucial for evaluation of the chemical composition. Results of the theoretical calculations will be compared with available experimental data.

1. K. Volz, O. Rubel, T. Torunski, S. D. Baranovskii and W. Stolz, Appl. Phys. Lett. **88** (2006), p. 081910.

Table 1. Strain ε of anion-cation bond lengths around isolated isovalent impurities in GaAs and in GaP cubic 64-atom supercells relative to the bond length of the host crystal calculated using density functional theory (DFT) and valence force field (VFF).

Method	GaAs:N	GaAs:B	GaAs:Sb	GaAs:P	GaP:Sb	GaP:As	GaP:N	GaP:B
DFT	−0.155	−0.113	+0.053	−0.025	+0.082	+0.025	−0.127	−0.122
VFF	−0.158	−0.117	+0.050	−0.025	+0.072	+0.023	−0.132	−0.124

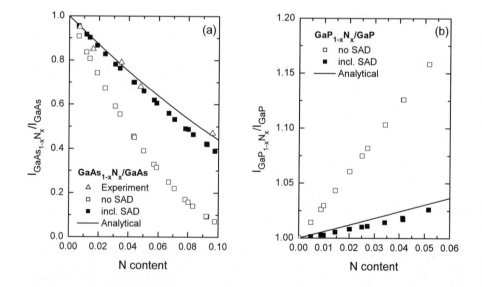

Figure 1. Compositional dependence of the (200) electron beam intensity in ternary GaAs(N) and GaP(N) semiconductor alloys calculated using the virtual crystal approximation (open boxes) and including static atomic displacements (SAD) for large supercells (filed boxes). Experimental results for GaAs(N)/GaAs quantum wells [1] are shown by open triangles on panel (a). The solid line indicates results of Eq. (1), which include the effect of atomic displacements in a simplified way.

Investigation of the local crystal lattice parameters in SiGe nanostructures by convergent-beam electron diffraction analysis

E. Ruh[1,2], E. Mueller[2], G. Mussler[3] and D. Gruetzmacher[3]

1. Labor for Micro-and Nanotechnology, PSI, 5232 Villigen PSI, Switzerland
2. Electron Microscopy, ETH Zürich, 8093 Zürich, Switzerland
3. Institut für Bio- and Nanosystems, FZ Jülich, 52425 Jülich, Germany

eruh@phys.ethz.ch

Keywords: CBED, STEM, SiGe, FEM calculations, strain and relaxation

Nanostructured SiGe material consisting of quantum wells, superlattices and quantum dots are promising candidates for fast optoelectronic devices. A main topic in the research of these nanostructures is the development of suitable configurations in order to tune the band structure for the desired optoelectronic applications. For that purpose strain conditions and local material composition within the nanostructures and especially at their interfaces have to be precisely known on an atomic scale.

The aim of the present work is to contribute to these researches by performing detailed layer characterisations with high spatial resolution based on convergent-beam electron diffraction (CBED) measurements and finite-element method (FEM) calculations on the example of Si/SiGe multi quantum well structures.

In scanning transmission electron microscopy (STEM) mode series of bright field CBED patterns were taken in the form of a line scan across the nanostructures using the [340]-zone axis. The higher order Laue zone (HOLZ) lines in each of these CBED pattern were fitted with the JEMS program [1] in order to deduce the local lattice parameters. Assuming equal strain conditions throughout the crystal column measured by the convergent electron beam the Ge concentration corresponding to the lattice parameters as deduced from CBED patterns was determined [2].

It has been shown that thin TEM samples consisting of Si/SiGe multi quantum well structures suffer from strong and asymmetric bending of the lattice planes near the upper and the lower end of the superlattice, which is confirmed by finite-element method (FEM) calculations [3]. These distortions of the crystal lattice are supposed to be responsible for the HOLZ line splitting [4,5] which makes the determination of the lattice parameters impossible in the affected regions of the Si substrate and SiGe layers. Figure 1 shows a strain-relaxed thin TEM foil consisting of a superlattice with 5 SiGe layers.

Furthermore the spatial resolution of this method was investigated on the interface of an AlN layer grown on Si(111). Since the plus and minus first order discs change position for Si and AlN, respectively, this is a relatively accurate measure to determine the spatial resolution of CBED, i.e. at which distance from the interface that the presence of the other material can still be detected. We found, that the resolution is not given by the beam geometry only (convergence angle, diameter), as the minimum

M. Luysberg, K. Tillmann, T. Weirich (Eds.): EMC 2008, Vol. 1: Instrumentation and Methods, pp. 225–226, DOI: 10.1007/978-3-540-85156-1_113, © Springer-Verlag Berlin Heidelberg 2008

226

resolvable distance is a factor of 2 to 3 larger than it should be due to geometrical reasons. We assume that multiple-diffraction is responsible for this phenomenon.

1. Electron microscopy software Java version (JEMS), P. Stadelmann, EPFL, Switzerland.
2. R. Balboni et al, Philosophical Magazine A **77(1)** (1998), p. 67 – 83
3. E. Ruh et al, Microscopy semiconducting materials XV (2007) proceedings, in press
4. A. Benedetti et al, Journal of Microscopy **223** (2006), p.249 – 252.
5. A. Chuvilin et al, Journal of Electron Microscopy **54(6)** (2005), p.515 – 517
6. We kindly acknowledge fruitful discussions with Karsten Tillmann, FZ Jülich, Hans Sigg, PSI and René Monnier, ETHZ

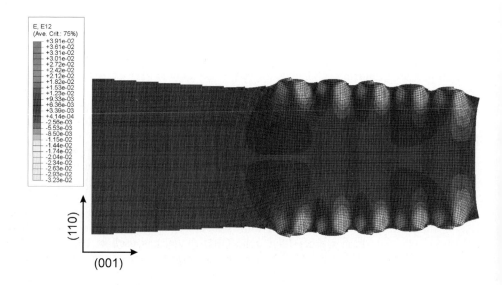

Figure 1. Contour plot of the off-diagonal strain component ε_{12} in a strain-relaxed TEM sample consisting of 5 SiGe layers as simulated by FEM calculations.

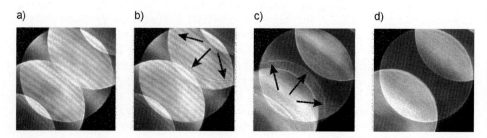

Figure 2. CBED patterns taken in a line scan across the interface of Si(111) and AlN: CBED pattern of pure Si a) and pure AlN d). In CBED patterns b), c) slight fringes marked by arrows indicate the change from Si to AlN.

Computer simulation of electron nanodiffraction from polycrystalline materials

K. Sugio[1] and X. Huang[2]

1. Mechanical System Engineering, Graduate School of Engineering, Hiroshima University, Higashi-Hiroshima 739-8527, Japan
2. Center for Fundamental Research: Metal Structures in Four Dimensions, Materials Research Department, Risø National Laboratory for Sustainable Energy, Technical University of Denmark, DK-4000 Roskilde, Denmark

ksugio@hiroshima-u.ac.jp

Keywords: electron nanodiffraction, polycrystalline materials, multiple diffraction, simulation

A recently developed X-Ray diffraction based technique, 3-Dimensional X-Ray Diffraction (3DXRD) microscope, has enabled to make 3D mapping of grain orientation and grain morphology in polycrystalline materials in a nondestructive way [1]. However, the spatial resolution of this technique is at the present limited to a few hundreds of nanometers. Inspired from the principle of 3DXRD microscope, we are exploring the possibility to develop an electron nanodiffraction based 3-Dimensional Transmission Electron Microscope (3DTEM) technique, with an aim to map the grain orientation and grain morphology for nanocrystalline materials. This technique will involve to index nanodiffraction patterns from overlapping grains and thus a major concern is the dynamic effect of electron diffraction. In this work, we demonstrate an example to analyze the nanodiffraction patterns from a twin boundary region (two overlapping twin-related grains), and the spot intensity change that will provide information about the boundary inclination.

Figure 1 (a) shows a twin boundary in an annealed copper sample. The twin boundary is inclined with respect to the foil plane and thus shows a wide projected width. The thickness fringes associated with the twin boundary are clearly seen. The grains separated by the twin boundary are labelled G1 and G2, as shown in Figure 1(a). A series of nanodiffraction patterns along the line indicated by the red arrow were acquired with a JEOL-3000F electron microscope operated at 300 kV. The beam size used to take the electron nanodiffraction patterns was about 5 nm. The nanodiffraction patterns taken within G1 or G2 show no clear change in the spot intensity, while the intensity of the nanodiffrcation spots varies from pattern to pattern in the twin boundary region. Figures 1 (b) and (c) show the nanodiffraction patterns from G1 and G2, respectively. Comparing the experimental patterns and the simulated ones (large red or small blue circles), the zone axes of the two patterns from G1 and G2 were determined to be $[1\bar{2}3]$ and $[\bar{1}\bar{2}3]$, respectively. Figure 1 (d) shows an example of the nanodiffraction patterns from the twin boundary region, which looks like from a single grain. Analysis shows that this pattern is a superimposition of the two patterns from G1 and G2, plus some extra spots, as shown in Figure 1 (d), where the extra spots are the ones not corresponding to large red (G1) or small blue (G2) circles. To understand the

M. Luysberg, K. Tillmann, T. Weirich (Eds.): EMC 2008, Vol. 1: Instrumentation and Methods, pp. 227–228, DOI: 10.1007/978-3-540-85156-1_114, © Springer-Verlag Berlin Heidelberg 2008

formation of the extra spots, a simulation of double diffraction was made: a diffracted electron beam from the first (top) grain acts as an incident beam for the second (bottom) grain [2], as shown in Figure 2 (a). In our simulation, the reciprocal vector diffracted by G1, $\mathbf{g}_1$, was calculated and then the reciprocal vector diffracted by G2, $\mathbf{g}_2$, was calculated, as shown in Figure 2 (b), where $\mathbf{k}_0$, $\mathbf{k}_1$ and $\mathbf{k}_2$ are wave vector of incident beam, first diffracted and second diffracted beam, respectively. If two-beam dynamical intensities in G1 were larger than a threshold, I_g^T, the second diffraction was calculated in G2. Figure 1 (e) shows the simulation results with $I_g^T = 0.1$. Transmitted and two diffracted beam from G1 (large red circles) were used as an incident beam into G2. The calculated positions of extra spots agree well with the experimental positions. Figure 1 (f) shows the simulation results with $I_g^T = 1.0 \times 10^{-5}$. Although many diffracted beam from G1 (large red circles) were used as an incident beam in G2, the simulation result is basically the same as that shown in Figure 1 (e). The boundary inclination, which is required for a 3D mapping of grain shapes, was analyzed based on spot intensity change for the nanodiffraction patterns taken from the twin boundary region.

1. H. F. Poulsen, "Three-Dimensional X-Ray Diffraction Microscopy", (Springer, Berlin) (2004)
2. B. Fultz and J. M. Howe, "Transmission Electron Microscopy and Diffractometry of Materials", (Springer, Berlin) (2004), p. 304.
3. We acknowledge E. Johnson, A. Godfrey, H. F. Poulsen and N. Hansen for fruitful discussions, and the Danish National Research Foundation for supporting the Center for Fundamental Research: Metal Structures in Four Dimensions, within which this work was performed. KS thanks the Ministry of Education, Culture, Sports, Science and Technology (MEXT), Japan, for supporting his visit to Risø National Laboratory by Overseas Advanced Educational Research Practice Support Program.

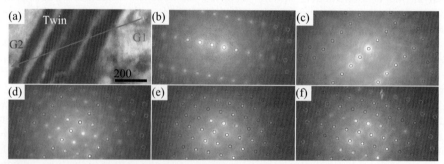

Figure 1. (a) Bright field image of a twin boundary separating two grains, G1 and G2, in copper. Nanodiffraction patterns from (b) G1, (c) G2 and (d) twin boundary region. Simulation with double diffraction (e) $I_g^T = 0.1$ and (f) $I_g^T = 1.0 \times 10^{-5}$.

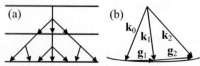

Figure 2. Schematics of double diffraction (a) in real space and (b) reciprocal space.

An analytical approach of the HOLZ lines splitting on relaxed samples

J. Thibault[1,#], C. Alfonso[1], L. Alexandre[1], G. Jurczak[1,*], C. Leroux[2], W. Saikaly[1,3], and A. Charaï[1]

1. CNRS, IM$_2$NP-UMR 6242, Aix-Marseille Université, Faculté des Sciences et Techniques, Campus de Saint-Jérôme, Avenue Escadrille Normandie Niémen - Case 262, F-13397 Marseille Cedex, France.
2. CNRS, IM$_2$NP-UMR 6242, Université Sud Toulon-Var, Bâtiment R, BP 132, F-83957 La Garde Cedex, France.
3. CP2M - Faculté des Sciences et Techniques, Campus de Saint-Jérôme, Avenue Escadrille Normandie Niémen - Case 221, F-13397 Marseille Cedex, France.

jany.thibault@univ-cezanne.fr
Keywords: CBED, HOLZ splitting, relaxed samples, Si$_{1-x}$Ge$_x$/Si heterostructures

For many years, Convergent Beam Electron Diffraction (CBED) has been intensively used for lattice parameter determination because of the extremely high sensitivity of High Order Laue Zone (HOLZ) lines position to small lattice parameter changes. But, for highly stressed systems, the sample thinning down to electron transparency induces a relaxation. In such cases, HOLZ lines are splitted due to an inhomogeneous variation of lattice parameter (Figure 1.a). Conventional measurement of lattice parameter using quasi-kinematical approach of HOLZ lines shift is then impossible and new models have to be developed. Recently, several approaches have been proposed in order to reproduce the experimental profile of split HOLZ lines also called "rocking curves" $I_g(s)$, and deduce the stress distribution in the sample [1, 2, 3, 4]. But, several questions are still open and relation between the rocking curve features and the atomic displacement profile is not completely understood.

In order to provide some clues to understand furthermore this relationship, we developed a new analytical approach of HOLZ splitting performed in the frame of the kinematical approximation of electron scattering. It will be shown that very simple analytical expressions of $I_g(s)$ can be obtained when the displacement profiles are approximated by simple continuous mathematical functions. This allows simple physical interpretation of the main features of the rocking curves. However, the displacement fields in a relaxed TEM sample are 3-dimensional, and this continuous approach can be a rather rough simplification. To take the 3D displacement field into account, we developed a second analytical "discrete model" using a slicing method similar to the Finite Element calculation. The main approximation of this slicing method is a constant displacement within a sub-slice. With this "discrete method", it will be shown that $I_g(s)$ have also an analytical expression that highlights the complex oscillations of the rocking curves. The physical interpretation of the different terms is less straightforward than in the case of the continuous model but this model can support complex expressions for the displacement or the results of the finite element calculation.

M. Luysberg, K. Tillmann, T. Weirich (Eds.): EMC 2008, Vol. 1: Instrumentation and Methods, pp. 229–230, DOI: 10.1007/978-3-540-85156-1_115, © Springer-Verlag Berlin Heidelberg 2008

230

In addition, this analytical approach provides a very simple method to extract a good value of the maximum atomic displacement. In fact, we will shown that, by measuring only the position and the intensity of two major peaks on the rocking curve, one is able to obtain this value with an uncertainty of about 10% with respect to the same value obtained after Finite Element Modelling (FEM). This error can even be lowered down to 2% with simple adjustment routine. This will be illustrated on $Si_{1-x}Ge_x/Si$ sample with x = 3 or 6%, as a model sample which exhibits a rather large relaxation (Figure 1)

1   L. Clément, R. Pantel, L.F.Tz. Kwakman, J.L. Rouvière 85(4), Applied Physics Letters (2004) 65.
2   Chuvilin A, Kaiser U, de Robillard Q, Engelmann HJ, Journal of Electron Microscopy, 54 (6), (2005) 515
3   F. Houdellier, C. Roucau, L. Clément, J.L. Rouvière, M.J. Casanove Ultramicroscopy, 106, (2006) 951
4   A. Spessot, S. Frabboni, R. Balboni, A. Armigliato Nuclear Instruments and Methods in Physics Research B 253, (2006) 149

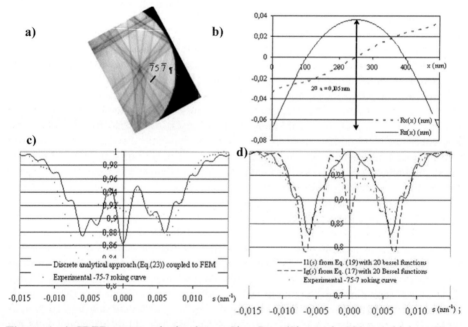

Figure 1. a) CBED pattern obtained on a $Si_{0.97}Ge_{0.03}/Si$ sample 500 nm thick at 80 nm from the interface. The $\overline{7}5\overline{7}$ HOLZ line is shown. b) Corresponding displacement profiles $R_x(x)$ and $R_z(x)$ calculated by FEM. The maximum atomic displacement is indicated. c) $\overline{7}5\overline{7}$ experimental rocking curve compared to the one calculated with the displacement profiles given on fig 2b and "discrete analytical model" d) Comparison with the curve calculated using the "continuous model" and a cosine model displacement profile.

ELDISCA C# – a new version of the program for identifying electron diffraction patterns

J. Thomas, T. Gemming

IFW Dresden, P.O. Box 27 01 16, 01171 Dresden, Germany

j.thomas@ifw-dresden.de
Keywords: electron diffraction, software

Electron diffraction is a well-established technique in transmission electron microscopy with regard to the instrumentation as well as to the simulation and identification of the diffraction patterns. E.g. the JEMS package by P. Stadelmann [1] is a well-known software. Nevertheless, special questions and requirements lead to wishes for special layouts and possibilities of the software.

Against this background we wrote our own software ELDISCA for the simulation and interpretation of point and ring diagrams obtained by selected area or nano-electron diffraction. The new version ELDISCA C# has been created with enhanced functionalities and additional possibilities running under Windows XP.

Beside the calculation of point and ring diagrams including the comparison with measurements the software allows the simulation of Kikuchi and CBED HOLZ patterns. For the determination of mixtures of phases an overlay of calculated ring diagrams of different phases with different grain sizes is possible. Their radii and intensities can be compared with the radial intensity distribution extracted from a measured pattern. The matching between measurement and simulation can be reached by a trial and error method varying the phase ratios and their grain sizes.

Additionally, the calculation of scattering curves of given atomic arrangements in amorphous materials using the Debye equation as well as the determination of radial density functions for measured scattering curves of amorphous materials are possible.

The program works on the basis of the kinematical theory with scattering factors by Doyle and Turner [2]. It needs files with the crystallographic data of the phases created with the help of the program (see Figure 1). These files can be used for all of the calculations. Measured results can be inserted by use of the windows clipboard function or by loading of an ASCII file or a bitmap, respectively.

Each option opens its own window, i.e. different patterns from different phases can be calculated and identified independently. The results remain in the windows. Figure 2 shows an example: the window for calculation of an overlay of maximal five ring diagrams as radial density distribution and the comparison with the measurement. It is possible to copy this window in the clipboard, to save it as a bitmap and to print it.

The program is supplemented by a help function with an introduction to the electron diffraction techniques, a detailed description of the algorithms, and a user guide.

1   JEMS, P. Stadelmann, CIME-EPFL, Lausanne,
     Switzerland:http://cimewww.epfl.ch/people/stadelmann/jemsWebSite/jems.html
2   P.A. Doyle, P.S. Turner, Acta Cryst. A24 (1968) 390

M. Luysberg, K. Tillmann, T. Weirich (Eds.): EMC 2008, Vol. 1: Instrumentation and Methods, pp. 231–232, DOI: 10.1007/978-3-540-85156-1_116, © Springer-Verlag Berlin Heidelberg 2008

232

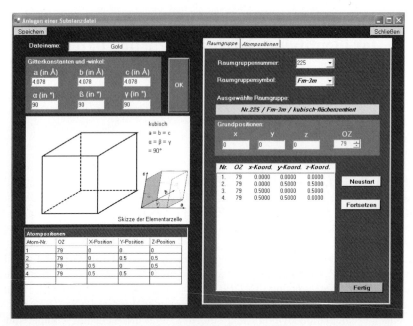

Figure 1. Window for input of a file with crystallographic data (example: gold).

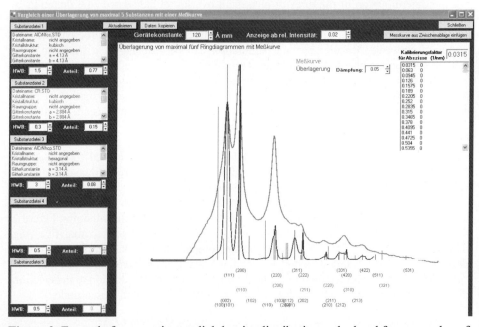

Figure 2. Example for an option: radial density distribution, calculated for an overlay of three phases und comparison with the measured curve.

Mixing Real and Reciprocal Space

R.D. Twesten[1], P.J. Thomas[1], H. Inada[2] and Y. Zhu[3]

1. Gatan Research and Development, Pleasanton, CA USA
2. Nanotech Systems Division, Hitachi High-Technologies Corp., Pleasanton, CA, USA
3. Center for Functional Nanomaterials, Brookhaven National Laboratory, Upton, NY USA

rtwesten@gatan.com

Keywords: Diffraction, Energy Filtering, Spectrum Imaging

Traditionally, TEM has been considered either an imaging or diffraction based technique; images are obtained to elucidate microstructure while diffraction patterns are used to determine crystal structure. This traditional separation, however, is completely arbitrary. A brief review of electron microscopy shows that nearly all TEM techniques rely on the mixing of real (r-space) and reciprocal space (k-space) to some degree [1]. For example, bright- and dark-field imaging relies on the filtering of k-space to produce contrast; large-angle CBED patterns directly mix real and reciprocal space to allow the measurement of strain centres. For some years, electron backscatter diffraction (EBSD) imaging in the SEM has allowed the direct combination of real and reciprocal space. This backscatter geometry has both advantages and limitations; a particular limitation, spatial resolution, can be greatly improved by moving to the transmission geometry.

We have recently developed tools to acquire and analyze diffraction image (DI) stacks within the Gatan spectrum image (SI) paradigm. The acquisition technique is completely generic, the only instrumental requirements are STEM mode and a digital camera. One particularly powerful mode is the use of an energy filter to record a DI stack of energy filtered diffraction patterns. Shown in figure 1 is data from a commercial Pt/Ru catalyst supported on carbon black recorded using a 200kV LaB6 (S)TEM and a post-column energy filter. Diffraction patterns can be extracted and analyzed from each particle to yield crystal structure information [2]. The filtered patterns (b-d) show higher signal-to-background ratio than the unfiltered data, especially true for the smallest particles, making diffraction analysis clearer.

Figure 2 shows data from the layered incommensurate thermoelectric material $Ca_3Co_4O_9$. Using a 200kV Cs corrected STEM probe, a series of 100 diffraction patterns were recorded with atomic resolution along a 3.5nm line. By acquiring a complete diffraction pattern at each point, we have the ability to vary the region of k-space producing the image. As seen in figure 2, varying the inner and outer ADF angles can have profound effects on image contrast and interpretation [3]. The method is particularly useful in mapping real space distribution of charge, orbital and spin orderings as well as lattice displacement that can be related to specific reciprocal signals [4].

1 see, for example, D.B. Williams and C.B. Carter (1996) Transmission Electron Microscopy: A Textbook For Material Science. Plenum, New York and references therein.
2 C.W. Hills, N.H. Mack, and R.G. Nuzzo (2003) J. Phys. Chem. **B 107**, 2626.

M. Luysberg, K. Tillmann, T. Weirich (Eds.): EMC 2008, Vol. 1: Instrumentation and Methods, pp. 233–234, DOI: 10.1007/978-3-540-85156-1_117, © Springer-Verlag Berlin Heidelberg 2008

234

3   D. E. Jesson and S. J. Pennycook (1993) Proc. Roy. Soc. Lond. **A 441**, 261.
4   Work at BNL was supported by US Department of Energy BES, under contract No. DE-AC02-98CH10886.

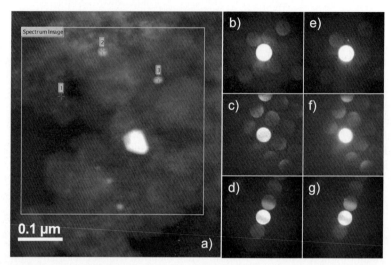

Figure 1. Analysis of filtered vs. unfiltered diffraction image (DI) stacks. a) A HAADF STEM survey image of a carbon black supported, Pt/Ru catalyst sample indicating the data collection region. From this area, a $(32x32)_{r\text{-space}}$ x $(256x256)_{k\text{-space}}$ DI stack was acquired onto GIF Tridiem CCD camera using a 10eV (filtered) or >100eV (unfiltered) slit. b-d) Filtered and e-g) unfiltered diffraction patterns corresponding to point 1-3 in a).

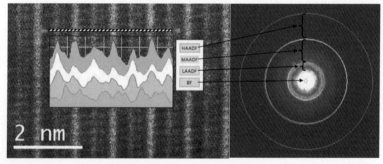

Figure 2. Atomic resolution diffraction image analysis: A series of 100 diffraction patterns at 35pm intervals were recorded along a line (indicated by red arrows) on a sample of $Ca_3Co_4O_9$ using the aberration corrected STEM probe of the Hitachi HD2700C STEM at Brookhaven National Laboratory. By post processing the resulting diffraction patterns, the angular dependence of the image contrast can be changed at will. Contrast reversals are evident as the scattering angle is reduced. The image collections angles are: 6mR, 30-59mR, 60-130mR, 133-210mR for the BF, LAADF, MAADF, and HAADF line traces, respectively.

Structure solution of intermediate tin oxide, SnO$_{2-x}$, by electron precession

T.A. White[1], S. Moreno[2] and P.A. Midgley[1]

1. Department of Materials Science and Metallurgy, University of Cambridge, Pembroke Street, Cambridge CB2 3QZ, UK
2. Centro Atomico Bariloche, 8400 - San Carlos de Bariloche, Argentina

taw27@cam.ac.uk

Keywords: tin oxide, precession electron diffraction, direct methods

Tin oxide is known to exist in two forms, SnO and SnO$_2$, but the existence of at least one additional form with a composition in between the two, i.e. SnO$_{2-x}$, is now generally accepted [1-3]. To date, no *ab initio* structural information has been extracted from this intermediate compound. A major reason for this is that the compound is usually found in the form of small (<1 μm), often defective, crystals which are attached to the metallic tin particles from which they grow (Figure 1a). Therefore, crystals large enough and of sufficient quality for single-crystal X-ray diffraction have not been available.

Electron diffraction allows the acquisition of single-crystal diffraction patterns from such small particles due to the ability of the electron microscope to form small focussed probes. The technique of precession electron diffraction [4] was applied to the intermediate tin oxide compound, and a typical diffraction pattern is shown in Figure 1b. The diffraction patterns from our specimens of the intermediate structure are consistent with the cell parameters given by Lawson [2] for Sn$_3$O$_4$.

Phasing the intensity data using the Tangent Formula [5] produced electrostatic potential maps which we interpret as showing the tin atoms as strong peaks, and possibly some semblance of the oxygen positions in the weaker background ripple. A map for the [010] projection according to Seko's choice of unit cell [3] is shown in Figure 2. Although dynamical effects are likely to be influencing the results quite strongly, the positions of the tin atoms in these maps are in agreement with models for the structure which are based on an oxygen-deficient supercell of the rutile structure of SnO$_2$, with small displacements of the tin atoms. In both our reconstructed map and Seko's proposed structure for Sn$_3$O$_4$, the tin atoms do not lie on a straight line as they would for an exact supercell of the rutile structure in this projection. A least-squares refinement to verify these subtle differences is the subject of ongoing work.

1. M. Batzill and U. Diebold, Progress in Surface Science **79** (2005) p. 47–154.
2. F. Lawson, Nature **215** (1967) p. 955.
3. A. Seko et. al., Physical Review Letters **100** (2008) 045702.
4. R. Vincent and P. Midgley, Ultramicroscopy **53** (1994) p. 271.
5. J. Karle and H. Hauptman, Acta Crystallographica **9** (1956) p. 635.
6. The authors thank the EPSRC and FEI Company for financial assistance.

M. Luysberg, K. Tillmann, T. Weirich (Eds.): EMC 2008, Vol. 1: Instrumentation and Methods, pp. 235–236, DOI: 10.1007/978-3-540-85156-1_118, © Springer-Verlag Berlin Heidelberg 2008

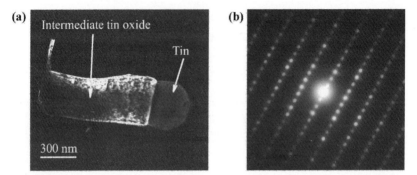

Figure 1. (a) Dark-field image of a cylindrical-shaped crystal of the intermediate tin oxide grown from a spherical particle of tin. **(b)** Typical precession electron diffraction pattern from the intermediate compound.

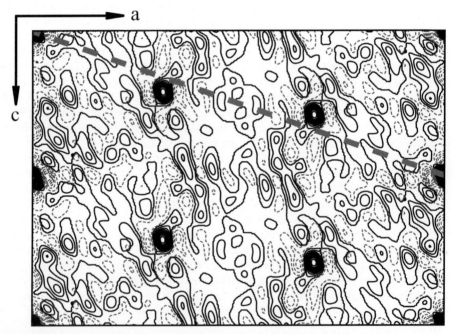

Figure 2. Electrostatic potential map reconstructed by direct methods showing the [010] projection according to Seko's choice of unit cell axes [3]. Dashed contours represent positive potential due to phase errors and Fourier truncation. The strong peaks reveal the tin positions, and do not lie on a common straight line as they would for an exact supercell of the rutile structure of SnO_2. The thick dashed line is a guide to the eye.

"Phase-scrambling" multislice simulations of precession electron diffraction

T.A. White[1], A.S. Eggeman[1] and P.A. Midgley[1]

1. Department of Materials Science and Metallurgy, University of Cambridge, Pembroke Street, Cambridge CB2 3QZ, UK

taw27@cam.ac.uk
Keywords: precession electron diffraction, multislice simulation, direct methods

Under exactly kinematical conditions, a particular crystal structure would produce the same diffraction pattern as a hypothetical crystal structure which had the same structure factor moduli but where their phases were different. This is simply a statement that solutions to the phase problem are not unique. However, if dynamical effects were present, the real and hypothetical crystal structures would produce different diffraction patterns since the results of re-interference between diffracted beams in the crystal depend on the phases of the beams which are interfering.

By using hollow cone illumination, precession electron diffraction enables patterns to be recorded whose intensities appear to be less sensitive to dynamical effects and thus makes the diffracted intensities more suitable for structure solution [1, 2]. By calculating dynamical precession patterns with a limited number of beams, Sinkler noted at a recent conference [3] that applying precession to diffraction data seemed to reduce their sensitivity to phase alterations of the type described above. We have carried out similar but more extensive simulations to verify this phenomenon using full multislice calculations [4, 5].

Simulations were performed using the silicon [110] projection with precession angles from 0 to 30 mrad, and with thicknesses up to a maximum of 100 nm. The calculated intensities were compared with those from an identical simulation performed using a projected potential where the phases of the structure factors had been randomly set to 0 or π just before the structure factors were transformed into the real-space potential. The two potential functions are shown in Figure 1. None of the symmetry of the projection was preserved other than its centrosymmetricity.

As shown in Figure 2, scatter plots were calculated showing the intensities from the phase-altered simulation against those from the original simulation. Little or no correlation is apparent when precession is not applied and where dynamical behaviour dominates. As the precession angle is increased, a clear correlation becomes evident at all thicknesses. The most dramatic increases in correlation appear to occur at lower precession angles. As a measure of the correlation, we calculated the unweighted intensity residual, R_2, between the two lists of reflections. This was 50% for 4 nm thickness with no precession, decreasing to 3.7% with 30 mrad of precession. At 100 nm thickness, R_2 was 59% with no precession and 13% when the precession angle was 30 mrad. These results support the assertion made by many researchers that precession electron diffraction produces intensities which can be treated as kinematical in nature.

M. Luysberg, K. Tillmann, T. Weirich (Eds.): EMC 2008, Vol. 1: Instrumentation and Methods, pp. 237–238, DOI: 10.1007/978-3-540-85156-1_119, © Springer-Verlag Berlin Heidelberg 2008

238

1. R. Vincent and P. Midgley, Ultramicroscopy **53** (1994) p. 271.
2. C. S. Own, L. D. Marks and W. Sinkler, Acta Crystallographica A **62** (2006) p. 434.
3. W. Sinkler, Microscopy and Microanalysis conference (2007).
4. E. J. Kirkland, Advanced Computing in Electron Microscopy, Plenum (1998).
5. The CamGrid environment was used: http://www.escience.cam.ac.uk/projects/camgrid/
6. The authors thank the EPSRC and FEI Company for financial assistance.

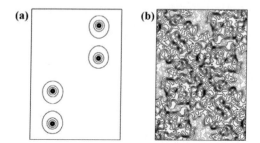

Figure 1. (a) Projected potential for silicon parallel to the <110> zone axis. **(b)** The same projected potential after scrambling the structure factor phases. Dashed contours represent positive electrostatic potential. The contour interval differs between the two plots.

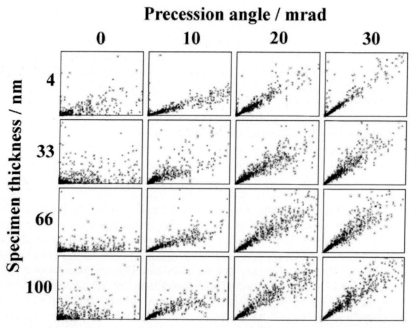

Figure 2. Montage of scatter plots for various thicknesses and precession angles. Each individual plot shows the true silicon intensities on the x-axis and phase-scrambled intensities on the y-axis.

High-Resolution Electron Holography on Ferroelectrics

M. Linck[1], H. Lichte[1], A. Rother[1], F. Röder[1], K. Honda[2]

1. Triebenberg Laboratory, Institut für Strukturphysik, Technische Universität Dresden,
Zum Triebenberg 50, 01328 Dresden, Germany
2. Fujitsu Laboratories Ltd., Device and Materials Lab, 10-1 Morinosato-Wakamiya,
Atsugi 243-0197, Japan

Martin.Linck@Triebenberg.de
Keywords: Holography, Ferroelectrics, Electric Fields

Within the continuing process of miniaturization in information technology analytical tools are required to characterize materials in terms of electric and magnetic fields on the nanometer scale. In this discipline, off-axis electron holography has proven to be a measuring talent. The holographic reconstruction offers an access to the complete complex object wave, i.e. amplitude and phase of the object-modulated electron wave with all details from largest area information up to the resolution limit of the microscope [1].

A lot of interest is currently focussed on ferroelectrics, especially because of technological promises such as use for non-volatile mass storage. Already, by evaluating the atomic displacements from Cs-corrected HRTEM micrographs, the polarization was estimated in direction and quantity on the nanometer scale [2]. This approach, however, represents an indirect determination of the polarization. Electron holography reveals the complete phase of the object wave, hence provides direct access to the projected electric potential of the ferroelectric material. Since ferroelectric distortions and nanoscopic potential variations can be measured at the same time, holographic investigations are essential for fundamental understanding of ferroelectrics.

Special care has to be taken on interpretation of experimental findings, because the measured projected potential gives rise to the projection of the electric field $E = (D - P)/\varepsilon_0$, which actually consists of both, the polarization P and the dielectric displacement D. ε_0 is the electric field constant. Ab-initio calculations of various perovskite structures, e.g. like in Figure 1, exhibit a strong compensation of polarization P by the dielectric displacement D. The remaining depolarization field E is usually small, but also strongly depending on the boundary conditions, i.e. charges at domain walls and interfaces to more or less conducting compositions, crystal defects, and size effects (e.g. [3]). In fact, for analysing these aspects, electron holography can benefit from its unique large area phase shift reconstruction to measure compensating charges.

In our work, we are systematically analysing different arrangements of ferroelectric materials in between conducting and non-conducting compositions. Figure 1 shows one of these arrangements. [4]

M. Luysberg, K. Tillmann, T. Weirich (Eds.): EMC 2008, Vol. 1: Instrumentation and Methods, pp. 239–240, DOI: 10.1007/978-3-540-85156-1_120, © Springer-Verlag Berlin Heidelberg 2008

240

1. H. Lichte, P. Formanek, A. Lenk, M. Linck, C. Matzeck, M. Lehmann and P. Simon, Annual Review of Materials Research, Vol. **37** (2007), 539-588, ISBN 978-0-8243-1737-9.
2. C.L. Jia, S.B. Mi, K. Urban, I. Vrejoiu, M. Alexe and D. Hesse, Nature Materials Vol. **7** Issue 1 (2008), 57-61.
3. J. Junqera, P. Ghosez, Nature Vol. **422**, 2003, 506-509.
4. The authors appreciate the enlightening discussions with all members of the Triebenberg group. Financial support from ESTEEM is gratefully acknowledged (http://esteem.ua.ac.be).

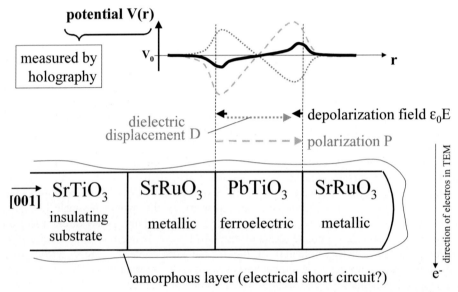

Figure 1. Schematic drawing of a cross-sectioned specimen with a ferroelectric layer of PbTiO₃ in between two conducting layers of SrRuO₃. The layers are grown epitaxially on a SrTiO₃-substrate in [001] direction. The hardly avoidable amorphous surface layer on top and bottom of the thin specimen may cause a short circuit between the SrRuO₃-electrodes. Above, the components of electric field E, polarization P and dielectric displacement D are indicated. The resulting electric potential difference V is the quantity, which finally is measured by means of electron holography.

Imaging parameters for optimized noise properties in high-resolution off-axis holograms in a Cs-corrected TEM

M. Linck[1]

1. Triebenberg Laboratory, Dresden University, Zum Triebenberg 50, 01328 Dresden

Martin.Linck@Triebenberg.de
Keywords: Holography, Cs-Corrector, Noise

The Cs-corrector nearly has become standard equipment for high-resolution imaging in Transmission Electron Microscopy [1]. Although it already provides an incredible tool for direct characterization on the atomic scale, there are still obstacles to be managed. One of them is the loss of phase information related to formation of intensity. Even though information transfer within the imaging system can be tuned to provide atomic phase contrast, the microscope still is blind for large area phase objects such as electric and magnetic fields and additionally suffers from residual aberrations.

A smart method to overcome the partial loss of information is off-axis electron holography, in that the complete complex image wave can be reconstructed and corrected for all coherent aberrations [2]. Also in electron holography the Cs-corrector has brought a set of remarkable improvements [3].

In conventional transmission electron microscopes the spherical aberration is fixed and for holography the defocus is used to enhance the spatial resolution of the image wave recorded in the hologram [4]. In the Cs-corrected TEM defocus Dz and spherical aberration Cs are free parameters within certain limits in that the improvements of the corrector are not recognizably violated. Because of holographic imaging these coherent aberrations can be corrected numerically afterwards. Therefore, Dz and Cs can be used to optimize other hologram properties.

The holographic reconstructed data sets are strongly affected by noise. In the empty hologram the uncertainty of fringe contrast V and phase shift φ are given as $\Delta V = \text{Sqrt}[(2-V)^2/N]$ and $\Delta\Phi = \text{Sqrt}[2/N]/V$, respectively, with N the number of electrons per fringe [5]. In fact, the object hologram is modulated in both, fringe contrast V and local electron dose N. Therefore, also noise becomes dependent on position: $\Delta V(r) = \text{Sqrt}[(2-V(r))^2/N(r)]$ and $\Delta\Phi(r) = \text{Sqrt}[2/N(r)]/V(r)$. To correct for the coherent aberrations, which were introduced before, the noisy amplitude (which is derived from the noisy fringe contrast) and the noisy phase are virtually back-propagated by numerical optics, leading to a noisy object wave.

A tuning of Dz and Cs within pre-defined limits changes $V(r)$ and $N(r)$ systematically. This offers a facility to obtain minimum noise in the reconstructed data. Figure 1 shows a scheme of the holographic procedure containing the aspects of noise.

By selecting certain optimization criteria for different applications, special sets of Dz and Cs can be found to provide low-noise results. In Figure 1 the low-noise measurement of the atomic phase shift was taken as the criterion for optimization. [6]

M. Luysberg, K. Tillmann, T. Weirich (Eds.): EMC 2008, Vol. 1: Instrumentation and Methods, pp. 241–242, DOI: 10.1007/978-3-540-85156-1_121, © Springer-Verlag Berlin Heidelberg 2008

242

1. M. Haider, H. Rose, S. Uhlemann, E. Schwan, B. Kabius, and K. Urban, Nature **392** (1998), 768-769.
2. H. Lichte and M. Lehmann, Rep. Prog. Phys. **71** (2008), 016102.
3. D. Geiger, H. Lichte, M. Linck and M. Lehmann, Microsc. Microanal. **14** (2008), 68-81.
4. H. Lichte, Ultramicroscopy **38** (1991), 13-22.
5. F. Lenz, Optik **79,** No.1 (1988), 13-14.
6. The fruitful discussions within the Triebenberg group are highly appreciated.

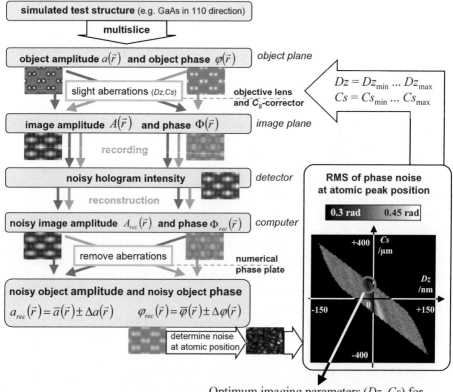

Optimum imaging parameters (Dz, Cs) for
low noise atomic phase shift reconstruction

Figure 1. Optimization scheme of the holographic procedure including effects of noise. In a Cs-corrected TEM the coherent aberrations (Dz, Cs) can be tuned within certain limits to optimize the noise properties in the reconstructed, aberration-corrected data. Here, as criterion the atomic phase shift has been chosen.

Partial coherence in inelastic holography

J. Verbeeck[1], G. Bertoni[1,2], D. Van Dyck[1], H. Lichte[3], P. Schattschneider[4]

1. EMAT, University of Antwerp, Groenenborgerlaan 171, 2020 Antwerp, Belgium
2. Italian Institute of Technology (IIT), Via Morego 30, I-16163 Genoa, Italy
3. Triebenberglabor, Dresden University, Zum Triebenberg 1, 01062 Dresden, Germany
4. Institut für Festkörperphysik, Technische Universität Wien, Wiedner Hauptstrasse 8-10, A-1040 WIEN, Austria

jo.verbeeck@ua.ac.be
Keywords: electron holography, mixed dynamic form factor, partial coherence

Inelastic holography is a technique in which electron energy loss spectroscopy (EELS) and electron holography are combined to study coherence effects in inelastic scattering. In typical inelastic cross section calculations it is assumed that we have one scattering atom and a single incoming plane wave. Although this setup seems to provide the answer to most of the practical inelastic scattering problems, there are several cases where it fails. The reason for this failure is the existence of interference effects in the scattered wave. This occurs when there is more than one incoming plane wave and/or when there is more than one scattering atom; like in almost every experiment in the electron microscope.

In this contribution we describe a set of different experiments where these effects can be clearly demonstrated. A typical example is the study of the partial coherence in the exit wave after inelastically exciting a plasmon. It turns out that the partial coherence is mainly generated by the fact that electrons interact with the object in a delocalised way through Coulomb interaction. This makes that even for scattering centers that are completely incoherent (viz. core excitations) partial coherence can still exist in the inelastic exit wave [1,2]. In view of the complexity and the counter intuitive aspects in partial coherence, it is important that we can numerically simulate these effects. It will be shown how these simulations are performed, and how they can be used to gain insight as well as good agreement with real life experiments [3,4].

We will show a set of applications that can only be understood with the described theories. Scanning transmission electron miscroscopy (STEM) EELS is probably the most important one, but also energy-loss magnetic chiral dichroism (EMCD) [5] and high resolution energy filtered TEM (EFTEM) require partial coherence to be included.

Current bottlenecks in simulations and experiments will be discussed and prospects for future development will be made [6].

1. Potapov P., Lichte H., Verbeeck J., Van Dyck D., Ultramicroscopy **106** (2006) p.1012-1018
2. Verbeeck J., van Dyck D., Lichte H., Potapov P., Schattschneider P., Ultramicroscopy **102** (2005) p.239-255
3. Schattschneider P., Verbeeck J., Ultramicroscopy (2008) doi:10.1016/j.ultramic.2007.05.011
4. Verbeeck J., Bertoni G., Schattschneider P., Ultramicroscopy **108-3** (2008) p. 263-269

M. Luysberg, K. Tillmann, T. Weirich (Eds.): EMC 2008, Vol. 1: Instrumentation and Methods, pp. 243–244, DOI: 10.1007/978-3-540-85156-1_122, © Springer-Verlag Berlin Heidelberg 2008

5.   Schattschneider P., Rubino S., Hébert C., Rusz J., Kunes J., Novák P., Carlino E., Fabrizioli M., Panaccione G., Rossi G., Nature **441** (2006) p. 486- 488
6.   J.V and G.B. want to thank the FWO-Vlaanderen for financial support under contract nr. G.0147.06 . J.V., H.L. and D.V.D are grateful for the financial support from the European Union under the Framework 6 program under a contract for an Integrated Infrastructure Initiative. Reference 026019 ESTEEM.

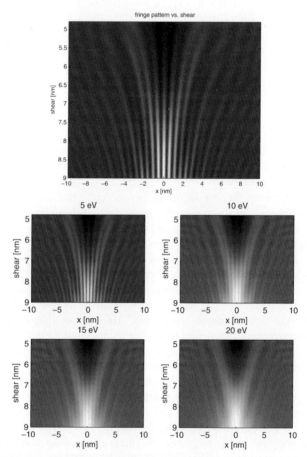

Figure 1. Simulated in-line holography fringe patterns for elastic (top) and inelastic scattering at 300 kV for different shear values (related to the biprism potential). The virtual biprism thickness in the image plane is taken to be 6 nm with a defocus of 5.1 μm. Note that fringes are still present for inelastic scattering although we assumed incoherent scattering centers. This can be understood due to delocalised interaction which leads to a reduction of the contrast with energy loss.

FIB prepared and Tripod polished prepared *p-n* junction specimens examined by off-axis electron holography

C. Ailliot, J.P. Barnes, F. Bertin, D. Cooper, J.M. Hartmann, P. Rivallin.

CEA-LETI, Minatec, 17 rue des Martyrs, 38054 Grenoble, Cedex 9, France.

cyril.ailliot@cea.fr
Keywords: off-axis electron holography, FIB, semiconductors, tripod polishing.

Off axis electron holography is a TEM-based technique which allows the amplitude and phase information that are carried in electrons that have passed through a specimen to be recovered. This is achieved by using an electron biprism that creates two virtual sources in the back focal plane which act to form an interference pattern, or hologram in the image plane. In the absence of magnetic fields, the phase shift of an electron as it passes through a specimen in the direction z is given by

$$\textbf{(1)} \qquad \varphi(x,y) = C_E \int_{sample} V(x,y,z)d(z),$$

where C_E is a constant depending on the energy of the electron beam, φ is the change in phase, and V is the potential of the specimen along the beam path [1].

If the potential of the specimen is invariant in the direction of the electron beam, then 2D phase maps can be directly calculated from the reconstructed phase images with nm-scale resolution [2].

One of the most critical issues is specimen preparation; an ideal TEM specimen should be representative of the bulk material being examined. Specimens can be prepared for electron holography using either Tripod polishing [2] or focused ion beam milling (FIB) [3].

FIB milling uses Ga ions typically with an energy of 30 kV to mill a thin parallel-sided membrane. FIB milling is easy, quick with a high success rate, however FIB milling introduces many artefacts into the specimens in the form of amorphous layers and defects introduced deep in the crystalline regions of the specimens [3].

Tripod polishing is a mechanical preparation technique. The sample is polished as a wedge until electron transparency is achieved. This technique is time consuming and the selectivity of a device is challenging, however, specimens prepared using this technique are not damaged to the extent that is observed in FIB-milling making the interpretation of the reconstructed phase images easier.

We observed a symmetrically 3×10^{18} cm^{-3} doped silicon *p-n* junction grown using reduced-chemical vapour deposition with boron and phosphorous used for the *p*-type and *n*-type dopants respectively. The SIMS profile of the junction is shown in Figure 1(a). The theoretical built in potential in this junction is 0.90 V.

A reconstructed phase image of a FIB prepared sample from this wafer is shown in Figure 1(b) clearly revealing the presence of the differently doped regions. The phase shift across the junction can be used to calculate the potential in the junction using formula **(1)**. The crystalline specimen thickness of a series of samples has been

M. Luysberg, K. Tillmann, T. Weirich (Eds.): EMC 2008, Vol. 1: Instrumentation and Methods, pp. 245–246, DOI: 10.1007/978-3-540-85156-1_123, © Springer-Verlag Berlin Heidelberg 2008

measured using convergent beam electron diffraction (CBED) [4]. Figure 1(c) shows the phase shift plotted as a function of crystalline specimen thickness for the series of junctions. By extrapolating the gradient to the x-axis we reveal an electrically 'inactive' thickness of 140 nm which does not contribute towards the phase measured across the junctions.

From the gradient in Figure 1(c), the built-in potential can be measured using an approach which is in theory independent of the electrically inactive thickness, however, using this method the measured value of the potential is 0.63 +/- 0.05 V which is much less than the 0.90 V expected from theory.

In this paper we will discuss the results obtained from p-n junctions prepared using a range of different specimen preparation techniques. We will highlight the influence of artefacts such as gallium implantation and defects in specimens prepared using FIB milling when examined using electron holography. We will also discuss the effects of electron irradiation on the phases measured by electron holography.

1. P.A. Midgley, Micron. **18**, 167184 (2001)
2. M.A. Gribelyuk, A.G. Domenicucci, P.A. Ronsheim, J.S. McMurray and O. Gluschenkov, J. Vac. Sci. Tech. B. **26**, 408414 (2008)
3. R.M. Langford and A.K. Petford-Long, J. Vac. Sci. Tech. A. **5**, 21862193 (2001)
4. D. Cooper, A.C. Twitchett-Harrison, P.A. Midgley and R.E. Dunin-Borkowski, J. Appl Phys. **101**, 094508 (2007)

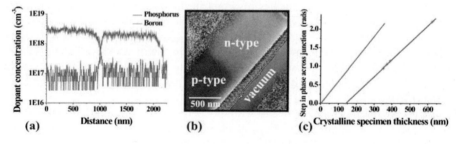

Figure 1. (a) shows SIMS profile of the p-n junction examined here. (b) Shows a reconstructed phase image of a 400-nm-thick FIB-prepared specimen containing a p-n junction clearly showing the presence of the p and n-doped regions. (c) Shows the step in phase measured across a series of junctions as a function of the crystalline specimen thickness measured using CBED. The x-intercept reveals the presence of an electrically 'inactive' thickness that does not contribute towards the phase measured across the junctions. The theoretical data for a sample with bulk properties is also shown.

Modelling kink vortices in high-T_c superconductors

M. Beleggia[1], G. Pozzi[2] and A. Tonomura[3]

1. Institute for Materials Research, University of Leeds, Leeds LS2 9JT, UK
2. Department of Physics, University of Bologna, V.le B. Pichat 6/2, Bologna, Italy
3. Advanced Reasearch Laboratory, Hitachi Ltd., Hatoyama, Saitama 350-0395, Japan

m.beleggia@leeds.ac.uk
Keywords: superconducting vortices, Lorentz microscopy, Josephson vortices

Transmission Electron Microscopy experiments on high-T_c superconducting (HTS) films where a magnetic field was applied at large angle with respect to the sample perpendicular showed novel and intriguing Fresnel-contrast features [1]. As the angle increases, the vortex contrast stretches along the field; for even larger angles (around 84°) the contrast seems to "split" and display a dumbbell-like appearance. With a model we developed for the interaction between the electron beam and the "pancake" vortices typical of layered HTS [2], we have demonstrated how these novel contrast features reveal a "kinked" structure of the vortex core [3].

Our model, however, did not include Josephson coupling between layers, and was derived within the assumption of infinite anisotropy (e.g. the penetration depth along the c-axis is infinitely large). For low-anisotropy materials such as YBCO, where the ratio between c-axis and ab-plane penetration depths is around 5, currents along the c-axis created by the interlayer Josephson coupling cannot be neglected. For this reason, we propose here an improvement of our model for layered HTS that takes into account piecewise-continuous deformations of the vortex core in finite-anisotropy HTS. To this purpose, we re-examine our earlier achievement of a continuous-anisotropic model [4] that proved successful in interpreting Fresnel images of vortices interacting with artificial columnar defects created in the sample, revealing a clear contrast difference between pinned and unpinned vortices.

We have extended the model by considering a sample divided into three superconducting (SC) regions and two vacuum regions (Figure 1). In the central layer, the vortex core is tilted around the chosen y-axis; in the top and bottom SC layers, the vortex core is straight and perpendicular to the surfaces; in the vacuum no vortex core exists and the field topography is dictated by the surface distribution of currents.

In order to obtain the magnetic vector potential associated to the kinked vortex we need to solve a system of 5 partial differential equations. The vector potential, once integrated along the electron, yields the electron optical phase shift responsible for the Fresnel-contrast intensity variations we observe in the out-of-focus image.

Results for the phase shift when electrons travel along the y-axis are shown in Figure 2, where we have simulated three tilt angles of the vortex core in the middle layer (kept at constant thickness) of 45, 75, 85 degrees. Projected induction lines reveal that the magnetic field is much smoother than the kinked core.

The model for kinked vortices enables us to carry out flexible simulations where relative thicknesses of the layers, tilt angles, as well as material parameters (anisotropy

M. Luysberg, K. Tillmann, T. Weirich (Eds.): EMC 2008, Vol. 1: Instrumentation and Methods, pp. 247–248, DOI: 10.1007/978-3-540-85156-1_124, © Springer-Verlag Berlin Heidelberg 2008

248

factor, penetration depth) can be varied arbitrarily. We are currently focussing our efforts on verifying that for very large tilt angles the phase shift converges to what we expect from a pure Josephson vortex [5,6].

1. A. Tonomura et. al., Phys. Rev. Lett. **88** (2002), p. 237001.
2. M. Beleggia and G. Pozzi, J. Electron Microsc. **51** (2002), p. S73.
3. M. Beleggia et al., Phys. Rev. B **70** (2004), p. 184518.
4. M. Beleggia et al., Phys. Rev. B **66** (2002), p. 174518.
5. M. Beleggia, Phys. Rev. B **69** (2004), p. 014518.
6. MB would like to acknowledge the Royal Society for financial support.

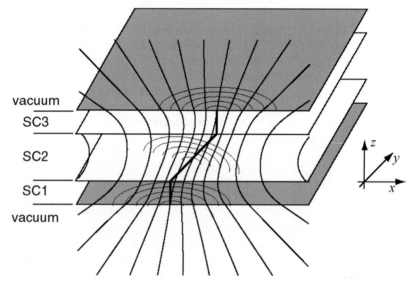

Figure 1. Scheme of the sample subdivided into 5 regions, and of the coordinate system employed. The kinked vortex is represented by the thick black line. Superimposed are representations of the projected induction lines and of the supercurrents.

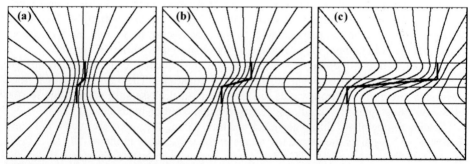

Figure 2. Phase shift when electrons are travelling along the y-axis. Each contour line represents a phase shift of $\pi/16$. The core tilt is 45° in (a), 75° in (b) and 85° in (c). The thickness of the middle SC layer is kept fixed as 1/5 of the total thickness.

Off-axis electron holography of FIB-prepared semiconductor specimens with mV sensitivity.

D. Cooper, R. Truche, P. Rivallin, J. Hartmann, F. Laugier, F. Bertin and A. Chabli

CEA LETI - Minatec, 17 rue des Martyrs, 38054 Grenoble, Cedex 9, France.

david.cooper@cea.fr
Keywords: off-axis electron holography, FIB, semiconductors

Off-axis electron holography is a TEM based technique that uses a biprism to interfere an object wave that has passed through a specimen with a reference wave that has passed through only vacuum. From the interference pattern, or hologram, both phase and amplitude images can be reconstructed. In the absence of magnetic fields, the phase change, $\Delta\Phi$ of an electron as it passes through a sample is given by the expression,

$$\Delta\Phi = C_E \int_0^t V(x,y,z)dz \, ,$$

where C_E is a constant dependent on the energy of the electron wave, V is the electrostatic potential and z is the direction of the electron beam [1]. As the phase of an electron is very sensitive to changes in potential in a specimen, such as from the presence of dopants, electron holography can in principle be used to fulfil the requirements of the semiconductor industry for a technique that can be used to quantitatively map dopants with nm-scale resolution [2].

Due to the continuous miniaturisation of state-of-the-art semiconductor devices it is desirable to improve the signal-to-noise ratio in the phase images as a doped region of interest can be less than 50-nm-thick in the direction of the electron beam, resulting in a small step in phase measured across the electrical junction. The theoretical phase resolution, $\delta\Phi$ in a reconstructed image is shown below, where μ is the contrast of the hologram and N, the number of electron counts [3].

$$\delta\Phi = \frac{\sqrt{2}}{\mu\sqrt{N}}$$

By using a probe corrected FEI Titan TEM with excellent electrical and mechanical stability; we have been able to record electron holograms for time periods of 128 seconds, with contrast levels of almost 40 % and an average signal on the charge-coupled device camera of 30,000 counts. We will show the latest results obtained from Si calibration specimens and from 45-nm-gate device specimens and highlight the improvement in the resolution of the reconstructed phase images.

An example is given in Figure 1. (a) which shows phase images of a Si calibration specimen prepared using 30 kV focused ion beam milling and cleaned using low energy Ar ion milling. The images have been acquired using different voltages applied to the biprism and different length acquisition times. The phase images reconstructed from the holograms formed using the biprism voltages of 100 and 140 V have spatial

M. Luysberg, K. Tillmann, T. Weirich (Eds.): EMC 2008, Vol. 1: Instrumentation and Methods, pp. 249–250, DOI: 10.1007/978-3-540-85156-1_125, © Springer-Verlag Berlin Heidelberg 2008

250

resolutions of 9 and 6 nm respectively. The improvement in the signal-to-noise ratio in the phase images is clear. Figure 1. (b) shows a SIMS profile of the calibration specimen examined, revealing the differently doped layers spaced by 30 and 60 nm. Figures 1 (c) and (d) shows profiles extracted from the phase images reconstructed with 9 and 6 nm spatial resolution. The inset of Figure 1. (d) shows the phase measured across the 30-nm-spaced doped layers extracted from a phase image reconstructed from a hologram acquired for 128 seconds. The specimen is 370-nm-thick, therefore the phase measured across the doped layers corresponds steps in potential of less than 0.030 +/- 0.003 V [3].

In this presentation we will discuss the microscope operating procedure that is used to achieve these results. We will discuss some of the current problems that are encountered when using long acquisition times. Finally we will show ways to improve specimen preparation which can further improve the signal-to-noise ratio in the reconstructed phase images.

1.   W.D. Rau, P. Schwander, F.H. Baumann, W. Hoppner and A. Ourmazd. Phys. Rev. Lett. **82**, 2614 (1999).
2.   International Technology Roadmap for Semiconductors, 2005 ed. http://public.itrs.net
3.   A. Harscher and H. Lichte, Ultramicroscopy, **64**, 57 (1994).
4.   D. Cooper, R. Truche, P. Rivallin, J. Hartmann, F. Laugier, F. Bertin and A. Chabli. Appl. Phys. Lett. **91**, 143501 (2007).

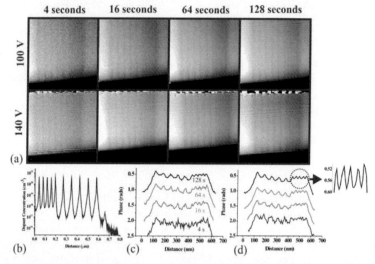

Figure 1. (a) Shows phase images acquired using different biprism voltages and acquisition times. (b) SIMS profile of specimen. (c) Profiles extracted from phase images acquired using 100 V biprism voltage (9 nm spatial resolution). (d) Profiles extracted from phase imaged acquired using 140 V biprism voltage (6 nm spatial resolution).

Off-axis electron holography and the FIB, a systematic study of the artefacts observed in semiconductor specimens.

D. Cooper, C. Ailliot, R. Truche, J. Hartmann, J. Barnes and F. Bertin

CEA LETI - Minatec, 17 rue des Martyrs, 38054 Grenoble, Cedex 9, France.

david.cooper@cea.fr
Keywords: off-axis electron holography, FIB, semiconductors

Focused ion beam (FIB) milling is a promising specimen preparation technique that is used to prepare specimens for examination using off-axis electron holography due to its ease of use and unprecedented site-specificity. However, the FIB introduces damage in the specimens in the form of amorphous surface layers and an electrically inactive thickness on the surfaces that results in the phase measured across the junctions that is much less than expected from theory [1].

In this paper we will show the effects of specimen charging, specimen geometry, FIB operating conditions and dopant concentration on the phases measured in Si calibration specimens and state-of-the-art devices. Figure 1(a). shows the phase measured across a 400-nm-thick specimen containing a p-n junction as a function of the intensity of the electron irradiation. The intensity of the electron irradiation was controlled by changing the C1 aperture (spot size) while maintaining a constant area of illumination on the specimen. A strong dependence on the electron beam intensity can be seen. If the electron beam intensity is kept below 1.4 mA then the phase measured across the junctions remains constant, however, this still results in a calculation of the potential across the junction that is much less than predicted by theory.

Figure 1(b). shows the step in phase measured across a series of FIB-prepared p-n junctions with different dopant concentrations as a function of the crystalline specimen thickness measured using convergent beam electron diffraction (CBED). From the x-intercept, an electrically 'inactive' thickness can be observed which is known to arise from the presence of defects introduced deep in the crystalline during FIB-milling [2]. The graph shows that the electrically 'inactive' thickness is strongly dependent on the dopant concentration in the specimens. Figure 1(c). shows the electrically 'inactive' thickness as a function of the dopant concentration in the p-n junctions. This result suggests that the electrically 'inactive' thickness is directly proportional to the dopant concentration and indicates that care must be taken when directly calculating the potential across p-n junctions from phase images. This result also suggests that thicker TEM specimens can be used to extend the detection limit for low dopant concentrations. For example by extrapolating the gradient plotted in Figure 1(c). it can be seen that to detect a p-n junction with a symmetrical dopant concentration of 1×10^{16} cm^{-3} a specimen thickness of over 500 nm will be required. However, using the latest generation microscopes, exceptionally long acquisition time periods can be used, permitting very thick TEM specimens to be examined [3].

M. Luysberg, K. Tillmann, T. Weirich (Eds.): EMC 2008, Vol. 1: Instrumentation and Methods, pp. 251–252, DOI: 10.1007/978-3-540-85156-1_126, © Springer-Verlag Berlin Heidelberg 2008

Even after accounting for the presence of the electrically 'inactive' thickness, the correct potential across the junction is not recovered from the phase. However, we will demonstrate that by attaching good electrical contacts to the specimens, the build-up of charge can be removed from the regions of interest and the theoretical built-in potential recovered.

Finally, by carefully controlling a range of parameters when performing off-axis electron holography on FIB-prepared specimens, we will show that a good reproducibility of results can be obtained.

1. A.C. Twitchett, R.E. Dunin-Borkowski, R.J. Hallifax, R.F. Broom and P.A. Midgley, Phys. Rev. Lett. **88**, 2383021 (2002).
2. D. Cooper, A.C. Twitchett, P.K Somodi, P.A. Midgley, R.E. Dunin-Borkowski, I. Farrer and D.A. Richie, Appl. Phys. Lett. **88**, 063510 (2006).
3. D. Cooper, R. Truche, P. Rivallin, J. Hartmann, F. Laugier, F. Bertin and A. Chabli, Appl. Phys. Lett. **91**, 143501 (2007).

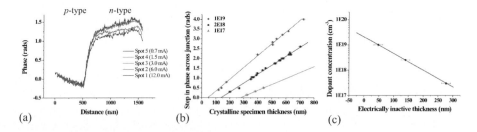

(a)　　　　　　　　　　　　(b)　　　　　　　　　　　　(c)

Figure 1. (a) Shows the unflattened phase measured across a 400-nm-thick specimen extracted from phase images reconstructed from holograms formed using different electron beam intensities. The current of the electron beam through the sample is displayed. (b). The step in phase measured across the symmetrical *p-n* junctions with different dopant concentrations as a function of crystalline specimen thickness measured using CBED. The electrically inactive layer can be determined from the x-intercept. (d) The electrically 'inactive' thickness as a function of the dopant concentration in the *p-n* junction specimens.

Analytical TEM and electron holography of magnetic field distribution in nanocrystalline Co layers deposited on Cu

B. Dubiel[1], D. Wolf[2], E. Stepniowska[1] and A. Czyrska-Filemonowicz[1]

1. Faculty of Metals Engineering and Industrial Computer Science, AGH University of Science and Technology, Al. A. Mickiewicza 30, PL-30059 Krakow, Poland
2. Triebenberg Laboratory for Electron Microscopy and Electron Holography, Technical University Dresden, Zum Triebenberg 50, D-01328 Dresden, Germany

bdubiel@agh.edu.pl

Keywords: nanostructure, magnetic domains, phase shift, electron holography

Magnetic field distribution in ferromagnetic nanolayers deposited on nonmagnetic substrate is of great interest due to their application in magnetic recording devices exhibiting giant magneto-resistance (GMR) effect [1].

In the present work, cobalt layers were deposited on copper substrate by molecular beam evaporation technique. The average thickness of the layers was 400 nm. Analytical TEM investigations were performed using JEM-2010 microscope using cross-section specimens prepared by ion beam milling.. The medium resolution off-axis electron holography was carried out using CM200FEG ST microscope with Lorentz lens and Möllenstedt biprism.

The TEM bright-field images of cobalt layer showed its nanocrystalline structure (Fig. 1). The size of the nano-grains was approximately 20 nm. To investigate the magnetic field distribution in nanocrystalline cobalt, the electron holography technique was used. In medium resolution holography of magnetic materials the lateral shift of the hologram fringes is caused by both the mean inner electrostatic potential of the analyzed material and by the magnetic field of the sample [2]. Then, the numerical reconstruction of the phase contains superposition of both the magnetic and electric information. The separation of the phase into its magnetic and electric contribution can be performed by turning the sample in the microscope, than the magnetic part inverts, but the electric part does not [3].

Because only the in-plane component of the magnetic field can be recovered, the specimen with nanocrystalline cobalt layer was tilted to a high tilt angle and 2 Tesla field was applied by the objective lens in the microscope. Thus, the specimen was magnetized in in-plane direction. The pairs of electron holograms were acquired at the same position with face-up and face-down of the specimen (Fig. 2).

The reconstruction of phases from the holograms was performed using Digital Micrograph software. After correction of residual displacements between the two phase images representing the same specimen position, the magnetic phase shift was gained by simple subtraction of each other. Finally, the projected magnetic field was calculated from the phase gradient.

Fig. 3 represents the projected in-plane component of the magnetic field with corresponding line profile, where the cobalt layer can easily be identified above the glue

M. Luysberg, K. Tillmann, T. Weirich (Eds.): EMC 2008, Vol. 1: Instrumentation and Methods, pp. 253–254, DOI: 10.1007/978-3-540-85156-1_127, © Springer-Verlag Berlin Heidelberg 2008

254

remaining from cross-section specimen preparation. Using a mean inner potential value of 30 Volt in order to determine the thickness of the specimen, it was calculated that averaged magnetic field in cobalt was about 0.6 T.

1.   B. Dubiel, H. Figiel, L. Gondek, F. Ciura, J. Chmist, A. Czyrska-Filemonowicz, Journal of Magnetism and Magnetic Materials, in press
2.   H. Lichte, P. Formanek, A. Lenk, M. Linck, Ch. Matzeck, M. Lehmann, P. Simon, Annu. Rev. Mater. Res. **37** (2007), p. 539
3.   H. Lichte, M. Lehmann, Rep. Prog. Phys. **71** (2008), p. 016102
4.   The authors acknowledge financial support from the European Union under the Framework 6 Program, contract for an Integrated Infrastructure Initiative. Reference 026019 ESTEEM. We also would like to thank T. Slezak (AGH-UST) for deposition of Co layers.

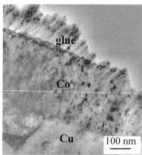
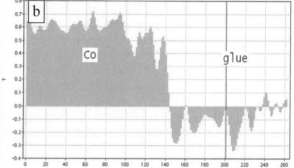

Figure 1. TEM bright-field image of Co layer

Figure 2. Pair of holograms taken at the same position with face-up (a) and face-down (b) of the specimen

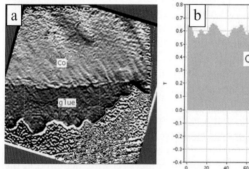

Figure 3. Reconstructed horizontal magnetic field component in cobalt (a) and corresponding line profile (b)

Electron Holography with Cs-corrected Tecnai F20 – elimination of the incoherent damping introduced by the biprism in conventional electron microscopes

D. Geiger[1], A. Rother[1], M. Linck[1], H. Lichte[1], M. Lehmann[2], M. Haider[3] and B. Freitag[4]

1. Triebenberg Laboratory, Institute of Structure Physics, Technische Universität Dresden, Germany
2. Institute for Optics and Atomic Physics, Technische Universität Berlin, Germany
3. CEOS GmbH, D-69126 Heidelberg, Germany
4. FEI Company, 5600KA Eindhoven, The Netherlands

Dorin.Geiger@Triebenberg.de

Keywords: Electron holography, Cs-corrected TEM, Signal damping

The exceptional progress achieved in the transmission electron microscopy (TEM) using Cs-corrector [1] allows now a true atomic resolution of 0.1 nm or better. Nevertheless, conventional TEM still suffers from the significant drawback that only a poor phase contrast can be obtained. The weaker phase structures like those produced by the meso- up to macroscopic electromagnetic fields are essential to understand solid-state properties but remain in conventional TEM virtually invisible.

Off-axis electron holography allows analysing also the electric structures of particular importance like in semiconductors and ferroelectrics fully quantitatively from the phase of the reconstructed object exit wave. Nevertheless, certain limitations due to the use of conventional, uncorrected TEM instruments still exert. Using Cs-corrected TEM, the performance of the off-axis electron holography has been significantly improved [2]. Main advantages are reduction of the quantum noise, enhancement of hologram contrast, and of the quality of residual aberration correction etc.

Besides the improvements, we already reported [2], an additional, previously not recognized incoherent aberration effect of the biprism has been analysed.

The damping of the fringe contrast in an empty off-axis electron hologram is produced by two factors: Firstly by the camera MTF and secondly by the incoherent superposition of plane-waves with incident angles distributed corresponding to the illumination system [3]. It amounts typically to values of 10% to 30%. The second factor is due to the slight distance of the biprism to the back image plane producing a defocus effect different for each incident angle. If an object is inserted into the beam, diffracted beams occur, which are modulated due to the well known phase plate describing the coherent aberrations of the objective lens. The incoherent spatial aberrations of the objective lens, i.e. the contrast damping of diffracted beams, are produced by an incoherent superposition of plane-waves with distributed incident angles in the same fashion like the fringe damping produced by the biprism. Consequently, in an off-axis holographic setup both incoherent contributions are not statistically independent, i.e. must not be described by a multiplication of two separate damping functions. A simultaneous incoherent averaging incorporating both the defocus effect

M. Luysberg, K. Tillmann, T. Weirich (Eds.): EMC 2008, Vol. 1: Instrumentation and Methods, pp. 255–256, DOI: 10.1007/978-3-540-85156-1_128, © Springer-Verlag Berlin Heidelberg 2008

due to the position of the biprism and the aberrations of the microscope yields a modification of the spatial envelope

$$K_S(q_x,q_y) = e^{-\frac{k^2}{2}\left([\ \nabla_{q_x}\chi(q_x,q_y)+\sqrt{2}\pi b\alpha/M\ ]^2\cdot\theta_x^2 + [\ \nabla_{q_y}\chi(q_x,q_y)\ ]^2\cdot\theta_y^2\right)}$$

(1)

in the direction perpendicular to the biprism. Here b is the biprism distance to the back image plane, M the objective lens magnification, α the superposition, k the electron wavenumber and θ the semiconvergence angle. The transmission cross-coefficient, TCC, valid for describing conventional TEM (centre band in diffractogram of the hologram) modulation remains unaffected. Using typical parameters valid when recoding holograms at CM 30 Special Tübingen (accelerating voltage $U = 300kV$, illumination ellipticity $\varepsilon = 20$, objective lens magnification $M = 50$, defocus $Dz = -80nm$, spherical coefficient Cs = 0.623mm, energy width of the electron beam 0.7eV, $\theta(\theta_x,\theta_y) = 0.5mrad$, biprism voltage $U_b = 600V$) a non-trivial modification of the total damping envelope occurs within the range of the recorded reflections, producing an asymmetry in that direction in the diffractogram. This previously not recognized effect is eliminated by Cs-correction through minimizing χ in (1).

1. M. Haider, H. Rose, St. Uhlemann, E. Schwan, B. Kabius, K. Urban, A spherical-aberration-corrected 200 kV transmission electron microscope, Ultramicroscopy **75** (1998), p. 53-60.
2. D. Geiger, H. Lichte, M. Linck and M. Lehmann: Electron Holography with Cs-corrected Transmission Electron Microscope, Microscopy and Microanalysis **14** (2008), p. 68–81.
3. H. Lichte, Parameters for high-resolution electron holography, Ultramicroscopy **51** (1993), p. 15-20.
4. The authors acknowledge financial support from Deutsche Forschungsgemeinschaft and from the European Union under the Framework 6 program under a contract for an Integrated Infrastructure Initiative. Reference 026019 ESTEEM.

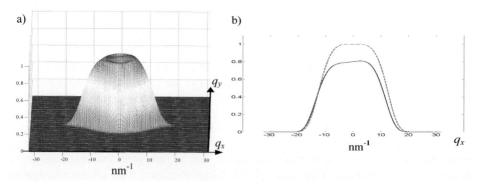

Figure 1. The integral damping envelope (a) occurs within the range of the recorded reflections (sideband centred at 0, center band on the right hand side outside the drawings) producing an asymmetry in the diffractogram perpendicular to the biprism, which overlaps with the asymmetry caused by the MTF. b) shows the projected profile of the envelope for a TEM like CM30 Special Tübingen (continuous line dark) and for a similar Cs-corrected TEM (dotted line).

Can the discontinuity in the polarity of the oxide layers at the interface SrTiO$_3$-LaAlO$_3$ be resolved by using Electron Holography with Cs-corrected TEM?

D. Geiger[1], S. Thiel[2], J. Mannhart[2] and H. Lichte[1]

1. Triebenberg Laboratory, Institute of Structure Physics,
Technische Universität Dresden, Germany
2. Experimentalphysics VI, Center of Electronic Correlations and Magnetism,
Institute of Physics, Universität Augsburg, Germany

Dorin.Geiger@Triebenberg.de

Keywords: Electron holography, oxide heterointerface, low-dimensional charge states

Intrinsic functionalities of complex oxides like ferroelectricity, magnetism, superconductivity or multiferroic behaviour can be combined in electronic devices based on epitaxially grown heterostructures. Searching for new materials with special properties in nanoscience implies also the control of interfaces down to the atomic level.

Interfaces between polar and nonpolar insulating oxide structures like those between the perovskites LaAlO$_3$ and SrTiO$_3$ show different behaviour in e.g. charge density and/or mobility depending among others on the sequence of the oxide layers in the interface. The charge distribution in the interface [1] depends also on the thickness of the upper oxide layer LaAlO$_3$. For thickness d_c up to 3 unit cells (uc), highly insulating interfaces are obtained. A tunable quasi-two dimensional electron gas (q2-DEG) can be generated at the interface of the oxide heterostructures. Similarly to the field effect transistor, by application of a gate voltage, a quantum phase transition from an insulating to a metallic state can be generated. Changes of the carrier concentration without modifying the level of disorder, as occurs when chemical composition is altered, can be controlled in a reversible way.

LaAlO$_3$ layers are grown epitaxially by pulsed laser deposition on TiO$_2$ terminated (001) SrTiO$_3$ surfaces. Reflection high-energy electron diffraction (RHEED) is applied to locally control the layer thickness d on a unit cell level. Additional scanning force microscopy (AFM) measurements show film thickness variations across the substrate of less than 5%.

Conventional transmission and scanning transmission electron microscopy (TEM and STEM), especially if equipped with Cs-correctors, have proven to belong to the most powerful investigation methods on atomic level. Unfortunately the phase of the image wave in coventional TEM is lost during image detection with the result that parts of the object phase information, which is the primary subject of quantitative interpretation in TEM, are definitely lost. The remaining phase contrast is generally very poor with the consequence that the phase structures like those produced by electric and magnetic fields, although decisive for solid-state properties, are virtually invisible.

M. Luysberg, K. Tillmann, T. Weirich (Eds.): EMC 2008, Vol. 1: Instrumentation and Methods, pp. 257–258, DOI: 10.1007/978-3-540-85156-1_129, © Springer-Verlag Berlin Heidelberg 2008

Using off-axis electron holography, it is possible to reconstruct the image wave and, after correction of the residual aberrations, to recover the object exit wave too. The object phase can be used for quantitative analysis of the investigated structure.

Is it possible to resolve the charge distribution in the oxide interface using the object phase reconstructed by means of off-axis electron holography [2]? Is the phase signal large enough to resolve the potential variations in the interface?

Using the recently achieved Cs-corrected TEM, the performance of off-axis electron holography has been significantly improved: quantum noise, information limit and spatial damping envelope, hologram contrast and accuracy of the residual aberration correction are the main improved aspects [2]. Especially the improvement of the phase detection limit opens new possibilities in crystal structure analysis of systems like $LaAlO_3/SrTiO_3$.

For the first step, two specimen were prepared in cross section procedure, a first one with 3 uc $LaAlO_3$-layer thickness, which is insulating, and a second one with 5 uc $LaAlO_3$-layer thickness, which has been shown to have a relatively highly conductive interface. A comparison of the phase signal at the interface between $SrTiO_3$ and $LaAlO_3$ for the two specimen could give more detailed quantitative information about the potential distribution in this heterointerface.

1. S. Thiel, G. Hammerl, A. Schmehl, C. W. Schneider, J. Mannharti, Tunable Quasi–Two-Dimensional Electron Gases in Oxide Heterostructures, Science **313** (2006) p. 1942 - 1945.
2. H. Lichte, Performance limits of electron holography, Ultramicroscopy **108** (2008) p. 256 – 262.
3. D. Geiger, H. Lichte, M. Linck and M. Lehmann: Electron Holography with Cs-corrected Transmission Electron Microscope, Microscopy and Microanalysis **14** (2008), p. 68 – 81.

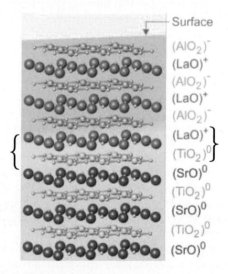

Figure 1. Show the layer structure of the epitaxial perovskite-heterosystem $LaAlO_3$ on $SrTiO_3$. In the interface layer of both oxide structures (marked), charge carriers with higher density can be generated.

Energy-filtered DBI/H

R.A. Herring

CAMTEC, Mechanical Engineering, University of Victoria, Canada V8W 3P6

rherring@uvic.ca

Keywords: electron holography, energy loss electrons, coherence

Energy-filtered Diffracted Beam Interferometry/Holography (DBI/H) [1, 2] is a method of interference that can measure the coherence properties (degree of coherence (γ) and lateral coherence) of elastically and inelastically scattered electrons, which requires a TEM with an electron source with good coherence, an electron biprism for holography and imaging energy filter (GIF) for separating the zero loss electrons and phonon loss electrons from the plasmon loss electrons, ionization edge loss electrons, etc. A condenser aperture focused on the diffraction plane is used to separate out the zero loss electrons from the phonon loss electrons, which self-interfere at the high scattering angles > 14 mrad depending on the material [3, 4]. At larger energy losses (10s eV), the bulk plasmons and surface plasmons can be self-interfered in a similar manner except a small convergent probe is used for the bulk plasmons, whereas, a variable-diameter planar beam is used for the surface plasmons [4]. At still higher energy losses (100s eV), the coherence properties of the ionization edge loss electrons can be measured as accomplished for Si, Ge and GaAs (e.g., Fig. 1) giving confidence that it will be possible for the magnetic materials of Co, Ni and Fe, which would enable a new method of magnetic microscopy to measure the magnetic potential of domains at the nanoscale, as well as, the coherence properties of the magnon (Fig. 2).

An important consideration in these measurements is the dynamic diffraction of the inelastically scattered electrons with themselves and the source electrons, which is accounted for in the simulations of DBI/H interferograms using the mixed dynamic form factor [5]. To help reduce dynamic diffraction, thin specimens are used having a thickness of ¼ extinction distance of the diffraction vector (~10 – 20 nm), which also provides for the highest contrast fringes in the DBI/H interferograms, simplifies the measurement of γ and provides for $\pi/2$ phase shift between the two beams. The energy loss electrons close to the zero loss electrons such as the phonon loss electrons and interband transition states loss electrons are most affected and thus require being measured at high scattering angles from the center of the DBI/H interferogram (e.g., > 14-16 mrad for Si [6]), whereas, those electrons having large energy losses such as the ionization loss electrons are not affected and can thus be measured in the central, high intensity part of the DBI/H interferogram at low scattering angles.

1. R.A. Herring, Ultramicroscopy 104 (2005) 261.
2. R.A. Herring, Ultramicroscopy (2007) doi:10.1016/j.ultramic.2007.11.001.
3. R.A. Herring, J. Appl. Phys. Submitted.
4. R.A. Herring, these proceedings.
5. P. Schattschneider and J. Verbeeck, Ultramicroscopy (2007)doi:10.1016/j.ultramic.2007.05.01.

M. Luysberg, K. Tillmann, T. Weirich (Eds.): EMC 2008, Vol. 1: Instrumentation and Methods, pp. 259–260, DOI: 10.1007/978-3-540-85156-1_130, © Springer-Verlag Berlin Heidelberg 2008

6. Z.L. Wang, Micron 34 (2003) 141.
7. Grants from UVic, NSERC and the Canadian Foundation for Innovation are gratefully appreciated.

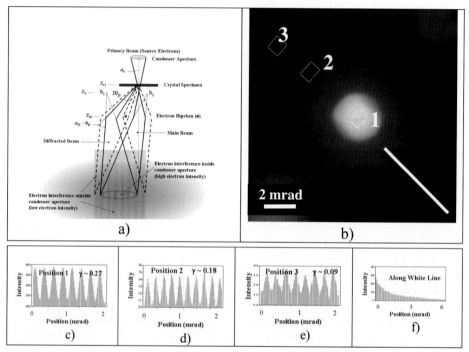

Figure 1. a) Schematic of energy-filtered DBI/H interferogram consisting of ionization loss electrons (30 eV) and 5 eV window using the main beam (000) and the 111 diffracted beam of Ge in b), having a fringe intensity profile in c) of $\gamma \sim 0.27$ at position 1, in d) $\gamma \sim 0.18$, in e) $\gamma \sim 0.09$ and in f) the intensity distribution along the white line, which falls off exponentially out to high scattering angles.

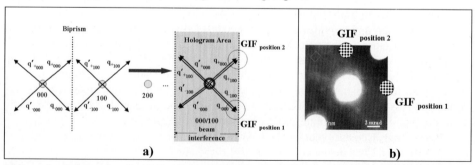

Figure 2. Schematic of energy-filtered DBI/H interferogram of ionization edge loss electrons showing the magnetic potentials, which can be determined from the intensities at position 1 and position 2, where the coherence properties of the magnon in the high intensity central region.

Strain determination by dark-field electron holography

F. Houdellier, M.J. Hÿtch, F. Hüe and E. Snoeck

CEMES-CNRS, 29 rue Jeanne Marvig, 31055 Toulouse, France

florent@cemes.fr

Keywords: electron holography, strain

Accurate determination of strain in electronic devices has been the subject of intense work during the last decades. Few techniques are able to provide highly localized and accurate information at the nanoscale. Among these, convergent-beam electron diffraction (CBED) combines the advantages of very small probes and remarkable sensitivity to small variations in the lattice parameter [1]. However, elastic relaxation effects make the analysis extremely difficult, necessitating time-consuming dynamical simulations combined with finite element modeling [2].

Rather than collecting data at isolated points, strain distributions can be mapped in a continuous fashion using high-resolution transmission electron microscopy (HRTEM) [3]. Unfortunately, HRTEM suffers from limitations due to specimen preparation, field of view and noise. These problems are minimised using a Cs-corrected microscope [4]. Whilst this technique is highly accurate at the nanometre scale, mapping strain in real devices like multilayers and transistors requires fields of view that are not easily accessible to HRTEM, even with a Cs-corrected microscope.

We have therefore sought to develop a new technique for measuring strain at lower magnification, for thicker samples and larger fields of view without sacrificing precision. The new method is a combination of the moiré technique and off-axis electron holography, called dark-field holography [5]. The experiments are carried out on the SACTEM-Toulouse, a spherical aberration corrected microscope fitted with a field emission gun (Schottky FEG) and rotatable biprism. The pseudo-Lorentz mode [6], corresponding to a special setting of the Cs-corrector, is essential to reach the desired field of view. Holograms are analysed using a modified version of GPA Phase 2.0 (HREM Research Inc.) for DigitalMicrograph (Gatan).

As an example, the method has been applied to Si(15.5nm)/$Si_{60}Ge_{40}$(13.5nm) multilayers grown on a virtual substrate of $Si_{80}Ge_{20}$. TEM specimens were prepared by a combinaison of tripod polishing and PIPS, though FIB preparation is preferable. The results using the (004) dark-field configuration are reported on Figure 1. In this case, the holographic fringes allow a spatial resolution of 2 nm for a precision of 0.07% strain over a half of a micron field of view.

1. J.C.H. Spence and J.M. Zuo in "Electron Microdiffraction" (Plenum Press, New York) (1992)
2. F. Houdellier, C. Roucau, L. Clément, J.-L. Rouvière and M.-J. Casanove, Ultramicroscopy **106** (2006), p. 951.
3. M.J. Hÿtch, J.-L. Putaux and J.-M. Pénisson, Nature **423** (2003), p. 270.

M. Luysberg, K. Tillmann, T. Weirich (Eds.): EMC 2008, Vol. 1: Instrumentation and Methods, pp. 261–262, DOI: 10.1007/978-3-540-85156-1_131, © Springer-Verlag Berlin Heidelberg 2008

262

4. F. Hüe, M.J. Hÿtch, H. Bender, F. Houdellier and A. Claverie, Phys Rev Lett (2008) accepted.
5. M.J. Hÿtch, F. Houdellier, F. Hüe and E. Snoeck. Patent Pending FR N° 07 06711.
6. E. Snoeck, P. Hartel, H. Müller, M. Haider and P.C. Tiemeijer, Proc. IMC16 International Microscopy Congress **2** (IMC, Sapporo) (2006), p. 730.
7. F. Hüe is grateful to the CEA-Leti for financial support. The authors thank the European Union for support under the IP3 project ESTEEM (Enabling Science and Technology through European Electron Microscopy, IP3: 0260019) and V. Destefanis and J.-M. Hartmann for the device material.

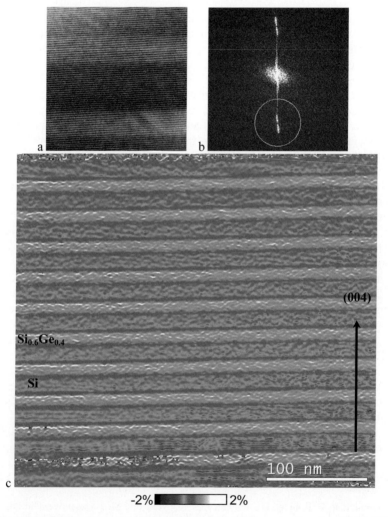

Figure 1. Dark-field holography of a $Si_{60}Ge_{40}/Si$ multilayer using the (004) diffracted beam: (a) holographic detail showing two layers; (b) FFT of hologram; (c) deformation map relative to the substrate of $Si_{80}Ge_{20}$.

Nonlinear Electron Inline Holography

C.T. Koch[1], B. Rahmati[1], P.A. van Aken[1]

1. Stuttgart Center for Electron Microscopy, Max Planck Institute for Metals Research, Heisenberstr. 3, 70569 Stuttgart, Germany

koch@mf.mpg.de

Keywords: holography, phase retrieval, potential mapping, focal series reconstruction

The phase of an electron wave passing through a specimen can only be measured by interfering it with an external reference wave (off-axis holography) or itself (inline holography). Some of the advantages of inline holography over off-axis holography are: (a) very simple experimental setup (works in any TEM), (b) possibility to record holograms far away from the specimen edge, (c) large fields of view because there is no need to oversample, and (d) specimen drift may easily be compensated, even during exposure. The disadvantage: complicated non-linear equations have to be solved, while off-axis holography is linear. Most focal series reconstruction algorithms available in the literature apply the quasi-coherent approximation to the image formation equation which reduces flux with defocus. This is unphysical and works only for small defoci. We will report on first experimental results obtained with a flux-preserving inline holography reconstruction algorithm [1] which allows the reconstruction of focal series recorded over a large focal range.

The diagram in Fig. 1 illustrates that (given a certain size of the objective aperture) the distance between points in the electron wave function whose relative phase difference may contribute to the image intensity is proportional to the defocus under which the image has been recorded (within the limits imposed by the spatial coherence of the illuminating electron beam). This means that very large defoci are necessary for recovering very low spatial frequency components of the complex exit face electron wave function. In a real microscope, images with such large differences in defocus differ also in magnification, image rotation, and possibly other distortions and aberrations.

We will report on the recent development of a software tool for the automated reconstruction of maps of the relative electron wave phase shift from general experimental focal series of TEM images, taking changes in image rotation and magnification into account. This method has been applied to the determination of potential profiles across interfaces in various ceramic materials. An example application is shown in Fig. 2, where the phase shift of the transmitted electron wave across an amorphous pocket in a SrTiO3 ceramic is shown.

Financial support from the European Commission under contract nr. NMP3-CT-2005-013862 (INCEMS) and the Integrated Infrastructure Initiative. Reference 026019 (ESTEEM) is acknowledged.

1. C.T. Koch, Ultramicroscopy **108** (2008) 141

M. Luysberg, K. Tillmann, T. Weirich (Eds.): EMC 2008, Vol. 1: Instrumentation and Methods, pp. 263–264, DOI: 10.1007/978-3-540-85156-1_132, © Springer-Verlag Berlin Heidelberg 2008

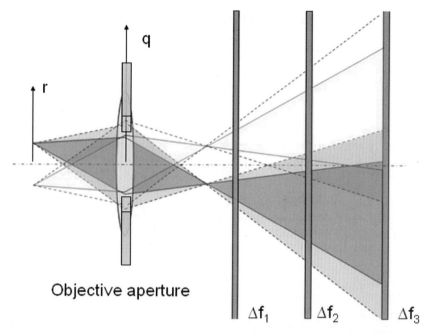

Figure 1. Diagram illustrating the need of a large defocus for reconstructing long-range phase relationships in the electron wave imaged by the objective lens of a TEM.

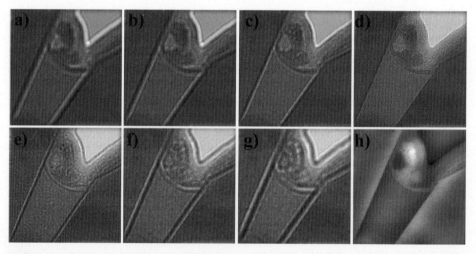

Figure 2. Focal series with images at Δf = -13.9 μm, -9.5 μm, -5.2 μm, -0.9 μm, 3.4 μm, 7.6 μm, and 11.6 μm. The image at the bottom right shows the phase reconstructed from this series. The size of each image is 135 x 135nm^2.

Electron Holography:
Performance and performance limits

Hannes Lichte

Triebenberg Laboratory, Institute of Structure Physics
Technische Universitaet Dresden, Germany

Hannes.Lichte@Triebenberg.de
Keywords: Electron holography, Resolution, Signal/Noise, Figures of merit

The most successful holographic method so far is off-axis electron holography, which uses the Moellenstedt electron biprism [1] as a beam splitter. It superposes an off-axis reference wave to the image wave hence gives rise to a cosinoidal interference pattern ("hologram"), which contains both amplitude and phase in the contrast and position of the interference fringes, respectively. By means of a reconstruction procedure, the distribution of amplitude and phase are determined quantitatively, i.e. the image wave is revived completely in the computer. Details and further references can be found in review [2].

At medium resolution with details larger than about 10 times Scherzer resolution, aberrations of the lens optics can be neglected, because the image wave agrees with the object wave. Thus, the reconstructed phase distribution can directly be interpreted in terms of the object:

- Mean Inner Potentials in solids
- Soft Matter: Phase contrast in focus without staining
- Functional potentials: pn-junctions in semiconductors
- Electric Fields, e.g. controlling growth in biominerals
- Ferroelectric polarization
- Electric potential distribution in charge-modulated structures
- Magnetic fields in and around magnetic structures down to a nanoscale

Examples of such applications are described in detail in [3].

At atomic resolution, amplitude and phase of the image wave cannot be interpreted directly. First, residual aberrations have to be corrected by appropriate processing. By this, both in amplitude and phase, a lateral resolution can be achieved as high as determined by the information limit (close to 0.1nm) offered by the TEM used for recording the hologram. Then, phase images allow revealing details, such as

- difference of atomic numbers of different constituents
- number of atoms in an atomic column
- interatomic electric potentials
- potentials across interfaces

Examples are given in [2].

The performance of electron holography mainly depends on aberrations and brightness of the TEM used for recording the hologram. The optimal reachable lateral resolution is given by the information limit of the microscope; we have reached about

M. Luysberg, K. Tillmann, T. Weirich (Eds.): EMC 2008, Vol. 1: Instrumentation and Methods, pp. 265–266, DOI: 10.1007/978-3-540-85156-1_133, © Springer-Verlag Berlin Heidelberg 2008

0.1nm so far using a Philips CM30 TEM. Besides lateral resolution, however, the signal resolution, i.e. the discernibility of details of the phase image, is often more essential to see the very weak object details. The two aspects, i.e. lateral and signal resolution, can be combined in one figure of merit, which we call the Information Content $InfoCont = n_\varphi n_{rec}$ [4]. It is the product of the number n_φ of phase values distinguishable in the range $(0, 2\pi)$, and the number n_{rec} of pixels reconstructed across the field of view, as shown in Fig.1. It turns out that *InfoCont* depends on the brightness of the electron emitter, stability of the TEM, and the quality of the electron detector used. In our case, $InfoCont = 7100$ is achievable. This means for example that in order to see details of $2\pi/50$ such as fields of single atoms, corresponding to $n_\varphi = 50$, the number of reconstructed pixels is limited to $n_{rec} = 140$. Reachable lateral resolution and field of view are related accordingly: aiming at a lateral resolution of 0.1nm, the reachable field of view is limited to 7nm, because one resolution element is made up by two pixels (Nyquist sampling theorem).

1. G. Möllenstedt and H. Düker, Z. Physik, 145 (1956) 377 - 397
2. H. Lichte and M. Lehmann, Rep. Prog. Phys. 71 (2008) 016102.
3. H. Lichte, P. Formánek, A. Lenk, M. Linck, Ch. Matzeck, M. Lehmann and P. Simon, Annual Review of Materials Research, Vol. 37 (2007) 539-588
4. H. Lichte, Ultramicroscopy 108 (2008) 256
5. Financial support obtained from DFG and EU (Framework 6, I3, Reference 026019 ESTEEM) is gratefully acknowledged.

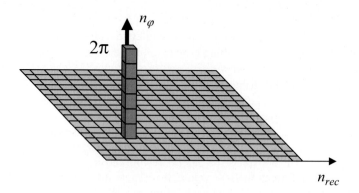

Figure 1. Definition of number of reconstructed pixels n_{rec} and number of discernible phase values n_φ. The squared hologram width gives the pixelated field of view.

Reconstruction methods
for in-line electron holography of nanoparticles

L. Livadaru[1,2], M. Malac[1,2] and R.A. Wolkow[1,2]

1. National Institute for Nanotechnology, 11421 Saskatchewan Drive, Edmonton, AB,T6G 2M9, Canada
2. Dept of Physics, University of Alberta, Edmonton, AB, T6G 2G7, Canada

Lucian@ualberta.ca

Keywords: Electron Holography, Digital reconstruction, Nanoparticles

Electron holography aims to retrieve the phase information about a sample, in addition to amplitude contrast. In-line holograms can be acquired in a TEM setup equivalent to that of Gabor's point-source holography[1]. For high electron energy the latter scheme is achieved if one uses the STEM mode to create a point-source to coherently illuminate the sample, as shown in inset (a) of Figure 1. However, in a TEM, one can realize several more equivalent holographic setups [2,3]. This can be done for instance if the waves propagate through exactly the same optical setup, but in the reverse direction as shown in Figure 1 inset (b), where multiple plane waves incident on the sample at different angles are shown. The hologram is the intensity of the resultant wave produced by the coherent interference of the reference and the scattered waves [4].

We explore several reconstruction methods of in-line holograms proposed for high or low energy electrons: (i) the Abbe imaging theory applied to electron microscopy[5]; (ii) Helmholtz-Kirchhoff formula, previously used in LEEPS holography[6] as well as in LEED and photoelectron holography[7] ; (iii) a new formula based on the Fresnel-Kirchhoff theory of diffraction; and (iv) iterative methods based on the above, which improve the quality of the reconstruction by reducing the distortion effect due to the twin image. According to (i), the reconstructed wavefront at the object exit plane is

$$\psi_{rec} = FT^{-1}\{T_z^* \cdot FT\{I_{holo}\}\} \tag{1}$$

where, T_z is transfer function or the Fourier transform (FT) of the spread function, t_z at location z. Within the small angle (parabolic) approximation of the a spherical wave, this is given by $T_z(\mathbf{u}) = \exp[-i\pi\lambda z u^2 + i\pi C_s \lambda^3 u^4 / 2]$. The method in (ii) employs

$$\psi_{rec}(\mathbf{r}) = \iint ds_x ds_y I_{holo}(\mathbf{s}) \exp(ik\mathbf{s} \cdot \mathbf{r}/s) , \tag{2}$$

while the reconstruction (iii) is done according to

$$\psi_{rec}(\mathbf{r}) = \iint ds_x ds_y I_{holo}(\mathbf{s}) \exp[ik(s - |\mathbf{r} - \mathbf{s}|)]/s|\mathbf{r} - \mathbf{s}|) , \tag{3}$$

where the integration in both formulas is performed over the recording screen.

The twin-image distortion is analyzed in TEM holograms, and we explore the use of Gerchberg-Saxton recursive reconstruction algorithms for its elimination.

1. D. Gabor, *Nature* **161** (1948) 777-778.
2. J.C.H. Spence and J.M. Cowley in "Introduction to Electron Holography", ed. Volkl E., Allard L.F., and Joy D.C., Kluwer (Academic/Plenum Publishers, New York) (1999), p. 311.

M. Luysberg, K. Tillmann, T. Weirich (Eds.): EMC 2008, Vol. 1: Instrumentation and Methods, pp. 267–268, DOI: 10.1007/978-3-540-85156-1_134, © Springer-Verlag Berlin Heidelberg 2008

268

3. J.M. Cowley, *Ultramicroscopy*, **41**, (1992) p.335-348.
4. M. Born and E. Wolf, *Principles of Optics*, Pergamon, Oxford, 1980.
5. J.C.H. Spence and J.M. Cowley, in "Introduction to Electron Holography", ed. Volkl E., Allard L.F., and Joy D.C.,(Kluwer Academic/Plenum, Publishers, New York) (1999) p. 17.
6. H.J. Kreuzer., K. Nakamura, A. Wierzbicki, H.W. Fink, and H. Schmid, *Ultramicroscopy* **45**, (1992) 381-403.
7. J.J. Barton *Phys. Rev. Lett.* **61**, (1998) 1356-1359.
8. We acknowledge NINT, NRC and NSERC for funding and Dr. J. Wang for an interesting sample.

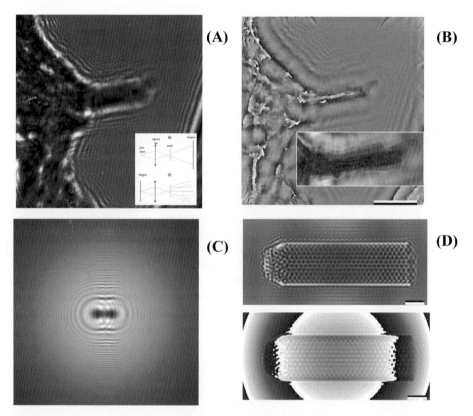

Figure 1. (A) In-line point projection hologram of titanium oxide nanowires taken with a defocus value of 215 μm. Inset: (a) Point-source in-line holographic setup as proposed by Gabor and (b) equivalent holographic TEM setup used in the current study. (B) Phase reconstruction of the hologram in (A) obtained with eq. (1). The scale bar is 100 nm and the inset is a crop of the amplitude reconstruction showing the fine structure of the central part of the sample. (C) Simulated hologram of a carbon nanotube segment for low energy electrons (100 eV). (D) Reconstruction (amplitude – top and phase -bottom) of the hologram in (C) with eq. (3) in the text; scale bar is 1nm

Electron holography of soot nanoparticles

M. Pawlyta[1], C.W. Tai[2], J.-N. Rouzaud[1] and Y. Lereah[2]

1. ENS Laboratoire de Géologie 24, rue Lhomond, 75231 Paris Cedex 3, France
2. Department of Physical Electronics, School of Electrical Engineering, Faculty of
Engineering, Tel Aviv University, Ramat Aviv 69978, Israel

mirka@pawlyta.pl

Keywords: soot, HRTEM, electron holography

Soot can be defined as the black solid product of an incomplete combustion or a pyrolysis of fossil fuels and other organic materials. It can be used as a filler and a pigment in industrial applications (carbon black). On the other hand, soot plays important roles as a traffic-related air pollutant (diesel soot) and as a significant component of the atmospheric aerosols. Despite all the efforts during the past years, the internal structure of soot aggregates is not well understood. High-resolution transmission electron microscopy (HRTEM) is a relevant tool to image directly the profile of the graphene layers forming the skeleton of agglomerated soot primary particles. This technique provides the unique information at high spatial resolution, and - what is important for inhomogenous environmental samples – it is free of interferences from the surrounding materials. The major advantage of electron holography is enhancing the differences in sample thickness, what facilitate the correct 3D reconstructions from 2D images. Two grades of industrial carbon black (Degussa: Printex 25 and FW200) and the soot sample obtained from wood burning were analysed in this study. Selected industrial CB grades are characterized by various morphologies and nanostructures [1]. Since heat-treatment strongly changes the shape and nanostructure of soot primary particles [2], CB samples heated at 2600°C were also examined. The TEM images and electron holograms were recorded on a FEI Tecnai F20 TEM operated at 200kV. The electron holography was implemented using a Möllenstedt biprism inserted into a selected area aperture. As the primary particles of CB Printex 25 and soot from wood burning are spherical, it was possible first to determine relationship between phase shift and thickness, next to reveal the thickness profile of a particle from the reconstructed phase images (Fig. 1). Measurements for several spherical particles with different diameters confirmed the results. Heat-treatment changes the shape and nanostructure of primary particles (Fig. 2a). A thickness profile from a hologram was determined. For particles with irregular shape, it was still possible to determine the height of individual elements (graphene layer stacks) (Fig. 3). This value could be obtained also for graphitic domains which are not parallel to electron beam and almost invisible on HRTEM images (point 3 on Fig. 3).

1. S. Duber, J.-N. Rouzaud, M. Pawlyta, H. Wistuba, B. Ptak, in: Carbonaceous and Catalytic Materials for Environment, Zakopane, 2004, pp.125–132
2. M. Pawlyta, J.-N. Rouzaud, S. Duber in: CESEP'07 : Carbon for Energy Storage and Environment Protection: the 2nd international conference: 2007, Krakow, Poland/ Krakow: WN „Akapit", p. 89
3. Acknowledgment: This study was supported by EC grant no. 29637.

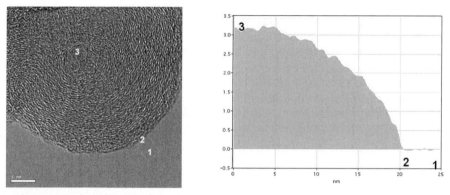

Figure 1. HRTEM image of a spherical industrial CB primary particle (Degussa: Printex 25) (left) and the thickness profile of line 1-2-3 from the phase image (right)

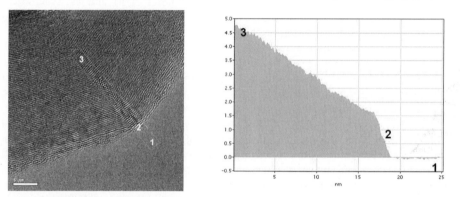

Figure 2. HRTEM image of a CB primary particle (Degussa: Printex 25, heated at 2600°C) (left) and the thickness profile of line 1-2-3 from the phase image (right)

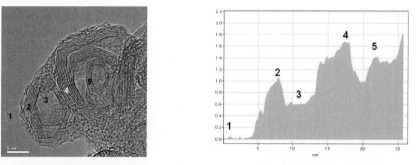

Figure 3. HRTEM image of the CB primary particles (Degussa: FW00, heated at 2600°C) (left) and the thickness profile of line 1-2-3-4-5 from the phase image (right)

Holographic tomography of electrostatic potentials in semiconductor devices

P.D. Robb[1] and A.C. Twitchett-Harrison[1]

1. Department of Materials, Imperial College London, Exhibition Road, London SW7 2AZ, UK

p.robb@imperial.ac.uk

Keywords: electron holography, tomography, semiconductors

With the continuing trend of semiconductor device miniaturisation, it is becoming ever more important that nanoscale devices are characterised in three dimensions. Surfaces and interfaces within device structures can greatly influence the overall functionality of a device if their electrical properties are significantly different from those of the bulk material [1]. Conventional two-dimensional techniques will be inadequate for the investigation of the inherently three-dimensional potential distribution in current and future semiconductor devices.

However, due to the combination of off-axis electron holography with electron tomography, it is now possible to map the three-dimensional (3-D) dopant distribution inside semiconductor devices at the nanometre scale [2]. Off-axis electron holography is a transmission electron microscopy based interference technique that allows the dopant-related phase change (introduced by the specimen) to be reconstructed [3]. An example of the phase change from a single electron hologram of a Si metal-oxide semiconductor field effect transistor (MOSFET) is shown in Figure 1. The 2D phase change allows the specimen's electrostatic potential, electric field and, ultimately, the charge density to be calculated [3].

The acquisition of a series of electron holograms over a range of specimen tilt angles, followed by tomographic reconstruction, permits the generation of 3-D phase information. Figure 2 presents a tomographic reconstruction that illustrates the 3-D electrostatic potential distribution across a Si p-n junction, which reveals quantitatively the electrostatic potential variation arising in the near-surface regions due to focused ion beam (FIB) specimen preparation and the presence of surfaces. By reconstructing the 3D potential the surface effects can be separated from 'bulk-like' effects in the centre of the membrane.

Holographic tomography has also been used to investigate more complex semiconductor device structures. A series of object and reference off-axis electron holograms of a B-doped Si MOSFET was acquired over a range of specimen tilts from -60° to +60° at 2° increments, and these were reconstructed tomographically using the SIRT algorithm [4]. Figure 3 shows the 3D tomographic reconstruction of the electrostatic potential. The p-type doped regions can be clearly delineated below the source and drain. Figure 4 shows a y-z plane view of the tomographic reconstruction of the MOSFET. It can be seen that there is a significant variation in the electrostatic potential along the edges of the reconstruction that are a result of surface effects. Further

M. Luysberg, K. Tillmann, T. Weirich (Eds.): EMC 2008, Vol. 1: Instrumentation and Methods, pp. 271–272, DOI: 10.1007/978-3-540-85156-1_136, © Springer-Verlag Berlin Heidelberg 2008

272

quantification and interpretation of these results will be presented. However, these results indicate that tomographic electron holography is a promising technique that can reveal the 3D electrostatic potential on the nanometre scale in semiconductor device structures, information that is not accessible with other characterization techniques.

1. J. H. Davies in "The physics of low-dimensional semiconductors", (Cambridge University Press, UK) (2005), p. 363.
2. A. C. Twitchett-Harrison, T. J. V. Yates, S. B. Newcomb, R. E. Dunin-Borkowski and P. A. Midgley, Nano Letters **Vol. 7** (2007), p. 2020-2023.
3. H. Lichte and M. Lechmann, Reports on Progress in Physics **Vol. 71** (2008), 016102.
4. P. Gilbert, Journal of Theoretical Biology **Vol. 36** (1972), p.105-117.

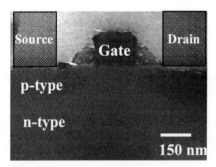

Figure 1. Reconstructed phase image of a Si B-doped MOSFET

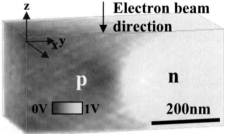

Figure 2. Tomographic reconstruction of the 3-D electrostatic potential within a Si p-n junction.

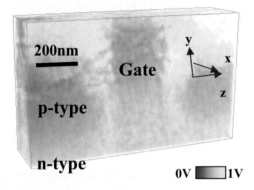

Figure 3. A tomographic reconstruction of the electrostatic potential from a Si MOSFET viewed along the z-direction. The gate, p-type and n-type regions can all be seen.

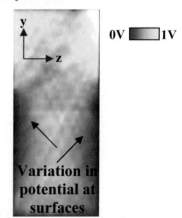

Figure 4. A tomographic reconstruction of the electrostatic potential from a Si MOSFET viewed along the x-direction. Surface regions show a reduced potential.

Electron Holography on the charge modulated structure $In_2O_3(ZnO)_m$ in comparison with DFT-calculations

Falk Röder [1], Axel Rother [1], Werner Mader [2], Thomas Bredow [3] and Hannes Lichte [1]

1. Triebenberg Laboratory, Institute of Structure Physics, University of Dresden, Zum Triebenberg 50, 01328 Dresden, Germany
2. Institute of Inorganic Chemistry, University of Bonn, Römerstr. 164, 53117 Bonn, Germany
3. Institute of Physical and Theoretical Chemistry, University of Bonn, Wegelerstr. 12, 53115 Bonn, Germany

Falk.Roeder@triebenberg.de

Keywords: charge modulation, electron holography, DFT, intrinsic fields, structure relaxation, indium-zinc-oxide

The structure - properties relation of modern materials requires fundamental understanding of the interaction between charge structure and relaxation processes e.g. induced by doping metal oxides. In this context, $In_2O_3(ZnO)_m$ provides a well defined arrangement of ZnO –like domains influenced by intersecting layers with fully occupied In^{3+}-ions and unoccupied metal sites. These layers are ordered periodically in a strictly alternating manner, with spacing controllable by the quantity m. An inversion of structural polarity of the domains [1] is forced by the octahedral coordination of In^{3+}-ions in so-called inversion domain boundaries (IDB) (Figure 1).

Correspondingly, X-ray diffraction studies of crystals with m=3 and m=5 [2,3] exhibit large displacements of the electron density around metal ions at lattice sites between the boundaries revealing a perturbed structure (Figure 1). This can be attributed to wide-spaced intrinsic electric fields stemming from charge zones at the boundaries. These are inducing structural and electronic relaxation processes leading to minimal total energy.

The result is a modulated charge structure, which is confirmed in this work by Mulliken population analysis of the electron density from DFT-calculations by means of the crystalline-orbital program package CRYSTAL06 [4]. Nevertheless, the calculated electrostatic potential in the case of a completely relaxed geometry indicates a remaining weak electric field in interatomic regions (Figure 2), which should be detectable experimentally by means of Electron Holography.

In a TEM, the phase of the electron wave is modulated sensitively by those electric fields but is unfortunately lost in the conventional acquisition of intensity images. Electron Holography provides a method to reconstruct also the phase information hence gives experimental access to electric field modulations at nm-scale [5]. First holograms of $In_2O_3(ZnO)_6$ were recorded at Triebenberg Laboratory in Dresden (Figure 3).

Although, the interpretation of phase images with respect to interatomic electric fields requires simulations including the whole self-consistent potential in the relativistic scattering process that has been performed by means of SEMI software package [6]. The investigations show the necessity of considering the influence of residual aberrations remaining in a Cs corrected electron microscope pretending field-induced phase modulations [7].

M. Luysberg, K. Tillmann, T. Weirich (Eds.): EMC 2008, Vol. 1: Instrumentation and Methods, pp. 273–274, DOI: 10.1007/978-3-540-85156-1_137, © Springer-Verlag Berlin Heidelberg 2008

274

1. C. Li, Y. Bando, M. Nakamura, N. Kimizuka, Journal of Electron Microscopy 46 (1997) 119.
2. C. Schinzer, F. Heyd, S. F. Matar., Journal of Materials Chemistry 9 (1999) 1569.
3. W. Pitschke, K. Koumoto, Powder Diffraction 14 (1999) 213.
4. R. Dovesi, V.R. Saunders, C. Roetti, R. Orlando, C.M. Zicovich-Wilson, F. Pascale, B. Civalleri, K. Doll, N.M. Harrison, I.J. Bush, Ph. D'Arco, M. Llunell, CRYSTAL06 User's Manual, University of Torino, Torino, 2006
5. H. Lichte, P. Formanek, A. Lenk, M. Linck, C. Matzeck, M. Lehmann, P. Simon; Electron Holography: Application to Materials Questions; Annual Review of Materials Research, Vol. 37 (2007), 539-588
6. Rother, SEMI software package, to be published
7. Financial support by ESTEEM and discussions within the Triebenberg group are gratefully acknowledged.

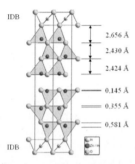

Figure 1. Structure model of m=5 showing displacements of ions between IDBs occupied with In.

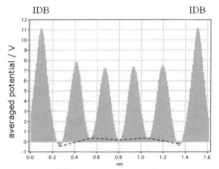

Figure 2. Weak modulation of calculated interatomic potential (dashed line) along hexagonal c-axis in m=5.

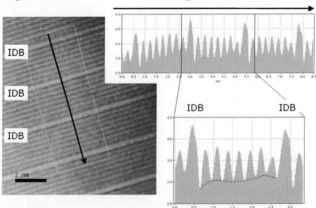

Figure 3. Experimentally reconstructed phase image of $In_2O_3(ZnO)_6$ revealing periodically arranged inversion domain boundaries (IDB). Slight phase modulations between atom columns are exhibited by averaged profiles (dashed line).

Correction of the object wave using iteratively reconstructed local object tilt and thickness

K. Scheerschmidt

Max Planck Institute of Microstructure Physics, D-06120 Halle, Germany

schee@mpi-halle.de

Keywords: inverse problems, object retrieval, dynamical theory, electron scattering

Object data can directly - without using trial-and-error matching - be retrieved from electron microscope exit waves, as e.g. reconstructed by holography [1, 2]. Such a inverse retrieval of local data, e.g. thickness, orientation and potential, as a basis of a general object reconstruction, can be gained by linearizing, regularizing and gene-ralizing the scattering problem. Figs. 1 shows the basic idea of trial-and-error versus inverse solution schemes and Fig. 2 the reconstruction of thickness and orientation at an Ge-CdTe interface applying different scattering potential for the Ge and the CdTe regions.

The inverse solution starts from moduli (A) and phases (P) of the set of plane waves Φ^{exp} at the exit surface (cf. Fig.1). Exit waves Φ^{th} are calculated using the dynamical scattering matrix M for an a priori model with a suitable scattering potential and by assuming a trial average beam orientation (K_{xo}, K_{yo}) predetermined by the experiment (upper arrow, direct trial-and-error formalism). With the sample thickness t_0 as a free parameter, a perturbation approximation yields both the moduli and phases of the plane wave amplitudes as linear functions of the object thickness t and orientation (K_x, K_y). The analytic form of the equations (cf. lower arrow, linearized inverse from scattering matrix M and its derivative δM) yields directly for each image pixel (i,j) the local thickness t_{ij} and local beam orientation $(K_x,K_y)_{ij}$. However, as pointed out in different previous analyses of the stability of the retrieval procedure (cf. the summaries in [3-5]) it requires the knowledge of the confidence region, conditions for stability, and restrictions due to modeling errors. The ill-posed inverse matrix is transformed by generalizing and regularizing to a well-posed but ill-conditioned one, which is equivalent to a least square (maximum-likelihood) minimization $\|\Phi^{exp}-\Phi^{th}\|+\gamma\|\Omega\|=Min$ or to pixel smoothing, controlled and optimized by the regularization parameter γ and the constraint Ω.

However, to reduce further the modeling errors and to extend the confidence region the start values and the assumptions for the underlying scattering model may be varied iteratively as demonstrated in Fig. 3, where the first subsequent iterative reconstruction yields noise reduction.

1. K. Scheerschmidt, Lecture Notes in Physics 486 (1997) 71-85.
2. K. Scheerschmidt, Journal of Microscopy 190 (1998) 238-248.
3. K. Scheerschmidt, Microsc. Microanal. 9, Suppl.6 (2003) 56-57.

M. Luysberg, K. Tillmann, T. Weirich (Eds.): EMC 2008, Vol. 1: Instrumentation and Methods, pp. 275–276, DOI: 10.1007/978-3-540-85156-1_138, © Springer-Verlag Berlin Heidelberg 2008

276

4. K. Scheerschmidt, WAVES 2003, Proc. 6th Int. Conf. Math. Num. Aspects Wave Prop., Eds.: G. C. Cohen and P. Joly, Springer 2003, 607-612.
5. K. Scheerschmidt, A. Rother, Microsc. Microanal. 13, Suppl.B (2007) 140-141.

Figure 1. Matrix formulation of the dynamical theory giving the basis for the iterative trial-and-error solution (arrow to the left) and of the local linearization and inversion (right arrow): The moduli and phases of the exit wave are the input to the generalized inverse yielding thickness t and orientation K_{xy} as function of the position (i,j) in the pixel space.

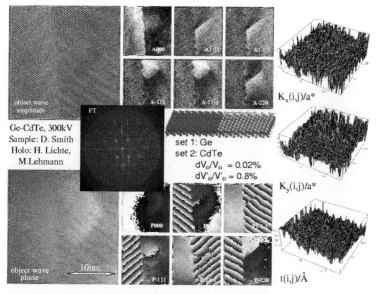

Figure 2. Object retrieval (thickness t, beam orientation K_{xy}) at an Ge-CdTe interface (moduli/phases of reconstructed single reflections in upper/lower rows, insets: sideband of Fourier transformed hologram and structural data used for exit wave calculations).

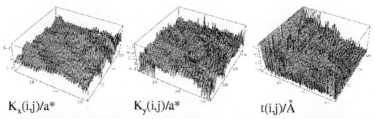

Figure 3. Smoothing of reconstructed thickness and orientation of Fig. 2 by using an iterative modification of the start values.

Extended field of view for medium resolution electron holography at Philips CM 200 Microscope

J. Sickmann, P. Formánek, M. Linck and H. Lichte

Triebenberg Laboratory, Institute of Structure Physics, Technische Universität Dresden, D-01062 Dresden, Germany

jan.sickmann@triebenberg.de

Keywords: electron holography, extended field of view

Electron holography has become a reliable tool for dopant profiling in semiconductor industry. Up to now, medium resolution and large field of view have been required, in order to adapt to object structures of interest [1]. With shrinking device dimensions, the resolution has to be improved, while preserving a large field of view.

A holographic field of view up to 500...1000 nm can easily be achieved by operating the microscope with the objective lens switched off and the Lorentz lens switched on instead. However, while the Lorentz lens provides a very large field of view, it is not able to reach the required resolution for the downscaled device structures. To use the objective lens instead will provide a better resolution, but a field of view of 30 nm only, which is to small for semiconductor devices. Thus the setup of the microscope has to be adjusted.

In order to obtain an optimized setup, in particular four critical parameters in electron holography have to be considered carefully: the hologram width, the fringe spacing, the contrast of the interference fringes and the coherent current over the area of interest. These parameters determine the holographic field of view, the lateral resolution and the quality of the detected phase signal [2].

Considering the holographic setup in Figure 1, with the biprism inserted between the back focal plane and and the first image plane, hologram width w_{hol} and fringe spacing s_{hol} related to the object are given as [3]:

$$w_{hol} = \frac{2\gamma\, b f}{a+b} - \frac{2 r f}{a} \qquad \text{and} \qquad s_{hol} = \frac{\lambda\, f}{2\gamma\, a} \,.$$

The distances a and b are given according to Figure 1, f is the focal length of the objective lens, r is the radius of the biprism and γ the deflection angle at a specific filament voltage. Unfortunately, as can be derived easily from the equations above, the hologram width and the fringe spacing have reciprocal dependence, which restricts the hologram width with respect to the desired resolution and the sampling rate of at least four pixels [4].

Since distance a and focal length f are almost fixed by design of the TEM column, apart from the deflection angle γ only distance b is adjustable. Thus we apply the free lens contol function of our microscope, to adjust the excitation of the diffraction lens hence change distance b. With the biprism positively or negatively biased, the first

M. Luysberg, K. Tillmann, T. Weirich (Eds.): EMC 2008, Vol. 1: Instrumentation and Methods, pp. 277–278, DOI: 10.1007/978-3-540-85156-1_139, © Springer-Verlag Berlin Heidelberg 2008

image plane can be shifted above the biprism plane [5]. The objective lens has to be refocused to perserve the imaging from the object to the detector plane.

Recording interference patterns for specific excitations of the diffraction lens and measuring the width of the holograms and the fringe spacing, an optimum setup for holography with an extended field of view has been derived. The largest field of view of 210 nm, was seven times larger than in the normal HRTEM-setup (Figure 2).

Special care has to be taken to preserve signal resolution.

1. W.D. Rau, P. Schwander, F.H. Baumann, W.Höppner, O. Ourmazd, Phys. Rev. Lett. **82** (1997) p. 2614-2617.
2. H. Lichte, Ultramicroscopy **108** (2008), p. 256-262.
3. H. Lichte, Ultramicroscopy **64** (1996), p. 79-86.
4. F. Lenz, E. Voelkl, Proc. ICEM **12** (1990) p. 228.
5. B.G. Frost, E. Voelkl, L.F. Allard, Ultramicroscopy **63** (1996), p. 15-20.

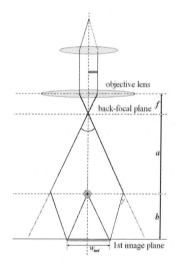

Figure 1. Scheme of the holographic setup. The biprism is inserted in the selected area diffraction aperture between the back-focal plane and the first image plane.

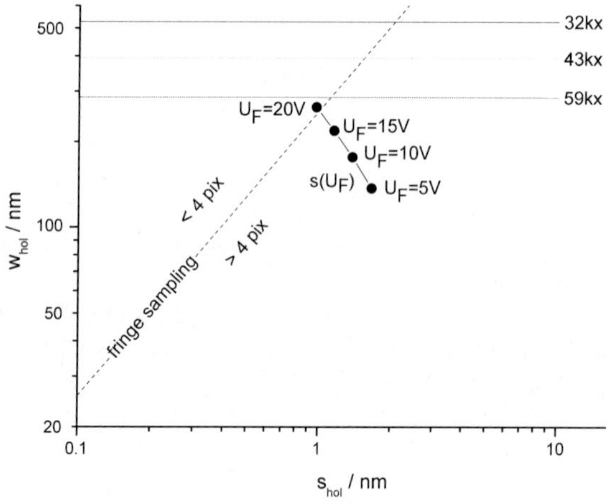

Figure 2. Hologram width w_{hol} vs. fringe spacing s_{hol} in double logarithmic scale taken at diffraction lens excitation of 30%. The straight horizontal lines indicate the fields of view at specific magnifications. The dashed line marks the four pixel sampling, with the criterion satisfied in the area below.

Electron holography of biological and organic objects

P. Simon[1]

[1]Max Planck Institute for Chemical Physics of Solids, Nöthnitzer Str. 40, 01187
Dresden, Germany

simon@cpfs.mpg.de
Keywords: Holography, biological samples, polymers, biominerals, organic solar cells, intrinsic
electric fields

Todate, there are only a few (~ 12) publications on electron holography of biological objects and much less of polymeric organic samples, thus this topic is quite small and concise. Unstained biological systems investigated by means of off-axis electron holography up to now are ferritin, tobacco mosaic virus, a bacterial flagellum, T5 bacteriophage virus, hexagonal packed intermediate layer of bacteria and the Semikli forest virus. Results on collagen fibers and surface-layer of bacteria, the so-called S-layer 2D crystal lattice are presented in this work and off-axis holography of materials related to biological systems, such as biomaterial composites.

Holographic investigations of the bacterial surface (S-layer) protein of Bacillus sphaericus NCTC 9602 are discussed [1]. The bacterial S-layer contains quasi 2D protein crystals, with different lattice symmetries of periodically arranged cell wall protein or glyco protein, which form the outermost part of the cell wall of various bacteria and archeae. We investigated air-dried, free-standing samples on a holey carbon film with a protective thin carbon layer evaporated onto the S-layer, to avoid charging.

Additionally, in this report we present results of electron holography of collagen fibrils at medium resolution [2]. Collagen is a common and important fiber protein, occurring in many living systems e.g. in glassy sponges, bone or tendon of vertebrates and invertebrate connective tissue. Collagen plays a key role in bone formation, thus, it would be helpful to image the morphology and topography of collagen fibers before and after mineralization. Electron holography enables us to observe this process without staining. Here, we report on collagen type-I from calf skin. In Fig. 1a, a hologram of a free-standing collagen type-I fiber is shown using Lorentz lens instead of the usual objective lenses. The fringe contrast of the hologram was 4 % with a resolution of the fringes of 8 nm, at a total dose of about 12 e $Å^{-2}$. The reconstructed phase image of the unstained crystal lattice of collagen type-I protein fiber shows a chessboard-like structure (Fig. 1b). Cross-striation with period of ~ 60 nm and even finer structural details down to ~ 8 nm are present as read out from the phase profile measured along the fibril long-axis (Fig. 1c, bottom). The height variation amounts to ~ 5 nm. Phase profile normal to the fiber axis (Fig. 1c top) indicates flattened fiber morphology.

Alongside, holography can be used to visualize electrical or magnetic fields produced either by the inorganic or organic component within a biological object. We found clear evidence for a direct correlation between intrinsic electrical fields and the self organized growth of the bio-composite system fluorapatite-gelatine [3,4]. In the case of the composite seeds, these fields are generated by a parallel orientation of triple-

M. Luysberg, K. Tillmann, T. Weirich (Eds.): EMC 2008, Vol. 1: Instrumentation and Methods,
pp. 279–280, DOI: 10.1007/978-3-540-85156-1_140, © Springer-Verlag Berlin Heidelberg 2008

helical protein fibers of gelatine and thus produce a mesoscopic dipole field as visualized by electron holography (see Fig. 2, left). We investigated also composite aggregates at the nanoscale. By conventional TEM the inner morphology cannot be imaged clearly since a certain defocus has to be applied to obtain contrast leading to a blurred pattern. By means of holography we obtain evidence that nanometer sized protein fibrils are integrated within the inorganic component as displayed in phase image (Fig. 2, right).

1. P. Simon, H. Lichte, R. Wahl, M. Mertig, W. Pompe, Biochim. Biophys. Act (Biomembr) **1663** (2004), 178-187.
2. P. Simon, H. Lichte, P. Formanek, M. Lehmann, R. Huhle, W. Carrillo-Cabrera, A. Harscher, H. Ehrlich, Micron **39** (2008), 229–256.
3. P. Simon, D. Zahn, H. Lichte, R. Kniep, Angew. Chem. **45** (2006), 1911-1915.
4. R. Kniep, P. Simon, Angew. Chem. **47** (2008), 1405-1409.
5. P.S. like to thank Prof. H. Lichte and Prof. M. Lehmann for their support on the TEM investigations at Triebenberg Laboratory at the Technical University of Dresden, Germany.

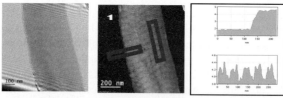

Figure 1. (a) Electron hologram of an unstained collagen type-I fiber at a magnification of ~61,000x, recorded with a dose of 12 e $Å^{-2}$. The resolution is 8 nm, given by the double of the interference fringe periodicity. The contrast of the fringes amounts to 4 %. (b) Reconstructed phase image of the unstained lattice of collagen type-I fiber, as shown in Fig. 1a. Cross-striations with period of ~ 60 nm and even finer structural details to ~ 8 nm are imaged. The phase profile drawn parallel to fiber direction is marked by the frame on the right: The periodicity corresponds to ~ 60 nm, height variation amounts, from maximum to minimum, to a value of ~ 5 nm.

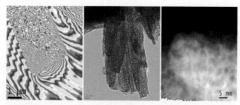

Figure 2. (left) Ideal hexagonal prismatic (appearing rectangular shaped in projection) fluorapatite–gelatine composite seed. Retrieved phase image of an electron hologram exhibit electric potential distribution around the seed. Colour code denotes from green to green a phase shift of 2 π. Fresnel fringes of the interferograms appear as striation pattern at the corners of the phase images. The observed projected potential corresponds to a mesoscopic dipole. (center) Conventional TEM of nano-sized composite aggregate. (right) High-resolution phase image indicating positions of integrated nano protein fibrils (dark areas) within the in the inorganic matrix (bright); dark area on top is representing vacuum.

Magnetic configurations of isolated and assemblies of iron 30 nm nanocubes studied by electron holography

E. Snoeck[1], C. Gatel[1], L.M. Lacroix[2], T. Blon[2], J. Carrey[2], M. Respaud[2], S. Lachaize[2], and B. Chaudret[2,3]

1. CEMES-CNRS, 29 rue J. Marvig, BP4347, 31055 Toulouse, France
2. Université de Toulouse; INSA, UPS; LPCNO, 135 av. de Rangueil, 31077 Toulouse,
3. LCC –CNRS, 205, route de Narbonne, 31077 Toulouse Cedex 04, France

snoeck@cemes.fr

Keywords: holography, magnetic, nanocubes

The magnetic remanent configurations of single and assemblies of crystalline 30 nm Fe nanocubes were studied using off axis electron holography (EH). Experiments were performed on the SACTEM-Toulouse (Tecnai-F20 microscope (FEI)) fitted with an aberration corrector (CEOS) in which we use the first transfer lens of the Cs corrector as a pseudo Lorentz lens to perform EH [1]. The magnetic contribution to the phase shift was obtained by recording two holograms between which the Fe cubes were magnetized in two opposite directions. This is achieved by tilting the sample by +45° and switching the microscope objective lens to apply a large field of about 2 Tesla to the sample. The objective lens was then switched off and the sample is tilt back to zero to record a first hologram. This procedure is done a second time but tilting the sample by -45°. This procedure relies on the ability to saturate the magnetization in the sample in two opposite directions before decreasing the magnetic field to zero. EH experiments performed on an isolated cube (Figure 1) evidence a vortex state which fulfilled the largest cylinder, fitting perfectly the central part of the cube. The magnetization continuously curls around the [001] axis with components in the (001) plane and turns progressively in the [001] direction when approaching its core. Additional features are visible along the [001] edges of the nanocube. In the four corners close to the border of the cube the magnetic flux cannot circulate around the vortex. The resulting magnetostatic energy is however reduced creating a flux closure around the [001] edges of the cube (see zoom in Figure 1d). This magnetic configuration was simulated by micromagnetic calculations carried on using the three-dimensional version of the micromagnetic code OOMMF [2]. The 3D nanocube shapes and sizes introduced for the calculations are deduced from the 2D TEM images. We used bulk magnetic parameters for iron: $M_S = 1.72 \times 10^6$ A/m, $A = 2.1 \times 10^{-11}$ J/m and a $K_1 = 4.8 \times 10^4$ J/m^3. We use in the simulations magnetic fields of same amplitude and direction than experimentally. Among different calculation outputs, OOMMF provides vectorial maps of the magnetization and of the total magnetic field. In electron holography the phase shift originates from the integration of the magnetic induction along the incident electron beam direction. To properly compare experiments and calculations we simulate the two-dimensional mapping of the induction similar to the holography result by summing along the z direction the calculated induction vectors. The simulation fits well with the

M. Luysberg, K. Tillmann, T. Weirich (Eds.): EMC 2008, Vol. 1: Instrumentation and Methods, pp. 281–282, DOI: 10.1007/978-3-540-85156-1_141, © Springer-Verlag Berlin Heidelberg 2008

282

EH results which confirms both the appearance of the vortex and the flux closure around the [001] edges.

Similar EH experiments and simulations were performed on four cubes in a square arrangement (Figure 2). In that case, the magnetic flux over the four cubes exhibits an axial symmetry, the centre of which is located in the middle of the square. The induction bends and circulates around a cylinder parallel to the [001] direction. The OOMMF simulation agrees well with the experimental map as shown in Figure 2d). Contrary to what was observed on the isolated cube, no vortex appears in that four cubes arrangement, i.e. the magnetisation is only curling around the middle of the square without any out-of-plane component. Moreover, as the magnetic induction cannot close in regions out of the flux cylinder i.e. at the outer corners of the four cubes, a flux closure around the external [001] edges of the four cubes is created to reduce the magnetostatic energy similar to what observed in the isolated cube case.

1. Snoeck E., Hartel P., Müller H., Haider M. & Tiemeijer P.C. Using a CEOS objective lens corrector as a pseudo Lorentz lens in a Tecnai F20 TEM. Proc. IMC16 International Microscopy Congress (Sapporo, 2006) Vol. 2, p. 730.
2. Donahue M. J. and Porter D. G. 1999 OOMMF User's Guide, Version 1.0, Interagency Report NISTIR 6376, National Institute of Standards and Technology, Gaithersburg, MD. We used version 1.2a3 of this public code that can be found at the URL http://math.nist.gov/oommf/.
3. The authors acknowledge financial support from the European Union under the Framework 6 program under a contract for an Integrated Infrastructure Initiative. Reference 026019 ESTEEM.

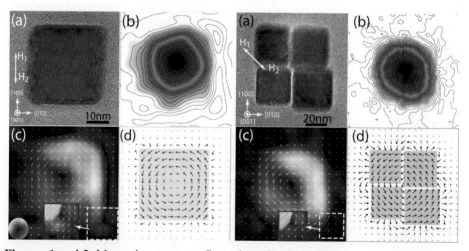

Figures 1 and 2. Magnetic vortex configuration of a single Fe nanocube (Figure 1) and curling dipolar coupling in four Fe nanocubes in a square arrangements (Figure 2) - a) TEM images, b) Phase image of the magnetic contribution to the phase shift with contours superimposed on it, c) vectorial map of the in-plane components of the magnetic induction, d) Micromagnetic simulation (OOMMF code).

Electron holography study of ferroelectric solid solutions

C.W. Tai and Y. Lereah

Department of Physical Electronics, School of Electrical Engineering, Faculty of Engineering, Tel Aviv University, Ramat Aviv, 69978, Israel

cwtai@eng.tau.ac.il

Keywords: electron holography, ferroelectric domains, ceramics

The configuration of domains, which is one of the determinant factors of the properties of ferroelectrics, has been investigated since the 1950s [1]. Transmission electron microscopy (TEM) is widely used to study ferroelectrics because the domains and walls can be characterized by imaging and diffraction mode simultaneously and the investigation can be performed at micron scale down to atomic resolution. In recent, off-axis electron holography has drawn considerable attention because further details of the domains and walls can be revealed [2].

In electron holography, the phase of the incident plane wave is modulated by the in-plane component of the projected polarization $\mathbf{P_{proj}}$. The phase shift can be expressed as Eq. (1) where σ is the interaction constant (0.0073 /V nm for 200 keV) and V_{proj} is the projected inner potential of the material. In a ferroelectric, the inner potential is substituted by the specific one, which is caused by the electric polarization in the material. Therefore, the specific projected inner potential is the integration of the projected polarization $\mathbf{P^{Ferro}}_{proj}$ over the in-plane volume where ε the dielectric constant of a ferroelectric [2]. Alternatively, the relationship between $\mathbf{P^{Ferro}}_{proj}$ and the phase gradient can be seen in Eq. (2).

$$\varphi(x,y) = \sigma \cdot V_{proj}(x,y) = \frac{\sigma}{\varepsilon} \int_{vacuum}^{\vec{r}(x,y)} \vec{P}_{proj}^{Ferro}(\vec{r})d\vec{r} \quad \cdots\cdots(1) \qquad \vec{P}_{proj}^{Ferro} = \frac{\varepsilon}{\sigma}\nabla\varphi(x,y) \quad \cdots\cdots(2)$$

Since the in-plane components of $\mathbf{P^{Ferro}}_{proj}$ contribute to the phase shift only, misleading interpretation of the phase image may be made. In the current study, various domain configurations in different phases of lead zirconate titanate (PZT) and the corresponding phase profiles are discussed.

A typical configuration of the ferroelectric domains in a tetragonal phase PZT grain is shown in Figure 1a as an example. The 90° domains configuration is characterized by the straight domain-walls. Domains cannot be observed in the regions closed to the sample edge due to surface relaxation in the very thin region. The reconstructed amplitude and phase images are shown in Figure 1b and c, respectively. The general profile of the intensity of a selected region does not show a noticeable trend though a few modulations are observed. Such changes of intensity are contributed by diffraction contrast that is commonly observed in ferroelectric ceramics. However, the phase profile of the selected region is different to the intensity profile. In Figure 1d, an increase of the phase is clearly seen, indicating the presence of the in-plane projected polarizations and their directions. In addition, a few steep phase changes are observed. We believe that the corresponding domain structure is rather complicated instead of the

M. Luysberg, K. Tillmann, T. Weirich (Eds.): EMC 2008, Vol. 1: Instrumentation and Methods, pp. 283–284, DOI: 10.1007/978-3-540-85156-1_142, © Springer-Verlag Berlin Heidelberg 2008

284

result of the head-to-head polarization, which is an improper configuration in a defect-free state. In this example, the difficulty in understanding the domain structures by the direct interpretation of the phase gradient has been demonstrated. Several domain configurations in different phases of PZT ceramics will be further given and discussed.

1. G. Shirane and A. Takeda, Journal of the Physical Society of Japan **7** (1952), p. 5.
2. H. Lichte, P. Formanek, A. Lenk, M. Linck, C. Matzeck, M. Lehmann and P Simon, Annu. Rev. Mater. Rev. **37** (2007), p. 539.
3. We would like to thank Mr. M. Linck and Prof. H. Lichte (Technische Universitaet, Dresden, Germany) for their support in the experimental setup and fruitful discussion. This study was supported by European Commission under Grant No. 29637.

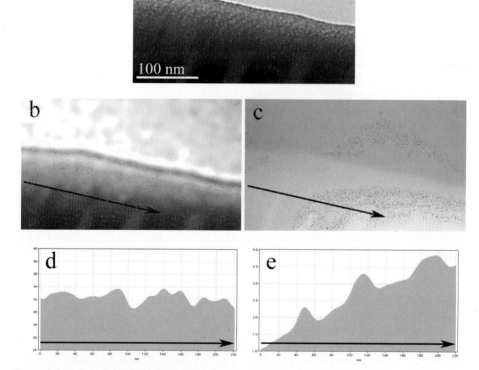

Figure 1. (a) Bright-field TEM image of a PZT grain. (b) Amplitude and (c) phase images of the grain after reconstruction. (d) & (e) The phase profiles taken in the amplitude and phase images of the same region, respectively.

Digital holographic interference microscopy of phase microscopic objects investigation

T.V. Tishko, D.N. Tishko and V.P. Titar

Lab of Holography, V.N. Karazin Kharkov National University, Svoboda sq. 4, 61077, Kharkov, Ukraine

Tatyana.V.Tishko@univer.kharkov.ua

Keywords: holography, microscopy, phase microscopic object, three-dimensional images, erythrocytes, thin films

Most biological micro-objects are phase objects that do not change the intensity of the radiation transmitted through them and are inaccessible to direct observation. Such methods of visualizing phase micro-objects are Zernike's phase contrast method and the interference contrast method. However the problem of 3D visualization of phase micro-objects has not been solved in classical microscopy. Electron microscopy makes it possible to obtain an image of a phase micro-object with high resolution. But this method requires special sample preparation and does not make it possible to study untreated micro-objects. The appearance of holography opened up new possibilities in the microscopy of phase micro-objects. The problem of the 3D visualization of phase micro-objects was solved by combining holographic methods with the methods of computerised image processing [1].

The first digital holographic interference microscope (DHIM), which allows the real-time 3D imaging of phase micro-objects and the quantitative measurements of their parameters, has been created at the Laboratory of Holography, Kharkov National University, Ukraine. The DHIM consists of three main units: holographic microinteferometer, digital video camera and computer. A He-Ne laser with a wavelength of 0.63 μ m serves as the radiation source. The interferograms of the micro-objects under study obtained using the holographic microinterferometer are recorded by the digital camera. The digital interferograms are computer processed using the mathematical algorithms that makes it possible to reconstruct the 3D images of micro-objects and to measure their geometrical parameters. The first 3D images of the native human blood erythrocytes were obtained using the microscope. The investigations, which were carried out in collaboration with different medical centres, have shown that a blood erythrocyte is a cell, which morphology reflects a state of a living organism and the level of its biological response on external factors influence [2]. It has been detected that erythrocytes morphology is changed not only in hematological diseases but in diseases of different genesis. The influence of ozone therapy on human and gamma radiation on rat blood erythrocytes has been investigated. For the first time the idea that morphological changes of blood erythrocytes reduce the oxygen capacity of blood and are the reason of different hypoxia pathologies, has been proposed.On Figure1 one can see the fragment of the blood sample with erythrocytes. The average diameter of the micro-objects is about 8 μ m.

M. Luysberg, K. Tillmann, T. Weirich (Eds.): EMC 2008, Vol. 1: Instrumentation and Methods, pp. 285–286, DOI: 10.1007/978-3-540-85156-1_143, © Springer-Verlag Berlin Heidelberg 2008

DHIM can be successfully used for thin films surface quality control, thickness and film defects parameters measurements. On Figure 2 one can see the fragment of AlN thin film deposited on acryl substrate damaged after UV-radiation influence. The film thickness is about 0,1 μ m.

1. T. V. Tishko, D. N. Tishko and V. P. Titar, J.of Opt.Tech. **72** (2005), p. 203.
2. T. V. Tishko, D.N. Tishko and V. P. Titar, SPIE **7006** (2007), p.9.

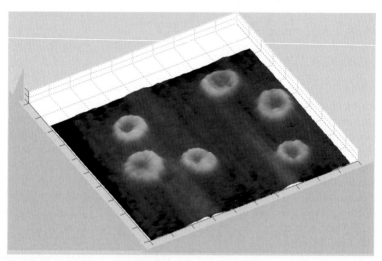

Figure 1. The fragment of the blood sample with human erythrocytes, obtained using the DHIM. The average diameter of the micro-objects is about 8 μ m.

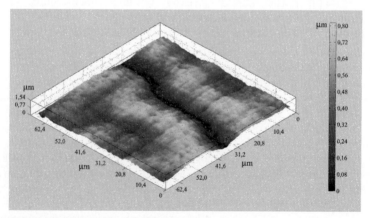

Figure 2. The fragment of the AlN thin film deposited on acryl substrate damaged after UV-radiation influence. The film thickness is about 0,1 μ m.

Reconstruction of 3D (Ge,Si) islands by 2D phase mapping

C.L. Zheng[1], H. Kirmse[1], I. Häusler[1], K. Scheerschmidt[2] and W. Neumann[1]

1.Humboldt-Universität zu Berlin, Institut für Physik, D-12489 Berlin, Germany
2. Max Planck Institut für Mikrostrukturphysik, Weinberg 2, D-06120 Halle, Germany

zcl@physik.hu-berlin.de
Keywords: Si-Ge islands, electron holography, mean inner potential

In the last few years, self-assembled (Ge,Si) islands grown on Si substrate have attracted much attention as promising application in electronic and photonics devices [1]. The physical properties due to quantum confinement of electrons in small space can be controlled by the size of islands [2]. TEM is an important tool for investigating structure and composition of such nanostructures. However, it is hard to get three dimensional size information from a single conventional TEM image.

Contrary to the CTEM imaging thechniques, electron holography records both amplitude and phase of the electron exit wave. The phase shift of the transmitted electron wave modulated by the inner potential of material is described by:

$$\Delta\varphi = C_E V_0 t \qquad [3],$$

where C_E is the interaction constant depending on the accelerating voltage, V_0 is the mean inner potential and t is the specimen thickness. If the mean inner potential is known, the 3D morphology can be reconstructed from two dimensional (2D) phase mapping.

$Ge_{0.4}Si_{0.6}$ islands were grown on an (001) oriented silicon substrate at 600°C by liquid phase epitaxy. The TEM specimen was prepared in plan-view. Electron holography investigations were performed on a JEOL JEM-2200FS transmission electron microscope operating at 200 kV (corresponding to an interaction constant of $0.00729 V^{-1} nm^{-1}$). The Lorentz mode (Magnification 50x to 15kx) is used for getting large field of view. The specimen was tilted by 3-5 degree from the Si [001] zone axis to reduce dynamical diffraction effects. Gatan Digital Micrograph and Holoworks were used for acquisition and reconstruction of the holograms. Further image processing and 3D reconstruction procedures were performed by Gwyddion package and Matlab software.

An electron hologram of $Ge_{0.4}Si_{0.6}$ islands on (001) Si is shown in Fig. 1a. The reconstructed amplitude and phase image is given in Fig. 1b and c. The displacement of interference fringes across the islands is due to thickness variations inside the island. Line profiles of amplitude and phase are compared in Fig. 2. An indication for the cross section of the island parallel to the growth direction could be recognized from the phase profile. Contrary to that, the amplitude oscillates along the line scan.

The mean inner potential of Ge-Si structure has been calculated using an isolated atom model [4]. The caculation provide 14.72 V for $Ge_{0.4}Si_{0.6}$. In order to improve the accuracy of the caculation of mean inner potential, density functional theory (DFT) calculations are in progress. The phase shift superimposed by the Si substrate has been

M. Luysberg, K. Tillmann, T. Weirich (Eds.): EMC 2008, Vol. 1: Instrumentation and Methods, pp. 287–288, DOI: 10.1007/978-3-540-85156-1_144, © Springer-Verlag Berlin Heidelberg 2008

subtracted as polynomial background before 3D reconstruction. The reconstructed 3D image of (Ge, Si) islands is shown in Fig. 3. The measured mean height of the islands is about 60 nm.

1. P. Schittenhelm, C. Engel and F. Findeis, et al., J. Vac. Sci. Technol. B, **16** (1998), 1575
2. C. Lang, D. J. H. Cockayne, and D. Nguyen-Manh. Phys. Rev. B, **72** (2005), 155328
3. L. Reimer, "Transmission Electron Microscopy" (Springer, Berlin, 1989).
4. P. Doyle and P. Turner, Acta Crystallogr, Sect. A: Cryst. Phys., Diffr, Theor. Gen. Crystallogr. **24**, 390 (1968)

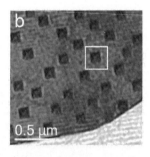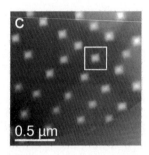

Figure 1: (a) Electron hologram of $Ge_{0.4}Si_{0.6}$ islands on (001) Si substrate (insert: enlarged island). (b) Reconstructued amplitude image and (c) unwrapped phase image.

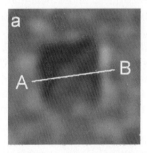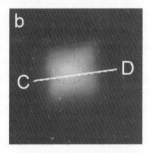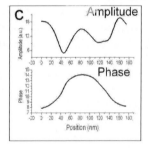

Figure 2: (a) Enlarged images of amplitude and (b) phase of a single $Ge_{0.4}Si_{0.6}$ island from the marked area in Fig. 2b and c. (c) Amplitude and phase profile along line AB and CD.

Figure 3: Reconstructed 3D image of $Ge_{0.4}Si_{0.6}$ islands on Si (001) substrate. The mean height of islands is about 60 nm.

Quantitative electron tomography of biological structures using elastic and inelastic scattering

R.D. Leapman[1], M.A. Aronova[1], A.A. Sousa[1], G. Zhang[1] and M.F. Hohmann-Marriott[1]

1. Laboratory of Bioengineering and Physical Science, National Institute of Biomedical Imaging and Bioengineering, National Institutes of Health, Bethesda MD 20892, USA

leapmanr@mail.nih.gov

Keywords: electron tomography, phosphorus, chromatin, nanoparticles, STEM, EFTEM

Conventional bright-field electron tomography (ET) provides valuable information about three-dimensional cellular architecture at 5-10 nm spatial resolution [1-3]. Reconstructed volumes can be obtained from heavy-atom stained plastic sections imaged with amplitude contrast, or from frozen-hydrated specimens imaged with phase contrast. Spatial resolution in conventional ET, as limited by stain artifacts and by radiation damage, is adequate for visualizing certain large macromolecular complexes, but smaller molecular assemblies often remain difficult to identify. We have investigated the use of alternative contrast mechanisms to help localize molecular structures of interest.

One approach is to use the elastic annular dark-field (ADF) signal in the scanning transmission electron microscope (STEM) to visualize heavy-element nanoparticles attached to antibodies or antibody fragments that can enter permeabilized cells to label specific proteins [4]. It is demonstrated that STEM tomography can be used to localize the three-dimensional distribution of Nanogold labels, each containing ~67 Au atoms, in 80-nm thick, lightly stained sections of rapidly-frozen, freeze-substituted and embedded cells, while simultaneously visualizing the surrounding biological structure [5-7]. Quantitative analysis of the experimental signal-to-noise ratio in ADF-STEM tomograms and comparisons with scattering theory provide a framework for establishing optimal conditions for localizing such heavy-atom nanoparticles [7].

Another approach that provides specific kinds of three-dimensional compositional information is quantitative electron spectroscopic tomography (QuEST), performed using energy-filtered transmission electron microscopy (EFTEM) [8-9]. By acquiring inelastic images at energy-losses above and below a characteristic core edge of interest, it is possible to determine distributions of specific elements within ~100-nm thick sections of rapidly-frozen, freeze-substituted and plastic-embedded cells. Such elements can act as intrinsic stains to localize cellular components, e.g., phosphorus reveals the distribution of RNA-containing ribosomes in the cytoplasm (Figure 1), and DNA within chomatin of the cell nucleus (Figure 2). Quantitative results are obtained by acquiring dual-axis tilt series, and by computing elemental distributions using the simultaneous iterative reconstruction technique (SIRT) [6].

Prospects are assessed for combining STEM tomography with EFTEM tomography to study the distribution of DNA and associated proteins in cell nuclei [10].

M. Luysberg, K. Tillmann, T. Weirich (Eds.): EMC 2008, Vol. 1: Instrumentation and Methods, pp. 289–290, DOI: 10.1007/978-3-540-85156-1_145, © Springer-Verlag Berlin Heidelberg 2008

1. J.R. Kremer, D.N., Mastronarde and J.R. McIntosh, J. Struct.Biol. **116** (1996), p. 71.
2. A.J. Koster et al., J. Struct. Biol. **120** (1997), p. 276.
3. W. Baumeister and A.C. Steven, Trends Biochem. Sci. **25** (2000), p. 624.
4. J.F. Hainfeld and F.R. Furuya, J. Histochem. Cytochem. **40** (1992), p. 177.
5. U. Ziese et al., J. Struct. Biol. **138** (2002), p. 58.
6. P.A. Midgley and M. Weyland, Ultramicroscopy **96** (2003), p. 413.
7. A.A. Sousa et al., J. Struct. Biol. **159** (2007), p. 507.
8. R.D. Leapman et al., Ultramicroscopy **100** (2004), p. 115.
9. M.A. Aronova et al., J. Struct. Biol. **160** (2007), p. 35.
10. This research was supported by the Intramural Research Program of the National Institute of Biomedical Imaging and Bioengineering, National Institutes of Health.

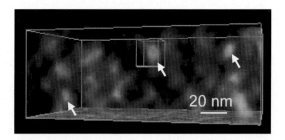

Figure 1. Volume rendered phosphorus distribution in cytoplasmic region of *Drosophila* larval cell, obtained using quantitative electron spectroscopic tomography (QuEST). Globular structures (arrows) are ribosomes; from analysis of subvolumes (indicated by box), the ribosomes are found to contain 7900±1900 P atoms (± st. dev.), in reasonable agreement with their known rRNA content.

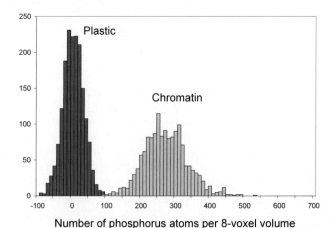

Figure 2. Histogram of the number of phosphorus atoms in 2x2x2 voxel (150 nm^3) regions of chromatin in nucleus of *Drosophila* larval cell, and from adjacent region of Epon embedding medium; no correction is applied for section shrinkage in z-direction.

Discrete tomography in materials science: less is more?

S. Bals, K.J. Batenburg, G. Van Tendeloo

1. EMAT, University of Antwerp, Groenenborgerlaan 171 B-2020 Antwerp, Belgium
2. Vision Lab, University of Antwerp, Universiteitsplein 1 B-2610 Wilrijk, Belgium

Sara.Bals@ua.ac.be
Keywords: discrete tomography, carbon nanotubes

Over the last decade, electron tomography has become a powerful tool in the investigation of nanostructures. In order to satisfy the projection requirement, different imaging techniques in the transmission electron microscope (TEM) have been successfully combined with tomography, whereas the development of dedicated tomography holders minimizes the missing wedge problem. Despite these efforts, quantitative interpretation of a 3D reconstruction is still not straightforward. Not only the remaining artifacts cause difficulties, also the segmentation step in which different grayscales in the reconstruction are linked to different compositions in the original structure is subjective and hampers a quantitative interpretation. Here, we therefore evaluate the use of so-called discrete tomography for quantitative electron tomography in materials science.

Discrete tomography is an alternative reconstruction technique that uses prior knowledge on the object that has to be reconstructed [1]. More specifically, an estimate of the number of different graylevels (corresponding to different compositions) that can be expected in the original object serves as an input. One of the advantages of this approach is an improvement of the quality of the reconstructions in comparison to reconstructions based on the conventional reconstruction algorithms. Moreover, if the number of possible grayvalues is small, it is often possible to obtain very accurate reconstructions based on a few projections. This is possible by exploiting the extra prior knowledge on the grayvalues in the image. Furthermore, missing wedge artifacts are strongly reduced, and segmentation is performed automatically during the reconstruction procedure.

In a first example, high angular annular dark field scanning TEM (HAADF-STEM) tomography is used to study the 3D inner structure and composition of catalytic nanoparticles used during growth of bamboo-like carbon nanotubes [2]. A 2D image from the tilt series is presented in Figure 1. The bamboo structure of the carbon nanotube as well as the catalyst particle can be clearly distinguished. In Figure 2, 3D reconstructions of the catalyst particle present at the tip of the carbon nanotube are shown. By comparing a conventional reconstruction algorithm (SIRT) and the discrete reconstruction, the advantages of using discrete tomography (DART algorithm) are obvious. Artifacts are strongly reduced and furthermore, the discrete reconstruction is already segmented, a major advantage when one wants to obtain quantitative information in 3D. In both the SIRT and DART reconstructions, it is obvious that the reconstructed catalyst particle is not homogeneous. Different grayscale values (labeled A, B, and C in Figure 2.f) are present inside the particle, suggesting that these areas

M. Luysberg, K. Tillmann, T. Weirich (Eds.): EMC 2008, Vol. 1: Instrumentation and Methods, pp. 291–292, DOI: 10.1007/978-3-540-85156-1_146, © Springer-Verlag Berlin Heidelberg 2008

have a different density and therefore, different compositions. The intensity of C is comparable to the intensity of the vacuum surrounding the carbon nanotube, suggesting that cavities are present inside the particle. Using EFTEM maps, the different compositions have been identified as Cu and CuO_2. The improved quality of the discrete tomography and its reliable segmentation (confirmed by simulations), allows one to obtain quantitative information in a straightforward manner by counting the number of voxels corresponding to each graylevel.

Other examples where the use of discrete tomography is beneficial include mesoporous structures and Au nanoparticles with sub nanometer dimensions.

1. K.J. Batenburg, S. Bals, J. Sijbers, C. Kübel, U. Kaiser, E. Gomez, N.P. Balsara, C. Kisielowski to be submitted to Ultramicroscopy.
2. S. Bals, K.J. Batenburg, J. Verbeeck, J. Sijbers, G. Van Tendeloo Nano Letters **7** (2007), p.3369.
3. S.B. and K.B. are grateful to the Fund for Scientific Research-Flanders (Contract No. G.0247.08).

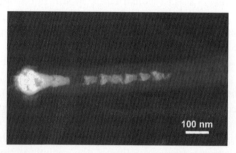

Figure 1. HAADF-STEM image from the acquired tilt series showing the catalyst particle and the bamboo-like carbon nanotube. The catalyst material is partially filling the hollow compartments.

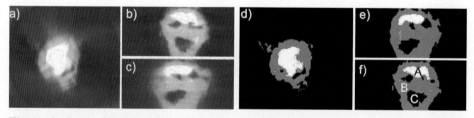

Figure 2. In a), b) and c) 3 orthogonal slice through the catalytic nanoparticle as reconstructed by SIRT are shown, whereas in d), e) and f), the corresponding slices obtained by discrete tomography are presented. In Figure 2.f, areas with different grayscale and possibly different composition are labeled A, B and C.

Towards atomic-scale bright-field electron tomography for the study of fullerene-like nanostructures

M. Bar Sadan[1,2], L. Houben[1], S.G. Wolf[3], A. Enyashin[4], G. Seifert[4], R. Tenne[2] and K. Urban[1]

1. Institute of Solid State Research and Ernst-Ruska Centre for Microscopy and Spectroscopy with Electrons, Research Centre Jülich GmbH, 52425 Jülich, Germany
2. Materials and Interfaces Dept., Weizmann Institute of Science, Rehovot 76100, Israel
3. Electron Microscopy Unit, Weizmann Institute of Science, Rehovot 76100, Israel
4. Physikalische Chemie, Technische Universität Dresden, 01062 Dresden, Germany.

m.bar-sadan@fz-juelich.de
Keywords: Tomography, Aberration Corrected Microscopy, Inorganic Fullerenes

Tomographic reconstruction from series of tilted objects is a widely used application in transmission electron microscopy (TEM), in particular in the biological sciences. In the field of physical sciences, the tomographic resolution attained so far has not surpassed the barrier of 1 nm^3. Yet, the knowledge of the three-dimensional structure and composition on the atomic scale hidden so far holds the information about the unique physical properties of nanomaterials compared with their bulk ancestors.

The basis for our approach to atomic-resolution tomography are delocalisation-reduced phase contrast in an aberration-corrected TEM in combination with low voltage operation [1]. The method is applicable to the analysis of the atomic architecture of a wide range of nanostructures where strong electron diffraction is absent, in particular carbon fullerenes and inorganic fullerenes. The benefit of using negative spherical aberration imaging (NCSI) is a close representation of the projected potential for such weakly scattering objects in the image and high sensitivity for light elements such as oxygen [3]. The point resolution of about 2 Å at 80 kV in the NCSI mode outperforms the point resolution of conventional microscopes at much higher acceleration voltages while the rate for knock-on radiation damage is significantly reduced.

Evidence for the feasibility of atomic-resolution electron tomography based on aberration-corrected imaging is given by first experimental data and a simulation study on the 3D reconstruction of inorganic MoS_2 nanooctahedra. Figure 1 shows the tomographic reconstruction using weighted backprojection of simulated images for a model structure established from a quantum mechanical approach [1,2]. The tomogram reconstitutes key features for the physical properties within the atomic structure: the coordination of Mo around the basal plane, the non-mirror symmetry in the seaming of three triangular MoS_2 faces and the square defect at the apices in the model structure on a scale better than 3 Å in all three dimensions.

Experimental tilt-series of MoS_2 nanooctahedra were acquired in a FEI Titan 80-300 transmission electron microscope equipped with a CEOS double-hexapole aberration-corrector. The tomogram in Figure 2 was reconstructed from a set of 22 experimental images with an angular step of 3°. The nested shells with their smallest separation of

M. Luysberg, K. Tillmann, T. Weirich (Eds.): EMC 2008, Vol. 1: Instrumentation and Methods, pp. 293–294, DOI: 10.1007/978-3-540-85156-1_147, © Springer-Verlag Berlin Heidelberg 2008

6.15 Å are reproduced in all of the slices. Internal structure features such as the vertex of the 2nd shell are uncovered. Details of the atomic coordination at the vertex are still hidden in this reconstruction due to a large missing wedge. However, the resolution of 2.8 Å obtained for the projected Mo distances in the xy-plane (Figure 2d) agrees with a predicted resolution better than 3 Å. The overall resolution in the experimental tomogram achieved so far is about $0.3 \times 0.6 \times 0.6$ nm^3 = 0.11 nm^3, an improvement of nearly one order of magnitude compared to state of the art electron tomography.

1. M. Bar-Sadan, L. Houben, S. G. Wolf, L. Enyashin, G. Seifert, R. Tenne, and K. Urban, Nanoletters, in press (2008).
2. M. Bar-Sadan, A. N. Enyashin, S. Gemming, R. Popovitz-Biro, S. Y. Hong, Y. Prior, R. Tenne, and G. Seifert, Journal of Physical Chemistry B **110** (2006), p. 25399-25410.
3. C. L. Jia, M. Lentzen, and K. Urban, Science **299** (2003), 870-873.

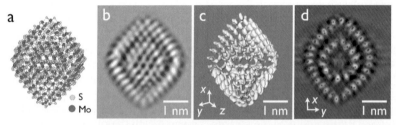

Figure 1. Simulation study: (a) A model structure of a two shell nanooctahedron. (b) A simulated 80 KV NCSI image for a single tilt of the particle. (c) Isosurface 3D visualization of a tomogram reconstructed from the simulated tilt series (120°, 1° step, tilt axis vertical). (d) Overlayed projected atom positions on a slice of the tomogram.

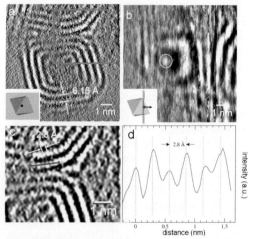

Figure 2. Experimental tomogram of a MoS$_2$ nanooctahedron: (a) A slice in the xy plane containing the tilt-axis (b) A slice in the xz plane orthogonal to the tilt-axis. (c) A magnified part of a, showing the resolved Mo-Mo distances, also presented in the line profile (d) along the marked arrow in c.

DART explained:
how to carry out a discrete tomography reconstruction

K.J. Batenburg[1], S. Bals[2], J. Sijbers[1], G. Van Tendeloo[2]

1. Vision Lab, University of Antwerp, Universiteitsplein 1 B-2610 Wilrijk, Belgium
2. EMAT, University of Antwerp, Groenenborgerlaan 171 B-2020 Antwerp, Belgium

Joost.batenburg@ua.ac.be

Keywords: discrete tomography, DART

Electron tomography is an important technique to investigate the three-dimensional structure of EM specimens in both micro-biology and materials science. Discrete Tomography is a relatively new computational technique for reconstructing images that consist of only a few different grey levels from their projection data [1]. This approach can be used effectively in electron tomography to reconstruct specimens that contain only a few different compositions.

Although the mathematical theory behind discrete tomography has been studied since the 1990s, the technique has not yet been used in many practical applications, due to the fact that no efficient and robust algorithm was available. Recently, a new algorithm for discrete intensity tomography, called DART (Discrete Algebraic Reconstruction Technique), was proposed [2,3,4] that is capable of reconstructing large 2D images (i.e., 1024x1024) in a few minutes on a standard PC. The same approach can be used for reconstructing 3D volumes, treating the volume as a stack of 2D slices. DART is an iterative reconstruction algorithm that alternates between steps of a continuous iterative algorithm, such as SIRT, and intermediate segmentation steps; see Figure 1.

For several reasons, applying DART to experimental datasets is currently not a straightforward task. After alignment of the tilt series, a sequence of steps must be carried out to compute a DART reconstruction.

First, after alignment of the projection images, further preprocessing of the projection data is often necessary to make it suitable for discrete tomography. This is a consequence of normalization steps commonly applied in the acquisition software, which are mathematically unsound. As DT aims at exploiting certain quantitative properties of the dataset, its requirements on processing of the projection data are more strict compared to conventional continuous tomography.

Subsequently, a continuous reconstruction must be computed to estimate the grey levels which will be used in the discrete reconstruction. Algebraic methods, such as SIRT are preferable to backprojection methods for this step, as they allow for a quantitative interpretation of the grey levels in the reconstructed image by direct comparison in the projection domain. Estimating the grey levels for each of the compositions in the sample is a relatively easy task if the materials are well-separated spatially, but can be difficult if the features of interest are close to the resolution of the microscope.

M. Luysberg, K. Tillmann, T. Weirich (Eds.): EMC 2008, Vol. 1: Instrumentation and Methods, pp. 295–296, DOI: 10.1007/978-3-540-85156-1_148, © Springer-Verlag Berlin Heidelberg 2008

Finally, the DART algorithm has several parameters which can affect the quality of the reconstruction. These parameters regulate the trade-off between the use of prior knowledge on one hand, and consistency with the measured projection data on the other hand. Their optimal values depend on acquisition characteristics, such as the noise level and the total number of projection image

In this presentation, the complete reconstruction procedure for DART will be demonstrated: preprocessing the aligned projection data, estimating the grey levels, choosing suitable algorithm parameters and carrying out the actual reconstruction. Particular attention will also be given to the various algorithmic steps that comprise DART, as shown in Figure 1.

1. G.T. Herman and A. Kuba, eds: Advances in Discrete Tomography and its Applications, Birkhäuser, Boston, 2007
2. K.J. Batenburg and J. Sijbers, Proc. of ICIP 2007, p.
3. K.J. Batenburg, S. Bals, J. Sijbers, C. Kübel, U. Kaiser, E. Gomez, N.P. Balsara, C. Kisielowski to be submitted to Ultramicroscopy
4. S. Bals, K.J. Batenburg, J. Verbeeck, J. Sijbers, G. Van Tendeloo Nano Letters 7 (2007), p. 3369
5. S.B., K.B. and J.S. are grateful to the Fund for Scientific Research-Flanders (Contract No. G.0247.08).

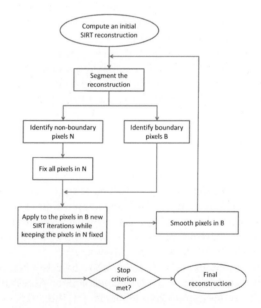

Figure 1. Flow chart of the DART algorithm.

Optical depth sectioning of metallic nanoparticles in the aberration-corrected scanning transmission electron microscope

G. Behan, A.I. Kirkland and P.D. Nellist

Department of Materials, University of Oxford, Parks Road, Oxford, OX1 3PH, United Kingdom

gavin.behan@materials.ox.ac.uk

Keywords: depth sectioning, scanning transmission electron microscopy, aberration correction

Aberration correctors have proved themselves an important addition to the electron microscope. The resulting increased numerical aperture of the objective lens has allowed sub-angstrom resolution [1]. The increased numerical aperture also reduces the depth of field [2], which is just a few nanometres for an aberration-corrected scanning transmission electron microscope (STEM). We can use this reduction to provide three-dimensional information by optically sectioning our sample much like in confocal optical microscopy, or probe specific depths in our sample. While we do not expect this technique will be able to compete with tomography for spatial resolution, the advantage is that data acquisition is a matter of minutes as opposed to hours and specific depths within a sample can be probed without the need for a lengthy reconstruction.

Optical sectioning has been demonstrated in previous work for individual Hf atoms in a gate oxide [3] and for heterogeneous catalyst nanoparticles [4]. In our own experiments, we examined Au and Pt nanoparticles of differing sizes. The data was recorded on the Oxford-JEOL 2200MCO at Oxford University, fitted with aberration correctors both pre- and post-sample, with an accelerating voltage of 200kV and semi-angle of convergence (α) of 22 mrad. Under these experimental conditions, we might expect a depth resolution of approximately 8.0 nm. Figure 1 shows the integrated intensity of a 3.1 nm Au particle plotted as a function of defocus, and the graph shows a much worse full-width half-maximum (FWHM) of approximately 200 nm. If we calculate the three-dimensional optical transfer function (OTF – see Figure 2) for our electron microscope, which is the Fourier transform of the three-dimensional probe intensity function we observe a missing cone region where the transfer is zero. While this result is well known in light optics [5] the size of the missing cone is controlled by the probe-forming beam convergence angle (Fig 2). For electron microscopy, the semi-angle of convergence is of the order of milliradians even with aberration correction, therefore most of Fourier space is set to zero. Extended objects consist predominantly of low lateral spatial frequencies, and therefore only explore regions of the OTF that provide poor depth resolution. Conversely, a single atom would have much stronger transfer in this direction because of the greater emphasis on higher lateral spatial frequencies. For an extended object, the depth resolution is given approximately by d/α where d is the size of our object, which agrees with the measured FWHM in Fig. 1.

The inherent poor depth resolution when optical sectioning in STEM provides a great motivation to look at other methods to improve the depth resolution, both

M. Luysberg, K. Tillmann, T. Weirich (Eds.): EMC 2008, Vol. 1: Instrumentation and Methods, pp. 297–298, DOI: 10.1007/978-3-540-85156-1_149, © Springer-Verlag Berlin Heidelberg 2008

experimental and computational. In light optics the use of confocal microscopy has solved the missing cone problem. Scanning confocal electron microscopy (SCEM) is one such technique [6,7] which provides an interesting challenge. Here we explore computational approaches. Maximum likelihood deconvolution methods are unable to recover this missing information, but we can make use of prior information by making assumptions about our sample to generate a simple 3D reconstruction (see Figure 3) of our sample. For our reconstructions, we have assumed that the particles are spherical, and that the particle is located at the defocus value where the edge of the particle is sharpest as determined by the use of the Sobel operator [8]

1. P. D. Nellist et al., *Science* **305** (2004) 1741
2. M. Born and E. Wolf, "Principles of Optics: Electromagnetic Theory of Propagation, Interference, and Diffraction of Light", (Pergamon, New York) (1989).
3. A.Y. Borisevich, A.R. Lupini, and S.J. Pennycook, *Proc. Natl. Acad. Sci.* **103** (2006) 3044.
4. K. Van Benthem, et al., *Appl. Phys. Lett.* **87** (2005) 034104.
5. B.R. Frieden, *J. Opt. Soc. Am.* **57** (1967) 36.
6. S.P. Frigo, Zachary H. Levine and Nestor J. Zaluzec, *Appl. Phys. Lett.* **81** (2002) 2122
7. P.D. Nellist, et al., *Appl. Phys. Lett.* **89** (2006) 124105.
8. J.C. Russ, "*The image processing handbook*", (CRC Press, Boca Raton) (2002)

9. The authors would like to acknowledge the support of Intel Ireland and the Department of Materials at the University of Oxford.

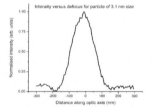 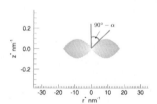

Figure 1. Plot of integrated HAADF intensity of a 3.1 nm particle as a function of defocus (see text for experimental conditions).

Figure 2. The logarithm of the 3D OTF of the STEM (see text for experimental conditions) plotted as a function of radial (r*) and longitudinal (z*) spatial frequencies. Note the large region of zero transfer preset

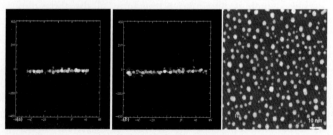

Figure 3. Projections of Au on C along one axis. The focal series was recorded between -300 nm and 294 nm in 6.0 nm steps. (a) shows a projection of the 3D volume. Sample was not tilted (b) shows the projection of the same sample which was tilted 20 degrees. Note the slight incline. (c) is an image of Au particles on C taken from the focal series.

3D-Geometrical and chemical quantification of Au@SiO$_x$ nano-composites in HAADF-STEM imaging mode

S. Benlekbir[1], T. Epicier[1], M. Martini[1,2], P. Perriat[1]

1. University of Lyon, INSA de Lyon, MATEIS, UMR CNRS 5510, Bât. B. Pascal, F-69621 Villeurbanne Cedex
2. Laboratory of Physico-Chemistry of Luminescent Materials, UMR CNRS 5620, University of Lyon 1, Berthollet Building, F-69622 Villeurbanne

samir.benlekbir@insa-lyon.fr

Keywords: STEM-HAADF, stereoscopy, nano-hybrids

Hybrid gold and Silica nano-composites (Au@SiO$_x$) can be used for photonic, optical and biomedical applications, for example for the development of novel contrast agents in magnetic resonance imaging (e.g. [1]). According to the synthesis conditions, these nano-hybrids can adopt various configurations, such as core-shell structures, distribution of Au nano-particles exclusively inside, or at the surface of the silica-based particles. The synthesis involves micro-emulsions; the native micellar structure (oil in water, presence of surfactant, co-surfactant) defines the final silica morphology, and the localization of gold particles with respect to the silica ones depends on the order of adding nano-reactors. An optimised feedback on the synthesis conditions requires a detailed geometrical and chemical analysis of the final products. According to the nanometric size of these objects (see figure 1), a stereoscopy approach in HAADF (High Angle Annular Dark Field) imaging in a Transmission Electron Microscope appears to be an elegant way for that purpose.

The method consists in acquiring few images but over a large tilt range, and to reconstruct the positions of all particles with the reasonable assumption that they can be described as spheres. It should be noted in that case that complete tilting tomography was unsuccessful in the present work, owing to the lack of stability of the particles during prolonged exposure under the electron beam.

Specific routines were developed as scripts of Digital Micrograph^{TM}, Gatan, in order to analyse precisely the stereoscopic series. They essentially consist in extracting the (x_i, y_i) positions of each particle at given tilts θ_i, and reconstructing their 3D-positions (X, Y, Z) according to simple trigonometric considerations. Taking into account the non-sphericity 'error' of the largest silica particles that can be measured from various projections, the localisation of the Au particles could be assessed with a relatively good confidence. In the case of figure 1, all gold particles appear to be at the surface of the silica, while other systems show a preferential localisation inside silica.

A further advantage of HAADF concerns the ability of quantifying the mass-thickness of the studied material in terms of atomic density and atomic number, as initially proposed by Treacy and Rice [2]. In figure 1, a core-shell structure of the silica particle is evidenced, which was produced by the use of various alcoxysilanes (such as APTES and TEOS) during the synthesis. Because HAADF imaging is insensitive to

M. Luysberg, K. Tillmann, T. Weirich (Eds.): EMC 2008, Vol. 1: Instrumentation and Methods, pp. 299–300, DOI: 10.1007/978-3-540-85156-1_150, © Springer-Verlag Berlin Heidelberg 2008

300

crystalline orientations effects, an accurate calibration could be done using the homogeneous gold particles, following the linear treatment proposed in [2] and shown in figure 2a). This procedure allows the experimental HAADF intensity to be scaled according to the basic relationship:

$$I_{HAADF} = c\,\rho_i\,Z_i^{\alpha}$$

which α measured to 1.85, ρ_i = atomic density (in number of atoms per unit volume), Z_i = atomic number of specie i, c = calibration constant.

From this calibration, it will be shown that the experimental intensity profile across the silica particle could be precisely matched with a core-shell model where the outer region is 25 % less-dense than the central one, as shown in figure 2b). [3]

1. P.J. Debouttière, S. Roux,F. Vocanson, C. Billotey, O. Beuf, A. Favre-Réguillon, Yi Lin, S. Pellet-Rostaing, R. Lamartine, P. Perriat and O. Tillement, *Adv. Funct. Mater.* 16 (2006), 2330.
2. M.M.J. Treacy, S. Rice, *J. Microsc.*, 156, (1989), 211.
3. The CLYM (Centre Lyonnais de Microscopie) is gratefully acknowledged for the access to the 2010F microscope.

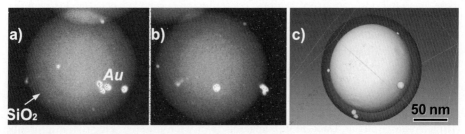

Figure 1: a-b): HAADF images of a gold and silica nano-composite at tilts equal to -25 and +25° respectively; c): surface rendering of its reconstruction: two large spheres are displayed, which correspond to the minimum and maximum diameters (respectively 128 and 140 nm) of the silica particle as measured from various projections.

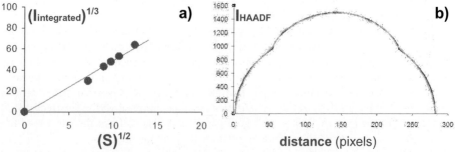

Figure 2: a): plot of integrated intensity $(I_{integrated})^{1/3}$ as a function of the square root of the projected surface $S^{1/2}$ of the gold nano-particles from figure 1; b): experimental intensity profile (grey line) across the SiO_x particle, and superimposed calculated profile (dark line) from a core-shell model where the silica shell has an atomic density equal to 75 % of that of the core region.

Electron tomography of mesostructured cellular foam silica

E. Biermans[1], S. Bals[1], E. Beyers[2], D. Wolf[3], J. Verbeeck[1], P. Cool[2] and G. Van Tendeloo[1]

1. EMAT, University of Antwerp, Groenenborgerlaan 171, B-2020 Antwerp, Belgium
2. Laboratory of Adsorption and Catalysis, University of Antwerp, Universiteitsplein 1, B-2610 Wilrijk, Belgium
3. Triebenberg Laboratory, Institute of Structure Physics, Technische Universität Dresden, D-01062 Dresden, Germany

Ellen.Biermans@ua.ac.be
Keywords: Electron tomography, mesoporous silica

In this study, the 3D structure of mesostructured cellular foam silica (MCF) [1] activated with titanium dioxide is investigated using bright field electron tomography. MCF is a structure yielding high surface area and large pore sizes which explains the interest in this material for catalyst applications. The MCF materials were activated following the acid-catalysed sol-gel method [2,3].

Electron tomography was performed in bright field TEM imaging mode. Therefore, a tilt series was recorded at 2° tilt-angle increments, over a tilt range of -68° to 56°, by a software package developed at the Triebenberg Laboratory. The software was originally developed for holographic tomography [4], e.g. on a Tecnai microscope, but was adapted to work on a Jeol 3000F. The software is written mainly in Digital Micrograph scripting language and can easily be adapted to any microscope with computer control. The 3D volume was reconstructed in Inspect3D with the SIRT algorithm using 30 iterations. Figure 1 shows a single 2D projection from the tilt series where the white rectangle indicates the area which was used for 3D reconstruction. A rendered isosurface together with an orthoslice through the structure is shown in Figure 2. From the reconstruction, the 3D spherical pore structure of the MCF material can clearly be recognized. The pore size is found to be approximately 20nm, which is in good agreement with the N_2-sorption results previously obtained for this structure. Furthermore, our results confirm that the pores are connected with each other. Further efforts will be made to locate the presence of titania nanoparticles in the structure.

1. P. Schmidt-Winkel, W.W. Lukens, D. Zhao, P. Yang, B.F. Chmelka, G.D. Stucky, J. Am. Chem. Soc., 121 (1999), 254-25. K. De Witte, V. Meynen, M. Mertens, O.I. Lebedev, A. Sepulveda-Escribano, F. Rodriguez-Reinoso, G. Van Tendeloo, E.F. Vansant, P. Cool, Appl. Catal. B - Env., In press (2008).
2. A. Battacharyya, S. Kawi, M.B. Ray, Catalysis Today, 98 (2004), 431-439.
3. K. De Witte, V. Meynen, M. Mertens, O.I. Lebedev, A. Sepulveda-Escribano, F. Rodriguez-Reinoso, G. Van Tendeloo, E.F. Vansant, P. Cool, Appl. Catal. B - Env., In press (2008).
4. D. Wolf, A. Lenk, H. Lichte contribution at this conference.
5. The authors acknowledge financial support form the European Union under the Framework 6 program under a contract for an Integrated Infrastructure Initiative.

M. Luysberg, K. Tillmann, T. Weirich (Eds.): EMC 2008, Vol. 1: Instrumentation and Methods, pp. 301–302, DOI: 10.1007/978-3-540-85156-1_151, © Springer-Verlag Berlin Heidelberg 2008

302

Reference 026019 ESTEEM. S.B. is grateful to the Fund for Scientific Research-flanders (Contract No. G. 0247.08). V. M. also acknowledges the Fund for Scientific Research-Flanders for financial support. The Deutsche Forschungsgemeinschaft is kindly acknowledged for financial support. The research was performed in the frame of a concerted research project, funded by the Special Research Fund from the University of Antwerp.

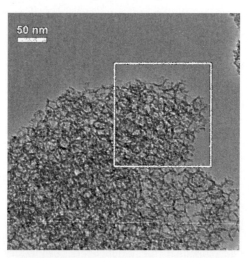

Figure 1. 2D projection obtained using bright field TEM. The area indicated by the white rectangle is used for 3D reconstruction.

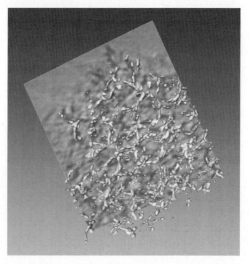

Figure 2. Rendered isosurface of the 3D reconstruction showing the spherical pore structure of the material.

A Study of Stacked Si Nanowire Devices
by Electron Tomography

P.D. Cherns[1], C. Dupré[1,2], D. Cooper[1], F. Aussenac[1], A. Chabli[1] and T. Ernst[1]

1. CEA-LETI, Minatec, 17 Avenue des Martyrs, 38054 Grenoble Cedex 9, France
2. IMEP, INPG-MINATEC, 3 Parvis Louis Ne´ el, 38016 Grenoble Cedex 1, France

peter.cherns@cea.fr
Keywords: tomography, nanowires, Gate-All-Around

In this study, we have investigated the use of electron tomography for the characterisation of novel stacked Si nanowire 3D devices. There has been significant interest in recent years in applying this technique to semiconductor device structures[1]. As devices become non-planar, and features become smaller than TEM specimen thickness, 3D analysis is required.

Stacked nanowire Gate-All-Around MOSFETs (GAAFET) offer excellent electrostatic control, combined with a high I_{ON} current [2]. During processing, a $(Si/Si_{0.8}Ge_{0.2}) \times n$ superlattice is epitaxially grown on top of a silicon-on-insulator (SOI) substrate and $(Si/SiGe) \times n$ fins are formed, by anisotropic plasma etching. The SiGe between the Si wires is removed and for a Gate-All-Around configuration (GAAFET) gate stack integration then takes place. Alternatively, spacers are introduced between the wires for a ΦFET configuration, which offers independent gate control [3]. A High Angle Annular Dark Field (HAADF) image and Energy Filtered TEM (EFTEM) bulk Si plasmon map of a typical ΦFET structure are shown in Figure 1. The surface morphology and roughness of the wires has a significant impact on the properties of the device. Our aim in this work has been to optimise sample preparation, series acquisition and reconstruction, to enable a quantitative 3D study of these samples.

The work presented has been carried out on an FEI Titan microscope, operating at 300kV with a CEOS probe corrector installed. For the tomographic series, the sample was mounted in a Fischione high tilt sample holder, allowing tilts up to +/- 80°.

It has been found during this study that conventional FIB sample preparation methods are not sufficient to obtain a region of interest that can be imaged successfully at tilts higher than +/-60°. We present and compare results obtained both by using novel methods of FIB preparation, most successfully with the preparation of cylindrical samples through annular milling, and also by the adaptation of typical in-situ lift out prepared lamellas. In either configuration, we find it possible to obtain samples with the nanowires in-plane (Figure 2) that can be imaged at +/- 70° without significant shadowing. Cylindrical samples are optimal due to constant thickness with tilt angle.

Figure 2 shows an example of the 0° HAADF image from a tilt series of a GAAFET device. The series consisted of 140 images acquired every 1° between +/-70°. Post acquisition alignment and reconstruction was carried out using FEI Inspect 3D software. Three perpendicular slices through the reconstructed volume are also shown. We demonstrate the use of reconstructions such as this one to investigate the roughness and

M. Luysberg, K. Tillmann, T. Weirich (Eds.): EMC 2008, Vol. 1: Instrumentation and Methods,
pp. 303–304, DOI: 10.1007/978-3-540-85156-1_152, © Springer-Verlag Berlin Heidelberg 2008

variation in thickness of the Si nanowires along the gate length, after processing for both GAAFET and ΦFET architectures. This information cannot be obtained from conventional 2D imaging techniques alone, due to projection effects.

1. C. Kubel, A. Voigt, R. Schoenmakers, M. Otten, D. Su, T.C. Lee, A. Carlsson and J. Bradley. Microscopy and Microanalysis, **11** (2005) p378.
2. C. Dupré, T. Ernst, V. Maffini-Alvaro, V. Delaye, J.-M. Hartmann, S. Borel, C. Vizioz, O. Faynot, G. Ghibaudo and S. Deleonibus Solid State Electron, Vol 52/4 pp 519-525
3. C. Dupré, T. Ernst, C. Arvet, F. Aussenac, S. Deleonibus, G. Ghibaudo, SOI Conference, 2007 IEEE International 1-4 Oct. 2007 p. 95 – 96
4. This work was supported by the French National Research Agency (ANR) through Carnot funding. The authors of this work would like to thank the different persons involved in LETI multiwires devices fabrication (V. Maffini-Alvaro, J.-M. Hartmann, S. Pauliac, C. Vizioz and C. Arvet).

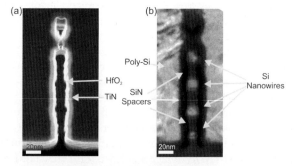

Figure 1. Images acquired from a 4 wire ΦFET structure : (a) HAADF, (b) EFTEM plasmon map acquired at 17eV, around the Si bulk plasmon.

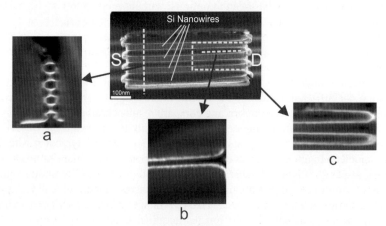

Figure 2. A tomographic series was acquired from a GAAFET structure. The 0° HAADF image is presented, along with three slices from the reconstruction.

Simulation of the electron radiation damage in an amorphous Ge sample

M.D. Croitoru[1], D. Van Dyck[1], S. Le Roux[2], and P. Jund[2]

1. EMAT, University of Antwerp, Groenenborgerlaan 171, B-2020, Antwerp, Belgium
2. Laboratoire de Physicochimie de la Matière Condensée-Institut Charles Gerhardt, Université Montpellier 2, Place E. Bataillon, Case 03, 34095 Montpellier, France

mihail.croitoru@ua.ac.be

Keywords: radiation damage, molecular dynamics, electron tomography

Transmission electron microscopy (TEM) with aberration correction provides one of the most powerful techniques for revealing atomic scale structure of solid samples. This technique has been widely used to solve the structure of crystalline samples. In practice, a lot of materials that are of key applications are either highly disordered or amorphous. The fact that the many interesting nanostructures are three dimensional and disordered makes them at the same time also much more difficult to study. Indeed in such structures many atoms are at non-crystallographic equilibrium positions and are thus much more prone to displacive radiation damage. This is essentially true in electron tomography [1], where the recording time is large, the accumulation of these defects during the tomographic acquisition process can lead to a significant degradation of the object, and hence to a significant influence of irradiation on the object reconstruction. Moreover, due to the improvement of resolution of TEM, the flux of electrons that will irradiate a local region in the experiment is rather large. Radiation damage is a serious problem also for structure analysis in nanomaterials that have many atoms on the surfaces, which tends to migrate under the illumination of electrons. This poses a fundamental limit on the ultimate resolution of the technique; especially for amorphous structures and it is questioned whether atomic resolution tomography is possible at all for such structures. Developing TEM technique for solving the structure of amorphous materials or highly disordered is of great challenge to today's microscopy.

The effects of change of physical, structural and chemical properties of the electron irradiated solids have been known for a long time and have provided a challenge to solid state physicists long before the advent of electron tomography in materials science and availability of aberration correctors in the new generation of electron microscopes [2]. In general, however, despite considerable efforts to shed light on the question of the radiation effects in solids, this problem has not been adequately considered for strongly disordered and amorphous structures. The investigation of these effects is indispensable for optimization of the TEM tomography experiment with amorphous objects so as to achieve the highest possible precision of the atomic positions in such structures against the minimal radiation damage [3].

A suitable tool to address the problem of sample modification due to electron irradiation beam is molecular dynamics simulation, in which an illuminated highly energetic electron is introduced in the structure and the phase trajectories of the damage-

M. Luysberg, K. Tillmann, T. Weirich (Eds.): EMC 2008, Vol. 1: Instrumentation and Methods, pp. 305–306, DOI: 10.1007/978-3-540-85156-1_153, © Springer-Verlag Berlin Heidelberg 2008

involved atoms are followed. For this purpose highly efficient, state-of-the-art computational code FIREBALL [4], designed to perform calculations using ab initio tight-binding molecular dynamics, having favorable accuracy/efficiency balance based on TB-DFT (Tight-Binding Density Functional Theory) is employed. Using a modified Rutherford differential cross section for modelling elastic scattering we predict the radiation damage in the amorphous Ge object as a function of electron energy and dose.

1. D. Van Dyck, S. Van Aert, and M. D. Croitoru, Int. J. Mat. Res. **97** (2006) 7.
2. M Haider et al., Ultramicroscopy 75 (1998) 53; M Haider et al., NATURE 392 (1998) 768; H. Rose, Optik 85 (1990) 19.
3. D. Van Dyck, S. Van Aert, A.J. den Dekker, A. van den Bos, Ultramicroscopy 98 (2003) 27.
4. http://fireball.phys.wvu.edu/LewisGroup/.

Observation of Three-dimensional Elemental Distribution by using EF-TEM Tomography

N. Endo[1], C. Hamamoto[1], H. Nishioka[1] and T. Oikawa[2]

1. JEOL. Ltd., Electron Optics Division, 1-2 Musashino 3-chome, Akisima, 196-8558, Tokyo, Japan
2. JEOL (EUROPE) SAS, Espace Claude Monet, 1, allee de Giverny 78290 Croissy-sur-Seine, France

nendo@jeol.co.jp

Keywords: Tomography, EF-TEM

The electron beam tomography draws recently considerable attention as a technique which visualizes the three-dimensional internal structure of various materials (a polymer, a catalyst, a semiconductor, etc.) by the spatial resolution of a nanometer scale [1-4]. In this paper, we reported three-dimensional elemental distribution using the energy filtering TEM (EF-TEM) tomography [5,6].

A specimen was a coating film which contained the particles of TiO_2. A thin film was prepared from the specimen by an ultra microtome. The EF-TEM tomography was carried out using a field-emission transmission electron microscope *JEM-2100F* (JEOL Ltd.) and an energy filter *GIF Tridiem* (GATAN Inc.). Energy filtered images of the thin film were recorded with the 4 degrees step over tilting angle from 68 degrees to -68 degrees. The total time of data acquisition was about one hour, and *3D Tomography Software* (GATAN Inc.) was used for data acquisition. Then, reconstruction of 3D tomography was performed by *Composer* (JEOL Ltd.) from 35 EF-TEM images. Visualization of that was performed by *Visualizer* (JEOL Ltd.).

A zero-loss image shows that the coating film contains the particles and the plates of various shapes as shown in Fig. 1. A titanium map by the Ti-L edge calculated by the three-window method is shown in Fig. 2. This map shows that large black particles with spherical shapes in Fig. 1 contain titanium and some of black particles dose not contain titanium. Figure 3 shows a three-dimensional elemental distribution of titanium which was reconstructed from the EF-TEM images. The three-dimensional elemental map clearly selects the target titanium particles. This report clearly shows the usefulness of this technique for observation of three-dimensional elemental distribution.

1. J. Frank (Ed.), Electron Tomography, Plenum Press, New York, (1992)
2. H. Sugimori, T. Nishi, and H. Jinnai, Macromolecules, **38**, 24 (2005), p.10226.
3. P. A. Midgley, J. M. Thomas, L. Laffont, M. Weyland, R. Raja, B.F.G. Johnson, and
4. T. Khimyak, J. Phys. Chem. B, **108** (2004), p.4590.
5. H. Bender, O. Richard, A. Kalio, and E. Sourty, Microelectronic Engineering, **84**, 11 (2007), p.2707.
6. P. A. Midgley, and M. Weyland, Ultramicroscopy, **96** (2003), p.413.
7. M. Weyland, and P. A. Midgley, Microscopy and Microanalysis, **9** (2003), p.542.

M. Luysberg, K. Tillmann, T. Weirich (Eds.): EMC 2008, Vol. 1: Instrumentation and Methods, pp. 307–308, DOI: 10.1007/978-3-540-85156-1_154, © Springer-Verlag Berlin Heidelberg 2008

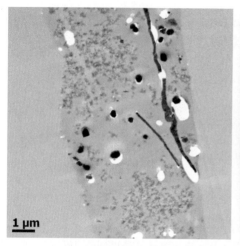

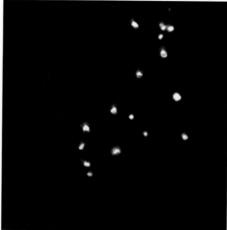

Figure 1. Zero-loss image of coating film   **Figure 2.** Titanium map by Ti-L edge

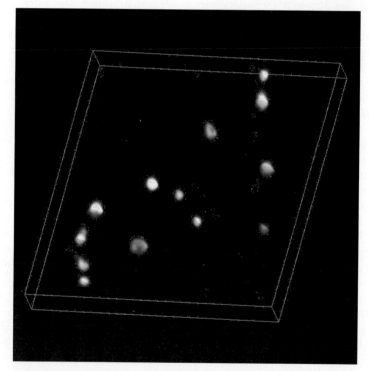

Figure 3. Three-dimensional elemental distribution of titanium

HAADF-TEM Tomography of the precipitation state in an Al-Zn-Mg alloy

T. Epicier[1], S. Benlekbir[1], F. Danoix[2]

1. Université de Lyon; INSA-Lyon, MATEIS, umr CNRS 5510, bât. B. Pascal, F-69621 Villeurbanne Cedex
2. Groupe de Physique des Matériaux, UMR CNRS 6634, University of Rouen, F-76801 Saint Etienne du Rouvray Cedex

Thierry.epicier@insa-lyon.fr
Keywords: TEM tomography, HAADF, precipitation, Al alloy, $MgZn_2$

Nano-tomography in Transmission Electron Microscopy (TEM) has been recently and successfully applied to the characterization of various precipitates embedded in a matrix (see for example [1,2]). In this context we have undertaken a study of the precipitation state of $MgZn_2$-based particles in the Al-Mg-Zn system. The material is a medium-strength commercial alloy extensively used for automotive applications, where weldability is an important feature. A previous characterization by means of TEM, Small-Angle X-ray Scattering (SAXS) and Atom Probe Tomography (APT) [3] has shown that, in the so-called T7 tempering state (6 hrs. at 170°C): (i) the stable η-$MgZn_2$ phase (hexagonal P63/mmc structure) mainly precipitates under the form of spheres, while minor variants, called η', adopt a platelet shape; (ii) the total precipitated volume fraction, as measured by SAXS, is 2.54 ± 0.3 %; (iii) the precipitates mean radius is 3.6 nm with a standard deviation of 1.0 nm. According to the above size and volume fraction, TEM tomography appears as an interesting alternative tool to SAXS, with which it can be delicate to discern the η and η' particles, and to APT, for which the density of precipitates makes any statistical approach hazardous.

Nano-tomography has thus been conducted in the STEM mode on a JEOL 2010F microscope, equipped with an HAADF (High Angle Annular Dark Field) detector. The material was prepared under the form of tips as for APT experiments. They were mounted on a home-made modified tip of the simple-tilt commercial specimen holder, compatible with a -80 to +80° tilting range with an Ultra-High Resolution pole-pieces.

Figure 1 shows a conventional TEM image and the corresponding HAADF micrograph extracted from a tilt series of 140 images acquired over a tilt range of 140° with a step of 1°, with the use of a home-made script developed under Digital Micrograph™. The viewing direction is here $[110]_{Al}$, and it will be shown that the projection of the 'elongated' precipitates does correspond to η' platelets lying in $\{111\}_{Al}$ planes. Although the HAADF contrast remains relatively poor owing to a small 'Z' difference between the Al matric and the $MgZn_2$ precipitates, 3D reconstruction could be performed with the algebraic reconstruction technique (ART) implemented in the TOMOJ software [4]. Figure 2 is a montage comparing the experimental HAADF image of the extremity of the tip with the re-projected tomogram at zero tilt. It clearly appears that the tomographic reconstruction improves the signal-to-noise ratio. From

M. Luysberg, K. Tillmann, T. Weirich (Eds.): EMC 2008, Vol. 1: Instrumentation and Methods, pp. 309–310, DOI: 10.1007/978-3-540-85156-1_155, © Springer-Verlag Berlin Heidelberg 2008

these experiments, a mean precipitate radius of 4.0 nm, and a volume fraction of 2.35 % were deduced, in perfect agreement with previously mentioned results. [5]

1. K. Kaneko, K. Inoke, K. Sato, K. Kitawaki, H. Higashida, I. Arslan, P.A. Midgley, *Ultramicroscopy*, 108, (2008), 210.
2. K. Inoke, K. Kaneko, M. Weyland, P.A. Midgley, K. Higashida, Z. Horita, *Acta Materialia*, 54, (2006), 2957.
3. M. Dumont, W. Lefebvre, B. Doisneau-Cottignies, A. Deschamps, *Acta Materialia*, 53, (2005), 2881.
4. C. Messaoudi, T. Boudier, C. OscarSanchez Sorzano, S. Marco, *BMC Bioinformatics*, (2007), doi:10.1186/1471-2105-8-288.
5. The CLYM (Centre Lyonnais de Microscopie) is gratefully acknowledged for the access to the 2010F microscope.

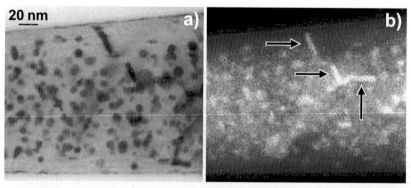

Figure 1. TEM and HAADF images of an almost cylindrical portion of the AlMgZn tip, showing equiaxed η precipitates and η' platelets (arrows).

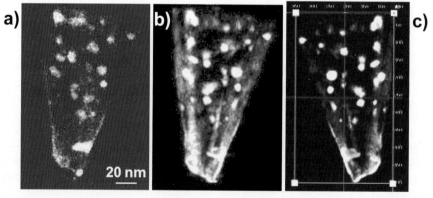

Figure 2. montage showing (a) the HAADF image of the extremity of a tip at zero tilt and (b) the re-projection of the reconstructed tomogram shown in (c) after a rotation of 90° along the vertical axis.

STEM electron tomography of gold nanostructures

J.C. Hernandez[1], M.S. Moreno[2], E.A. Coronado[3], P.A. Midgley[1]

1. Department of Materials Science and Metallurgy, University of Cambridge, Pembroke Street, Cambridge, CB2 3QZ, UK
2. Centro Atomico Bariloche, 8400 - San Carlos de Bariloche, Argentina
3. INFIC, Centro Laser de Ciencias Moleculares. Dpto de Fisico-quimica, Facultad de Ciencias, Universidad Nacional de Cordoba, Argentina

jch65@cam.ac.uk
Keywords: electron tomography, STEM, core-shell nanoparticles

The electronic and optical properties of many nanoparticles can depend critically on their size and morphology [1]. As such, the morphological control of noble metal nanoparticles, technologically important in many fields such as catalysis, optics and surface enhanced Raman spectroscopy, has become of great interest. Metallic core-shell nanostructures have attracted attention since it is possible to enhance physical and chemical properties that cannot be obtained in single-component nanoparticles. Their plasmonic properties can be tuned with nanoparticle geometry (size and shape) and dielectric environment to such extent that a very small change in their geometry leads to dramatic changes in their optical properties [2]. Knowledge of the 3D morphology as well as the precise thickness of the shell is needed in order to understand the unusual physico-chemical properties of these core-shell nanostructures. Here we apply STEM-based electron tomography, a technique well suited to materials science applications [3], to achieve a full 3D analysis of gold and gold-silver core-shell nanostructures. Tomographic reconstruction allows 3D visualization of the nanoparticle shape and enables measured opto-electronic properties to be related to the particle morphology. In particular, we would like to understand variation in the surface plasmon resonance as a function of core-shell dimension and composition [4].

HAADF-STEM tilt series were acquired on a FEI Tecnai F20 microscope, using a Fischione ultrahigh-tilt tomography holder. The images in the tilt series were aligned and reconstructed using Inspect3D software. Our first example is taken from a gold nanoparticle synthesized by chemical seed-mediated growth. Figure 1 shows a tomographic reconstruction, using the simultaneous iterative reconstruction technique (SIRT), from a preliminary STEM-HAADF tilt series recorded in 10° steps. Even with such limited data, an approximate decahedral morphology is apparent from the tomographic reconstruction. The five dominant facets, delineated by the dashed lines in the figure, are likely to be {111} planes. However, for accurate metrology (surface area, volume, angles between facets, etc), finer tilt steps are needed and/or the use of novel reconstruction algorithms such as discrete tomography [5]. Figure 2 shows a SIRT tomographic reconstruction of a cluster of Au-Ag core-shell nanoparticles. HAADF-STEM images were recorded every 2° over +/-74°. Here we highlight the sub-volume near the uppermost nanoparticle as an example of nanoscale quantitative analysis: the mean diameter of the Au core is 23 nm and the Ag shell has a mean thickness of 2.3 nm.

M. Luysberg, K. Tillmann, T. Weirich (Eds.): EMC 2008, Vol. 1: Instrumentation and Methods, pp. 311–312, DOI: 10.1007/978-3-540-85156-1_156, © Springer-Verlag Berlin Heidelberg 2008

312

1. A. Sanchez-Iglesias, I. Pastoriza-Santos, J. Perez-Bustamante, B. Rodriguez-Gonzalez, F.J. Garcia de Abajo and L.M. Liz-Marzan, Advanced Materials, **18** (2006), 2529.
2. J. Zhu, Y. Wang, L. Huang and Y. Lu, Physics Letters A, **323** (2004), 455.
3. P.A. Midgley and M. Weyland, Ultramicroscopy, **96**(3-4) (2003), 413.
4. K.L. Kelly, E. Coronado, L.L. Zhao and G.C. Schatz, J. Phys. Chem. B, **107** (2003), 669.
5. S. Bals, K.J. Batenburg, J. Verbeeck, J. Sijbers and G. Van Tendeloo, Nano Letters, **7**(12) (2007), 3669.
6. The authors acknowledge financial support from the European Union under the Framework 6 program for an Integrated Infrastructure Initiative, Ref.: 026019 ESTEEM. Support of CONICET, ANPCyT, Agencia Córdoba Ciencia and SECyT-UNC, Argentina, is also acknowledged. E.R. Encina and H.E. Butt are thanked for the preparation of the samples.

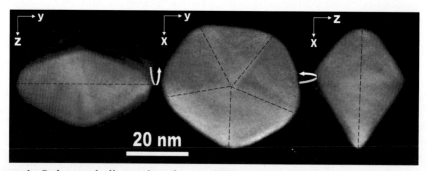

Figure 1. Orthogonal slices taken from a SIRT reconstruction of a gold nanoparticle showing an approximate decahedral morphology. Dashed lines are a guide to the eye.

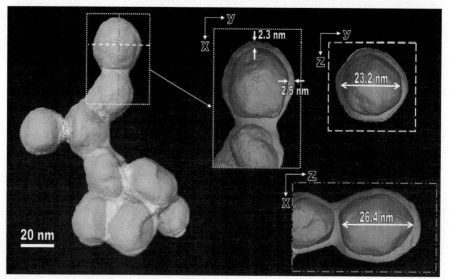

Figure 2. (Left) Tomographic reconstruction of a cluster of Au-Ag core-shell nanoparticles. (Right) Orthogonal views of the uppermost nanoparticle. The particle has been cut to reveal the Ag shell thickness in cross-section and the Au interior.

Three-dimensional imaging of semiconductor nanostructures by compositional-sensitive diffraction contrast electron tomography studies

J.C. Hernandez[1], A.M. Sanchez[2], R. Beanland[1] and P.A. Midgley[1]

1. Department of Materials Science and Metallurgy, University of Cambridge, Pembroke Street, Cambridge, CB2 3QZ, UK
2. Dpto. CC. de los Materiales, Ing. Metalurgica y Q. Inorganica, Facultad de Ciencias, Universidad de Cadiz, Campus Rio San Pedro s/n, Puerto Real 11510(Cadiz), Spain

jch65@cam.ac.uk
Keywords: electron tomography, dark-field TEM, semiconductor quantum dots

New heterostructures in semiconductor materials like self-assembled quantum dots (QDs) are ones of the most studied systems for their potential application as a new generation of opto-electronic devices applications [1]. Because device properties now depend on size and shape as much as they depend on the traditional parameters of structure and composition, accurate morphological and metrological measurements of these complex structures are required. A full 3D analysis can be achieved using electron tomography which is very well suited to materials science applications [2]. A tomographic tilt series of embedded quantum dots has been carried out using HAADF-STEM images [3,4]. In this imaging mode, the intensity is approximately proportional to the square of the atomic number, and in a ternary system such as $In_xGa_{1-x}As$ it is possible to obtain a relatively unambiguous measure of composition. For more complex structures – such as those containing $Al_y(In_xGa_{1-x})_{1-y}As$ material – other imaging modes may give complementary information on composition.

Here we report a study performed using compositionally sensitive diffraction contrast in semiconductor materials. For sphalerite structures, dark-field (DF) 002 diffraction conditions are compositionally sensitive, with the diffracted intensity being approximately proportional to the difference in group III and group V atomic number, integrated through the thickness of the TEM specimen [5]. Electron tomography of such structures using 002 imaging conditions may therefore give further insights into the 3D structure of these nanoscale structures.These methods are applicable to any system, providing a unique and versatile three-dimensional visualization tool.

An [010] TEM specimen of a three-layer $Al_yIn_{1-y}As/InAs/Al_yIn_{1-y}As$ QD heterostructure on (001) GaAs was prepared as described in [6]. Electron tomography experiments were performed on a Philips CM30 operating at 300kV; TEM images were recorded every 2° from -44° to +46°. Keeping the diffraction conditions approximately constant across the tilt series, the use of photographic plates allowed detailed analysis of a relatively large area of material with high pixel resolution. The images forming the tilt series were aligned using Inspect3D software and a reconstruction carried out using the SIRT algorithm. Figure 1(a) shows an orthogonal slice through the reconstructed DF002 data set with a single dot and surrounding layers at close to zero tilt, i.e. close to [010].

M. Luysberg, K. Tillmann, T. Weirich (Eds.): EMC 2008, Vol. 1: Instrumentation and Methods, pp. 313–314, DOI: 10.1007/978-3-540-85156-1_157, © Springer-Verlag Berlin Heidelberg 2008

Al$_y$In$_{1-y}$As layers appear bright, whereas the InAs QDs appear dark with a brighter core. Figure 1(b) and 1(c) shows the other otrhogonal slices through the reconstructed data set with reconstructed 002 sections looking along [001] and [100] – i.e. perpendicular to Figure 1(a) – directions which are inaccessible in the microscope. Such images allow a complete description of the distribution of the QDs as well as their location within the TEM specimen. Due to the limited range of tilts used in the reconstruction (missing wedge), some loss of resolution has occurred along [010]; however several interesting features can be seen. For example there is a clear alignment of the basal edges of the dot along <110> directions and an asymmetry in the composition of the capping layer. We hope that further analysis using the mechanics of the contrast mechanism may allow a full reconstruction of composition in three dimensions.

1. Y. Masumoto in "Semiconductor Quantum Dots", Springer, Berlin (2002).
2. P.A. Midgley and M. Weyland, Ultramicroscopy, **96**(3-4) (2003), 413.
3. T. Inoue, T. Kita, O. Wada, M. Konno, T. Yaguchi and T. Kamino, Appl. Phys. Lett. **92** (2008) 031902.
4. I. Arslan, T. J. V. Yates, N. D. Browning and P. A. Midgley, Science **309** (2005), 2195.
5. R. Beanland, Ultramicroscopy **102** (2005), 115.
6. R. Beanland, Microscopy Today **11** (2003), 29.
7. The authors acknowledge financial support from the European Union under the Framework 6 program for an Integrated Infrastructure Initiative. Ref.: 026019 ESTEEM.

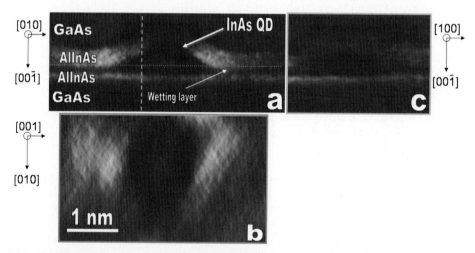

Figure 1. Orthogonal slices through the reconstructed DF002 data set. (a) a slice corresponding to zero tilt, close to [010]; (b) and (c) show cross-section and plan view slices along perpendicular directions marked by the orange and red lines in (a).

A full tilt range goniometer inside a TEM goniometer

X.J. Xu, A. Lockwood, R. Gay, J.J. Wang, Y. Peng, B.J. Inkson, and G. Möbus

Dept. of Eng. Materials, University of Sheffield, Mappin Street, Sheffield, S1 3JD, UK.

beverley.inkson@sheffield.ac.uk

Keywords: goniometer, specimen holder, electron tomography, tungsten nanotip

Over the last decade, electron tomography has developed into a powerful and widely applicable tool for three-dimensional (3D) characterisation problems in materials science [1-3]. However, the limited tilt range available from the commercial goniometers in modern TEMs of materials science laboratories remains a major problem to avoid missing wedge artefacts. Often the narrowest of available pole piece gaps is preferred in materials science for maximising lattice resolution which allows only size reduced specimens to be rotated to high angles [4].

We have developed a piezoelectric goniometer which can operate in a normal TEM without any modification to the TEM and its standard goniometer to allow direct observation of the specimen under unlimited tilt angles. A complete 180 degrees tomographic acquisition series can be easily achieved with the tilt increment as fine as affordable with respect to the duration of acquisition, specimen contamination and drift problems. The capability to even tilt further than 180° will allow for applications beyond the field of tomography, e.g. for studies of specimens by pairs of opposite viewing directions, top-bottom scattering asymmetries and for magnetic holography.

A tilt series with a tilt range of ~ 260 degrees was acquired using ADF-STEM on a JEM 2010F-URP TEM (JEOL, Japan) with 2mm pole piece gap as shown in Fig. 1. A home-made tungsten tip by the electrochemical etching method is embedded in a WO_3 oxide shell and a second shell of carbon contamination. The tip mount is perfectly rotationally symmetric as reported earlier [4-6]. Only the images at about every 20 degrees are displayed in this figure. An example pair observed in two opposite viewing directions (180°) is selected as Fig 2.

Earlier reconstruction of W-tips by ADF-STEM and EDX tomography [6-7] has clearly confirmed the importance of high tilt range. Quantitative postprocessing of tomograms into numerical quantities, such as volumes or surface areas will further push the need to avoid any missing wedge [8].

1.   G. Möbus and B.J. Inkson, Applied Physics Letters **79** (2001), p.1369.
2.   Midgley, P. A.; Weyland, M. Ultramicroscopy, **96** (2003), p. 413.
3.   X. Xu, Z. Saghi, R. Gay and G. Möbus, Nanotechnology, **18,** 22, (2007), p. 225501.
4.   X. Xu, Z. Saghi, Y. Peng, R. Gay, B. J. Inkson, G. Möbus, Microscopy and Microanalysis **12** **S2** (2006), p. 648.
5.   X. Xu, G. Yang, Z. Saghi, Y. Peng, R. Gay, G. Möbus, Mater.Res.Soc.Symp.Proc., **928E** (2007) 0982-KK02-04.
6.   X. Xu, Y. Peng, Z. Saghi, B. J. Inkson, G. Möbus, J. Phys.: Conf. Ser. **61** (2007), p. 810.
7.   Z. Saghi, X. Xu, Y. Peng, B. Inkson, G Möbus, Appl. Phys. Lett. **91** (2007), p. 251906.
8.   This work was supported by EPSRC under grant number GR/S85689/01.

M. Luysberg, K. Tillmann, T. Weirich (Eds.): EMC 2008, Vol. 1: Instrumentation and Methods, pp. 315–316, DOI: 10.1007/978-3-540-85156-1_158, © Springer-Verlag Berlin Heidelberg 2008

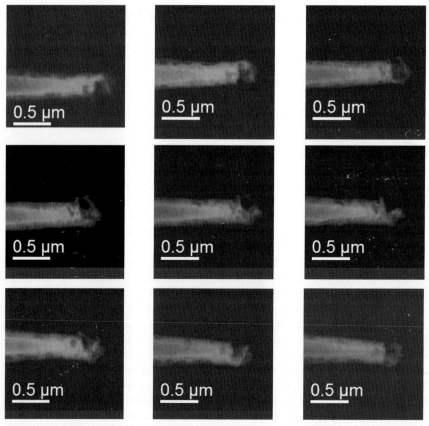

Figure 1: ADF-STEM tilt series of a tungsten nanoindentation tip with total tilt range of 260°. Shown are 9 selected images every 29°. The sharply defined inner conical W-core is surrounded by amorphous oxide and carbon contamination (JEM 2010F, 200kV).

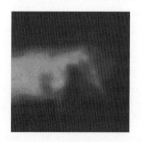
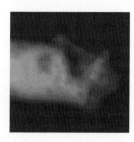

Figure 2: Two members of above tilt series with 180° relative angle (opposite views along the same line)

Four-dimensional STEM-EELS Tomography

K. Jarausch[1], D. Leonard[2] , R. Twesten[3] and P. Thomas[3]

1. Hitachi High-Technologies Corp., 5100 Franklin Dr, Pleasanton, CA, USA
2. Dept. of Physics & Astronomy, Appalachian State University, Boone, NC, USA
3. Gatan, Inc., 5794 W. Las Positas Blvd, Pleasanton, CA USA

konrad.jarausch@hitachi-hta.com
Keywords: Tomography, Spectrum Imaging, EELS

Advances in electron based instrumentation are enabling multi-dimensional data acquisition to explore the unique structure-property relationships of nano-structured materials [1-3]. Here we report a technique for directly probing and analyzing a material's three-dimensional (3D) electronic structure. A rotation holder [4] is used to vary specimen orientation and record STEM EELS spectrum images at regular angular intervals (using a Hitachi HD2300A FEG-STEM equipped with a Gatan ENFINA spectrometer). Experimental conditions were optimized to facilitate acquisition rates of over 100 spectra/second and thereby make the acquisition of such large data sets practical. By using a pillar shaped sample, and a rotation holder the electronic properties are sampled with a constant projection thickness and over a complete 180 degree rotation to minimize artefacts. Analysis software was developed to align the four-dimensional (4D) data volume, and extract the spectral properties of interest [5]. By combining energy loss information from such a series of spectrum images it is possible to map not only the microstructure, but also the elemental, physical and chemical state information of a material in three dimensions. Details and limitations of the 4D STEM-EELS acquisition will be discussed here, while the options and requirements for analysis of 4D EELS data sets will be discussed separately [5]. This technique has been applied to map the 3D properties of a variety of samples and two examples are reported here.

4D STEM-EELS was used to directly probe a W to Si contact from a semiconductor device and a ZnO film with embedded Au nanodots. The 4D data sets were analyzed to map the composition (Fig. 1a), bonding and phases of the W to Si contact in three dimensions [5]. Providing the chosen EELS signal meets the projection criterion, tomographic reconstruction can be used to obtain a "volumetric" map of the property of interest (Fig. 1b) [5]. The ZnO thin film was specially prepared to allow sampling of the electronic properties over a range of crystallographic orientations using a single axis of rotation. The film was rotated from e-beam parallel to the c-axis (plan-view, Fig. 2a) to e-beam perpendicular to c-axis (cross-section, Fig. 2b), and then back again. Subtle, but systematic changes in low-loss structure were observed as a function of sampling angle, similar to those reported elsewhere [6,7]. These examples illustrate how the 4D STEM EELS technique reported here can be used to directly probe the electronic structure of complex nano-structures in three dimensions.

1. C. Jeanguillaume and C. Colliex, Ultramicroscopy **28** (1989) p. 252.
2. P.A. Midgley and M Weyland, Ultramicroscopy **96** (2003) p.413.

M. Luysberg, K. Tillmann, T. Weirich (Eds.): EMC 2008, Vol. 1: Instrumentation and Methods,
pp. 317–318, DOI: 10.1007/978-3-540-85156-1_159, © Springer-Verlag Berlin Heidelberg 2008

318

3. M H. Gass et al, Nano Letters **6** (2006) 376.
4. T. Kamino et. al, J. Electron Microscopy **53** (2004) p.583.
5. P. Thomas et al, submitted to EMC2008 Proceedings
6. Juan Wang, et. al., Applied Physics Letters **86** (2005) p.201911.
7. Yong Ding, et. al., Journal of Electron Microscopy, **54** (2005) p.287.

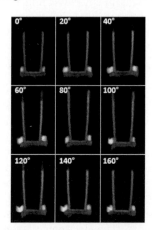 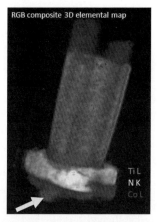

Figure 1. (a) Computed RGB composite elemental distribution maps from the 4D STEM-EELS rotation series (key: Ti L red, N K green, Co L blue). Images are shown at 20° steps for clarity. (b) RGB composite image showing the volumetric elemental distribution maps obtained by tomographic reconstruction from the corresponding elemental map rotation-series. The "leaky cobalt pipe" defect highlighted with an arrow

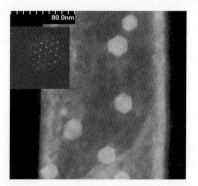 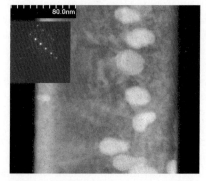

Figure 2. (a) Planview (c-axis) and (b) cross-section (a-xis) HAADF images and diffraction patterns of the ZnO film with embedded Au Nanodots. The electronic structure of this sample was systematically probed as a function of orientation using 4D STEM-EELS to explore the effects of crystal anisotropy on the electronic structure [6,7].

Embedment-free section electron microscopy (EM): a highly potential advantage in application to EM tomography

Hisatake Kondo[1] and Tetsuo Oikawa[2]

1. Division of Histology, Department of Rehabilitation, Faculty of Medical Sciences and Welfare, Tohoku Bunka Gakuen University, Sendai, Japan
2. JEOL(Europe) SAS, 1 Allée de Giverny, 78209 Croissy-sur-Seine, France

hkondo@mail.tains.tohoku.ac.jp

Keywords: embedment-free, PEG, stereo-view, tomography

Embedment-free section electron microscopy (EM) is successfully made possible by employing polyethylene glycol as a transient embedding media and by critical-point drying after de-embedding of sections by immersion into warm water [1,2,3,4]. Because of the section thickness several times thicker than epoxy sections and the high contrast without urany and lead staining together with high efficiency in the electron beam penetration through sectioned specimens due to the absence of any embedding media, this method presents remarkable advantages in the three-dimensional observation of biological structures by stereo-viewing of two tilted sections as well as EM tomography.

The first example of the advantages is the translucency to neural myelins. In contrast to the general view based on epoxy section images that the myelins are highly electron-opaque, this method makes intra- and extra-axonal elements visible through myelins from exterior and interior of myelinated axons, respectively, by stereo-viewing (Figure 1). The second is much more clear and accurate demonstration of the renal glomerular slit than its epoxy section image by means of EM tomography as well as stereo- viewing of two tilted sections (Figure 2). By this method, the glomerular slit is revealed to represent not diaphragm but ladder-structures in which ladder-strands are aligned with irregular intervals and may terminate intermittenedly. Although the EM tomography image of the slit has been reported by other authors [5], their result was based on images from epoxy sections which make vague the appearance of such non-plasma membrane-formed structures as the slit, and it is possible that the present image in embedment-free section EM represents a much more real structure of the glomerular slit than that in conventional epoxy section EM.

In addition, the present authors propose in this paper a new idea about the biological significance of the microtrabecular lattices in the cytoplasmic matrix which was originally revealed by the late KR Porter and his group [6] in whole-mount EM whose methodological essence is the same as the present embedment-free section EM. Based on our observation in this method of various protein solutions at different concentrations and consistency as well as various tissue cells, the microtrabecular lattice with a higher compactness in a cell represents a higher concentration of cytosolic proteins in the cell than the other, while the lattice with a higher compactness in a portion of a cell represents the gelated state of cytosolic proteins in the cell portion than the remainder of

M. Luysberg, K. Tillmann, T. Weirich (Eds.): EMC 2008, Vol. 1: Instrumentation and Methods, pp. 319–320, DOI: 10.1007/978-3-540-85156-1_160, © Springer-Verlag Berlin Heidelberg 2008

320

the cell. However, the microtrabecular strands themselves are regarded as unavoidable and histological preparation consequence of any proteins and thus biologically non-significant.

1. Kondo H, Wolosewick JJ & Pappas GD (1982) J Neurosc 2:57-65
2. H. Kondo, J Electr Microsc (2006) 55:231-243
3. H. Kondo, Tohoku U Exp Med 214 (2008) 167-174
4. H. Kondo, Microsc Res Tech (2008) in press
5. J. Wartiovaara et al, J Clin Invest (2004) 114:1475-1483
6. JJ. Wolosewick & KR. Porter, J Cell Biol (1979) 82:114-139

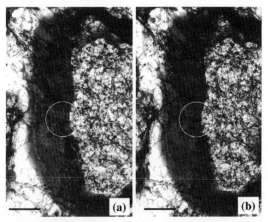

Figure 1. Two embedment-free EMs of same myelin with tilting angle of 7°. Note visibility of intraxonal cytoskeletons through myelin, indicating electron translucency of myelin.

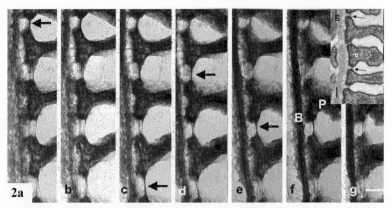

Figure 2. Stereo-pairs of embedment-free section EMs of renal glomerular slits in comparison with single conventional epoxy EM (insert). Tilting angle is 7° between any two adjacent EMs. Note ladder-alignment of strands (arrows) with irregular intervals but not diaphragms. B:basement lamina, P:podocyte pedicles. EM tomography of this feature will be demonstrated on PC in front of poster.

Quantification and Segmentation of Electron Tomography Data– Exemplified at ErSi$_2$ Nanocrystals in SiC

J. Leschner, J. Biskupek, A. Chuvilin and U. Kaiser

Central Facility of Electron Microscopy, Group of Materials Science, Albert-Einstein-Allee 11, 89069 Ulm, Germany

Jens.Leschner@Uni-Ulm.de

Keywords: electron tomography, nanocrystals in SiC, quantification, segmentation

Small transition metal crystals embedded in a semiconducting matrix are nanostructured systems with potential applications in nanotechnology. As their size distribution plays a key role for the detailed understanding of the device properties, the precise three-dimensional quantification is a must. However, the quantification process of the observed three-dimensional object is still under big discussion [1], due to subjective (manual) segmentation and desired ease of data-processing. However, before application related questions can be addressed, the nanocrystal formation as well as basic questions in the field of quantification have to be answered and robustly solved.

In the past we studied the nanocrystal formation process of hill-like shaped ErSi$_2$ nanocrystals formed after high dose Er ion implantation in 4H-SiC by HAADF-STEM imaging and spectroscopy and we suggested that the nanocrystals start to form at matrix defects, rather than to grow spontaneously [2,3]. However, the crystal shape within the volume could not be derived from two-dimensional imaging, but by HAADF-STEM tomography only. Using iterative reconstruction techniques, it was possible to visualize not only facets of the nanoparticles (Figure 1), but also Er decorated voids and channels filled with Er atoms [4,5].

As a next step the obtained data has to be quantified: facets, nanocrystal size and distribution with respect to implantation depth. The crucial point is to guarantee that the visualized object is in one-to-one correspondence to the original object. In order to decouple the quantified results from the adjustments during reconstruction and – as the method of choice – segmentation by thresholding, the contribution to the object contrast has to be investigated and an objective measure for thresholding has to be derived. Different imaging modes and the influence of the reconstruction algorithm as well as its parameters have been already discussed for a specific case of material and home-made reconstruction algorithm [6], but a quantitative way for easy segmentation is still not established.

In this work we evaluate how to perform robust segmentation for our material system and reconstruction algorithm (SIRT, FEI Inspect3D®) in order to implement it into an objective segmentation algorithm. Therefore, we model phantoms of the nanocrystals, reconstruct them and evaluate the influence of the developed algorithm. Afterwards we derive the quantified values from the experimental nanocrystals with higher reliability (Figure 2).

M. Luysberg, K. Tillmann, T. Weirich (Eds.): EMC 2008, Vol. 1: Instrumentation and Methods, pp. 321–322, DOI: 10.1007/978-3-540-85156-1_161, © Springer-Verlag Berlin Heidelberg 2008

322

1. Discussion at the Tomography Workshop at the Hahn-Meitner-Institute in Berlin, Oct. 2007.
2. U. Kaiser, D.A. Muller, J.L. Grazul, A. Chuvilin, M. Kawasaki, Nature Materials **1** (2002), p. 102.
3. U. Kaiser, D.A. Muller, A. Chuvilin, G. Pasold, W. Witthuhn, Microscopy and Microanalysis **9** (2003), p. 36.
4. C. Kübel and U. Kaiser, Microscopy and Microanalysis **12** (2006), p. 1546.
5. J. Leschner, U. Kaiser, J. Biskupek, A. Chuvilin and C. Kübel, Microscopy and Microanalysis **13** (2007), p. 116.
6. H. Friedrich, M. R. McCartney and P. R. Buseck, Ultramicroscopy **106** (2005), p. 18.
7. This work has been supported by the German Research Foundation (DFG) – project KA1295/7-1.

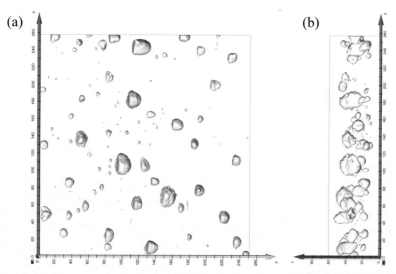

Figure 1. Reconstructed volume of ErSi$_2$ nanocrystals embedded in a crystalline SiC matrix (scale is nm): (a) three-dimensional view of the hill-like shaped particles in the xy-plane, and (b) along the implantation direction [11-20] (xz-plane) showing facets

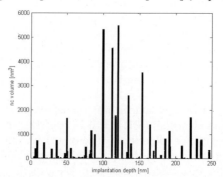

Figure 2. Quantification of the nanocrystals according to size and distribution w.r.t. (relative) implantation depth: small nanocrystals can be found at the surface and deep within the sample, whereas big grown particles are found dominantly in the middle

3-D TEM observation of xenon nano-precipitates in aluminium crystals

M. Song[1], H. Matsumoto[2], M. Shimojo[1,3], and K. Furuya[1]

1. National Institute for Materials Science, Sakura 3-13, Tsukuba, Ibaraki, 3050003 Japan.
2. Yokohama Lab., Mitsubishi chemical group science and research center, INC, 1000 Kamoshida-cho, Aoba-ku, Yokohama, 2278502, Japan.
3. Saitama Institute of Technology, Fusaiji 1690, Fukaya, Saitama, 3690293 Japan

6503883@cc.m-kagaku.co.jp
Keywords: TEM, tomography, nano-particle

It is known that the inert gas, such as xenon (Xe), is not soluble in typical metals and tends to precipitate in the metals. The Xe-precipitates in Al have been observed with TEM in usual modes, such as the bright field (BF) and high-resolution (HREM) modes [1]. However, almost only 2-dimensional (2-D) information on the structure and distribution of Xe precipitates has been studied. On the other hand, tomography observation is a powerful method to analyze the 3-D information, such as the distribution, shape, and structure of the small objects embedded in a matrix. In the present work, we carried out the 3-D observation of Xe-precipitates in Xe-implanted Al crystals using a new developed tomography specimen holder for JEM-ARM1000. One of the features of the holder is that it utilizes an in-holder-rotation system, with which the holder can be rotated over +/- 90 degrees not dependent on the rotation system of the microscope. The distribution, size and density of the Xe-precipitates in Al crystals were analyzed.

A rod-shaped Al specimen in diameter of about 100 nm was prepared with FIB (focused-ion-beam) and Ar^+ ion milling. The Xe-ion implantation was carried out with the ion-implanter-interfaced TEM, JEM-ARM1000 [2]. The energy of the ions is 50 keV, which corresponds to the peak stop range of 26 nm and the distribution straggling of 7 nm. The flux and fluence of the Xe-ion are 1.3 x 10^{13} ions cm^{-2} s^{-1} and up to 4.6 x 10^{15} ions cm^{-2}, respectively. The TEM observations were carried out in BF mode and in rotation range from -90 degrees to +90 degrees with an increasing step of 2.16 degrees. The TEM images recorded with CCD camera were reconstructed using the IMOD program developed by Mastronarde et al [3]. All the experiments were performed at room temperature.

Fig. 1 schematically shows the direction-relationships of the ion-implantation and the TEM observation to the specimen. The ion-implantations were performed in directions A and B to fluences of 4.6 x 10^{15} ions cm^{-2} and 1.5 x 10^{15} ions cm^{-2}, respectively. Fig. 2 shows a TEM BF image taken at a rotation near 0 degree. The area with larger precipitates shown by the arrows is near the side implanted to the larger ion fluence. Fig. 3 shows a set of cross-section images from the reconstructed tomogram. The nanosized Xe-precipitates, some of them are as small as about 2 nm, and the 3-D

M. Luysberg, K. Tillmann, T. Weirich (Eds.): EMC 2008, Vol. 1: Instrumentation and Methods, pp. 323–324, DOI: 10.1007/978-3-540-85156-1_162, © Springer-Verlag Berlin Heidelberg 2008

324

distribution of the precipitates are well identified. The distances between the precipitates and the sizes of the precipitates were measured directly in several XZ planes. They are useful for analyzing the growth process of the nano-precipitates. The precipitates in the side implanted with larger ion fluence have larger size but lower density as observed in Fig. 3a. Also, there are few Xe-precipitates in the centre part of the specimen. These results suggest that the implanted-Xe atoms diffuse locally to precipitate, and that the coarsening of the precipitates happens during the ion-implantation.

1. M. Song, K. Mitsuishi and K. Furuya, Mat. Sci. and Eng. A 304-306 (2001) p. 135.
2. N. Ishikawa and K. Furuya, Ultramicroscopy 56 (1994) p. 211.
3. J.R. Kremer, D.N. Mastronarde and J.R. McIntosh, J. Struct. Biol. 116 (1996) p. 71; D.N. Mastronarde, J. Struct. Biol. 120 (1997) p. 343; Also: http://bio3d.colorado.edu/imod/.

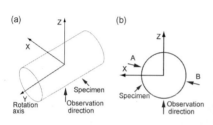

Figure 1. A schematic drawing showing the configuration of the ion-implantation and the observation. A and B show the directions of the ion-implantation

Figure 2. A typical TEM BF micrograph of Xe-implanted Al specimen.

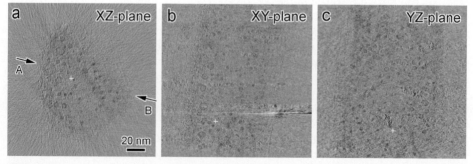

Figure 3. A set of cross-section images of the reconstructed tomograpm of Xe-implanted Al specimen. A and B show the directions of the ion-implantation

Dark-field TEM tomography of ordered domain morphology in a Ni₄Mo alloy

K. Kimura[1], K. Matsuyama[1], S. Hata[2] and S. Matsumura[1]

1. Department of Applied Quantum Physics and Nuclear Engineering, Kyushu University, Fukuoka 819-0395, Japan
2. Department of Engineering Sciences for Electronics and Materials, Kyushu University, Kasuga 816-8580, Japan

syo@nucl.kyushu-u.ac.jp
Keywords: electron tomography, dark-field imaging, ordered domains

Various TEM imaging modes, such as bright-field imaging, STEM and energy-filtering and so on, have been being utilized in electron tomography [1]. In contrast to the above imaging modes, dark-field (DF) TEM imaging is useful for observing crystalline microstructures such as lattice defects, grain boundaries, ordered domain structures, etc. However, it is not generally considered that dark-field (DF) TEM imaging is applicable to electron tomography, since the image contrast is not a simple function of thickness and density. One has to control the diffraction condition precisely to get well-defined DF images, because the image contrast is quite sensitive also to the direction of the incident electron beam. Contrary to the general understanding, we have shown a technique to obtain a DF TEM tilt series and its applicability to electron tomography for crystalline specimens [2]. In this paper, we discuss our latest results of the tomographic DF TEM observation of variant structures in a Ni₄Mo ordering alloy from the view point of 3D morphology of different variants.

Single-phase microstructures with the $D1_a$ superstructure was obtained in a Ni-19.5 at% Mo alloy by annealing at 1073 K for 24h. A TEM specimen was set so that a [420] crystallographic direction, or a systematic row including a (420)/5 superlattice reflection, was exactly parallel to the tilt axis of the specimen holder (Fig. 1(a)). A tilt series of DF images was taken with the (420)/5 superlattice reflection over the tilt angle from -60° to +60°, with a step interval of 2° (Fig. 1(b)). To keep the diffraction condition, the incident beam direction was precisely adjusted at every step of the specimen tilt. In order to observe another variant phase in the same area, a DF TEM tilt series was taken with a (620)/5 superlattice reflection (Fig. 2) in the same way after replacing the specimen on the holder. Three-dimensional images were reconstructed from the DF TEM tilt series using Composer™ (JEOL Ltd.).

Figures 3(a) and 3(b) show tomographic reconstruction images taken with (420)/5 and (620)/5 superlattice reflections, respectively. Bright areas correspond to $D1_a$ domains of a specific orientation variant. Figure 3(c) was obtained by superposition of Fig. 3 (a) and (b). One may recognize that these two pictures are almost fit in each other. It clearly shows that only the two variants mostly prevail in the 3D volume observed, though there are six equivalent variants of $D1_a$. It is suggested that there is a selected growth of the two orientation variants with a common C-axis along a <001> direction.

M. Luysberg, K. Tillmann, T. Weirich (Eds.): EMC 2008, Vol. 1: Instrumentation and Methods, pp. 325–326, DOI: 10.1007/978-3-540-85156-1_163, © Springer-Verlag Berlin Heidelberg 2008

Thus DF TEM tomography opens up a research field of real morphologies of crystal domains in three dimensions, for which no experimental technique has existed so far.

1.   P. A. Midgley and M. Weyland, Ultramicroscopy, **96** (2003) 413.
2.   K. Kimura, S. Hata, S. Matsumura and T. Horiuchi, J. Electron Micros , **54** (2005) 373.

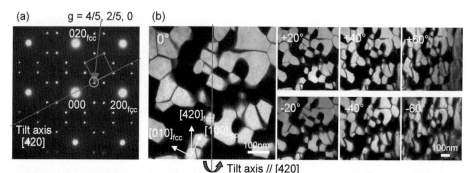

Figure 1. [001] Electron diffraction pattern (a), and DF images of the $D1_a$-Ni$_4$Mo domains taken with the (420)/5 superlattice reflection (b).

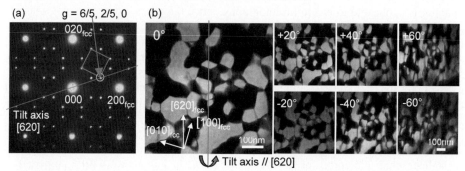

Figure 2. [001] Electron diffraction pattern (a), and (620)/5 DF images of the same specimen area as in Fig. 1.

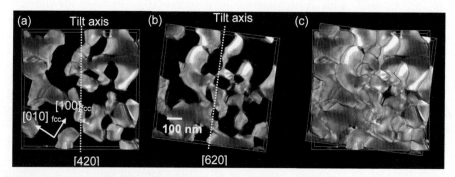

Figure 3. $D1_a$-Ni$_4$Mo domains of an orientation variant (a) and of another variant (b). (c) is superimposition of (a) and (b).

Optimum optical condition of Tomography for thick samples

S. Motoki[1], C. Hamamoto[1], H. Nishioka[1], Y. Okura[1], Y. Kondo[1] and H. Jinnai[2]

1. JEOL Ltd. 1-2 Musashino, 3-chome Akishima Tokyo 196-8558 Japan
2. Kyoto Institute of Technology Matsugasaki, Sankyo-ku, Kyoto 606-8585 Japan

smotoki@jeol.co.jp

Keywords: Tomography, focal depth, soft materials

In the soft materials science, it is known that structures of the order of 10 ~ 100 nm significantly influence various properties of materials. Transmission Electron Microtomography is one of the most powerful tools for structural analysis of three-dimensional entities in such materials. Structural analysis of those materials requires thicker samples than those examined so far due to the larger internal structures.

In the observation of the thick samples with TEM, results of 3D reconstruction depend on the electron optical conditions, such as focal depth. In this paper, we compared long focal length optical system with conventional one in order to clarify the influence of focal depth in topographic reconstruction.

Figure 1 shows optical systems of (A) long focal length optical system and (B) conventional one. Table 1 shows optical conditions of (A) and (B). In general, focal depth is given by eq. (1). l, d and α are the focal depth, the resolution at the specimen and the detection angle, respectively. When the detector resolution is larger than the resolution of the microscope, d can be replaced by the detector resolution d_{pix} [see eq. (2)]. With weak lens approximation, α is determined by the radius of objective aperture r and focal length f_o. Accordingly, eq. (2) can be rewritten as eq. (3) with effective focal depth: l_{ef}. The focal depth in condition (A) is 5 times larger than the one in the condition (B). The large value of spherical aberration: Cs in condition (A) has no practical effect in imaging at the magnification in our study, because the point resolution of the microscope d estimated from the Cs value is smaller than d_{pix}.

Figure 2 shows the TEM images of ABS resin/OsO_4 sample with the thickness of 500 nm that was prepared by Cryo-microtomy. These images were taken with the 200 kV TEM (JEM-2100), the magnification is 6,000 x and specimen tilt angle is 0 degree. The left image [Fig. 2(a)] was taken under the condition (A), while the right one was taken under the condition (B) [Fig. 2(b)]. It is obvious that Fig. 2(a) appeared sharper than Fig. 2(b). This may be attributed to the deeper focal depth and effective aperture size determined by the focal length of the objective lens and also the limiting aperture size. These results demonstrate that the long focal optical condition is of great advantage for tomography.

Equations of depth of field and effective depth of field

$$\ell = \frac{d}{\alpha} \quad \text{Eq. (1)} \qquad \ell_{ef} = \frac{d_{pix}}{\alpha} \quad \text{Eq. (2)} \qquad \ell_{ef} = \frac{d_{pix} f_o}{Mr} \quad \text{Eq. (3)}$$

M. Luysberg, K. Tillmann, T. Weirich (Eds.): EMC 2008, Vol. 1: Instrumentation and Methods, pp. 327–328, DOI: 10.1007/978-3-540-85156-1_164, © Springer-Verlag Berlin Heidelberg 2008

Table 1. The optical conditions of (A) long focal length optical system and (B) conventional one.

	A Long focal depth	B Conventional optical system
focal length fo	52 mm	2.3 mm
Spherical avaration constant Cs	460 mm	1 mm
theoretical resolution d	1 nm	0.23 nm
Pixel size of ccd (x6000 d_{pix})	18 μm x 18 μm (2 nm x 2 nm)	
detection angel α (effective aperture size)	1.4 mrad. (6 μm)	6.5 mrad. (30 μm)
effective focal length l_{ef}	1400 nm	300 nm

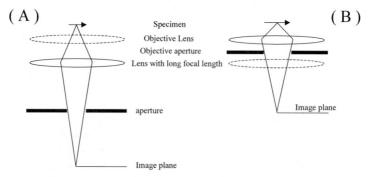

Figure 1. The optical-ray diagrams of (A): long focal length optical system and (B): conventional one.

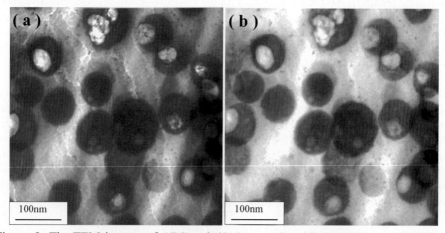

Figure 2. The TEM images of ABS resin/OsO$_4$ sample with the thickness of 500 nm, which is prepared by Cryo-microtomy. Images in part (a) and part (b) were taken with the condition (A) and (B) as listed in the Table 1.

3-dimensional nanoparticle analysis using electron tomography

T. Oikawa[1, 2], D. Alloyeau[2, 3], C. Ricolleau[2], C. Langlois[2], Y. Le Bouar[3] and A. Loiseau[3]

1. JEOL(Europe) SAS, 1 Allée de Giverny, 78290 Croissy-sur-Seine, France
2. Université Paris 7 / CNRS, UMR 7162, 2 Place Jussieu, 75251 Paris, France
3. ONERA / CNRS, UMR 104, B.P. 72, 92322 Châtillon, France

oikawa@jeol.fr
Keywords: CoPt nanoparticle, Electron tomography, Nanoparticle thickness

It is now well known that physical and chemical properties of nanoparticles depend not only on their size but also on three-dimensional (3D) morphology. Therefore, the knowledge of those parameters is crucial and their accurate measurement highly desirable. Recently, electron tomography has made huge progress both in the acquisition and the reconstruction processes [1, 2]. The acquisition is now automated and the reconstruction processes are improved by the development of new algorithms and also by the computers. In this paper, 3D morphologies of CoPt nanoparticles deposited on amorphous carbon film were analysed using electron tomography.

CoPt nanoparticles on carbon film were produced by Pulsed Laser Deposition (PLD) in a high vacuum chamber. Co and Pt were alternatively deposited by PLD using pure Co and Pt targets. As the metallic species do not wet amorphous carbon substrate, nanoparticles were formed instead of a continuous film. Two samples were prepared with the same CoPt nominal thickness and were annealed in a vacuum furnace during 1 hour at 650°C (sample A) and at 750°C (sample B). Tomography experiments were performed on the *JEM-2100* (200 kV TEM). The basic principles of tomography are well described in a recent paper [3]. 3D reconstruction and 3D particle analysis were carried out with *"TEMography II"* and *"TRI/3D-PRT"* softaware.

Figures 1(a) and 1(b) show a bright field (BF) TEM image and its 3D representation (tomogram) of the selected area in sample B. Figure 2 shows a representation of colored segmentation obtained after processing the 3D particle analysis. The colours correspond to particle 3D size (*i.e.* thickness). From the result of the analysis the particle size were measured automatically. Figure 3 shows a histogram of 3D diameter measured by segmentation automatically. Thus, t/d ratios (t: particle thickness, d: diameter in the substrate plane) are compared in both samples. For a same projected size (d), the t/d ratio of the sample A is most often smaller then the one of the sample B. It is concluded that nanoparticles annealed at 650°C are flatter than those annealed at 750°C. The electron tomography and 3D particle analysis can be used successively to determine the 3D morphology of CoPt nanoparticles deposited on a substrate.

1. H. Furukawa, M. Shimizu, Y. Suzuki and H. Nishioka, JEOL News **36** (2001) 12 – 15..
2. N. Kawase, M. Kato, H. Nishioka and H. Jinnai, Ultramicroscopy **107** (2007) 8 – 15.
3. P. A. Midgley and M. Weyland, Ultramicroscopy **96** (2003) 413 – 431.

M. Luysberg, K. Tillmann, T. Weirich (Eds.): EMC 2008, Vol. 1: Instrumentation and Methods, pp. 329–330, DOI: 10.1007/978-3-540-85156-1_165, © Springer-Verlag Berlin Heidelberg 2008

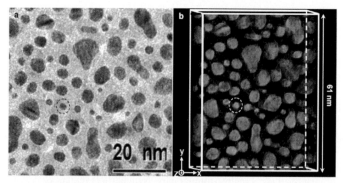

Figure 1. (a) BF-TEM image of CoPt nanoparticles of sample B (1 hour annealing at 750°C)   (b) Tomogram of the same area.

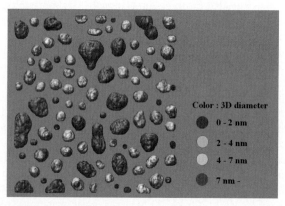

Figure 2. Representation of colored segmentation from the tomogram (Fig. 1(b)). The colours correspond to particle 3D size

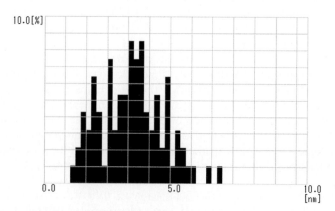

Figure 3. A histogram of 3D particle diameter of CoPt nanoparticles measured automatically by segmentation shown in Fig. 2.

Electron Tomography of ZnO Nanocones with Secondary Signals in TEM

V. Ortalan[1], Y. Li[1], E.J. Lavernia[1], N.D. Browning[1,2]

1. Department of Chemical Engineering and Materials Science, University of California-Davis, One Shields Ave, Davis, CA 95616, USA
2. Chemistry, Materials and Life Sciences Directorate, Lawrence Livermore National Laboratory, 7000 East Avenue, Livermore, CA 94550, USA

vortalan@ucdavis.edu

Keywords: tomography, secondary electrons, nanocone

Electron tomography, which has been successfully utilized in life sciences, has attracted attention in materials science recently. Z-contrast imaging using a High Angle Annular Dark Field (HAADF) detector and Energy Filtered Transmission Electron Microscopy (EFTEM) imaging have been proved to be applicable in tomographic studies provided that the contrast in the image is only due to a change in mass-thickness of the investigated sample but not the crystallographic orientation [1]. However, for these imaging modes, the thickness of the sample is the main constraint in determining the reliability of the reconstructions. Especially HAADF Scanning TEM imaging is highly sensitive to thickness variations which makes it almost impossible to use this technique for materials thicker than a couple of hundreds of nanometres. As the thickness increases, the number of electrons transmitted decreases, therefore, using the backscattered and secondary electrons become more effective.

Due to their unique properties and potential to be used in electronic and optoelectronic nanodevices, one dimensional nanoscale materials has received great attention lately. Especially, nanocones have been used in many application areas from field emitter devices to the probes for scanning microscopy owing to their sharp tip and unique morphological characteristics. ZnO has been generally used for its electrical, optical and photochemical properties [2]. In this work, the tomographic reconstructions of ZnO nanocones were performed by using secondary signals in order to understand their growth mechanism.

Figure 1.A illustrates the significance of thickness effect in transmission imaging. The thickness of the nanocone reaches up to about 1micron in certain regions and the TEM imaging failed in terms of projecting the density of the sample. On the other hand, for a thick sample, there exists more secondary signal compared to transmitted electrons that can be used to form the image. Moreover, the addition of high energy electrons in a TEM makes secondary and backscattered electron imaging a strong candidate in terms of forming projection images as in Figure 1.B. The tilt series using secondary signals was collected by using a 1° tilt-angle interval over the range -71° to +71°. The volume rendering of the tomograms reconstructed using these images were shown in Figure 2. The application of this imaging technique to tomography reveals the 3D structure of the

M. Luysberg, K. Tillmann, T. Weirich (Eds.): EMC 2008, Vol. 1: Instrumentation and Methods, pp. 331–332, DOI: 10.1007/978-3-540-85156-1_166, © Springer-Verlag Berlin Heidelberg 2008

332

nanocones effectively. Therefore, utilization of secondary and backscattered electrons in tomography is very promising for the 3D morphologic analysis of thick samples.

1. P. A. Midgley and M. Weyland, Ultramicroscopy 96, 413 (2003).
2. L. Vayssieres et. al, Chem. Mater. 13, 4386 (2001).

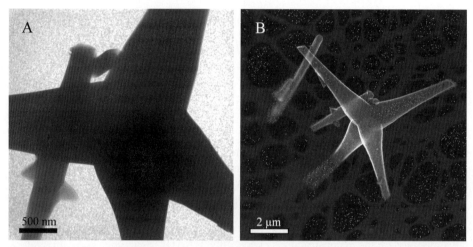

Figure 1. A. Bright Field TEM image of the ZnO nanocone. B. Secondary and backscattered electron image of the same nanocone.

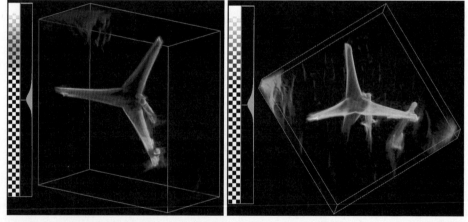

Figure 2. Volume rendering of tomograms reconstructed using secondary and backscattered electrons.

Quantification of Nanoparticle Tomograms

Z. Saghi, X. Xu and G. Möbus

Dept. of Eng. Materials, University of Sheffield, Mappin Street, Sheffield, S1 3JD, UK.

z.saghi@sheffield.ac.uk

Keywords: electron tomography, quantification, nanoparticles, fractal dimension

The reconstruction of nanoparticles embedded in a matrix or free-standing has recently become a major aim of electron tomography [1-4]. The generation of iso-threshold displays or 3D video-animations is the standard method of communicating results of tomography studies. Beyond qualitative visualisation, there is also a significant amount of quantifiable information to be gained from post-processing of tomograms, such as the estimation of volumes, surface areas, shape asymmetry, or surface roughness. Three examples of such processing are presented:

(i) Nanoparticle volume estimation: Octahedrally facetted CeO_2 nanoparticles with constant density and chemistry were reconstructed by EFTEM [5] (Figure 1) and binarised after backprojection so as to extract the particle (voxel value: 1) from the background (voxel value: 0). The sum of all voxels is then a measure of the volume. For regularly shaped nanoparticles, the error evaluation can be done by comparing the "reconstructed volume" V_r with the "geometric volume" V_g as calculated from edges and basal planes directly from the micrograph.

(ii) Particle volume fraction of a nanocomposite: Dendritic CeO_2 precipitates inside a borosilicate glass [6] appear with a high contrast ratio in ADF-STEM due to Z-contrast (Figure 2). A tomogram of a micrometre-sized glass fragment with a dozen of nanoparticles can therefore be thresholded and binarised twice: at a lower level to reveal the glass fragment total shape, and at a higher level to isolate all CeO_2 particles above a dark background. The total voxel count of the two binary tomograms then quantify glass fragment volume and precipitate volumes, and the ratio estimates the essential particle volume fraction V_v. This method is more accurate than counting knots on a mesh lying inside or outside a particle, and requires sections instead of projections.

(iii) Volume quantification of a fractal particle: One dendrite from the same glass composite can be tomographically reconstructed [7] and evaluated for its actual volume and fractal dimension. A surface estimate (and surface/volume ratio) can be gained by edge enhancement (Laplace filter) and binarisation. The sum of all non-zero surface voxels then gives the surface area. According to Figure 3, the ratio of a selected cross-sectional area of the dendrite relative to the chosen support envelope is 0.53 in 2D (depending on threshold) while this ratio for the entire 3D dendrite is 0.48. The ratio of its edge contour length relative to the contour length of the support envelope is 3.64 (depending on shape of extracted support) and amounts to 2.80 in 3D. The fractal dimension d [8] estimated from a box counting approach is d=1.61 on the 2D section, while d=2.39 for the 3D dendrite.

Errors/ambiguities arise from noise on the input images, threshold selection, choice of reference for the ratios and also the missing wedge artefact [9].

M. Luysberg, K. Tillmann, T. Weirich (Eds.): EMC 2008, Vol. 1: Instrumentation and Methods, pp. 333–334, DOI: 10.1007/978-3-540-85156-1_167, © Springer-Verlag Berlin Heidelberg 2008

334

1. A.J. Koster, U. Ziese, A.J. Verkleij, A.H. Janssen and K.P. De Jong, Journal of Physical Chemistry B **104** (2000), p. 9368.
2. G. Möbus and B.J. Inkson, Applied Physics Letters **79** (2001), p.1369.
3. M. Weyland and P.A. Midgley, Materials Today **79** (2004), p.32.
4. G. Möbus and B.J. Inkson, Materials Today **10** (2007), p.18.
5. X. Xu, Z. Saghi, R. Gay and G. Möbus, Nanotechnology **18** (2007), art. no. 225501.
6. G. Yang, G. Möbus and R.J. Hand, Micron **37** (2006), p.433.
7. X. Xu, Z. Saghi, G. Yang, R.J. Hand and G. Möbus, Crystal Growth & Design (2008).
8. E.P.W. Ward, T.J.V. Yates, J-J. Fernandez, D.E.W. Vaughan and P.A. Midgley, The Journal of Physical Chemistry C **111** (2007), p.11501.
9. We acknowledge funding by EPSRC, UK under the grant number GR/S85689/01.

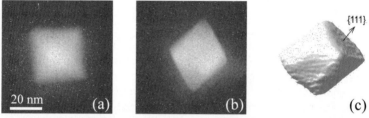

Figure 1. (a-b) EFTEM images of a CeO2 nanoparticle at -25° and +55°. (c) isosurface view of the tomogram. Quantification: (i) 2D measurements: a = b = 31.7 nm, c = 45.7 nm; geometric volume V_g = 15300 nm^3 (based on $a^2 * c/3$); (ii) Reconstructed tomogram volume: voxel size = 0.26nm^3; V_r = 17300nm^3; Systematic error $(V_r - V_g)/V_g$ = 13.2%.

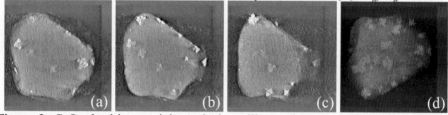

Figure 2. CeO$_2$ dendrite precipitates in borosilicate glass, (a-c): 3 sections through tomogram, (d) re-projection of all slices to show overestimation of particle density without tomography. Quantification: (i) Measured particle volume fraction from tomogram: V_v = 0.030; (ii) From mesh counting (inside/outside particle hits): Np/N = 10/266 = 0.037 (±0.003); (iii) Similar mesh counting on (2d): Np/N = 0.33.

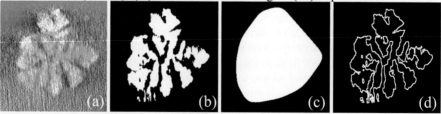

Figure 3. CeO$_2$ dendrite selected from glass of Figure 2: (a) section through tomogram [5]; (b) volume segmentation for surface and volume estimates; (c) support of the particle; (d) edge enhancement of (b) for the evaluation of edge contour length.

Tomographic imaging ultra-thick specimens with nanometer resolution

E. Sourty[1,2], B. Freitag[1], D. Wall[1], D. Tang[1], K. Lu[2,4], J. Loos[2,3,4]

1. FEI Company, Achtseweg Noord 5, Building AAE, 5600 KA Eindhoven/Acht, The Netherlands.
2. Laboratory of Materials and Interface Chemistry and Soft-Matter CryoTEM Research Unit, and
3. Laboratory of Polymer Technology, Eindhoven University of Technology, PO Box 513, NL-5600 MB Eindhoven, The Netherlands.
4. Dutch Polymer Institute, Eindhoven University of Technology, PO Box 902, NL-5600 AX Eindhoven, The Netherlands.

Erwan.Sourty@fei.com, j.loos@tue.nl
Keywords: STEM, Tomography, ultra-thick specimen

Transmission electron microscopy (TEM) is a well established and powerful technique to explore matter down to the atomic scale in two dimensions. Since real world objects have a three dimensional character the need to obtain volume information on nanometer scale is increasing especially on the mesoscopic dimensions of devices or complex mixtures of multiphase objects. TEM and scanning TEM (STEM) tomography methods have been developed in the past to fulfil this need, but these methods are limited to relatively thin specimens. In TEM this is caused mainly by the chromatic aberration and the dramatic increase of inelastic scattering in the sample, which leads ultimately to a strong resolution loss. Only high acceleration voltage TEM (above 1 million volts) allows for investigation of thicker specimens; this technique is, however, extremely expensive, far from being routine and highly destructive, in particular for organic matter. STEM tomography can minimize the effect of chromatic aberration, but is limited by the focal depth when large convergence angles are used. In this paper we establish a novel method based on dark field STEM tomography that pushes the resolution in all three dimensions down to a few nanometers for ultra-thick specimens of several micrometers at 300kV acceleration voltage.

In conventional STEM the electron beam is focused at the specimen and the illumination is convergent. The convergence angle used will determine the beam crossover i.e. the probe size at the specimen; the smallest probe size is obtained at a convergence where the superimposed contributions from electron source size, spherical aberrations, and diffraction effects at the beam limiting aperture are minimal. Due to its convergence, the beam will be focused at a specific height within the specimen, the beam size being broader at higher or lower positions. This results in a relatively shallow depth-of-field; hence high resolution only can be obtained for specimens with few 100-nm thicknesses at maximum.

Instead, using the flexibility of state-of-the-art three condenser illumination systems in a modern TEM, the convergence angle can be set to low values allowing the best

M. Luysberg, K. Tillmann, T. Weirich (Eds.): EMC 2008, Vol. 1: Instrumentation and Methods, pp. 335–336, DOI: 10.1007/978-3-540-85156-1_168, © Springer-Verlag Berlin Heidelberg 2008

compromise between probe size at the crossover (a few nanometers) and depth-of-field (a few micrometers). Figure 1 (a) represents the volume morphology of a polymer specimens filled with nanoparticles. The specimen has a thickness of about 1.0 µm. Obviously, a micrometer scale networks of the nanoparticles is observed within the entire volume of the specimens. Figure 1 (b) shows in top view the 5 nm gold beads deposited on both sides of a similar micrometer thick specimen: the brighter ones correspond to the top, the greyish gold beads to the bottom surface of the specimen. All beads are in focus, which evidently demonstrates the resolution, at least, that can be achieved by our approach.

In summary, low-convergence angle STEM allows volume analyzing and imaging of several micrometer thick specimens with few nanometer local resolution in all three dimensions; such information is not accessible with any other available technique. This unique approach introduced is applicable to a wide range of materials covering inorganic or condensed matter as well as soft matter like synthetic polymers and biomaterials. Potential challenging targets for our unique approach are characterization of the active site and substrate porosity in real bulk catalyst structures, identifying micrometer-scale networks and phase separation in polymer systems, or volume imaging of whole cells, to name but a few.

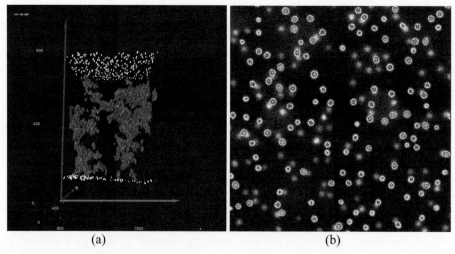

(a)   (b)

Figure 1. (a) 3D volume reconstruction visualizing the morphology of (a) ~ 1.0 µm thick polymer nanocomposite specimen filled with nanoparticles. In the 3D representation a network path of the nanoparticles throughout the whole specimen volume is highlighted; (b) image of 5-nm gold markers visible on top (bright) and bottom (reddish) surfaces of a similar micrometer thick specimen as achieved by applying low convergence-angle STEM.

Three-dimensional imaging at the mesoscopic scale using STEM-in-SEM

P. Jornsanoh[1], G. Thollet[1], C. Gauthier[1] and K. Masenelli-Varlot[1]

1. Université de Lyon, MATEIS UMR 5510, INSA-Lyon, 7 avenue Jean Capelle, 69621 Villeurbanne, France

Karine.Masenelli-Varlot@insa-lyon.fr

Keywords: tomography, STEM, SEM, ESEM

Tomography is a very powerful tool to investigate the three-dimensional structures of biological or materials specimens. However, the mesoscopic scale is covered neither by X-Ray tomography (insufficient spatial resolution) nor by electron tomography in a transmission electron microscope (too small reconstructed volume). We developed an electron tomography stage in a scanning electron microscope in order to obtain the missing information at the mesoscopic scale [1].

Controlled-pressure (or environmental) scanning electron microscopy indeed has a lot of advantages: it enables the observation of uncoated non-conductive samples, e.g. biological materials, ceramics or polymers. Moreover, several detectors are available in the microscope chamber, allowing the observation of sample surface. Observations in transmission mode (STEM-in-SEM) are also possible if thin enough samples are used. Interestingly, a previous work has shown that a signal could be obtained in STEM-in-SEM even for a few microns thick samples [2]. Finally, the 360°-rotation of a sample is made possible by the great dimensions of the microscope chamber.

Figure 1 presents the reconstructed volume of a PVC matrix filled with large inorganic particles (several hundreds of nm) and shock modifiers (diameter of about 80 nm). The images were acquired using STEM-in-SEM in a HAADF mode. Interestingly, a very good contrast could be obtained in the images, even between the shock modifiers and the PVC matrix, although the sample was not stained. This allowed the volume reconstruction. A very good compromise could be found between the size of the analysed volume (in that case $5x5x2$ μm^3) and the spatial resolution (estimated to a few tens of nm, since the shock modifiers could be detected).

A image of a polyurethane foam, acquired under the same conditions, is displayed in Figure 2. The foam is only partially electron transparent: the struts are too thick to give rise to a signal and appear in black, whereas the walls are thin enough and appear in light grey. Although a reconstructed volume cannot be computed from this kind of image series, a lot of pieces information regarding the foam structure can be derived from the tilted images series : cells geometry, walls thickness and curvature as well as the struts topography.

1. P. Jornsanoh, G. Thollet, K. Masenelli-Varlot, C.Gauthier, FR Patent 06-09-708, 2006.
2. A. Bogner, G. Thollet, D. Basset, P.H. Jouneau, C. Gauthier, Ultramicroscopy **104** (2005), pp. 290-301.

M. Luysberg, K. Tillmann, T. Weirich (Eds.): EMC 2008, Vol. 1: Instrumentation and Methods, pp. 337–338, DOI: 10.1007/978-3-540-85156-1_169, © Springer-Verlag Berlin Heidelberg 2008

Figure 1. 3D reconstruction of the inorganic fillers and shock modifiers dispersion in a PVC matrix (size of the reconstructed volume : 5x5x2 μm^3).

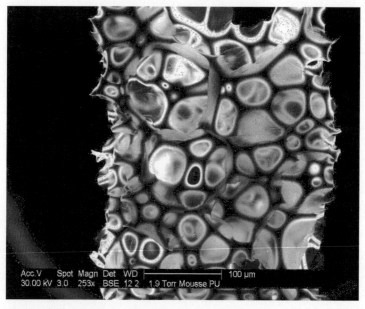

Figure 2. Transmission image of a polyurethane foam, with non electron-transparent struts and transparent walls.

Three-dimensional potential mapping of nanostructures with electron-holographic tomography

Daniel Wolf, Andreas Lenk, Hannes Lichte

Triebenberg Laboratory, Institute of Structure Physics, Technische Universität Dresden, D-01062 Dresden, Germany

Daniel.Wolf@Triebenberg.de

Keywords: electron tomography, electron holography, mean inner potential

The physical properties of solids are strongly correlated with the intrinsic electrostatic potentials. Therefore, the knowledge of the potential distribution is of high interest in many fields of research, e.g. solid-state physics, materials science or semiconductors industry. Electron-holographic tomography, i.e. combining off-axis electron holography with tomography, uses a series of electron holograms over a large range of object tilt angles to reconstruct the electrostatic potential in three dimensions [1]. The holographically reconstructed phase signal is - in the absence of diffraction effects - proportional to the projected potential, hence satisfies the projection requirement for tomography.

However, like in other tomographic methods, a high degree of automation for recording and alignment is essential for an applicable and quantitative tool for 3-D characterization [2]. For this reason, a software package within Digital Micrograph™ was developed that provides a complete solution for holographic tomography. Furthermore, interfaces to IMOD-software [3] and MATLAB TOM toolbox [4], offering sophisticated procedures for tomography, are created. The acquisition of the tilt series is performed by a software-tool with user-friendly graphical interface. This enables the operator to control the whole recording process of the electron holograms. Then, the tilt series of phase images is reconstructed. Since the phase shift in the phase images, obtained from the holographic reconstruction, often exceeds 2π, the phase images must be unwrapped. After alignment of the phase tilt series, i.e. shifting all projections to the common tilt-axis, simultaneous iterative reconstruction (SIRT) as well as weighted back-projection is used to reconstruct the 3-D electrostatic potential of the specimen. However, phase-unwrapping artifacts, the limited number of projections (missing wedge) and dynamical diffraction reduce the accuracy of this method. Thus, simulations are made in order to quantify the influence of these problems.

The two examples below demonstrate the capability of electron-holographic tomography as a measurement tool for 3-D electrostatic potentials. The first example, a 3-D reconstruction of a zinc oxide nano-rod (Figure 1), recovers a homogeneous potential distribution, where a mean inner potential value of 15.1 V can be determined by histogram analysis. The hexagonal cross-section indicates the [0001] growth direction. The second example, a FIB-prepared specimen with a p-n junction in silicon (Figure 2), exhibits that the analysis of doped semiconductors can be successfully extended into three dimensions using electron-holographic tomography [5]. The reconstructed 3-D potential can be evaluated quantitatively and locally, i.e.

M. Luysberg, K. Tillmann, T. Weirich (Eds.): EMC 2008, Vol. 1: Instrumentation and Methods, pp. 339–340, DOI: 10.1007/978-3-540-85156-1_170, © Springer-Verlag Berlin Heidelberg 2008

measurements of mean inner potential, built-in voltage across p-n junction and dead layer can be performed inside the reconstructed volume by accordingly positioned slices. It is a decisive advantage of tomography that it measures the potential locally voxel by voxel, instead of projected (averaged) across the whole object resulting from a single hologram.

1. G. Lai, T. Hirayama, K. Ishizuka, T. Tanji, A. Tonomura, J. Appl. Optics **33** (1994), p. 829.
2. D. Wolf, Microscopy and Microanalysis, Volume **13**, Suppl. 3 (2007), p. 112.
3. J.R. Kremer, D.N. Mastronarde, J.R. McIntosh, J. Struct. Biology **116** (1996), p. 71.
4. S. Nickell, F. Förster, A. Linaroudis, W. Del Net, F. Beck, R. Hegerl, W. Baumeister, J.M. Plitzko, J. Struct. Biology **149**, (2005), p. 227.
5. A.C. Twitched, T. J. V. Yates, R. E. Dunin-Borkowski, S.B. Newcomb, P.A. Midgley, Journal of Physics, Conference Series **26** (2006), p. 29.
6. The authors acknowledge financial support from the Deutsche Forschungsgemeinschaft and the European Union under the Framework 6 program under a contract for an Integrated Infrastructure Initiative. Reference 026019 ESTEEM

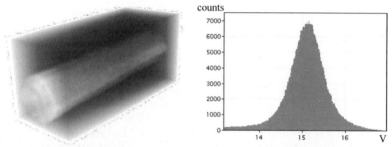

Figure 1. Reconstructed electrostatic 3-D potential of a grown zinc oxid rod (diameter: 100 nm) with corresponding histogram. The reconstruction recovers the hexagonal cross-section of the rod. The mean inner potential value of 15.1 V can be gained from the histogram. (Specimen kindly provided by Z. L. Wang, Georgia Institute, Atlanta)

Figure 2. Central slice through the reconstructed electrostatic 3-D potential of a needle-shaped FIB-prepared silicon sample with a p-n junction. The brighter upper part represents the n-doped substrate, the darker lower part represents the p-doped layer. Right: The profile-scan corresponding to the white arrow shows the expected absolute values of about 12 V (=MIP of silicon) as well as the expected potential variation of about 1 V (= built-in voltage across a silicon p-n junction).

Design of high-speed tomography with the 3MV ultrahigh voltage electron microscope

Kiyokazu Yoshida, Ryuji Nishi and Hirotaro Mori

Research Center for Ultra High Voltage Electron Microscopy, Osaka University, 7-1, Mihogaoka, Ibaraki, Osaka 567-0047, JAPAN

yoshida@uhvem.osaka-u.ac.jp

Keywords: tomography, electron microscope, high-speed

Since ultrahigh voltage electron microscopes (UHVEM) make it possible to observe thick specimens, they are widely used in electron tomography experiments. In order to take photographs of as many samples as possible in the limited machine time available, there is a need to shorten the time required to take photos. One of the reasons why so much time is required is that one must correct for variation in sample position and focus for each photo when the sample is tilted. In addition, CCD-camera exposure time and reading and recording time are required as well. Although practical automation of these correction and photography tasks has been achieved for conventional EMs, it seems no substantial effects have been achieved in shortening the total time required.

In this study, we employ a high-precision specimen tilting apparatus that can eliminate the need for the above correction. The result for this has been reported previously [1]. In the present work, we made further improvements that realized movement of no more than 3 μm in sample position when tilted by ±60°. Since, in this condition, the sample will remain within the field of view at magnification of 10,000 times or less, correction mentioned above becomes unnecessary.

Another cause of the long time requirements of electron tomography is the time needed to take photos. The present cooled CCD camera for the UHVEM requires one minute per photo for optical switching, exposure, reading, and recording. To shorten this time requirement, we developed a method of using an SLR digital camera and a Super HARP high-definition camera to record photos. Fig. 1 shows a comparison of images taken with each camera. The specimen is an LSI cross section. The direct magnification is 20,000 times. From left to right, the photos shown are for the cooled CCD, the digital SLR, and the high-definition camera, in that order. The lower photos have been magnified five times. Exposure times were four seconds, two seconds, and 1/30 second, respectively. Pixels, after trimming to match the field of view of the high-definition camera, numbered 3960×3170, 2680×2150, and 1280×1020, respectively.

Since, in general, tomography reconstruction is processed by resizing images to 1000×1000 pixels, any of the above resolutions is sufficient. However, comparison of details after magnifying the images shows that the signal-to-noise ratio for the digital camera and the resolution for the high-definition camera were inferior to those of the cooled CCD.

M. Luysberg, K. Tillmann, T. Weirich (Eds.): EMC 2008, Vol. 1: Instrumentation and Methods, pp. 341–342, DOI: 10.1007/978-3-540-85156-1_171, © Springer-Verlag Berlin Heidelberg 2008

In order to increase the efficiency of overall operations, we developed a software that links and automates the tilting and photographing operations. As a result, the entire series of photographic operations now can be conducted automatically.

The combination of the high-precision tilting apparatus with a digital camera or a high-definition camera and with the automated photography software made it possible to take photographs suited to their purposes of use in a short period of time at magnifications of 10,000 times or less — completing in approximately 20 minutes when using a digital camera and in approximately 75 seconds when using a high-definition camera what previously would have required 2-5 hours to take a series of 121 photos.

This work was mainly supported by the Ministry of Education, Culture, Sports, Science and Technology (MEXT), Japan, under a Grant-in-Aid for Scientific Research (Grant No. 19560663).

1. Design of a High Precision Tilting Apparatus for Electron Tomography with the 3MV HVEM, Kiyokazu Yoshida, Tadao Furutsu, Shuichi Mamishin and Hirotaro Mori, Proc. 16th Inter. Conf. on Electron Microscopy, 2006, Sapporo, 1146CD

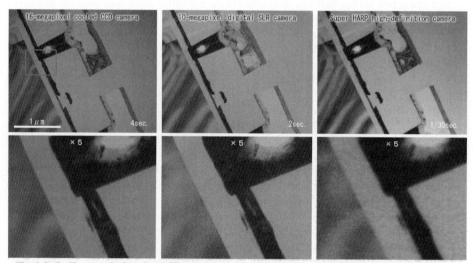

Fig. 1: Left: 16-megapixel cooled CCD camera, center: 10-megapixel digital SLR camera, right: Super HARP high-definition camera. In the above fields of view, pixel counts numbered 3960×3170, 2680×2150, and 1280×1020, respectively. Exposure times were 4, 2, and 1/30 seconds, respectively. Lower photos represent the area indicated by the square and magnified five times. The specimen used was an LSI cross section. Direct magnification: 20,000 times.

The point spread function assessment of MeV electron imaging quality for thick specimens

Fang Wang[1], Hai-Bo Zhang[1], Meng Cao[1],
Ryuji Nishi[2], Kiyokazu Yoshida[2] and Akio Takaoka[2]

1. Department of Electronic Science and Technology, Xi'an Jiaotong University, Xi'an 710049, People's Republic of China
2. Research Centre for Ultrahigh Voltage Electron Microscopy, Osaka University, 7-1 Mihogaoka, Ibaraki, Osaka 567-0047, Japan

hbzhang@mail.xjtu.edu.cn

Keywords: point spread function, thick specimen, image quality

Electron tomography (ET) with the high voltage electron microscope (HVEM) or the ultra-HVEM is becoming particularly useful for relatively thick amorphous specimens. This may extend utilization of ET to both biology and physical science [1]. However, tilting a thick specimen in ET results in the increase of its effective thickness, and thus may degrade the image quality due to multiple scattering. In this presentation, we investigate the dependence of the image quality on the accelerating voltage and the effective thickness of the tilted thick specimen in the ultra-HVEM at Osaka University.

The image quality of the thick specimen can be evaluated by the calculation of the standard deviation of the point spread function (PSF) from gold-particle images [2]. Under different accelerating voltages, a series of images of gold particles with a mean diameter of 100 nm on the bottom surface of the tilted 5 μm thick epoxy-resin film was recorded as the blurred images $g(x,y)$. The effective thickness H of the specimen with tilting angle θ is $h/\cos\theta$, where h is the real thickness of the specimen. Based on the assumption that the imaging system in the experiment is linear, $g(x,y)$ can be expressed by the convolution between the ideal image $f(x,y)$ of a 100 nm gold particle and the PSF $h(x,y)$ as

$$g(x,y) = f(x,y) \otimes h(x,y).$$

$h(x,y)$ can be acquired by deconvolution using the increment Wiener filter [2], and then be approximated to a Gaussian function with a standard deviation σ_θ. The narrower PSF, or the smaller σ_θ corresponds to the higher image quality for the thick specimen.

Figure 1 illustrates the variation of σ_θ with the effective thickness H at the accelerating voltages of 0.5, 1, and 3 MV for electrons. The image quality is improved with the increase of the accelerating voltage, and degraded with the increase of H. Further, Figure 2 shows σ_θ as a function of the accelerating voltage, at the tilting angle of 70° and H of 14.6 μm. Here, the image quality is considerably improved at 1−3 MV, compared with the situation of 0.5 MV.

The deterioration of image quality for the tilted thick specimen is mainly due to electron beam broadening caused by multiple elastic scattering [3]. According to Bothe's theory, we have calculated the angular distribution of 3 MeV electrons

M. Luysberg, K. Tillmann, T. Weirich (Eds.): EMC 2008, Vol. 1: Instrumentation and Methods, pp. 343–344, DOI: 10.1007/978-3-540-85156-1_172, © Springer-Verlag Berlin Heidelberg 2008

penetrating specimens of different thicknesses. Figure 3 shows that with the increase of specimen thickness, transmitted electrons distribute to a wider angular range. Only electrons scattered within the half angle of the objective aperture α_0 can be collected to image. Figure 4 describes the calculated angular distribution of electrons under different accelerating voltages. In short, we can obtain a more intensive signal of transmitted electrons and higher image quality by increasing the accelerating voltage for electrons.

1. A. Takaoka, T. Hasegawa, K. Yoshida and H. Mori, Ultramicroscopy **108** (2008), p. 230.
2. C. Yang, H.B. Zhang, J.J. Li and A. Takaoka, J. Electron Microsc. **54** (2005), p. 367.
3. H.B. Zhang, F. Wang and A. Takaoka, J. Electron Microsc. **56** (2007), p. 51.
4. This work was supported by "Specialized Research Fund for the Doctoral Program of Higher Education" and "Nanotechnology Network Project of the Ministry of Education, Culture, Sports, Science and Technology (MEXT), Japan" at the Research Centre for Ultrahigh Voltage Electron Microscopy, Osaka University (Handai multi-functional Nano-Foundry).

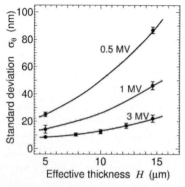

Figure 1. The PSF standard deviation σ_θ vs. the effective specimen thickness under different accelerating voltages.

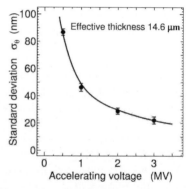

Figure 2. The variation of σ_θ vs. the accelerating voltage, at the effective specimen thickness of 14.6 μm.

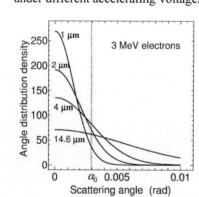

Figure 3. The angular distribution of 3 MeV electrons that have penetrated specimens of different thicknesses.

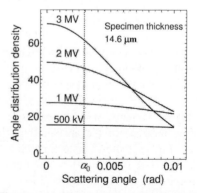

Figure 4. The angular distribution of electrons at different accelerating voltages. The specimen thickness is 14.6 μm.

Relevance of the minimum projection number to specimen structures for high-quality electron tomography

Hai-Bo Zhang[1], Meng Cao[1], Yong Lu[1], Chao Li[1], Ryuji Nishi[2] and Akio Takaoka[2]

1. Department of Electronic Science and Technology, Xi'an Jiaotong University, Xi'an 710049, People's Republic of China
2. Research Centre for Ultrahigh Voltage Electron Microscopy (Ultra-HVEM), Osaka University, 7-1 Mihogaoka, Ibaraki, Osaka 567-0047, Japan

hbzhang@mail.xjtu.edu.cn

Keywords: specimen structure, projection number, reconstruction quality

Electron Tomography (ET) is becoming a powerful 3D analysis approach in many fields such as biology, life science, materials science and physics. However, some factors that influence the ET quality have been poorly understood, and the ET quality of the integrated circuit (IC) specimen with a complex structure needs to be improved [1]. Therefore, from the point of view of signal processing, we have investigated, for the first time, the relationship between the structural characteristic of a specimen and the minimum projection number required for the perfect or high-quality reconstruction.

We first consider the case of a 2D image with the discrete spectrum. For terms $-N$ to N of the Fourier series of the image along the projection angle, we have demonstrated that the perfect reconstruction of the image is available from $N+1$ projections at least, which are distributed homogeneously over the projection angle range of $0-180°$. The relevance of the projection number to the image structure is equivalent to the Nyquist criterion in the discrete-time system. In other words, the Nyquist criterion can be extended to evaluate the condition of the perfect reconstruction for an image.

Further, if above term number N turns to infinite, the frequency spectrum of an image will become continuous. As the original image to be reconstructed, we chose a tomographic section that was taken from 3D reconstruction results of a thick IC specimen by the ultra-HVEM tomography [2]. The inverse Fourier transform algorithm for the reconstruction here was used to reduce the algorithm error. Figure 1 illustrates the reconstruction quality in the mean square error (MSE) and the calculated frequency spectrum of the image. According to Figure 1(a), the abscissa value of the convergent point in the MSE curve represents the minimum number of projections for high-quality reconstruction, 40 projections. Evidently, the number is just equal to the frequency bandwidth in Figure 1(b) of the original image, satisfying the extended Nyquist criterion again. The results in Figures 2(a)–(c) were reconstructed from 20, 40 and 60 projections, respectively. When the projection number is less than 40, there is an obvious information distortion in Figure 2(a). When the projection number is 40, the reconstructed image shown in Figure 2(b) contains most characteristics of the original structure and is, therefore, of high quality. Nevertheless, the reconstruction quality is not so improved if the projection number is more than 40, as seen in Figure 2(c). Note that the original image in Figure 2(d) is also a 2D distribution of the practical specimen.

M. Luysberg, K. Tillmann, T. Weirich (Eds.): EMC 2008, Vol. 1: Instrumentation and Methods, pp. 345–346, DOI: 10.1007/978-3-540-85156-1_173, © Springer-Verlag Berlin Heidelberg 2008

It can thus be concluded that in both image cases of discrete and continuous frequency spectrum, the extended Nyquist criterion is valid for image reconstruction, and also applicable to 3D reconstruction. Practically, high-quality ET will be realized if the frequency bandwidth of a sampling system is not less than that of the specimen structure, which should be considered for further ET applications in complex specimens.

1. H.B. Zhang, X.L. Zhang, Y. Wang and A. Takaoka, Rev. Sci. Instrum. **78** (2007), 013701.
2. A. Takaoka, T. Hasegawa, K. Yoshida and H. Mori, Ultramicroscopy **108** (2008), p. 230.
3. This work was supported by "Specialized Research Fund for the Doctoral Program of Higher Education" (No. 20070698013) and "Nanotechnology Network Project of the Ministry of Education, Culture, Sports, Science and Technology (MEXT), Japan" at the Research Centre for Ultrahigh Voltage Electron Microscopy, Osaka University (Handai multi-functional Nano-Foundry).

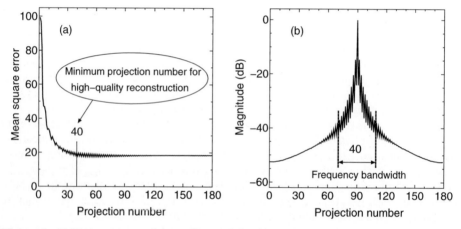

Figure 1. (a) The reconstruction quality and (b) the equivalent frequency spectrum of a 2D specimen distribution in the domain of projection number. Here, the extended Nyquist criterion is valid for 2D reconstruction of the image.

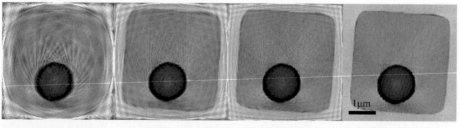

(a) 20 projections   (b) 40 projections   (c) 60 projections   (d) original image

Figure 2. Reconstructed images (a)–(c) at different projection numbers and the original image (d) as one of tomographic sections by the ultra-HVEM tomography.

Atomic-Scale Chemical Imaging of Composition and Bonding by Aberration-Corrected Microscopy

D.A. Muller[1], L. Fitting Kourkoutis[1], M. Murfitt[2], J.H. Song[3], H.Y. Hwang[4], J. Silcox[1], N. Dellby[2], O.L. Krivanek[2]

1. Applied and Engineering Physics, Cornell University, Ithaca, NY, 14853, USA
2. Nion Co., Kirkland, WA 98033, USA
3. Department of Physics, Chungnam National University, Daejeon 305-764, Korea
4. Department of Advanced Materials Science, University of Tokyo, Kashiwa, 277-8561, Japan

dm24@cornell.edu

Keywords: STEM, EELS, Aberration Corrector

The unique chemical identification of all atoms in a sample, with atomic resolution, has been a long-standing goal of analytical microscopy. The first step towards atomic-resolution EELS imaging was the point-by-point identification of the bonding states across an interface, using a scanning transmission electron microscope (STEM)[1]. Without the addition of a corrector, small chemical images are possible, but can take over an hour to record [2].The recent correction of third order electron-optical aberrations by multipole correctors has allowed the illumination aperture size to be increased, giving a six-fold increase in the beam current for the same probe size[3]. Using a fifth-order aberration-corrected scanning transmission electron microscope, which provides a factor of 100x increase in signal over an uncorrected instrument, we demonstrate two-dimensional elemental and valence-sensitive imaging at atomic resolution using electron energy loss spectroscopy, with acquisition times for small images (4096 pixels) of well under a minute [4].

There is more to EELS chemical imaging than simply increasing the beam current and reducing the probe size. Unless the collection optics are improved as well, it is difficult to take advantage of the increased beam current, most of which may fall at angles outside the entrance aperture of the spectrometer. Worse, as has been recently demonstrated, when the collection angle is much smaller than the incident probe angles, elastic scattering artifacts can dominate the EELS spectrum image[5]. Strong elastic scattering on heavy atoms columns can prevent inelastic electrons from entering the spectrometer, resulting in contrast reversals and inelastic images that reflect mostly the elastic scattering and not the elemental distribution in the sample. A common symptom of this elastic scattering artifact is that all inelastic maps peak at the same spatial positions independent of the elements actual location. The effect can be suppressed by increasing the collection angle so that almost all of the elastically scattered electrons enter the spectrometer. However at such large angles, the energy resolution of the spectrum can be degraded. With proper design and corrective optics, the energy resolution can be preserved and incoherent EELS maps free of diffraction and other elastic artifacts can be obtained[6].

M. Luysberg, K. Tillmann, T. Weirich (Eds.): EMC 2008, Vol. 1: Instrumentation and Methods, pp. 347–348, DOI: 10.1007/978-3-540-85156-1_174, © Springer-Verlag Berlin Heidelberg 2008

The spectrum-image of a $La_{0.7}Sr_{0.3}MnO_3/SrTiO_3$ superlattice (Figs 1&2) was recorded with an incident beam current of ~780 pA with ~600 pA collected in the spectrometer and a dwell time of 7 ms per pixel. The entire image was recorded in less than a minute (28 seconds live time, 10 seconds readout). In addition to the high-brightness source and improved probe formation, the post specimen optics were matched to collect over a much wider range of scattering angles than in an uncorrected instrument [6]. As a consequence, the elastic scattering artifacts that masked the chemical signal in earlier EELS work has been suppressed, and the image acquisition times were reduced by a factor of 10-100x.Ultimately, the signal obtained is sufficient that the EELS fine structure indicative of chemical bonding is visible in individual spectra[4,7].

1. P. E. Batson, Nature **366**, 727 (1993).
2. K. Kimoto, *et al.*, Nature **450**, 702 (2007).
3. P. E. Batson, N. Dellby, O. L. Krivanek, Nature **418**, 617 (2002).
4. D. A. Muller, *et al.*, Science, in press Feb 22 (2008).
5. M. Bosman, *et al.*, Phys. Rev. Lett. **99**, 086102 (2007).
6. O. L. Krivanek, *et al.*, Ultramicroscopy (2007).
7. This work supported by NSF grants DMR-9977547, EEC-0117770, DMR-0520404 and the Japan Society for the Promotion of Science. In-kind contributions were made by Gatan, Inc., Sandia National Laboratory, Pacific Northwest Laboratory and Bell Laboratories.

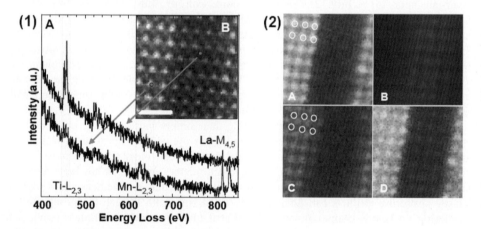

Figure 1. A) Individual EELS spectra from a 64 x 64 spectrum image series across a $La_{0.7}Sr_{0.3}MnO_3/SrTiO_3$ multilayer. Each spectrum is recorded with a 7 ms dwell time. (B) The simultaneously recorded annular dark field (ADF) image. The brightest atoms are La, then Sr, then Ti/Mn. 1 nm scale bar.

Figure 2. Spectroscopic-imaging of a multilayer, showing the different chemical sublattices in a 64x64 pixel spectrum-image extracted from the data of Fig 1. (A) La-M edge (B) Ti-L edge (C) Mn-L edge (D) Red/Green/Blue false color image obtained by combining the rescaled Mn,La, and Ti images. 3.1 nm field of view.

EMCD with nm Resolution and Below:
Experiments, Proposals, and a Paradox

P. Schattschneider[1], M. Stöger-Pollach[2], F. Tian[1] and J. Verbeeck[3]

1. Institut für Festkörperphysik, Technische Universität Wien, Wiedner Hauptstrasse 8-10, A-1040 Wien, Austria
2. Univ. Service Center for Transmission Electron Microscopy, Technische Universität Wien, Wiedner Hauptstrasse 8-10, A-1040 Wien, Austria
3. EMAT, University of Antwerp, Groenenborgerlaan 171, B-2020 Antwerp, Belgium

schattschneider@ifp.tuwien.ac.at
Keywords: magnetic dichroism, EELS, coherence

In the two years after the discovery of the EMCD effect [1] (energy loss magnetic chiral dichroism) an impressive improvement in signal and spatial resolution has been achieved. Several labs are now beginning to use this novel technique with the aim to investigate nano-magnetic systems used in spintronics applications.

The particular attraction of EMCD rests in the possibility to detect atom specific magnetic moments in combination with spin and orbital sum rules with nm resolution.

A review of the most important results achieved so far will be presented. The dependence of the EMCD on several experimental conditions (such as thickness, relative orientation of beam and sample, collection and convergence angle) is investigated. Different scattering geometries are illustrated; their advantages and disadvantages are detailed, together with their actual limitations.

A careful analysis of the inelastic scattering process based on the kinetic equation [2] reveals new aspects of EMCD. In particular, it appears feasible to perform EMCD in image mode instead of the standard diffraction setting. Simulations of a simple model system suggest detection of the EMCD signal in a combined mode - (x, q) or (x, E) - where x is a one-dimensional spatial coordinate, q is the scattering vector perpendicular to x, and E is the energy loss. Thus it should be possible to obtain a one-dimensional spatial map, directly visualising chiral electronic transitions with atomic resolution. A demonstration example is given in Fig. 1. The scale corresponds to a $2p \rightarrow 3d$ dipole allowed transition in Si and is purely hypothetical, giving an order of magnitude of the expected effect. Simulations for more realistic systems are under way.

An (x, q_y) map can be realized using a cylinder lens as sketched in Fig. 2. (e.g. a quadrupole of a Cs corrector). The detector is placed in the image plane. These maps would allow measuring the angular momentum of the transition from that of the probe

$$\langle L_z \rangle = \hbar \iint dx \, dq_y \; \rho(x, q_y) \, x \, q_y \, .$$

It is shown that the crystal supplies additional angular momentum to the probe during its propagation through the lattice. This must be taken into account when determining $< L_z >$. We discuss also the paradox that in the standard geometry (incident

M. Luysberg, K. Tillmann, T. Weirich (Eds.): EMC 2008, Vol. 1: Instrumentation and Methods,
pp. 349–350, DOI: 10.1007/978-3-540-85156-1_175, © Springer-Verlag Berlin Heidelberg 2008

plane wave, detection of a plane wave) no angular momentum seems to be transferred to the specimen whereas the atom's angular momentum changes in the chiral transition.

1. Schattschneider P., Rubino S., Hébert C., Rusz J., Kunes J., Novák P., Carlino E., Fabrizioli M., Panaccione G., Rossi G., Nature **441** (2006) 486- 488
2. S. Dudarev, L. Peng, M. Whelan, Phys. Rev.B 48 (1993) 13408
3. J.V. acknowledges financial support from the European Union under a contract for an Integrated Infrastructure Initiative. Reference 026019 ESTEEM.

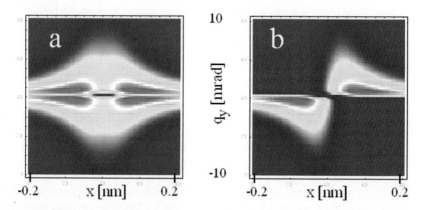

Figure 1. Schematic map of a 2p → 3d transition for atomic Si in the (x,q_y) plane. The color coding (rainbow chart) represents the probability to find the probe electron in the respective (x,q_y) state. The incident electron was assumed to be a plane wave with k=2500/nm (corresponding to 200 keV energy) impinging along the z axis.

  a) $|p> \rightarrow |d>$ transition with degenerate magnetic levels;

  b) $|p> \rightarrow |d,m=-1>$ transition. The scattering vector q_y is given in mrad.

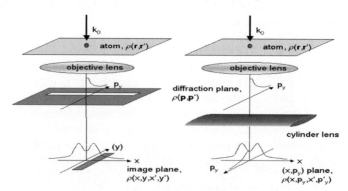

Figure 2. Scattering geometries for EMCD and density matrix ρ.
Left: Selection of $q_y > 0$ and $x > 0$. This is equivalent to the standard geometry.
Right: a cylinder lens maps the q_y coordinate onto the image plane.

Combining electronic and optical spectroscopy at the nanometer scale in a STEM

S. Mazzucco[1], R. Bernard[1], M. Kociak[1], O. Stéphan[1], M. Tencé[1],
L.F. Zagonel[1], F.J. Garcia de Abajo[2] and C. Colliex[1]

1. Laboratoire de Physique des Solides, CNRS, Université Paris Sud 11,
91405 Orsay, France
2. Instituto de Optica, CSIC, Serrano 121, 28006 Madrid, Spain

bernard@lps.u-psud.fr
Keywords: STEM, electronic and optical spectroscopy, metallic nanoparticles

Low loss EELS has proved to be a very useful tool to study plasmons in metallic nanoparticles. Combined with the spatial resolution of a Scanning Transmission Electronic Microscope (STEM), it has been possible to map plasmon modes on a single metallic nanoparticle [1,2]. The limitations of this spectroscopic method are the energy resolution that can be achieved due to energy dispersion of the electrons and the detection of very low energy plasmonic modes (< 3 eV) which are masked by the Zero Loss Peak tail. Cathodoluminescence is a way to overcome these difficulties [3]. In fact, coupling optical spectroscopy with the STEM electron beam as an excitation source presents several advantages: energy of visible and near IR photons lie in the 1-3 eV range which is difficult to access in EELS, the energy resolution is much better and the polarization of the emitted light constitutes one additionnal information about the physical phenomena occuring inside the sample.

Our project aims to design a versatile system allowing both cathodoluminescence measurements in our STEM (see Figure 1) and the opposite process, which means using photons as the excitation source and electrons as a probe. In this poster, we will discuss the theoretical background of this project and the interest to perform electron spectroscopy on optically active systems under light illumination. Furthermore, we will describe the experimental setup that we are currently implementing to achieve this goal. Hopefully we will be able to present our first results of metallic nanoparticle spectroscopy under both electron and photon excitation.

1. J. Nelayah, M. Kociak, O. Stephan, F. J. G. de Abajo, M. Tence, L. Henrard, D. Taverna, I. Pastoriza-Santos, L. M. Liz-Marzan and C. Colliex, Nature Physics 3 (2007), p. 348.
2. M. Bosman, V. J. Keast, M. Watanabe, A. I. Maaroof and M. B. Cortie, Nanotechnology 18 (2007), p. 165505.
3. N. Yamamoto, K. Araya and F. J. Garcia de Abajo, Phys. Rev. B 64 (2001), p. 205419.

M. Luysberg, K. Tillmann, T. Weirich (Eds.): EMC 2008, Vol. 1: Instrumentation and Methods, pp. 351–352, DOI: 10.1007/978-3-540-85156-1_176, © Springer-Verlag Berlin Heidelberg 2008

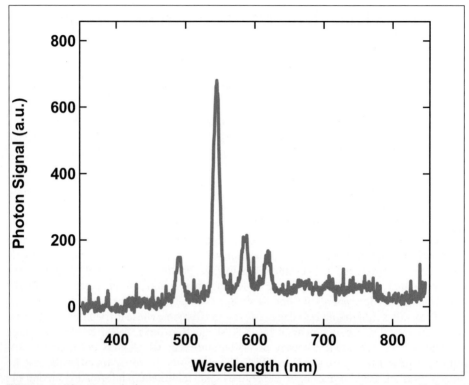

Figure 1. Photon emission spectrum from YAG powder under electron excitation at 100kV in a VG-HB501 STEM

Deconvolution of core loss electron energy loss spectra

G. Bertoni[1,2], J. Verbeeck[1]

1. Electron Microscopy for Materials Science (EMAT), University of Antwerp, Groenenborgerlaan 171, B-2020 Antwerp, Belgium
2. Italian Institute of Technology (IIT), Via Morego 30, I-16163 Genoa, Italy

Giovanni.Bertoni@ua.ac.be
Keywords: Electron energy loss, deconvolution, model based

Experimental electron energy loss spectra (EELS) are related to inelastic scattering events in the sample by the so called single scattering distribution $S(E)$ [1] (the differential cross section after integration in the detector acceptance angle). The relationship between the experimental spectrum and its $S(E)$ is complicated because of several effects, for instance:

- Multiple scattering occurs (and it is dependent on the thickness).
- The energy resolution of the instrument (TEM) is finite (due to a non monochromatic gun).
- Aberrations are present in the spectrometer.
- Instabilities and drift.

For core loss spectra, it is common to deconvolve the experimental spectrum with a low loss spectrum acquired at the same conditions. In a noise free system this could remove, for instance, the effect of the resolution of the system, making possible to replace expensive monochromators with software. Unfortunately, every experiment contains noise, and the amount of information that can be recovered depends on this noise.

In this contribution we present a way to retrieve an approximation of the single scattering distribution $S(E)$ for core loss ionization edges, making use of a model based technique. The experimental spectrum is modelled using a set of parameters, and then fitted to the data. This is the procedure we already used successfully in EELS quantification [2], and implemented in the EELSMODEL program [3]. The results of the retrieved single scattering term are shown in Figure 1 for Anatase TiO_2 and Rutile TiO_2 respectively. The spectra were measured on a Philips CM30 with a native energy resolution of 0.7eV. The model based results are compared with the results from a standard Fourier ratio deconvolution with a Gaussian modifier, from maximum entropy method based on Richardson-Lucy algorithm (RLA), and with X-ray results from literature [4,5].

1. R.F. Egerton in "Electron Energy-Loss Spectroscopy in the Electron microscope", Second Edition, (Plenum Press, New York and London) (1996), pp. 158-169.
2. G. Bertoni, J. Verbeeck, Ultramicroscopy (2008), doi:10.1016/j.ultramic.2008.01.004
3. The program is freely available under the GNU public license and it can be downloaded from http://webhost.ua.ac.be/eelsmod/eelsmodel.htm)

M. Luysberg, K. Tillmann, T. Weirich (Eds.): EMC 2008, Vol. 1: Instrumentation and Methods, pp. 353–354, DOI: 10.1007/978-3-540-85156-1_177, © Springer-Verlag Berlin Heidelberg 2008

354

4.   R. Ruus, A. Kikas, A. Saar, A. Ausmees, E. Nõmmiste, J. Aarik, A. Aidla, T. Uustare, and I. Martinson, Solid State Commun. 104 (1997), pp. 199-203.

5.   G. Bertoni acknowledges the financial support from FWO-Vlaanderen (project nr. G.0147.06). J. Verbeeck thanks the financial support from the European Union under Framework 6 program for Integrated Infrastructure Initiative, Reference 026019 ESTEEM.

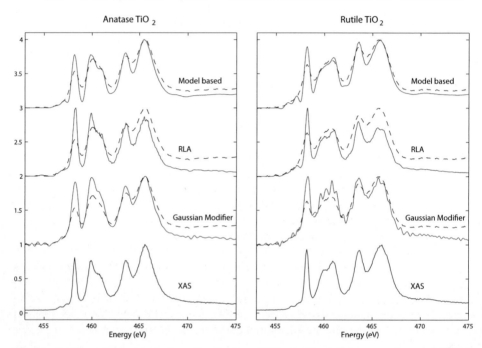

Figure 1. The single scattering distribution *S(E)* obtained from experimental EELS spectra (*dashed lines*) for TiO$_2$ using different deconvolution techniques (*full lines*), and compared with published results from X-ray absorption (XAS)[4], for TiO$_2$ Anatase (*left*), and TiO$_2$ Rutile (*right*). A Gaussian modifier of 0.5eV was used for Fourier ratio, and 200 iterations for the RLA. 40 points were used for the model based function.

Obtaining the loss function from angle resolved electron energy loss spectra

G. Bertoni[1,2], J. Verbeeck[1] and F. Brosens[3]

1. Electron Microscopy for Materials Science (EMAT), University of Antwerp, Groenenborgerlaan 171, B-2020 Antwerp, Belgium
2. Italian Institute of Technology (IIT), Via Morego 30, I-16163 Genoa, Italy
3. Department of Physics, University of Antwerp, Groenenborgerlaan 171, B-2020 Antwerp, Belgium

Giovanni.Bertoni@ua.ac.be
Keywords: Electron energy loss, plasmon, model based

In this communication we show results on the angular dependence of scattering for plasmons excitations (bulk) in the case of an interacting electron gas, which can serve as a model system for simple metals. We make use of energy filtered diffraction pattern from TEM to collect the signal as a function of momentum transfer. Figure 1 shows the scattering profile at energy loss centered on the bulk plasmon in aluminum. A first cut-off at θ_E, due to the dominant Lorentzian term in the angular part, and a second smooth cut-off at θ_C are visible. The second cut-off is due to the damping of the plasmon by particle-hole excitations [1]. Figure 1 shows the enhancement in the visibility of θ_C by removing the quasi-elastic scattering by means of a deconvolution with a diffraction pattern filtered around the zero loss peak.

Furthermore, we propose a way to retrieve the angle-resolved bulk loss function from the experiment, using model based technique, as alternative to deconvolution techniques. A starting guess for the loss function in the (q,E) plane is obtained according to the Random Phase Approximation (RPA), or to the Dynamical Exchange Decoupling (DED) [2]. This function, defined in a $(n{\times}m)$ mesh of points, is used to build the single inelastic scattering term $S(q,E)$. From this the full multiple scattering terms are calculated as:

$$\tilde{J}(q,E) = I_t \tilde{Q}(q,E) \exp(-t/\lambda) \exp\left[(t/\lambda)\tilde{S}(q,E)\right],$$

Where $\tilde{Y}$ denotes the Fourier transform in energy plus Hankel transform in momentum of Y (assuming circular symmetry in q) [3]. $Q(q,E)$ is the quasi-elastic scattering term (thermal diffuse scattering), t/λ is the thickness, and I_t the total electronic intensity in the spectra. The model is then fitted to the experiment to find the best estimates for the $(n{\times}m)$ parameters. Figure 2 illustrates the construction of the final spectra, in the case of a $(41{\times}41)$ mesh for the loss function. The model obtained from DED present a lower dispersion and a broader peak in the continuum, in better agreement with the experiment, but both RPA and DED predict a higher intensity for the onset of the continuum of particle-hole excitations with respect to the experiment [4].

M. Luysberg, K. Tillmann, T. Weirich (Eds.): EMC 2008, Vol. 1: Instrumentation and Methods, pp. 355–356, DOI: 10.1007/978-3-540-85156-1_178, © Springer-Verlag Berlin Heidelberg 2008

1. R.F. Egerton in "Electron Energy-Loss Spectroscopy in the Electron microscope", Second Edition, (Plenum Press, New York and London) (1996), pp. 158-169.
2. J.T. Devreese and F. Brosens, in "Electron Correlations in Solids, Molecules, and Atom", vol. 81 (Plenum Publishing Corporation, New York) (1983).
3. D. Su and P. Schattschneider, Philosophical Magazine A 65 (1992), pp. 1127-1140.
4. G. Bertoni acknowledges the financial support from FWO-Vlaanderen (project nr. G.0147.06). J. Verbeeck thanks the financial support from the European Union under Framework 6 program for Integrated Infrastructure Initiative, Reference 026019 ESTEEM.

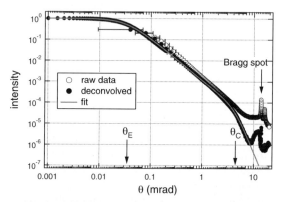

Figure 1. Radial intensity distribution for plasmon scattering in Aluminum, resulting from integration of a filtered diffraction pattern.

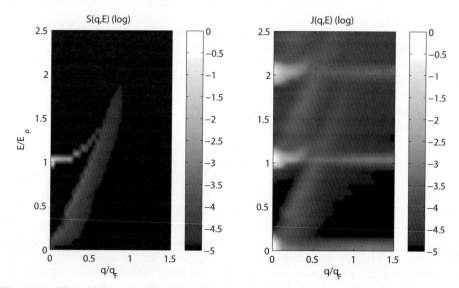

Figure 2. The single scattering term $S(q,E)$ obtained from the bulk loss function for aluminum in the DED approximation (*left*), and the spectrum $J(q,E)$ obtained after multiple inelastic scattering and quasi-elastic scattering (*right*).

Revisiting the determination of carbon sp^2/sp^3 ratios via analysis of the EELS carbon K-edge

R. Brydson[1], Z. Zhili[1,2], A. Brown[1]

1. Institute for Materials Research, University of Leeds, Leeds LS2 9JT, U.K.
2. Institute of Materials, Beijing Jiaotong University, Beijing 100044, China.

mtlrmdb@leeds.ac.uk
Keywords: EELS, carbon, bonding, sp^2 fraction

Carbon materials are highly versatile owing to the strong dependence of their physical properties on the ratio of sp^2 (graphitic) to sp^3 (adamantine) bonds. Graphitic carbons have a variety of forms with various degrees of ordering ranging from microcrystalline graphite to glassy carbons. Amorphous carbons can have any mixture of sp^3, sp^2 and, in some cases, sp^1 bonded carbon sites with the additional possible presence of hydrogen, nitrogen or boron. The structure of amorphous or poorly crystalline carbon materials are difficult to study using conventional characterisation tools; techniques which have been successfully applied include neutron scattering, nuclear magnetic resonance, (*uv*) Raman spectroscopy and EELS.

EELS analysis is enabled by consideration of the intensity of the spectral signals specific to π bonding which are associated with sp^2 and sp^1 bonding of carbon. These are contained in both the low loss region: the π plasmon and/or π to π * transition at approximately 6 eV), as well as the 1s to π * transition at 285 eV at the carbon K-edge. The low loss approach has been investigated by Fink et al. [1] and more recently by Daniels et al. [2]. Here we concentrate on analysis of the C K-edge.

A typical analysis procedure is as follows. The 1s to π * peak intensity at the C K-edge (background stripped and deconvoluted to remove plural scattering) is measured either using an energy window of typically 5 eV or by fitting a Gaussian under the peak (I_π). The ratio of this intensity to the total C K-edge intensity is then taken – the total intensity being usually measured in a window of up to 20 eV extending from the edge onset and covering both the π * peak and the majority of the subsequent 1s to σ* structure (I_σ). This intensity ratio represents the fraction of C atoms in the unknown (U) involved in π* bonding, however, this needs to be normalized to a reference spectrum (R). Typically a 100% sp^2 carbon reference material is employed, e.g. microcrystalline graphite or in some cases C_{60} fullerite crystals or even a thin foil of glassy carbon. The normalized intensity ratio is given by $N_{int\ ratio} = \dfrac{I_\pi^U / I_\sigma^U}{I_\pi^R / I_\sigma^R}$ and the sp^2 fraction (x) is

given $N_{int\ ratio} = \dfrac{3x}{(4-x)}$. Note x is the sp^2 fraction [3] and not the sp^2/sp^3 ratio as quoted in some texts [4].

M. Luysberg, K. Tillmann, T. Weirich (Eds.): EMC 2008, Vol. 1: Instrumentation and Methods, pp. 357–358, DOI: 10.1007/978-3-540-85156-1_179, © Springer-Verlag Berlin Heidelberg 2008

358

There are a number of issues surrounding this analytical procedure. Firstly, if a crystalline, anisotropic reference (e.g. graphite) or unknown material is measured then the relative intensities of features in the ELNES is dependent on sample orientation and this will affect the analysis results. The only way to avoid this is to record all spectra using an experimentally determined *magic* collection semi-angle of approximately twice the characteristic angle [5]; for the carbon K-edge this is ca. 1.7 mrads. Secondly, the presence of additional species (e.g. hydrogen). C-H bonds give rise to transitions to C-H σ* states centred at around 287 eV. Figure 1 shows a fit to a carbon K-edge spectrum, which has employed unconstrained Gaussian fits to C=C π* (at ca. 285 eV), C-C σ* (ca. 292 eV) and C=C σ* (ca. 300 eV). The residuals of the fit clearly show a peak at 287 eV which we associate with C-H σ* states. The intensity of this peak can be normalized (in a number of different ways) and used to extract the fraction of C-H bonds and hence estimate hydrogen contents in carbon materials. Thirdly, based on the fit in figure 1 we are investigating if it is possible to simply use the normalized intensity ratio of the fitted Gaussians for C=C π* (at ca. 285 eV), C-C σ* (ca. 292 eV) to determine sp^2/sp^3 ratios. Finally, there is a need for a systematic correlation of sp^2 contents in differing carbon materials as determined by EELS, Raman and NMR, if we are to improve the reliability of such methods particularly in the light of radiation induced damage in graphite at TEM voltages > 80 kV and the increasing use of FIB for cross-sectional TEM/EELS analysis of carbon coatings, which can also lead to sample alteration and damage.

1. J. Fink et al., Phys. Rev. B **30** (1984), p. 4713
2. H. Daniels et al., Phil. Mag. **87** (2007), p. 4073.
3. J. Bruley et al., J. Microscopy **180** (1995), p. 22.
4. J. Yuan and L.M. Brown, Micron **31** (2000), p. 515.
5. H. Daniels et al., Ultramicroscopy. **96** (2003), p. 523; C. Hebert et al., **101** (2004), p. 271.

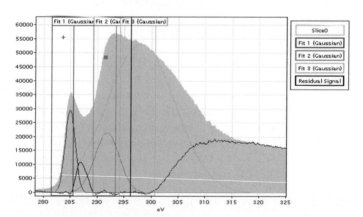

Figure 1. Example fit to the carbon K-edge of a graphitizable petroleum pitch heat treated at 600 °C (see [2]). The three Gaussian fit for C=C π*, C-C σ* and C=C σ* states leave a residual with intensity peaked at 287 eV due to C-H σ* states.

Orbital and spin sum rules for electron energy loss magnetic chiral dichroism: Application to metals and oxides

L. Calmels[1], B. Warot[1], F. Houdellier[1], P. Schattschneider[2], C. Gatel[1], V. Serin[1], and E. Snoeck[1]

1. CEMES-CNRS, 29 rue Jeanne Marvig, BP 94347, Toulouse Cedex 4, France
2. Institute for Solid State Physics, Vienna University of Technology, Wiedner Hauptrasse 8-10/138, A-1040 Vienna, Austria

calmels@cemes.fr

Keywords: EELS, sum rules, magnetic moment, transition metals

The possibility of using electron energy loss spectroscopy (EELS) to measure a magnetic chiral dichroic spectrum analogous to the x-ray magnetic circular dichroism (XMCD) signal has been suggested a few years ago [1]. This effect has been demonstrated recently for metallic samples [2]. The energy loss magnetic chiral dicroism (EMCD) signal is obtained by subtracting the EELS signal recorded at two symmetrical positions in the diffraction pattern, see figure 1. These two positions can be reached by moving the spectrometer aperture in the diffraction pattern [2], or by using the energy spectrum imaging (ESI) technique which consists in recording the whole diffraction pattern for successive energy windows [3].

We describe spin and orbital sum rules for EMCD spectra which are analogous to the XMCD sum rules [4, 5]. Their application allows the spin and the orbital magnetic moments to be determined from the experimental spectra and from products of Bloch wave coefficients of the incident and scattered probe electrons, which can be calculated within dynamical diffraction theory [6, 7]. Even without any knowledge of the Bloch wave coefficients the ratio of the spin to orbital magnetic moments can be extracted from the experimental spectra. The sum-rules have been applied to spectra recorded for metallic samples (Fe, Fe_xCo_{1-x}) as well as for ferromagnetic oxides (Fe_3O_4).

1. C. Hébert and P. Schattschneider, Ultramicroscopy **96**, 463 (2003).
2. P. Schattschneider, S. Rubino, C. Hébert, J. Rusz, J. Kunes, P. Novak, E. Carlino, M. Fabrizioli, G. Panaccione, and G. Rossi, Nature (London) **441**, 486 (2006).
3. B. Warot-Fonrose, F. Houdellier, M. J. Hÿtch, L. Calmels, V. Serin, and E. Snoeck, Ultramicroscopy **109**, 393 (2008).
4. L. Calmels, F. Houdellier, B. Warot-Fonrose, C. Gatel, M. J. Hÿtch, V. Serin, and E. Snoeck, Phys. Rev. B **76**, 060409(R) (2007).
5. J. Rusz, O. Eriksson, P. Novák, P. M. Oppeneer, Phys. Rev. B **76**, 060408(R) (2007).
6. J. C. H. Spence and J. M. Zuo, Electron Microdiffraction (Plenum Press, New-York, 1992).
7. J. Rusz, et al., Phys. Rev. B **75**, 214425 (2007).

M. Luysberg, K. Tillmann, T. Weirich (Eds.): EMC 2008, Vol. 1: Instrumentation and Methods, pp. 359–360, DOI: 10.1007/978-3-540-85156-1_180, © Springer-Verlag Berlin Heidelberg 2008

360

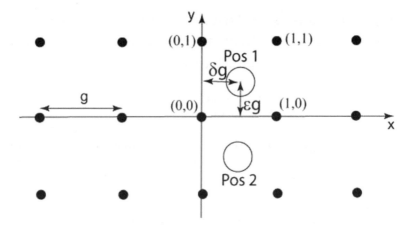

Figure 1. Four-fold diffraction pattern which has been used to calculate the differential cross section. The two symmetrical positions of the spectrometer aperture are labelled *pos1* and *pos2*. In the experiment, the systematic row along x is selected, by tilting the zone axis [001] a few degrees off the optical axis along the y direction.

Probing bright and dark surface plasmon modes in individual and coupled Au nanoparticles using a fast electron beam

Ming-Wen Chu[1], Viktor Myroshnychenko[2], F. Javier García de Abajo[2] and Cheng Hsuan Chen[1]

1. Center for Condensed Matter Sciences, National Taiwan University, Taipei 106, Taiwan
2. Instituto de Óptica - CSIC, Serrano 121, 28006 Madrid, Spain

chumingwen@ntu.edu.tw
Keywords: surface plasmons, STEM, EELS, nanoparticles, monochromator

Surface plasmons (SPs) are quantized collective plasma oscillations of conduction electrons propagating at the surface of metals. The excitations of SPs in the visible spectral regime dictate the color of noble metals. In noble metal nanoparticles (NPs) with only a few tens or hundreds of nanometers, the SP resonances are further tailored by the size, shape, and coupling of the NPs, determining their optical properties ranging from near-infrared to UV spectral regimes as demonstrated in nanoprisms, nanorings, nanostars, (coupled) nanorods, and (coupled/arrayed-) nanospheres.[1-3] Using individual and/or coupled NPs with proper designed geometrical constraints, the SPs can, therefore, be tuned to the proximity of the laser energies (wavelengths) conveniently available. The associated resonant excitations have been shown to amplify the electric near-field of the SPs by astronomical orders of magnitudes, opening up the opportunities for several fascinating applications of the NPs such as guiding light beyond the subwavelength limit — photonics and plasmonics, surface enhanced Raman scattering, and scanning near-field optical microscopy (SNOM) with a spatial resolution around 20~30 nm.[4-6]

The novel applications mentioned above are not possible using bulk noble metals and strongly correlated with the spatial variations of SP modes on the nanomaterials. There is, therefore, a rising interest in spectrally *mapping* the SPs on noble metal NPs by electron energy-loss spectroscopy (EELS) in conjunction with scanning transmission electron microscopy (STEM) due to the unmatched *nanometer-scale* spatial resolution with respect to SNOM and the extended spectral range from near-infrared to soft X-rays of this combined technique. In STEM-EELS, the evanescent electromagnetic field of the nanometer- or atomic-scale incident fast electron beam (close to the speed of light) behaves as high-energy white light, polarizing the NPs at given SP energy with the corresponding inelastic electron energy loss for this excitation.

Using a 2-nm ***monochromatized*** fast electron beam in a STEM, we have studied the SP modes of individual and coupled Au NPs with various sizes (10~40 nm) and shapes. The STEM-EELS spectra were acquired with the electron beam in grazing incidence to the surface of NPs, similar to the optical near-field setup, and the so-called *bright* SP

M. Luysberg, K. Tillmann, T. Weirich (Eds.): EMC 2008, Vol. 1: Instrumentation and Methods, pp. 361–362, DOI: 10.1007/978-3-540-85156-1_181, © Springer-Verlag Berlin Heidelberg 2008

modes (dipole-active) were identified in both individual and coupled NPs ranging from *near-IR* to *visible* spectral regimes. In an individual long nanorod with an aspect ratio of ~6, surprisingly a *dark* SP mode invisible by light (thus dipole-inactive) was excited and the theoretical calculations of the corresponding macroscopic dielectric responses in the framework of Maxwell's equations confirms this experimental observation. Further with the NPs coupling, rich bright and dark SP modes emerge and the suppression of one mode over the other is dictated by the electron-beam position. The electron scattering geometry thus plays the role of the SP-mode selection, unexpected and never documented before. Our results surpass the conventional knowledge in SP excitations by light and may stimulate future interests beyond optics.

1. J. Nelayah, M. Kociak, O. Stéphan, F. J. García de Abajo, M. Tencé, L. Henrard, D. Taverna, I. Pastoriza-Santos, L. M. Liz-Marzán, and C. Colliex, Nature Physics **3**, 348 (2007).
2. J. Aizpurua, P. Hanarp, D. S. Sutherland, M. Käll, G. W. Bryant, and F. J. García de Abajo, Phys. Rev. Lett. **90**, 057401 (2003).
3. F. Hao, C. L. Nehl, J. H. Hafner, and P. Nordlander, Nano Lett. **7**, 729 (2007).
4. W. L. Barnes, A. Dereux, and T. W. Ebbesen, Nature **424**, 824 (2003).
5. E. Ozbay, Science **311**, 189 (2006).
6. H.-H. Wang, C.-Y. Liu, S.-B. Wu, N.-W. Liu, C.-Y. Peng, T.-H. Chan, C.-F. Hsu, J.-K. Wang, and Y.-L. Wang, Adv. Mater. **18**, 491 (2006).

Dual energy range EELS spectrum imaging using a fast beam switch

A.J. Craven, M. MacKenzie and S. McFadzean

Department of Physics & Astronomy, University of Glasgow, Glasgow, G12 8QQ, UK

a.craven@physics.gla.ac.uk
Keywords: EELS, fast beam switch

Electron energy loss spectroscopy (EELS) is used to examine the structural and chemical properties of materials on a sub-nanometre scale. For full data processing, it is essential to collect both core and low loss spectra at each pixel under the same electron-optical conditions. The Glasgow FEI Tecnai F20 TEM/STEM is equipped with a field emission gun, a Gatan ENFINA electron spectrometer and an EDAX X-ray spectrometer. Mounted on one of the two 35 mm camera ports is a retractable fast beam switch (FBS) [1,2]. The other 35mm port is used for a Fischione high angle annular dark field (HAADF) STEM detector. An additional annular dark field (DF) STEM detector is located immediately before the ENFINA entrance aperture. The FBS has a sub-microsecond response time which allows rapid shuttering of the beam and so is capable of handling the large range of exposure times required to cover both core and low loss spectra.

In the case of spectrum images (SIs), we collect a core loss spectrum followed by a low loss spectrum at each pixel. As the datasets have "exact" pixel correlation, it is possible to correctly remove multiple scattering from the EELS core edges, either by Fourier-ratio deconvolution or by splicing the EELS SIs together and using Fourier-Log deconvolution. The spectra can also be normalized for intensity allowing absolute quantification. Accurate energy calibration allows energy shifts to be investigated. The EDX signal can be acquired in parallel with the EELS acquisitions providing additional analytical information. In addition the electron signals from the STEM imaging detectors can be acquired at the same pixel allowing accurate correlation with image features. Absolute thickness mapping and determination of numbers of atoms per unit area and per unit volume are also possible and the effects of diffraction contrast and variation of specimen thickness can be removed.

Figure 1 contains an example of a dual EELS dataset taken from a high-k dielectric stack sample. The stack was deposited by IMEC and has the structure: Si substrate/SiO_2/HfSiO/TiN/poly-Si. Figure 1(a) is the DF STEM image of the wafer; the white vertical line indicates the region from which the dual energy range EELS SI was acquired. The data was acquired using an energy dispersion of 0.3eV/channel and energy offsets of 300eV and -290eV for the core and low loss parts of the spectrum, respectively and 5 reads of 1sec. For the low loss part of the spectrum, the fast beam switch was used to limit the integration time to 3msec per read. The as-acquired SI is shown in Figure 1(b). Figure 1(c) shows the DF STEM signal acquired simultaneously with the core loss acquisition of the EELS SI. The low loss part of the spectrum was

M. Luysberg, K. Tillmann, T. Weirich (Eds.): EMC 2008, Vol. 1: Instrumentation and Methods, pp. 363–364, DOI: 10.1007/978-3-540-85156-1_182, © Springer-Verlag Berlin Heidelberg 2008

used to correct for energy drift during the acquisition and to perform Fourier-ratio deconvolution. Figure 1(d) compares the background subtracted N K and Ti L-edges from the TiN layer before and after deconvolution.

This demonstrates an extremely flexible and powerful tool for EELS data collection. It is essential for the full processing of EELS datasets and for studies involving more subtle energy shift changes.

1. A.J. Craven, J. Wilson and W.A.P. Nicholson, Ultramicroscopy **92** (2002), p. 165.
2. J. Scott, P.J. Thomas, M. MacKenzie, S. McFadzean, J. Wilbrink, A.J. Craven and W.A.P. Nicholson, submitted to Ultramicroscopy (2008).
3. We kindly acknowledge IMEC for supplying the wafer, B. Miller (University of Glasgow) for preparing the TEM specimen, Paul Thomas of Gatan Research and Development for help in developing the processing and EPSRC for financial support under grant GR/S44280.

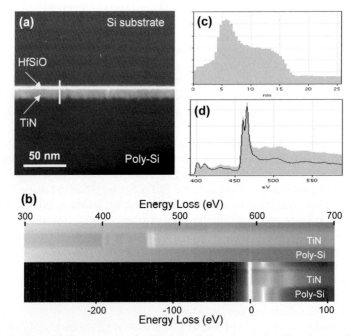

Figure 1. (a) DF STEM image of a high-k dielectric stack: Si substrate/SiO_2/HfSiO/ TiN/poly-Si. The white vertical line indicates the region from which the dual EELS dataset was acquired. (b) The as-acquired dual EELS spectrum image. The core loss is in the upper half and the low loss in the lower half. The spatial axis is vertical with the substrate on the top and corresponds to the line in (a). The energy loss axis is horizontal with the energy losses shown (c) DF STEM signal during the core loss acquisition with the substrate on the left. (d) Comparison between the background subtracted spectra before (solid) and after (line) Fourier-ratio deconvolution.

Determination of local composition of Li-Si alloys by Electron Energy-Loss Spectroscopy

J. Danet[1,2], D. Guyomard[1], T. Brousse[2] and P. Moreau[1]

1. Institut des Matériaux Jean Rouxel (IMN), Université de Nantes, CNRS, 2 rue de la Houssinière, BP 32229, 44322 Nantes cedex 3, France
2. Laboratoire de Génie des Matériaux et Procédés Associés, Polytech Nantes, Rue Christian Pauc, BP 50609, 44306 Nantes cedex 3, France

juliendanet@cnrs-imn.fr

Keywords: Silicon lithium alloys, EELS, local composition, lithium ion battery

Most commercially available Li-ion batteries use graphite as a negative electrode which provides a theoretical capacity of 372 mAh/g. With a maximum capacity of 3579 mAh/g ($Li_{15}Si_4$), silicon seems to be an excellent substitute for this graphite electrode [1]. Due to a large volume expansion (400%), the important degradation of silicon particles during repeated charge/discharge of the battery, followed by a quick fade in capacity, is the main obstacle to its development [2]. Several solutions have been proposed to improve the actual capacity and long term cyclability of silicon-based negative electrodes [3,4]. These solutions usually based on large scale electronic and structural characterisations [1].

Very few studies present results obtained at a nanometer scale allowing to analyse local compositions of the Si/Li alloys formed during or after cycling [5,6]. Electron energy-loss spectroscopy (EELS) in a Transmission Electron Microscope is well suited for this purpose. Low energy-loss spectra, although not often used in literature, can provide information much like high energy losses but within a few milliseconds, a decisive factor to be considered knowing the large beam sensitivity of lithium based compounds.

From all published atomic structures of alloys [7], it is possible to determine theoretical positions for the plasmon energies considering a simple Drude model [8] (Figure 1). Values span from 16.7 eV (pure Si) to 7.5 eV (pure Li). The variation of the plasmon energy is not linearly dependent on the percentage of lithium as observed in standard alloys [8]. The non metallic nature of silicon modifies this usual picture. However, a linear relation for amorphous Li-Si compounds is expected.

Pure alloys were prepared at high temperature (550°C) in stainless steel tubes and characterized by X-ray diffraction with an INEL diffractometer. Since these compounds are air sensitive, all samples were prepared and handled under a controlled argon atmosphere and a vacuum sample holder was used to perform the TEM/EELS studies.

A typical EELS spectrum for a $Li_{12}Si_7$ composition is presented in Figure 2, showing the large plasmon peak (≈ 15 eV). Spectra were gain, dark count, detector and plural scattering corrected before plasmon energies, E_p, were obtained by fitting spectra with a Drude-Lorentz model. The ratio between Si $L_{3,2}$ edges (≈ 99 eV) and Li K edges (≈ 55 eV) provide another information to validate the composition of the analyzed alloys. The

M. Luysberg, K. Tillmann, T. Weirich (Eds.): EMC 2008, Vol. 1: Instrumentation and Methods, pp. 365–366, DOI: 10.1007/978-3-540-85156-1_183, © Springer-Verlag Berlin Heidelberg 2008

Si $L_{3,2}$ edge features are very similar to those for pure silicon, implying that the electronic structure of silicon does not significantly change in the alloys. This data bank of EELS spectra for different Si-Li alloys is a very useful tool to quickly determine the local composition of silicon anodes upon galvanostactic cycling.

1. M.N. Obrovac and L. Christensen, Electrochem. Solid-State Lett. **7** (2004), A93-A96.
2. U. Kasavajjula, C. Wang, and A.J. Appleby, J. Power Sources **163** (2007) 1003-1039.
3. B. Lestriez, S. Bahri, I. Sandu, L. Roue, and D. Guyomard, Electrochem. Com. **9** (2007), 2801-2806.
4. C. K. Chan *et al.* Nat. Nano **3** (2008), 31-35.
5. H.Li *et al.* Solid State Ionics **135** (2000), 181-191.
6. G. W. Zhou *et al..* Appl. Phys. Lett. **75** (1999), 2447-2449.
7. Inorganic Crystal Structure Database (2007). NIST/FIZ
8. R. F. Egerton in "Electron Energy-Loss Spectroscopy in the Electron Microscope", 2nd edition, ed. Plenum Press (1996).

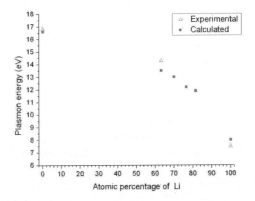

Figure 1. Calculated and experimental plasmon energies, E_p, for different Li-Si alloys.

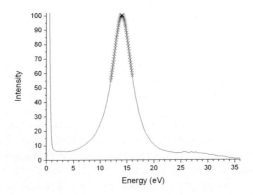

Figure 2. Experimental EELS low loss spectrum for the $Li_{12}Si_7$ alloy (full line) with the associated Drude-Lorentz fit (crosses).

Elemental and bond mapping of complex nanostructures by MLS analysis of EELS spectrum-imaging data

F. de la Peña[1], R. Arenal[2], O. Stephan[1], M. Walls[1], A. Loiseau[2], C. Colliex[1]

1. Laboratoire de Physique Solides, Université Paris-Sud, 91405 Orsay, France
2. Laboratoire d'Etude des Microstructures, CNRS-ONERA, 92322 Châtillon, France

delapena@lps.u-psud.fr

Keywords:EELS, Spectrum Imaging, MLS, Chemical Bonding Maps

The huge and increasing interest in nano-structures has stimulated the development of appropriate characterisation methods in order better to understand (and thus exploit) their properties in possible applications. Electron Energy Loss Spectroscopy (EELS) is an exceptionally useful analytical technique in this regard. In particular, the spectrum-imaging (SPIM) mode [1], which retains the full energy and spatial resolution of the technique, permits the mapping of a material's electronic properties at a resolution of less than 1 nm [2].

Although conventional data analysis methods developed for a single spectrum can directly be applied to extract the large amount of information contained in a SPIM, there exist techniques specially designed to take full advantage of the information available [3]. In parallel, Multiple Least Squares (MLS) [4] for elemental quantification has recently received new attention [5] due to its higher accuracy and reliability [6] compared to the widespread integration window method.

In this work, we combine the high energy resolution of the SPIM acquisition mode and the accuracy of multiple least squares to map simultaneously both the oxidation state of boron and the chemical composition of a complex boron nitride nanotubes sample by fitting the experimental data to a linear model consisting of simulated edges and experimental boron K-edge fine structure profiles [7]. The fine structure profiles can be obtained from reference samples [8] but due to the difficulty of exactly reproducing the experimental conditions, extracting them from regions of pure compounds in the sample under investigation is preferred when possible. We also explore the possibility of extracting the components of the fine structure when no pure references are available in the sample by using independent component analysis.

Fig. 1 (a) shows an HADF image recorded in parallel with the EELS signal and corresponding to a SPIM (64x64 recorded spectra) acquired on a region of several nano-cages in the BNNTs sample. Fig. 2 (b-d) displays the elemental maps for boron, nitrogen and oxygen. As mentioned above, further analysis can be developed thus Fig. 2 shows maps of the fitting weight of each edge type over a region containing pure boron, boron oxide and boron nitride. The anisotropic boron nitride is mapped as a function of c-axis orientation with respect to the beam direction (the pi and sigma excitations' relative strengths are inverted in passing from horizontal to vertical). The BN is in the form of flat foils which are mostly roughly horizontal but are curled up at the edges, presenting the hexagonal planes vertically. For the most part, the regions showing

M. Luysberg, K. Tillmann, T. Weirich (Eds.): EMC 2008, Vol. 1: Instrumentation and Methods, pp. 367–368, DOI: 10.1007/978-3-540-85156-1_184, © Springer-Verlag Berlin Heidelberg 2008

368

strong bonding of a particular type correlate closely with the associated elemental map - However some apparently anomalous effects have been detected and their meaning will be discussed in detail.

For the present work, we have developed a new programme capable of MLS analysis of SPIMs, EELSLab [9], that will soon be made publicly available.

1. C. Jeanguillaume, C. Colliex, Ultramic. **28,** 252 (1989)
2. V.J. Keast, A.J. Scott, R. Brydson, D.B. Williams, J. Bruley, Microsc **203,** 135 (01)
3. M. Bosman, M. Watanabe, D.T. Alexander, Keast, Ultramic. **106,** 1024 (06)
4. H. Schuman, A.P. Somlyo, Ultramicroscopy **21**, 23 (1987)
5. J. Verbeeck, S.V. Aert, Ultramicroscopy **10** , 207 (2004)
6. T. Manoubi, M.Tencé, M. Walls, C. Colliex, Micros. Microanal. Microstr.**1** 23 (90)
7. R. Arenal, O. Stephan, F. de la Peña, M. Walls, A. Loiseau, C. Colliex, Ultram. To be submitted (2008)
8. L.A. Garvie, P.R.B. Peter, Nature **396,** 667 (1998)
9. EELSLab. http://www.eelslab.org

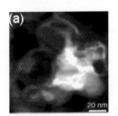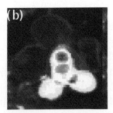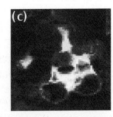

Figure 1. HADF image of a cumul of nanocages. (b-d) Boron, nitrogen and oxygen elemental maps obtained from the SPIM after background subtraction.

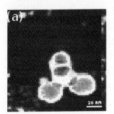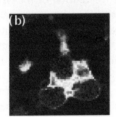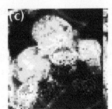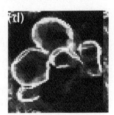

Figure 2. MLS maps extracted from the SPIM (Fig. 1) and corresponding to: pure boron, boron oxide, h-BN for two extreme orientations (see text).

EELS analysis of plasmon resonance in the UV-vis energy range of metal alloy nanoparticles

J.W.L. Eccles[1], U. Bangert[1] and P. Christian[2]

1. School of Materials, University of Manchester, Manchester, M1 7HS, UK
2. School of Chemistry, University of Manchester, Manchester, M1 7HS, UK

James.Eccles@manchester.ac.uk

Keywords: Silver and Gold alloy nanoparticles, Plasmon resonance, EELS

Gold and Silver nanoparticles are used in a variety of optical applications. Of particular consequence to the scattering/absorption of light is the surface plasmon resonance (SPR); a collective oscillation of the conduction electrons induced by a passing electromagnetic field. The nanoparticle size, shape, orientation and permittivity determine the frequency of this SPR. Alloy particles of Au and Ag (Figure 1a,b) are also significant as they can provide a greater frequency range than their pure metal equivalents [1]. We have attempted to characterise the composition of Au/Ag alloy nanoparticles utilising a variety of analytical techniques. Initial EDS analysis suggests that the alloy nanoparticles do not have a homogeneous mixture of the two elements (see Figure 2a). If discrete regions of Au and Ag exist within an alloy particle, we would expect two separate resonance peaks when exciting the SPR with an electron beam [2]. A spatially resolved EELS analysis was conducted to verify this. To support our conclusions, the behaviour of the SPR in the pure metal equivalent Au and Ag nanoparticles will also be presented. Figure 2b highlights spectroscopic detail such as the increase in the amplitude of a bulk resonance with increasing Ag particle diameter.

The EELS and EDS analysis were performed on a VG HB-601 FEG-STEM (scanning transmission electron microscope). The EELS spectrometer system was a Gatan Enfina instrument, enabling the acquisition of a spectrum image. The system operates with a HT voltage of 100 kV, has an energy resolution of 0.35 eV (FWHM of the ZLP) and is capable of high dispersion (0.01 eV per channel). These factors enable detailed scrutiny of the valence band spectra and in particular, the plasmon resonances in the UV energy-loss range (Figure 2a). The EDX spectrometer was an Oxford Instrument device. This EDX technique uses a small probe size (0.8 nm) which gives high spatial X-ray resolution. The 0.8 nm probe has a current of 500 pA allowing the acquisition of single point spectra, linescans (Figure 2b) and image maps. The data were collected using a convergence angle of 11 mrad and dwell times of 15,000 ms for linescans. Initial HR-TEM analysis was carried out on a Tecnai F30 electron microscope operating at 300 kV.

The iterative Richardson-Lucy deconvolution algorithm has been applied to improve the energy resolution in the visible energy range of the EEL spectra [3, 4]. The routine has previously been applied effectively to both core-loss and low-loss EEL spectra [5, 6]. The amplification of the background noise after each iteration loop is compensated for by terminating the iteration before it develops false features in the spectrum.

M. Luysberg, K. Tillmann, T. Weirich (Eds.): EMC 2008, Vol. 1: Instrumentation and Methods, pp. 369–370, DOI: 10.1007/978-3-540-85156-1_185, © Springer-Verlag Berlin Heidelberg 2008

1. Y.Sun and Y. Xia, Analyst, **128** (2003) p. 686
2. R.G. Freeman et al, J. Phys. Chem, **100** (1996) p. 718.
3. W.H Richardson, J. Opt. Soc. Am, **62** (1972), p. 55.
4. L.B. Lucy, Astrophys. J, **79** (1974), p. 745.
5. A. Gloter, A. Douriri, M. Tencé, C. Colliex, Ultramicoscopy, **96** (2003), p. 385.
6. J. Nelayah et al, Nature Physics, **3**, p. 348.
7. We kindly acknowledge the support of Dr P. Thomas of Gatan Inc. and Dr S. Romani at the STEM facility at the University of Liverpool.

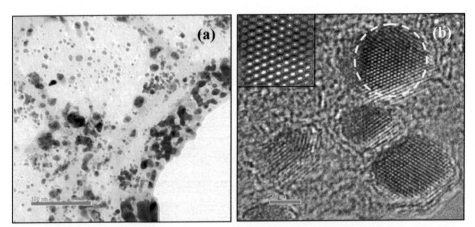

Figure 1. (a) STEM Bright-field image (200kx) of Au/Ag alloy nanoparticles (b) High-resolution TEM Bright-field image (1mx) of Au/Ag alloy nanoparticles with inset Fourier enhanced lattice image of a single nanoparticle (dashed line).

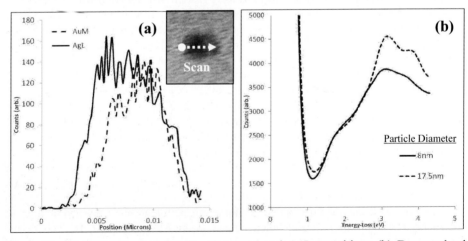

Figure 2. (a) EDS linescan showing the AuM and AgL transitions (b) Deconvolved EEL spectra measured in the centre of Ag nanoparticles. The peaks at 3.1 and 3.8 eV are the surface and bulk plasmon respectively.

Anisotropic effects in ELNES of the O-K edge in rutile: a case of trichroism

V. Mauchamp[1], T. Epicier[2], J.C. Le Bossé[2]

1. PhyMAT (ex LMP) SP2MI - Boulevard 3, Téléport 2 - BP 30179, F-86962 Futuroscope Chasseneuil Cedex
2. Université de Lyon; INSA-Lyon, MATEIS, umr CNRS 5510, bât. B. Pascal, F-69621 Villeurbanne Cedex

Thierry.epicier@insa-lyon.fr
Keywords: anisotropy, ELNES, dichroism, trichroism, O-K edge TiO_2, FEFF, Wien2k

When probing atomic species with non-isotropic atomic environments in low-symmetry materials by electron energy loss spectroscopy (EELS), simulation of ionization edges, as measured in a Transmission Electron Microscope (TEM), requires to consider experimental parameters that can affect the near-edge structures (ELNES): the energy, orientation and convergence of the electron beam, and the collecting angle aperture. The well-known case of dichroism (a situation where a threefold, fourfold or sixfold main rotation axis exists through the probed atom) has been extensively studied and elucidated (see [1-2] and references within). Within the dipole approximation, the current of inelastic electrons emitted from a particular atomic core level can be expressed as a linear combination of two particular electron inelastic scattering cross sections (the intrinsic components), respectively calculated for orientations of the transferred wave vector parallel and perpendicular to the main rotation axis **c**. In the situation of lower symmetries around the probed atom, different cases of trichroism [3] have to be considered. It has recently been shown that the collected current can then be expressed as a linear combination of three, four or six intrinsic components with adequate weights [4]. This analytical theory provides the mean of calculating the collected current for any beam orientation and any collector aperture from these intrinsic components.

The present contribution aims at investigating the case of the O-K edge in the rutile form of TiO_2, where the local symmetry around the oxygen site is described with the C2v point group. This corresponds to a situation of "a" trichroism [3], where the collected current depends on three intrinsic components. These contributions are calculated using either the FEFF8.2 code [5] or the TELNES subroutine of the WIEN2k code [6]. The simulations will be compared with experimental O-K near-edge structures recorded in a 200 kV Jeol 2010F microscope equipped with a Gatan DigiPeels spectrometer, for different orientations and collection conditions. Calculations are in a good agreement with the experimental results, and orientation as well as collection angle effects on the ELNES are evidenced (see figure 1). Moreover, it is clearly shown that, in such a case, the relativistic velocity of the incident electrons has to be considered: neglecting these relativistic effects induces significant changes in the ELNES, leading to a markedly deterioration of the agreement between experiment and

372

theory. These results will be discussed in relation with recent available data from the literature [7].

1. P. Schattschneider, C. Hébert, H. Franco, and B. Jouffrey, *Phys. Rev.*, B72, (2005), 045142.
2. J.C. Le Bossé, T. Epicier, and B. Jouffrey, *Ultramicrosc.*, 106, (2006), 449.
3. C. Brouder, *J. Phys.: Condens. Matter*, 2, 701 (1990).
4. J.C. Le Bossé, T. Epicier, and H. Chermette, *Phys. Rev.*, B76, (2007), 075127.
5. P. Blaha, K. Schwarz, G. Madsen, D. Kvaniscka, and J. Luitz, *WIEN2k, An Augmented Plane Wave + Local Orbitals Program for Calculating Crystal Properties* (edited by K. Schwarz, Techn. Universität Wien, Austria, 2001).
6. A. Andukinov, B. Ravel, J. Rehr, and S. Conradson, *Phys. Rev.*, B58, (1998), 7565.
7. The CLYM (Centre Lyonnais de Microscopie) is gratefully acknowledged for the access to the 2010F microscope.

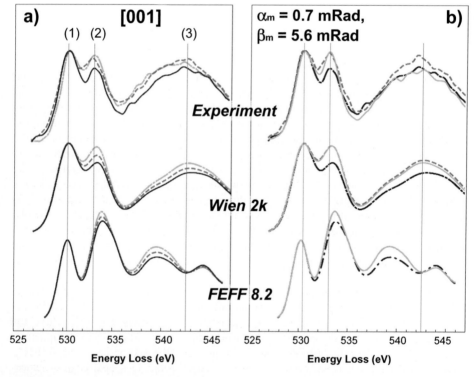

Figure 1. ELNES of the O-K edge in rutile TiO_2 showing 3 main peaks labelled (1), (2) and (3). a): incident beam along [001] for various collection conditions: $\beta_m = 1.86$ mRad (dot-dashed *green* line), $\beta_m = 2.8$ mRad (dashed *red* line), $\beta_m = 5.6$ mRad (solid *dark* line). b): variation of the incident beam orientation from [001] (dot-dashed *black* line) to [100] (dashed *red* line) and [110] (solid *green* line) with fixed convergence and collection angles ($\alpha_m = 0.7$ and $\beta_m = 5.6$ mRad respectively). In each case all spectra have been normalized in such a way that the intensity of the first peak is equal to unity.

Dissimilar cation migration in (001) and (110) La$_{2/3}$Ca$_{1/3}$MnO$_3$ thin films

S. Estrade[1], I.C. Infante[2], F.Sanchez[2], J.Fontcuberta[2], F. de la Peña[3], M. Walls[3], C. Colliex[3], J. Arbiol[1,4], F. Peiró[1]

1. EME/CeRMAE/IN2UB, Dept. d'Electrònica, Universitat de Barcelona, c/ Marti Franques 1, 08028 Barcelona, CAT, Spain
2. Institut de Ciència de Materials de Barcelona -CSIC, 08193 Bellaterra, CAT, Spain
3. Laboratoire de Physique des Solides, (UMR CNRS 8502), Bâtiment 510, Université Paris Sud, 91405 Orsay (France)
4. TEM-MAT, Serveis Cientificotecnics, Universitat de Barcelona, 08028, Barcelona, CAT, Spain

sestrade@el.ub.es
Keywords: LCMO, STO, EELS

Mixed valence manganites, such as La$_{2/3}$Ca$_{1/3}$MnO$_3$ (LCMO), have been the focus of a great deal of research efforts due their ferromagnetic and metallic character that can be useful as electrodes in magnetic tunnel junctions. These devices ar based in naometric layers of ferromagnetic electrodes separated by an insulating barrier, and it is of major relevance the control of the thin film properties for the final device functionality. When grown as thin films, it has been reported that film magnetic properties are much poorer than those of their bulk counterparts [1,2]. The origin of this decrcase in the magnetic properties has been ascribed to the loss of the ferromagnetic metallic mixed valence state of the Mn ions (Mn$^{3+/4+}$) through the development of electronic phase separated states (ferromagnetic insulating Mn^{3+} and Mn^{4+} or non-magnetic Mn ions) along the layer [8]. Electron energy loss spectroscopy is a tool to determine the local chemistry and Mn oxidation state at the nanoscale. In this sense, several studies have been carried out in mixed valence manganites grown with (001) orientation [3-7]. Yet, very little attention has been paid to mixed valence manganites grown with other orientation, although recently (110) oriented manganite films have been reported to display enhanced magnetic properties [8].

In the present work, we present transmission electron microscopy (TEM) and electron energy loss spectroscopy (EELS) comparative results of LCMO films grown on (001) and (110) SrTiO$_3$ (STO) substrates. Thin films of nominal thickness of 85 nm were simultaneously grown by rf-sputtering. Structural characterization using X-ray diffraction has indicated that the film lattice parameters are differently adapted to the substrate parameter depending on the film orientation: (001) film presents fully strained state whereas (110) film is less strained, showing that the in-plane lattice parameter along the [1-10] direction is less strained that that along the [001] one [8]. In agreement with these diffraction results, TEM analysis reveals that plastic strain relaxation was observed for (110) LCMO.

M. Luysberg, K. Tillmann, T. Weirich (Eds.): EMC 2008, Vol. 1: Instrumentation and Methods, pp. 373–374, DOI: 10.1007/978-3-540-85156-1_187, © Springer-Verlag Berlin Heidelberg 2008

374

Ca/La relative concentrations obtained from EELS experiments (Figure 1a) indicated that a Ca enrichment towards free surface occurs for (001) orientation, while no local variations were found for (110). EEL spectra also showed a shift of the L_3 Mn peak onset towards higher energies and a decreasing of the L_3/L_2 intensity ratio when moving from the interface to the top of the layer for the (001) oriented layer, while for the (110) layer no variation was found (Figure 1b). These data suggest that a cationic migration and concomitant Mn oxidation state variation take place in (001) LCMO as a strain accommodation mechanism. As the magnetic properties are improved for (110) films when compared to those of (001) ones [8], it seems that strain accommodation via cation migration is more detrimental to the magnetic performance than accommodation via defect creation, because it is less effective (i.e: layers remain in higher strain conditions) and because of local Mn oxidation state deviations from its nominal value.

1. J. S. Moodera and G. Mathon, J. Magn. Magn. Mater. **200** (1999) 248
2. S. S. P. Parkin, K. P. Roche, M. G. Samant, P. M. Rice, R. B. Byers, R. E.Scheuerlein, E. J. O'Sullivan, S. L. Brown, J. Bucchigano, D. W. Abraham,Yu Lu, M. Rooks, P. L. Trouilloud, R. A. Wanner, and W. J. Gallagher, J. Appl. Phys. **85** (1999) 5828
3. J. Simon, T. Walther, W. Mader, J. Klein, D. Reisinger, L. Alff, R. Gross, App. Phys. Lett. **84** (2004) 3882
4. J.-L. Maurice, D. Imhoff, J.-P. Contour, C. Colliex, Phil. Mag. **15** (2006) 2127
5. L. Samet, D. Imhoff, J.-L. Maurice, J.-P. Contour, A. Gloter, T. Manoubi, A. Fert, and C. Colliex, Eur. Phys. J. B **34** (2003) 179
6. F. Pailloux, D. Imhoff, T. Sikora, A. Barthélémy, J.-L. Maurice, J.-P. Contour, C. Colliex , A. Fert, Phys.Rev. B **66** (2002) 14417
7. S. Estradé, J. Arbiol, F. Peiró, Ll. Abad, V. Laukhin, Ll. Balcells, B. Martínez, App. Phys. Lett. **91** (2007) 252503
8. Infante, I.C.; Sánchez, F.; Fontcuberta, J.; Wojcik, M.; Jedryka, E.; Estrade, S.; Peiro, F.;Arbiol, J.; Laukhin, V.; Espinós, Phys. Rev. B.. **76** (2006) 224415-1
9. Support by Support from the European Union project ESTEEM allowed us to perform STEM measurements in the LPS (France). Financial support by the MEC of the Spanish Government (Project No. NAN2004-9094-C03, MAT2005-5656-C04) is also acknowledged. We are very thankful to Dr. A. Romano-Rodríguez for FIB sample preparation.

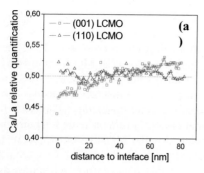
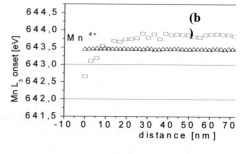

Figure 1. Ca/La quantification (a) and Mn L_3 onset position (b) for (001) and (110) LCMO layers.

Energy-loss near edge structures of Cr_2O_3, CrO_2 and $YCrO_4$ phases

M.S. Moreno[1], E. Urones-Garrote[2] and L.C. Otero-Díaz[2,3]

1. Centro Atómico Bariloche, 8400 S.C. de Bariloche, Argentina
2. Centro de Microscopía y Citometría, Universidad Complutense de Madrid, 28040 Madrid, Spain
3. Departamento de Química Inorgánica, Facultad de Ciencias Químicas, Universidad Complutense de Madrid, 28040 Madrid, Spain

esteban.urones@pdi.ucm.es

Keywords: chromium oxides, EELS, simulation

It is known that the valence of Cr strongly affects the solubility and toxicity of Cr complexes. The oxides Cr_2O_3 and CrO_2, with 3+ and 4+ oxidation state respectively, are technologically important materials with several applications [1]. It is worth to mention that the Cr^{5+} oxidation state is highly unstable and its coordination polyhedron in $YCrO_4$ is a bisdisphenoid (triangular dodecahedron: 8-fold coordination), which is not very abundant in crystal chemistry [2,3]. The aim of this work is to study the O-K fine structure of three different chromium oxides Cr_2O_3, CrO_2 and $YCrO_4$.

The employed Cr_2O_3 and $YCrO_4$ samples, have been prepared by solid state reaction using the metal nitrates as precursors, and CrO_2 was obtained with high pressure techniques [6]. Electron Energy-Loss spectra were acquired with a GIF 200 attached to a Philips CM200 FEG microscope operated at 200 kV. The energy resolution, as determined from the full-width at half maximum of the zero loss peak, was ~1 eV. Collection angle β was 0.83 mrad.

Typical spectra acquired with the incident beam along the [100] direction are shown in Figure 1. The spectra were aligned at the energy-loss value of Cr-L_3 edge. It can be seen that for the binary oxides the O-K edge consists of two broad peaks while for the $YCrO_4$ phase the fine structure consists of 3 peaks. In the case of CrO_2 a small peak below 535 eV is present. These features allow us to identify and distinguish between the three phases. With the exception of $YCrO_4$ a clear anisotropy in the ELNES was not observed.

To correlate ELNES spectral features with cation and O contributions, we have carried out full multiple scattering calculations using the FEFF 8.20 program [4] as indicated in [5] (see Figure 2). The calculated density of states reveals a covalent nature of these phases. The structures appearing within the first 10 eV above the threshold arise from a covalent mixing of mainly O $2p$ and Cr s-p (plus Y in $YCrO_4$) states.

1. T.L. Daulton and B.J. Little, Ultramicroscopy **106** (2006), p. 561.
2. B.G. Hyde and S. Andersson, "Inorganic Crystal Structures", (John Wiley, New York) (1989)
3. J.M. Hanchar and P.W.O. Hoskin, "Reviews in Mineralogy & Geochemistry: Vol. 53 Zircon", (MSA, Washington) (2003).

M. Luysberg, K. Tillmann, T. Weirich (Eds.): EMC 2008, Vol. 1: Instrumentation and Methods, pp. 375–376, DOI: 10.1007/978-3-540-85156-1_188, © Springer-Verlag Berlin Heidelberg 2008

376

4. A.L. Ankudinov, B. Ravel, J.J. Rehr and S.D. Conradson, Phys.Rev.B **58** (1998), p. 7565.
5. M.S. Moreno, K. Jorissen and J.J. Rehr, Micron **38** (2007), p. 1.
6. Financial support from Conicet (Argentina) and from MAT-2007-63497 research project (Spain) is acknowledged. We would like to thank Prof. M. A. Alario-Franco for providing the CrO_2 sample.

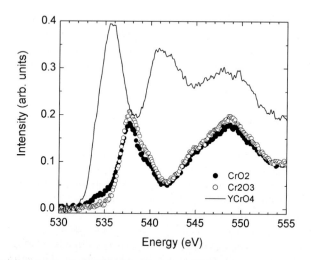

Figure 1. Comparison of EELS data for the different Cr oxides. The spectra were acquired with the incident beam along the [100] direction.

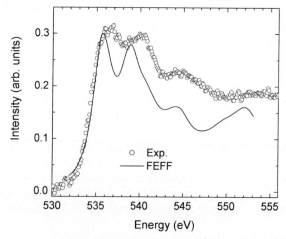

Figure 2. Comparison of the EELS data (open circles) and calculations (solid line) for $YCrO_4$ for the [001] direction.

Distortion corrections of ESI data cubes for magnetic studies

C. Gatel[1], B. Warot-Fonrose[1], F. Houdellier[1] and P. Schattschneider[2]

1. CEMES-CNRS, 29 rue J. Marvig, 31055 Toulouse, France
2. Institute for Solid State Physics, Vienna University of Technology, Wiedner Hauptrasse 8-10/138, A-1040 Vienna, Austria

gatel@cemes.fr
Keywords: datacube, EFTEM, aberrations

Electron energy loss spectroscopy in a transmission electron microscope (TEM) is widely used to investigate chemical and electronic properties of materials with the main advantage of being a local technique. Most results are reported on chemical composition at a local scale but EELS can also be a powerful tool to investigate physical properties from the reciprocal space. A property which can be obtained from diffraction patterns is the electron magnetic chiral dichroism[1]. This technique makes it possible to get magnetic information at the scale determined by the electron probe. The knowledge of local magnetic properties is an exciting challenge to clear up issues arising from the reduction of magnetic devices, for which interfaces and surfaces play a dominant role.

The measurement of this magnetic property needs the recording of electron energy loss spectra at precise locations in the diffraction pattern. Different experimental configurations are available to combine spatial and energy data[2,3]. One of these methods is the energy spectrum imaging (ESI) method applied in the diffraction plane i.e. the whole diffraction pattern is recorded for each energy slice and a three dimensional data set, called data cube, is obtained[4] (Figure 1).

Recording EELS data is only one stage of a complete process which includes 3 steps: the experimental set-up for acquisition, the method chosen for recording spectra and the post-treatments necessary to get reliable information from the collected data. The combination of optimised solutions for the 3 steps is needed if nanometer information is aimed and quantitative data are required. In this presentation, we will briefly explain the experimental set-up used to get reliable signal and the recording process and we will focus on post-treatments developed to ensure that the dichroic signal is meaningful. These post-treatment can be applied on classical ESI data cubes. In a typical EELS experiment, we consider two origins for deviations for perfect data collection: the microscope and the spectrometer. We will assume that the microscope is correctly aligned for the experiment but we will consider that the microscope stage can move for the duration of the acquisition, this is the stage drift, resulting in a sample drift. The drift of the diffraction pattern originates too from instabilities of the image shift coils that are used to move the pattern in the spectrometer entrance axis. These coils are inside the aberration corrector and need time to be stabilised. The second source of deviation from theoretical data is the spectrometer aberrations that are

M. Luysberg, K. Tillmann, T. Weirich (Eds.): EMC 2008, Vol. 1: Instrumentation and Methods, pp. 377–378, DOI: 10.1007/978-3-540-85156-1_189, © Springer-Verlag Berlin Heidelberg 2008

378

adjusted by lenses but never perfectly compensated. The aberration we will consider here is the non isochromaticity which corresponds to the difference in energy from position to position in an image. This aberration needs to be corrected to make sure we compare pixels recorded at the same energy inside a single energy slit. The measurements of sample drift and non isochromaticity will be detailed as well as their corrections. We will point out treatment differences between our method and previous studies[5,6].

1. P. Schattschneider, S. Rubino, C. Hébert, J. Rusz, J. Kunes, P. Novak, E. Carlino, M. Fabrizioli, G. Panaccione, G. Rossi, Nature **441** (2006), p.486

2. C. Hébert, P. Schattschneider, S. Rubino, P. Novak, J. Rusz and M. Stöger-Pollach, Ultramicroscopy (2007), doi:10.1016/j.ultramic.2007.07.011

3. P. Schattschneider , Cécile Hébert, Stefano Rubino, Michael Stöger-Pollach, Ján Rusz and Pavel Novák, Ultramicroscopy (2007), doi:10.1016/j.ultramic.2007.07.002

4. B. Warot-Fonrose, F. Houdellier, M.J. Hÿtch, L. Calmels, V. Serin and E. Snoeck, Ultramicroscopy (2007), doi:10.1016/j.ultramic.2007.05.013

5. B. Schaffer,G. Kothleitner,W. Grogger, Ultramicroscopy 106 (2006) p.1129

6. J.C. Russ, The Image Processing Handbook, third ed., CRC, Boca Raton, FA, 1998

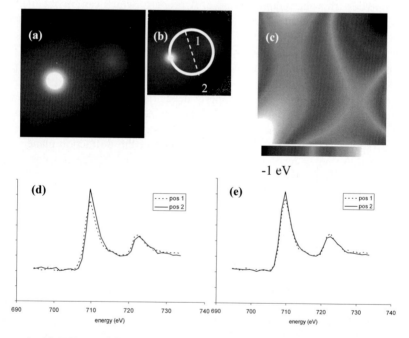

Figure 1. (a) ESI acquisition of the intensity of the diffraction pattern on a Fe single film; (b) diffraction pattern and drawing of the Thales circle with the 1 and 2 positions used to measure the dichroic signa; (c) Fit of the non isochromaticity measured at the iron edge; (d) Spectra acquired in position 1 and 2 on the rawdata cube; (e) Spectra acquired in position 1 and 2 on the corrected data cube

Optimisation of the Positions and the Width of the Energy Windows for the Recording of EFTEM Elemental Maps

Benedikt Gralla and Helmut Kohl

Physikalisches Institut and Interdisziplinäres Centrum für Elektronenmikroskopie und Mikroanalyse (ICEM), Westfälische Wilhelms-Universität Münster, Wilhelm-Klemm-Str. 10, 48149 Münster, Germany

lexx.matrix@uni-muenster.de

Keywords: Elemental Maps, SNR, EFTEM

Due to the small cross-sections for inner-shell excitations the detection limit in elemental maps is determined by the signal-to-noise ratio (SNR). Therefore it is vital to choose instrumental parameters leading to an optimized SNR. We have investigated the influence of the positions and widths of the energy windows for recording elemental maps following earlier works by Kothleitner [1] and Berger [2]. To be able to easily obtain reliable data for arbitrary energy windows, we acquired experimental electron energy loss spectra (EEL spectra) and determined the values for energy windows of larger widths by summing the spectral data over the corresponding energy losses. For the optimization procedure of the SNR we wrote a computer program to deal with an experimentally measured electron energy loss spectrum and varied the energy loss window positions in the pre-edge as well as in the ionization edge region, the energy slit width and the width of the pre-edge region, using one energy loss window above and at least three windows below the characteristic energy loss. The number of pre-edge windows can be selected by the user. The calculation performed by this program yields the energy loss window positions and the window width referring to the maximum SNR as well as the corresponding SNR value. The window positions and window width may then be applied to record an inelastically filtered image series from which an elemental map can be computed.

Due to the short calculation time of about 1 minute when using three pre-edge images the imaging parameters can be calculated on-line. Therefore it is possible to acquire the series of inelastically filtered images from the same sample position as the experimentally taken EEL spectrum.

Additionally, we wrote a computer program to correct the sample drift and the background signal in elemental maps. The background corrected elemental maps have been calculated from three pre-edge images at 605-630 eV, 654-679 eV and 680-705 eV. This leads to an elemental map with an improved EFTEM SNR.

1. G. Kothleitner and F. Hofer, Micron **29** (1998), p. 349-357.
2. A. Berger and H. Kohl, Optik **92** (1993), p. 175-193.
3. Images courtesy of Tobias Heil

M. Luysberg, K. Tillmann, T. Weirich (Eds.): EMC 2008, Vol. 1: Instrumentation and Methods, pp. 379–380, DOI: 10.1007/978-3-540-85156-1_190, © Springer-Verlag Berlin Heidelberg 2008

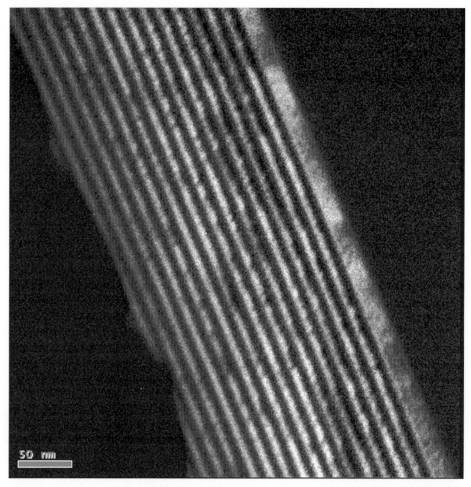

Figure 1. Elemental map of iron (708-733 eV energy loss) with an optimized SNR and the new sample drift correction. The elemental map has been calculated from three pre-edge images [3].

Band gap mapping using monochromated electrons

L. Gu, W. Sigle, C.T. Koch, V. Srot, J. Nelayah and P.A. van Aken

Stuttgart Center for Electron Microscopy, Max Planck Institute for Metals Research, Heisenbergstr. 3, 70569 Stuttgart, Germany

gu@mf.mpg.de

Keywords: Semiconductor, Band gap, Energy filter, Monochromator

The recent development of monochromators for transmission electron microscopes has made valence electron energy-loss spectroscopy (VEELS) a powerful technique to study the semiconductor band structure with high spatial resolution. Albeit difficulties of the band structure measurements were encountered [1], solutions have been demonstrated for several material systems [2]. Taking advantage of the Zeiss SESAM microscope, an energy resolution below 100 meV is achieved routinely [3], which is highly appreciated for band structure measurements using low-loss EELS.

Energy-filtered transmission electron microscopy (EFTEM) series were acquired on a sapphire/GaN/$Al_{45}Ga_{55}N$ layered structure for band gap mapping. The sum of all images of an EFTEM series (1 eV to 32 eV) is shown in Figure 1(a). Spectra of different materials were extracted from the corresponding regions marked in Figure 1(a). A blue-shift of the band gap onset energy and the bulk plasmon energy is readily observed from GaN to $Al_{45}Ga_{55}N$ (Figure 1(b)). Several GaN interband transitions are clearly revealed from the inset spectrum which was integrated from a large region to enhance the signal-to-noise ratio. These details were found to be comparable with previous VEELS band structure measurements [2].

One of the major concerns upon using low-loss EELS is the delocalised inelastic scattering event. To cope with delocalisation effects, collection of the off-axis inelastic electrons was proposed [4]. The MANDOLINE in-column energy filter with a post-filter annular-dark-field (ADF) detector provides access to a novel energy-filtered scanning transmission electron microscopy (EFSTEM), which is of great promise for band gap mapping with high spatial resolution. A schematic ray diagram of EFSTEM is shown in Figure 2(a). Spectra of GaN and $Al_{45}Ga_{55}N$ were extracted from the EFSTEM series (Figure 2(b)). Comparable results to EFTEM were observed on band gap onset and plasmon energy. Note that the EFSTEM has less energy sampling steps than EFTEM. Both the effective atomic number and the collection angle are the main factors responsible for the portion of inelastic contribution to the EFSTEM contrast; therefore the EFSTEM spatial resolution can only be improved at the cost of losing inelastic signal which is mostly diminished at large collection angles.

1. M. Stöger-Pollach, H. Franco, P. Schattschneider, S. Lazar, B. Schaffer, W. Grogger and H.W. Zandbergen, Micron **37** (2006), 396.
2. L. Gu, V. Srot, W. Sigle, C. Koch, P. van Aken, F. Scholz, S.B. Thapa, C. Kirchner, M. Jetter and M. Rühle, Phys. Rev. B **75** (2007), 195214.

M. Luysberg, K. Tillmann, T. Weirich (Eds.): EMC 2008, Vol. 1: Instrumentation and Methods, pp. 381–382, DOI: 10.1007/978-3-540-85156-1_191, © Springer-Verlag Berlin Heidelberg 2008

382

3. C.T. Koch, W. Sigle, R. Höschen, M. Rühle, E. Essers, G. Benner and M. Matijevic, Microsc. Microanal. **12** (2006) 1.
4. D.A. Muller and J. Silcox, Ultramicroscopy **59** (1995), 195.
5. We acknowledge financial support from the European Union under the Framework 6 program under a contract for an Integrated Infrastructure Initiative. Reference 026019

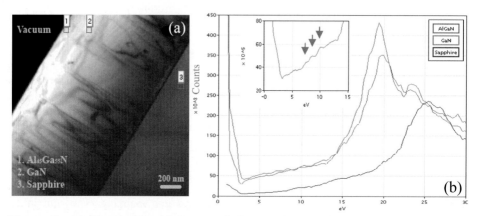

Figure 1. (a) Energy-filtered image of the sapphire/GaN/Al$_{45}$Ga$_{55}$N structure; (b) spectra of the corresponding materials extracted from the EFTEM series, inset shows a GaN spectrum integrated over a large region for better statistics.

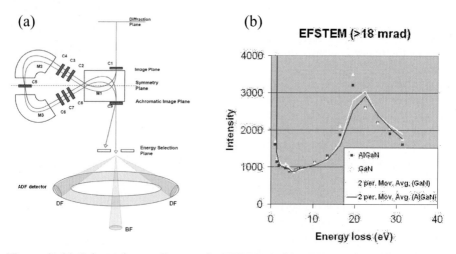

Figure 2. (a) Schematic ray diagram for EFSTEM; (b) EFSTEM intensity of GaN and Al$_{45}$Ga$_{55}$N for the collection full-angle above 18 mrad with an illumination full-angle of about 15 mrad; moving average trendlines were added for corresponding materials.

StripeSTEM, a new method for the isochronous acquisition of HAADF images and monolayer resolved EELS

M. Heidelmann[1], L. Houben, J. Barthel and K. Urban

Ernst Ruska-Centre for Microscopy and Spectroscopy with Electrons, Institute of Solid State Research, Research Centre Jülich, 52428 Jülich, Germany

m.heidelmann@fz-juelich.de
Keywords: STEM, EELS, StripeTEM, interfaces

A scanning transmission electron microscope (STEM) offers the advantage to simultaneously collect spatially resolved high-angle annular dark-field (HAADF) signal and electron energy loss spectroscopy (EELS) data. It is however state of the art that HAADF images and EELS spectra are recorded sequentially, meaning that the image acquisition is interrupted for spectrum acquisition. The disadvantage of such sequential recording techniques is that, there is frequently no control of the measurement location on an atomic scale, regardless of the beam residing at one spot or scanning along a line during EELS data acquisition. Loss of the measurement location at high resolution is in particular inflicted by inherent specimen and beam drift due to the high dwell time required for attaining a reasonable signal-to-noise ratio in EELS spectra.

In order to overcome obstacles due to specimen and beam drift we propose a simultaneaous acquisition of a series of spectra and an HAADF-image with horizontal scan lines parallel to an interface or a crystal plane. In this way, an image is recorded containing spectral data vs. spatial distance, similar to images recorded with the StripeTEM (or ELSP) method in TEM [1]. Shutter times are set such that every spectrum corresponds to several scan lines of the HAADF-image. Thereby the spectra are recorded with high SNR, while distributing the electron dose on the sample. In addition to the spectral image, the HAADF-image supplies coincident information about the measurement location. Distortions in the HAADF-image are used to detect the combined drift of the specimen and the beam.

StripeSTEM measurements were performed on a probe side aberration-corrected FEI Titan 80-300 equipped with a Gatan GIF. Fig. 1 shows a series of Ti-edge spectra and the coincident HAADF-image of bulk $SrTiO_3$. The intensity modulation in y-direction corresponds to alternating monolayers in $SrTiO_3$ as recorded in the coincident HAADF-image. Our StripeSTEM method is in particular attractive for interface analysis. Fig. 2 shows the application of the method to a $La_2CuO_4/SrTiO_3$ interface. The Ti-L, the O-K and the La-L edges are recorded with monolayer resolution across the interface.

The StripeSTEM method is a powerful tool for the investigation of interfaces as it enables doubtless correlation of spectra and atomic planes. The electron dose is spread and the resolution along the spatial coordinate is limited only by the localization of the energy loss signal [2].

M. Luysberg, K. Tillmann, T. Weirich (Eds.): EMC 2008, Vol. 1: Instrumentation and Methods, pp. 383–384, DOI: 10.1007/978-3-540-85156-1_192, © Springer-Verlag Berlin Heidelberg 2008

1. T. Walther, Ultramicroscopy **96** (2003), 401
2. R.F. Egerton, in "Electron Energy Loss Spectroscopy in the Electron Microscope" (Plenum Press, 2nd ed., 1996)

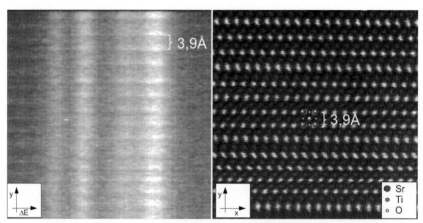

Figure 1. Series of EELS spectra of bulk $SrTiO_3$ (100) showing the monolayer modulation of the Ti-L_{23} edge signal and the coincident high-resolution HAADF-image.

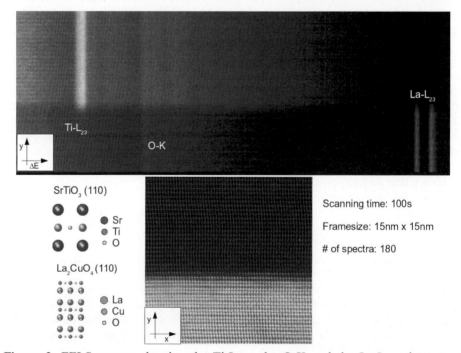

Figure 2. EELS spectra showing the Ti-L_{23} , the O-K and the La-L_{23} edges at an interface between La_2CuO_4 and $SrTiO_3$ as well as the coincident high-resolution HAADF image.

Comparing Transmission Electron Microscopy (TEM) and Tomographic Atom Probe (TAP) through Measurements of Thin Multilayers

Tobias Heil[1,3], Patrick Stender[2,3], Guido Schmitz[2,3] and Helmut Kohl[1,3]

1. Physikalisches Institut,
2. Institut für Materialphysik und
3. Interdisziplinäres Centrum für Elektronenmikroskopie und Mikroanalyse (ICEM)
Universität Münster, Wilhelm-Klemm-Str. 10, 48149 Münster, Germany

tobiheil@uni-muenster.de

Keywords: EFTEM, Elemental Mapping, Tomographic Atom Probe (TAP)

While transmission electron microscopy (TEM) is well established, the tomographic atom probe (TAP) is a rather new instrument for three-dimensional analysis of structures at the nanometre scale. The TAP has been developed from the field ion microscopy (FIM) that is used to obtain information about the crystalline structure of sharp (<50 nm tip radius) metal tips, by attaching a spatial detector and performing time of flight measurements. This allows the chemical analysis of individual atoms extracted from the tip, resulting in a complete chemical analysis over the whole volume [1]. To determine the accuracy and reliability of both methods for sub-nanometre chemical analysis of the interface between the layers of film systems we compare their respective results.

For our analysis we have evaporated Fe/Cr multilayers with layer thicknesses of 1-2 nm. The samples were prepared through alternate sputtering of the materials onto a plane silicon substrate (TEM) and a tungsten tip (TAP) simultaneously. Ten chromium and nine iron layers were deposited between two thick iron layers. The TEM image and the elemental map obtained through energy-filtered imaging (EFTEM) with our Zeiss Libra 200FE TEM can be seen in Figure 1. The TAP image of the chromium-iron mulitlayer system and its concentration profile is shown in Figure 2.

To compare the TEM and the TAP measurements we created a numerical reconstruction based on the TAP results, which should be a direct image of the sample. Starting from this model, we calculated the influence of the instrumental parameters and the drift of the sample. The resulting theoretical elemental signals were compared with the experimental TEM results (Figure 3). A good correlation for the iron signal can be seen, minor deviations can be explained by the inaccurate layer thickness. The chromium signal does not match quite as well, especially the background signal is higher than calculated. This could be caused by errors in the drift value that we estimated from the measurable drift between the different images.

In conclusion through the good correlation between the TEM and the TAP images of the Fe/Cr multilayer system the comparability of both instruments can be shown. Further measurements of different samples are planned to confirm these results.

M. Luysberg, K. Tillmann, T. Weirich (Eds.): EMC 2008, Vol. 1: Instrumentation and Methods,
pp. 385–386, DOI: 10.1007/978-3-540-85156-1_193, © Springer-Verlag Berlin Heidelberg 2008

386

1. D. Blavette, B. Deconihout, A. Bostel, J. Sarrau, M. Bouet, A. Menand, Rev. Sci. Instrum. **64** (1993), p. 2911.
2. We wish to express our gratitude to Dr. D. Baither, C. Reinke and L. Rettich (Institut für Materialphysik) for invaluable discussions and their help with specimen preparation.

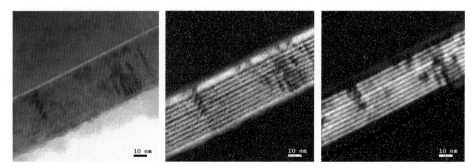

Figure 1. Unfiltered TEM image (left) and elemental map of the iron (middle) and the chromium (left) signal from the prepared multilayer system.

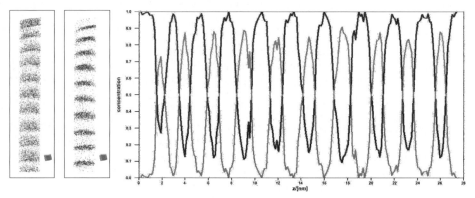

Figure 2. TAP image of the iron (left) and the chromium (middle) of the multilayer system. The grey cube has an edge length of 1 nm. The profile of iron (black) and chromium (grey) is shown on the right side.

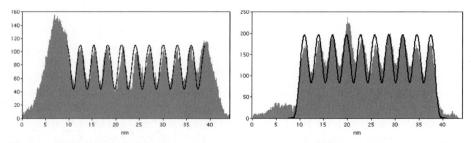

Figure 3. Theoretical (black curve) and experimental TEM measurements of the iron (left) and chromium (right) layers.

Development of a Process for Cleaning a TEM Column by Chemical Etching of Oxygen Radicals

Shin Horiuchi[1], Takeshi Hanada[2], and Masaharu Ebisawa[2]

1. Nanotechnology Research Institute, AIST, 1-1-1, Higashi, Tsukuba, Ibaraki 305-8565, Japan
2. Consult Zeroloss Imaging Inc., 1-33-31, Shinkoji, Machida, Tokyo 195-0057, Japan

s.horiuchi@aist.go.jp

Keywords: TEM, contamination, cleaning, plasma, EELS

Contamination deposition on a specimen during the irradiation of the electron beam in a TEM column interferes accurate elemental analysis and high-resolution imaging. Such specimen contaminations are mainly caused by the interactions of an electron beam with hydrocarbons existing in a high-vacuum TEM chamber introduced by high-vacuum pumping system, specimen damage, and leaks. Therefore, an effective process for the cleaning and removal of the contaminants in a TEM column has been desired

It has been know that hydrocarbon contamination in a SEM chamber can be cleaned by oxygen radicals generated in a RF-plasma device [1]. A SEM chamber is so large that reactive oxygen radicals easily distributed throughout the chamber. On the other hand, most part of a TEM chamber is narrow tubes, and thus it is difficult to achieve high conductance of activated radical flows in a column. In order to clean a TEM column, a process to achieve high conductance of oxygen radicals under a relatively low vacuum level for the stable generation of plasma should be developed.

Figure 1 shows the specimen contaminations created on a 10 nm thick carbon foil by irradiating the beam with 5.6×10^4 el/nm·s at a 200 kV accelerating voltage in a 5 years-used LEO922 energy-filtering transmission electron microscope. The EELS spectra acquired from the contaminated regions indicate that the thickness of the contaminated region increases significantly with the irradiation time. Evactron 45 (XEI Scientific, Inc.), used as an oxygen radical generator, was directly mounted on one of the ports close to the specimen chamber. The generated oxygen radicals were distributed into the chamber by introducing a vacuum pumping system at a position closed to the plasma chamber. The plasma generation (3 min at 0.4 torr) and N_2 purge (3 min at 0.9 torr) were repeated 20cycles. The upper side of Figure 2 shows typical locations of the plasma chamber and the vacuum system and the lower side shows the contaminations produced on a carbon film after the cleaning by the corresponding set-ups, which indicates the significant reduction of the specimen contaminations. The contaminations were barely seen even for the 5 min irradiation. Although the 10 min irradiation seems to produce heavy contamination, they were ten times thinner than the contamination before the cleaning as estimated by EELS. The results clearly indicate that the developed process successfully brings the active oxygen radicals into the TEM chamber and the contaminations were chemically etched away outside the chamber.

1. A. E. Vladár, M. T. Postek and R. Vane, Proc. SPIE, **835**, (2001), p. 4344.; R. Vane and V. Carlino, Microsc. Microsanal. **11 (Suppl 2)**, (2005), p. 900.

M. Luysberg, K. Tillmann, T. Weirich (Eds.): EMC 2008, Vol. 1: Instrumentation and Methods, pp. 387–388, DOI: 10.1007/978-3-540-85156-1_194, © Springer-Verlag Berlin Heidelberg 2008

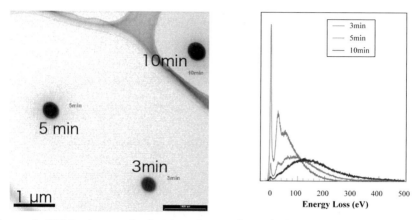

Figure 1. TEM micrograph showing electron-beam-induced contaminations deposited on a carbon thin foil created by the irradiation of 5.6 x 10^4 el/nm·s at a 200 kV accelerating voltage for 3, 5 and 10 min.

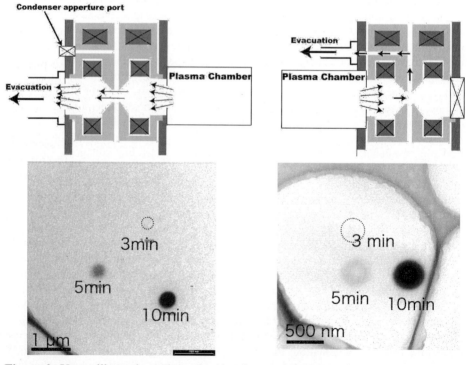

Figure 2. Upper illustrations show the cleaning system set-ups into a TEM specimen chamber, and lower TEM micrographs show the reduction of the specimen contaminations after the cleaning by the corresponding system set-ups shown above.

Low loss EELS study of gold nanoparticles using a monochromated TEM

S. Irsen, N.P. Pasoz and M. Giersigr

caesar, center of advanced European studies and research,
Ludwig Erhardt Alle 2, 53175 Bonn

irsen@caesar.de

Keywords: monochromated TEM, low loss EELS, gold nanoparticles

The investigation of low losses using EELS has become popular since the last generation of TEMs with monochromators and higher order corrected spectrometers have become more easily accessible. Several examples mainly focusing on semi conductors can be found in recent literature [1], [2]. The advantage in using EELS for the analysis of surface plasmons at very low energy losses is the combination of obtaining information about optical properties and band structures in combination with a spatial resolution which normally cannot be achieved using different techniques, e.g. HREELS, Raman or optical spectroscopy [3]. The wide application range of gold nanoparticles, e.g. in tumor therapy [4], brings the need for a reliable method studying these particles with both high spatial and high energy resolution. However, for gold particles, the surface plasmon resonances occur at very low energies, covering the visible range of the electromagnetic spectrum. So EELS analysis is challenging in spite of the excellent energy resolution of modern MC-TEMs. Nevertheless working at an energy resolution below 200 meV, low losses in the 1 eV regime should be visible in raw data.

In this preliminary study, we   show the possibilities of analyzing EEL-spectra at very low energy losses in single particles as well as clusters of gold nanoparticles. The final aim, a correlation of EELS results to other spectroscopic methods will be topic of further investigations.

Experiments were carried out on a 200 kV FEG (S)TEM system with monochromator ,a CS-corrector for the illumination system (ZEISS Libra200MC CRISP) and a 90° in column energy filter. An energy resolution of the system of 0.14 eV (zero loss half width) is achievable [5]. Spectra were recorded in CTEM mode at a convergence angle of 1.6 mrad and an acceptance angle of 5 mrad respectively. The dispersions of the recorded EELS spectra were 0.026 eV/channel.

Figure 1 shows examples of spherical gold nanoparticles as used for EELS analysis. All images were taken after the spectrum acquisition. The particles show high crystallinity as can easily be seen in the HRTEM images. The Spectra in figure 2 are the raw data of the EELS-scans corresponding to the particles in figure 1. For each spectrum 10 single spectra were recorded aligned and summed. The graphs in figure 2 are normalized to the integrated zero loss intensity and shifted for a better visualization. The black line derived from the dimer particle shown in figure 1a represents a low loss EELS signal similar to a bulk specimen. The grey graph in figure 2 on the other hand was

M. Luysberg, K. Tillmann, T. Weirich (Eds.): EMC 2008, Vol. 1: Instrumentation and Methods, pp. 389–390, DOI: 10.1007/978-3-540-85156-1_195, © Springer-Verlag Berlin Heidelberg 2008

390

derived from a single gold particle (see figure 1 b). Here a significant shift of the low loss signal to lower energies into the range of visible light can be observed (see arrows in figure 2). This is a promising result for the investigation of low loss spectra of nobel metals.

For a better understanding of these low loss shifts especially using different shaped particles e.g. gold nano-wires, more detailed investigations are required.

1. I.R. Khan, D. Cunningham, S. Lazar, D. Graham, W.E. Smith, and D.W. McComb, FARADAY DISCUSSIONS **132**, 171 (2006).
2. [K.A. Mkhoyan, T. Babinec, S.E. Maccagnano, E.J. Kirkland, and J. Silcox, Ultramicroscopy **107**, 345 (2007).
3. Y. Chen, R.E. Palmer, E.J. Shelley, and J.A. Preece, Surf. Sci. **502**, 208 (2002).
4. [R.K. Visaria, R.J. Griffin, B.W. Williams, E.S. Ebbini, G.F. Paciotti, C.W. Song, and J.C. Bischof, MOLECULAR CANCER THERAPEUTICS **5**, 1014 (2006).
5. E.Q. T. Walther, H. Stegmann, A. Thesen, G. Benner, Ultramicroscopy **106**, 963 (2006).

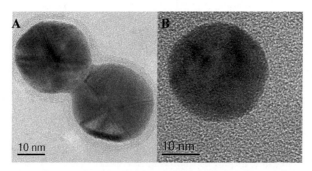

Figure 1. F- images of the gold nanoparticles used for EELS low loss investigation. All images were taken after recording the EEL spectra. The monochromator aperture was removed prior to the scans.

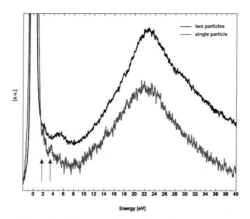

Figure 2. Low loss EEL-spectra of gold nanoparticles. The black graph represents a scan of two particles while the grey line was derived fro a single gold particle.

Analytical RPA response of Carbon and BN single-walled nanotubes: Application to EELS and wave loss spectra

P. Joyes, O. Stéphan, M. Kociak, A. Zobelli, and C. Colliex

Laboratoire de Physique des Solides, CNRS, UMR8502, Université Paris-Sud, 91405 Orsay, France

joyes@lps.u-psud.fr

Keywords:tubule, RPA, photon and electron spectroscopy

We calculate the local RPA (Random Phase Approximation) polarizability of single-walled carbon and BN nanotubes and apply our results to the loss spectra of electrons (EELS: Electron Energy Loss Spectrum) and photons in the case of □ excitations. We take into account crystal local field effects (CLFE) by developing a two-order matricial treatment. For spatially periodic tubules, the whole calculation (one-electron and RPA steps) is analytical and leads to simple formulas. We compare our results to other RPA-type ones where CLFE are neglected (fig.1). We also examine opened tubule where the analytical derivation is only possible for the one-electron polarizability, the RPA step being made numerically. We operate a characterization of the excitation modes by determining the local phase-shifts and the displaced charges on each site. The general conclusion is that EELS and electromagnetic wave spectra exhibit common excitation modes with differences in intensities depending on the conditions of the interaction. An interesting result is given by EELS with an incident beam direction parallel to the tubule axis when the impact parameter is varied: for a central interaction or zero impact parameter (respectively, for a large impact parameter) the spectrum is closely similar to the wave spectrum with parallel polarization (respectively, with perpendicular polarization). The evolution of the various modes with tubule size is also studied.

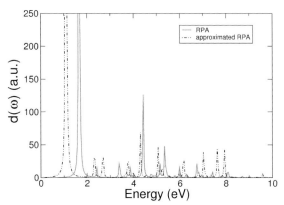

Figure 1. Carbon nanotube dipole moment for an electric field direction parallel to the nanotube axis; broken line: CLFE neglected, full line: this work.

M. Luysberg, K. Tillmann, T. Weirich (Eds.): EMC 2008, Vol. 1: Instrumentation and Methods, pp. 391–392, DOI: 10.1007/978-3-540-85156-1_196, © Springer-Verlag Berlin Heidelberg 2008

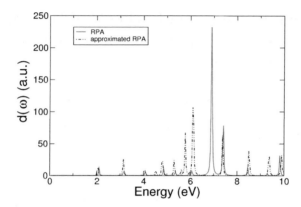

Figure 2. Carbon nanotube dipole moment for an electric field direction perpendicular to the nanotube axis; broken line: CLFE neglected, full line: this work.

Improvement of energy resolution of VEELS spectra with deconvolution method for electronic and optical properties analysis on ferroelectric oxides in nano-scale

T. Kiguchi[1], N. Wakiya[2], K. Shinozaki[3] and T.J. Konno[1]

1. Institute for Materials Research, Tohoku University, 2-1-1 Katahira, Aoba-ku, Sendai, Miyagi, 980-8577, Japan
2. Department of Materials Science and Chemical Engineering, Shizuoka University, 3-5-1 Johoku, Hamamatsu, Shizuoka, 432-8561, Japan
3. Department of Metallurgy and Ceramics Science, Graduate School of Science and Engineering, Tokyo Institute of Technology, 2-12-1 O-okayama, Meguro-ku, Tokyo, 152-8550, Japan

tkiguchi@imr.tohoku.ac.jp

Keywords: Valence-EELS, Deconvolution, Optical properties, Ferroelectric oxide film

Electron Energy Loss Spectra of low-energy loss regions below 50eV, which is called VEELS, is the signature of valence band structures, displaying the inter-band transition (single excitation), plasma oscillation (collective excitation), etc. It can also be employed to reveal electronic and optical properties of materials [1]. VEELS analysis can elucidate a local variation of electronic and optical properties of materials in nano-scale using the transmission electron microscope (TEM) in the nano-probe mode. TEM-VEELS method has several advantages over conventional optical spectroscopy, such as UV-VIS-IR spectrometry, because it could measure a wider energy (wave length) range and local electronic and optical properties from a small volume. However, the energy resolution of TEM-EELS spectrum is less than that of high resolution EELS (HREELS) or reflective EELS methods used with a lower incident energy, or conventional optical measurements such as UV/VIS/IR spectrometry. In TEM-EELS methods operated under a high voltage, the monochromater has been applied for improving the energy resolution. However, this method is restricted for the limited number of instruments, and there exist the cost of loss of intensity. On the othr hand, a software besed monochromater, which was a deconvolution method, is free from these difficulties [2]. This technique has been applied to improve the energy resolution of core-loss spectra by eliminating the energy spread of incident electrons and shown to be effective. However, it has not been reported for application of VEELS.

In this study, we applied this technieque to the investigation of the local electrical and optical properties of $Pb(Mg_{0.33}Nb_{0.67})$-0.34mol%$PbTiO_3$ solid solution (PMN-PT) film stacked on Si(001) wafer via $(La,Sr)CoO_{3-x}/CeO_2/YSZ$ buffer layers prepared by PLD (Pulsed-Laser Deposition) method [4]. VEELS spectra have been obtained with TEM-EELS method in the CBED-Diffraction mode (JEM-2010F, JEOL / ENFENA model776, Gatan). The full width of half maximum (FWHM) of the nano-probe was c.a. 1nmφ. The convergence and the collective semi-angles were 0.9 and 5.7mrad, respectively, and the energy dispersion used was 0.1eV/ch. The deconvolution method

394

using Richardson-Lucy algorithms (DeConvEELS, HREM Research Inc.) was employed to improve the energy resolution.

Figure 1 (a) shows a low-loss spectrum obtained from PMN-PT layer. The dashed line shows the as-obtained spectrum, the solid line the deconvoluted spectrum, demonstrating improved energy-resolution spectrum. The full width of half maximum (FWHM) of the as-obtained spectrum is 1.2eV, which can be improved to 0.4eV after processing. An ELF (energy-loss function) has then been derived as shown in (b) by deconvoluting an elastic (and semi-elastic) scattering and multiple scattering components from the original spectrum (a). The ELF indicates some shoulders and peaks corresponding to inter-band transitions, collective excitations (plasmon) and core-loss edges from the shallow core level. It is clearly seen that the fine structure can be reconstructed by the deconvolution. A dielectric function ε_1 (real part), ε_2 (imaginary part) (c) has been derived from an ELF (b) by Kramers-Kronig analysis. The fine structure of dielectric function can also be reconstructed. ε_2 shows optical absorption arising from inter-band transitions, and the onset energy shows an optical band gap of PMN-PT. The value is about 3.5eV in this study, which is comparable to the reported values obtained from ellipsometry [4-6]. Another optical properties, such as refractive index, extinction coefficient, optical absorption coefficient etc., have also been considered.

1. R.F.Egerton in "Electron Energy Loss Spectroscopy in the Electron Microscope 2nd.ed.", (PLENUM PRESS, NEW YORK) (1996), p. 142, p. 312, p. 398
2. K.Ishizuka, K.Kimoto, and Y.Bando, Proc. Micros. Microanal. 2003, San Antonio (2003) p.832
3. K.Shinozaki, M.Kasahara, T.Kiguchi, N.Mizutani and N.Wakiya, Jpn. J. Appl. Phys. **46** (2007), p. 657
4. K.Y. Chan, W.S. Tsang, C.L. Mak and K.H. Wong, J. Eur. Ceram. Soc **25** (2005) p.2313
5. Xinming Wan, H. L. W. Chan and C. L. Choy, Xiangyong Zhao and Haosu Luo, J. Appl. Phys. **96** (2004) p.1387
6. W.S.Tsang, K. Y. Chan, C. L. Mak, and K. H. Wong, Appl. Phys. Lett. **83** (2003) p.1599

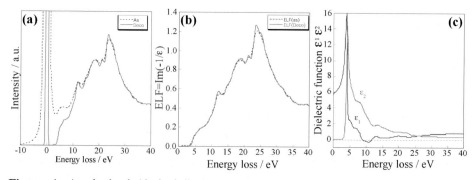

Figure 1. As-obtained (dashed line) and decovoluted (solid line) (a) valence-EEL spectrum and (b) ELFs of PMN-PT thin film. (c) The dielectric function of the real part ε_1 and the imaginary part ε_2.

Low Loss Electron Energy Spectroscopy on LiFePO₄ for Li ion Battery Applications

M. Kinyanjui[1,2], U. Kaiser[1], M. Wohlfahrt-Mehrens[2], J. Li[3], and D. Vainkin[4]

1. Electron Microscopy Group of Materials Science, University of Ulm, Albert Einstein Allee 11, 89081 Ulm, Germany
2. Centre for Solar Energy and Hydrogen Research, Helmholtzstr. 8, 89081 Ulm, Germany
3. Department of Materials Science and Engineering, University of Maryland, College Park, MD 20742, USA
4. Ames Laboratory and Department of Physics and Astronomy, Iowa State University, Ames, IA 50011, USA.

michael.kinyanjui@uni-ulm.de

Keywords: EELS, DFT, LiFePO₄, Li ion batteries

$LiFePO_4$ is a potential cathode material for high energy Li ion battery applications. The potential of this material in Li ion battery applications is hindered by its low electronic and ionic conductivity [1]. Doping $LiFePO_4$ with multivalent cations has been one of the proposed methods to improve $LiFePO_4$ electron properties [2]. This requires a fundamental understanding of the electronic structure and conduction properties of $LiFePO_4$. Band structure calculations derived from Density Functional Theory (DFT) calculations have produced controversial results that predict $LiFePO_4$ as being semi-metallic with a small band gap of ~ 0.1 eV or an insulator characterized by a large band gap ~ 4 eV band gap [3, 4, 5]. There is also a lack of conclusive experimental studies on the electronic properties of $LiFePO_4$.

The low loss region (0-50eV) of an Electron Energy Loss Spectrum contains information about valence electron excitations and thus the electronic structure of a specimen. Here we present results from a low-loss electron energy-loss spectroscopic (EELS) study of $LiFePO_4$. The low loss EELS spectra were acquired at 80 KV (with an energy resolution of 0.5 eV) in order to reduce the influence from the relativistic Cerenkov losses. Zero loss peak de-convolution and Kramers-Kronig transformations have been used to analyze the low-loss EELS spectrum from which the real part and imaginary parts of the energy loss spectrum, ε_1, ε_2 respectively, were extracted. Figure 1(a) shows the derived real ε_1 and imaginary ε_2 parts of the low loss EELS spectrum while Figure 1(b) shows plots of the calculated optical absorption spectrum and imaginary ε_2 of the energy loss spectrum. The obtained results are in good agreement with first principles calculations and Ultraviolet-Visible-Infrared (UV-Vis-IR) optical measurements [3] supporting the argument of $LiFePO_4$ as a wide gap insulator. We discuss these results vis-à-vis, the EELS methods used and *ab initio* modelling of the low loss properties. We put these results into context of $LiFePO_4$ electronic properties and its application as a cathode material for Li ion batteries.

M. Luysberg, K. Tillmann, T. Weirich (Eds.): EMC 2008, Vol. 1: Instrumentation and Methods, pp. 395–396, DOI: 10.1007/978-3-540-85156-1_198, © Springer-Verlag Berlin Heidelberg 2008

396

1. A.K. Padhi, K.S Nanjundaswamy and J.B.Goodenough, Journal Electrochem. Soc. 144 (1997) 1188
2. S.Y. Chung, J.T. Bloking, and Y.M. Chiang, Nature Materials, **1**, (2002) 123
3. F.Zhou, K.Kang, T.Maxisch, G.Ceder, D.Morgan, Solid State Communications, 132, 3-4, (2004), 181
4. A. Yamada and S.-C. Chung, Journal of the Electrochemical Society **148** (2001), A960
5. Y.N. Xu, S.Y. Chung, J.T. Bloking, Y.M.Chiang, W.Y. Ching, Electrochemical and Solid-State Letters **7** (2004) A131
6. The authors acknowledge the BMBF for the partial funding of the project REALIBATT

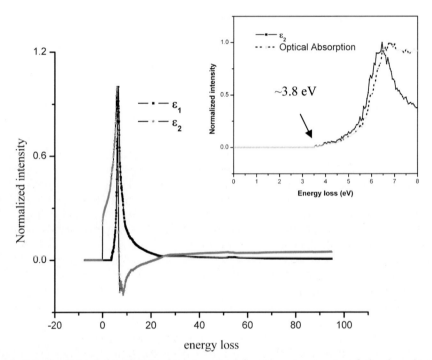

Figure1. (a) A plot of the obtained real ε_1 and imaginary ε_2 parts of the low loss LiFePO$_4$ EELS spectrum (b) Plots of the calculated optical absorption spectrum and imaginary ε_2 of the energy loss spectrum showing the onset of the obtained spectral intensity at ~ 3.8 eV.

Atomic-resolution studies of complex oxide materials using in-situ scanning transmission electron microscopy

G. Yang, Y. Zhao, and R.F. Klie

Nanoscale Physics Group, Department of Physics, University of Illinois at Chicago, 845 W. Taylor Str., Chicago, IL, 60616

rfklie@uic.edu

Keywords: STEM, EELS, $LaCoO_3$, $Ca_3Co_4O_9$

Layered cobaltate materials have been the focus of many recent studies due to the wide variety of electrical, magnetic, and structural properties they exhibit. These phenomena have been commonly associated with the presence of mixed Co-valence, spin transitions and the presence of tilted octahedral CoO_2 layers.[1] In this presentation, we will use aberration-corrected Z-contrast imaging and EELS in combination with in-situ heating/cooling experiments to study the spin-state transition in perovskite oxide $LaCoO_3$ and the effects of charge transfer in misfit-layered $Ca_3Co_4O_9$.

Figure 1 shows a temperature study of the perovskite oxide $LaCoO_3$ in the range between 10 K and 300 K. It has been previously shown that $LaCoO_3$ exhibits an anomaly in its magnetic susceptibility at around 80 K which is commonly associated with a thermally excited transition of the Co^{3+}-ion spin. Here, we show the effect of this spin-state transition on the O K-edge fine-structure in $LaCoO_3$. While the crystal structure of $LaCoO_3$ does not change during the in-situ cooling experiment to 10K (Figure 1b), the O K-edge pre-peak intensity decreases for T>80 K. A detailed analysis of the O K-edge fine-structure and comparative DFT calculations will be presented.[2]

Figure 2a) shows an atomic-resolution Z-contrast image of $Ca_3Co_4O_9$ ([010]). This misfit-layered structure has been of great interest due to its high thermopower rivaling that of more conventional (non-oxide) thermoelectric materials.[1] The structure of $Ca_3Co_4O_9$ consists of five distinct layers; three rock salt-type layers Ca_2CoO_3 sandwiched between two CdI_2-type CoO_2 layers along the c-direction.[3] Figure 1b) shows atomic-column resolved EELS spectra of the O K-edge from the individual layers within the $Ca_3Co_4O_9$ unitcell. It can be clearly seen that the O K-edge pre-peak, which is associated with the Co-$3d$ O-$2d$ hybridization, decreases in the CoO layer.

Of particular interest, is the abrupt change of the electrical resistivity in $Ca_3Co_4O_9$ at 420 K, which is assumed to be due to the spin-state transitions of Co ions.[3] Using in-situ heating experiments, we examine the structural and electronic changes of $Ca_3Co_4O_9$ at different temperatures (Figure 3). The most obvious change in the O K-edge near-edge fine-structure at 500 K compared to room-temperature is the decrease in the pre-peak intensity. Since the O K-edge pre-peak intensity can be used to quantify the spin-state (see Figure 1)[2], the change at 500K could indicate a spin-state transition in $Ca_3Co_4O_9$ that is responsible for the change in the resistivity at elevated temperature.

M. Luysberg, K. Tillmann, T. Weirich (Eds.): EMC 2008, Vol. 1: Instrumentation and Methods, pp. 397–398, DOI: 10.1007/978-3-540-85156-1_199, © Springer-Verlag Berlin Heidelberg 2008

398

1. S. Li et al., J. Mater. Chem. **9** (1999) 1659
2. R. F. Klie et al., Phys. Rev. Lett. **99** (2007) 047203
3. C. Masset et al., Phys. Rev. B **62** (2000) 166

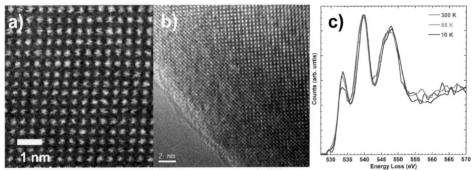

Figure 1. a) Z-contrast image of LaCoO₃ (221) at 85K; b) HRTEM image of LaCoO₃ (221) at 10 K; c) EELS-spectra of the O K-edge at 300K, 86K and 10K.

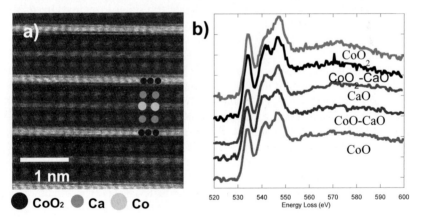

Figure 2. a) Z-contrast image of Ca₃Co₄O₉ [010]; b) ELNES of O K-edge in Ca₃Co₄O₉.

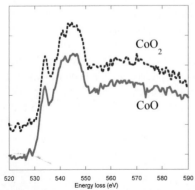

Figure 3. ELNES of O K-edge in Ca₃Co₄O₉ at 500 K.

Low-loss EELS measurements on an oxide multilayer system using monochrome electrons

G. Kothleitner[1], B. Schaffer[1] and M. Dienstleder[1]

1. Institute for Electron Microscopy, Graz University of Technology, Graz, Austria

gerald.kothleitner@felmi-zfe.at
Keywords: EELS, EFTEM SI, monochromator, bandgap measurements

The analysis of TEM low-loss electron energy-loss spectra i.e. the range up to 50 eV, has recently become attractive and feasible with the advent of advanced TEM equipment and analysis instrumentation, such as monochromators and improved energy-filters / spectrometers [1,2]. The better energy resolution opens up new possibilities for a more accurate measurement of bandgaps and optical properties via the dielectric function. However, a crucial step in the analysis of valence EELS data is the separation of the zero-loss peak [3], ranging into the spectral regime between 0-5 eV. A monochromator can help shortening the tail by narrowing the peak itself, and secondly it makes the shape symmetric, facilitating the study of the band-gap region considerably, even with narrow band-gap semiconductors. On the other hand, physical effects like Čerenkov excitation may pose a limit to the accuracy of band structure investigations, which are further influenced by experimental parameters, bandgap character and / or specimen thickness. Despite the instrumental improvements, the handling of dynamically changing intensities in the low-loss area and the correction of artifacts caused by energy- and spatial drift as well as residual aberrations require an elaborated data acquisition and post-processing scheme, including extra interactivity, robustness and flexibility [4].

Low-loss EFTEM measurements involving optimized acquisition and processing as well as ultra-narrow slits in combination with monochrome illumination have been carried out on a multilayered oxide system (MgO, ZrO, ZnO), fabricated by pulsed laser PVD, in order to extract bandgap information. The achieved symmetric ZLP width, resulting from the used slit width (370meV), the residual aberrations (100eV), the intrinsic spread (160meV) and the parasitic magnetic fields (50meV), amounted to 420meV, and was sufficient to reveal the relevant low-loss features, which match the monochrome EELS reference spectra, taken on the same system (Figure 1). As the sample was rather uniform in thickness and did not exceed 0.5λ , systematic bandgap shifts could not be observed and the true direct bandgaps of the respective layers could be mapped (Figure 2).

1. P.C. Tiemeijer, Ultramicroscopy **78** (1999), p.53.
2. T. Walther, E. Quandt, H. Stegmann, A. Thesen & G. Benner, Ultramicroscopy **106** (2006) p.963.
3. R. Erni and N.D. Browning, Ultramicroscopy **107** (2007) p.267.
4. B. Schaffer, W. Grogger, G. Kothleitner & F. Hofer, Anal Bioanal Chem 390 (2008) p.1439.

M. Luysberg, K. Tillmann, T. Weirich (Eds.): EMC 2008, Vol. 1: Instrumentation and Methods, pp. 399–400, DOI: 10.1007/978-3-540-85156-1_200, © Springer-Verlag Berlin Heidelberg 2008

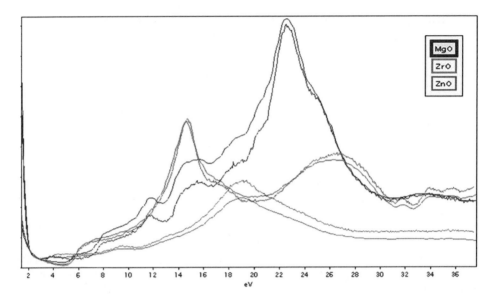

Figure 1. High-resolution low-loss EELS spectra from an oxide multilayer system. The two curves per colour represent spectra reconstructed from a monochrome EFTEM SI (~0.4eV) in comparison to monochrome reference spectra (0.2eV).

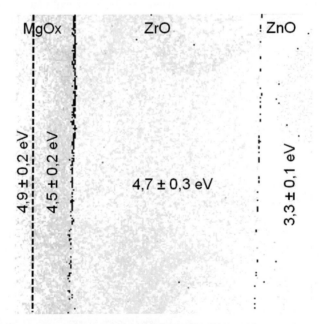

Figure 2. Direct bandgap map from the MgO/ZrO/ZnO multilayer stack. For the dark lines at the phase boundaries no clear values can be given.

White noise subtraction for calculating the two-particle-structure factor from inelastic diffractograms

C. Kreyenschulte and H. Kohl

1. Physikalisches Institut and Interdisziplinäres Centrum für Elektronenmikroskopie und 2. Microanalyse (ICEM), Universität Münster, Wilhelm-Klemm-Straße 10, 48149 Münster, Germany

carsten.kreyenschulte@uni-muenster.de

Keywords: elemental maps, EFTEM, diffractograms

A method to determine the two-particle-structure factor from inelastic diffractograms has been described in previous publications [1,2,3,4]. To obtain quantitative results it is vital to subtract the noise in the diffractogram correctly.

In Fourier-space the image signal $I_{CCD}(u)$ recorded with a CCD-camera consists of the modulation-transfer-function $MTF(u)$, the image-drift during image acquisition $D(u)$, the inelastic transfer function $ITF(u)$, the white noise $B(u)$ and a factor $S_{aa,bb}(u)$ proportional to the two-particle-structure factor,

$$\left| I_{CCD}(u) \right|^2 = MTF(u)^2 \cdot D(u)^2 \cdot ITF(u)^2 \cdot S_{aa,bb}(u) + B(u).$$

We used a sample of an aged iron-chromium-alloy displaying spinodal decomposition. Therefore we could work under the assumption of an approximately isotropic material allowing us to use the circular average of the squared modulus of the recorded signal.

The $MTF(u)$ of the CCD-camera is usually supplied by the manufacturer or can be measured using the knife-edge method [5,6]. The image drift correction $D(u)$ can be calculated from the image itself and the $ITF(u)$ by the methods described in [1,2,3].

Using the approach in [6,7] the white noise $B(u)$ can be calculated using a series of evenly illuminated empty images. This works quite well for high signal intensities and short acquisition times. In our case the sample drift and the inelastic filtering at energy losses of several hundred eVs limits our acquisition times to a couple of minutes resulting in low signal intensities in the recorded images. Due to these rather extreme conditions the camera shows a different behaviour than in usual conditions and the white noise subtraction had to be modified by assuming an additional additive term.

Figure 1 shows an elemental map of iron from the mentioned sample and figure 2 the corresponding two-particle-structure factors for iron and chromium calculated with our new noise correction.

1. H. Kohl, H. Rose, Adv. Electron El. Phys. **65** (1985), 173-227.
2. R. Knippelmeyer and H. Kohl, J. Microscopy **194** (1999), 30-41.
3. A. Thesing, PhD Thesis, University of Münster.
4. C. Kreyenschulte and H. Kohl, Microsc Microanal **13** Suppl 3 (2007), 284-285.
5. A. Weickenmeier, W. Nüchter and J. Mayer, Optik **99** (1999), 147-154.

M. Luysberg, K. Tillmann, T. Weirich (Eds.): EMC 2008, Vol. 1: Instrumentation and Methods, pp. 401–402, DOI: 10.1007/978-3-540-85156-1_201, © Springer-Verlag Berlin Heidelberg 2008

402

6. C. Hülk, PhD Thesis, University of Münster.
7. R. Meyer and Angus Kirkland, Microsc Research and Technique **49** (2000), 269-280.

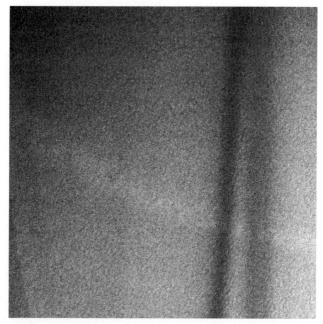

Figure 1. Elemental map of iron, taken at 718eV, Fe(45at%)Cr aged for three weeks

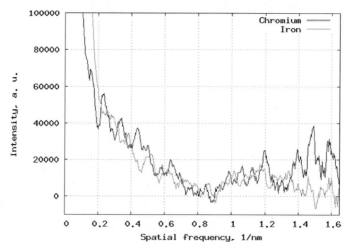

Figure 2. Two-particle-structure-factor for iron and chromium, calculated by using our new noise subtraction

Local Analysis of BaTiO₃/SrTiO₃ interfaces by STEM-EELS

H. Kurata, R. Kozawa, M. Kawai, Y. Shimakawa and S. Isoda

Institute for Chemical Research, Kyoto University, Uji, Kyoto 611-0011, Japan

kurata@eels.kucir.kyoto-u.ac.jp

Keywords: HAADF-STEM, VEELS, strained interface

Electron energy-loss spectroscopy (EELS) combined with a scanning transmission electron microscopy (STEM) is a powerful tool to analyze electronic structure in local area such as interfaces and defects. Here we report the study on local analysis of BaTiO₃ (BTO)/SrTiO₃ (STO) interfaces using STEM-EELS method. The influence of residual strain due to the lattice mismatch between the film and the substrate on the electronic structure is investigated.

Epitaxial BTO thin films were grown on STO (001) substrates at 973 K in an oxygen partial pressure of 0.75 Pa by a pulsed laser deposition technique [1]. High-resolution HAADF imaging and STEM-EELS measurement were performed using a 0.1 nm probe in a JEM-9980TKP1 equipped with a Cs corrector for illumination lens and a cold field-emission (nano-tip) gun [2] operated at a 200 kV. The spatially resolved valence EELS (VEELS) were measured by an omega type of filter with an energy resolution of 0.5 eV.

Figure 1 shows a high-angle annular dark-field (HAADF) image of a BTO/STO interface. In the vicinity of the interface misfit dislocations are observed within the BTO side. These dislocations are edge type with Burgers vectors b = a<100>. The dislocation (A) is about 4 nm away from the interface, so the region between the interface and the dislocation (A) should be strained. Figure 2 shows the variation of lattice parameters in BTO film measured from the cation positions in the HAADF image using a STO lattice parameter as the calibration standard. They reveal that the in-plane lattice parameter in region from the interface to the dislocation (A) is coincident with that of STO (a_{STO} = 0.3905 nm), while the out-of-plane lattice parameter is larger than that of a bulk BTO (c_{BTO} = 0.4036 nm). Therefore, this interfacial layer is in a strained state, while the region below the dislocation is almost relaxed to the bulk BTO structure. Figure 3 shows spatially resolved VEELS measured at eleven cation sites across the interface and the dislocation core (A). From the shift of volume plasmon peak shown by broken lines the position of interface can be determined accurately. In order to extract the change of electronic structure at strained BTO, a Kramers-Kronig analysis was carried out for each VEELS to obtain dielectric function. Figure 4 shows the spectra of imaginary part of dielectric function (ε_2) obtained from the strained and relaxed-BTO areas and dislocation core. The peak (b) in the ε_2 spectrum of strained-BTO is slightly shifted to high energy compared to that of relaxed-BTO. This peak is mainly assigned to the interband transitions from oxygen 2p to barium 5d bands by the first principles band structure calculation. The shift of peak (b) is attributed to the change of local electronic

404

structure caused by a compressive misfit strain along in-plane directions in the strained-BTO layer.

1. M. Kawai, D. Kan, S. Isojima, H. Kurata, S. Isoda and Y. Shimakawa, J. Appl. Phys. 102 (2007) 114311.
2. K. Kurata, S. Isoda and T. Tomita, Proc. of IMC16, p.583, 2006.
3. We acknowledge to Mrs. Kawasaki, Otsuka and Miss. Nishimura of TRC for preparing the cross-section samples.

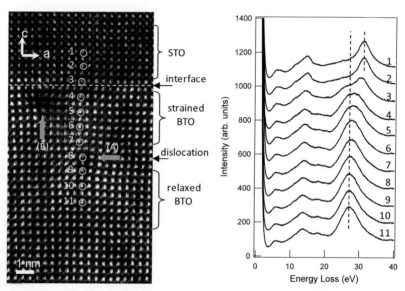

Figure 1. HAADF image of BTO/STO interface.

Figure 3. Spatially resolved VEELS across the interface and the dislocation core (A) in Fig. 1.

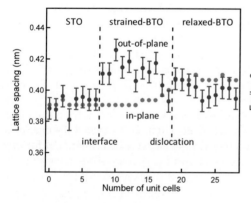

Figure 2. Variation of the in-plane and out-of-plane lattice parameters.

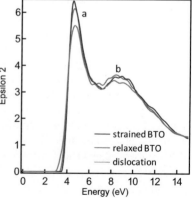

Figure 4. ε_2 spectra obtained from the strained and relaxed BTO layers and dislocation core (A) in Fig. 1.

Experimental conditions and data evaluation for quantitative EMCD measurements in the TEM

H. Lidbaum[1], J. Rusz[2,3], A. Liebig[2], B. Hjörvarsson[2], P.M. Oppeneer[2], E. Coronel[1], O. Eriksson[2], K. Leifer[1]

1. Department of Engineering Sciences, Uppsala University, Box 534, SE-75121 Uppsala, Sweden
2. Department of Physics, Uppsala University, Box 530, SE-75121 Uppsala, Sweden
3. Institute of Physics, Academy of Sciences of the Czech Republic, Na Slovance 2, CZ-182 21 Prague, Czech Republic

hans.lidbaum@angstrom.uu.se

Keywords: TEM, magnetic circular dichroism, EELS

The recently demonstrated technique electron magnetic circular dichroism (EMCD) opens new routes for characterization of magnetic materials using transmission electron microscopy [1]. The technique enables quantitative measurements of orbital to spin magnetic moments with element specificity, according to the recently derived sum rules [2]. Electron energy-loss spectra is obtained at well defined scattering geometries, see figure 1, from where the EMCD-signal is obtained.

The principle of the technique has been demonstrated, further progress is required to obtain reliable quantitative information about the magnetic properties of the sample. By using energy filtered diffraction patterns, the distribution of the EMCD signal in reciprocal space is obtained. We study the influence of experimental geometries on the EMCD signal and optimize the data analysis of the probed reciprocal plane. This is essential to obtain correct and reliable magnetic information. Especially normalization, signal to noise optimization and consideration of the entire edge intensities are important. The data cubes consisting of the reciprocal plane and energy-loss were acquired using a FEI Tecnai F30ST microscope equipped with a Gatan GIF2002 spectrometer [3]. In figure 2, two spectra that were extracted at the P+ and P- positions are shown, revealing an EMCD-signal. The experimental results are compared with calculations of the EMCD signal for a thin Fe film, showing very good agreement.

1. P. Schattschneider, S. Rubino, C. Hébert, J. Rusz, J. Kuneš, P. Novák, E. Carlino, M. Fabrizioli, G. Panaccione and G. Rossi, Nature **441** (2006), p. 486-488.
2. J. Rusz, O. Eriksson, P. Novák and P.M. Oppeneer, Phys. Rev. B **76** (2007), 060408(R).
3. H. Lidbaum, J. Rusz, A. Liebig, B. Hjörvarsson, P.M. Oppeneer, E. Coronel, O. Eriksson and K. Leifer, submitted.

M. Luysberg, K. Tillmann, T. Weirich (Eds.): EMC 2008, Vol. 1: Instrumentation and Methods, pp. 405–406, DOI: 10.1007/978-3-540-85156-1_203, © Springer-Verlag Berlin Heidelberg 2008

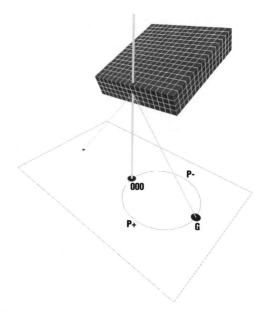

Figure 1. A sketch of the sample oriented in a two beam case geometry where the transmitted and a Bragg scattered beam G are strongly excited. The circle indicates the so called Thales circle positions P+ and P-.

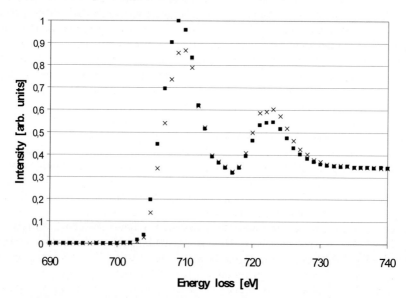

Figure 2. EELS spectra of Fe $L_{2/3}$ edges measured on a thin film of Fe. The spectra were extracted from energy filtered diffraction patterns at the P+ and P- positions in reciprocal space, as indicated in figure 1.

Investigation of the valency distribution in $Cu_{1.2}Mn_{1.8}O_4$ using quantitative EELS near-edge structures analysis

C. Maunders[1,2], B.E. Martin[1], P. Wei[1], A. Petric[1] and G.A. Botton[1,2]

1. Department of Materials Science and Engineering, McMaster University, 1280 Main St. West, Hamilton L8S 4M1, Canada
2. Brockhouse Institute for Materials Research, McMaster University, 1280 Main St. West, Hamilton L8S 4M1, Canada

maunder@mcmaster.ca

Keywords: Cu-Mn-O Spinel System, EELS, Valency Distribution, Solid Oxide Fuel Cells (S.O.F.C)

Electron energy loss spectroscopy (EELS) is a well-established technique used to probe the electronic structure of complex materials. Improvements in spectrometer design and the introduction of monochromators combined with aberration correctors in modern transmission electron microscopes provide spectra of ultrahigh energy resolution (equivalent to synchrotron data) at the atomic level.

In this work we employ EELS to investigate the electronic structure of the spinel compound $Cu_{1.2}Mn_{1.8}O_4$ which is being investigated as a new candidate for components of solid oxide fuel cells (S.O.F.C.). Since the late 1950's there has been significant interest in the atomic and electronic structure of the $Cu_xMn_{3-x}O_4$ spinel system (where $1 < x < 1.5$). Many studies have been conducted on their thermal and electrical properties in order to facilitate their use in industry [1,2] and in general these properties are well understood, however, some important aspects of the electronic structure, such as the distribution and valence states of the Cu and Mn cations have not been directly determined. Understanding the cation valence distribution in this system is therefore critical to the understanding the conductivity behaviour and advancing their use in industrial applications.

The Mn valency distribution in $Cu_{1.2}Mn_{1.8}O_4$ was extracted using a quantitative fitting procedure of the Mn $L_{2,3}$ core loss edge with Mn $L_{2,3}$ edges from reference compounds containing Mn^{2+} (MnO), Mn^{3+} (Mn_2O_3) and Mn^{4+} (MnO_2) normalised according to the number of holes in the $3d$ band. The fitting procedure proceeds by varying the relative amounts of each of the fitting components until both the L_3/L_2 ratio and the fine structure of the Mn $L_{2,3}$ edge are simultaneously reproduced. The relative amounts of each component then represent the Mn valency distribution. Using this fitting procedure we have determined the valency distribution to be 55% Mn^{4+}, 37% Mn^{3+} and 8% Mn^{2+}. Analysis of the Cu $L_{2,3}$ edge demonstrates that all Cu is present as Cu^{2+} and is coordinated both octahedrally and tetrahedrally. The O-K edge presents a pre-peak of the EELS Near-edge Structure (ELNES) consistent with the predominant presence of Mn^{4+}, whilst the features in the ELNES are identical to those observed in other spinel compounds. The effect of site symmetry of the various edges, the charge balance as well as the nature of the defects necessary to maintain charge neutrality are discussed. The importance of high-energy resolution as well as the application of this

M. Luysberg, K. Tillmann, T. Weirich (Eds.): EMC 2008, Vol. 1: Instrumentation and Methods, pp. 407–408, DOI: 10.1007/978-3-540-85156-1_204, © Springer-Verlag Berlin Heidelberg 2008

technique to detect local changes in valency at boundaries and surfaces of nanoparticles used in energy storage applications is discussed. Examples related to the determination of valency in lithiated and delithiated $LiFePO_4$ and $LiFeMnPO_4$ compounds used in rechargeable batteries are presented as examples to demonstrate the advantages of high spatial resolution and high-energy resolution measurements for mapping the valence distribution in complex oxides.

1. R.E. Vandenberghe, G.G. Robbrecht, V.A.M. Brabers, Mat. Res. Bull. **8** (1973), p. 571.
2. A.D.D. Broemme, V.A.M. Brabers, Solid State Ionics **16** (1985), p. 171.

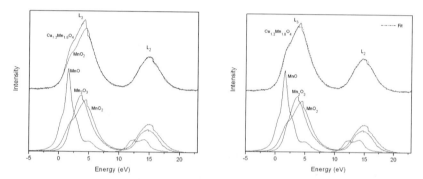

Figure 1. Comparison of the Mn $L_{2,3}$ edge from $Cu_{1.2}Mn_{1.8}O_4$ and the reference spectra MnO (Mn^{2+}), Mn_2O_3 (Mn^{3+}) and MnO_2 (Mn^{4+}) (a) and fit with a combination of the reference spectra (b).

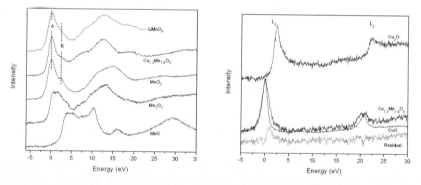

Figure 2. O-K edge from $Cu_{1.2}Mn_{1.8}O_4$ and the reference materials (a) and the Cu $L_{2,3}$ edge and reference CuO and Cu_2O (b).

EELS mapping of surface plasmons in star-shaped gold nanoparticles: morphological behaviour of optical properties from star to sphere

S. Mazzucco, O. Stéphan, M. Kociak, C. Colliex

Laboratoire de Physique des Solides, Univ. Paris-Sud, CNRS, UMR 8502, F-91405 Orsay Cedex, France.

mazzucco@lps.u-psud.fr

Keywords: STEM, EELS mapping, surface plasmons, gold nanoparticles, optical properties.

Surface plasmons (SPs) were discovered almost one century ago [1] but a first theoretical explanation was given only around 60 years later [2]. In the past 20 years a renewed interest in the study of their properties has come about, so that a wide range of applications and tools is now available to the scientific community [3, 4].

SPs can be viewed either as collective excitations of (valence/conductive) electrons at the surface of a material or as electromagnetic waves coupled to the surface itself and induced by an external electromagnetic source (photons or electrons). SPs are strongly influenced by the dielectric environment as well as by the shape and size of the involved nano-object. In the case of metals such as gold or silver, SPs take place at optical energies, so that they are particularly useful in sub-wavelength devices where classical optics cannot be applied.

The most used technique for spatially resolved SP analysis of nano-objects is the Scanning Near-field Optical Microscopy (SNOM) that can achieve a lateral resolution of $10 - 30$ *nm* [5], and this without spectral analysis. Electron Energy Loss Spectroscopy (EELS) in a Scanning TEM improves the lateral resolution down to 1 *nm* and thus allows to map SPs' local features on a single nanoparticle [6, 7]. Rastering the STEM electron probe over an area with a defined grid size we can collect a data set of EEL spectra at each pixel and then study the SP local (e.g. eigenstates) and global (e.g. energy and intensity) behaviour on the considered nano-object.

This work deals with the SP mapping of isolated star-shaped gold nanoparticles (nanostars). Our aim is to understand how SPs build up in such nanostructures when their morphology changes from a simple shape to a more complex one.

Different shapes have been taken into account to find out how SPs are modulated by the morphology of the particles. Passing from a star-like to a sphere-like particle, the SP features change from the tips to the core of the nanostars shifting from "intense red" to "soft blue" and coming close to the typical optical properties of a perfect sphere.

SP energy varies at the extremities of the particle from $\sim$1.7 *eV* to $\sim$2.0 *eV* depending on the roundness of the tips and the length of the branches while at the core its value is $\sim$2.2 *eV*, very close to the SP of a sphere (2.3 *eV*). The tip SP is much more intense than the core SP (three to four times stronger) showing that the electric field enhancement is more important far from the particle centre (Figure 1).

M. Luysberg, K. Tillmann, T. Weirich (Eds.): EMC 2008, Vol. 1: Instrumentation and Methods, pp. 409–410, DOI: 10.1007/978-3-540-85156-1_205, © Springer-Verlag Berlin Heidelberg 2008

410

These first EELS measurements for this kind of nanoparticles are in good agreement with theoretical predictions [8] and encourage us to proceed to furher investigations. A systematic study of the influence of the branch lenght and the tip roundness as well as the analysis of nanoparticles of different shapes could clarify the role played by morphology in the physics of surface plasmons.

1. R.W. Wood, Proc. Phys. Soc. of London **18** (1902) p. 269-275.
2. R A. Ferrel, Phys. Rev. **111** (1958) p. 1214-1222.
3. W.A. Murray and W. L. Barnes, Advanced Materials **19** (2007) p. 3771-3782.
4. W. L. Barnes, A. Dereux and T.W. Ebbesen, Nature **424** (2003) p. 824-830.
5. S. Ducourtieux, V.A. Podolskiy, S. Grésillon and S. Buil, Phys. Rev. B **64** (2001) p. 165403.
6. J. Nelayah, M. Kociak, O. Stephan, F.J. García de Abajo, M. Tencé, L. Henrard, D. Taverna, I. Pastoriza-Santos, L.M. Liz-Marzán and C. Colliex, Nat. Phys. **3** (2007) p. 348-353.
7. M. Bosman, V.J. Keast, M. Watanabe, A.I. Maaroof and M.B. Cortie, Nanotech. **18** (2007) p. 165505.
8. P.S. Kumar, I. Pastoriza-Santos, B. Rodríguez-González, F.J. García de Abajo and L.M. Liz-Marzán, Nanotech. **19** (2008) p. 15606.
9. The authors want to acknowledge the Colloid Chemistry Group of Vigo University led by Prof. Luis M. Liz-Marzán for providing the samples and Marcel Tencé of the STEM Group at the Laboratoire de Physique des Solides of Paris-Sud University for the technical support during the STEM – EELS experiments.

This research has been partly financed by the EU under the Pierre et Marie Curie contract "MEST CT 2004 514307".

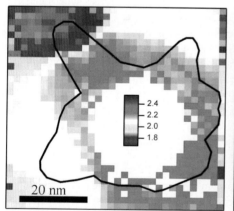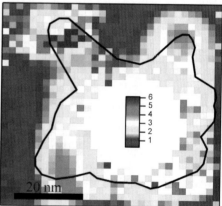

Figure 1. Surface plasmon energy (left) and intensity (right) map for a gold nanostar, the colour bar units are respectively *eV* and *a.u.*; the pixel size is 2x2 *nm²*.

Fast local determination of phases in Li$_x$FePO$_4$

P. Moreau[1], V. Mauchamp[2] and F. Boucher[1]

1. Institut des Matériaux Jean Rouxel (IMN), Université de Nantes – CNRS, 2 rue de la Houssinière, B.P. 32229, 44322 NANTES Cedex 3 FRANCE
2. PhyMAT (ex LMP) SP2MI – Boulevard 3, Téléport 2 – BP 30179 - 86962 FUTUROSCOPE, CHASSENEUIL CEDEX FRANCE

Philippe.Moreau@cnrs-imn.fr
Keywords: lithium batteries, electron energy loss spectroscopy, DFT

There is a renewed interest for the simulation of low-losses in Electron Energy-Loss Spectroscopy experiments (VEELS, "Valence" EELS). In particular, due to the increasing research on nanomaterials, finite size effects are carefully included in order to interpret the subtle changes observed in the spectra. Interface and surface plasmons, Cherenkov radiations, must be calculated [1, 2]. Simultaneously, remarkable improvements have been achieved in the calculation of dielectric functions of various compounds [3]. VEELS is chiefly applied to compounds for the semiconductor industry, nanocarbons, but hardly any example can be found in the lithium battery research field. Since the lithium battery community is increasingly considering nanosized active materials in the latest electrodes in order to improve the power behaviour of the electrochemical system, it is only natural to precisely study these materials with VEELS. Due to its cheap and non toxic nature, LiFePO$_4$ has received much attention, and we chose to concentrate our study on that compound [4, 5].

Samples were synthesized by the nitrate method [4] so that no carbon coating was present. Chemically delithiated compounds were obtained by using proper amounts of NO$_2$BF$_4$. Experiments were carried out on a H2000 Hitachi Transmission Electron Microscope at low temperature to avoid beam damages. A modified Gatan 666 spectrometer was used giving a 0.65 eV energy resolution. Experimental spectra were gain, dark count, detector and plural scattering corrected. First principle calculations were performed with WIEN2k [6], a Full potential Linearized Augmented Plane Wave (FLAPW) code, based on the Density Functional Theory (DFT). A spin polarized calculation is necessary and an antiferromagnetic order was introduced as previously done by Zhou et al [7]. The GGA+U approximation was used with a mean U$_{eff}$ value (4.3 eV) for both LiFePO$_4$ and FePO$_4$.

In Figure 1, experimental VEELS spectra for both compounds are strongly different. In particular a strong double peak is observed around 5 eV, fine structures are modified between 8 and 13 eV, and the plasmon shape in the 20-30 eV is also altered. All these differences are extremely well reproduced in the calculated spectra in Figure 2. Furthermore, a spectrum was acquired on a interface between LiFePO$_4$ and FePO$_4$ in a sample with global composition Li$_{0.6}$FePO$_4$. The shape of the peak around 5 eV is modified and an increase in intensity is observed around 13 and 30 eV. These observations are well reproduced in the interface simulated spectrum considering the standard formula Im $(-1/(\varepsilon_{LiFePO4}+\varepsilon_{FePO4}))$. First principles calculations are essential in

M. Luysberg, K. Tillmann, T. Weirich (Eds.): EMC 2008, Vol. 1: Instrumentation and Methods, pp. 411–412, DOI: 10.1007/978-3-540-85156-1_206, © Springer-Verlag Berlin Heidelberg 2008

order to avoid unreliable Kramers-Kronig analyses. Let us note that experimental spectra were acquired in a total time of 20 ms, somehow two order of magnitude less than high energy losses. This could prove an extremely valuable advantage to study the dynamics of lithium insertion in lithium battery compounds.

1. P. Moreau, and M. C. Cheynet, Ultramicroscopy **94** (2003) 293 .
2. R. Erni, S. Lazar, and N. D. Browning, Ultramicroscopy **108** (2008) 270 .
3. L. Reining *et al.*, Phys. Rev. Lett. **88** (2002) 066404 .
4. L. Laffont *et al.*, Chem. Mater. **18** (2006) 5520.
5. S. Miao *et al.*, J. Phys. Chem. A **111** (2007) 4242
6. P. Blaha et al. WIEN2K, An Augmented Plane Wave + Local Orbitals Program for Calculating Crystal Properties, K. Schwarz, Techn. Universität Wien, Austria, 2001.
7. Zhou F. *et al.*, Solid State Comm. **132** (2004) 181.

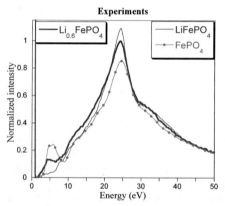

Figure 1. Comparison of experimental spectra for $LiFePO_4$ (thin line), $FePO_4$ (line with dots) and $Li_{0.6}FePO_4$ obtained on an interface between $LiFePO_4$ and $FePO_4$ (thick line).

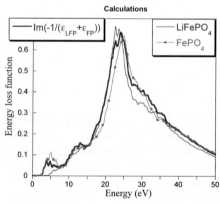

Figure 2. Comparison of calculated spectra for $LiFePO_4$ (thin line) and $FePO_4$ (line with dots) in the GGA+U. A simulation of an interface plasmon spectrum is also given when using the calculated dielectric functions of $LiFePO_4$ and $FePO_4$ (thick line).

EELS/EFTEM in life science: proof of the presence of H$_2$O$_2$ in human skin by Ce deposition in melanosomes

Elisabeth Müller[1], Miriam Droste[1], Katja Gläser[2] and Roger Wepf[1]

1. Electron Microscopy ETH Zürich (EMEZ), ETH Zürich, 8093 Zürich, Switzerland
2. Beiersdorf AG, 20253 Hamburg, Germany

elisabeth.mueller@emez.ethz.ch

Keywords: EELS, EFTEM, skin, life science, melanin degradation

The skin of humans has the ability to protect itself against UV irradiation of the sun by production and storage of melanin. While this complex is produced in the skin of the members of all human races, its quantity and distribution within the skin strongly differ. Humans with dark skin exhibit a melanin content in the skin, which is up to four times higher than that of humans with bright skin. But all have in common, that the melanin is degraded in the melanosomes during the final differentiation of keratinocytes in the epidermis. For humans with bright skin it might be an advantage to impede this process in order to increase the protection against damage by the solar UV-radiation. On the other side, people with dark skin might prefer to enhance the degradation process, since a brighter skin would imply social advantages, as e.g. in India. An understanding of the degradation of melanin in the skin is therefore of high interest.

Several degradation mechanisms of melanin were discussed so far: disintegration by treatment with alkaline pH or oxidation by potassium permanganate or hydrolysis with hydrogen iodide – all extremely unlikely in living cells [1, 2]. The only remaining possible degradation pathway consistent with the structure of the melanin oligomer as supposed by Clancy and Simon [3] points to oxidative degradation. A promising candidate is hydrogen peroxide capable of degrading melanin *in vitro* [4], and also known to arise in several cell types present in the skin [5]. NADPH oxidase, a membrane-associated, multi-subunit enzyme complex, is indicated as a possible source for hydrogen peroxide [1].

In order to investigate the localisation of the H$_2$O$_2$ production in the epidermis, the principle of cerium capture cytochemistry [6] was adapted to skin biopsies, i.e. Ce-chloride was added during the TEM-preparation of the skin biopsy. A fine electron dense precipitation is expected as a result from a reaction of H$_2$O$_2$ with Ce ions.

In the investigations with CLSM (confocal laser scanning microscopy) a dense staining was only observed along the plasma membranes and in the intercellular space. For the TEM investigation the resin sections were not stained with heavy metals, in order to avoid staining artefacts and to preserve the visibility of the cerium precipitates contrasting against the ultrastructure of the specimen. Precipitates appeared as fine dotted enhancement of specific structures - predominantly in the lower layers of the epidermis, along cell membranes, in membrane bound vesicles and frequently also in the intercellular space. Melanocyte dendrites demonstrated particularly distinct accumulation of the dark precipitates. Less dense accumulation of precipitate granules were frequently observed around, and even located directly on melanosomes and

M. Luysberg, K. Tillmann, T. Weirich (Eds.): EMC 2008, Vol. 1: Instrumentation and Methods, pp. 413–414, DOI: 10.1007/978-3-540-85156-1_207, © Springer-Verlag Berlin Heidelberg 2008

414

melanosome clusters. In the case of the latter ones, the precipitates were evenly distributed within the membrane bound complex (Figure 1).

In order to confirm the presence of cerium in the biopsies and to verify its localisation on or near melanosomes, energy filtered TEM as well as electron energy loss spectroscopy (in STEM mode) were applied. Since the atomic number of Ce atoms is very high compared to typical constituents of biological material, HAADF STEM (high angular annular dark field scanning transmission electron microscopy) was used additionally, in order to localize areas where heavy atoms were accumulated. EELS measurements proved that the areas containing melanosomes were characterized by a high accumulation of Ce as also confirmed by the high intensity in the HAADF STEM image (Figure 1). With EFTEM it could be shown that the Ce distribution is identical with the dark TEM-contrast of the melanosomes (Figure 2). H_2O_2 is therefore demonstrated to play an important role in the degradation of melanin in human skin.

1. J. Borovansky and M. Elleder, Pigment Cell Research 16 (2003), p. 280.
2. S. Ito and K. Wakamatsu, Pigment Cell Research 16 (2003), p. 523.
3. C.M. Clancy and J.D. Simon, Biochemistry 40 (2001), p. 13353.
4. W. Korytowski and T. Sarna, J. Biol. Chem. 265 (1990), p. 12410.
5. G.M. Bokoch and U.G. Knaus, Trends Biochem. Sci. 28(3) (2003), p. 502.
6. E.A. Ellis and M.B. Grant in "Oxidants and Antioxidants", ed. D. Armstrong, (Humana Press, Totowa NJ) (2002), p. 3.
7. We would like to thank Prof. Dr. St. Förster, Institute of Physical Chemistry, University of Hamburg, Germany, for his kind support of this work.

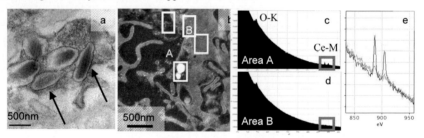

Figure 1. BF TEM image of melanosomes framed by precipitates (a), HAADF STEM image of a skin biopsy (b), EELS spectra taken from the bright area A representing the melanosomes (c) prove an accumulation of Ce in this area compared to the surrounding regions, e.g. area B (d); enlarged spectra of the Ce M-edge of areas A and B (e).

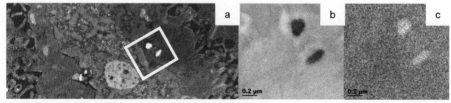

Figure 2. HAADF STEM image of a skin biopsy (a), BF image of the melanosomes marked by a frame (b), Ce elemental map of the same area (c).

Phase separation study of annealed SiOx films through energy filtered scanning transmission electron microscope

G. Nicotra[1], C. Bongiorno[1], C. Spinella[1], and E. Rimini[1,2]

1. Istituto per la Microelettronica e Microsistemi–CNR, Stradale Primosole 50, I-95121 Catania, Italy
2. Dipartimento di Fisica ed Astronomia, Università di Catania, via S. Sofia 64, I-95123

giuseppe.nicotra@imm.cnr.it

Keywords: STEM, EFTEM, EELS, SRO, Si QDs

Different physical and chemical methods can produce Si nanocomposite films, and most of them start with the deposition of silicon rich oxide (SRO) followed by high-temperature annealing [1].

Although much work has been published concerning this topic, results appear often to be discordant as far as annealing temperature crystallization time and amount of phase separation. The differences have to be ascribed by the preparation techniques for the starting material whom produce different results even at the same stoichiometry. A systematic study on the formation of Silicon nanoclusters formed in substoichiometric silicon oxide films by annealing at 1100 °C for 1/2 h as a function of two different deposition techniques, has been performed. The depositions were carried out by two different methods: a plasma enhanced chemical vapour deposition (PECVD) and a radio frequency (RF) magnetron sputtering. The samples were analyzed by energy filtered transmission electron microscopy (EFTEM), scanning transmission electron microscopy (STEM), Rutherford backscattering (RBS) and electron energy loss spectroscopy (EELS). We quantified the clustered silicon concentration in annealed substoichiometric silicon oxide layers deposited through PECVD by using an analytical methodology based on electron energy loss spectroscopy and energy-filtered transmission electron microscopy [2]. As far as the PECVD specimen, within our experimental accuracy, at any deposition condition the clustered silicon concentration is significantly lower than the initial silicon excess concentration, demonstrating that high temperature anneal of PECVD SiOx films induces only partially the phase separation between Si and SiO2. This behaviour is explained by taking into account the free energy difference between the metastable SiO and stable SiO2 phase and the strain energy associated with the different atomic densities of Si and SiO2 [3]. The Si nanoclusters remain embedded in a substoichiometric silicon oxide whose final composition strongly depends on the initial composition of the as-deposited homogeneous film. A further confirmation of this theory comes by the experimental results obtained on the layers deposited by RF magnetron sputtering, where we have obtained an almost complete phase separation, over 90%, at the same composition, the same annealing temperature, and the same annealing time. A detailed study of the host material, performed by series of line scan of a 2 Angstroms probe of a scanning transmission electron microscope (STEM), acquiring a series of valence electron energy loss spectra (VEELS) among

M. Luysberg, K. Tillmann, T. Weirich (Eds.): EMC 2008, Vol. 1: Instrumentation and Methods, pp. 415–416, DOI: 10.1007/978-3-540-85156-1_208, © Springer-Verlag Berlin Heidelberg 2008

agglomerated Silicon dots, demonstrates that the host material in the PECVD layers is still a substochiometric silicon oxide film whose excess of silicon is the one after to be subtracted by the agglomerated one "Figure 1a". On the contrary, the VEELS spectra acquired by line scan the STEM beam among agglomerated Silicon nanodots, on the layers by RF magnetron sputtering, show the presence of a stoichiometric SiO2 matrix "Figure 1b".

1. A.Meldrum, Rec. Res. Dev. Nucl. Phys. **1** (2004) 93
2. C. Spinella, C. Bongiorno, G. Nicotra, E. Rimini, A. Muscarà, and S. Coffa, Appl. Phys. Lett. **87**, 044102 (2005)
3. A. La Magna, G. Nicotra, C. Bongiorno, C. Spinella and E. Rimini Applied physics letters **90** (2007) 183101
4. G. Nicotra, C. Bongiorno, C. Spinella, and E. Rimini to be published Journal of Applied Physics

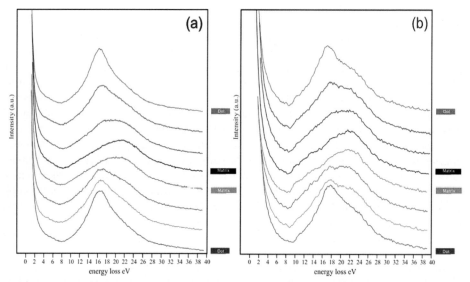

Figure 1. VEELS spectra obtained by linescan the STEM beam across two agglomerated Si Qds, respectively (a) on the PECVD layer and (b) on the layer obtained by RF magnetron sputtering. The labelled spectra are the ones in correspondence of the matrices and dots bulk, whose show that the PECVD specimen has a still substochiometric silicon oxide matrix whereas the sputtered ones shows a clear SiO2 spectrum, underlining a complete phase separation.

Phase Identification of Aluminium Oxide Phases by Analysis of the Electron Energy-loss Near Edge Structure

Daesung Park, Thomas E. Weirich and Joachim Mayer

Gemeinschaftslabor für Elektronenmikroskopie, RWTH Aachen, Ahornstr. 55, D-52074 Aachen, Germany

park@gfe.rwth-aachen.de

Keywords: aluminium oxides, ELNES, phase identification

Over the past decade, electron energy-loss spectroscopy (EELS) has become an increasingly important tool for characterising the chemical and electronic structure of materials on the nanometre scale [1]. In particular the characteristic fine structure beyond the onset of the core-loss ionization edges (the Energy-loss Near Edge Structure, ELNES) provides detailed insight into the coordination environment around the excited atom. Hence this feature can also be exploited by comparison with spectra of known structures to identify the material under investigation [2]. However, in practise one faces often the problem that spectra of the relevant reference materials are not readily available or cannot be obtained due to the lack of suitable samples. A promising way out of this quandary is provided by artificial core-loss spectra that can be calculated by quantum mechanical calculations from crystallographic data as input.

A challenging material system for such phase via ELNES analysis are the aluminium oxides that exist in various forms (θ, δ, γ, α, σ, τ, ε modification depending on the history of synthesis). So it was shown for example that multiple scattering and first-principles calculations of the Al-L$_{23}$ ELNES are in good agreement with experimental spectra and thus allow to distinguish between the prominent α- and γ-phase [3, 4]. In the present work we have picked up this idea and started to simulate the Al-L$_{23}$ and O-K ELNES for various alumina phases using the multiple scattering approach (FEFF 8.5 code [5]).

As first results we show here our calculation of the Al-L$_{23}$ and O-K ELNES for α- and γ-alumina in Figure 1 and Figure 2 respectly. In α-alumina all aluminium atoms are coordinated by six oxygen atoms, whereas in γ-alumina the aluminium atoms are surrounded by four and six oxygen atoms. The different local structure in the two phases is also reflected by the differences in the ELNES that can serve to discriminate between the two modifications. Similar calculations for the remaining aluminium oxide phases with known structure are in progress and will be compared with experimental ELNES spectra for verification.

1. R.F. Egerton, Electron Energy-Loss Spectroscopy in the Electron Microscope, 2nd ed. (plenum, New york, 1996)
2. R.Brydson, H.Sauer, and W. Engel in Transmission Electron Energy Loss Spectrometry in Material Science, edited by M. M. Disko, C. C. Ahn, and B. Fultz (The Minerals, Metals & Material Society, Warrendale, PA, 1992)

M. Luysberg, K. Tillmann, T. Weirich (Eds.): EMC 2008, Vol. 1: Instrumentation and Methods, pp. 417–418, DOI: 10.1007/978-3-540-85156-1_209, © Springer-Verlag Berlin Heidelberg 2008

418

3. Rik Brydson, Multiple scattering theory applied to ELNES of interfaces, Appl. Phys. **29** (1996) 1699-1708
4. Koji Kimoto, Kazuo Ishizuka, Teruyasu Mizoguchi, Isao Tanaka and Yoshio Matsui, Journal of Electron Microscopy **52(3)** (2003) 299 - 303
5. M. S. Moreno, K. Jorrissen, J.J. Rehr, Practical aspects of electron energy-loss spectroscopy (EELS) calculations using FEFF8, Micron **38** (2007), 1-11

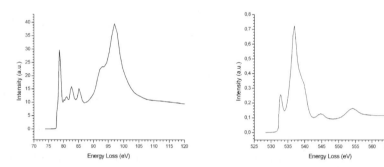

Figure 1. Al-L$_{23}$ (left) and O-K (right) ELNES of α-alumina calculated using FEFF 8.5.

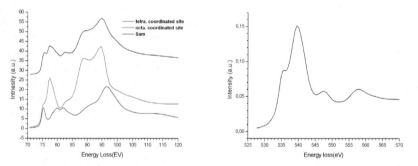

Figure 2. Al-L$_{23}$ (left) and O-K ELNES (right) of γ-alumina calculated using FEFF 8.5. Two different crystallographic sites were calculated and weighted by crystallographic multiplicity for Al-L$_{23}$ ELNES. The blue one is the tetrahedrally coordinated aluminium site and the red one is the octahedrally coordinated aluminium site.

Valence sensitivity of Fe-L$_{2,3}$ white-line ratios extracted from EELS

T. Riedl[1], R. Serra[1], L. Calmels[1] and V. Serin[1]

1. Centre d'Elaboration de Matériaux et d'Etudes Structurales, 29, rue Jeanne Marvig, BP 94347, F-31055 Toulouse Cedex 4, France

riedl@cemes.fr

Keywords: EELS, white-line ratio, valence sensitivity

Several authors have shown that the EELS white-line intensity ratio (WLR) of 3d transition metals can be used to measure the metal valency [1,2]. For the WLR extraction from EEL edge spectra various procedures have been proposed [2,3], including fitting and 2nd derivative methods. The aim of the present study is to evaluate quantitatively the influence of the extraction method on the differentiability of noisy Fe-L$_{2,3}$ white-line spectra belonging to two different valencies.

For that, three different types of WLR extraction procedures have been analyzed (Figure 1): The first one, after eventual smoothing of the original spectrum, integrates the L$_3$ and L$_2$ intensities in intervals of width u centred around the white-line maxima. The second group calculates the 1st derivative of the smoothed original spectrum, and evaluates either the areas in intervals of width u or the amplitudes of the derivative. In an analogous way the third group determines either the areas or the amplitudes of an averaged 2nd derivative of the original spectrum.

These procedures have been applied to two sets of 50 artificial spectra corresponding to experimental spectra of Fe metal and Fe$_2$O$_3$, respectively. The artificial spectra have been constructed using a simple model composed of two Lorentzians $L_2(E)$ and $L_3(E)$, and Gaussian random numbers $r(E)$ as noise:

$$I(E) = \alpha_2 \cdot L_2(E) + \alpha_3 \cdot L_3(E) + \beta \cdot r(E) \qquad (\alpha_2, \alpha_3, \beta: \text{real factors}).$$

Within each set the spectra differ only in the noise content, whereas the noise amplitude remains constant. Since the measurement of EEL edge spectra at high spatial resolution often involves poor edge signals, we have selected a low signal-to-noise ratio of ≈30. Height, width, and energy position of the Lorentzians have been chosen such to match experimental spectra of Fe metal and Fe$_2$O$_3$. In addition, the WLR has been determined of two reference spectra calculated for the same compounds, which were obtained by leaving out the noise component. We will discuss how the extraction method affects the resulting WLR, its statistical error (Figure 2), and thus the valence sensitivity of the procedure.

1. D.H. Pearson, C.C. Ahn and B. Fultz, Phys. Rev. B **47** (1993), p. 8471.
2. V. Stolojan, Dissertation University of Cambridge (2000).
3. G.A. Botton, C.C. Appel, A. Horsewell and W.M. Stobbs, J. Microsc. **180** (1995), p. 211.
4. We kindly acknowledge the European Union for support under the IP3 project ESTEEM (Enabling Science and Technology through European Electron Microscopy, IP3: 0260019).

M. Luysberg, K. Tillmann, T. Weirich (Eds.): EMC 2008, Vol. 1: Instrumentation and Methods, pp. 419–420, DOI: 10.1007/978-3-540-85156-1_210, © Springer-Verlag Berlin Heidelberg 2008

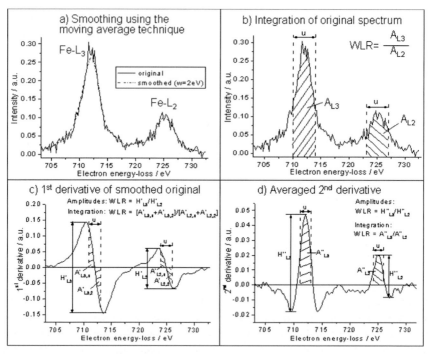

Figure 1. a) Smoothing; b-d) Different types of WLR extraction methods, applied to an artificial Fe-$L_{2,3}$ white-line spectrum.

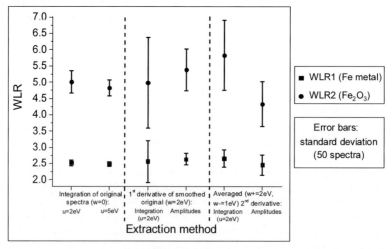

Figure 2. Fe-$L_{2,3}$ WLR and WLR standard deviation obtained for two sets of 50 artificial spectra using different extraction methods. w denotes the width for spectrum smoothing, w+ and w- the averaging widths used for calculation of the 2nd derivative, and u is the integration interval width.

Calculation of inelastic scattering events within second order QED – Implications of fully relativistic scattering

A. Rother

Institute of Structure Physics, Technische Universitaet Dresden, Germany

Axel.Rother@Triebenberg.de
Keywords: EELS, EMCD, QED

Inelastically scattered electrons carry a wealth of information about the specimen, thus are extensively explored through a variety of TEM techniques (EELS, ELNES, EFTEM, etc.). The standard interpretation of the data is based upon the calculation of scattering cross sections derived from non-relativistic QED electron-electron scattering. It was shown that relativistic effects are not negligible though [1]. In this work a fully relativistic formalism and explicit numerical calculations are presented. The following inelastic scattering events are considered (Fig. 1): 1. Electron-electron scattering, 2. Electron-core scattering, 3. Bremsstrahlung. The scattering matrix elements will be calculated by both Feynman diagram techniques and a fully relativistic version of the Yoshioka equations [2]. It is shown that both formalisms are equivalent within second-order fully relativistic QED. The numerical evaluation of the fully relativistic Yoshioka equations is however computationally advantageous in the case of electron-electron and electron-core scattering since both, the dimension and number of the integrals can be drastically reduced. Moreover an accurate and fast implementation into a fully relativistic version of the Multislice formalism is possible, additionally reducing the computation time and automatically taking into account preceding and succeeding elastic scattering events. The electronic states of the specimen electron are calculated within a fully relativistic Density Functional approach. Results on atomic shell excitation show the explicit influence of retarded 4-potentials, different radial dependency in the large and small component of the Kohn-Sham state and the influence of spin in inelastic scattering. The prospects of magnetic material characterization through spin polarized TEM will be discussed. Electron-core scattering and Bremsstrahlung start at very low energy transfers and occur over a wide energy range. We calculate single-ion and phonon excitations as a superposition of single ion excitations. The interaction radius of electron-core scattering with a low energy transfer shows, that inelastic events occur in vacuum regions, thus contributing to off-axis electron holographical contrast. The evaluation of the Bremsstrahlung is carried out analytically in the low energy transfer regime. The results show, that significant magnitudes occur at large reflection angles only (Fig. 2). The superposition of single Bremsstrahlung events can however lead to a more significant influence. An extension of the formalism to quasi-particle excitation (Plasmons, etc.) will be discussed.

1. B. Jouffrey, P. Schattschneider, C. Hebert, Ultramicroscopy **102**, 61-66 (2004)
2. H. Yoshioka, Journal of the Physical Society of Japan 12, 618 (1957)

M. Luysberg, K. Tillmann, T. Weirich (Eds.): EMC 2008, Vol. 1: Instrumentation and Methods, pp. 421–422, DOI: 10.1007/978-3-540-85156-1_211, © Springer-Verlag Berlin Heidelberg 2008

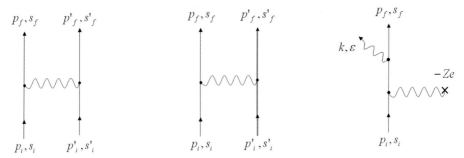

Figure 1. Feynman diagrams of electron-electron, electron-core and Bremsstrahlung scattering

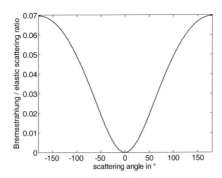

Figure 2. Angular depended magnitude of Bremsstrahlung

Role of asymmetries for EMCD sum rules

J. Rusz[1,2], H. Lidbaum[3], A. Liebig[1], P. Novák[2],
P. M. Oppeneer[1], O. Eriksson[1], K. Leifer[3], B. Hjörvarsson[1]

1. Department of Physics, Uppsala University, Box 530, S-751 21 Uppsala, Sweden
2. Institute of Physics, Academy of Sciences of the Czech Republic, Na Slovance 2,
CZ-182 21 Prague, Czech Republic
3. Department of Engineering Sciences, Uppsala University, Box 530, S-751 21
Uppsala, Sweden

jan.rusz@fysik.uu.se

Keywords: EELS, Magnetic circular dichroism, Dynamical diffraction theory

The recent derivation of sum rules for electron energy loss near edge spectra (ELNES) [1,2] provided the theoretical foundation for extracting spin and orbital magnetic moments from ELNES spectra that are measured under particular scattering conditions – a phenomenon named electron magnetic chiral dichroism (EMCD) [3]. This theory predicts that the ratio of orbital to spin magnetic moments can directly be obtained from EMCD spectra, when these are constructed as a difference from two symmetrical measurements.

We present a detailed computational study of the applicability of sum rules in several scattering geometries – two-beam case, three-beam case (Figures 1 and 2), general systematic row geometry and zone axis geometry. We study influences of different sorts of deviations, such as asymmetry due to the beam tilt in systematic row geometries or slight misorientations within a three-beam case and zone axis geometries [4].

Our results indicate that when operating the transmission electron microscope (TEM) in spectroscopic mode, the two-beam case geometry is the preferred choice. On the other hand, when acquiring energy filtered diffraction patterns, the three-beam case allows more accurate L/S extraction.

We support these results by our recent experiments performed in both two-beam case and three-beam case geometries [5].

1. Rusz J., Eriksson O., Novák P., Oppeneer P.M., Phys. Rev. B **76**, 060408 (2007)
2. Calmels L., Houdellier F., Warot-Fonrose B., Gatel Ch., Hÿtch M.J., Serin V., Snoeck E., Schattschneider P., Phys. Rev. B **76**, 060409 (2007)
3. Schattschneider P., Rubino S., Hébert C., Rusz J., Kuneš J., Novák P., Carlino E., Fabrizioli M., Panaccione G., Rossi G., Nature **441**, 486 (2006)
4. Rusz J., submitted
5. Lidbaum H., Rusz J., Liebig A., Hjörvarsson B., Oppeneer P.M., Coronell E., Eriksson O., Leifer K., submitted
6. This work was supported by STINT and Swedish Research Council.

M. Luysberg, K. Tillmann, T. Weirich (Eds.): EMC 2008, Vol. 1: Instrumentation and Methods, pp. 423–424, DOI: 10.1007/978-3-540-85156-1_212, © Springer-Verlag Berlin Heidelberg 2008

424

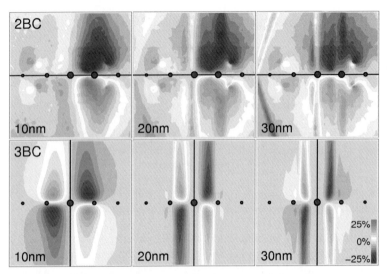

Figure 1. Relative difference maps at L_3 edge in two- and three-beam case geometries calculated for bcc iron at 300keV, **G**=(200), incoming beam direction (016). Blue circles indicate positions of Bragg spots on the systematic row, transmitted beam is in the center.

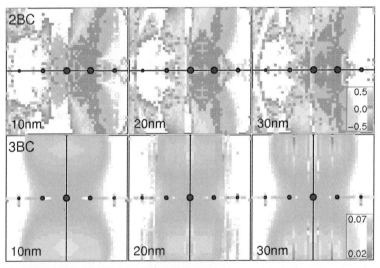

Figure 2. Apparent L/S maps constructed from difference maps shown in Fig.1 by applying the sum rules. Grey color in the 2BC maps denotes the area, where an accurate L/S extraction is possible.

Smart acquisition EELS

K. Sader[1,2*], P. Wang[1], A.L. Bleloch[1], A. Brown[2], R. Brydson[2]

1. SuperSTEM, Daresbury Laboratory, Warrington, WA4 4AD, UK
2. Institute for Materials Research, University of Leeds, LS2 9JT, UK

k.sader@leeds.ac.uk

Keywords: EELS, beam damage, low dose

Electron energy loss (EEL) spectroscopy and high angle annular dark field (HAADF) imaging in aberration-corrected electron microscopes are powerful techniques to determine the chemical composition and structure of materials at atomic resolution. Aberration correction enables the formation of angstrom sized electron probes that contain enough current for both low loss and near-edge core excitation spectroscopy to be performed. Therefore, spatially distributed EEL spectra can be collected and correlated with atomic resolution HAADF images.

However, the electron doses required for reasonable signal core loss spectra are high because of the small scattering cross section, causing electron beam induced chemical and structural changes of beam sensitive materials [1-3], such as organic and some inorganic materials [4].

One possibility to avoid artefacts is to sum the core loss signal collected from different areas of the same specimen, irradiating each with less than the critical dose [5] for damage. Current detection systems are limited in their ability to record very low dose EEL spectra by their read-out noise, which prevents the binning or arbitrary combination of "spectrum imaging" pixels. The other solution is to raster scan over similar areas of interest, while collecting a single spectrum, thus averaging the signal into one spectrum with a single read-out. This latter concept is not entirely new, with very similar ideas proposed by Hunt et al. [6] in relation to the collection of reference spectra of beam sensitive polymers. However, with the successful implementation of aberration correction, atomically resolved areas can now be defined and analysed making this concept worth revisiting.

Digital Micrograph (DM, Gatan) scripts have been written to utilize existing beam control commands to control the position of the electron probe on a VG HB501 cold field emission dedicated STEM equipped with a Nion quadrupole-octupole probe aberration corrector, an Enfina spectrometer, and a Digiscan beam control unit (SuperSTEM). Shapes to be scanned can be defined in a HAADF image from any combination of regions of interest (ROIs) in DM, or from binary image masks created by thresholding the image itself (Figure 1). This has allowed the averaging of the EEL signal over defined features with a minimum dwell time of currently 5ms at any one position. This dwell time will likely be decreased by streamlining the script or compiling in C. The potential benefits of this methodology include the ability to perform low dose, core loss spectroscopy from both beam sensitive materials in pristine condition or an array of similar but dispersed objects at high spatial resolution with high statistical accuracy. The latter could facilitate the study of spatially resolved features

M. Luysberg, K. Tillmann, T. Weirich (Eds.): EMC 2008, Vol. 1: Instrumentation and Methods, pp. 425–426, DOI: 10.1007/978-3-540-85156-1_213, © Springer-Verlag Berlin Heidelberg 2008

426

such as surfaces and internal interfaces and allow an averaged, statistically representative measurement to be made of such features (a task which is often difficult in TEM).

1. Hitchcock, A.P., et al., Comparison of NEXAFS microscopy and TEM-EELS for studies of soft matter. Micron, 2008. **39**(3): p. 311-319.
2. Rightor, E.G., et al., Spectromicroscopy of Poly(ethylene terephthalate): Comparison of Spectra and Radiation Damage Rates in X-ray Absorption and Electron Energy Loss. J. Phys. Chem. B, 1997. **101**(11): p. 1950-1960.
3. Yakovlev, S. and M. Libera, Dose-limited spectroscopic imaging of soft materials by low-loss EELS in the scanning transmission electron microscope. Micron. **In Press**.
4. Pan, Y., et al., Electron beam damage studies of synthetic 6-line ferrihydrite and ferritin molecule cores within a human liver biopsy. Micron, 2006. **37**(5): p. 403-11.
5. Egerton, R.F., P. Li, and M. Malac, *Radiation damage in the TEM and SEM.* Micron, 2004. **35**(6): p. 399-409.
6. Hunt, J.A., R. Leapman, and D.B. Williams, *Low-dose EELS and Imaging Strategies in the STEM*, in *Microbeam Analysis*. 1993

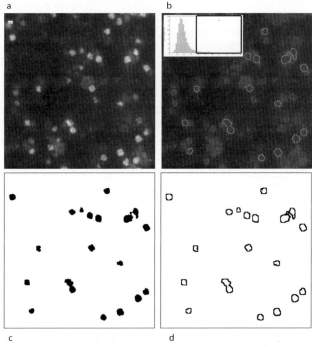

Figure 1. a) HAADF image of ferritin molecules (the iron oxyhydroxide mineral cores of the molecules have strong white contrast). b) Image thresholded in DM, with histogram of pixel intensities inset (black box showing pixel intensities). c) Binary pixel mask created from thresholded image for collecting EEL spectra from several particles. d) Binary pixel mask containing only the outline of particles for collecting EEL spectra from the projected surfaces. Image dimensions are 150nm by 150 nm.

EELS fine structure tomography using spectrum imaging

Z. Saghi, X. Xu and G. Möbus

Dept. of Eng. Materials, University of Sheffield, Mappin Street, Sheffield, S1 3JD, UK.

z.saghi@sheffield.ac.uk

Keywords: precipitates, cerium oxide, tomography, spectrum imaging, EELS, ELNES

Electron energy loss spectroscopy fine structure (EELS/ELNES) enables to measure with high spatial resolution not only chemical composition but also changes in element valence (oxidation state) [1-2]. A particularly sensitive relationship exists between the ratio of the double-M-edge-lines for lanthanide elements and their bivalent oxidation state (+III versus +IV). In earlier work we discovered that Ce can coexist in two different valences in one glass composite material, adopting mostly +III dissolved in the glass matrix while switching to +IV inside precipitates, which form as nano-dendrites of pure CeO_2 upon exceeding the total solubility of Ce in the particular alkali-boro-silicate glass described elsewhere [3]. The glass nanocomposite as a whole and its individual particles have been found highly suitable for tomographic reconstruction including quantitative post-processing [4,5].

We combine in this study our earlier work on ELNES spectrum imaging (SI) across a particle/matrix cross-section [3] with our work on tomographic reconstruction of glasses [4] to achieve EELS fine structure tomograms of cross-sections through the material perpendicular to the specimen holder axis. By ratio-imaging of the two tomograms corresponding to the energies of the M5 and M4 Ce-lines, we finally compute a chemical valence tomogram.

Data were acquired on a JEM 2010F + GIF instrument, the EELS interval was selected to include O-K to Ce-M-edge and exponential background subtraction (BG) was applied using Gatan SI+EELS software. A tomographic tilt series over 120 degrees total angular range with 10 degrees increment was acquired, including one spectrum image plus one ADF-STEM image (for assisting in alignment) for every angle (Figure 1). Reconstruction by filtered backprojection was applied to the BG-subtracted SI files after alignment in the single spatial dimension (Figure 2). For the ratio images a sum of 10 tomogram slices was used to estimate peak integration as a first approximation.

The result in Figure 2(c) shows the right jump in M_5/M_4 ratio corresponding to our earlier findings of valence in glass and matrix. Our method introduces a new type of EELS data cube, formed by $(y,\Theta,\Delta E)$ and backprojected into $(y,z,\Delta E)$ instead of the normal SI and EFTEM data cubes formed by $x,y,\Delta E$ (z= beam direction, x= holder direction, y=perpendicular, Θ = tilt angle) and allows for the first time to "look inside" a material for inner shell quantitative analysis. The method complements earlier spectroscopic tomography on a 4D space $(x,y,z,\Delta E)$ by EFTEM [6] and EDX [7], with only a few ΔE values pre-selected. Since one M_5/M_4 ratio is measured in every voxel $(\Delta x, \Delta y, \Delta z)$, thickness is no longer a parameter in the final data cube. [8]

M. Luysberg, K. Tillmann, T. Weirich (Eds.): EMC 2008, Vol. 1: Instrumentation and Methods, pp. 427–428, DOI: 10.1007/978-3-540-85156-1_214, © Springer-Verlag Berlin Heidelberg 2008

428

1. C. Jeanguillaume, and C. Colliex, Ultramicroscopy **28** (1989), p. 252.
2. J.A. Fortner, E.C. Buck, A.J.G. Ellison and J.K. Bates, Ultramicroscopy **67** (1997), p. 77.
3. G. Yang, G. Möbus and R.J. Hand, Micron **37** (2006), p.433.
4. X. Xu, G. Yang, Z. Saghi, Y. Peng, R. Gay and G. Möbus, Materials Research Society Symposium Proceedings **928E** (2007), 0982-KK02-04.
5. Z. Saghi, X. Xu and G. Möbus, this conference (2008).
6. G. Möbus and B.J. Inkson, Applied Physics Letters **79** (2001), p.1369.
7. G. Möbus, R.D. Doole and B.J. Inkson, Ultramicroscopy **96** (2003), p. 433.
8. This work was supported by EPSRC under grant number GR/S85689/01.

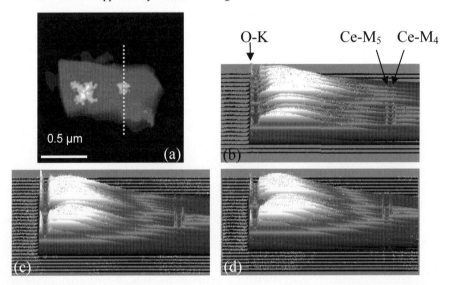

Figure 1. (a) ADF-STEM image showing the location of the embedded CeO2 nanoparticle inside a glass fragment; the Spectrum Image tilt series were acquired along the yellow line (20 points); (b-d) -40°, 0° and +30° background subtracted spectrum images, showing the O K-edge and Ce $M_{5,4}$ edges.

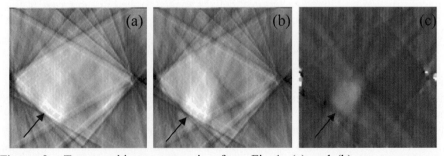

Figure 2. Tomographic reconstruction from Fig 1: (a) and (b) are two y-z cross-sections perpendicular to the energy axis of the 3D data cube and show the distribution of Ce M_5 and M_4 edges respectively. (c) shows the ratio M_4/M_5 between Ce M_5 and Ce M_4 signals, with a value ~1.15 inside the CeO_2 particle and ~0.95 in the matrix.

Shift in electron energy loss compared for different nickel silicides in a Pt alloyed thin film

M. Falke[1], T. Schaarschmidt[1], H. Schletter[1], R. Jelitzki[1], S. Schulze[1], G. Beddies[1], M. Hietschold[1], M. MacKenzie[2], A.J. Craven[2], A. Bleloch[3]

1. Institute of Physics, Chemnitz University of Technology, 09107 Chemnitz, Germany
2. Dept. of Physics & Astronomy, University of Glasgow, Glasgow G12 8QQ, U.K.
3. SuperSTEM Laboratory, Daresbury, Cheshire, WA4 4AD, U.K.

thomas.schaarschmidt@s2002.tu-chemnitz.de
Keywords: silicides, STEM, EELS

Nickel monosilicide (NiSi) is a promising material for electrical contacts and interconnects in the latest generation of CMOS devices [1]. Despite already being used in certain industrial applications, there is a lack of information, especially about chemical or structural variations in nanometer scale compounds. With continuing miniaturisation, knowledge of such variations is crucial for successful application. Another important aspect is the thermal stability of the low-resistivity NiSi phase which generally changes into the disilicide ($NiSi_2$) at a temperature of 700°C. This phase transformation can be shifted towards higher temperatures by alloying with Pt [2].

Ni films were sputtered together with Pt onto a Si(001) substrate with varying Pt concentrations and subsequently annealed at selected temperatures. The resulting Ni-Pt-Si layers were investigated and it was found that a platinum concentration of approx. 8% was sufficient to avoid the formation of $NiSi_2$ at a temperature of 900°C.

Layers with a lower Pt concentration were examined by energy dispersive X-ray spectroscopy (EDX) and electron backscattering diffraction (EBSD) to study their composition. A case, in which NiSi and $NiSi_2$ coexisted, as seen in Figure 1, was chosen for electron energy loss spectroscopy (EELS). Recent work [3] has shown that it is possible to distinguish between different pure nickel silicide phases because of a shift of the Ni-$L_{2,3}$ edge as well as a shift of the plasmon peak in EELS spectra. Using the so-called Dual EELS, with a fast beam switch [4] to allow a nearly parallel collection of low and core loss spectra, a similar edge shift could be observed for the Ni-$L_{2,3}$ edge in the Pt alloyed thin films at hand. However, the observed shift was smaller than it had been reported for the pure phases (Figure 2). Plasmon shifts were investigated as well. Finally, high angle annular dark field contrast at superSTEM 1, an abberation corrected dedicated STEM, revealed the position of Pt within the alloy on atomic scale.

1. C. Lavoie, F.M. D'Heurle, C. Detavernier, C. Cabral, Microelectron. Eng. 70 (2003), 144
2. D. Mangelinck, J.Y. Dai, J.S. Pan, S.K. Lahiri, Appl. Phys. Lett. 75 (1999), 12
3. M.C. Cheynet, R. Pantel, Micron 37 (2006), 377
4. J. Scott, P.J. Thomas, M. MacKenzie, S. McFadzean, J. Wilbrink, A.J. Craven and W.A.P. Nicholson, submitted to Ultramicroscopy (2008)
5. We kindly acknowledge the financial support by the DAAD; project D/07/09995.

M. Luysberg, K. Tillmann, T. Weirich (Eds.): EMC 2008, Vol. 1: Instrumentation and Methods, pp. 429–430, DOI: 10.1007/978-3-540-85156-1_215, © Springer-Verlag Berlin Heidelberg 2008

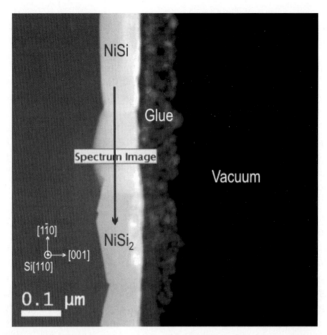

Figure 1. High angle annular dark field (HAADF) image of a silicide film on the Si(001) substrate in Si[110] zone axis. NiSi forms a wave like interface with Si and appears bright in HAADF because of its higher metal content. The less bright disilicide forms a sharper interface with silicon. In the middle of the line scan a nickel monosilicide and a disilicide crystallite overlap.

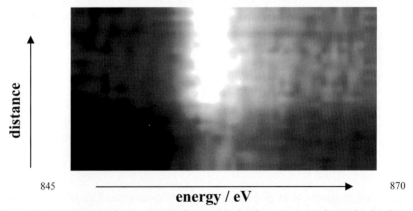

Figure 2. The image shows EELS data with the loss energy along the horizontal axis, while the vertical axis represents the real space location along the line indicated in Figure 1. The energy range around the Ni-$L_{2,3}$ edge is displayed and a peak shift of less than 1eV upon transition from the NiSi (above) into the NiSi$_2$ (below) can be observed.

Distribution of Fe and In dopants in ZnO: A combined EELS/EDS analysis

H. Schmid and W. Mader

Inst. Inorg. Chemistry, University of Bonn, Römerstr. 164, 53117 Bonn, Germany

herbert.schmid@uni-bonn.de

Keywords: EELS, EDS Spectroscopic Imaging, ZnO

Zinc oxide (ZnO) is a highly versatile material with unique combinations of optical, electronic and piezoelectric properties which can be controlled by the addition of Fe or In. Undoped ZnO crystallizes in the non-centrosymmetric wurtzite structure with alternating (0002) layers of tetrahedrally coordinated O^{2-} and Zn^{2+} stacked along the c axis. In the present investigation the distribution of dopants in ZnO is studied by a combined quantitative EELS/EDS analysis. In ZnO doped with Fe^{3+} or In^{3+} a characteristic inversion domain structure with planar inversion domain boundaries (IDBs) is observed on (0001) planes (basal IDBs) and $\{2\,\bar{1}\,\bar{1}\,5\}$ planes (pyramidal IDBs), respectively (Fig.1). The number of IDBs is directly correlated to the local dopant concentration; quasi-periodic structures are observed at dopant concentrations ≥ 5 at.% of cations (Fig.2). The (0002) lattice planes are well resolved, albeit severely distorted in the vicinity of IDBs in Fe-ZnO. Elemental mapping in EFTEM indicates that dopants are essentially located within the IDBs (Fig.3), however, it does not allow for a quantitative assessment of dopant concentrations [1]. EEL spectra were acquired with high spatial resolution in diffraction/nanoprobe mode. Regions analyzed by EELS measurements are indicated in Fig.3a (open circles). Quantitative measurements, corrected for O-K EXELFS oscillations [2], yielded a solid-solubility < 0.4 at.% Fe in unaffected ZnO domains (Z_{SS}), whereas Fe is depleted in inverted domains (Z_0). The Fe content in single basal IDBs was measured by EELS using the variable beam diameter method [1], yielding an effective boundary thickness $\delta \approx 0.27$ nm, corresponding to one close-packed monolayer of Fe^{3+}. The trivalent oxidation state was confirmed by Fe-L ELNES analysis, utilizing the valence-sensitive L_3/L_2 white-line intensity ratio [3].

These results were verified by EDS measurements in STEM Spectroscopic Imaging by X-rays (SIX). Sample areas containing IDBs in edge-on orientation were scanned and analyzed by SIX, yielding a quantitative correlation between the ratio of (0002) cation layers in IDBs and ZnO domains (dopant : Zn ratio) and corresponding dopant concentration within the scanned region as measured by EDS (Fig.4). SIX measurements in sintered Fe-ZnO and In-ZnO confirmed that both types of IDBs contain one monolayer of dopant ions, whereas measurements in In-ZnO nanorods grown by the VLS mechanism yielded indium concentrations equivalent to two monolayes per IDB.

1. O. Köster-Scherger, H. Schmid et al., J. Am. Ceram. Soc., **90** (2007) 3984.
2. H. Schmid and W. Mader, Proc. IMC 16, Sapporo/Japan (2006) 829.
3. H. Schmid and W. Mader, Micron, **37** (2006) 426.

M. Luysberg, K. Tillmann, T. Weirich (Eds.): EMC 2008, Vol. 1: Instrumentation and Methods, pp. 431–432, DOI: 10.1007/978-3-540-85156-1_216, © Springer-Verlag Berlin Heidelberg 2008

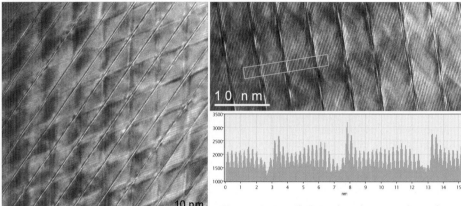

Figure 1. Basal and pyramidal inversion domain boundaries in Fe doped ZnO.

Figure 2. Basal IDBs in edge-on orientation (Z.A. [1 $\bar{1}$ 0 0]); (0002) lattice planes are distorted at IDBs. Profile shows (0002) fringe contrasts according to cation stacking.

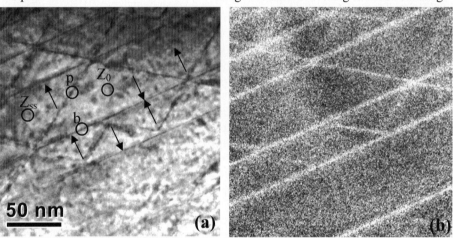

Figure 3. Inversion domain structure in Fe-ZnO: (a) TEM-BF image, Z.A. [2 $\bar{1}$ $\bar{1}$ 0]. Regions analyzed by EELS are indicated; c-axis orientations are indicated by arrows. (b) Iron distribution shown qualitatively in Fe jump-ratio map of same sample region.

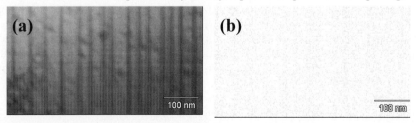

Figure 4. Inversion domain structure in In-ZnO: (a) STEM-BF image showing basal IDBs; (Z.A. [1 $\bar{1}$ 0 0]). (b) In L-mapping by spectroscopic imaging with X-rays (SIX).

Changes in the Soot Microstructure during Combustion studied by SEM, TEM, Raman and EELS

M.E. Schuster[1], M. Knauer[2], N.P. Ivleva[2], R. Niessner[2], D.S. Su[1], and R. Schlögl[1]

1. Fritz-Haber Institut der Max-Planck Gesellschaft, Faradayweg 4-6, 14195 Berlin, Germany
2. Institute of Hydrochemistry, Technische Universität München, Marchioninistr. 17, 81377 Munich, Germany

manfred@fhi-berlin.mpg.de
Keywords: soot formation, TEM, EELS, Raman, NEXAFS

The aim of this work is the understanding of the changes in the structure of soot during the combustion process. Soot formation in the combustion process of diesel engines and changes of structure during after-treatment are important topics because of the risk of penetration of the soot particles into the lungs and harmful interference with the human body. The oxidation behaviour is therefore an important issue for the elimination of soot in the after treatment process.

The soot will be investigated at various oxidation states to make the changes in the combustion of the soot visible. These changes in the structure will be important to understand how the soot has to be modified to prevent it from interfering with the human body.

In the project we investigate the combustion process of different carbon based materials. In particular we compare Euro4 and Euro6 soot with GfG (spark discharge generator) soot, SWCNTs, MWCNTs and carbon black.

The SEM images of Euro4 soot in Figure 1 show significant difference in the shape of the secondary soot agglomerates. Whereas the unheated sample has a spherical shape with big holes inside, the temperature treated one looks more directly connected without any spaces at this scale.

Preliminary TEM investigations of Euro4 soot already show a difference between the samples at different heating stages as seen in Figure 2. The sample on the right side was heated up in a 5% O_2 flow in N_2 with a heating rate of 5K/min to a final temperature of 773 K.

It can already be seen in Figure 2 that the graphene layers are broken on the right sample compared to the left sample which indicates that the heat treatment has already an effect on the structure of the soot particles.

Furthermore the magic angle approach will be used to investigate the C-K edge in the samples. These special settings will be used to compensate the anisotropy effects that occur due to different orientation between incident electron and sample and therefore it will be possible to determine the sp^2/sp^3 ratio. The results of these measurements will be correlated to Raman measurements, which could, according to [1, 2], give information about the bonding structure.

M. Luysberg, K. Tillmann, T. Weirich (Eds.): EMC 2008, Vol. 1: Instrumentation and Methods, pp. 433–434, DOI: 10.1007/978-3-540-85156-1_217, © Springer-Verlag Berlin Heidelberg 2008

434

The origin of these differences is the presence of PAH (polycyclic aromatic hydrocarbons) in the untreated sample which makes it hydrophobic. Due to heating up to 773 K and therefore oxidation of the PAH the sample becomes hydrophilic. This is seen as the reason for the change of the shape of the secondary agglomerate size.

Parallel to that we will also do a heating slope at the synchrotron beamline (BESSY) which gives the possibility to obtain information from NEXAFS about the oxygen groups on the surface of the samples.

1. A. Sadezky, H. Muckenhuber, H. Grothe, R. Niessner, U. Pöschl Carbon 43 (2005) 1731-1742.
2. N. P. Ivleva, A. Messerer, X. Yang, R. Niessner, U. Pöschl, Environ. Sci. Technol., 41 (10), 2007, 3702 -3707.
3. The support for this work by DFG grant is kindly acknowledged.
4. MAN is kindly acknowledged for providing the Euro4/6 samples

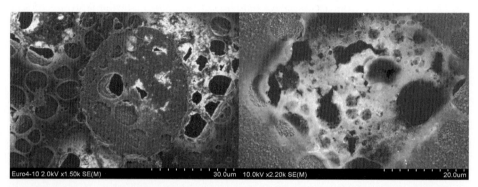

Figure 1. SEM micrograph of an Euro4 sample; left graph untreated, right graph heated up to 773 K

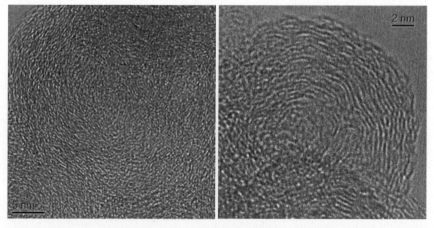

Figure 2. High-resolution electron micrograph of an Euro4 sample; left image untreated; in the right image the sample has been heated up to 773 K.

EELS and EFTEM-investigations of aluminum alloy 6016 concerning the elements Al, Si and Mg

S. Schwarz[1], M. Stöger-Pollach[1]

1. University Service Centre for Transmission Electron Microscopy, Vienna University of Technology, Wiedner Hauptstraße 8-10/137, A-1040 Vienna, Austria

schwarz@ustem.tuwien.ac.at

Keywords: Aluminum alloy, EELS, EFTEM

Aluminum alloys are one of the most promising materials concerning high hardness and low weight (important for the use in automotive engineering) and consequently low cost in operation (low fuel consumption), and comparatively low costs in fabrication. For the investigations done in this work, the aluminum alloy 6016 containing Mg and Si as major elemental additions, have been taken into account. The precipitation sequence in Al-Mg-Si alloys consists of: Al (SSS) – clusters of Si atoms, clusters of Mg atoms – dissolution of Mg clusters – formation of Mg/Si co-clusters – small precipitates of unknown structure - Beta″ precipitates – B′ and Beta″ precipitates - Beta(Mg2Si) precipitates [1].

For microstructural investigation, transmission electron microscopy (TEM) is a powerful tool both for phase analysis (e.g. by electron diffraction patterns and high resolution (HR) TEM) and for chemical analysis (e.g. by energy dispersive x-ray (EDX) analysis, electron energy loss spectroscopy (EELS) and energy filtered (EF) TEM).

The Mg-Si-rich precipitates of an AA6016 alloy with 2 different heat treatments (one sample as received + 400°C/1h, and the other sample heat treatment T6) have been investigated by EELS and EFTEM. The sample thickness for the EFTEM investigation was 0.7 mean free path; the Al-L and Si-L edge were used for analysis. The TEM samples have been prepared by electropolishing described in [2].

In the first step, the EELS spectra of the edges for Al (matrix), Mg and Si (precipitates) have been recorded and analysed. For Al and Si the L-edges could have been taken for the EFTEM picture (see Fig. 1), whereas for Mg only the K-edge could be taken into account. The Al and Si maps are shown in Fig. 1. The L-edge of Mg coincides with the Al-plasmons in the lower energy range, so for Mg only the K-edge could be used for mapping Mg concentrations (see Fig.2).

It was shown that EELS/EFTEM is a powerful tool for the investigation of Mg-Si-rich precipitates in Al matrix (here AA6016) for the size and distribution of precipitates.

1. G.A. Edwards, K. Stiller, G.L. Dunlop, Acta mater. Vol. 46 No.11 (1998), pp. 3893-3904.
2. A.K. Gupta, D.J. Lloyd, S.A. Court, Materials Science and Engineering A301 (2001), pp. 140-146.
3. This work was supported by the Hochschul-Jubiläumsstiftung der Stadt Wien, grant No. H-01585/2007, and the Institut für Werkstoffwissenschaft und Werkstofftechnologie of the Vienna University of Technology.

M. Luysberg, K. Tillmann, T. Weirich (Eds.): EMC 2008, Vol. 1: Instrumentation and Methods, pp. 435–436, DOI: 10.1007/978-3-540-85156-1_218, © Springer-Verlag Berlin Heidelberg 2008

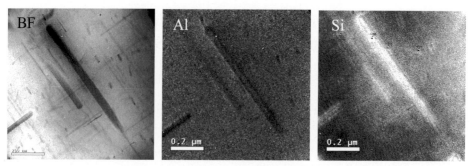

Figure 1. TEM bright field image and EFTEM images of the Al matrix and the precipitation Si of AA6016 as received + 400°C/1h.

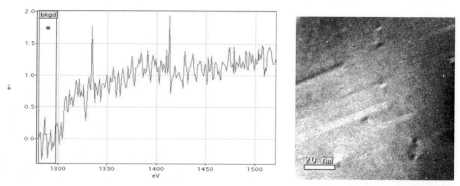

Figure 2. EELS spectrum and EFTEM-image of AA6016 -heat treatment T6- of the Mg-rich precipitates (Mg K edge at 1305 eV).

EELS modelling using a pseudopotential DFT code

C.R. Seabourne, A.J. Scott, R. Brydson

Institute for Materials Research, School of Process, Environmental and Materials Engineering, University of Leeds, LS2 9JT, United Kingdom.

chmcrs@leeds.ac.uk

Keywords: Electron energy loss spectroscopy (EELS), ELNES, density functional theory (DFT), CASTEP, TEM

EELS, when performed in the (S)TEM, enables chemical state and coordination information to be obtained with a potential spatial resolution on the atomic scale [1].

Good agreement between experiment and theory has been achieved utilising both multiple scattering (FEFF) and *all-electron* DFT (WIEN2k) based methods [2]. CASTEP is a *pseudopotential* DFT code. Pseudopotential codes consider only valence electronic states thereby reducing the calculation time and making the study of more complex systems possible [3]. However, ignoring the core electronic states means that approximating for the core hole is difficult.

Recent additions to CASTEP have been made that allow EELS spectra to be calculated for a specific atom in a unit cell, using an on-the-fly (OTF) pseudopotential where we can specify exactly the *full* electronic configuration. Therefore, the code can now be used to effectively simulate the core-hole [4].

We have been used this new code, CASTEP-EELS, to reproduce results obtained for group IV carbides. We have further recognised the significance of the correct treatment of broadening in theoretical EELS. It is common to account for just instrumental and core state broadening, by applying an energy independent numerical function; this can lead to simulated peaks high above the edge onset (>20 eV) being too intense as compared to experiment. Therefore, we have added an energy-dependent final state lifetime broadening function, as reported by Moreau et al. [5].

Figure 1 compares calculated and experimental carbon K-edges from ZrC. It shows there is considerable potential for the code, particularly when combined with the Moreau et al broadening model. We have also investigated the significance of parameter selection in obtaining reliable, converged spectra. Figure 2 depicts the comparative significance of k-point mesh and energy cut-off selection. k-point selection appears to be the key consideration.

1. V.J. Keast, A.J. Scott, R. Brydson, D.B. Williams, J. Bruley, J. Microscopy **203** (2001), p135
2. A.J. Scott, R. Brydson, M. MacKenzie, A.J. Craven, Phys. Rev. B. **63** (2001), p.245105.
3. S.J. Clark, M.D. Segall, C.J. Pickard, P.J. Hasnip, M.J. Probert, K. Refson, M.C. Payne, Zeitschrift für Kristallographie **220** (2005), p.567.
4. S.P Gao, C.J. Pickard, M. C. Payne, J. Zhu, J. Yuan, Phys. Rev. B. **77** (2008), p. 115122.
5. P. Moreau, F. Boucher, G. Goglio, D. Foy, V. Mauchamp, G. Ouvrard, Phys. Rev. B. **73** (2006), p.195111.

M. Luysberg, K. Tillmann, T. Weirich (Eds.): EMC 2008, Vol. 1: Instrumentation and Methods, pp. 437–438, DOI: 10.1007/978-3-540-85156-1_219, © Springer-Verlag Berlin Heidelberg 2008

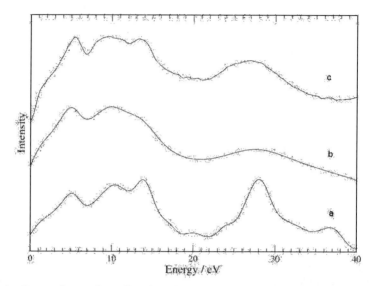

Figure 1. Comparison of predicted and experimental carbon K-edges for zirconium carbide. (a) CASTEP-EELS with numerical broadenings (0.4 Lorentzian, 0.8 Gaussian), (b) CASTEP-EELS with additional energy-dependent final state lifetime broadening, (c) experimental data (VG HB5 cold FEG STEM, 0.3 eV energy resolution).

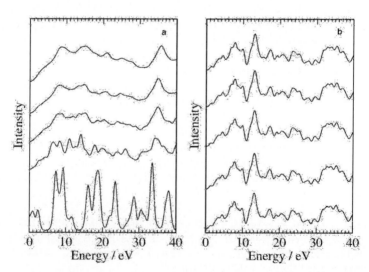

Figure 2. Unbroadened, theoretical aluminium metal K-edge spectra, with a single electron core-hole. (a) investigation of the significance of *k*-point grid selection, with increasing numbers of *k*-points used from bottom to top. (b) variation with kinetic energy cut-off, increasing in value from bottom to top.

The EELS spectrum database

Thierry Sikora[1] and Virginie Serin[2]

1. CEMES-CNRS, 29 rue Jeanne Marvig, 31055 Toulouse, France
2. Savantic AB, Rosenlundsgatan 50, Stockholm, Sweden, and LPS-CNRS, Université Paris-Sud, 91405 Orsay Cédex France.

serin@cemes.fr
Keywords: EELS, Materials Science

The EELS database is a public interactive consultable web repository of outer-shell and inner-shell excitation spectra from Electron Energy Loss Spectroscopy and X-Ray experiments, which forms a reference catalog of fine structures for materials. Each spectrum is available with a full set of recording parameters providing a complete overview of the working conditions. The database must also be seen as a research tool for EEL spectroscopists, theoreticians, students, or private firms and a central "location" for the growing EELS community.

Contributions are made on a voluntary basis. Once registered, the user can easily search, compare, download or submit spectra directly via web forms. As in a scientific journal, a system of referees examines the submitted data prior to their public availability to ensure quality. As of today the database contains more than 200 spectra, while more than 1000 people involving all known EELS institutions worldwide are registered in the database.

In this communication, the latest improvements to the database are presented together with the future projects.

1. http://www.eelsdatabase.net and eelsdb@cemes.fr
2. T.S. would like to thank the EC, ESTEEM Contract number 0260019, for funding. The authors would like to thank Christian Colliex for supporting and promoting the EELS database. The authors warmly acknowledge the EELS labs that recently contributed to the database.

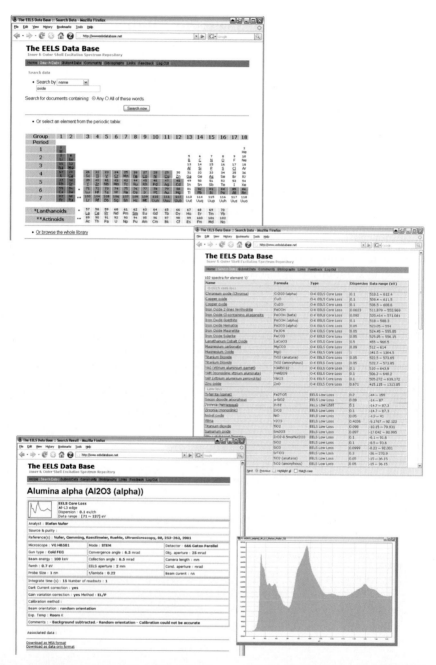

Figure 1. Figures extracted from the http://www.eelsdatabase.net showing a list of some data available in the database, and example of the AlL$_{2,3}$ spectrum in αAl$_2$O$_3$ (Courtesy to Stefan Nufer).

STEM-EELS analysis of interface magnetic moments in Fe(100)/Co(bcc) superlattices

R. Serra[1], L. Calmels[1], V. Serin[1], B. Warot-Fonrose[1], S. Andrieu[2]

1. CEMES-CNRS, 29 rue Jeanne Marvig, BP 94347, 31055 Toulouse Cedex 4
2. LPM, Université Henri Poincaré, Bld des Aiguillettes, BP 239, 54506 Vandoeuvre les Nancy

rserra@cemes.fr

Keywords: EELS, white-line ratio $I(L_3)/I(L_2)$, Iron $L_{2,3}$ edges

Information on the magnetic moment can be obtained experimentally by Electron Energy Loss Spectroscopy (EELS) at the $L_{2,3}$ edge of iron, since the moment depends upon the Fe-$3d$ band occupancy [1,2]. Experimental works have for instance shown that the Fe-$I(L_3)/I(L_2)$ ratio is directly correlated to the Fe magnetic moment for an important number of compounds and alloys containing iron [2,3,4,5,6]. In this contribution, we focus on the magnetic moment modification induced by the interfaces of a thin iron layer. Local and averaged values of the $I(L_3)/I(L_2)$ ratio have been measured in cross sectional sample, and for several Fe(n atomic layers)/Co(3 atomic layers) *bcc* superlattices, with $n=3,5,10,20$, grown by Molecular Beam Epitaxy (MBE) on a single crystal Fe(100) buffer, see Fig. 1. The samples differ one from the others by the iron layer thickness. The Fe-$I(L_3)/I(L_2)$ ratio measured in the superlattices is higher than the Fe buffer value and increases when the number of atomic planes in the thin iron layer of the Fe/Co superlattice decreases. This can be clearly seen in Fig. 2. These variations of the $L_{2,3}$ ratio can be interpreted in term of an interface-induced enhancement of the Fe spin magnetic moment: This has been confirmed by ab-inito calculations of the magnetic moment for the superlattices corresponding to our samples. The relative modification of the $L_{2,3}$ ratio and of the calculated spin magnetic moment can be compared. They are of the same order of magnitude and increase when the proportion of iron atoms which are located at the interface increases.

1. B. T. Thole, G. Van Der Laan, Phys. Rev. B 38 (1988), p. 3158
2. D. M. Pease, A. Fasihuddin, M. Daniel, J. I. Budnick, Ultramicroscopy, vol. 88 (2001), p. 1-16
3. H.Kurata, N.Tanaka, *Microsc. Microanal. Microstruct.*, 2 (1991), p. 183
4. T. I. Morrison, M. B. Brodsky, N. J. Zaluzec, L. R. Sill, Phys. Rev. B 32 (1985), p. 3107
5. T. I. Morrison, C. L. Foiles, D. M. Pease, N. J. Zaluzec, Phys. Rev B 36 (1987), p. 3739
6. Falqui, V.Serin, L.Calmels, E.Snoeck, A.Corrias, G.Ennas, *Jourrnal of microscopy*, 210 (2003), p. 80-88

M. Luysberg, K. Tillmann, T. Weirich (Eds.): EMC 2008, Vol. 1: Instrumentation and Methods, pp. 441–442, DOI: 10.1007/978-3-540-85156-1_221, © Springer-Verlag Berlin Heidelberg 2008

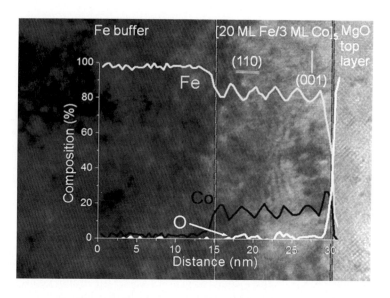

Figure. 1. Composition profile along the growth axis and corresponding HREM micrograph of the analyzed zone.

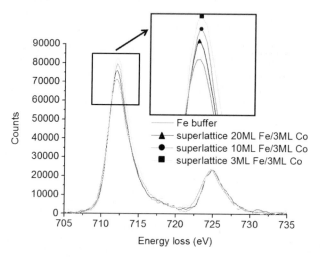

Figure. 2: Iron L_3, L_2 thresholds of the Fe buffer and in the Fe(n atomic planes)/Co(3 atomic planes) *bcc* superlattices with n=20, 10 and 3. The data shown for the superlattice are single spectra extracted from STEM line-spectra. These spectra correspond to an averaged information over a probe size of 1,5 nm of the Fe/Co interface.

EMCD at high spatial resolution: comparison of STEM with EELS profiling

M. Stöger-Pollach[1], P. Schattschneider[1,2], J. Perkins[3], D. McComb[3]

1. Univ. Service Center for Transmission Electron Microscopy, Technische Universität Wien, Wiedner Hauptstrasse 8-10, A-1040 Wien, Austria
2. Institut für Festkörperphysik, Technische Universität Wien, Wiedner Hauptstrasse 8-10, A-1040 Wien, Austria
3. Department of Materials, Imperial College, South Kensington Campus, London SW7 2AZ, United Kingdom

stoeger@ustem.tuwien.ac.at

Keywords: magnetic dichroism, EELS, STEM

Soon after the discovery of the EMCD effect [1] (energy loss magnetic chiral dichroism) we achieved a considerable improvement in signal strength and spatial resolution [2].

In the present work we propose a new scattering geometry based on energy loss spectroscopic profiling (ELSP) [3] and compare it with the previous (STEM) approach for nm-resolved EMCD.

In the STEM approach two EELS line scans are recorded at the two different chiral positions [2] in the three-beam case still having the interfaces parallel to the beam direction. The spatial resolution is then limited by the beam diameter, which is 1.7 nm in the present experiment. The investigated sample is a Fe/Au multilayer stack shown in fig. 1, together with a simulation of the expected signals with superimposed noise (Fig. 1 C).

In the ELSP approach the image of a multilayer sample is projected onto the SEA with the interfaces parallel to the energy dispersive axis using a double tilt rotation holder. The chiral electronic transitions are selected by placing the objective aperture at the respective positions in the diffraction plane. The intrinsic advantage of ELSP is that the whole stack can be investigated within two EELS recordings. These advantages are counterbalanced by technical problems like spectrometer aberrations, which become dominant due to selecting the chiral positions being off the optical axis, and drift, which is increased due to heating of the sample within the intense electron beam. Fig. 2 shows a HAADF image of the Fe/Au stack and an ELSP at the Fe $L_{2,3}$ edge. The results have to be carefully interpreted.

1. Schattschneider P., Rubino S., Hébert C., Rusz J., Kunes J., Novák P., Carlino E., Fabrizioli M., Panaccione G., Rossi G., Nature **441** (2006) 486
2. Schattschneider P., Stöger-Pollach M., Rubino S., Sperl M., Hurm C., Zweck J., Rusz J., Ultramicroscopy, *submitted*
3. Reimer L., Fromm I., Hirsch P., Plate U., Rennekamp R., Ultramicroscopy **46** (1992) 33 5
4. The Hochschuljubiläumsstiftung der Stadt Wien, contract Nbr. H-01585/2007, is kindly acknowledged for funding.

M. Luysberg, K. Tillmann, T. Weirich (Eds.): EMC 2008, Vol. 1: Instrumentation and Methods, pp. 443–444, DOI: 10.1007/978-3-540-85156-1_222, © Springer-Verlag Berlin Heidelberg 2008

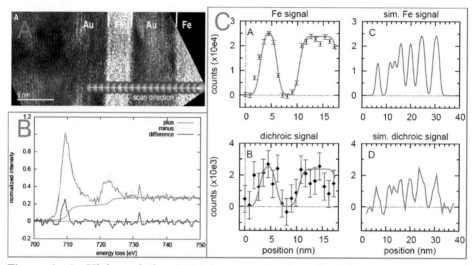

Figure 1: A: High-resolution TEM image of the investigated area; the shape and position of the beam during the scan are indicated by the superimposed circles.
B: Spectra from the middle of the 3 nm Fe layer, taken at the position marked with a cross. The difference is the dichroic signal.
C: (A) Profile of the line scanned in Fig. 2. The experimental points are the integrated Fe signal (sum of the "+" and "-" spectra) at the L3 edge (707.9-713.9 eV). The best fit with a Gaussian spot shape gives a FWHM of 1.66 nm for the spot. The error bars are $3\sigma = 855$ counts. (B) Corresponding line profile of the dichroic signal (difference of the "+" and "-" spectra) integrated at the L3 edge. (C, D): Simulated Fe (C) and EMCD (D) profiles for a stack of 4 nm Au, 1 nm Fe, 4 nm Au, followed by double layers of 1 nm Fe/Au, 2 nm Fe/Au and 3 nm Fe/Au. The EMCD profile is shown with simulated Poissonian noise corresponding to the present experimental conditions, demonstrating a resolution of ≈ 2 nm.

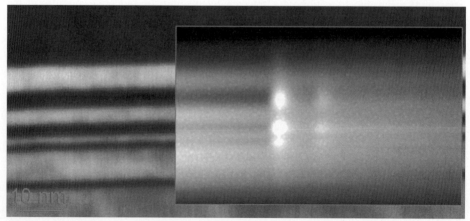

Figure 2: HAADF image of the layer stack and ELSP of the Fe-$L_{2,3}$ edge.

Elemental, Chemical and Physical State Mapping in Three-Dimensions using EELS-SI Tomography

P.J. Thomas[1], C. Booth[1], R. Harmon[1], S. Markovic[1], R.D. Twesten[1] and K. Jarausch[2]

1. Gatan Inc., 5794 West Las Positas Blvd, Pleasanton, CA 94588 USA
2. Hitachi High Technologies America 5100 Franklin Drive Pleasanton, CA 94588 USA

pthomas@gatan.com

Keywords: EELS, tomography, spectrum-imaging

The EELS spectrum provides a wealth of information regarding the elemental, chemical and physical state of the material under investigation with typically nanometre resolution. When coupled with the STEM Spectrum-Imaging mode of acquisition, this information can be spatially resolved allowing properties to be calculated as two dimensional maps [1]. The ability to do this in a fast, automated manner has resulted in the EELS STEM-SI technique to become the method of choice for high resolution microanalysis in the transmission electron microscope. However, information acquired in this way is always projected in the direction parallel to the electron beam.

The recent development of the tilt-series tomography technique allows this lost dimension to be resolved. By acquiring a series of data-sets over a range of sample tilts, reconstruction methods can be applied to retrieve the sample information that is usually integrated. Typically, the HAADF or BF signals are used for tomography since they satisfy the "projection" criterion to a reasonable approximation [2]. While these signals contain contrast arising from the change in atomic number (or Z-contrast) of the sample at a local scale, they do not provide the additional information offered by EELS. Use of the EELS signal for tomographic reconstruction has been explored before via EFTEM (e.g. [3]). However, plural scattering effects make the EELS spectrum highly sensitive to changes in projected thickness, typically encountered during a tilt series acquisition, which makes the use of EELS problematic for tomography reconstruction.

In this work, the feasibility of three-dimensional mapping using various signals in the EELS spectrum is investigated using STEM EELS-SI tomography. The problem of varying sample thickness with tilt angle is overcome by using a micro-pillar sample, providing approximately constant projected thickness with varying tilt angle. Also, the ability to rotate through 360° avoids the "missing wedge" effect encountered in regular tilt-limited tomography series [4]. Practical aspects of the sample preparation and data-acquisition are reported in detail elsewhere [5]. By acquiring a low- and core-loss EELS-SI datasets at regular tilt steps over full rotation, a four dimensional data-set is generated containing electron energy-loss, spatial and tilt angle information. Once acquired, the 4D data-set can be analyzed using standard EELS analysis methods to provide a tilt-series map of any feature in the acquired EELS spectrum; for example, performing MLLS fingerprinting of a core-loss edge found in multiple chemical states (Fig. 2). Providing the chosen signal satisfies the projection criterion, tomographic reconstruction can then be applied to obtain a "volumetric" map of the property of

M. Luysberg, K. Tillmann, T. Weirich (Eds.): EMC 2008, Vol. 1: Instrumentation and Methods, pp. 445–446, DOI: 10.1007/978-3-540-85156-1_223, © Springer-Verlag Berlin Heidelberg 2008

interest. The potential, and also the validity, of this approach will be discussed with respect to the wide range of properties available from the EELS spectrum.

1. J.A. Hunt and D. B. Williams, Ultramicroscopy **38** (1991) p. 47.
2. P.A. Midgley and M Weyland, Ultramicroscopy **96** (2003) p.413.
3. M H. Gass et al, Nano Letters **6** (2006) 376.
4. K. F. Jarausch and D. N. Leonard, Imaging & Microscopy **9** (2007) p.24.
5. K. F. Jarausch et al, submitted to EMC 2008 Proceedings.

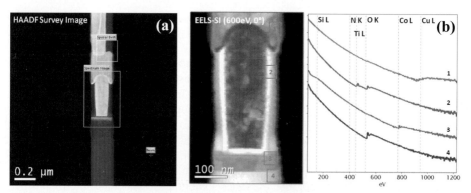

Figure 1. (a) HAADF survey image of the micro-pillar semiconductor sample. The STEM-EELS data was acquired from the region as shown. (b) An acquired EELS core-loss spectrum-image tilt series data-set (total range 60-1400eV loss, 0-180° tilt) with spectra extracted from the regions marked (offset intensities with log scale for clarity).

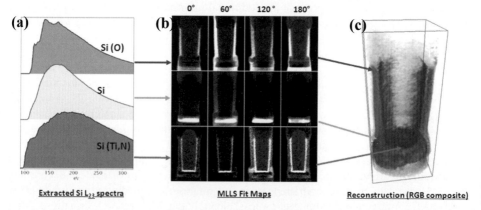

Figure 2. Volumetric chemical-state mapping of Si from the STEM-EELS SI tilt series. (a) Spectra extracted from the EELS core-loss data show chemical state specific variations in the Si L_{23} edge shape. (b) MLLS fingerprint mapping over 90-130eV loss range yields a distribution map tilt-series for each silicon phase (maps shown with 60° steps for clarity). (c) Tomographic reconstruction of each map tilt series produces a 3D chemical state map, shown as an RGB composite using the colour key from (a).

Sub-0.5 eV EFTEM Mapping using the Zeiss SESAM

C.T. Koch, W. Sigle, J. Nelayah, L. Gu, V. Srot, and P.A. van Aken

Stuttgart Center for Electron Microscopy, Max Planck Institute for Metals Research, Heisenbergstraße 3, 70569 Stuttgart, Germany

vanaken@mf.mpg.de

Keywords: EELS, low-loss EELS, interfaces, plasmons

Being equipped with a monochromator and a high-transmissivity energy filter (MANDOLINE filter, $T_{0.5eV}$ = 3250 nm^2/sr) the Sub-Electron-Volt-Sub-Ångstrom-Microscope (SESAM) [1] recently developed by Zeiss SMT is an ideal tool for EFTEM and EELS applications requiring high energy and spatial resolution. Figure 1 shows the instrument itself and the shape of the zero-loss peak, having a width of 77 meV. Although an energy-resolution of < 50 meV has already been achieved [2], the spectrum shown here is more representative for the instruments everyday performance.

Examples of recent EFTEM results obtained on the SESAM will be presented. These include band-gap mapping of semiconductor structures, mapping of surface plasmon resonances (SPRs) in metallic nanostructures (as first demonstrated by [3]), and EELS profiling of interfaces and grain boundaries.

Figure 2a shows a single image from an EFTEM series of a Σ17 grain boundary in SrTiO$_3$ producing a data stack of 1024 x 1024 x 100 pixels and spanning an energy range of 0 to 39.6 eV. The spatial sampling in this image is 0.64 nm, the mono-chromator and energy filter slit widths and energy step were all set to 0.4 eV. Integrating spectrum profiles across the grain boundary along the axis of symmetry produces spectra with very good statistics, despite an average of only 20 – 60 counts per pixel (for ΔE > 3 eV) (see Figure 2 b and c). The arrow in Fig. 2c points to an additional peak at 19 eV found only at the grain boundary, and not in either of the 2 grains.

Probing optical properties by low-loss EFTEM requires particularly good energy resolution. Figure 3 shows selected members of an EFTEM series of a silver nano-prism revealing the real-space shape of 3 different SPRs. This EFTEM series has been recorded with an energy resolution of 0.2 eV exceeding that of comparable STEM-EELS experiments [3].

1. C.T. Koch, W. Sigle, R. Höschen, M. Rühle, E. Essers, G. Benner and M. Matijevic, Microscopy and Microanalysis **12** (2006) 506.
2. E. Essers, M. Matijevic, G. Benner, R. Höschen, W. Sigle, C.T. Koch, Microscopy and Microanalysis **13** (2007) 18.
3. J. Nelayah, M. Kociak, O. Stephan, F.J.G. de Abajo, M. Tencé, L. Henrard, D. Taverna, I. Pastoriza-Santos, L.M. Liz-Marzan, and C. Colliex., Nature Physics 3 (2007) 348.
4. We acknowledge financial support from the European Union under the Framework 6 program under a contract for an Integrated Infrastructure Initiative. Reference 026019. The instrument has been funded by the DFG within its "Großgeräteinitiative Hochauflösende Elektronenmikroskopie" (Ru 342/16-1) as well as the Federal State of Baden-Württemberg and the Max Planck Society (MPG). Proposal by Manfred Rühle, Joachim Mayer, Frank Ernst and Erich Plies.

M. Luysberg, K. Tillmann, T. Weirich (Eds.): EMC 2008, Vol. 1: Instrumentation and Methods, pp. 447–448, DOI: 10.1007/978-3-540-85156-1_224, © Springer-Verlag Berlin Heidelberg 2008

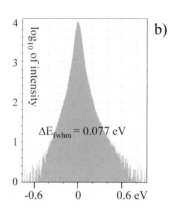

Figure 1. a) The Zeiss SESAM 200 kV FEG-TEM, b) Log_{10} of a typical zero-loss peak (ΔE_{FWHM} = 77 meV, sum of 300 spectra with 50 ms exposure time per spectrum)

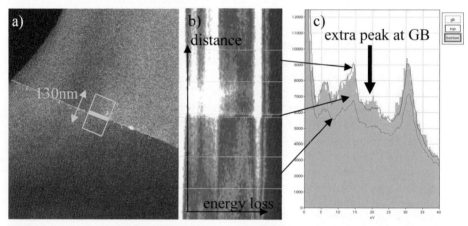

Figure 2. a) 1k x 1k EFTEM map of a $\Sigma17$ grain boundary in $SrTiO_3$ recorded with 0.4 eV energy resolution (the plane at 30 eV energy loss is shown). b) Spectrum profile across grain boundary integrated along the thick line in a). c) Spectra extracted from b).

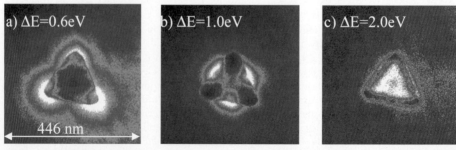

Figure 3. EFTEM images of an Ag triangle. MC and filter were set to ~ 0.2 eV slit width.

Acquisition of the EELS data cube by tomographic spectroscopic imaging

W. Van den Broek[1], J. Verbeeck[1], S. De Backer[2], D. Schryvers[1] and P. Scheunders[2]

1. EMAT, University of Antwerp, Groenenborgerlaan 171, 2020 Antwerp, Belgium
2. Vision Lab, University of Antwerp, Universiteitsplein 1, 2610 Wilrijk, Belgium

wouter.vandenbroek@ua.ac.be
Keywords: Tomography, EELS data cube, energy filter

The EELS data cube combines spatial and spectral information because it has an EELS spectrum in each pixel of a spatial image. The two major methods of acquiring it are image-spectroscopy (or an Energy Filtered Series or EFS) and spectrum-imaging. In [1] we proposed a new method: Tomographic Spectroscopic Imaging (TSI). A proper defocus of the energy filter induces a mixing of spatial and spectral information that is equivalent to a projection of the data cube onto the CCD camera. The projection angles are directly related to the defocus value and all angles between 0° and 180° are reached. The projections are used to reconstruct the data cube by tomographic techniques. In [1] we concluded that TSI needs a lower electron dose than EFS to acquire a data cube with the same resolution and mean signal-to-noise ratio.

Experiments are conducted on a Jeol 3000F TEM equipped with a Gatan post-column energy filter (GIF). We aim at recording a data cube measuring 50 pixels in the energy direction (~1 eV/pixel) and 90 by 90 pixels in the spatial directions (1.3 nm/pixel). By defocusing the GIF in imaging mode, projections through angles from 0° to 73° and from 110° to 180° are possible; with the focused image being a projection through 0°. Defocusing in spectroscopy mode yields angles ranging from 77° to 105°; with the EELS spectrum being a projection through 90°. To obey the projection requirement the data cube needs to be bounded in both the spatial and the spectral directions, so the 0.6 mm entrance aperture of the GIF is inserted as well as a 50 eV wide energy slit. 46 projections in imaging mode and 10 in spectroscopy mode were made, and then fed into an algebraic reconstruction technique.

The sample is a TiO_2 particle with a $L_{2,3}$ peak at 460 eV energy loss. Its data cube is recorded with TSI and with EFS with a very high dose that serves as an approximation to the true data cube, see Figure 1. Visual comparison of both data cubes shows that TSI reconstructs the true data cube faithfully, see Figure 2. This is the first direct proof that TSI is capable of recovering the real data cube. It also shows that in the whole data processing chain, from measurements to reconstruction, all steps are carried out at sufficient accuracy and precision. [2,3]

1. W. Van den Broek, J. Verbeeck, S. De Backer, P. Scheunders and D. Schryvers, Ultramicroscopy **106** (2006), p. 269.
2. W. Van den Broek is supported by a Concerted Action project, University of Antwerp, 2002/1, "Characterisation of nanostructures by means of advanced electron energy spectroscopy and filtering"

M. Luysberg, K. Tillmann, T. Weirich (Eds.): EMC 2008, Vol. 1: Instrumentation and Methods, pp. 449–450, DOI: 10.1007/978-3-540-85156-1_225, © Springer-Verlag Berlin Heidelberg 2008

3. The authors acknowledge financial support from the European Union under the Framework 6 program under a contract for an Integrated Infrastructure Initiative. Reference 026019 ESTEEM.

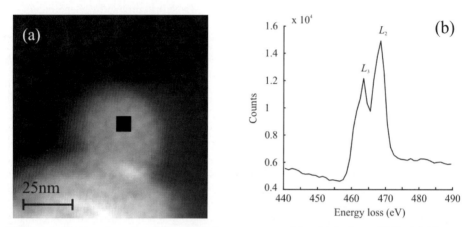

Figure 1. Data cube of a TiO$_2$ crystal, as measured by high dose EFS. (a) Energy filtered image at 468.5 eV; (b) EELS spectrum, the average of the spectra in the dark square in (a).

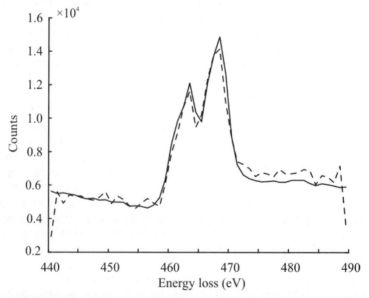

Figure 2. Spectra averaged over the black square in Figure 1(a). The full line stems from the EFS data, the broken line from TSI.

A low electron fluence EELS study of Fe-coordination within ferrihydrite and phosphorous doped ferrihydrite nanoparticles

G. Vaughan[1], A.P. Brown[1], R. Brydson[1], K. Sader[2]

1. Institute for Materials Research, University of Leeds, UK
2. SuperSTEM, Daresbury Laboratories, UK

g.m.vaughan07@leeds.ac.uk

Keywords: Ferritin, Ferrihydrite, Phosphorous, EELS, TEM

Ferritin is a major iron storage molecule which acts as a reservoir for excess iron, storing it in a non-toxic form[1]. Without ferritin (and several other dedicated proteins and chelating molecules) the oxygen rich environments of biological systems would result in the formation of insoluble iron oxides, rendering this iron inaccessible. The ferritin molecule is known to consist of a spherical protein shell (~12nm in diameter) which forms a central cavity (~6nm in diameter) within which iron is stored as an oxyhydroxide mineral core with a structure similar to that of the nano-crystalline mineral ferrihydrite (FHY)[1, 2]. It is known that the iron rich mineral cores of ferritins have varying amounts of phosphorous incorporated into their structure[3, 4]. It is thought that the majority of this phosphorus is bound to surface sites on the core with the remainder being located at discontinuities and stacking faults located within the crystalline cores[1].

The relative ease with which FHY can be synthesised[5] makes it a readily available analogue for the structure of ferritin mineral cores. We propose to synthesise a series of phosphorous doped FHYs by solution precipitation with a range of doping levels (atomic: 5%, 10% 15%) similar to those seen in human liver biopsies[6] which will allow us to investigate where phosphorus is incorporated in the mineral's structure (surface/bulk) and what effect this has on its internal atomic arrangement.

One way to investigate the exact position and co-ordination of the phosphorous and iron in these mineral cores is to use analytical electron microscopy. As an experimental precursor to the work to be carried out on Ferritin mineral cores we are developing a novel low electron fluence EELS acquisition method as applied to FHY nano-particles. All work will be carried out using a VG HB501 STEM retro-fitted with a Nion spherical aberration corrector and a Gatan Enfina electron energy loss spectrometer at the Daresbury Laboratories, UK (SuperSTEM1)

The fine probe of the STEM (routinely ~1.1 angstroms with the possibility of sub-angstrom in the SuperSTEM) opens up the possibility of acquiring EELS spectra from individual atomic columns within a material. However to achieve a sufficient signal to noise ratio for, say, analysis of the energy loss near edge structure (ELNES), large fluences are required. These are usually achieved by integrating over time and this can potentially damage the specimen. Certainly it is known that FHY is an electron beam sensitive material and will undergo internal atomic rearrangement when exposed to the

M. Luysberg, K. Tillmann, T. Weirich (Eds.): EMC 2008, Vol. 1: Instrumentation and Methods, pp. 451–452, DOI: 10.1007/978-3-540-85156-1_226, © Springer-Verlag Berlin Heidelberg 2008

electron beam of a TEM or STEM [7]. Our work seeks to acquire a range of low fluence EELS spectra whilst simultaneously maintaining a signal-to-noise ratio of sufficiently high level to be able to perform analysis of the fine structure present in the Fe L_{2-3} energy loss edge from spatially resolved regions of FHY. We propose to achieve this by averaging the energy loss signal from a large specimen area. By allowing the electron beam to scan a predefined total area of specimen while continuously acquiring an EEL spectrum such that beam dwell time (and so fluence) on a given area of sample is minimised ($\leq 10^4$-10^5 electrons nm^{-2}) the data is acquired over total integration times sufficient to optimise the signal-to-noise in the spectrum (Figure 1 and K. Sader this proceedings). By first capturing an image of the area (particle(s)) to be analysed and predefining a scan pattern for the acquisition (Figure 2) we will be able to acquire and distinguish low fluence spectra averaged over the surface or the bulk of the particles. Thus specifically we can determine the true co-ordination of Fe specifically at the surface and in the bulk of pristine and phosphorous modified FHY and develop this "smart acquisition" EELS technique for more general use.

1. Massover, W.H., Ultrastructure of ferritin and apoferritin: a review. Micron, 1993. **24**(4): p. 389-437.
2. Cowley, J.M., et al., The structure of ferritin cores determined by electron nanodiffraction. J Struct Biol, 2000. **131**(3): p. 210-6.
3. Treffry, A., et al., *A note on the composition and properties of ferritin iron cores.* Journal of Inorganic Biochemistry, 1987. **31**(1): p. 1-6.
4. Desilva, D., J.H. Guo, and S.D. Aust, *Relationship between Iron and Phosphate in Mammalian Ferritins.* Archives of Biochemistry and Biophysics, 1993. **303**(2): p. 451-455.
5. U.Schwertmann, R.M.C., *Iron Oxides in the Laboratory. Preparation and Characterization,* ed. D.C. Dyllick-Brenzinger. 1991: VCH Verlagsgesellschaft mbH, Weinheim.
6. Y.H.Pan, Electron Microscopy of Mineral Cores in Ferritin and Haemosiderin, Ph.D 2006, University of Leeds.
7. Pan, Y., et al., Electron beam damage studies of synthetic 6-line ferrihydrite and ferritin molecule cores within a human liver biopsy. Micron, 2006. **37**(5): p. 403-411.

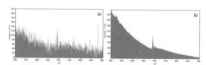

Figure 1. (a) – 100 summed spectral each taken from a single pixel of a square area (spectrum image) over the bulk of a FHY particle each with an acquisition time of 0.05s per pixel, (b) – "smart acquisition" EELS spectra of the same area as in (a) with a dwell time of 0.05s per pixel.

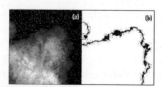

Figure 2. (a) - HAADF micrograph of a FHY nano-particle, (b) – "smart acquisition" scan mask defining the scan pattern (black) for the surface of the particle in (a).

Optimal aperture sizes and positions for EMCD experiments

J. Verbeeck[1], C. Hébert[2], S. Rubino[3], P. Novák[4], J. Rusz[4,5], F. Houdellier[6], C. Gatel[6], P. Schattschneider[3]

1. EMAT, University of Antwerp, Groenenborgerlaan 171, 2020 Antwerp, Belgium
2. CIME & LSME, EPFL, Station 12, 1015 Lausanne, Switzerland
3. Institut für FestkörperPhysik, Technische Universität Wien, Wiedner Hauptstrasse 8-10, A-1040 WIEN, Austria
4. Institute of Physics, Academy of Sciences of the Czech Republic, Na Slovance 2, CZ-182 21 Prague, Czech Republic
5. Department of Physics, Uppsala University, Box 530, S-751 21 Uppsala, Sweden
6. CEMES, 29 rue Jeanne Marvig, BP 94347 31055 Toulouse cedex 4, France

jo.verbeeck@ua.ac.be

Keywords: EELS, Circular dichroism, Magnetic

The signal-to-noise ratio (SNR) in Energy-loss Magnetic Chiral Dichroism (EMCD)[1,2]- the equivalent of X-ray Magnetic Circular Dichroism (XMCD) in the electron microscope- is optimized with respect to the detector shape, size and position. We show that an important increase in SNR can be obtained when taking much larger detector collection angles. Using an analytical formula for the SNR, one can even derive the ideal shape of the collection aperture by optimisation. Figure 1. shows such an optimal shape for α-Fe close to a three-beam case for t=16 nm. The ideal shape is unfortunately heavily dependent on the thickness and on the exact orientation of the specimen and it is rather intricate.

Round apertures are on the other hand readily available in any microscope and simulations show that they are a good compromise if position and radius are chosen to optimise the SNR. Figure 2 shows a set of optimal radii and positions for circular apertures for a simulation of α-Fe in three-beam conditions for a range of thicknesses t=2-40 nm. This figure shows that at all thicknesses a rather large circular aperture with a radius around $R=g_{110}$ is preferred and that the position is closer to the $\overline{1}\,\overline{1}\,0$ beam.

Figure 3 shows the relative SNR as a function of thickness and aperture radius. This figure demonstrates the gain in SNR when going to aperture radii of around $R=g_{110}$ and also shows the dependence on thickness[3,4].

1. J. Verbeeck, C. Hébert, S. Rubino, P Novák, J. Rusz, F. Houdellier, C. Gatel, P. Schattschneider, Ultramicroscopy (2008) in press.
2. Hébert C., Schattschneider P., Ultramicroscopy, Vol. 96, Issues 3-4, (2003), 463-468
3. Schattschneider P., Rubino S., Hébert C., Rusz J., Kunes J., Novák P., Carlino E., Fabrizioli M., Panaccione G., Rossi G., Nature 441 (7092): 486- 488 (2006)
4. This work was supported by the European Commission under contract Nr. 508971 CHIRALTEM. J.V. and F.H. thank the financial support from the European Union under the Framework 6 program under a contract for an Integrated Infrastructure Initiative. Reference 026019 ESTEEM.

M. Luysberg, K. Tillmann, T. Weirich (Eds.): EMC 2008, Vol. 1: Instrumentation and Methods, pp. 453–454, DOI: 10.1007/978-3-540-85156-1_227, © Springer-Verlag Berlin Heidelberg 2008

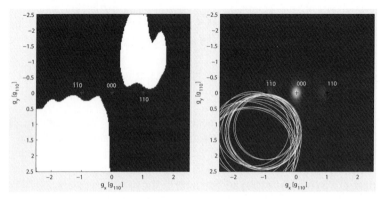

Figure 1. Ideal aperture shape for α-Fe in a 9 beam Bloch wave calculation with three excited beams for a thickness of 16 nm.

Figure 2. Optimal position and radius for a circular aperture for t=2 to 40 nm. Note that the ideal position is closer to the $\bar{1}\,\bar{1}\,0$ beam and the optimal radius is around R=g_{110}.

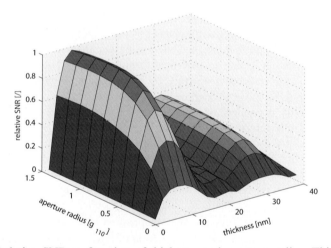

Figure 3. Relative SNR as function of thickness and aperture radius. This plot shows that although the signal is dependent on thickness, considerable gains in SNR can be made by choosing R~g_{110}.

Chemical analysis of nickel silicides with high spatial resolution by combined EDS and EELS (ELNES)

E. Verleysen[1,2], O. Richard[1], H. Bender[1], D. Schryvers[3], and W. Vandervorst[1,2]

1. IMEC, Kapeldreef 75, 3001 Leuven, Belgium
2. K.U. Leuven, IKS, Celestijnenlaan 200D, 3001 Leuven, Belgium
3. Universiteit Antwerpen, EMAT, Groenenborgerlaan 171, 2020 Antwerpen, Belgium

Eveline.Verleysen@imec.be

Keywords: EELS, ELNES, nickel silicides

The continuous scaling in semiconductor technology has made accurate characterization of transistor compounds more challenging. Often, the main difficulty is to obtain both high spatial resolution and information on the chemical composition of the material. In this context, Ni-silicides are selected as a test-vehicle to examine the combination of quantification and spatial resolution, by using different analytical TEM techniques in parallel.

In the 45 nm technology, Ni-silicides are integrated in CMOS devices as contact layers to lower electrical resistivity, and as gate electrodes to reduce sheet resistance and to overcome the poly-silicon depletion problem [1]. The nickel silicide system consists of several phases that differ in composition ratio. Since these phases have comparable interplanar distances and are formed in small-grained polycrystalline layers, they are difficult to distinguish based on their lattice spacings in HREM. As a consequence, it is essential for correct phase determination to combine HREM studies with chemical analysis.

Ni-silicide layers of about 100 to 150 nm thick are formed by rapid thermal annealing of a Ni-layer and a thin polycrystalline Si film deposited on top of a SiO_2/Si substrate. The temperature and thickness ratio of the Ni and Si-layer are varied to obtain different phases. Figure 1 shows a TEM and HAADF-STEM image of such a Ni-silicide layer containing two phases ($Ni_{31}Si_{12}$ and Ni_2Si). EDS and EELS analyses are carried out on a TECNAI F30 operating at 300keV, equipped with an EDAX detector and a Tridiem GIF energy filter.

Initial examination of the layers is done by EDS. A Ni-silicide reference layer is used as a standard for accurate quantification. The analysis allows to obtain the atomic composition of the Ni-silicide phases with a high accuracy. Table 1 shows the results of the EDS analysis on the sample displayed in figure 1.

Further research is done by EELS and associated energy loss near edge structure (ELNES). The atomic composition is obtained using conventional quantification based on background subtraction and integration of spectra from the core excitations of Ni and Si. Also, the use of model based quantification is investigated [2]. In parallel with the EELS analysis, the ELNES of the different phases is examined. Since the ELNES contains information about chemical bonding in the material, it should exhibit small differences in the fine structure of each phase. The use of the ELNES as a fingerprint,

M. Luysberg, K. Tillmann, T. Weirich (Eds.): EMC 2008, Vol. 1: Instrumentation and Methods, pp. 455–456, DOI: 10.1007/978-3-540-85156-1_228, © Springer-Verlag Berlin Heidelberg 2008

specific for each phase, is tested by examining the Ni-$L_{2,3}$ and Si-$L_{2,3}$ peaks. Figure 2 shows the Ni-$L_{2,3}$ peak of the phases displayed in figure 1.

The methodology is applied to the full range of available Ni-silicide phases. This allows to investigate the composition dependence of the accuracy of the results. Also the dependence on the specimen thickness is systematically examined.

The complementary information that can be obtained by combining these analytical TEM techniques allows to accurately distinguish the observed Ni-silicide phases with a high spatial resolution. The procedure will be further extended by combining the EDS/EELS information with the HAADF-STEM intensities.

1.   J. A. Kittl, A. Lauwers, M. A. Pawlak, Electrochem. Soc. Proc. **2005-05** (2005), p. 225-232.
2.   J. Verbeeck, S. Van Aert, Ultramicroscopy **101** (2004), p. 207-224.

Table 1. Results of EDS analysis on the sample shown in figure 1. N is the number of repeated measurements. The obtained ratios are presented, together with the standard deviation (st. dev.). Between parentheses the expected value is given. The relative precision and the relative accuracy with respect to the expected values are listed in the last two columns.

Phase	N	Ratio Si/Ni (expected)	st. dev.	rel. prec.	rel. acc.
Top layer: $Ni_{31}Si_{12}$	15	0.389 (0.387)	0.027	6.9%	0.5%
Bottom layer: Ni_2Si	15	0.501 (0.500)	0.028	5.5%	0.2%

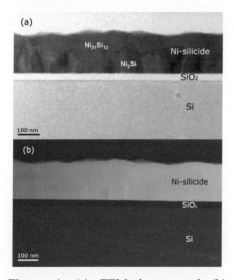

Figure 1. (a) TEM image and (b) HAADF-STEM image of a Ni-silicide layer containing $Ni_{31}Si_{12}$ (top layer) and Ni_2Si (bottom layer).

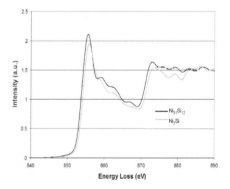

Figure 2. Ni-$L_{2,3}$ edge of $Ni_{31}Si_{12}$ and Ni_2Si (figure 1). The background contribution and the plural scattering effect have been removed. The intensity was normalized to eliminate difference in the number of Ni atoms in the measured area.

Retrieving dielectric function by VEELS

L. Zhang[1], J. Verbeeck[1], R. Erni[2], G. Van Tendeloo[1]

1. EMAT, University of Antwerp, Groenenborgerlaan 171, B2020 Antwerp, Belgium
2. National Center for Electron Microscopy, Lawrence Berkeley National Laboratory, One Cyclotron Rd., Berkeley, CA 94720, USA

Liang.zhang@ua.ac.be

Keywords: VEELS, retardation effect, dielectric function, TEM

The properties of materials are largely determined by their electronic structure [1]. With the increasing use of nanomaterials, it becomes therefore important to probe this electronic structure on the nanoscale. A good candidate for this is Valence electron energy loss spectroscopy (VEELS) in (scanning) transmission electron microscopy ((S)TEM) since regions of atomic scale can be investigated, although the technique needs further development.

Aiming at the determination of the dielectric function of semiconductors, a detailed VEELS study is presented using a method that can also be applied to nanodiamond materials. Several contributions like plural scattering, retardation and surface losses [2] can considerably influence the measured energy-loss signal of diamond and therefore we need to process the raw VEELS data.

We propose a procedure which consists of three basic steps. The first step is the acquisition of the VEELS data under well defined experimental conditions (figure 1). In the second step the scattering contributions discussed above are removed to obtain a single scattered loss function. It is only after this procedure that in the third step a Kramers-Kronig analysis [3] can be applied in order to derive the complex dielectric function (figure 2). In this contribution, the experimental results as well as the different required processing step will be discussed and the applicability to nanoscale problems will be demonstrated.

1. H. Müllejans, R. H. French, Microscopy and Microanalysis (2000), 6: 297-306 Cambridge University Press.
2. R. Erni, N. D. Browning, Ultramicroscopy, doi:10.1016/j.ultramic.2007.03.005. 1.
3. R.F.Egerton, Electron Energy-loss spectroscopy in the Electron Microscope, second edition, Plenum Press, New York, 1996.
4. Authors would thank the financial support from the European Union under the Framework 6 program under a contract for an Integrated Infrastructure Initiative. Reference 026019 ESTEEM. Authors also acknowledge financial support from IAP-VI

458

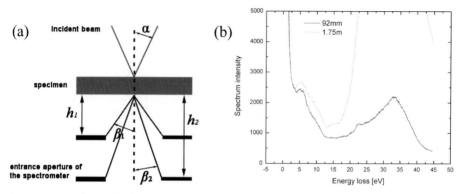

Figure 1. **a** Schematic drawing of two camera length setting, **b** two spectra recorded with two different camera lengths (92mm and 1.75m). Normalized by ZLP. h_1 and h_2 are two different camera length, β_1 and β_2 are the corresponding collection angles.

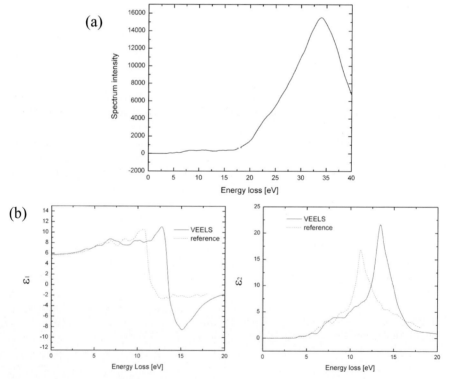

Figure 2. **a** Single scattering signal, subtracted from the spectra with two different collection angles, **b** the corresponding dielectric function ε_1 and ε_2. The (black) solid line is the VEELS result and the (red) dashed line is the theoretical calculation as the reference.

Some Recent Materials Applications of In Situ High Resolution Electron Microscopy

R. Sinclair[1], S.K. Kang[2], K.H. Kim[3] and J.S. Park[1]

1. Department of Materials Science and Engineering, Stanford University, Stanford, CA 94305, USA
2. Samsung Electronics Co., Ltd., Suwon, Republic of Korea
3. Korea Institute of Ceramic Engineering and Technology, Seoul, Republic of Korea

bobsinc@stanford.edu

Keywords: In Situ, High Resolution TEM, Semiconductor Processing

Since the realization that high resolution electron microscopy (HREM) imaging could be carried out under non-ambient conditions [1, 2], in situ HREM has broadened significantly and is now a well-accepted technique [3]. Our work has largely focused on heating materials to observe at atomic resolution the reactions which occur at elevated temperatures. The extension to environmental studies is likewise a significant achievement [4-6].

One area that has received our attention concerns the changes which occur upon processing of the materials for future field effect transistors. As the dimensions are made increasingly small, replacement of the standard silicon channel, SiO_2 gate oxide, polysilicon gate combination is required. Suitable materials include higher mobility semiconductor (e.g. Si-Ge, Ge or GaAs), higher dielectric constant gate oxides (e.g. metal oxides or silicates) and higher conductivity metals (e.g. silicides). We focus here on the latter two.

Metal oxides, deposited by such methods as atomic layer deposition (ALD), are normally structurally amorphous, like their predecessor thermally grown silicon oxide. However unlike the latter they readily crystallize upon heating, which introduces crystals of different orientations and their associated grain boundaries. A comprehensive study of the crystallization of ALD tantalum oxide by in situ HREM has been described elsewhere [7] and this is really a model of our approach. Basic kinetic studies of nucleation and growth are accompanied by high-resolution observations of the growth mechanisms. We have applied this to crystallization and phase separation studies of ALD hafnium silicates of varying combination, whereby spinodal decomposition accompanies the crystallization process. The temperatures involved are often above 800°C, which is a notable extension for HREM work.

Nickel silicide is likewise being considered as a candidate gate material. The preferred phase is NiSi which can be fabricated by deposition of the appropriate proportion of nickel and silicon, followed by their reaction upon heating. In situ HREM shows that nickel diffusion through the NiSi layer is the dominant reaction mechanism [8], consistent with prior bulk studies. However, as NiSi is not thermodynamically stable in contact with silicon (the disilicide is, of course), penetration through the thin gate oxide results in further silicidation at high temperatures (e.g. 950°C) with silicide

M. Luysberg, K. Tillmann, T. Weirich (Eds.): EMC 2008, Vol. 1: Instrumentation and Methods, pp. 459–460, DOI: 10.1007/978-3-540-85156-1_230, © Springer-Verlag Berlin Heidelberg 2008

460

"spikes" into the silicon substrate, clearly an undesirable eventuality. We have recorded this reaction also by in situ HREM (e.g. Figure 1) with the NiSi$_2$-NiSi interface sweeping through the silicide film fed by silicon atom diffusion from the substrate.

In summary, in situ HREM continues to be a very powerful means of studying reactions in solids. In the present case we have applied this method successfully to future field effect transistor materials.

1. R. Sinclair and M.A. Parker, Nature **322** (1986), p.531.
2. R. Sinclair, T. Yamashita, M.A. Parker, K.B. Kim, K. Holloway and A.F. Schwartzman, Acta Crystallogr. Sec. A **44** (1988), p. 965.
3. D.B. Williams and C.B. Carter, Transmission Electron Microscopy (1996), p. 546.
4. P.L. Gai, Topics in Catalysis 21 (2002), p.161.
5. S. Helveg, C. Lopez-Cartes, J. Sehested, P.L. Hansen, B.S. Clausen, J.R. Rostrup-Nielsen, F. Abid-Pedersen and J.K. Norskov, Nature 427 (2004), p. 426.
6. R. Sharma and K. Weiss, Microsc. Res. and Tech. **42** (1998), p. 270.
7. K.H. Min, R. Sinclair, I.S. Park, S.T. Kim and U.I. Chung, Philos. Mag. **85** (2005), p. 2049.
8. R. Sinclair, IMC 16 Proc. (2006), p. 1322.
9. This work was initially funded by the Stanford University INMP program and the Stanford Nanocharacterization Laboratory, and receipt of KOSEF scholarships (Kang and Kim) are appreciated.

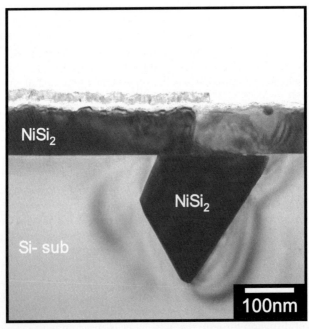

Figure 1. Bright field TEM image of "spiking" of nickel disilicide into the silicon substrate after reaction with the prior nickel monosilicide gate layer.

Melting and solidification of alloys embedded in a matrix at nanoscale

K. Chattopadhyay, V. Bhattacharya, K. Biswas

1. Indian Institute of Science, Bangalore 560 012, India
2. Indian Institute of Technology, Kanpur, India

kamanio@materials .iisc.ernet.in

Keywords: nanoparticles, alloy, eutectics, melting, solidification, phase transformation, insitu studies

The transformation behaviour of nanometer sized particles of pure metals has been studied extensively in recent time [1,2]. It has been shown that in the case of melting transition of embedded metal particles, the transformation temperature can increase or decrease depending on the nature of the interfaces. This is in contrast to the free particles where the melting temperature generally depresses. These transitions can be very effectively studied by insitu heating and cooling inside the microscope. Such a study also revealed that a planar defect free interface between the particles and the embedding matrix promotes superheating.

In contrast to the work on pure metals, our understanding of the behaviour of the alloy particles is less well understood. For example, in the case of an embedded Pb-In alloy particle, the mobility of different interfaces during melting as well as the roughening transition itself differs [3]. There exists report of thermodynamic calculations which suggest a significance increase in the segregation of solutes near the boundaries of the nano particles [4]. Probably the most interesting cases pertain to multiphase alloy nanoparticles and their transformation behaviour. In the presentation, we shall summaries our recent results obtained through insitu electron microscopy of embedded alloy nanoparticles with particular emphasis on melting and solidification transformation. Although we shall touch several low melting systems, a detailed analysis will be presented on our recent results on two phase eutectic nanoparticles of lead and tin. We shall show that at small length scales their behaviors are significantly different. Formation of a solid solution precedes both melting to a single phase liquid and solidification to a two phase microstructure. These results will be analysed in the light of some recent suggestions about the phase transformation behaviour of the small particles.

1. K. Chattopadhyay and R. Goswami, Prog Mater Sci 42 (1997), p. 287
2. Q.S. Mei, K. Lu Prog Mater Sci 52 (2007) 1175–1262
3.P. Bhattacharya and K.Chattopadhyay, International Journal of Nanoscience ,4(2005) 909-920
4. S.Acharya, Ph.D thesis,Indian Institute of Science, 2005

M. Luysberg, K. Tillmann, T. Weirich (Eds.): EMC 2008, Vol. 1: Instrumentation and Methods, pp. 461–462, DOI: 10.1007/978-3-540-85156-1_231, © Springer-Verlag Berlin Heidelberg 2008

462

a)

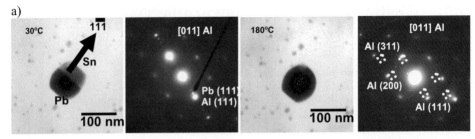

Figure 1a. A heating sequencesshowing single phase solid solution of Pb-Sn eutectic particle prior to melting

b)

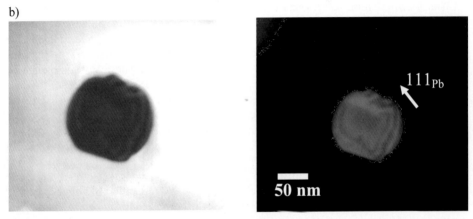

Figure 1b. Pb-Sn eutectic particle just after solidification showing single phase solid solution of Sn in Pb at 150°celsius.

Advances in transmission electron microscopy: in situ nanoindentation and in situ straining experiments

Jeff Th.M. De Hosson

Department of Applied Physics, the Netherlands Institute for Metals Research and Zernike Institute for Advanced Materials, University of Groningen, Nijenborgh 4, 9747 AG Groningen, the Netherlands

j.t.m.de.hosson@rug.nl

Keywords: in situ nano-indentation, in-situ straining, shear bands

Undisputedly microscopy plays a predominant role in unraveling the underpinning mechanisms in plastic deformation of materials. There are at least two reasons that hamper a straightforward correlation between microscopic structural information and mechanical properties: one fundamental and one practical reason. First, the defects affecting these properties, like dislocations, are in fact not in thermodynamic equilibrium and their behavior is very much non-linear. Second, a quantitative evaluation of the structure-property relationship can be rather difficult because of statistics. In particular, situations where there is only a small volume fraction of defects present or a very inhomogeneous distribution statistical sampling may be a problem. A major drawback of experimental research in the field of crystalline defects is that most of the microscopy work has been concentrated on static structures.

Direct observation of dislocation behavior during indentation has recently become possible through in-situ nanoindentation in a transmission electron microscope. In this contribution we will concentrate on the dynamic effects of dislocations and cracks in crystalline and amorphous metals observed with in-situ TEM nanoindentations and in-situ TEM straining experiments. The objective of this contribution is not to address all the various deformation mechanisms in metallic systems but rather to discuss the various recent advances in in-situ TEM techniques that can be helpful in attaining a more quantitative understanding of the dynamics of dislocations, see Figure 1 [1,2].

Besides moving dislocations in crystalline materials, significant progress has been made in recent years in the understanding of the associated deformation and crack propagation in amorphous metals, together with possible control of shear band propagation by virtue of (nano-)crystalline additions in order to suppress the tendency for instantaneous catastrophic failure. However, it is also apparent that there is still much inconsistency, and whilst many sound hypotheses and proofs abound, clarity is often lacking when comparing published results. The shear band thickness lies in the range to be 10 – 20 nm for several BMG compositions. TEM should be a suitable tool for this kind of analysis of shear band formation in metallic glasses, since their thicknesses are very low and it may be expected that shear bands may lead to (nano-scale) structural changes in amorphous materials, see Figure 2 [3].

1. W.A. Soer, J.T.M. De Hosson, A.M. Minor, Z. Shan, S.A.S. Asif , O.L. Warren ,Appl. Phys. Lett. **90** (2007), 181924.

M. Luysberg, K. Tillmann, T. Weirich (Eds.): EMC 2008, Vol. 1: Instrumentation and Methods, pp. 463–464, DOI: 10.1007/978-3-540-85156-1_232, © Springer-Verlag Berlin Heidelberg 2008

464

2. J. T.M. De Hosson, W.A. Soer, A.M. Minor, Z. Shan, E.Stach, S.A.S. Asif, O. Warren, J. Mater.Sci. **41**(2006), 7704.
3. D.T.A. Matthews, V. Ocelík, P.M. Bronsveld, J.Th.M. De Hosson, Acta Materialia, (2008), in press and online: doi:10.1016/j.actamat.2007.12.029.

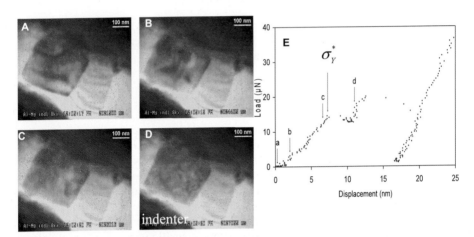

Figure 1. TEM bright- field Image sequence (a-d) from the initial loading portion (e) of the indentation on Al-2.6%Mg. The first dislocations are nucleated between (a) and (b), i.e. prior to the apparent yield point. The nucleation is evidenced by an abrupt change in image contrast: before nucleation, only thickness fringes can be seen, whereas more complex contrast features become visible at the instant of nucleation.
see http://www.dehosson.fmns.rug.nl/.

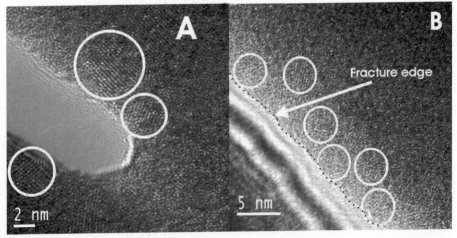

Figure 2. HRTEM micrographs revealing (a) a rapid-propagation induced meniscus at the crack-tip for $Cu_{47}Ti_{33}Zr_{11}Ni_6Sn_2Si_1$ ribbon (b) HRTEM image revealing nano-crystallization close to the fracture surface edge for $Zr_{50}Cu_{30}Ni_{10}Al_{10}$

Observing Nanosecond Phenomena at the Nanoscale with the Dynamic Transmission Electron Microscope

G.H. Campbell[1], N.D. Browning[1,2], J.S. Kim[1,2], W.E. King[1], T. LaGrange[1], B.W. Reed[1], and M.L. Taheri[1]

1. Lawrence Livermore National Laboratory, Livermore, CA, 94550, USA
2. University of California, Davis, CA, 95616, USA

ghcampbell@llnl.com

Keywords: DTEM, martensite, reactive foils, nanosecond microscopy

The dynamic transmission electron microscope (DTEM) [1] is a standard TEM that has been modified such that the electron beam can be operated with a single intense pulse of electrons ($>10^8$ e$^-$) with a pulse duration of just 15 ns. The short pulse of electrons is created via photoemission at the microscope cathode. Figure 1 shows a schematic layout of the instrument with the pulsed cathode laser operating in the ultraviolet (211 nm) brought into the column through a specially made additional section. This column section has a mirror to direct the laser up to the microscope gun. The usual microscope filament is replaced by a Ta disk photocathode, however it can also be operated in thermionic emission mode for normal operation of the microscope for alignment and experiment setup. Additional modifications have also been made to the optical design of the condenser lens system.

The goal of the instrument is to enable *in situ* experiments that can be observed with high time resolution. The experiments currently use a second laser to initiate the dynamic response of interest and this laser is brought through an x-ray port. The relative timing of the pulses from the two laser systems sets the time of the observation relative to the initiation of the event and can be set from 10 ns to 100 μs. The specimen drive pulse duration is comparable to the cathode laser pulse duration and the wavelength can be chosen as 1064 nm, 532 nm, or 355 nm, depending on experimental requirements.

The DTEM has been used to investigate a number of rapid phenomena in materials. It has characterized the rate of the martensitic transformation of Ti [2] as it transforms from α to β phase at elevated temperature. The data have created the first time-temperature-transformation (TTT) diagram with nanosecond time resolution and have illuminated a mechanism changes that depends on temperature. DTEM has also been used to study reactive multilayer films [3] that sustain a reaction front speed greater than 10 m/s (Fig. 2). A transient cellular morphology to the reaction front was identified [4].

1. T. LaGrange, et al., Appl. Phys. Lett. **89** (2006), p. 044105
2. T. LaGrange, G.H. Campbell, P.E.A. Turchi, and W.E. King, Acta Mater. **55** (2007), p. 5211
3. J.S. Kim, B.W. Reed, N.D. Browning, and G.H. Campbell, Microscopy and Microanalysis **12** (Suppl. 2) (2006) p. 148
4. Work performed under the auspices of the US Dept. of Energy, Office of Basic Energy Sciences by Lawrence Livermore National Lab under contract DE-AC52-07NA27344.

M. Luysberg, K. Tillmann, T. Weirich (Eds.): EMC 2008, Vol. 1: Instrumentation and Methods, pp. 465–466, DOI: 10.1007/978-3-540-85156-1_233, © Springer-Verlag Berlin Heidelberg 2008

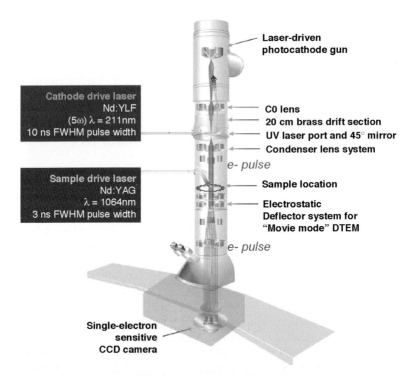

Figure 1. A Schematic representation of the DTEM at LLNL.

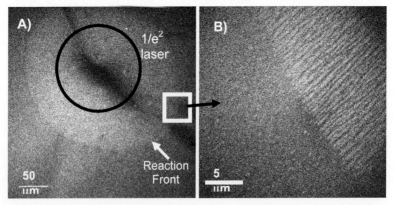

Figure 2. 15ns exposure images of NiAl reaction front morphology, A) low magnification image. Note the sharp transition between unreacted (dark) and reacted (light), indicating the position of reaction front. B) higher magnification pulsed image of cellular microstructure which forms behind reaction front.

TEM characterization of nanostructures formed from SiGeO films: effect of electron beam irradiation

C. Ballesteros[1], M. I. Ortiz[1], B. Morana[2], A. Rodríguez[2], and T. Rodríguez[2]

1. Departamento de Física, E.P.S, Universidad Carlos III de Madrid, Avda. Universidad 30, 28911 Leganés Madrid, Spain.
2. Departamento de Tecnología Electrónica, E.T.S.I. de Telecomunicación, Universidad Politécnica de Madrid. 28040 Madrid, Spain

balleste@fis.uc3m.es

Keywords: Ge-nanoparticles, electron beam irradiation in TEM, chemical vapour deposition

Group IV (Si or Ge) nanocrystals embedded in a dielectric medium have potential applications in electronic and optoelectronic devices. Previous results indicate that low pressure chemical vapour deposition (LPCVD) of amorphous discontinuous SiGe films in SiO_2 on top of Si-substrates and their subsequent crystallization permit to obtain a homogeneous distribution of SiGe nanocrystals, the process being fully compatible with CMOS technology [1]. In this work, Si-Ge-O films were deposited by LPCVD and furnace annealed at temperatures from 600 to 1100 °C for times up to 1 hour in a N_2 atmosphere to segregate the expected excess of Si and/or Ge in the form of nanocrystals embedded in an oxide matrix. High-resolution transmission electron microscopy (HREM) and electron-diffraction pattern simulation by fast Fourier transform was used to analyze the crystal structure and distribution of the nanoparticles and their stability under in situ electron irradiation. TEM observation and electron beam irradiation were performed in a Philips Tecnai 20F FEG microscope operating at 200 kV.

In the as-deposited samples an amorphous layer containing Si, Ge, and O is observed. After annealing at 600°C a distribution of Ge-rich nanocrystals embedded in SiO_2 is obtained. In Figure 1 a cross-sectional multibeam image of a sample annealed at 600°C along the <110> direction of the Si-substrate is shown.

As a general fact, electron beam irradiation induces heating and/or generation of defects, from either direct atoms displacement or dangling bonds reorganization. TEM "in situ" electron irradiation modifies the crystal structure and gives useful information on the stability of SiGe nanostructures [2]. We have study the structure modifications induced by TEM electron irradiation on the Ge-rich nanostructures. Figure 1b,c shows the effect of 1 min electron beam irradiation on sample annealed at 600°C. Electron beam irradiation induces structural alternate order-disorder transitions in annealed crystallized Ge-rich nanoparticles and the formation of nanocrystals in the as-deposited amorphous layers

1. M. I. Ortiz, J. Sangrador, A. Rodríguez, T. Rodríguez, A. Kling, N,. Franco, N.P. Barradas and C.Ballesteros. Phys. Stat. Sol. (a) **203**, (2006),1284
2. M.I.Ortiz, A.Rodríguez, J.Sangrador, T.Rodríguez and C.Ballesteros. J.Phys.Conf.Ser. (2008)

M. Luysberg, K. Tillmann, T. Weirich (Eds.): EMC 2008, Vol. 1: Instrumentation and Methods, pp. 467–468, DOI: 10.1007/978-3-540-85156-1_234, © Springer-Verlag Berlin Heidelberg 2008

468

3. TEM work has been carried out at the LABMET (Red de Laboratorios CAM) funded by the Spanish government, CICYT Project MEC-MAT2007-66181-C03-01.

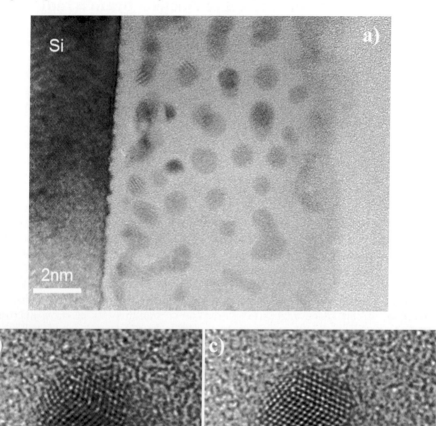

Figure 1. a) Multibeam cross-sectional micrograph along the <110> direction of the Si-substrate of a sample annealed at 600°C, a distribution of Ge-rich nanocrystals embedded in SiO_2 are imaged. b) High-resolution micrograph of a nanocrystal. c) Same area after 1min irradiation.

In situ Lorentz microscopy in an alternating current magnetic field

Z. Akase[1], H. Kakinuma[1], D. Shindo[1], M. Inoue[2]

1. Institute of Multidisciplinary Research for Advanced Materials, Tohoku University, Sendai 980-8577, Japan
2. JEOL Ltd. Akishima 196-8558, Japan

akase@tagen.tohoku.ac.jp

Keywords: Lorentz microscopy, in situ observation, alternating current magnetic field

By utilizing an alternating current (AC) magnetizing system recently developed on our transmission electron microscope (TEM), in-situ Lorentz microscope observations of soft magnetic materials were carried out in an AC magnetic field [1]. The domain walls moved smoothly in a low frequency AC magnetic field. It was demonstrated that in-situ Lorentz microscopy in an AC magnetic field is very useful for the investigation of interactions between the microstructure and the motion of magnetic domain walls in soft magnetic materials.

Figure 1 shows a schematic illustration of the magnetizing system. A specimen is set on the magnetizing stage which has an electromagnet and yokes on the tip of the specimen holder in order to apply horizontal magnetic field on the specimen. The object lens on the TEM is the so-called "Lorentz lens" which does not disturb the intrinsic magnetic domain structure of the specimen. Above the specimen, there are two deflectors additionally. AC is divided into the deflector coils and the magnetizing stage coil synchronously. The two deflectors control the incident direction of electron beam in order to control the defocus value of objective (mini) lens. Figure 2 shows how the defocus is controlled. The electron trajectory below the specimen has curvature because the magnetic field from the magnetizing stage exists not only on the specimen but also below the specimen. This curvature makes "a pivot image". If the object mini lens is focused on the pivot image, the image on the screen does not vibrate when the function generator provides sine wave current in the magnetizing stage (Fig.2 (a)). The height of the pivot image can be controlled by the incident beam direction (Fig.2 (b)). Figure 3 shows the Lorentz micrographs of a Sendust in the AC magnetic field (8.0 kA/m, 1Hz) captured from a videotape. Fig. 3 (a,b) were taken in over focus, and Fig. 3 (c,d) were taken in under focus. Figure 4 shows Lorentz micrographs of an electrical steel sheet in an AC magnetic field (2.4 kA/m, 1 Hz). In the pictures on the right, a domain wall contrast and two AlN precipitates are indicated by dotted line and arrowheads respectively. By using diffraction contrast, interactions between precipitates and the motion of the magnetic domain walls were visualized and clarified. Eventually, the magnetic domain walls were found to be pinned at strain fields around precipitates.

1.   Z.Akase, D. Shindo, M. Inoue and A. Taniyama, *Mater. Trans.* **48** (2007) 2626-2630

M. Luysberg, K. Tillmann, T. Weirich (Eds.): EMC 2008, Vol. 1: Instrumentation and Methods, pp. 469–470, DOI: 10.1007/978-3-540-85156-1_235, © Springer-Verlag Berlin Heidelberg 2008

470

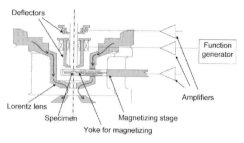

Figure 1. A schematic illustration of the magnetizing system.

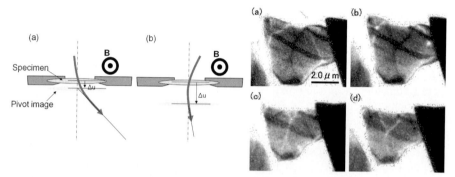

Figure 2. The electron trajectory near the specimen. The height of pivot image depends on the incident direction of electron beam.

Figure 3. Lorentz micrographs of Sendust in an AC magnetic field (8.0 kA/m, 1Hz). (a,b) over focus, (c,d) under focus.

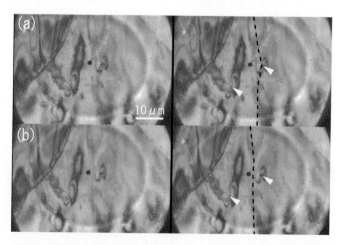

Figure 4. Lorentz micrographs of an electrical steel sheet in an AC magnetic field (1.0 Hz, 2.4 kA/m). In the pictures on the right, a domain wall contrast and two AlN precipitates are indicated by dotted line and arrowheads, respectively.

In-situ transmission electron microscopy investigation of TiO islands nucleating on SrTiO₃ (100) and (110) surfaces at high temperature

P.J. Bellina, F. Phillipp and P.A. van Aken

Stuttgart Center for Electron Microscopy, Max Planck Institute for Metals Research, Heisenbergstr. 3, 70569 Stuttgart, Germany

paul@mf.mpg.de

Keywords: $SrTiO_3$, HRTEM, SAED

Oxide surfaces play an important role in technological applications, being used in catalysis and thin films growth. In particular strontium titanate ($SrTiO_3$, STO) is a perovskite with a high dielectric constant, which is applied as a substrate material for the growth of high-temperature superconductor thin films and is becoming a promising gate dielectric in electronic technology. Several factors are known to affect the surface of STO; these include annealing (time and temperature), sample's environment (reducing or oxidizing), and Ar^+-ion thinning.

In this work the behaviour of (100) and (110) STO surfaces has been investigated in situ by high-resolution transmission electron microscopy (HRTEM) and selected area electron diffraction (SAED) from room temperature up to temperatures higher than 900 °C. TEM samples were obtained from STO single crystals and thinned by Ar^+-ion milling.

At room temperature the samples show superlattice reflections (SLR) in the <111> direction, supposedly as a consequence of the Ar^+-ion thinning. This superlattice changes its orientation reversibly at high temperature, rearranging in the <110> direction on the (110) surface and in the <100> direction on the (100) surface (Fig. 1). The latter was already observed by TEM, and attributed to the reconstruction of a surface monolayer [1]. The SLR are interpreted to arise from ordered arrays of oxygen and strontium vacancies in the surface region. In fact the depletion of these elements after annealing has already been observed by Auger electron spectroscopy [2].

At temperatures higher than 900 °C the surfaces are faceted into the (100) and (110) planes. The faceting allows us to observe atomically flat surfaces in cross section and in plan view. Such condition enables the investigation in HRTEM of the nucleation of the TiO islands (Fig. 2), which were already observed previously [3].

It seems reasonable that the depletion of Sr and O leads to the observed superstructure in the near surface region and to a phase separation at the surface, which forms the TiO termination layer. It is thought that the so formed epitaxial TiO layer evolves in the formation of the TiO islands in order to reduce the epitaxial stresses, following a Stranski-Krastanov mode of growth.

M. Luysberg, K. Tillmann, T. Weirich (Eds.): EMC 2008, Vol. 1: Instrumentation and Methods, pp. 471–472, DOI: 10.1007/978-3-540-85156-1_236, © Springer-Verlag Berlin Heidelberg 2008

472

1. N. Erdman, K.R. Poeppelmeier, M. Asta, O. Warschkow, D.E. Ellis, and L.D. Marks, Nature **419** (2002), p. 55.
2. D.T. Newell, A. Harrison, F. Silly, and M.R. Castell, Physical Review B **75** (2007), p. 205429.
3. S.B. Lee, F. Phillipp, W. Sigle, and M. Rühle, Ultramicroscopy **104** (2005), p. 30.

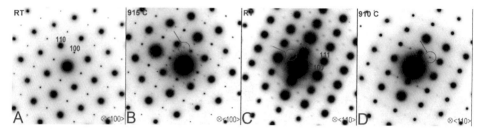

Figure 1. SAED patterns acquired at the Jeol ARM 1250: A) <100> zone axis (ZA) at room temperature; B) <100> ZA at 915 °C SLR <100> direction; C) <110> ZA at room temperature, SLR in <111> direction; D) <110> ZA at 910 °C, SLR in <110> direction.

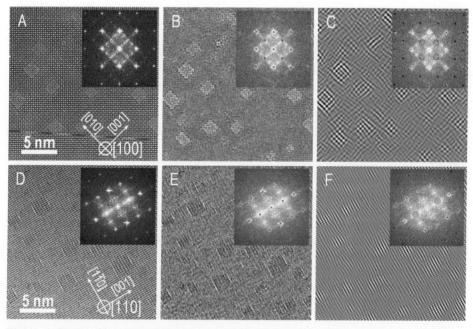

Figure 2. Processing of HRTEM images acquired at the Jeol ARM 1250, each image is presented with FFT in the inset: A) and D) unfiltered images acquired at 980 °C showing TiO islands on the STO surface; B) and E) images obtained from A and D removing the spatial frequencies (highlighted in blue) of STO by Bragg filtering; C and F) images obtained from B and E selecting the frequencies highlighted by red circles.

Probing integration strength of colloidal spheres self-assembled from TiO₂ nanocrystals by in-situ TEM indentation

C.Q. Chen, Y.T. Pei, J.Th.M. De Hosson

Department of Applied Physics, Netherlands Materials Innovation Institute, University of Groningen, Nijenborgh 4, 9747 AG Groningen, the Netherlands

c.chen@rug.nl

Keywords: self assembly, integration, strength, in-situ TEM, nanoindentation

Small building blocks such as molecules and nanoparticles, with controlled size, shape, and properties, have been recently utilized as artificial building blocks to assemble two- or three-dimensional structures via "bottom up" processes. Unlike the well known ionic, metallic, or covalent bonds combining "real" atoms, the nature of the bonding between these small aggregates, and the mechanical stability of the self-assemblies, are far from clear. Recent development of in-situ TEM nanoindentation technique enables simultaneous monitoring of mechanical and structural responses of a nanosized cluster under externally applied load at real time in high resolution TEM.

In this context, in-situ TEM indentation is performed on nanoscale colloidal spheres self-assembled from TiO_2 nanocrystals, with the size of the clusters ranging from 80-350 nm, while the composing nanocrystals are 6-8 nm in diameter.

Our results show that these self-assemblies are capable of supporting significant contact stress up to 3 GPa. The load responses of these self assemblies indented with a flat diamond punch show a load drop under displacement control (Fig. 1), or displacement burst under load control. This kind of instability is correlated to the surface crack opening that occurs under indentation.

Further, the colloidal particles show a time-dependent behavior and stress relaxation when keeping the indenter tip at an applied indentation depth. Detailed analysis of these indentation behaviors provides an insight into the interparticle bonding inside the self assemblies. Possible size effects on the load and structural response are also investigated.

M. Luysberg, K. Tillmann, T. Weirich (Eds.): EMC 2008, Vol. 1: Instrumentation and Methods, pp. 473–474, DOI: 10.1007/978-3-540-85156-1_237, © Springer-Verlag Berlin Heidelberg 2008

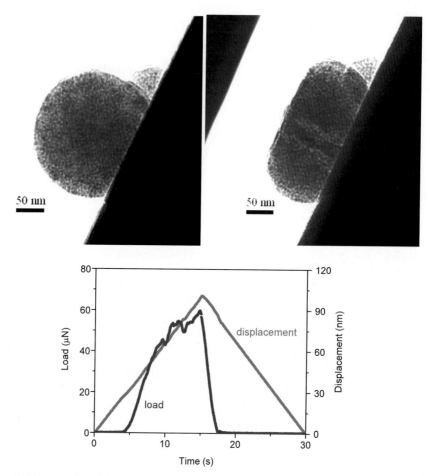

Figure 1. Structural and load response of a colloidal sphere of 223 nm diameter self-assembled from TiO_2 nanocrystals of 6-8 nm size: bright field images showing morphology of a colloidal particle (a) before and (b) after indentation with a flat punch indenter. The two bands of lighter contrast in (b) correspond to the appearance of two obvious load drops in the load curve in (c).

Installation and operation of an *in situ* electron microscopy facility

M.W. Fay[1], H.K. Edwards[2], M. Zong[3], K.J. Thurecht[3], S.M. Howdle[3] and P.D. Brown[2]

1. Nottingham Nanotechnology and Nanoscience Centre, University of Nottingham, Nottingham NG7 2RD, UK
2. School of M3, University of Nottingham, Nottingham NG7 2RD, UK
3. School of Chemistry, University of Nottingham, Nottingham NG7 2RD, UK

michael.fay@nottingham.ac.uk
Keywords: in-situ TEM, cryomicroscopy

The Nottingham Nanotechnology and Nanoscience Centre has recently commissioned an electron microscopy facility comprising a FIB-SEM with cryo transfer system and a field emission gun transmission electron microscope equipped with an extensive range of sample holders that allow *in situ* analysis of a wide range of materials to be carried out (figure 1). These instruments are configured to serve a wide range of interests across the fields of the life sciences and physical sciences.

The TEM is configured to operate fully at both 200keV and 100keV, allowing beam sensitive materials to be analysed, such as single walled carbon nanotubes as described in the presentation by Khlobystov [1].

The cryo-transfer system on the FIB-SEM, in conjunction with cryotransfer/tomography stage and low-dose imaging on the TEM, allow a wide range of biological materials to be characterised. By combining the facilities available in the centre, cryotransfer from cryo-FIB-SEM has been successfully carried out using this equipment, as described in the presentation by Edwards et al [2].

Digital video acquisition allows the recording of dynamic *in situ* experiments. Electrical and STM-TEM holders allow the direct correlation of structure and properties to be carried out in the microscope column.

The TEM is also equipped with an environmental cell and heating holder, allowing the investigation of temperature sensitive materials. By way of example, the analysis of phase separation of fluorine in a polymer material at room temperature and after heating to near Tg is described (figure 2).

These holders are backed with the expected array of detectors; i.e. EDX, EFTEM/EELS, BF/DF STEM and conventional imaging. Focal and Tilt Series Reconstruction software allows the further improvement of the resolution.

An overview of the microscopy suite installation and initial experimental results will be presented.

1. A.N. Khlobystov, M.W. Fay and P.D. Brown, European Microscopy Congress 2008
2. H.K. Edwards, M.W. Fay, P.D. Brown, et al European Microscopy Congress 2008
3. We kindly acknowledge the University of Nottingham and the four partner Schools of Chemistry, M3, Pharmacy and Physics and Astronomy for supporting this initiative through the SRIF3 funding scheme.

M. Luysberg, K. Tillmann, T. Weirich (Eds.): EMC 2008, Vol. 1: Instrumentation and Methods, pp. 475–476, DOI: 10.1007/978-3-540-85156-1_238, © Springer-Verlag Berlin Heidelberg 2008

476

Figure 1. (left) The FEI Quanta200 3D FIB-SEM with Quorum Technologies PT2000 cryo-transfer unit and (right) the JEOL 2100F FEGTEM in the Nottingham Nanotechnology and Nanoscience Centre

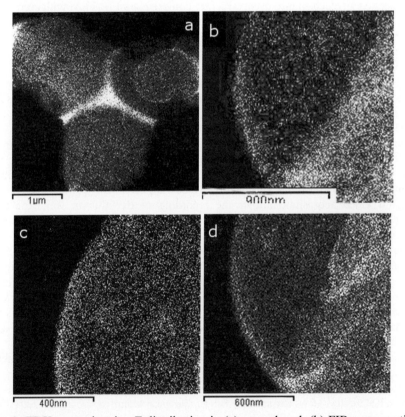

Figure 2. EDX maps showing F distribution in (a) as analysed, (b) FIB cross sectioned and (c,d) cross sectioned polymer structures after *in situ* heat treatment for 1.5 hours at 80 and 110°C

Bringing chemical reactions to life: environmental transmission electron microscopy (E-TEM)

B. Freitag[1], S.M. Kim[2,3], D.N. Zakharov[3], E.A. Stach[2,3] and D.J. Stokes[1]

1. FEI Company, PO Box 80066, 5600 KA Eindhoven, The Netherlands
2. School of Materials Engineering, Neil Armstrong Hall of Engineering, 701 West Stadium Avenue, West Lafayette, IN 47907-2045, USA
3. Birck Nanotechnology Center, Purdue University, 1205 W State Street, West Lafayette, IN 47907-2057, USA

Bert.Freitag@fei.com

Keywords: Environmental TEM, *in situ*, chemical reactions

Electron microscopy can provide more than just static observations and characterization of materials. For example, the environmental transmission electron microscope (E-TEM) enables the synthesis of materials in the TEM, and allows us to study dynamic behaviour under the influence of gases and temperature excursions, while maintaining atomic resolution. By varying the temperature, pressure and composition of the gaseous environment, it becomes possible to directly interrogate chemical processes using both imaging and spectroscopic techniques. This allows a deep understanding of both the mechanisms and kinetics of reactions at the nanoscale, as evidenced by the growing body of literature (see, for example [1-6]).

As a specific example, we will present STEM and TEM analyses of the role of catalysts in both nanotube synthesis and in NOx conversion. In the case of nanotube synthesis, we will demonstrate how E-TEM studies can be used to directly investigate the shape of the catalysts responsible for nanotube nucleation, and will correlate this with quantification of nanotube chirality. In the case of NOx conversion, real time observations of oxidation and reduction are used to investigate the role of Pt particle size, structure and electronic structure on NOx conversion. These investigations highlight the utility of real time observations of both structural changes in electronic structure effects during nanoscale observations of reaction processes.

We also introduce a new member of the Titan family specifically designed for E-TEM studies, with a pressure range of up to 40mbar (4 kPa) of gas at the specimen area and a wide range of temperatures. So far, this has been tested for nitrogen gas, with the ultimate aim of allowing *in situ* experiments involving water vapour. This special microscope is equipped with a `gas sniffer` and a plasma cleaner in the latest standard versions and can be operated in non-E-TEM mode with the standard sub-Angstrom specifications of a conventional Titan. The performance at different gas pressures and gaseous species will be discussed and the influence on resolution will be demonstrated. Specifically, we will demonstrate the effect of gas scattering on the coherence envelope of the contrast transfer function of the microscope, and correlate this with existing models of electron scattering from gaseous species. Figure 1 shows an information transfer of 1.2 Å, obtained for a gold specimen in an aberration -corrected E-TEM.

M. Luysberg, K. Tillmann, T. Weirich (Eds.): EMC 2008, Vol. 1: Instrumentation and Methods, pp. 477–478, DOI: 10.1007/978-3-540-85156-1_239, © Springer-Verlag Berlin Heidelberg 2008

478

1. Sharma, R. and Weiss, K., Microscopy Research and Technique **42**(4) (1998), p. 270-280.
2. P.L. Hansen, J.B. Wagner, S. Helveg, J.R. Rostrup-Nielsen, B.S. Clausen and Topsoe, H., Science **295**(5562) (2002) p. 2053-2055.
3. S. Helveg, C. Lopez-Cartes, J. Sehested, P.L. Hansen, B.S. Clausen, J.R. Rostrup-Nielsen, F. Abild-Pedersen and Norskov, J.K., Nature **427** (2004) p. 426-429.
4. R. Sharma, P. Rez, M. Brown, G.H. Du, and M.M.J. Treacy, Nanotechnology **18**(12) (2007)
5. S. Hofmann, R. Sharma, C. Ducati, G. Du, C. Mattevi, C. Cepek, M. Cantoro, S. Pisana, A. Parvez, F. Cervantes-Sodi, A.C. Ferrari, R. Dunin-Borkowski, S. Lizzit, L. Petaccia, A. Goldoni and J. Robertson, Nano Letters **7**(3) (2007) p. 602-608.
6. P.L. Gai, R. Sharma and F.M. Ross, MRS Bulletin **33**(2) (2008) p. 107-114.

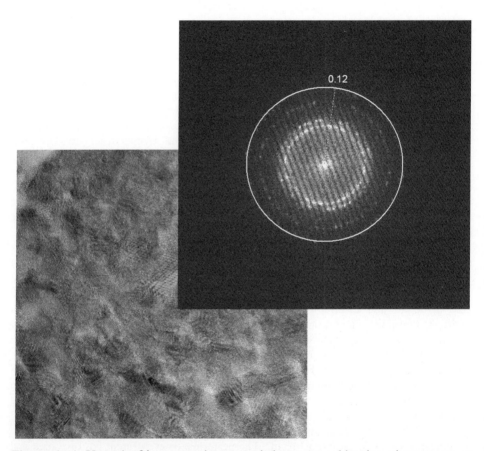

Figure 1. A Young's fringe experiment carried out on gold using nitrogen gas at pressure p = 5 mbar (500 Pa) with an aberration-corrected E-TEM. The information limit in this case is 1.2 Å.

Dynamic *in situ* experiments in a 1Å double aberration corrected environment

Pratibha L. Gai[*] and Edward D. Boyes[#]

1. Departments of Chemistry and Physics* and Departments of Physics and Electronics#
2. The University of York, The York JEOL NanoCentre, Science Park, Heslington, York YO10 5BR, UK

pgb500@york.ac.uk and eb520@york.ac.uk

Keywords: *in situ*, aberration-correction, nanoparticle

Dynamic in-situ experiments access metastable states and the mechanisms of reactions, phase transformations and nanostructure development (and in some cases destruction), as well as contributing key information to synthesis, processing and stability studies. They have proved particularly informative in catalysis [1, 2, 3], semiconductor studies [4] and metallurgy. To be useful they need to be conducted under controlled conditions of temperature, and where appropriate environment, and of course always with a minimal, mechanistically non-invasive influence of the electron beam. This means the microscope must be able to accept a hot stage, if heating is the driving force for change being investigated, and with all other conditions held constant.

The need to accommodate a hot stage, perhaps for the first time in an aberration corrected machine, has been one of the more important criteria driving the specification of the JEOL 2200FS double aberration corrected FEG TEM/STEM in the new York JEOL Nanocentre at the University of York. Since aligning the sample into a zone axis orientation is a prerequisite for atomic resolution electron microscopy of crystalline materials, an increased specimen tilt range was also desired. Both these conditions of course benefit from a larger gap (HRP) objective lens polepiece, and in this case the Cs (C3) aberration correctors on both the STEM probe forming and TEM image sides of the instrument were used to provide the desired expanded specimen geometry with minimal effect on the 1Å imaging performance of the system.

The advantages of the configuration we have adopted, include, promoting a contrast transfer function (CTF) extending to higher spatial frequencies and imaging resolution; allowing image recording at close to zero defocus, including to strengthen information at internal interfaces and nanoparticle surfaces; analysing small (<2nm) nanoparticles and clusters on supports, using high resolution TEM as well as HR STEM; facilitating HAADF STEM and extending HAADF STEM resolution to 1Å (0.1nm) and below.

Figure 1 shows in-situ data from Pt-Pd nanoparticles on carbon at 500°C, with the FFT/optical diffractogram illustrating achievable resolution of 0.1nm with a hot stage in the JEOL 2200 FS 2AC TEM/STEM at the University of York.

Figure 2 illustrates gold nanoparticle sintering also at 500°C and Figure 3 shows image detail from a performance critical size range of gold nanoparticle at 25°C.

M. Luysberg, K. Tillmann, T. Weirich (Eds.): EMC 2008, Vol. 1: Instrumentation and Methods, pp. 479–480, DOI: 10.1007/978-3-540-85156-1_240, © Springer-Verlag Berlin Heidelberg 2008

1. P L Gai, Phil Mag 43 (1981) 841
2. E D Boyes and P L Gai, Ultramicroscopy 67 (1997) 219
3. P.L. Gai, E.D. Boyes, S. Helveg, P.Hansen, S.Giorgio, C.Henry, MRS Bull. 32 (2007) 1044
4. R Sinclair, T Yamashita and F Ponce, Nature 290 (1981) 386
5. This work and the new facilities are supported by funding from the regional development agency, Yorkshire Forward, the European Union through the European Regional Development Fund, JEOL (UK) Ltd and The University of York. Ian Wright assisted.

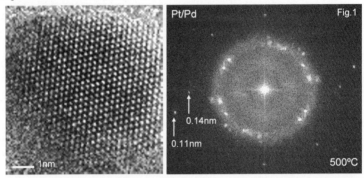

Figure 1. shows in situ data from Pt-Pd nanoparticles on carbon at 500°C, with the FFT/optical diffractogram with achievable image resolution of 0.1nm using a hot stage in the JEOL 2200 FS 2AC TEM/STEM at the University of York.

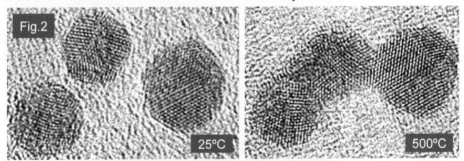

Figure 2. illustrates gold nanoparticle sintering also at 500°C

Figure 3. shows image detail from a performance critical size range of gold nanoparticle at 25°C

A very high temperature (2000°C) stage for atomic resolution *in situ* ETEM

Pratibha L. Gai[1,2] and Edward D. Boyes[2,3]

Departments of Chemistry[1], Physics[2] and Electronics[3]
The York JEOL Nanocentre, University of York, York YO10, 5DD, U.K.

pgb500@york.ac.uk
Keywords: *in situ*, hot-stage, graphite

Some important applications in the chemical and materials sciences require very high temperatures under gaseous atmospheres. For example, the formation of ceramics, carbon nanotubes, the conversion of carbon based systems, and phase transformations in white pigments require temperatures in the region of 1100 °C to 2000 °C in reactive gases. Commercial holders for ETEM are limited to about 1000 °C [1-3]. Therefore, there is the need to develop a very high temperature hot stage (about 2000 °C) for applications in the atomic resolution-*in situ* environmental transmission electron microscopy (ETEM).

We have designed a very high temperature (VHT) holder to meet these demands. The environmental holder design is based on the early work in vacuum by Kamino and Saka [4]. We have used stainless steel and brass components for the holder and tungsten (W) filaments. The diameter of the filament was selected after testing filaments with different sizes. The holder was extensively safety tested including for x-ray radiation and it has electrostatic screening.

Figure 1 (a) and (b) show the holder tip at different magnifications. We calibrated temperatures using an optical pyrometer in a test rig. Temperatures were recorded from 800°C up to 2200°C and up to 2300°C in a few cases. Temperature versus milliamp (mA) calibration data with optical pyrometer readings through glass are shown in the table. In the test rig, experiments were performed on the VHT holder in different atmospheres, namely nitrogen, hydrogen and hydrocarbons.

Samples of cellulose were then attached to the filament and the holder was transferred to the ETEM. Nitrogen gas was introduced at a pressure of 1mbar. The transformation of cellulose into graphite was observed at about 2000 °C (Figure 2). The design and initial applications of the VHT holder have been carried out and more applications are planned.

1. P.L. Gai, Catal. Rev. Sci. Eng. 34 (1992) 1.
2. E.D. Boyes and P.L. Gai, Ultramicroscopy. 67 (1997) 219.
3. P.L. Gai, E.D. Boyes, S. Helveg, P.Hansen, S.Giorgio, C.Henry, MRS Bull. 32 (2007) 1044.
4. T. Kamino and H. Saka, Microsc. Microanal. Microstruct. 4 (1993) 127.

M. Luysberg, K. Tillmann, T. Weirich (Eds.): EMC 2008, Vol. 1: Instrumentation and Methods, pp. 481–482, DOI: 10.1007/978-3-540-85156-1_241, © Springer-Verlag Berlin Heidelberg 2008

482

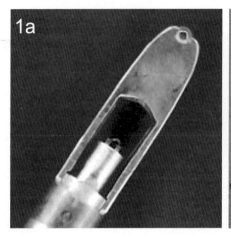 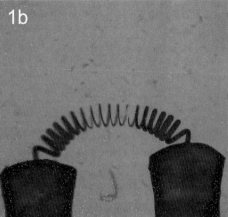

mA	Temp °C
151.3	859
157.6	1015
159.6	1067
170.5	1245
180.4	1355
196.1	1532
221.0	1720
237.0	1850
247.0	1922
255.0	1990
285.0	2020

Figure 1. High temperature hot stage (above) with calibration (left)

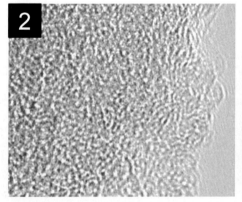 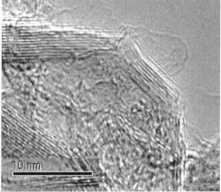

Figure 2. Transformation of cellulose precursor to graphitic structure at ~2000°C (below)

Environmental High Resolution Electron Microscopy With a Closed Ecell: Application to Catalysts

S. Giorgio, M. Cabié, C.R. Henry

CINAM- CNRS, Campus de Luminy, Case 913, 13288 Marseille Cedex 9   France

giorgio@crmcn.univ-mrs.fr

Keywords: environmental electron microscopy, catalysts

Metal nano-clusters have catalytic properties related to their size, their shape and crystalline structure. The most important advances for catalysis in electron microscopy, was the possibility to insert gas in the microscope during observations with the aim to close simultaneously the material and the pressure gap in catalysis [1-3].

In a standard high resolution electron microscope (Jeol 3010), an environmental sample holder with carbon windows designed by Jeol, has been used for in situ observations at the atomic scale of Au and Pd catalysts, during oxydo reduction cycles [4]. The gas pressure at the level of the sample is lower than 10 mbar and the temperature can be adjusted between room temperature and 350 °C. A mass spectrometer is connected at the exit of the sample holder to analyse the products of the catalytic reaction on the sample.

In figures 1, the same Au particle supported on amorphous carbon is in situ observed under pure H_2 and pure O_2 at a pressure P = 4 mbar.

The Au particle contaminated after the air transfer (1), is facetted by adsorption of H_2 (2-3), then rounded after circulation of O_2 (4-5).

Hydrogen induces faceting by (001) and (111) faces which corresponds to the Wulff shape while O_2 induces rounding of the shape resulting for a strong interaction with oxygen.

This result is surprising whereas it is admitted that O_2 is adsorbed on the edges, corners and defects of Au clusters and has little interaction with low index facets [5-8].

New fields of discussion are opened for the understanding of catalytic properties of small Au clusters.

1. R. Sharma, P. Crozier in "Transmission Electron Microscopy in Nanotechnology", Nan Yao & ZL Wang, Eds., Springer – Verlag and Tsinghua University Press, (2005) 531- 565
2. E.D. Boyes and P.L.Gai, Ultramic. 67, 219, 1997
3. PL. Hansen, J. B. Wagner, Stig Helveg, J.R.Rostrup-Nielsen, B.S. Clausen, H. Topsoe, Science, 295 (2002) 2053
4. S. Giorgio, S. Sao Joao, S. Nitsche, G. Sitja, CR. Henry, Ultramicroscopy 106 (6) (2006) 503
5. V.A. Bondzie, S.C. Parker and C.T. Campbell, J. Vac. Sci. Technol. A 17 (1999) 1717
6. J.M. Gottfried, K.J. Schidt, S.L..M. Schroeder, K. Christmann, Surf. Sci. 525 (2003) 197
7. J.D. Stiehl, T.S. Kim, S.M. McClure and C.B. Mullins, J.Am.Chem. Soc., 126 (2004) 1606
8. I.N. Remediakis, N. Lopez, J.K. Norskov, Angew Chem. Int. Ed. 44 (2005) 1824

M. Luysberg, K. Tillmann, T. Weirich (Eds.): EMC 2008, Vol. 1: Instrumentation and Methods, pp. 483–484, DOI: 10.1007/978-3-540-85156-1_242, © Springer-Verlag Berlin Heidelberg 2008

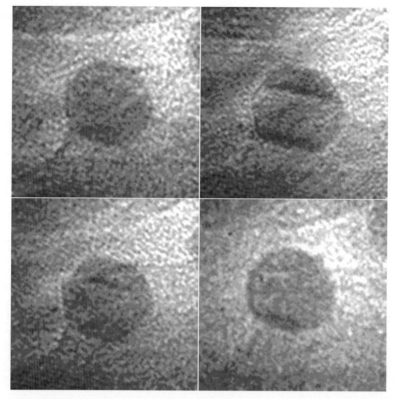

Figure 1. Au cluster supported on TiO$_2$ during an oxido- reduction cycle

Pulsed-mode photon and electron microscopy surveyed

A. Howie

Cavendish Laboratory, University of Cambridge, Cambridge CB3 0HE, UK

ah30@cam.ac.uk

Keywords: pulsed-mode microscopy, pump-probe microscopy, electron-photon microscopy

Photons are superior to electrons for microscopy except at highest spatial resolution. For ancillary spectroscopy they cover an enormous energy range with great precision and selectivity in excitation. Highly coherent, tuneable photon beams with sufficient intensity can also exploit nonlinear interactions such as multi-photon excitation, sum-difference or Raman spectroscopy reaching sub-wavelength resolution via the tip field enhancement effect. Thus subsurface single spins in silica were imaged at 25 nm in AFM magnetic resonance using microwave field with cantilever frequency interruptions [1]. Raman peaks from individual adenine molecules were identified with 50 nm spatial resolution in the CARS technique with three optical frequency pump fields [2].

Pulsed mode excitation, again readily exploited with photons, offers a Fourier complement to spectroscopy in the linear response regime but more significantly opens a window to ultra-fast, time-resolved microscopy particularly for reversible, re-settable phenomena. Thus in fluorescence lifetime microscopy, an ultra-fast gating image intensifier allows ns timing in protein interaction studies [3]. The randomly pulsed nature of low current electron beams can be used for 10 ns timing measurements with coincidence detection techniques [4]. Far better electron pulse control is achieved with a photocathode driven with a regular train of fs photon pulses from a mode-locked laser at 100 MHz rate. In this way 10 ps pulses of 10 or less electrons were generated in an SEM study of InGaAS quantum dot structures [5]. Using a streak camera detector in photon counting mode, the evolution of 5 different CDL peaks coming from different parts of the 30 nm pyramidal quantum dot was followed with a time resolution of 10 ps. Gated detector timing over a whole image plane has also been achieved in pulsed photoelectron emission microscopy (PEEM) using a delay line detector [6].

Pump-probe microscopy is an attractive alternative to gated detector timing of dynamic events even if it is generally necessary to repeat the whole process as many as 10^8 times for each time delay τ between the pump and probe pulses in order to acquire an acceptable image. Purely photon-based operation is so far the best developed option here though in some cases the pump pulse lies in the RF or microwave region. In time-resolved scanning optical microscopy (TRSOM) for instance the pulse train from a mode-locked laser was used to trigger (in the 2-4 GHz frequency range) an EM driving field for a ferroelectric. Reflection images at submicron resolution were then obtained from the probe illumination [7, 8]. An X-ray synchrotron probe pulse following a lower frequency pump pulse triggered by the synchrotron electronics was used to observe the vibrations of a dislocation in lithium niobate under the influence of a 0.6 GHz surface acoustic wave [9]. Similarly magnetic domain switching processes in response to a 400 ps magnetic pump pulse were probed using the circular dichroism effect to yield a

M. Luysberg, K. Tillmann, T. Weirich (Eds.): EMC 2008, Vol. 1: Instrumentation and Methods, pp. 485–486, DOI: 10.1007/978-3-540-85156-1_243, © Springer-Verlag Berlin Heidelberg 2008

pulsed X-ray PEEM image at 20 nm spatial resolution [10]. In purely optical work, chirped 775 nm, 150 fs, 800 μJ laser pulses were used in ablation studies in Si with a frequency-doubled probe pulse for reflection image recording in a CCD camera [11]. Growth of molten Si regions was followed with 0.5 μm resolution. In a photolysis study, subdivision of the probe entrance pupil yielded an impressive sequence of 400 absorption spectra at 25 fs intervals, though inevitably at poor spatial resolution [12].

Photon-pump, electron-probe pulse microscopy has been pioneered in recent years so far yielding a spatial resolution of 1.46 nm and sub-100 fs timing by adding the results of many single electron pulses in a few seconds [13,14]. Dynamic studies of melting and other phase transitions have been made on the 2 ps time-scale, although so far the diffraction rather than the TEM imaging facility has been more useful [15,16]. STEM imaging with the electron probe pre-positioned on a defect or other feature of potential interest in the phase transition could perhaps be efficient for real space work.

Other noteworthy developments include two-photon PEEM imaging at 60 nm and 10 fs timing with movies of surface plasmon evolution [17,18]. For phenomena that do not repeat themselves so exactly at the image resolution employed, single shot operation is the only option. Despite Coulomb repulsion in the large pulses required for electron imaging, 20 nm spatial resolution has been obtained with 1.5 ns pulses [19]. Diffraction data good enough to resolve long-range order have however been collected with a single 700 fs pulse of just 4000 electrons [20]. Even shorter 25 fs pulses of 30 nm soft X-rays have been used for single shot diffraction and imaging at 90 nm spatial resolution using phase retrieval techniques [21,22]. A new frontier in ultra-fast imaging may be opened up if this approach can be pushed to shorter wavelengths and higher resolution.

1. D. Rugar, R. Budakian, H. J. Mamin and B. W. Chui, Nature 430 (2004), p. 329.
2. H. Watanabe, Y. Ishida, N. Hazayawa, Y. Inoue et al, Phys. Rev. B 69 (2004), p. 155418.
3. M. Elangovan, R. N. Day and A. Periasamy, J. Microsc., 205 (2002) p. 3.
4. H. Mullejans, A. L. Bleloch, A. Howie and M. Tomita, Ultramicrosc. 52 (1993) p. 360.
5. M. Merano, S. Sonderegger, A. Crottini, S. Collin et al, Nature 438 (2005), p. 479.
6. G. Schonhense and H. J. Elmers, Surf. and Interface Anal., 38 (2006) p. 1578.
7. C. Hubert, J. Levy E. Cukauskas and S,. W. Kirchoefer, Phys. Rev. Lett., 85 (2000) p. 1998.
8. J. Haenl, P. Irvin, W. Chang, R. Uecker, P.Reiche et al., Nature 430 (2004) p. 758.
9. D. Shilo and E. Zolotoyabko, Phys. Rev. Lett., 91 (2003) p. 115506.
10. H. Stoll, A. Puzik, B. van Waeyenberg et al., Appl. Phys. Lett., 84 (2004) p. 3328.
11. J. P.McDonald, J. Nees and S. M. Yasilove, J. Appl. Phys. 102 (2007) 063109.
12. P.R. Poulin and K. A. Nelson, Science 313 (2006) p. 1756.
13. V. A. Lobastov, R. Srinivasan and A. H. Zewail, Proc. N.A.S. 102 (2005) p. 7069.
14. H. S. Park, J. S. Baskin, O. H. Kwon and A. H. Zewail, Nanolett., 7 (2007) p. 2545.
15. N. Gedik, D-S. Yang, G. Logvenov, I. Bozovic and A. H. Zewail, Science 316 (2007) p. 425.
16. C-Y. Ruan, Y. Murooka, R. K. Ramani and R. A. Murdick, Nanolett., 7 (2007) p. 1290.
17. Kubo, Y. S. Jung, H.K. Kim and H. Petek, J. Phys. B 40 (2007) p. 470.
18. F.-J. Meyer zu Heringdorf, L. I. Chelaru, S. Mollenbeck et al., Surf. Sci., 601 (2007) p. 4700.
19. W. E. King, G. H. Campbell, A. Frank, B. Reed et al., J. Appl. Phys. 97 (2005) 111101.
20. J. Cao, H. Zao, H. Park, C. Tao, D. Kau et al., Appl. Phys. Lett., 83 (2003) p. 1044.
21. R. L. Sandberg, A. Paul, D. A. Raymondson et al., Phys. Rev. Lett., 99 (2007) 011301.
22. H. Chapman, A. Barty, M. J. Bogan, S. Boutet et al., Nature Phys., 2 (2006) p. 839.

In-situ Observation of Nano-particulate Gold Catalysts during Reaction by Closed-type Environmental-cell Transmission Electron Microscope

T. Kawasaki[1,2], H. Hasegawa[1], K. Ueda[1] and T. Tanji[3]

1. Dept. Electrical Eng. and Computer Sci., Nagoya Univ., Furo-cho, Chikusa-ku, Nagoya, 464-8603, Japan
2. PREST, Japan science and Technology Agency, 4-1-8 Honcho Kawaguchi, Saitama
3. EcoTopia Science Inst., Nagoya Univ., Furo-cho, Chikusa-ku, Nagoya, 464-8603

kawasaki@nuee.nagoya-u.ac.jp

Keywords: Environmental-cell TEM, Nano-particle, Gold catalyst

Gold exhibits catalytic activity when it is in the form of fine particles having a size of less than 10 nm and is tightly supported on specific metal oxides such as TiO_2, etc [1]. To reveal its mechanism, dynamic observation of the sample structures during the reaction by the transmission electron microscope (TEM) is quite essential. "Environmental-cell (E-cell) TEM" technique [2], which enables gas introduction around specimens, is one of the most powerful methods for the purpose. The authors also have developed the closed-type E-cell TEM [3], and had revealed that the shape of catalytic gold nano-particles are changing during reaction [4]. In the paper, we report about difference of the shape change of gold nano-particles between under reaction gas and non-reaction gas conditions.

Fig. 1 shows a schematic diagram of our E-cell TEM system. This consists of two developed apparatuses equipped with a conventional 200kV-TEM (H-8000, Hitachi). One is a closed-type E-cell specimen holder. This has a small gas room, called the E-cell, at its top and two pipes for the gas in/out. The gas in the E-cell is separated from vacuum with ultra-thin carbon films. The films used are specially developed with high toughness and anti-oxidative property. About 10nm thick films, less than half of conventional ones, enabling to withstand more than atmospheric pressure were achieved. The other developed apparatus is a gas control unit. This is the hand-made equipment but enables fine tuning of introducing gas pressure and so on. The completed system realizes the "real" catalytic reaction condition in the TEM.

In the present experiment, a specimen was nano-size gold particle supported on TiO_2. The gas introduced were 1% CO in dry air as a reaction gas and dry N_2 as a non-reaction gas. Their pressure around the specimen was set at about 750 Pa. Fig. 2 shows in-situ TEM images of a gold nano-particle picked up from among those recorded sequentially. In this case, the introduced gas was CO in dry air. Therefore, catalytic reaction happened on the catalyst surface; CO was oxidized into CO_2. As shown in these images, the shape of gold particle was dramatically changed. Various facets were appeared and disappeared. On the other hand, almost no change was observed in the case of non-reaction gas N_2 condition, as shown in Fig. 3. Although slight shape changes might occur due to electron beam irradiation in the case of Fig. 3, the difference of al-

M. Luysberg, K. Tillmann, T. Weirich (Eds.): EMC 2008, Vol. 1: Instrumentation and Methods, pp. 487–488, DOI: 10.1007/978-3-540-85156-1_244, © Springer-Verlag Berlin Heidelberg 2008

terations of the particle shape can be clearly shown by comparing with Fig. 2. There results prove that the catalytic reaction causes the shape changes of the gold nano-particles.

1   M. Haruta, Catalysis Today **36** (1997) p. 153.
2   P. L. Gai, Topics in Catalysis **21** (2002) p. 16.
3   T. Kawasaki *et al.*, Proc. of M and M 07 (2007) p. 644.
4   T. Kawasaki *et al.*, Proc. of 9th Inter-American Congress of Electron Microscopy 2007 (2007) p. 79F.

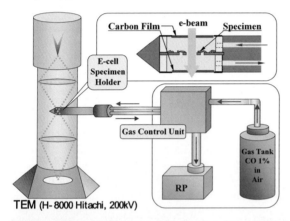

Figure 1. Schematic diagram of the developed E-cell TEM

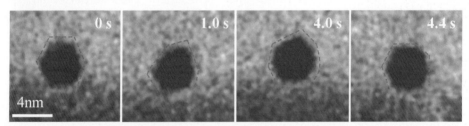

Figure 2. In-situ TEM images of a catalytic gold nano-particle during CO oxidation (Reaction gas is CO with dry air; Pressure is about 750Pa)

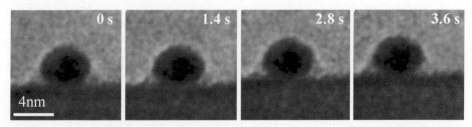

Figure 3. In-situ TEM images of a catalytic gold nano-particle in the condition of non-reaction gas environment (Non-reaction gas is dry N_2; Pressure is about 750Pa)

In situ transmission electron microscopy on leadzirconate-titanate under electrical field

J. Kling[1], L. Schmitt[1], H.-J. Kleebe[2] and H. Fuess[1]

1. Structure research, Institute for materials science, Petersenstr. 23, 64287 Darmstadt, Germany
2. Geo-material-science, Institute of applied geoscience, Schnittspahnstr. 9, 64287 Darmstadt, Germany

j_kling@st.tu-darmstadt.de

Keywords: TEM, in situ, ferroelectrics

Ferroelectric materials play an important role in today's functional material discussion and are widely used in industrial applications as injection systems, actuators or sensors. A detailed knowledge of the microstructure is prerequisite to optimise these materials for further applications.

Leadzirconate-titanate (PZT) is one of the most frequently used ferroelectric materials in industry, although the microstructure and the behaviour under continuous cycling are not well understood. It shows the best performance around the morphotropic phase boundary (MPB), where the composition is between $PbZr_{0.54}Ti_{0.46}O_3$ and $PbZr_{0.52}Ti_{0.48}O_3$.

The high strain in this region can not only be explained by an intrinsic effect of the material. The behaviour of the microstructure, like domain evolution and switching, the extrinsic effect, seems to be of importance. In addition, these materials show significant fatigue under continuous cycling which degrade their properties such as strain or polarizability. Doping may improve the stability of the properties although this influence is not fully understood yet.

In order to obtain a detailed insight in the microstructure and the domain evolution under electrical field, an *in situ* transmission electron microscopy (TEM) experiment was performed. A modified double-tilt holder with insulated feedthroughs was used. As electrodes two Cu-apertures, one above and one below the sample, were mounted and fixed at the holder by Cu-cables (see figure 1). Applying a voltage, an electrical field nearly parallel to the electron beam on the observable sample region is produced. The achievable field strength was approximately 2 kV/mm, which is around the coercive field of PZT.

The experiments showed a considerable change of the contrast within the microdomains but the microdomain walls remained visibly unchanged in the observed voltage range [1]. In the sample $PbZr_{0.54}Ti_{0.46}O_3$ this contrast change can be related to the change in nanodomain configuration. These nanodomains were already observed in several compositions around the MPB [2]. Their mobility seems to be much higher than that of the microdomains. Therefore the nanodomains seem to be an important feature of the microstructure which also contribute to the piezoelectric properties of the material.

M. Luysberg, K. Tillmann, T. Weirich (Eds.): EMC 2008, Vol. 1: Instrumentation and Methods, pp. 489–490, DOI: 10.1007/978-3-540-85156-1_245, © Springer-Verlag Berlin Heidelberg 2008

A new electrode geometry was realised for additional investigations. The electrodes were sputtered directly to the flat surface of the sample. They are half circled shaped with a slit in between. An applied voltage results in an electrical field perpendicular to the electron beam and a higher field strength as compared to the former geometry. Preliminary results with this new geometry will be presented.

1. R. Theissmann et al, JAP **102** (2007), p. 024111.
2. L.A. Schmitt et al, JAP **101** (2007), p. 074107.

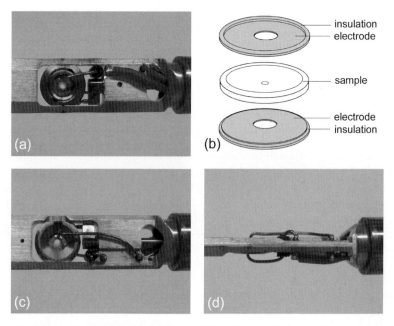

Figure 1. Electrode geometry parallel to the electron beam. (a), (c), (d) show a contacted specimen at the holder tip from different viewing angles. (b) is a schematically sketch of the geometry.

Elongation of Atomic-size Wires:
Atomistic Aspects and Quantum Conductance Studies

M.Lagos, V.Rodrigues , D.Ugarte,

1. Laboratório Nacional de Luz Síncrotron (LNLS), Campinas-Brazil
2. Universidade Estadual de Campinas (UNICAMP), Campinas-Brazil

mlagos@lnls.br

Keywords: time-resolved HRTEM, atomic-size nanowires, quantum conductance

The study of atomic-size metal nanowires (NW´s) is attracting a great interest due to occurrence a novel physical and chemical phenomena. Among these new phenomena, we can mention conductance quantization that will certainly influence the design of nanodevices. NW´s are usually generated by mechanical deformation and the conductance is measured during the wire elongation. The interpretation of the results is troublesome, because conductance is measured during the modification of the atomic structure. This kind of experimental study has been performed by many research groups and, a quite wide range of temperatures (4 - 300 K) and vacuum conditions have been used (from ambient to UHV). In fact, the results display significant variation, what has generated several controversial interpretations. It must be emphasized that many models have been derived without taking into account that the NW structural deformation should be significantly dependent on temperature.

In this work, we have studied how thermal effects influence the atomistic aspects of the gold NW deformation and the influences the quantum conductance behavior. The structure of NW´s has been studied by means of time-resolved high resolution transmission electron microscopy; the NWs transport measurements were based on a mechanically controlled break junction operated in ultra-high-vacuum. The experiments were performed at ~150 K (LT) and 300 K (RT).

Our results have shown that at RT gold NW´s are always crystalline and free of defects, and the atomic structure is deformed such that one of the [111]/[100]/[110] crystallographic axis becomes approximately parallel to the stretching direction [1]. LT observations revealed important differences: i) Au NWs show extended defects, mainly stacking faults and twinning (Figure 1). ii) NWs elongated along the [110] axis evolve to suspended atomic chains (ATC), while at RT they break abruptly (Figure 2). (iii) Formation of ATC is enhanced at LT [2]. The global histograms of conductance at LT showed that: i) a increase of the 1 Go peak intensity due to enhancing of the ATC; ii) slight reduction of the NWs conductance due to scattering at defects and; iii) the peak at ~2 Go shows a sub structure, what is due to the occurrence of two different atomic arrangements with similar conductance.

1.   V.Rodrigues, T.Fuhrer and D.Ugarte, PRL **85**, 4124 (2000).
2.   M.Lagos, F.Sato, V.Rodrigues, D.Galvão and D.Ugarte, manuscript in preparation.
3.   We acknowledge P.C.Silva and J.Bettini for assistance during experiments. We also thank LNLS, FAPESP and CNPq for financial support.

M. Luysberg, K. Tillmann, T. Weirich (Eds.): EMC 2008, Vol. 1: Instrumentation and Methods, pp. 491–492, DOI: 10.1007/978-3-540-85156-1_246, © Springer-Verlag Berlin Heidelberg 2008

492

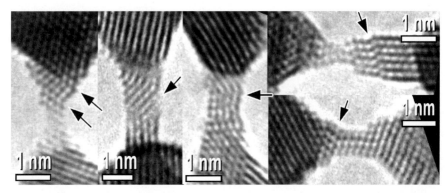

Figure 1. High-resolution images of gold atomic-size nanowires (NW) deformed mechanically at low temperature (~150 K). Note that arrows indicate planar defects (stacking faults and twins) in the NW atomic structures. Atomic positions appear dark.

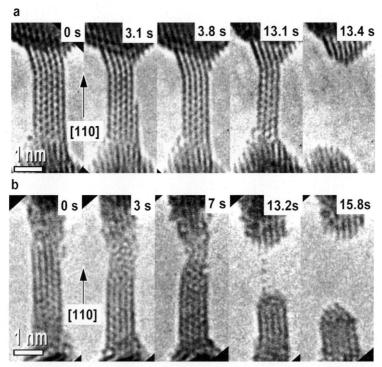

Figure 2. Time sequences in the elongation and rupture of gold rod-like NW deformed mechanically along the [110] direction at **(a)** ~ 300 K and **(b)** ~150 K; atomic positions appear dark. Note that during stretching at room temperature NW keeps its rod morphology and to breaks abruptly when still is formed by 4 atomic planes in width, meanwhile, at low temperature NW evolves into atomic chain and finally breaks.

Atomic-size Silver Nanotube

M. Lagos[1,2], F. Sato[2], J. Bettini[1], V. Rdrigues[2], D. Galvão[2] and D. Ugarte[1,2]

1. Laboratório Nacional de Luz Síncrotron (LNLS), Campinas-Brazil
2. Universidade Estadual de Campinas (UNICAMP), Campinas-Brazil

mlagos@lnls.br
Keywords: time-resolved HRTEM, atomic-size nanowires, metal nanotube

The atomic arrangement of nanosystems may be quite different from the traditional materials; surface energy minimization plays a dominant role in this size range, and accounts for many of these new structures. Graphitic nanotubes [1] represent the best example, being fromed by a rolled the graphitic layer, which is tradionally flat. Subsequently the rolling of the compact (111) atomic planes was reported for gold nanowires (NW) generated by mechanical stretching [2]. But, we may expect many more surprises from the interplay between atomic and electronic structure.

Herein, we report the spontaneous formation of a square cross-section hollow metal wires during the elongation of silver nano-contacts along [001] direction. Pure Silver shows a face centered cubic (fcc) metals with almost identical lattice parameter that Au, but subtle changes of the surface energy (cubic facets (100) gain importance) generates clear differences of structural and mechanical behavior as revealed by real time atomic resolution transmission electron microscope (HRTEM) [3].

We have determinant the structure of the silver nanotube (Figure 1) by associating time-resolved atomic resolution HRTEM, and also image simulations. Our results revealed that: (i) the hollow NW atomic structure is formed by 2 different atomic planes (A,B), each one containing four atoms, keeping the stacking sequence $4_A/4_B$, instead of the $5_A/4_B$ stacking present in the perfect fcc [001] wire (Figure 2).

1. S. Iijima, Nature **354**, 56 (1991).
2. Y. Kondo and K. Takayanagi Science **289**, 606 (2000).
3. V.Rodrigues, J.Bettini, A.R.Rocha, L.G.C.Rego and D.Ugarte, PRB **65**, 153402 (2002).
4. We acknowledge P.C.Silva for support during sample preparation. We also thank LNLS, FAPESP and CNPq for financial support.

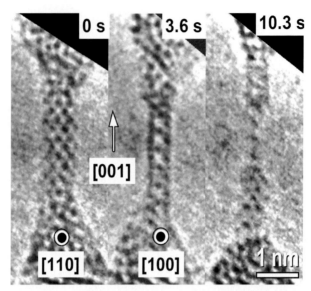

Figure 1. Thinning of a Ag NW being elongated along the [001] axis. Note that significant changes of the NW image (from 0 to 3.6 s) occurs during elongation; finally the wire forms an atom chain (10.3 s) before breaking (atomic positions appear dark).

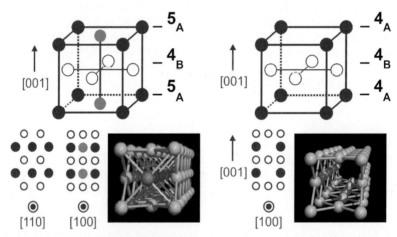

Figure 2. Left: Schematic drawing of a fcc unit cell, and the expected image contrast when a [100] wire of width a is projected along [110] and [100] directions; Right: fcc unit cell and expected contrast patter along [100] direction after eliminating the atom located at center at the cube [001] facet), what generates a one-lattice wide Ag hollow NW (see three dimensional ball/stick schema).

In-situ TEM mechanical testing of a Si MEMS nanobridge

A.J. Lockwood[1], R.J.T. Bunyan[2] and B.J. Inkson[1]

1. Department of Engineering Materials, University of Sheffield, Sheffield, S1 3JD. UK
2. QinetiQ, St. Andrews Road, Malvern, Worcestershire, WR14 3PS. UK

a.lockwood@sheffield.ac.uk

Keywords: in-situ TEM, nanoindentation, MEMS, plastic deformation

The failure mechanisms of micro-electromechanical systems (MEMS) are an area of current interest as many aspects of MEMS failure are somewhat uncertain despite MEMS usage and applications dramatically increasing. Current understanding of Si MEMS failure is generally based on oxide thickening and stress corrosion cracking (SCC) [1], leading to catastrophic failure. Testing and observing MEMS structures within a vacuum is becoming more widespread, for example actuating the MEMS devices over many thousands of cycles inside scanning electron microscopes (SEM) [2]. This SEM technique is useful in understanding cycles to failure, however does not address the effects that microstructure has upon failure. One method to understand how failure occurs with respect to the microstructure is to analyse the failure post mortem using a transmission electron microscope (TEM). This technique does not address dynamic microstructural effects during operation and failure. Recent advancements in TEM are exploring dynamic testing using various in-situ techniques such as atomic force microscopy (AFM)-TEM, scanning probe microscopy (SPM)-TEM and TEM-nanoindentation. Real-time observations made during mechanical probing of nanosized structures are currently being undertaken using in-situ TEM nanoindentation.

Here we have investigated the mechanical properties of Si nanobridges (figure 1). The nanobridges were machined from 320nm thick poly-Si MEMS cantilevers produced by QinetiQ, UK. Focused ion beam (FIB) milling was used to post-pattern nanostructures along the ends of the cantilevers to form the bridges with a crossbeam thickness of 115nm and side support thicknesses of 180nm separated by a gap of over 3μm. Initial experiments demonstrate how flexible and elastic the material becomes at the nanoscale. Figure 1 shows an indenter tip brought into contact with a Si nanobridge and loaded at a rate of 20nm/s. Initially, over penetration depths <200nm the structure elastically regains its original undeformed shape after the tip is removed. Increasing the penetration depth to over 500nm introduces plastic deformation into the structures, a characteristic of single crystal silicon under high indentation loading [3] but unseen in previous work performed on poly-Si micro/nanoscale structures [4, 5]. At a peak deflection of 550nm, the applied force is calculated to be around 634μN and a contacting pressure of over 20GPa. After the tip is fully withdrawn, the structure remains plastically deformed by a residual deflection of ~94nm.

On a second structure seen in Figure 2, after deflecting the crossbeam by 535nm it fractures fully through the beam resulting in failure. The crack initiation point does not occur at the point of tip contact (and maximum contact pressure) but ~165nm to the side where a grain boundary intersects at the beam surface. As seen in the first example, the

M. Luysberg, K. Tillmann, T. Weirich (Eds.): EMC 2008, Vol. 1: Instrumentation and Methods, pp. 495–496, DOI: 10.1007/978-3-540-85156-1_248, © Springer-Verlag Berlin Heidelberg 2008

structure remains plastically deformed after the tip is fully removed. Imaging the fracture point at high magnification indicates that the fracture is not cleanly broken, but material is lost from both the inner and outer surfaces, with some fracture debris collecting on the indenter tip.

In summary, in-situ TEM nanoindentation is beginning to realise mechanical testing of more complex functional structures. The technique is very flexible and transferable to other more active devices. It also demonstrates that polycrystalline silicon at dimensions of less than 320nm can retain some plastic strain after unloading at room temperature [6].

1. W. Merlijn van Spengen, Microelectronics Reliability, 43 (2003) 1049.
2. C.L. Muhlstein, R.T. Howe, R.O. Ritchie, Mechanics of Materials, 36 (2004) 13.
3. A.M. Minor, E.T. Lilleodden, M. Jin, E.A. Stach, D.C. Chrzan, J.W. Morris, Philosophical Magazine, 85(2-3) (2005) 323.
4. J.N. Ding, Y.G. Meng, S.Z. Wen, Mat. Sci. Eng, B83 (2001) 42.
5. X. Li, B. Bhushan, K. Takashima, C.-W. Baek, Y.-K. Kim, Ultramicroscopy, 97 (2003) 481.
6. This work was support by a grant from EPSRC, GR/S85689/01, UK.

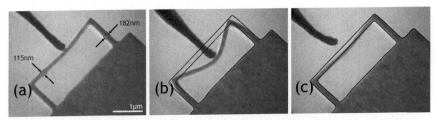

Figure 1. 3 frames extracted from a video sequence showing a W-tip being used to deform a FIB milled poly-Si MEMS nanobridge, a) the tip is brought into contact with the 115nm wide cross-beam, b) the tip is used to deform the beam to a penetration depth of ~550nm and after the tip is fully withdrawn, c) shows some residual deformation that remains within the structure. (The original shape is denoted by a black outline).

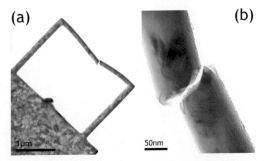

Figure 2. TEM images showing the fracture of a Si nanobridge which occurred during loading. a) An overview of the entire structure demonstrating plastic deformation and b) magnified view of the fracture point.

In-situ TEM nanoindentation and deformation
of Si-nanoparticle clusters

A.J. Lockwood and B.J. Inkson

Department of Engineering Materials, University of Sheffield, Sheffield, S1 3JD. UK

a.lockwood@sheffield.ac.uk
Keywords: in-situ TEM, nanoindentation, nanoparticle deformation

Understanding the mechanical properties and interaction of individual nanoparticles within a cluster, and other small volume nanoobject interactions, is becoming increasingly important with the development of devices with sub-100nm components. At the nanometre scale, materials tend to behave differently compared to bulk structures, and object-object interactions, adhesion and strain induced at nano-contact sites can be quite significant [1]. Recent advances in transmission electron microscopy (TEM) technology, by the combination of TEM with scanning probes, have enabled the application of electrical and mechanical driving forces to samples whilst simultaneously imaging them at TEM resolutions [2-4].

A custom made in-situ TEM nanoindentation holder designed for use in a JEOL (Japan) 2010/3010 series instrument was used in this study to probe small clusters of crystalline silicon nanospheres (MTI Corp, USA) agglomerated within an amorphous silicon matrix.

Figure 1 shows a small cluster of ~50nm Si-nanoparticles on the edge of a Ti substrate. A W-tip with end radius ~110nm was brought into contact with a pair of particles in the cluster. The two particles were loaded in a four-stage sequence. Each stage consisted of ~21nm displacement of the tip at a loading rate of approximately 20nms^{-1}, followed by a hold phase to enable microstructural observation. Figure 1 shows (a) the tip first brought into contact with the Si nanoparticle cluster, (b-d) microstructure during holding after the 1st-3rd loading stages, and (e-f) images before and after fracture during the 4th loading stage. After the cluster of particles fracture, the tip loses contact with the two closest particles in the group and a large jump of the tip towards the substrate is observed.

Figure 2 shows the load-displacement curve of the four stage indentation. The load applied to the Si cluster increases through each loading phase, until fracture occurs at t=68.8s. Two small load drops of 3.5 and 1.3µN are observed to occur at the beginning of the first and second holding stages (points I and II in Figure 2). This movement is due to time-dependant microstructural deformation to relieve stress within the Si cluster. At fracture, contact is lost between the tip and the sample, the force immediately falls to zero, and a large jump in displacement of the loaded tip occurs towards the Ti substrate (Figure 1). Fracture occurs at the a-silicon matrix between two Si particles and the applied force at fracture is 55µN.

In summary, the development of a high stiffness TEM nanomanipulation and indentation system has enabled the quantitative characterisation of the dynamical

M. Luysberg, K. Tillmann, T. Weirich (Eds.): EMC 2008, Vol. 1: Instrumentation and Methods, pp. 497–498, DOI: 10.1007/978-3-540-85156-1_249, © Springer-Verlag Berlin Heidelberg 2008

498

mechanical properties of a cluster of interacting 50nm Si nanoparticles. This study clearly demonstrates significance and flexibility of in-situ TEM nanoindentation for investigating the dynamical mechanical properties of nanostructures [5].

1. A. Erts, R. Lohmus, H. Olin, A.V. Pokropivny, L. Ryen, K. Svensson, Appl. Surf. Sci., 188 (2002) 460.
2. K. Svensson, Y. Jompol, H. Olin, E. Olsson, Rev. Sci. Instrum., 74 (1995) 4945-4947.
3. L. Oleg, M. Yasuji, S. Naoya, Ceramics Japan, 40(11) (2005) 953-957.
4. M.S. Bobji, C.S. Ramanujan, J.B. Pethica, B.J. Inkson, Meas. Sci. Technol., 17 (2006) 1324.
5. This work was support by a grant from EPSRC, GR/S85689/01, UK.

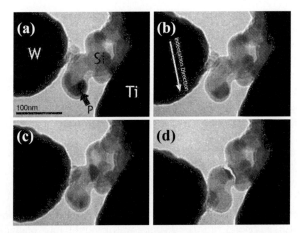

Figure 1. A Si particle cluster observed by TEM during multi-stage loading with a W-tip, (a) just prior to contact with strong Bragg contrast at point P, (b) holding position after 1st loading stage with tilt of the nanoparticles clearly changing Bragg contrast (c) 0.04s prior fracture and (d) first observed frame with visible fracture during the 4th loading stage.

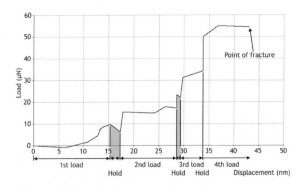

Figure 2. Load-displacement curve for the 4-stage loading scheme applied to the cluster of silicon nanoparticles.

Electron Holography of in-situ ferroelectric polarisation switching

Ch. Matzeck, B. Einenkel, H. Müller and H. Lichte

Triebenberg Laboratory, Institute of Structure Physics, Technische Universität Dresden, 01062 Dresden Germany

christopher.matzeck@triebenberg.de

Keywords: Electron Holography, Ferroelectrics, in-situ switching

Below Curie temperature, Barium Titanate (BTO) has a slightly tetragonally distorted unit cell with separated positive and negative centres of charge. This generates an electric dipole moment in each unit cell, which yields a spontaneous electric polarisation in absence of an external electric field. But ferroelectricity requires at least two stable states of a crystal that are equivalent in their crystallographic structure and, additionally, differ in their state of spontaneous electric polarisation. These two states can be toggled by application of an external electric field. The behaviour of this polarisation switching is essential for many applications of ferroelectric crystals, e.g. non-volatile memories.

Corresponding in-situ TEM-investigations were done by several groups with bright field imaging [1]-[3]. This method, however, implicates the loss of the phase information of the electron wave [4]. Since the electric structure of crystals influences the phase of electron waves in the TEM, electron holography is capable to measure this modulation [4]. It was shown a powerful tool to investigate ferroelectricity in crystals [5].

Theoretical calculations predicted an external electric field of ~1kV/mm needed to switch the polarisation in BTO. But these estimations disregard the creeping behaviour of ferroelectric domains. Early measurements show that already an electric field of 0.26kV/mm reverses the polarisation of BTO "Figure 1" [6].

To achieve these field strengths for an in-situ experiment in the TEM, specially prepared specimens were developed and a holder with an electrical feedthrough was used. The specimens are built of a small piece of BTO between two semicircular metal electrodes, kept inside a ceramic ring "Figure 2". First experiments are shown in "Figure 3".

1. E. Snoeck, L. Normand, A. Thorel, C. Roucau, Phase Transitions **46** (1993), p. 77
2. Z. Xu, X. Tan, P. Han, J.K. Shang, Appl. Phys. Lett. **76** (2000), p. 3732
3. X.Y. Qi, H.H. Liu, X.F. Duan, Appl. Phys. Lett. **89** (2006), p. 092908
4. H. Lichte, M. Lehmann, Rep. Prog. Phys. **71** (2008), p. 016102
5. H. Lichte, M. Reibold, K. Brand, M. Lehmann, Ultramicroscopy **93** (2002) 199
6. E.A. Little, Phys. Rev. **98** (1955), p. 978
7. The financial support from the Deutsche Forschungsgemeinschaft for the Research Group on Ferroic Functional Components FOR 520 is gratefully acknowledged.

M. Luysberg, K. Tillmann, T. Weirich (Eds.): EMC 2008, Vol. 1: Instrumentation and Methods, pp. 499–500, DOI: 10.1007/978-3-540-85156-1_250, © Springer-Verlag Berlin Heidelberg 2008

500

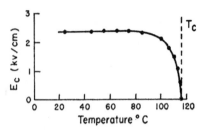

Figure 1. Minimum field needed to switch the ferroelectric polarisation of BTO depends on temperature. At room temperature it is ~260V/mm [6]. T_c is the Curie temperature.

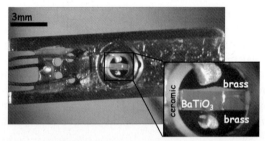

Figure 2. The specially prepared specimen consists of BTO embedded between two brass electrodes for the electric field. The outer ceramic ring [1]-[3] insulates electrodes and holder.

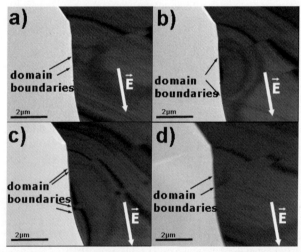

Figure 3. Ferroelectric domains in BTO under application of an external electric field. a) no field ; b) 650V/mm; c) 850V/mm; d) back to zero field. The change of the domain structure with increasing electric field is evidently visible. After one cycle the remaining domain in d) is slightly larger than the initial one in a). The main difference is the bending caused by the induced strain.

Development of fast CCD Cameras
for *in-situ* Electron Microscopy

Bill Mollon, Lancy Tsung, Ming Pan, Yan Jia, Paul Mooney, and Chengye Mao

Gatan, Inc. 5794 W. Las Positas Blvd., Pleasanton, CA 94588, USA

bmollon@gatan.com
Keywords: CCD, TV rate, *in-situ*

In-situ electron microscopy has shown resurgence in recent years [1, 2]. This resurgence has been largely enabled by recent advances in electron microscope design and new CCD cameras utilizing the latest digital imaging technologies. However, challenges still remain in designing a suitable digital imaging system that allows *in-situ* experiments to be observed and/or recorded in real time and at atomic resolution. In this presentation, we will review the current technology in high-speed CCD camera development and practical examples from *in-situ* experiments.

Traditionally, *in-situ* electron microscopy experiments are recorded using analog technology via a TV-rate camera installed on the microscope. The correction of camera defects and gain non-uniformity from this analog signal is difficult. Furthermore the recorded data is not easily shared or transferred to another system that uses different video format. On the other hand, digital CCD cameras have the advantage of improved image quality, shareable data formats, and the ability to correct for cosmetic defects and non-uniformity that may be present in the image. With a moderately configured computer (sufficient RAM memory, fast hard drives), suitable acquisition hardware and software it is possible to create digital videos from a CCD camera. However, such a system has not been widely adopted for *in-situ* electron microscopy mainly because of the slow speed of large-format fiber-optically coupled CCD cameras or the low sensitivity of small, fast CCD's coupled with lenses.

Our recent effort in developing a system suitable for *in-situ* applications has resulted in a totally digital CCD imaging system capable of acquiring images at true TV rate, i.e. 30 frames per second (fps), with a large field of view. High speed and low noise are achieved by optimizing CCD read out timing within a flexible 30MHz camera controller. Digital streaming video (DSV) can be generated simultaneously by DigitalMicrograph software and recorded into a digital movie via video authoring software. The camera has adequate sensitivity for each captured individual frame to show atomic resolution details. Figure 1 shows 5 individual frames extracted from a digital movie of a small Au nano-particle in a 400kV TEM. The high speed imaging allows the user to observe rapid structural changes (single and multiple twinning, moving of the twinning plane, creation and disappearance of multiple twinning, etc.) of this nanometer-scale Au particle under the electron beam. If desired, a full resolution, single frame image can be acquired at any time using the camera's high quality mode of operation (Figure 2). In addition to the CCD chip's anti-blooming capability, enhanced handling of high-dynamic range images allows users to acquire streak-free diffraction

M. Luysberg, K. Tillmann, T. Weirich (Eds.): EMC 2008, Vol. 1: Instrumentation and Methods, pp. 501–502, DOI: 10.1007/978-3-540-85156-1_251, © Springer-Verlag Berlin Heidelberg 2008

pattern (Figure 3) both in high-quality single frame mode and in high-speed video mode suitable for observing structure changes during *in-situ* experiments.

1. ASU/NSF workshop "Dynamic *in-situ* electron microscopy as a tool to meet the challenges of the nanoworld", Jan 3-6, 2006, Tempe, AZ
2. FEMMS conference "Ultrafast & *In-situ* Electron Microscopy" sessions, Sept 23-28, 2007, Sonoma, CA

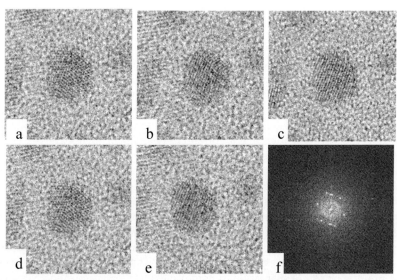

Figure 1. Individual frames extracted from a 30fps digital video; (a) frame #467 (0.00sec); (b) Frame #506 (1.30sec); (c) frame #663 (6.53 sec); (d) frame #776 (10.30 sec); (e) frame #951 (16.13 sec); (f) FFT of a single video frame showing single twinning in an Au particle

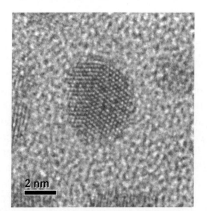

Figure 2. Acquired image with full CCD, 1x binning, 0.5 seconds exposure time

Figure 3. Si [110] diffraction pattern

In situ characterization of the mechanical properties of nanoparticles and nanoscale structures

J. Deneen Nowak, Z.W. Shan and O.L. Warren

Hysitron, Inc., 10025 Valley View Road, Minneapolis, MN 55344 USA

jnowak@hysitron.com

Keywords: in situ, TEM, indentation

One major difficulty in characterizing the mechanical properties of nanoscale materials is the inherently small size of the structures of interest. With traditional nanomechanical-testing techniques, the inability to see nanoscale structures directly can leave a number of unknowns. For example, the crystallographic orientation of the sample, the presence of any pre-existing defects, and even the certainty that contact was both made and maintained during deformation must necessarily be assumed. While the transmission electron microscope (TEM) is uniquely well-suited for determining these "unknowns," mechanical testing inside the TEM has consisted of, until quite recently, predominantly *qualitative* studies [1-3]. Over the past decade a number of advances in *in situ* indentation specimen holders, such as the integration of not only a piezoelectric actuator, but also a capacitive transducer, have allowed for the development of a holder which is capable of quantitative load–displacement measurements [4].

This holder is particularly useful for characterizing the mechanical response of nanoscale volumes, which tend to exhibit enhanced mechanical properties compared to their bulk counterparts. These properties have been a topic of interest for decades, notably in a 1956 study of the tensile strength of metal whiskers [5], but have gained increased attention in recent years in studies of nanocrystals and nanoparticles [6, 7]. The mechanisms proposed to explain these enhanced behaviours, however, typically lack direct-observational support. Recent work using the newly developed quantitative sample holder has illustrated the benefits of this *in situ* technique in determining the mechanisms responsible for the observed behaviours. For example, the role of dislocation-source starvation in the strengthening of Ni nanopillars, such as the pillar shown in Figure 1, has been ascertained through an *in situ* compression study [8].

In the present work, careful design of the sample geometry allows for the compression of individual nanostructures *in situ,* where the phenomena can be observed directly by the electron microscope. Further, by coupling this setup with a video recording device it is possible to view and record deformation events as they happen in the TEM. This study investigates the mechanical response of a number of different types of nanostructures with a variety of different chemistries and morphologies. Current *in situ* indentation, compression, and bending studies of nanoscale structures will be discussed at length and the future of *in situ* mechanical testing will be considered.

M. Luysberg, K. Tillmann, T. Weirich (Eds.): EMC 2008, Vol. 1: Instrumentation and Methods, pp. 503–504, DOI: 10.1007/978-3-540-85156-1_252, © Springer-Verlag Berlin Heidelberg 2008

504

1. A.M. Minor, E.T. Lilleodden, E.A. Stach and J.W. Morris, Jr., J. Electron. Mater. **31** (2002), p. 958

2. J. Deneen Nowak, W.M. Mook, A.M. Minor, W.W. Gerberich and C.B. Carter, Philos. Mag. **87** (2007), p. 29

3. A.M. Minor, E.T. Lilleodden, M. Jin, E.A. Stach, D.C. Chrzan and J.W. Morris, Jr., Philos. Mag. **85** (2005), p. 323

4. O.L. Warren, Z. Shan, S.A. Syed Asif, E.A. Stach, J.W. Morris Jr. and A.M. Minor, Materials Today **10** (2007), p. 59

5. S.S. Brenner, J. Appl. Phys. **27** (1956), p. 1484

6. V.G. Gryaznov and L.I. Trusov, Prog. Mater. Sci. **37** (1993), p. 289

7. W.W. Gerberich, W.M. Mook, C.R. Perrey, C.B. Carter, M.I. Baskes, R. Mukherjee, A. Gidwani, J.V.R. Heberlein, P.H. McMurry and S.L. Girshick, J. Mech. Phys. Solids **51** (2003), p. 979

8. Z.W. Shan, R.K. Mishra, S.A.S. Asif, O.L. Warren and A.M. Minor, Nature Materials **7** (2008), p. 115

9. This work was supported in part by a DOE SBIR Phase II grant (DE-FG02-04ER83979) awarded to Hysitron, Inc., which does not constitute an endorsement by DOE of the views expressed in this article. This work was also supported by the Director, Office of Science, Office of Basic Energy Sciences, of the US Department of Energy under Contract No. DE-AC02-05CH11231.

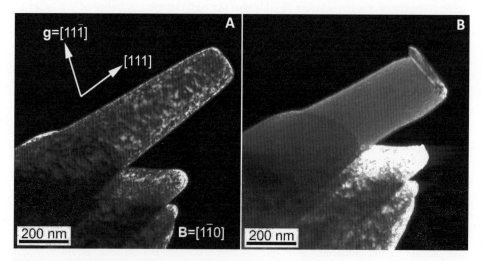

Figure 1. Dark-field TEM images of a Ni nanopillar before (A) and after (B) *in situ* compression. Upon compression the initially high dislocation density disappears [8].

In-situ engineering of nanostructures with near atomic precision and property measurements

L.-M. Peng, M.S. Wang, Y. Liu and Q. Chen

Key Laboratory for the Physics and Chemistry of Nanodevices and Department of Electronics, Peking University, Beijing 100871, China

lmpeng@pku.edu.cn

Keywords: Nanostructure, Manipulation, Property

Carbon nanotube (CNT) and nanowire materials are important building materials for nanotechnology. These materials may be synthesized via a range of physical and chemical methods, and new nanotube and nanowire materials are produced every day. Measurements on individual nanostructures remain, however, difficult and it is even more challenging to control the property of these nanomaterials via structure modification at near atomic resolution. A very promising and perhaps the best method to tackle this problem is to combine the scanning tunnelling microscope (STM) with electron microscope (EM) so that manipulation and structure modification may be made via a highly controllable fashion on individual nanostructure [1-2].

Nanostructures can be fabricated, manipulated and engineered with high precision (e.g. Fig. 1), and their real time electrical and mechanical properties can be measured in-situ inside the EM (e.g. Fig. 2). While it is very convenient to carry out manipulation and measurement on nanostructure in a scanning electron microscope (SEM) [3], the resolution of the SEM is limited and the vacuum level is typically not as good as in a transmission electron microscope (TEM). The higher resolution and vacuum level in a TEM has been utilized for revealing the importance of the CNT tip structure on its electron field emission characteristics and effects of deformation on the conductance of the CNT. These experiments show clearly that the conductance of the large diameter multi-walled CNT is not easily affected by deformation and these CNTs may in principle be used in the fabrication of novel nanoelectronic circuit as interconnects [4].

A quantitative analysis of the electric transport property of the semiconducting nanowire also requires the detailed structure of the contact and nanowire. Two terminal I-V characteritics may be measured inside TEM as shown in Fig. 3, and the diameter, length etc. of the nanowire may readily be obtained from TEM imaging and varied during experiments providing valuable input for the quantitative analysis of the transport property of the nanowires [5].

1. L.-M. Peng et al., MICRON 35 (2004) 495
2. M.S. Wang, Q. Chen, and L.-M. Peng, Adv. Mater. 20 (2008) 724; Small (2008) in press
3. Y. Liu et al., Appl. Phys. Lett. 92 (2008) 033102; L. Shi et al., Nano Letters 7 (2007) 3559
4. M.S. Wang et al., Adv. Func. Mater. 15, (2005) 1825; 16 (2006) 1462
5. Z.Y. Zhang et al., Appl. Phys. Lett. 88 (2006) 073102; Adv. Func. Mater. 17 (2007) 2478

M. Luysberg, K. Tillmann, T. Weirich (Eds.): EMC 2008, Vol. 1: Instrumentation and Methods, pp. 505–506, DOI: 10.1007/978-3-540-85156-1_253, © Springer-Verlag Berlin Heidelberg 2008

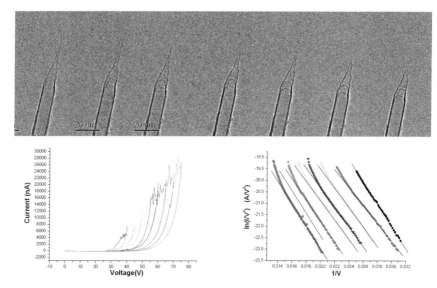

Figure 1. TEM images showing an extremely sharp tip of the CNT being modified under the electron field emission experiments and corresponding I-V curves and F-N plots [2].

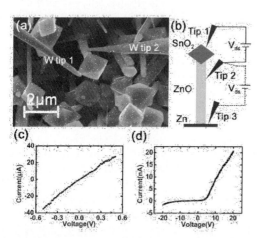

Figure 2. (a) SEM image showing a typical I-V measurement setup using three probe system. (b) Schematic showing the measurement circuit for Zn(sub)–ZnO(NW) junction and ZnO(NW)-SnO2(cap) junction. Corresponding experimental I-V curves for (c) Zn(sub)-ZnO(NW) and (d) ZnO(NW)-SnO2(cap) junctions [3].

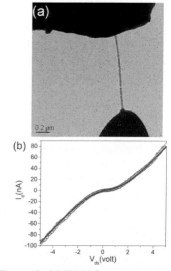

Figure 3. (a) TEM image showing a ZnO nanowire based two terminal device and (b) experimentally measured and fitted I-V curves [5].

In-situ TEM investigation of the contrast of nanocrystals embedded in an amorphous matrix

M. Peterlechner, T. Waitz and H.P. Karnthaler

Physics of Nanostructured Materials, University of Vienna, Boltzmanngasse 5, 1090 Vienna, Austria

martin.peterlechner@univie.ac.at

Keywords: TEM, in-situ heating, nanocrystals, amorphous, bright-field contrast

Bulk amorphous NiTi alloys can be obtained by severe plastic deformation [1]. Upon heating, the kinetics of the nanocrystallization were analysed by in-situ heating experiments in the transmission electron microscope (TEM) [2]. The present investigation is focused on the contrast of nanocrystals embedded in the amorphous matrix.

Figure 1 shows a diffraction pattern of diffuse rings caused by the amorphous phase of NiTi. Strong diffraction occurs at a reciprocal distance of 4.6 nm^{-1}. Figure 2 shows a TEM bright field image taken during the in-situ heating. The diameter of the objective aperture was 2.7 nm^{-1} and therefore excludes the strongly diffracted intensity of the amorphous phase. During in-situ heating, embedded in the amorphous matrix (showing uniform contrast) nanocrystals nucleate and grow until they impinge upon each other. With respect to the amorphous matrix, nanocrystals of various sizes show bright contrast (the contrast is weak when the nanocrystals have a size of 5 to 10 nm and nanocrystals of bright contrast that have a size of less than about 5 nm are hardly detected). Some of the nanocrystals show a marked change of contrast when their size increases (marked by circles in Fig. 2). The contrast of the nanocrystal denoted 1 changes from dark to bright when its diameter increases by about 12 nm. At specific diameters, even vanishing contrast can occur (cf. the nanocrystals denoted 3 and 2 in Fig. 2 a and b, respectively).

Since the crystallization is polymorphous (i.e. without change of the chemical composition [3]) it is concluded that the bright, dark and vanishing contrast of nanocrystals arises by a diffraction effect. Regarding the orientation of the crystals with respect to the incident beam, two different cases can be distinguished: Firstly, crystals off a Bragg orientation. For each projected thickness, these crystals scatter less than the amorphous matrix of equal mass thickness and, therefore, show bright contrast [4]. Secondly, crystals near a Bragg orientation. In this case, strong dynamical diffraction effects of the growing crystals are expected to lead to oscillations of their contrast as a function of size: With respect to the amorphous matrix that shows an exponential decrease of the transmitted intensity, dark, bright and vanishing contrast will occur with increasing projected thickness of the nanocrystals in agreement with the experiment.

Finally, it should be noted that caused by the very weak contrast of nucleating nanocrystals that are off an Bragg orientation these crystals are hardly detected during the in-situ experiment unless they have grown to a size exceeding 5 to 10 nm.

M. Luysberg, K. Tillmann, T. Weirich (Eds.): EMC 2008, Vol. 1: Instrumentation and Methods, pp. 507–508, DOI: 10.1007/978-3-540-85156-1_254, © Springer-Verlag Berlin Heidelberg 2008

508

1. T.Waitz, V.Kazykhanov, H.P.Karnthaler, Acta Mat. 52 (2004) 137
2. M. Peterlechner, T. Waitz and H.P. Karnthaler, Proc. Intern. Symp. Bulk Nanostr. Mater., Ufa, Russia (2007) p. 57.
3. H. Ni, H-J. Lee, A. G. Ramirez, J. Mater. Res., 20 (2005) 1728
4. L. Riemer, Z. Angew. Physik. 22, (1967) 287
5. The authors acknowledge the support by the research project "Bulk Nanostructured Materials" within the research focus "Materials Science" of the University of Vienna. M.P. acknowledges the support by the I.K. "Experimental Materials Science – Nanostructured Materials", a college for PhD Students at the University of Vienna.

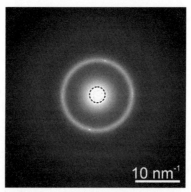

Figure 1. Amorphous NiTi. Selected area diffraction showing a pattern of diffuse rings that are excluded by the objective aperture (indicated by a dashed circle). Weak spots arise from nanocrystals already present in the amorphous matrix prior to in-situ heating.

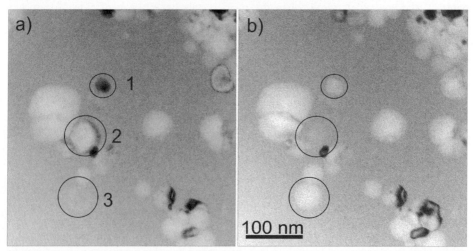

Figure 2. Nanocrystallization of NiTi. TEM bright field images taken during in-situ heating. Most of the nanocrystals show bright contrast. Some crystals change contrast with size (marked by circles). (a) Crystal 1, 2 and 3 show dark, bright and vanishing contrast (b) Same area as (a) after heating for 12 min at 352°C. Crystal 1, 2 and 3 now show bright, vanishing and bright contrast, respectively.

In situ HRTEM – Image corrected and monochromated Titan equipped with environmental cell

J.B. Wagner[1], J.R. Jinschek[2], T.W. Hansen[1], C.B. Boothroyd[1], R.E. Dunin-Borkowski[1]

1. DTU-CEN, Center for Electron Nanoscopy, Technical University of Denmark, Fysikvej, Building 307, 2800 Kgs. Lyngby, Denmark
2. FEI Company, Achtseweg Noord 5, 5600 KA Eindhoven, Netherlands

jakob.wagner@cen.dtu.dk

Keywords: Environmental HRTEM, Titan, C_s image corrector, monochromator

High-resolution environmental TEM has become available within the last decade. This includes atomic resolution imaging of catalyst nanoparticles under controlled gas atmospheres up to 20mbar pressure [1,2,3] as well as dynamical growth studies of semiconductor nanowires and of carbon containing nanowires [4,5].

At the newly inaugurated Center for Electron Nanoscopy (CEN) at the Technical University of Denmark the next step for improved resolution in HRTEM under non-vacuum conditions has been taken. A monochromated Titan TEM with a spherical aberration (C_s) image corrector has been installed and equipped with an environmental cell. The microscope achieves a resolution of 1.1 Ångström with a controlled gas environment around the specimen. Besides the improved spatial resolution for environmental TEM the monochromated microscope is equipped with a post-column energy filter providing the opportunity for high energy resolution EELS of both gas phase and solid samples.

Accurate control of sample temperature, gas flow and gas composition is essential to prevent thermal drift in the specimen for stable long term experiments. A gas inlet system has been specially designed to allow maximum control over the gas mixing and flow into the microscope column.

Under optimum conditions it will be feasible to directly study phenomena such as surface (re)constructions of catalytic nanoparticles under working conditions, crystal twinning and grain boundaries. Operating in a dynamic mode, allows for the investigation of phenomena like diffusion and nanowire growth mechanisms.

Figure 1 demonstrates the easily achievable resolution of at least 1.1Å at 0.5mbar N_2 pressure by a Youngs fringe experiment. In Figure 2, images of an Au/Al$_2$O$_3$ based catalyst for methanol synthesis acquired in situ with working pressures up to 5mbar N_2 are shown. The {111} fringes of the Au particle are easily recognized and C_s correction means no delocalization.

First results from CEN's environmental C_s image corrected Titan will be presented.

1. P. L. Gai and E. Boyes, Ultramicroscopy 67 (1997), p. 219.
2. P. L. Hansen, J.B. Wagner, S. Helveg, J.R. Rostrup-Nielsen, B.S. Clausen and H. Topsøe, Science 295 (2002), p. 2053.
3. T. W. Hansen, J. B. Wagner, P.L. Hansen, S. Dahl, H.Topsøe, and C.J.H. Jacobsen, Science, 294 (2001), p 1508.

M. Luysberg, K. Tillmann, T. Weirich (Eds.): EMC 2008, Vol. 1: Instrumentation and Methods, pp. 509–510, DOI: 10.1007/978-3-540-85156-1_255, © Springer-Verlag Berlin Heidelberg 2008

510

4. S. Kodambaka, J. Tersoff, M. C. Reuter and F. M. Ross, Science **316** (2007), p. 729.
5. S. Helveg, C. López-Cartes, J. Sehested, P .L. Hansen, B. S. Clausen, J. R. Rostrup-Nielsen, F. Abild-Pedersen and J. K. Nørskov, Nature **427** (2004), p. 426.

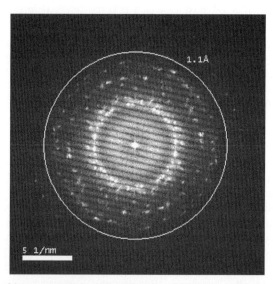

Figure 1. Youngs fringe experiment showing the resolution of the environmental Titan under 0.5mbar N_2 pressure. The fringes are easily recognized up to 9.2 nm^{-1} corresponding to 1.1 Å.

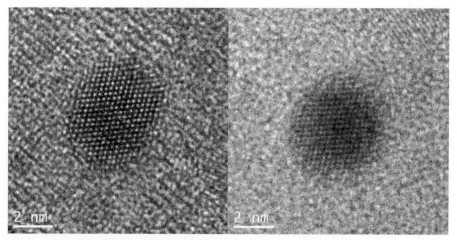

Figure 2. Image of Au particle supported on Al_2O_3 recorded at 0.5mbar (left) and 5mbar (right) pressure of N_2. The sample is a catalyst for methanol synthesis.

The surface dynamics of the transient oxidation stages of Cu and Cu binary alloys

J. C. Yang[1], Z. Li[1], L. Sun[1], G.W. Zhou[2], J.E. Pearson[3], J.A. Eastman[3], D.D. Fong[3], P.H. Fuoss[3], P.M. Baldo[3], L.E. Rehn[3]

1. Mechanical Eng. & Materials Sci., University of Pittsburgh, 848 Benedum Hall, Pittsburgh, PA, USA
2. Mechanical Eng., Binghamton University, Binghamton, NY, USA
3. Materials Science Division, Argonne National Laboratory, 9700 S. Cass Ave Argonne, IL, USA

jyang@engr.pitt.edu

Keywords: oxidation, in situ, copper alloy

The transient stages of oxidation — from the nucleation of the metal oxide to the formation of the thermodynamically stable oxide — represent a scientifically challenging and technologically important *terra incognita*. These issues can only be understood through detailed study of the relevant microscopic processes at the nanoscale *in situ*. We are studying the dynamics of the initial and transient oxidation stages of a metal and alloys with *in situ* methods, including ultra-high vacuum (UHV) transmission electron microscopy (TEM) as well as synchrotron X-ray diffraction. *In situ* methods permits direct visualization of the nucleation, growth and morphological evolution of oxides at the nanoscale under reaction environments, where the UHV environment provides controlled surface conditions necessary for quantitative insights.

These experiments were carried out in a modified JEOL 200CX TEM. This microscope is equipped with an ultra-high vacuum (UHV) chamber with base pressure $\sim 10^{-8}$ Torr. A controlled leak valve attached to the column permits the introduction of oxygen gas directly into the microscope at a partial pressure (pO_2) between 5×10^{-5} and $\sim 5 \times 10^{-4}$ Torr. Cu(100), Cu(110) and Cu(111), CuAu(100) and CuNi(100) single crystal films with 700-1000Å thickness were grown on single crystal NaCl by sputter deposition. The metal films were removed from the substrate by flotation in deionized water, washed and mounted on a specially prepared sample holder that allows for resistive heating to a maximum temperature of 1000°C.

We have previously demonstrated that the formation of epitaxial Cu_2O islands during the transient oxidation of Cu(100), (110) and (111) films bear a striking resemblance to heteroepitaxy, where the initial stages of growth are dominated by oxygen surface diffusion and strain impacts the evolution of the oxide morphologies. Furthermore, we noted that temperature and surface orientation have a dramatic effect on the oxide morphology (Figure 1). We are presently investigating the early stages of oxidation of Cu-Au and Cu-Ni as a function of oxygen partial pressures and temperatures. For Cu-Au oxidation, the addition of the second element reduces the strain between the metal and oxide, and thereby provides a second method, besides temperature, to control the oxide shape. The addition of Au also led to a self-limiting

M. Luysberg, K. Tillmann, T. Weirich (Eds.): EMC 2008, Vol. 1: Instrumentation and Methods, pp. 511–512, DOI: 10.1007/978-3-540-85156-1_256, © Springer-Verlag Berlin Heidelberg 2008

512

growth of the oxide due to the Au build-up around the oxide island, which led to a dendritic oxide growth. For Cu-Ni oxidation, the addition of Ni causes the formation Cu_2O and/or NiO where the oxide type(s) and the relative orientation with the film depend on the Ni concentration, oxygen partial pressure and temperature. Figure 2 is a cross-sectional TEM image of the CuNi film grown on STO oxidized in O_2 *ex situ*, where an irregular-shaped NiO island is seen. This research program is funded by the National Science Foundation and the Department of Energy.

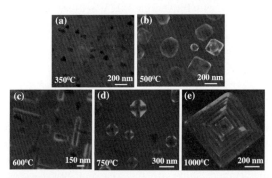

Figure 1. Different Cu_2O nanostructures formed during oxidation of Cu(001) at different temperatures at $P(O_2) = 10^{-4}$ torr.

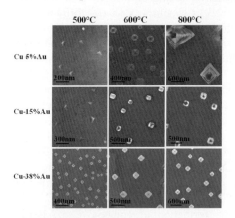

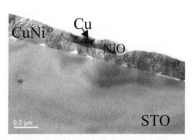

Figure 2. The effect of Au concentration and temperature on the Cu_2O morphologies due to oxidation of Cu-Au alloys *in situ*.

Figure 3. Cross-sectional TEM images of oxidized Cu-Ni alloy reveals Cu on the surface of a NiO island (courtesy L. Wang, UIUC)

In-situ TEM for altering nanostructures and recording the changes at an atomic resolution

X.F. Zhang[1], and T. Kamino[2]

1. Hitachi High Technologies America, Inc, 5100 Franklin Dr, Pleasanton, California 94588, USA
2. Hitachi High Technologies Corp., Hitachinaka, Ibaraki, Japan

xiao.zhang@hitachi-hta.com
Keywords: in-situ TEM, atomic resolution, nanostructure

In recent years, progresses in in-situ transmission electron microscopy (TEM) provided unique imaging and analytical capabilities for studying structural evolutions in versatile environments. Aiming at atomic resolution in-situ TEM capability, various sample holders have been developed by a group led by Takeo Kamino in Hitachi High Technologies Corporation, including gas injection-heating holder, single- and double-tilt heating holders, and double-heater sample holder [1-2]. Using these sample holders, in-situ heating TEM studies in vacuum or in a gas environment, and in-situ evaporation deposition can be done in a standard 300 kV H-9500 high-resolution transmission electron microscope [3], true atomic resolution can be achieved at elevated temperatures for example at 1500°C, and digital recording of the dynamic structural evolutions is realized using a high speed CCD camera.

Various nanomaterials have been studied using the aforementioned in-situ TEM system to study solid-solid, solid-liquid, and solid-gas interactions. It has been found that 300 kV electron beam could alter some nanostructures at room temperature even though the nanomaterials were composed of 'robust' materials such as carbon and metals. However, when heating samples to elevated temperatures, electron beam irradiation helped in-situ TEM study in many ways that it might minimize knock-on damages, burn off amorphous surface layers, or trigger structural changes in nanostructures. Figure 1 shows beam damage on multiple wall carbon nanotubes, but the similar beam irradiation caused no visible damage on the multiple carbon nanotubes heated to 600°C as shown in Figure 2. In study of metallic nanoparticles, atomic layer-by-atomic layer structural changes at various temperatures have been observed directly, the changes in structure would be impossible to be explained without in-situ atomic resolution TEM. These data provide insights into the structural processes in the middle stage before the environmental impacts became catastrophic to materials, therefore can help to elucidate puzzled phenomena often encountered in ex-situ experiments or in in-situ TEM experiments at low resolution or with too long time intervals for image recording.

1   T. Kamino and H. Saka, Microsc. Microanal. Microstruct. **4** (1993) p. 127.
2   T. Kamino, **T**. Yaguchi, M. Konno, A. Watabe, T. Marukawa, T. Mima, K. Kuroda, H. Saka, S. Arai, H. Makino, Y. Suzuki and K. Kishita, J. of Electron Microscopy **54** (2005) p. 497.
3   X.F. Zhang and T. Kamino, Microscopy Today **9** (2006) p. 16.

M. Luysberg, K. Tillmann, T. Weirich (Eds.): EMC 2008, Vol. 1: Instrumentation and Methods, pp. 513–514, DOI: 10.1007/978-3-540-85156-1_257, © Springer-Verlag Berlin Heidelberg 2008

514

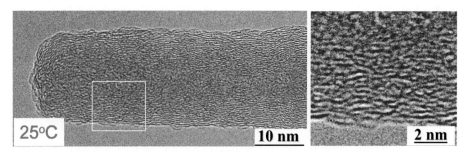

Figure 1. A multiple wall carbon nanotube was damaged by electron beam irradiation for 20 min at room temperature.

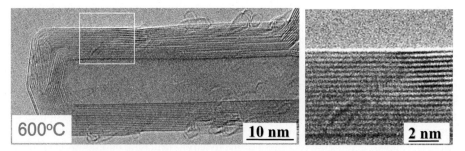

Figure 2. A multiple wall carbon nanotube was irradiated by electron beam for 20 min at 600°C, no structural damage is recognized.

Aberration correction in SEM: Relaunching an old project

J. Zach

CEOS GmbH, Englerstr. 28, D-69126 Heidelberg, Germany

zach@ceos-gmbh.de

Keywords: aberration correction, SEM, chromatic aberration

After the basic correction concepts of chromatic aberration correctors had been developed in the early 1970s for Transmission Electron Microscopes (TEMs) [1], the idea to use such correctors also for Scanning Microscopes (SEMs) came up soon: It seemed to be obvious that the best improvements by chromatic correction could be achieved, where the beam energy was low and, consequently, the relative energy width was high.

After the theoretical basis for corrected Low Voltage Scanning Electron Microscopy had been laid in the late 1980's [2], the development of a prototype corrector started at the European Molecular Biology Laboratory resulting in an instrument with significantly improved resolution in 1995 [3]. This corrector was adapted for a commercial instrument in a collaboration between CEOS and JEOL. A commercial corrected SEM, the JSM-7700F, is available since a couple of years.

However, corrected SEM still is a niche application with very few installations, whereas correction in TEM and Scanning TEM has become a standard technique in high resolution transmission microscopy. What are the reasons for this?

- Are the uncorrected instruments good enough to see, what people would like to see in their micrographs?
- Are there limitations other than probe size, which limit the resolution like specimen preparation or beam-specimen interaction?
- Is the operation of correctors too complicated for the ordinary SEM user?
- Are correctors too expensive for the SEM market?

We will discuss these questions in some detail in our talk. But two aspects shall already be mentioned here:

Resolution is typically estimated by looking at some small details in the image and the users are satisfied, if they can see the details, they are interested in. There is nothing wrong with that as long as noise, dose or radiation damage are of no importance. However, if a certain image quality should be achieved with the minimum number of incident electrons, it becomes important that most of the electrons are confined to a region corresponding to the target resolution. Unfortunately, for many uncorrected SEMs this is not the case. Figure 1 shows the probe current distribution for 1 kV in an uncorrected high resolution SEM. One could probably see 1.8 nm details with this instrument, but over 90% of the primary electrons are not confined to a disc with 1.8 nm diameter! In the corrected case the number of electrons within this disc increases by almost a factor of 10!

For the same reason in all analytical applications, where rather high beam currents are used, the analytical resolution is much worse compared to the pure image resolution.

M. Luysberg, K. Tillmann, T. Weirich (Eds.): EMC 2008, Vol. 1: Instrumentation and Methods, pp. 515–516, DOI: 10.1007/978-3-540-85156-1_258, © Springer-Verlag Berlin Heidelberg 2008

516

Due to the high beam current the central peak of the intensity distribution gives enough signal to generate a crisp image. However, the analytical signal is coming from an area, which has a diameter, which is more than 10 times that of the central peak.

These are the reasons, why we think that aberration correction has some strong potential also in SEM. We have invented a new corrector set-up together with a transfer lens concept. The system is axially fully corrected up to fifth order. The corrector part is shown in figure 3. We are convinced that this design together with a new alignment strategy will overcome most of the obstacles, which have until now prevent the success of correction in SEM.

1. H. Rose, Optik **33** (1971), p. 1.
2. J. Zach, Optik **83** (1989), p. 30.
3. J. Zach, M. Haider, Nucl.Inst..Meth. A **363** (1995) p. 316.

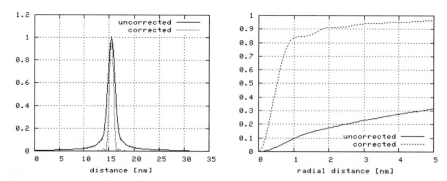

Figure 1. Left: Central peak of the scanning spot. The FWHM is not very different for the corrected and uncorrected case. Right: Integrated intensity of the same spot showing that most of the electrons are not in the central peak in the uncorrected case.

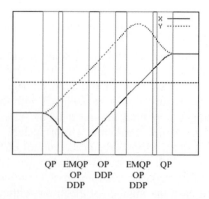

Figure 2. The optical elements and the Gaussian beam path in the new corrector. The elements are: QP=quadrupole, EMQP=electric/magnetic quadrupole, OP=octupole, and DDP=duodecipole.

Changes and reversals of contrasts in SEM

J. Cazaux

Department of Physics, Faculty of Sciences, BP 1039, 51687 Reims, France.

jacques.cazaux@univ-reims.fr
Keywords: SEE, SEM metrology, image simulation, material & topographic contrasts

The material and topographic contrasts of SEM images may be investigated by following the various steps of the image formation from the SEE yields, $\delta=f(E°)$, of the specimen's components up to the image simulation via the role of the detector: position and angle of collection passing by the influence of beam energy, $E°$, and angle of tilt, i. For such an investigation, a specimen of known dimension and composition (similar to that obtained from a lithographic process in the semiconductor industry) is postulated.

The two components of the binary specimen are a heavy metal, Pt, and a light insulator, quartz, for which the yield curves, Fig. 1a, are available in the literature. The crossing of the yields at around $E°(inv)$ ~2 keV leads to expect not only a contrast change but also a contrast reversal with the beam energy $E°$. From the δ values at 1 keV and at 4 keV combined to the use of linear scale from $\delta= 0.8$ to $\delta= 2.5$ for the grey levels, a simulation of the corresponding images is shown in Fig. 1b and it is next compared to true SEM images, fig. 1c, of a top view of a Cr/quartz integrated circuit taken at normal incidence [1]. Despite deviations in the grey levels (to be discussed), the experimental images support the contrast reversal predicted from fig. 1a.

The calculated angular distribution of the emitted SE, $\partial\delta/\partial\alpha$, for their most probable energy into the vacuum, $E_{k,}$, illustrates the role of the detection system via the value of the mean detection angle, α: Fig. 2c. The physical reason of the large difference between metals and insulators is the large difference between the corresponding inner kinetic energies, E_S, measured with respect to the bottom of the conduction band, Fig. 2a, and then to difference in the refraction effect at the specimen/vacuum interface [2]. A consequence is the change of the contrast inversion energy with the detector position (or α) via the change of the ratio, I(quartz)/I(Pt): fig. 2b. The topographic contrast is influenced by $\partial\delta/\partial\alpha$ in combination to the change of δ with the angle of tilt but only for energies at or above the maximum yield where $\delta(i)$~$\delta°/\cos i,$. Fig.2d shows the material dependence of the topographic contrast (note the possible contrast reversals with angle i) and its significant change with $E°$ [2].

There are only a few circumstances where a contrast reversal may be observed in SEM but, besides this spectacular effect, the role of various parameters on the observed contrasts remains. Another example contrast reversal is given in [3] and the limitation of the use of published yield curves will be discussed in a work in progress.

1. Courtesy of Dr P. Buffat, EPFL, Lausanne (Ch), with author's acknowledgments.
2. J. Cazaux, J. of Microsc. **217** (2005) 16 & Nucl. Instr. & Meth. in Phys. B. **244** (2006) 307
3. J. Cazaux, these proceedings.

M. Luysberg, K. Tillmann, T. Weirich (Eds.): EMC 2008, Vol. 1: Instrumentation and Methods, pp. 517–518, DOI: 10.1007/978-3-540-85156-1_259, © Springer-Verlag Berlin Heidelberg 2008

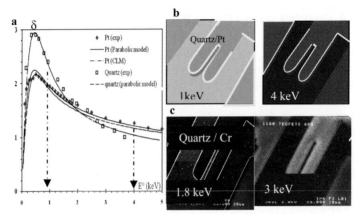

Figure 1. Simulation of the contrast reversal of a quartz/Pt specimen between 1keV and 4 keV, b, deduced from experimental yield curves, $\delta=f(E°)$ shown in: a. c: experimental contrast reversal of a similar quartz/Cr specimen obtained with a lateral detector [1].

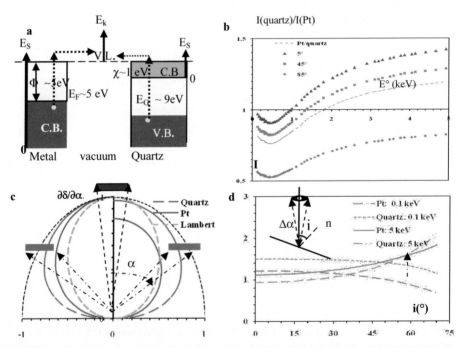

Figure 2. a: Band structure scheme of a metal and of an insulator. c: radial distribution of the emitted SE's compared to a Lambert distribution. b: change of the ratio of the collected intensities, I(quartz)/I(Pt), with E° for different mean angles of collection, α. d: change of the collected signals, I, with the angle of tilt, i°. Note the opposite evolution at low and high energies and the expected contrast reversal at i~60° for E°= 5keV(arrow).

Surface potential and SE detection in the SEM

J. Cazaux

Department of Physics, Faculty of Sciences, BP 1039, 51687 Reims, France.

jacques.cazaux@univ-reims.fr
Keywords: Angular SE emission, contrast in SEM, charging

In SEM there are many situations where the surface potential of the specimen, V_S, differs from zero:negative biasing of the specimen holder in cathode lens systems [1]; positive biasing in the SE energy filtering [2]; investigation of non-conductive materials where charging effects may lead to >0 or <0 surface potential [3]. Some consequences of such a potential are deduced from the calculation of the SE trajectories of initial energy E_K and angle of emission α (to the normal) for a specimen surface of potential V_S, at $z=0$, and grounded plate in front of it at $z=z^\circ$. Fig. 1a shows the examples of such SE trajectories derived from: $z/z^\circ = - (E_S/4E_K) (\sin^{-2}\alpha)(x/z^\circ)^2 + (z/z^\circ) \cot \alpha$ with $E_S=qV_S$. Fig 1a also shows three groups of emitted electrons: SE_D being detected for a in-lens detection defined by $\tan \alpha^\circ = x^\circ/z^\circ$; SE_S returning back to the specimen surface; and the remaining, SE_{SH} and SE_{PP}, evacuated to the ground. The initial energy distribution, $\partial\delta/\partial E_K$, has been derived from Chung & Everhart expression for metals (with $\Phi \sim 5eV$) and the initial angular distribution, $\partial\delta/\partial\alpha$, includes the refraction effects at the specimen/vacuum interface (see Fig 2c in J. Cazaux: these proceedings).

For a positive bias, the intensity of the detected SE decreases with the increase of V_S, as expected, but the SE spectral distribution does not present a sharp cut-off at $E_K = qV_S$ when x°/z° is as low as 0.1 or $\alpha^\circ = 5.7^\circ$ and the true spectral distribution is distorted for E_K slightly above qV_S (Fig. 2a). In addition, a significant number of SE are returning back to the specimen or the specimen holder where they arrive at a rather large distance, a few z° (working distance) or several millimeters, from their emission point (Fig. 1a). Then, for insulators, the steady state (where $\delta+\eta=1$) is delayed by the delay in the charge compensation: neutralization of the holes left by the emitted SEs by the returning SEs. This delay increases with the decrease of the dimension of the specimen.

For a large negative bias, the collection efficiency of the detector rapidly increases, as expected, with the increase of $|V_S|$. It approaches the unity for $V_S>1000$volts and $x^\circ/z^\circ \sim 0.1$ and it is two orders of magnitude larger than that of normal conditions. The spectral distribution of the collected SEs is shifted upwards by qV_S [3] but the tail of this distribution may be also distorted because the energetic SEs emitted at large oblique emission angles cannot be collected (Fig. 2b). This increase in the collection efficiency explains why the insulating parts of a specimen appear very bright when there are imaged with a in-lens detection system and at large beam energies: $E^\circ > 5keV$. This point is illustrated in Fig.3a) for SiC particles embedded in a Si matrix. Compared to Fig. 3b obtained with a lateral detector, this pair of images offers another example of contrast reversal in SEM: J. Cazaux, these proceedings.

M. Luysberg, K. Tillmann, T. Weirich (Eds.): EMC 2008, Vol. 1: Instrumentation and Methods, pp. 519–520, DOI: 10.1007/978-3-540-85156-1_260, © Springer-Verlag Berlin Heidelberg 2008

520

1. Müllerova & L. Frank in Adv. In Imaging and Electron Physics **128**, (2003) 309
2. P. Kazemian, et al., Ultramicroscopy, **107**, (2007) 140
3. J. Cazaux, J. of Microsc. **217** (2005) 16 & Nucl. Instr.& Meth. in Phys. B. **244** (2006) 307.
4. F.Grillon. & J. Cazaux, Proceedings of EUREM XII, Brno, July 2000, **3** (2000) 229

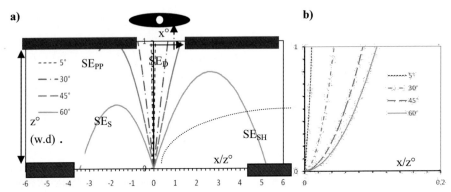

Figure 1. Calculated SE trajectories. **a)**-positive bias-: $E_K=2E_S$ (left) and for $E_K=3E_S$ (right). **b)**- negative bias-: $E_K=-E_S/1000$.

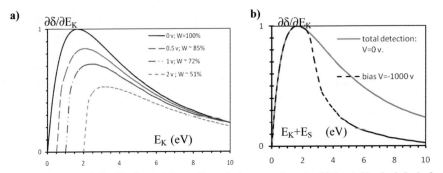

Figure 2. Spectral distribution of the collected electrons for $x°=z°/10$. **a)** $V_S=0$; 0.5; 1; 2 volts. 'w': relative weight of the integrated SE intensity refereed to the collected intensity at 0 v.
b) $V_S=-1000$ v. The reference is the total emission spectrum (note the shift in the energy scale).

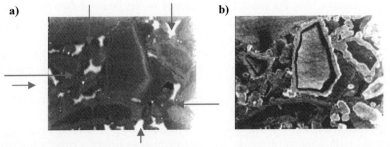

Figure 3. Images of SiC particles embedded in Si for a working distance $z°=16$ mm and $E_0 = 5$ k.e.V (from [4]). **a)** In-lens detection. **b)** Lateral detection. Note the contrast reversal between the two, in particular for the white particles- arrows on the left-.

On the Spatial Resolution and Nanoscale Features Visibility in Scanning Electron Microscopy and Low-Energy Scanning Transmission Electron Microscopy.

V. Morandi[1], A. Migliori[1], F. Corticelli[1], M. Ferroni[2]

1. CNR-IMM Section of Bologna, via Gobetti 101, 40129 Bologna, ITALY
2. INFM-CNR SENSOR and Dept. of Chemistry and Physics, Brescia University, via Valotti 9, 25133 Brescia, ITALY.

morandi@bo.imm.cnr.it

Keywords: SEM, Backscattered Electrons, Secondary Electrons, Low Energy STEM

Lecture in honour of P.G. Merli.

In the last fifteen years a significant work on the interpretation of Backscattered Electron (BSE) and Secondary Electron (SE) compositional imaging has been performed. The starting point of the research was represented by the experimental evidence of nanometric resolution of compositional features in BSE and SE imaging at relatively high energy (20-30 keV). The interpretation of these results required a reconsideration of the resolution definition, usually associated to the Full Width Half Maximum (FWHM) of the collected signal and to the local variations of the SE yield. In fact, this approach was not able to justify the experimental results with BSE because an interaction volume of the order of a μm and an exit profile of the BSE signal having a comparable FWHM did not appear compatible with the observed resolution of a few nm. Also the SE images were hard to interpret because the yield is practically independent of the layer composition [1-3].

The key point for the explanation of the results has been the following: the physical observable is not represented by the electron trajectories but by the number of electrons counted by the detector. If we consider a sequence of equidistant layers A,B,A,B... with different composition their visibility will depend on the difference in the number of the detected electrons when the beam is positioned in A or B. This difference allows one to compute the contrast, and consequently to deduce the critical current and the spot size that defines the resolution. If we use the same Monte Carlo modelization and we focus the attention on the exit profile of the electron trajectories we are not able to explain the experimental results whereas image interpretation is straightforward, when, according to the previous scheme, the output of the simulation is expressed in terms of the observable quantity, i.e. the number of electrons reaching the detector, independently of their origin [1-4].

The consequences of this new methodological approach are several and only partially investigated.

One of these concerns the definition of the output signal in BSE imaging as: $O(x) = S(x) \otimes I(x)$, where $I(x)$ is the Gaussian beam intensity distribution, whose FWHM still defines the resolution, while $S(x)$ is the specimen response that determines the contrast and depends on the beam energy [5]. From this relation it is possible to deduce the

M. Luysberg, K. Tillmann, T. Weirich (Eds.): EMC 2008, Vol. 1: Instrumentation and Methods, pp. 521–522, DOI: 10.1007/978-3-540-85156-1_261, © Springer-Verlag Berlin Heidelberg 2008

conditions providing the maximum contrast for nanoparticles or specimen details, as well as the conditions for nanoanalysis with BSE. Moreover the previous relation is formally equal to the one providing the signal in Scanning Transmission Electron Microscopy (STEM) [5]. This formal analogy has a clear physical meaning: the rules governing the image formation with Forward Scattered Electrons (FSE), obviously in a thin specimen, and BSE are the same. The only difference is related to the intensity of the signal that is higher for scattering angles smaller than 90° as in the transmission mode. An immediate consequence of this analogy is that the beam broadening and the energy loss do not affect the resolution again defined by the probe size, while the beam broadening affects the contrast.

Following the continuous demand of improvement in spatial resolution and sensitivity of the characterization techniques, due to the increasing miniaturization processes of the semiconductor technology, this approach has been validated through the BSE, SE, and FSE observation of layers of different materials with a well defined composition and width (semiconductors multilayers), as well as regions with a gradual and well controlled composition variation (doped regions) which are the base elements of any electron device [6-8].

The results allowed a quantitative description of the mechanism of the contrast formation based on a careful analysis of the scattering of the electrons inside the specimen as well as on the detailed study of the portion of the emitted signal collected in the different imaging modes [6-8]. At the same time the fundamental role played by the specimen features, as the thickness and the boundary conditions of the observed detail [8-11] has been highlighted, showing the potentialities of the SEM not only in the field of microelectronics but also for the observations of specimens with a more complex structure as the biological ones [12]. Moreover, following this approach, the importance of a proper design of the SEM camera and of the availability of properly designed tools suitable to a signal handling and collecting has been demonstrated, and an improved STEM detection system has been proposed (Patent pending, n. BO2007A000409) [13].

1. P.G. Merli and M. Nacucchi, Ultramicroscopy **50** (1993), p. 83.
2. P.G. Merli, A. Migliori, M. Nacucchi, D. Govoni, G. Mattei, Ultramicroscopy **60** (1995), p. 229.
3. P.G. Merli, A. Migliori, M. Nacucchi, M. Vittori-Antisari, Ultramicroscopy **63** (1996), p. 23.
4. P.G. Merli, A. Migliori, V. Morandi, R. Rosa, Ultramicroscopy **88** (2001), p. 139.
5. P.G. Merli, V. Morandi, F. Corticelli, Ultramicroscopy **94** (2003), p.89.
6. P.G. Merli, F. Corticelli, V. Morandi, Applied Physics Letters **81** (2002), p. 4535.
7. P.G. Merli and V. Morandi, Microscopy and Microanalysis **11** (2005), p. 97.
8. P.G. Merli, V. Morandi, G. Savini, M. Ferroni, G. Sberveglieri, Applied Physics Letters **86** (2005), p. 101916.
9. P.G. Merli, V. Morandi, F. Corticelli, Journal of Microscopy **218** (2005), p. 180.
10. V. Morandi, P.G. Merli, M. Ferroni, Journal of Applied Physics **90** (2006), p. 043512.
11. V. Morandi and P.G. Merli, Journal of Applied Physics **101** (2007), p. 114917.
12. V. Morandi, P.G. Merli, D. Quaglino, Applied Physics Letters **90** (2007), p. 163113.
13. V. Morandi, A. Migliori, P. Maccagnani, M. Ferroni, F. Tamarri, abstract submitted to EMC 2008 (2008).

Scanning electron microscopy techniques for cross-sectional analyses of thin-film solar cells

D. Abou-Ras[1], U. Jahn[2], J. Bundesmann[1], R. Caballero[1], C.A. Kaufmann[1], J. Klaer[1], M. Nichterwitz[1], T. Unold[1], H.W. Schock[1]

1. Hahn-Meitner-Institut Berlin, Glienicker Strasse 100, Berlin, Germany
2. Paul-Drude-Institut für Festkörperelektronik, Hausvogteiplatz 5-7, Berlin, Germany

daniel.abou-ras@hmi.de
Keywords: chalcopyrite-type, thin-film solar cells, EDX, EBSD, EBIC, cathodoluminescence

Thin-film solar cells with chalcopyrite-type absorbers such as $Cu(In,Ga)Se_2$ or $CuInS_2$ provide a technology with low-cost perspective at high photovoltaic performance. These solar cells generally consist of $ZnO/CdS/Cu(In,Ga)Se_2/Mo$ or $ZnO/CdS/CuInS_2/Mo$ thin-film stacks on, e.g., glass substrates (see Figure 1). Since $Cu(In,Ga)Se_2$ and $CuInS_2$ are the photoactive layers in these solar cells, they are subjected to refined analyses. For research and development, it is essential to determine the influences of process parameters such as the substrate temperature during the absorber deposition or the Cu concentration within the absorber on the microstructural, compositional and electrical properties of these thin-film stacks.

Recently, it has been presented how chalcopyrite-type layers in completed solar cells may be prepared for cross-sectional analyses by electron backscatter diffraction (EBSD) [2]. This approach allows not only for precise measurements of grain sizes but also provides access to local orientations and grain boundaries. In addition, it is possible to combine this technique with various other scanning electron microscopy (SEM) methods.

In Figure 2, an SEM image, an EBSD map and an electron beam-induced current (EBIC) image from the same area on a $Cu(In,Ga)Se_2$ layer is shown. The grain areas can be identified unambiguously by EBSD, whereas EBIC gives information on the carrier collection. It is possible to estimate the recombination velocities at the grain boundaries from the collection profiles by appropriate modelling.

Combined studies applying EBSD and energy-dispersive X-ray spectrometry (EDX, used at 7 kV) allow for phase identification, e.g., of Cu-S islands found on top of $CuInS_2$ produced by rapid sulfurisation of Cu-In precursors under Cu-rich conditions (Figure 3). By the use of low acceleration voltages for EDX, lateral resolutions down to 100-150 nm may be achieved even in a scanning electron microscope.

For a spatially resolved, optoelectronic analysis of such $CuInS_2$ layers, EBSD was combined with cathodoluminescence (CL), see Figure 4. The EBSD and CL results show how important it is to identify grains in order to be able to relate areas of decreased CL intensities to either grain boundaries or defects within the grains.

1. The authors are grateful to N. Blau, B. Bunn, C. Kelch, M. Kirsch, P. Körber, and T. Münchenberg for solar-cell processing.

2. D. Abou-Ras, S. Schorr, and H.-W. Schock, J. Appl. Cryst. **40** (2007), p. 841.

M. Luysberg, K. Tillmann, T. Weirich (Eds.): EMC 2008, Vol. 1: Instrumentation and Methods, pp. 523–524, DOI: 10.1007/978-3-540-85156-1_262, © Springer-Verlag Berlin Heidelberg 2008

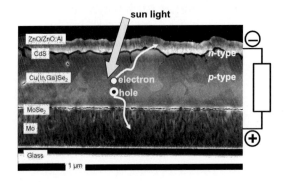

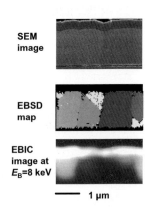

Figure 1. Cross-sectional SEM image of a Cu(In,Ga)Se$_2$ thin-film solar cell and its mode of operation.

Figure 2. SEM image, EBSD map and EBIC signal of a Cu(In,Ga)Se$_2$ solar cell.

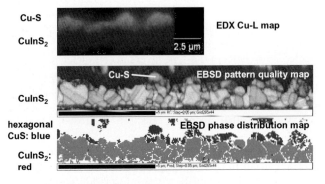

Figure 3. EDX Cu-L and EBSD maps of a Cu-S/CuInS$_2$/Mo/glass stack. The Cu-S phase can be identified from these EDX and EBSD results as hexagonal CuS.

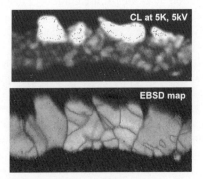

Figure 4. CL (820 nm) and EBSD maps with Σ3 grain boundaries highlighted by red lines from the identical region of a CuInS$_2$ layer in a completed solar cell. Within several grains, the CL signal is not uniform and also exhibits areas of reduced intensity, probably caused by planar defects.

Maximising EBSD acquisition speed and indexing rate

Shunsuke Asahina[1], Franck Charles[1], Keith Dicks[2], and Natasha Erdman[3]

1. European Application Group, JEOL(Europe) SAS Espace Claude Monet, 1, allée de Giverny, Croissy sur Seine 78290 France
2. Oxford Instruments Analytical Halifax Road, High Wycombe Bucks HP12 3SE UK
3. JEOL USA, INC. 11 Dearborn Road Peabody, MA 01960 USA

asahina@jeol.fr

Keywords: EBSD, FE-SEM, Sample preparation, High probe current,

Electron Backscatter Diffraction (EBSD) is a powerful technique capable of characterising extremely fine grained microstructures in a Scanning Electron Microscopy (SEM). Electron Back Scatter Patterns (EBSPs) are generated near the sample surface, typically from a depth in the range 10 – 50nm. Consequently, EBSD requires that the surface is adequately damage free on a crystallographic scale, in order to generate useable EBSP's. Thus sample preparation is critical in achieving good EBSP quality, which in turn, is an important prerequisite to be able to maximise acquisition speed.

Recently, JEOL have developed a new cross sectioning apparatus based on milling with an Ar ion beam using a shield plate [1], which can produce a very good quality cross section of specimens for high spatial resolution microscopy and microanalysis. The apparatus can also produce surfaces suitable for EBSD directly,. The principle of this method is shown schematically in Fig.1. This apparatus is called the Cross Section Polisher (CP) [2]. In this paper, the CP instrument was used to prepare samples for EBSD analysis.

SEM performance is also a critical factor, especially the ability to deliver high probe current coupled with small probe diameter, which is advantageous for high acquisition speed EBSD analysis, whilst maintaining high spatial resolution. For conventional EBSD analysis, beam current in the range 1nA to 10nA is typical.

Recently, Oxford Instruments NanoAnalysis have introduced a new EBSD system [3] capable of producing 400 analysed points per second. This new system uses a newly developed EBSD detector (The Nordlys F400) and s/w package running multithreaded programming on a multiple processor PC. It can achieve 400 pixels per second, analysed in real time, with a 99% solution rate for Ni prepared by electropolishing.. The indexing rate is linked to pattern intensity and quality, amongst other factors, which in turn (ignoring the sample condition for a moment) is dependant on the available beam current and probe diameter.

JEOL have developed a New FE-SEM called the JSM-7001F. This FE-SEM has an 'In-Lens' Shottky FE-Gun [4] which is capable of producing high probe current (up to 200nA at 15kV guaranteed). In this article, we have investigated high speed EBSD acquisition for differing beam currents using a strained Fe BCC sample, cross-sectioned using the CP polisher. The sample was analyzed without any further preparation using the HKL Nordlys F400 running with Fast Acquisition Software, fitted to a JSM 7001F.

M. Luysberg, K. Tillmann, T. Weirich (Eds.): EMC 2008, Vol. 1: Instrumentation and Methods, pp. 525–526, DOI: 10.1007/978-3-540-85156-1_263, © Springer-Verlag Berlin Heidelberg 2008

526

Fig.2 shows the EBSD results for different probe currents acquired at high acquisition speed. The EBSD result Fig. 2 (a) indicates a 85.48% indexed map, acquired using a 20nA beam current. The EBSD result Fig.2 (b) shows the same area mapped using a beam current of 80nA. This achieved a substantially increased indexing rate (92.49%), whilst maintaining the same acquisition speed.

As a result, it is clear that high probe current combined with small probe diameter can be advantageous in achieving high acquisition speed, combined with high indexing rate, as in the case of ion milled, strained specimens shown here.

1. W. Hauffe: *Sputtering by Particle Bombardment III*; R. Behrish and K. Wittmaack, ed., Springer-Verlarg, Berlin (1991) p.305
2. M. Shibata: JEOL News, [39] (2004) p.28
3. Oxford Instruments NanoAnalysis Nordly F400 brochure, OIA/152/A/0607
4. Patent No; US6,753,533 B2 (22 June.2004)

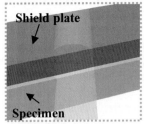 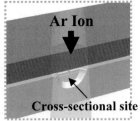

The specimen on the unmasked front side below a shield plate is milled to produce a cross section of the specimen by an Ar ion beam. Because the ion beam is irradiated parallel to the surface of the cross section, milling rate is almost independent of constituents in the specimen and radiation damage due to the ion beam is minimized, to produce a very high quality smooth section with minimum artifacts.

Figure 1. Principle of Cross Section Polisher

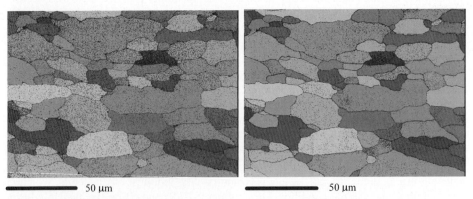

Figure 2. EBSD results from different probe current at high acquisition speed.

(a) 20nA 20kV indexing 85.48% Iron BCC (b) 80nA 20kV indexing 92.49% Iron BCC

EBSD acquisition speed; 364points/sec   Step size; 0.1um   Points; 1216×912   Sample; Iron BCC

Helium ion microscope: advanced contrast mechanisms for imaging and analysis of nanomaterials

David C. Bell[1], L.A. Stern[2], L. Farkas[2] and J.A. Notte

1. SEAS, Harvard University, Cambridge, MA, USA
2. ALIS Corp., Carl Zeiss SMT Company, Peabody, MA, USA

dcb@seas.harvard.edu
Keywords: ion microscopy, secondary electron imaging, nanomaterials, nanofabrication

Since 1960s when the first combination of FIM and mass spectrometry allowed single atom detection and imaging, it took nearly 40 years to develop stable He ion imaging source and commercialize this technology. The Helium ion microscope developed by ALIS Corporation (Orion microscope) presents new and compelling prospects in terms of imaging and characterization of nanomaterials and composites.

The image formation principle in the helium ion microscope involves interaction of a relatively low energy He ion with the specimen that produces secondary electron signal that is mainly confined to the surface of the material [1]. The resulting secondary signal exhibits superb resolution and image fidelity. Minimal contribution from high energy backscattered electrons to the collected signal enhances image fidelity even further. Use of ions as an imaging source, as opposed to electrons, allows to image non-conductive materials (Fig. 1) that would otherwise require coating or use of low vacuum environment in SEM. Optional RBS (Rutherford Backscatter) type detector provides additional imaging capabilities. RBS imaging is sensitive to atomic number difference (z-contrast) and the image in essence becoming a real time spectral map of materials constituents.

In this paper we will show our results using He ion microscope to image a wide variety of nanomaterials including: nanotubes, nanowires, nanoparticles, composite materials and structures such as polymers and multi-layer fabricated devices and biological materials. One of the goals of our research is to understand the underlying physics of the image formation, sample/ion interactions and subsequent image contrast interpretation. We will show direct comparison between the type of information that can be obtained from FEG-SEM operated at various conditions (different kVs and different detector systems, including SE energy filtering) and the information that is obtained from the Orion microscope. We will discuss the possible image formation mechanisms and image interpretation and how they can be related to the familiar realm of SE imaging in a scanning electron microscope.

We will also show one of the unique applications of the Orion microscope – milling of beam sensitive materials using He ions. While using He ion source the sputtered target atoms occur at a rate of 100 times less than Ga ions in a conventional FIB system, meaning long imaging times without sample alteration and no sample-altering ion implantation. Materials that would be termed 'difficult' in a Ga FIB and would possibly require cryo conditions for stabilization can be gently milled or sputtered in the He ion

M. Luysberg, K. Tillmann, T. Weirich (Eds.): EMC 2008, Vol. 1: Instrumentation and Methods, pp. 527–528, DOI: 10.1007/978-3-540-85156-1_264, © Springer-Verlag Berlin Heidelberg 2008

528

microscope without contamination or thermal effects as normally seen in FIB system.

The helium ion source has proven to be very stable and offers high brightness, low virtual size and low energy spread as compared to a conventional FEG type SEM, and an ion source with low contamination issues [2]. The Orion microscope offers unique imaging capabilities - remarkable depth of field, compositional analysis ability via RBS system, no charging effects and excellent surface information on low Z materials. The Helium Ion Microscope (Orion type microscope) will evolve into the surface imaging instrument for the 21st Century.

1. J. Notte and B. Ward, Scanning Vol 28 (2006)
2. V.N. Tondare, J. Vac. Sci. Vol 23, 6 p1498-1507 (2005)

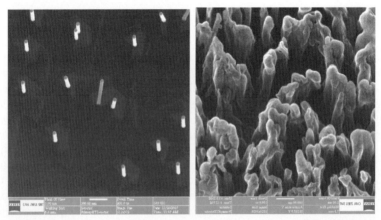

Figure 1. Images of InAs nanowires showing SE contrast (left); Mo whiskers on substrate (right).

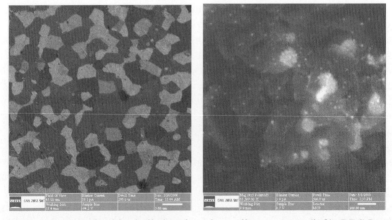

Figure 2. RBS Image of solder ball showing channeling contrast (left), RBS Imaging of nanoparticles indicating size and contrast variations due to compositional variations.

Hygroscopic properties of individual aerosol particles from aluminum smelter potrooms determined by environmental scanning electron microscopy

N. Benker[1], M. Ebert[1], P.A. Drabløs[2], D.G. Ellingsen[3], Y. Thomassen[3] and S. Weinbruch[1]

1. Institute of Applied Geosciences, Technical University Darmstadt, Schnittspahnstraße 9, 64287 Darmstadt, Germany
2. Karmøy Plant, Norsk Hydro, 4265 Håvik, Norway
3. National Institute of Occupational Health, P.O. Box 8149 Dep., 0033 Oslo, Norway

benker@geo.tu-darmstadt.de

Keywords: workplace aerosol, aluminum, hygroscopic properties, ESEM

Aluminum is produced by electrolysis of alumina (Al_2O_3). To decrease the melting point cryolite (Na_3AlF_6) is added. At present, the Norwegian aluminum industry uses two different technologies to produce aluminum: the so called Søderberg and Prebake processes [1]. In both cases, particles with variable composition, several gases (for example hydrogen fluorides) and aromatic hydrocarbons (PAHs) are formed. Inhalation of these pollutants leads to adverse health effects for the workers. For better understanding of the toxicological relevance of the particulate matter present in the workrooms, individual particle analysis was performed. Special emphasis was placed on the hygroscopic properties of the different particles sampled in the potrooms, as the presence of thin water films or small droplets may provide an opportunity for hydrogen fluoride (HF) to be transported deep into the lung.

Aerosol particles with aerodynamic diameters between 0.1 and 10 μm were collected in two aluminum smelter potrooms. Particles were sampled on Cu foils with a two stage cascade impactor. Previous work by Höflich et al. [2] determined the size, morphology and chemical composition of approximately 1000 particles in both potrooms by scanning electron microscopy and energy-dispersive X-ray microanalysis. According to these authors, soot, aluminum oxides (predominantly β-alumina, $NaAl_{11}O_{17}$), cryolite and a mixture of aluminum oxides and cryolite are the most abundant particle groups. The hygroscopic properties of these most abundant particle groups were studied in the present work.

Hygroscopic properties of aerosol particles are important for exposure assessment, as they influence the deposition in the human respiratory tract. In contrast to particles consisting of a single phase, the hygroscopic behavior of complex agglomerates (mixtures of various phases on a nanometer scale) cannot be predicted from their chemical composition. Therefore, the hygroscopic behavior of individual particles was studied in situ by ESEM with a Quanta 200 F instrument (FEI, The Netherlands). Technical details of the hygroscopicity experiments are described by Ebert et al. [3]. All experiments were carried out at a temperature of 5 °C and relative humidities (RH) up to 100 %.

M. Luysberg, K. Tillmann, T. Weirich (Eds.): EMC 2008, Vol. 1: Instrumentation and Methods, pp. 529–530, DOI: 10.1007/978-3-540-85156-1_265, © Springer-Verlag Berlin Heidelberg 2008

The hygroscopic behavior of two cryolite particles is shown in Figure 1. At a RH of approximately 65 % the particles begin to grow, presumably due to the formation of a water film. At a high RH of approximately 99 %, partial deliquescence (formation of droplets of a saturated solution) is observed.

Aluminum oxide particles show a different behavior. First adsorption of water (indicated by a slight modification of the particle morphology) is already observed at a RH of 35 %. After reducing the relative RH, the original particle morphology is observed again. Soot particles develop a water film at a high RH of approximately 99 %.

In summary, most particles present in the potrooms develop a small water film or show partial deliquescence indicating the ubiquitous presence of surface coatings of soluble material. Due to the high solubility of hydrogen fluoride in water, these observations provide an opportunity to transport HF deep into the lung.

1. G. Nechev, S. Tsymbalov, L. E. Swartling and G. E. Volfson, Proceedings TMS Annual Meeting, Orlando, FL, USA (1997), p. 201.
2. B. L. W. Höflich, S. Weinbruch, R. Theissmann, H. Gorzawski, M. Ebert, H. M. Ortner, A. Skogstad, D. G. Ellingsen, P. A. Drabløsd and Y. Thomassen, J. Environ. Monit. 7 (2005), p. 419.
3. M. Ebert, M. Inerle-Hof, and S. Weinbruch, Atmos. Environ. 36 (2002), p. 5909.

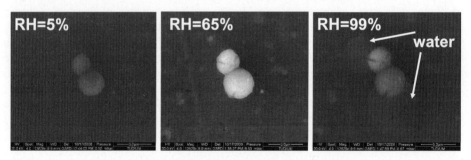

Figure 1. Secondary electron images of cryolite at low and high RH

Analysis of individual aerosol particles by automated scanning electron microscopy

N. Benker, K. Kandler, M. Ebert, and S. Weinbruch

Institute of Applied Geosciences, Technical University Darmstadt, Schnittspahnstraße 9, 64287 Darmstadt, Germany

benker@geo.tu-darmstadt.de

Keywords: automation, SEM, aerosol, particle analysis

Characterisation of individual particles is very time consuming if carried out manually. In the case of automated measurements, a large number of particles can be investigated offering the possibility of monitoring the atmospheric aerosol composition by scanning electron microscopy (SEM). The reliability of automated particle analysis is shown in the present contribution with two examples from recent field campaigns: (a) desert dust collected at the island of Tenerife (Spain), and (b) aerosol monitoring at an industrial site in the city of Duisburg (Germany).

In both examples, aerosol particles were collected with a two stage cascade impactor. The samples were characterized automatically by environmental scanning electron microscopy (ESEM) with a Quanta 200 F instrument (FEI, The Netherlands) coupled with an energy-dispersive X-ray detector (EDX). Using an ESEM has the advantage of avoiding substantial loss of volatile components due to the low vacuum in the sample chamber. The automated measurements require an optimized particle sampling (sampling time, substrate) [1].

The chemical composition of more than 22000 particles of Saharan mineral dust (collected on the island of Tenerife) was analysed by EDX. From the chemical composition, various particle groups were defined. The average relative abundance of the different particle groups as function of particle size is shown in Figure 1. Based on calcium concentrations, it is possible to determine the source regions of the aerosol particles. High calcium concentrations in the aerosol are related to high calcite concentrations in the soil of the source region. In addition, the presence of thin sulphate surface coatings on a large number of particles (more than 16000) was detected by investigating the scaling behaviour of various element signals with the particle size. There is no other technique which allows the detection of thin surface coatings on such a large number of particles.

In the second example, more than 40000 aerosol particles collected near a steel smelter were investigated. Based on chemical composition, morphology (Figure 2) and beam stability different particle groups including metallic and oxidic fly ashes, soil particles (e.g. silicates, oxides), sea salt, soot and secondary aerosol particles were defined.

Based on the automated particle characterization together with the analysis of meteorological parameters it was possible to quantify the contribution of various sources (e.g. industry, traffic, natural sources) to the aerosol composition.

M. Luysberg, K. Tillmann, T. Weirich (Eds.): EMC 2008, Vol. 1: Instrumentation and Methods, pp. 531–532, DOI: 10.1007/978-3-540-85156-1_266, © Springer-Verlag Berlin Heidelberg 2008

Both examples clearly demonstrate that a large number of particles can be characterized (chemical composition, size and morphology) with scanning electron microscopy. Thus, individual particle analysis by SEM is feasible for monitoring purposes.

1. K. Kandler, N. Benker, U. Bundke, E. Cuevas, M. Ebert, P. Knippertz, S. Rodríguez, L. Schütz, S. Weinbruch, Atmospheric Environment 41 (2007), 8058-8074.

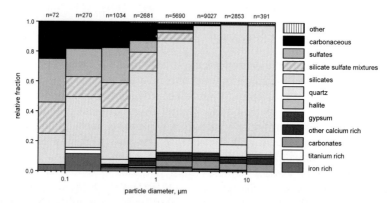

Figure 1. Relative number abundance of different particle groups of Saharan mineral dust collected at Tenerife.

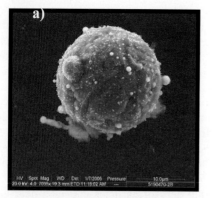
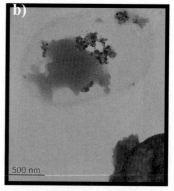

Figure 2. SEM image of an iron oxide fly ash (a), and TEM image of a secondary aerosol particle with soot inclusions (b).

A new quantitative height standard for the routine calibration of a 4-quadrant-large-angles-BSE-detector

D. Berger[1], M. Ritter[2], M. Hemmleb[2], G. Dai[3] and T. Dziomba[3]

1. Zentraleinrichtung Elektronenmikroskopie (ZELMI), Technische Universität Berlin, Straße des 17. Juni 135, 10623 Berlin, Germany
2. m2c microscopy measurement & calibration GbR, Alt Nowawes 83a, 14482 Potsdam, Germany
3. Physikalisch Technische Bundesanstalt (PTB), Bundesallee 100, 38116 Braunschweig, Germany

dirk.berger@tu-berlin.de

Keywords: sample morphology, quantitative height analysis, 3D calibration standard, nanomarker

The 4-quadrant-large-angles-BSE-analysis (4Q-analysis) is well known for the fast analysis of the 3D sample morphology of catalysts, fracture surfaces and semiconductor devices in the scanning electron microscope (SEM) using backscattered electrons (BSE). As reported previously [1-3] the 4Q-analysis works well for the **qualitative** reconstruction of the sample morphology, i.e. the relative heights of surface structures are measured properly with respect to each other. The method can resolve heights down to 10 nm.

However, the **quantitative** height analysis is often not reproducible and depends strongly on many parameters of detector- and SEM-settings. Therefore, a calibration of the detector is necessary for each individual measurement! For this purpose a new fast exchangeable 3D calibration standard was developed on the basis of a commercially[2] available calibration structure for scanning probe microscopy (SPM). The new standard (Figure 1) consists of three pyramids (edge length 10 μm, height 1.5 μm), with 263 nanomarkers, i.e. landmarks that serve as reference points [4]. Additionally, a half sphere which is used for the essential balancing of the four BSE-diodes of the detector, i.e. the offset and the gain of the four amplifiers are adjusted to identical brightness and contrast during imaging of the half sphere with each diode. The new calibration structure was produced by an automated focused ion beam (FIB) patterning process.

It was calibrated at PTB by the Metrological Large-Range Scanning Probe Microscope [5] based on the NanoMeasuringMachine (SIOS GmbH). The positions of all three translation axes of this instrument are monitored by laser interferometers and thereby ensure direct traceability to the SI unit metre. In this way, a high-quality determination of the nanomarker coordinates is accomplished, thus turning the sample into a high-quality standard.

For a quantitative height analysis of the sample morphology, the diodes are balanced using the half sphere first. Then, the four 4Q-images with different take off angles both of the calibration standard and of the sample under investigation are recorded. The relative heights of the surface morphology on both samples are evaluated using the reconstruction software MEX 4.2 (Alicona Imaging GmbH). Then, the newly developed soft-

M. Luysberg, K. Tillmann, T. Weirich (Eds.): EMC 2008, Vol. 1: Instrumentation and Methods, pp. 533–534, DOI: 10.1007/978-3-540-85156-1_267, © Springer-Verlag Berlin Heidelberg 2008

534

ware microCal (m_2c GbR) calculates the height calibration factor by comparing the nanomarker coordinates in the 4Q-image with the PTB reference values. The included software microCorrect applies the resulting height calibration factor on the 4Q-image of the sample. Another advantage of the m_2c software bundle is the immanent compensation of inhomogeneous magnification of the SEM in x- and y-direction and their shear. Figure 2 shows a typical example of a scale and shear corrected 4Q-measurement.

1. D. Berger, Microscopy Conference MC 2007, Microsc. Microanal. 13 (2007) 74-75.
2. U. Wendt, H. Heyse, O. Kisel, Microscopy Conference 2005, Davos.
3. W. Drzazga, J. Paluszynski, W. Slowko, Meas. Sci. Technol. 17 (2006) 28-31.
4. M. Ritter, T. Dziomba, A. Kranzmann, L. Koenders, Meas. Sci. Technol. 18 (2007) 404-414
5. G. Dai, F. Pohlenz, H.-U. Danzebrink, Rev. Sci. Inst. 75 (2004) 962-969

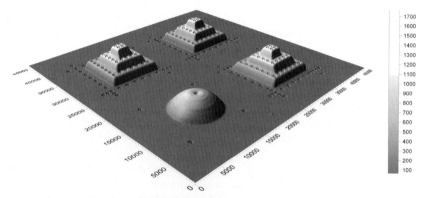

Figure 1. 3D view of the new calibration standard measured by the Metrological Large-Range Scanning Probe Microscope at PTB; edge length of the pyramids is 10 μm.

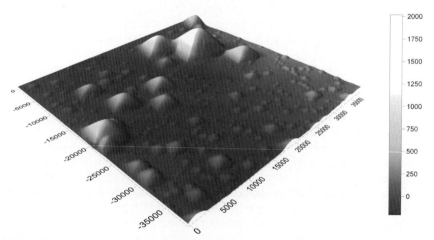

Figure 2. Measured quantitative 3D surface morphology of etched Si, heights are two-times amplified (scaling in nm).

SEM-EDS for effective surface science and as a next generation defect review tool for nanoparticle analysis?

Edward D. Boyes

Departments of Physics and Electronics, University of York, York YO10 5DD, UK

eb520@york.ac.uk
Keywords: SEM, EDS, Surface-Analysis

The development of high resolution low voltage (towards 1nm at 1kV) SEM systems and medium to low voltage (5kV) SEM-EDS [1,2] has demonstrated new capabilities of SEM-based methods in surface analysis. They appear to be particularly important in the analysis of crystallographically specific or other patch field coatings on sub-micron substrate particles, including nanoparticles, with complex three dimensional geometries. These are considered to be not well matched to wider area surface analyses or scanning probe approaches; especially where re-entrant or otherwise steep surfaces are involved.

The data have been calibrated with wide area thin films on planar substrates with thicknesses monitored by accepted methods such as quartz crystal thickness monitors. The classical crystallographic symmetry of some substrate particles has provided an elegant cross-check with related and chemically deposited (and therefore we expect not deposition direction dependent) surface patches accessible in both plan view and profile with gratifying consistency in results.

It is necessary to have high sensitivity of analysis under minimally invasive conditions with respect to a clean dry high vacuum system and low beam currents. The proof-of-principle system [1,2] combines superior imaging capabilities at low beam energies with greatly improved access for a custom EDS detector to provide both imaging (to <0.4nm and more importantly close to 1nm at 1kV) and chemical analysis sensitivity (0.3sr) at the highest levels so far attained in an SEM accepting a bulk sample, defined in this context as a piece of unthinned Si wafer, and up to 7.5mm wide. The EDS detection geometry is at least 15x more effective than in regular SEM geometries and some are much worse than that (>50x).

The high detection efficiency (50kcps per nA for Al at 5kV with a 2nm imaging probe) can be exploited, as here, for high sensitivity analysis (~1nm) of thin films of C or Al, or for very high speed analysis (especially for spectral imaging maps), or for low intensity and minimally invasive data collection from delicate samples; or some suitable application specific compromise between these various attributes. One way in which the new capabilities can be focused is for the detection and analysis of small particles on bulk sample surfaces; both as intentional nanoparticles and as inadvertent ones, such as those which are defect review tool targets. The demonstrated successful detection and analysis of 28nm SiO_2 particles on an Al substrate by the Si k-line fingerprint translates into a sensitivity for <20nm particles for the single element variety and meeting with an extension of SEM-EDS the current published ITRS Roadmap [3] requirements for next

M. Luysberg, K. Tillmann, T. Weirich (Eds.): EMC 2008, Vol. 1: Instrumentation and Methods, pp. 535–536, DOI: 10.1007/978-3-540-85156-1_268, © Springer-Verlag Berlin Heidelberg 2008

536

generation semiconductor process defect review tools. And doing it in a minimally invasive way which could be engineered to extend to full wafer analyses; although then, as so often, target discovery and acquisition would inevitably be an issue. The test targets used here are much simpler to analyse in this regard.

Further improvements in analytical sensitivity and thereby in minimum detectable mass can be predicted by using x-ray detectors with higher P/B, e.g. conventional wavelength dispersive spectrometers, parallel beam devices or microcalorimeters [4]; although none of these devices would be easy to integrate efficiently into a high spatial resolution instrument and the methods might then prove slower and more invasive. It seems the immediate goals can be achieved with imaginative incremental improvements (I^3) to established SEM-EDS methods and that target discovery and acquisition is likely to be the more substantial problem than basic detection efficiency.

1. E D Boyes, Inst of Physics (UK) Conference Series 186 (2001) 115
2. E D Boyes, Mikrochimica Acta, 138 (2002) 225.
3. International Technology Roadmap for Semiconductors, http://www.itrs.net/ 2007
4. D Newbury, D Wollman, K Irwin, G Hilton, and J Martinis, Ultramicroscopy, 78 (1999) 73
5. Hitachi S5000SPX FESEM with Thermo Noran 30mm^2 ATW EDS detector at 0.3sr

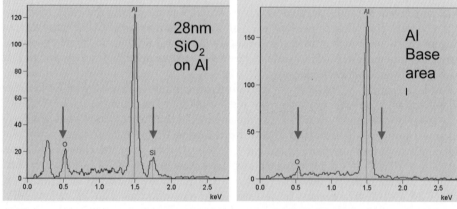

Figure (a) (b). EDS spectra (a) acquired from an identified silica particle 28±1nm in diameter using 3kV beam energy at 100,000x and the comparable data (b) from the related Al base area showing sensitivity for nanoparticle analysis to ITRS next generation defect review tool specification with sensitivity equivalent to <20nm single element particle detection using current technologies in custom form [5].

Detection of Signal Electrons
by Segmental Ionization Detector

P. Cernoch[1], J. Jirak[1,2]

1. Dept. of Electrotechnology FEEC BUT, Udolni 53, 60200, Czech Republic
2. Institute of Scientific Instruments of ASCR, Kralovopolska 147, 61264, Czech Rep.

xcerno07@stud.feec.vutbr.cz

Keywords: segmental ionization detector, side detection

As the variable pressure scanning electron microscopes and the environmental scanning electron microscopes are capable to operate with pressures up to several thousand Pa in the specimen chamber, they are frequently utilized for observations and diagnostics of a wide scale of specimens without modifications, including wet and non-conducting ones, and a wide scale of experiments, such as changes of a structure and a composition of a specimen caused by changes of specimen temperature and ambient in the specimen chamber. For information acquisition about the specimens in these microscopes, detectors that takes advantages of the process of the impact ionization in the gas ambient in the specimen chamber are often utilized [1, 2, 3].

A segmental ionization detector (SID) [4], that is a modification of an ionization detector [1], consists of several electrodes from which anyone can be selected for a signal detection. It is possible to attach varied voltages in a range from -500 V to +500 V on any of the electrodes of the SID when a specimen holder is grounded. A version of the SID consisting of four circular electrodes divided into a left and a right half placed above the specimen concentrically to the microscope axis is shown in Figure 1a, b. This halved SID enables a mode of the signal electrons detection from the electrodes placed on the side towards the imaging area. If the same voltages are attached on the electrodes D_L and D_R and the signal is detected only by one of these electrodes D_L or D_R, respectively, then detection from the left or right side, respectively, is achieved. Other electrodes of the SID must be grounded. Two examples of such side detection acquired by the SID are shown in Figure 2a, b, c, d.

As presented in the images, the halved SID enables to accomplish similar results as scintillation and semiconductor detectors positioned on the side toward the specimen. Moreover in the case of the halved SID, it is possible to achieve a change of shadow and diffusion contrasts in specimen images not only by a selection of the detection electrodes but also by a change of the voltages on the non-detection electrodes.

1. G. D. Danilatos, Sydney: Academic Press (1990), 103 p. (ISBN 0-12-014678-9).
2. S. W. Morgan, M. R. Phillips, J. of App. Physics **Vol. 100** (2006), 16 pages.
3. B. L. Thiel, et al., Rev. of Sci. Ins. **Vol. 77** (2006), 7 pages.
4. P. Cernoch, J. Jirak, R. Autrata, Proceedings of the 7th MCM (2005), p. 367 – 368.
5. This research is supported by project MSM 0021630516 and by the Academy of Sciences of the Czech Republic, the Grant No. KJB 200650602.

M. Luysberg, K. Tillmann, T. Weirich (Eds.): EMC 2008, Vol. 1: Instrumentation and Methods, pp. 537–538, DOI: 10.1007/978-3-540-85156-1_269, © Springer-Verlag Berlin Heidelberg 2008

538

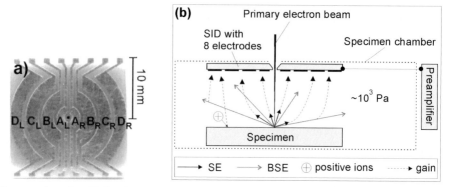

Figure 1a, b. Halved segmental ionization detector (halved SID) consisting of 8 electrodes: (a) photograph of SID, (b) position of SID in specimen chamber.

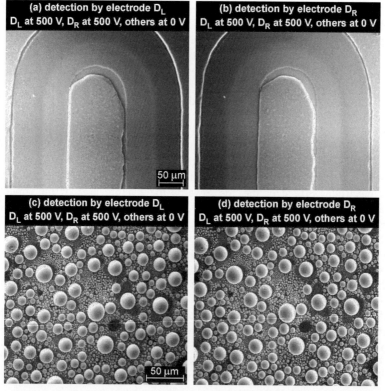

Figure 2a, b, c, d. Images of specimens acquired by side detection by left electrode D_L and right electrode D_R: (a), (b) passivated transistor with aluminium metallization, (c), (d) resolution test specimen "tin on carbon" (Agar S1937U).

Low-voltage Scanning Transmission Electron Microscopy of InGaAs nanowires.

L. Felisari[1], V. Grillo[1], F. Jabeen[1,2], S. Rubini[1] and F. Martelli[1]

1. CNR-INFM TASC Area, Science Park, S.S. 14 Km 163.5, 34012 Trieste, Italy
2. Sincrotrone Trieste S.C.p.A., Trieste, Italy

felisari@tasc.infm.it

Keywords: low energy STEM, nanowires

Scanning Transmission Electron Microscope (STEM), using High Angle Annular Dark Field (HAADF) detectors is a powerful technique for material characterisation. In high voltage STEM (200-300kV), the probe size that defines spatial resolution has atomic size, while in low voltage approach (20-30kV) the probe is larger. In this latter case channelling effects are reduced and the specimen response will be more directly interpretable as due to mass-thickness contrast.

A conventional SEM can be easily converted into a hybrid SEM-STEM instrument using a dedicated specimen holder able to convert transmitted electrons into Secondary Electrons (SE) that can be collected by an Everhart-Thornley detector [1]. The low voltage STEM approach is inexpensive and suitable for rapid screening of large number of samples, for example nanowires, nanotubes with spatial resolution of few nanometers. Moreover, Z-contrast imaging in low voltage STEM could be used to give complementary information to Energy Dispersive X-Rays Spectroscopy (EDX).

Nanowires are 1D nanostructures that can be exploited as building block for nanoscale device applications in several fields, ranging from sensors to optoelectronics and solar cells. Nanowires (NWs) are generally produced in large numbers per growth, therefore it is extremely important to find a versatile experimental tool that allows a reliable and fast characterization of their intimate structure with statistical relevance. While some NWs features as length, diameter and morphology can be obtained by SEM imaging, other important aspects as chemical interfaces (e.g. in nanowires heterostructures) nature of the catalyst particle and defects detection require a transmission electron microscopy. STEM in SEM would allows these characterizations in a fast and simple way.

We applied low voltage STEM to the characterization of Au-catalyzed InGaAs nanowires (NWs), grown by Molecular Beam Epitaxy. InGaAs NWs could find applications in long wavelength optical transmission and integrated photonics.

In figure1a a low voltage STEM image of a few InGaAs NWs, grown on GaAs substrate at 580° C for 45 minutes, is reported. This image has been collected using a Zeiss Supra 40 with field emission gun operated at 20kV and with measured spatial resolution of about 4nm. The detection angle was set between 20° and 60°. The diameter of the nanowires close to the tip is about 20nm. The brighter dots on top of the NWs are Au-rich particles, with hemispherical shape and the same diameter of the wire at its end.

In order to test the reliability of chemical mapping based on this imaging mode a line-profile of a catalyst particle has been obtained (figure 1b). The intensity profile can

M. Luysberg, K. Tillmann, T. Weirich (Eds.): EMC 2008, Vol. 1: Instrumentation and Methods, pp. 539–540, DOI: 10.1007/978-3-540-85156-1_270, © Springer-Verlag Berlin Heidelberg 2008

be satisfactorily fitted with a mass thickness based model function $I = A\sqrt{R^2 - x^2} \exp(-\mu\sqrt{R^2 - x^2})$ where R is the particle radius, x the radial coordinate, A is the intensity scaling factor and μ is an absorption factor. The best fit values for these quantities are indicated.

The intensity contrast $C = (I_{Catalyst} - I_{wire})/(I_{wire})$ between the catalyst and the InGaAs wire body has been also considered for about 40 wires. The histogram of the obtained contrast is shown in figure 1c. The contrast values are peaked at C~0.7 with tails up to 1. These data have been compared with high voltage (200KeV) HAADF experiments performed on the same sample (experimental details are described elsewhere [2]). A typical high voltage HAADF image is shown in figure 1d. For high voltage experiments the contrast was C~2.5, much higher then in the low voltage approach. A possible qualitative account of this large difference can be given in the hypothesis that large absorption affects take place in the Au-rich particle at low voltages as already indicated by the fitting in figure 1b.

Our data are encouraging results towards the application of STEM-SEM on NWs and indicate that a parametric study of low voltage STEM is necessary for a correct image interpretation. In an earlier phase, a comparison with high voltage STEM will be helpful to the development of the technique.

1. P.G. Merli, A. Migliori, M. Nabucchi, M. Vittori Antisari, Ultramicroscopy, **65**, (1996), p.23
2. E. Carlino, V.Grillo, Phys. Rev. B. **71** (2005), p. 235303

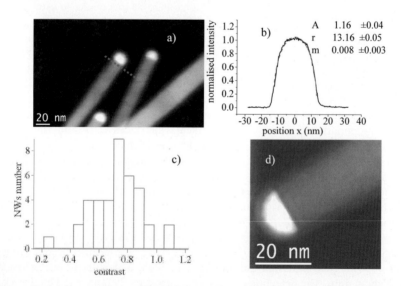

Figure 1. 1a) Low voltage STEM image of Au catalyzed InGaAs nanowires deposited on C coated Cu grid. b) intensity line profile across a catalyst particle c) low voltage SEM contrast distribution at different wires. d) High voltage STEM-HAADF image of a nanowire along with its catalyst.

Secondary Electrons Characterization
of Hydrogenated Dilute Nitrides

L. Felisari[1], V. Grillo[1], S. Rubini[1], F. Martelli[1], R. Trotta[2], A. Polimeni[2], M. Capizzi[2],
L. Mariucci[3]

1. Laboratorio Nazionale INFM-CNR TASC Basovizza, Trieste, Italy
2. CNISM-Dipartimento di Fisica, Universita` di Roma "La Sapienza,", Roma, Italy
3. IFN-CNR, via Cineto Romano, Roma, Italy

felisari@tasc.infm.it

Keywords: dilute nitrides, secondary electrons, H irradiation

Dilute nitrides alloys, as $GaAs_xN_{1-x}$, have been extensively studied in the last few years since the presence of small amounts of N in III-V semiconductors leads to major changes in the physical properties of the host materials. In particular, the strong reduction of band gap induced by N enables important technological applications in the field of active optical devices and microelectronics.

The epitaxial growth of GaAsN based heterostructures on GaAs substrates can then be used to achieve a control over electronic and optical properties in the growth direction. A recent work [1] has demonstrated the possibility to modulate electronic properties also in the plane of growth by means of spatially controlled hydrogen bombardment. Indeed, hydrogen irradiation of diluted nitrides can reverse the changes induced by N incorporation, leading to partial or complete recovering of the band gap and lattice parameter of GaAs. Using simple lithographical techniques, it is possible to design patterns of GaAsN embedded inGaAsN:H matrix.

Characterization of planar patterned structure has been previously carried out by means of microphotoluminescence experiments, but the spatial resolution of this technique is not sufficiently high for a satisfactory nanoscale characterization of patterning and interface sharpness. We demonstrate that such an investigation can successfully be performed by Secondary Electron (SE) imaging in a Scanning Electron Microscope (SEM). This approach allows a high spatial resolution, down to tens of nanometers, and the simplicity of the analysis makes this technique suitable for routine characterization. Indeed SE images are collected at room temperature, and a fast screening of samples can be performed in short time.

SE emission is sensitive to different factors as compositional variations, sample morphology and local variation in surface barrier height (work function). However, we expect that, in the case of the patterned structure formed by two adjacent regions of hydrogenated and not hydrogenated GaAsN, the secondary electron emission will be dominated by the local variation of band structure, since the variations in composition or morphology between the two regions can be assumed to be of minor importance [2].

For our experiments a 200 nm-thick $GaAs_{1-x}N_x$ layer with x=0.9 has been grown by Molecular Beam Epitaxy on a GaAs substrate [001]. A Ti mask with parallel lines 500nm wide and 5μm spaced has been applied to the sample surface. The successive

M. Luysberg, K. Tillmann, T. Weirich (Eds.): EMC 2008, Vol. 1: Instrumentation and Methods,
pp. 541–542, DOI: 10.1007/978-3-540-85156-1_271, © Springer-Verlag Berlin Heidelberg 2008

542

hydrogen irradiation has been performed by low energy (110eV) and low current (a tens of μA/cm^2) Kaufman source with the sample at 200°C; H impinging dose was equal to $d_H=3\times10^{18}$cm^{-2}. After H irradiation, the Ti mask has been removed. SE images have been collected by means of Zeiss Supra SEM. An in-lens detector has been used, thus achieving higher contrast than provided by the Everhart Thornley detector.

Figure 1a shows a SEM image of the planar heterostructure. Consistently with an intensity prediction based on the band-gap values, the darker regions correspond to hydrogenated areas with wider bandgap while the narrow (nominally 500nm thick) brighter areas correspond to the not hydrogenated GaAsN with smaller bandgap. Defining I_{GaAsNH} and I_{GaAsN} the SE intensity in the hydrogenated and not hydrogenated areas, the contrast between the two regions can be defined as $C=(I_{GaAsNH}-I_{GaAsN})/I_{GaAsN}$.

The maximum intensity contrast has been obtained at 5 kV and working distance of 6mm; this contrast can be further enhanced by a small (5°) tilt of the sample. Figure 1b shows an intensity profile obtained in the marked region of Figure. 1a. The slow rise of the intensity profile on the scale of 100 nm at the two sides of the curve is most probably due to an effective slow variation of the implantation profile.

Preliminary experiments have been performed applying an external electric field in order to produce an energy-angle filtering of the secondary electrons. These data seem to confirm also numerically that the observed contrast is a due to genuine work function variation between the two materials. However a more accurate modelling of the effect of the filtering electric field is required. On going experiments on samples with higher N concentration, and higher contrast, will be reported at the conference, together with data on samples hydrogenated at different temperatures to investigate how the H diffusion profile depends on the implantation temperature.

1.   M. Felici, A. Polimeni, G. Salviati, L. Lazzarini, N. Armani, F. Masia, M. Capizzi, M. Lazzarino, G. Bias, M. Piccin, S. Rubini and A. Franciosi, Adv. Mat. **18**, (2006), p. 1993.
2.   A.A. Suvorova, S. Samarin, Surface Science **601** (2007) p. 4428

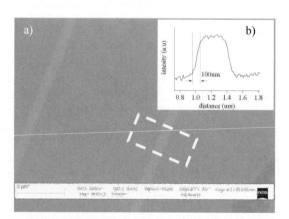

Figure 1. a) SE image of GaAs$_x$N$_{1-x}$/GaAs planar heterostructures. Brighter, thinner lines are GaAsN separated by darker GaAs region. b) Linescan of white dotted box, spatial resolution is about 100nm.

Mapping of the local density of states with very slow electrons in SEM

Z. Pokorná, L. Frank

Institute of Scientific Instruments ASCR, Královopolská 147, 61264 Brno, Czech Rep.

ludek@isibrno.cz

Keywords: density of states, SEM, very low energy electron microscopy

The local density of electron states is important characteristics of solids, which can be utilized in materials science when distinguishing between different crystal orientations in a polycrystalline matter, mapping the local density of dopants in semiconductors, etc. Non-imaging application of the reflection of slow electrons (VLEED – very low energy electron diffraction) knows so called energy band structure region on the intensity vs. energy curve for the specularly reflected (00) spot, which appears below the threshold where the first nonspecular diffracted beam appears [1]. The incoming electron wave impinging on the sample surface has to "convert" into electron waves of the crystal periodicity, i.e. into Bloch states. This condition makes penetration of incident electrons proportional to the local density of states coupled to the incident wave. Electron reflection becomes then inversely proportional to the local density of states and the reflected flux can measure the density. Crucial condition is sufficiently low absorption of injected hot electrons, which is available below about 20 to 30 eV above the Fermi level.

The only way of preserving the resolution down to arbitrarily low energy is to use a zero working distance electrostatic lens, i.e. the cathode lens (CL) for retarding the primary beam [2]. The retarding field is then applied directly to the specimen and the energy dependence of the spot size improves to $(energy)^{-1/4}$ in a worst case but can even disappear at certain combinations of parameters [2]. Thus, the CL equipped SEM allows us to enter the region where the contrast of the local density of states is expected.

Demonstration experiments have been performed on polycrystalline Al samples prepared by extruding to a rod, cutting to discs, annealing at 500°C and quenching in chilled water. The surfaces were finished by conventional electrolytic polishing method in a mixture of perchloric acid and ethanol. The grains grew to a size in tens of μm and exhibited various orientations with respect to the final surface. SEM used was of a thermo-emission gun type with nominal resolution of 3 nm at 30 keV and about 7 nm at 10 keV, the primary energy used for the experiment. When retarded to 10 eV in the CL mode, the spot size grows to some 20 nm at low beam currents.

Figure 1 shows the reflectance curves $R(E)$ acquired by measuring average image signals S_i over two grains, together with their contrast $(S_1 - S_2)/(S_1 + S_2)$. $R(E)$ look similar to that measured for the Al (111) surface [1] but the peaks are shifted by 1 and 7 eV, respectively, to higher energies. Experiment combined with the EBSD mode on another couple of grains (see Figure 2) provided the reflectance curves mutually shifted by appr. 0.3 eV only, but again the maximum contrast exceeds the 10% level.

M. Luysberg, K. Tillmann, T. Weirich (Eds.): EMC 2008, Vol. 1: Instrumentation and Methods, pp. 543–544, DOI: 10.1007/978-3-540-85156-1_272, © Springer-Verlag Berlin Heidelberg 2008

544

The experiments have been performed under standard vacuum conditions at about 10^{-3} to 10^{-4} Pa so the signal variations connected with the changes in reflectance are "sunk" in the background generated owing to presence of amorphous surface contamination layers. These layers become more transparent at lowest energies so the 10 % contrast level can be exceeded. The main advantage of the proposed method lies in very high signal to noise ratio, achieved by collimating the signal electrons in the cathode lens field to a bundle the major part of which impinges on the detector. The grain maps can be so obtained in seconds, as distinct from the EBSD methods with frame times at least in minutes.

1. R.C. Jaklevic and L.C. Davis, Phys. Rev. B **26** (1982), p. 5391.
2. I. Müllerová and L. Frank, Adv. Imaging Electron Phys. **128** (2003), p. 309.
3. The Al specimen was prepared by Mr. J. Tsukiyama, Faculty of Engineering, University of Toyama, Japan.

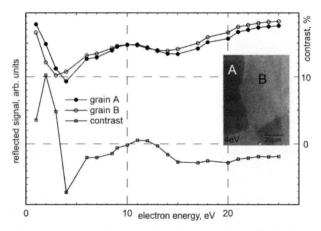

Figure 1. The reflectance curves for two Al grains, together with their contrast.

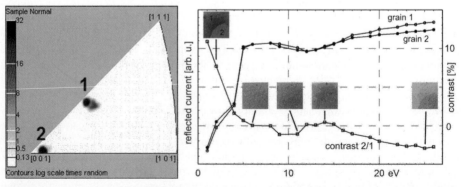

Figure 2. The inverse pole plot from the EBSD method for two Al grains (left), and the $R(E)$ and contrast curves (right).

Thickness and composition measurement of thin TEM samples with EPMA and the thin film analysis software STRATAGem

F. Galbert[1], D. Berger[1]

1. Zentraleinrichtung Elektronenmikroskopie, Technische Universität Berlin, Straße des 17 Juni 135, 10623 Berlin, Germany.

francois.galbert@tu-berlin.de
Keywords: thin sample, composition, thickness

The SAM´x Software Company [1] offers the STRATAGem software for the determination of the mass thickness (in unit $\mu g/cm^2$) and the composition of thin layers on bulk substrates from only one measurement. Good results are already been obtained for thin C-layers (10 – 50 nm) on different substrates [2]. If the density of the layer is known, the mass thickness can be converted into the geometrical thickness in nm. If the thickness of the layer is known, the density of this material can be determined. This method is absolutely non-destructive for the sample.

In this paper, we deal with the question: Can we extend the STRATAGem / Electron Probe Microanalyser (EPMA) [4] analysis on thin self-supporting TEM samples without substrate?

The first step is a conventional quantitative element analysis [3] with the wave length dispersive spectrometer (WDS) of the EPMA for the elements in the sample. Then, the ratio of the x-ray counts I_x of the elements in the sample to the x-ray counts I_{std} in pure reference standard is imported into STRATAGem graphic users interface.

The element ionization depends on the energy attenuation of the incident electrons in the sample. The "Monte Carlo Simulation" method shows the electron distribution depending on the depth in the target. For a given acceleration voltage, the function $\Phi(\rho z)$ is a measure for the distribution of the characteristic x-ray emission from the element in units of mass thickness. Using the Monte Carlo Simulation, the function $\Phi(\rho z)$ and the ratio I_x/I_{std}, STRATAGem calculates the corresponding mass thickness and the composition of the sample or layers. Since the STRATAGem algorithm expects bulk substrates, we assume Helium as a virtual substrate, which does not influence the calculation.

We test STRATAGem by measuring the thickness of different aluminium films evaporated in high vacuum. The expected thicknesses of 50 nm, 150 nm and 500 nm are considerable smaller than the penetration depth of 20kV electrons in aluminium (about 3.5 µm). STRATAGem delivers respectively 46.4 nm, 165.4 nm and 587 nm for the aluminium thicknesses. Using the same method, the thickness of the aluminium oxide layer is found to be less than 7 nm. All results were compared successfully with Field Emission Scanning Electron Microscope images on cross sections.

M. Luysberg, K. Tillmann, T. Weirich (Eds.): EMC 2008, Vol. 1: Instrumentation and Methods, pp. 545–546, DOI: 10.1007/978-3-540-85156-1_273, © Springer-Verlag Berlin Heidelberg 2008

Our new method is even applicable for the analysis of the composition of thin films. This has been checked on very small fragments of olivine mineral. The homogeneity and the composition (O 43.73 %, Ni 0.31 %, Mg 30.46 %, Si 19.09 % and Fe 6.42 %) of polished bulk olivine is well known by EPMA measurements; the density is measured to 3.4 g/ cm^3. Thin cleaved fragments (thickness about 500 nm) were deposited on a conventional TEM copper grid with thin Formvar film. The thickness determined by STRATAGem corresponds to results from SEM cross section images shown in "Figure1".

The results of the quantitative analysis "Figure 2" from STRATAGem agree very well with the results of bulk olivine.

1. STRATAGem programme (version 4.3.0): developed in cooperation with Jean-Louis Pouchou (ONERA). STRATAGem is included in SAM´x package of J.F. Thiot.
2. F.Galbert. Microsc Microanal 13 (Suppl. 3), 2007
3. Electron microprobe: Hardware CAMEBAX MICROBEAM, Software SAM´x
4. Xmas Plus-Analysis in SAM´x software package.

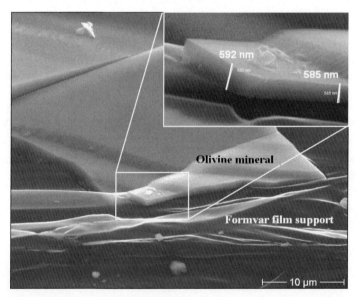

Figure 1. SE cross section image of olivine sample, recorded at SEM HITACHI 2700, Image with courtesy of J. Nissen, ZELMI

Figure 2.	Measurement of olivine cleaved fragment					
	thickness	O wt%	Ni wt%	Fe wt%	Si wt%	Mg wt%
STRATAGem results	547.4 nm	43.53	0.32	6.57	18.57	31.01
Reference composition of bulk	density 3.4 g/cm³	43.73	0.31	6.42	19.09	30.46

Automatic acquisition of large amounts of 3D data at the ultrastructural level, using serial block face scanning electron microscopy

C. Genoud[1], J. Mancuso[2], S. Monteith[1], B. Kraus[3]

1. GATAN UK, 25 Nuffield Way, Abingdon OX14 1RL, United Kingdom
2. GATAN Inc, 5794 W. Las Positas Blvd, Pleasanton, CA 95488
3. GATAN Gmbh, Ingolstaedterstr. 12 80807 Muenchen, Germany

cgenoud@gatan.com
Keywords: electron microscopy, reconstruction, SBFSEM

Serial block face scanning electron microscopy (SBFSEM) is a new technique designed to obtain image slices through specimens inside a scanning electron microscope (SEM) in an automated process. Biological tissue in particular, is prepared by classical en-block staining and resin embedding as used for many years for transmission electron microscopy.

Obtaining meaningful 3D data is becoming critically important in the life sciences. Understanding the traffic of cell components, the organisation of cells in different tissues and the modification of cells following environmental constraints are key studies in many domains. The techniques of confocal microscopy including multiphoton imaging have made large progress in this field. However, they are limited by the use of fluorochromes as well as by the resolution of the light microscopy. Higher resolution studies require using electron microscopy.

This is done by cutting an entire block of tissue inside a SEM in order to image each freshly exposed surface, sequentially after each cut. This technique was first developed by Denk and Horstmann [1]. We show now that SBFSEM is suitable for numerous tissue samples, as long they are processed by en-block staining and embedded in epoxy. We also show that not only animal tissue, but also plant tissue and cell in cultures can be studied in 3D.

A neuron first visualised by fluorescence microscopy in a mouse brain is revealed with diaminobenzidine (DAB) and then processed for SBFSEM. The resolution allows synapses to be distinguished and visualisation of the environment of the dendrite. Numerous studies have shown the necessity to combine EM with fluorescence microscopy [2,3]

Studies have also been made of different histological samples in order to visualise the 3D structure of tissues. For example, mouse skin has been processed using SBFSEM Figure 1 shows the type of images that can be obtained with SBFSEM We also present image stacks from the kidney (Figure2) as well as 3D rerndering obtained with Gatan Visualisation toolTM, All these examples are unique observation of these tissues at an ultrastructural level but also on a large field of view. 3D details of the cells forming the tissue, their organisations as well as their interactions to form a functional tissue provide new insights for histological observation.

M. Luysberg, K. Tillmann, T. Weirich (Eds.): EMC 2008, Vol. 1: Instrumentation and Methods, pp. 547–548, DOI: 10.1007/978-3-540-85156-1_274, © Springer-Verlag Berlin Heidelberg 2008

548

All the examples illustrate the intimate link between the 3D ultrastructure and the functions of tissue. In some cases it is envisaged that 3D ultrastructure from SBFSEM will provide new evidence which will question present biological models of healthy and pathological tissue, or else provide answers to known problems in life sciences.

1. W. Denk and H. Horsmann, PLOS 2(11) 2004; e329
2. G. W. Knott, A. Holtmaat, L. Wilbrecht, E. Welker, K. Svoboda, Nature Neuroscience Sep 9(9) 2004
3. V Lucić, A.H. Kossel, T. Yang, T. Bonhoeffer, W. Baumeister, A. Sartori. J Struct Biol. 2007 Nov;160(2) (2007) :146-56.

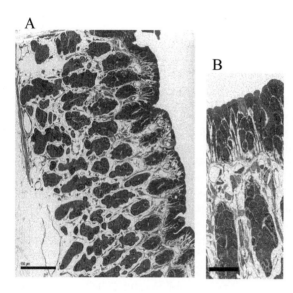

Figure 1 Example of the large field of view that we can obtain with 3View based on a sample showing the mouse skin. The original field of view (A) has been scanned with the resolution of 4k x 4k and represents 500x500 micrometers. The extract shown in B represents a subfield of view. The total volume observed is 12.5×10^6 μm^3 as 999 sections of 50 nanometers have been shaved of the surface overnight. Scale bar: 20 microns

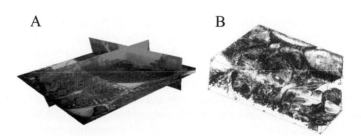

Figure 2. Example of 3D reconstruction obtained from a healthy mouse kidney, imaged using GATAN 3View SBFSEM system. A cut through the raw stack of images to show a renal tubule with the microvilli in the medulla. B 3D rendering of the same area (GATAN visualization tool) in order to understand the organization of

MCSEM- a modular Monte Carlo simulation program for various applications in SEM metrology and SEM photogrammetry

D. Gnieser[1,2], C.G. Frase[1], H. Bosse[1] and R. Tutsch[2]

1. Physikalisch-Technische Bundesanstalt, Bundesallee 100, 38116 Braunschweig
2. Technische Universität Braunschweig, Institut für Produktionsmeßtechnik, Schleinitzstraße 20, 38106 Braunschweig

Dominic.Gnieser@PTB.de
Keywords: Monte Carlo Simulation, SEM, metrology, photogrammetry

MCSEM is a Monte Carlo simulation program for the modeling of image formation in scanning electron microscopy. The program is written entirely in C++ and uses object-orientated programming techniques. Different aspects of the image formation process, as the probe forming, the geometric specimen model, the electron-specimen-interaction, and the electron detection are modeled in separate program modules. Due to this modular structure, the program can easily be enhanced and adapted to new simulation tasks by integration of new modules.

The electron-specimen-interaction module simulates the electron diffusion in solid state and the generation and emission of secondary electrons. Elastic scattering is based on tabulated Mott scattering cross sections, calculated by Salvat and Mayol [1], inelastic scattering is modeled by the Bethe formula in the modification of Joy and Luo [2]. Different specimen model modules are available, ranging from simple 2D specimen to complex 3D structures, offering a flexible system of specimen definition. Specimen definition files can be transformed to VRML (virtual reality modeling language) format for 3D visualization. Figure 1 shows the 3D model of a silicon specimen structure and the trajectories of 20 keV electrons. The detector module enables the detection of backscattered electrons (BSE), transmitted electrons (TE), and secondary electrons (SE). This module calculates individual SE trajectories in the close-up range above specimen surface and uses precalculated lookup-tables based on electron raytracing calculations by Konvalina and Müllerová [3] for macroscopic SE movement.

MCSEM is tested in various application examples like TE imaging of nanoparticles and SE and BSE imaging of complex 3D semiconductor and metal structures. The testing is performed by comparing the simulation with measurement results and with results of other Monte Carlo simulation programs [4,5]. Moreover, 3D structures are scanned from different observation angles to generate simulated stereopair images which can be used to test SEM photogrammetry algorithms.

1. F. Salvat, R. Mayol: "Elastic Scattering of Electrons and Positrons by Atoms. Schrödinger and Dirac Partial Wave Analysis." Comp. Phys. Comm. 74, pp.358-374, (1993)
2. D.C. Joy, S. Luo, "An Empirical Stopping Power Relationship for Low-Energy Electrons", SCANNING 11(4), pp.176-180, (1989)

M. Luysberg, K. Tillmann, T. Weirich (Eds.): EMC 2008, Vol. 1: Instrumentation and Methods, pp. 549–550, DOI: 10.1007/978-3-540-85156-1_275, © Springer-Verlag Berlin Heidelberg 2008

550

3. I. Konvalina, I. Müllerová: "The Trajectories of Secondary Electrons in the Scanning Electron Microscope", SCANNING VOL. 28, pp. 245-256, (2006)
4. L. Reimer, M. Kässens, L. Wiese, "Monte Carlo Program with free Configuration of Specimen Geometry and Detector Signals", Microchim. Acta 13, pp.485-492, (1996)
5. J.R. Lowney: "MONSEL-II: Monte Carlo Simulation of SEM Signals for Linewidth Metrology", Microbeam Analysis 4, pp. 131-136, (1995)

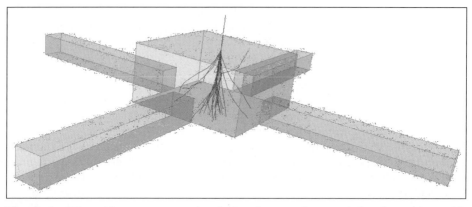

Figure 1: 3D specimen structure (total size 4 µm x 4 µm x 0.5 µm) together with simulated trajectories of 20 keV probe electrons, visualized in VRML.

Wien filter electron optical characteristics determining using shadow projection method

I. Vlček, M. Horáček and M. Zobač

Institute of Scientific Instruments, Academy of Sciences of the Czech Republic, Královopolská 147, 61264 Brno, Czech Republic

mih@isibrno.cz
Keywords: Wien filter, shadow method, very low energy electrons

Wien filter is suitable for the separation of the primary and the signal electron beams in very low energy scanning electron microscope with cathode lens [1]. We have modified the *two-grid shadow method* [2] to determine experimentally electron optical properties (cardinal elements and aberrations) of the Wien filter, which is not a rotationally symmetric element. We call the modified method the *shadow method with grid and moving screen*. The advantage of the shadow method is its geometrical simplicity allowing the comparison of the experimentally obtained and numerically computed trajectories.

The arrangement of the *two-grid shadow method* is shown schematically in Figure 1(a). The electron beam from the point source passes through horizontally oriented linear grid, next the beam passes through the measured optical element, makes an image of the source, passes through a vertically oriented second linear grid, and casts a shadow upon a fluorescent screen. The inclination tangent of input trajectories can be determined from the point source position and from the known dimension of the first grid. The inclination tangent of the output trajectory can be determined from the shadow image of the first grid and the distance between the screen and the intersection of the trajectory with the optical axis. This distance can be determined from the second grid and its shadow image. Angular magnification can be calculated from these trajectories. Hence, with knowledge of potentials in image and object planes, we can calculate linear magnification. Two measurements at two different geometrical configurations are necessary to calculate the remaining cardinal elements, i.e. the positions of principal planes and focal distances [3]. The calculation of angular magnification from shadow image of two grids is possible only for rotationally symmetric element. Therefore, our modification of the *two-grid shadow method* is necessary to measure electron optical elements without rotational symmetry.

The arrangement of the modified *shadow method with grid and moving screen* is shown in Figure 1(b). The rectangular grid situated in front of the measured Wien filter (outside the filter fields) together with the point source position fully determines the input trajectories. Output trajectories can be determined from the two shadow images taken at two fluorescent screen positions. The advantage of the method remains the use of the grid as a scale, thanks to this the coordinates of the characteristics points are sufficiently precisely determined before the measurement. The grids can be used to calibrate distances on the screen. Additionally, the information about transverse

M. Luysberg, K. Tillmann, T. Weirich (Eds.): EMC 2008, Vol. 1: Instrumentation and Methods, pp. 551–552, DOI: 10.1007/978-3-540-85156-1_276, © Springer-Verlag Berlin Heidelberg 2008

552

coordinates of the trajectories is contained in one image, so image processing methods can be used to read the coordinates of the characteristic points from the image. Only the determination of the screen position is important.

All the types of aberrations will manifest in the shadow image: geometrical, chromatic, and diffractive, as well as blurring due to finite dimension of the point source. The method is based on reading the coordinates of the characteristic points from the shadow image. The edge blurring caused by the chromatic and diffractive aberrations and the finite dimension of the point source makes the precise reading of the points coordinates difficult. Therefore, deviations in the image caused by the above mentioned aberrations have to be minimized and the deviations in the image caused by the geometrical aberrations have to be maximized. Simultaneously the aberrations must be large enough to be measured. The calculation of the aberration coefficients will be made by fitting of the measured trajectory using the least squares method. The method can be also used for calculation of chromatic aberrations coefficients if the measurements for different electron energies would be made.

The experimental UHV apparatus for measurement of our Wien filter properties have been designed and made. The Wien filter uses eight combined poles-electrodes to produce nearly identical magnetic and electrical field. Schottky thermal field cathode is used as a point source of electrons.

1. I. Vlček, B. Lencová, M. Horáček in "Proc. of the 13th EMC" (Antwerp, 2004) Vol. I, p.337
2. K. Spangenberg and L.M. Field, Electrical Communications **20** (1942), p. 305.
3. L.A. Baranova and S.YA. Yavor in "Advances in Electronics and Electron Physics", ed. P.W. Hawkes (Academic Press, Boston) (1989), p. 1.
4. Supported by Grant Agency of the Academy of Sciences of the Czech Republic, grant no. IAA100650803 and Academy of Sciences of the Czech Republic, grant no. AV0Z20650511.

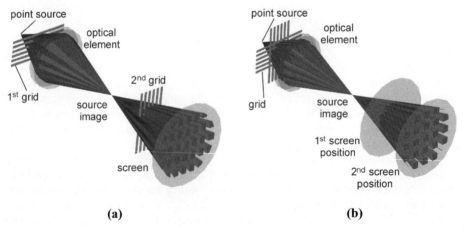

(a)　　　　　　　　　　(b)

Figure 1. (a) The arrangement of the two-grid shadow method. **(b)** The arrangement of the modified shadow method with grid and moving screen.

Strain related Contrast mechanisms in crystalline materials imaged with AsB detection

Heiner Jaksch

Carl Zeiss SMT–Nano Technology System Div., Carl-Zeiss-Str. 56, 73447 Oberkochen, Germany

Jaksch@smt.zeiss.com

Keywords: Backscattered elctrons, angle contrast, Z contrast, AsB detector, EsB detector, strain, deformation

At high landing energies the conventional BSE images are showing a contrast mechanism, dominated by multiple elastic scattering processes in the sample. These processes are described in the Rutherford scattering equation. To see crystallographic information we have to "remove" this blurring multiple elastic scattering contrast – known as Z-contrast - and optimize the image formation by "selecting" more single elastic scattered electrons. These electrons carry mainly crystallographic contrast, also known as channelling contrast coming from Mott scattering. This contrast is one of the oldest contrast mechanisms described in the literature of electron microscopy.

In "normal" BSE images you have a combination of both Rutherford and Mott scattering, depending on the density (d-spacing) and the landing energy (λ) according to Bragg´s law: n x λ = 2d x sin ϑ.

New improvements come from an integrated GEMINI lens detector, selecting and separating the BSE signal via Z contrast or *angle* contrast. While the high angle BSE electrons, which are detected in the unique in-column energy selective *E*sB detector, the large and very large angles, coming from different scattering processes, are collected in the *A*sB detector, or *A*ngular selective BSE detection system.

In homogeneous crystalline bulk materials, we have strong demands to characterize these materials and determine the treatment history. Chemically etching the polished surface is a common technology to describe grain structure, but sub grain information is lost. The channelling contrast, coming from mainly Mott scattered electrons, highlights the mechanism, which is used in this detector. The GEMINI lens separates the *very* large angle single elastic scattered electrons from large angle *multiple* elastic scattered BSE electrons. As a result we detect unmatched crystalline contrast (Figure 1). Very low angle changes ($< 1^0$) of the lattice orientation changes can be imaged.

Seeing the crystalline grain orientation in very high contrast, we can also see the *deformation* of the crystalline "lattice", coming from strain. Production processes of metals, or any other crystalline materials, cause deformation of the lattice. Deformation in general, causes problems in the material (stability, corrosion etc.). With the AsB detector, we are able to detect these deformations on bulk polished samples in high resolution. Contrast and image quality is comparable to images acquired on TEM

M. Luysberg, K. Tillmann, T. Weirich (Eds.): EMC 2008, Vol. 1: Instrumentation and Methods, pp. 553–554, DOI: 10.1007/978-3-540-85156-1_277, © Springer-Verlag Berlin Heidelberg 2008

554

1. Jaksch H. et. al. Microsc. Microanal. (2004) 1372 CD
2. Jaksch H. et. al. Microsc. Microanal. 9 (Suppl) (2003) 106
3. Jaksch H. Field emission SEM for true surface imaging and analysis, Materials World, Oct. 1996

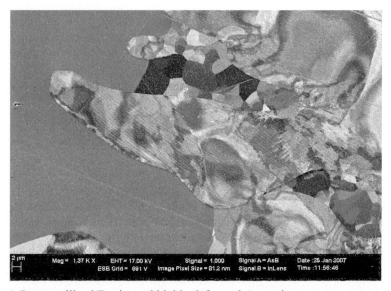

Figure 1. Recrystallized Ferrite and highly deformed Austenite

Low Loss BSE imaging with the EsB Detection system on the Gemini Ultra FE-SEM

Heiner Jaksch

Carl Zeiss SMT–Nano Technology System Div., Carl-Zeiss-Str. 56, 73447 Oberkochen, Germany

jaksch@smt.zeiss.com

Keywords: Low-loss BSE, EsB detector, In-lens BSE detector

The sensitivity of the GEMINI Inlens detection systems is worldwide known. Especially the EsB detection provides compositional contrast down to the ppm level of concentration. Due to direct detection of the boosted low voltage BSE electrons, a very strong signal is generated on the on **axis Inlens BSE** detector. Additional a filtering of the BSE electrons can be applied. The filtered information can be reduced down to the low loss BSE region where only a few 10eV to 100eV of

BSE energy are used for the imaging. With this technology we are able to visualize smallest density differences of compositions such as different oxidation states of metals (Figure 1), hybridisation states ($sp^2 - sp^3$ hybrids) or ligand bondings. Due to the effect of filtering the low loss BSE electrons, we have an extreme sensitive from the real surface coming signal. At landing energies far below 1kV, the scatter volume becomes very small and the filtered LL-BSE signal – reduced to a few 10 or 20eV energy loss – shows finest compositional differences. Different polymers and even proteins can be separated and characterized at lowest landing energies with the EsB detector. Examples of different oxidisation states from metal oxides or nitrides will also be shown in the presentation.

1.   Jaksch H. et. al. Microsc. Microanal. (2004) 1372 CD

M. Luysberg, K. Tillmann, T. Weirich (Eds.): EMC 2008, Vol. 1: Instrumentation and Methods, pp. 555–556, DOI: 10.1007/978-3-540-85156-1_278, © Springer-Verlag Berlin Heidelberg 2008

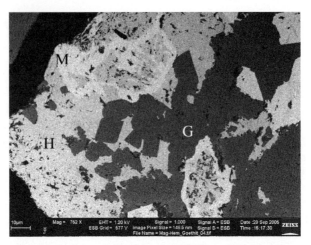

Figure 1. Magnetite, Hematite and Goethite at 1.2kV

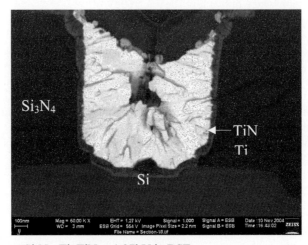

Figure 2. Silicon, Si$_3$N$_4$, Ti, TiN at 1.27kV in BSE

Accurate calculations of thermionic electron gun properties

P. Jánský[1], B. Lencová[1] and J. Zlámal[2]

1. Institute of Scientific Instruments of the ASCR, v.v.i., Královopolská 147, 612 64 Brno, Czech Republic
2. Institute of Physical Engineering, Faculty of Mechanical Engineering, Brno University of Technology, Technická 2, 616 96 Brno, Czech Republic

jansky@isibrno.cz

Keywords: electron emission, electron gun, space charge

Thermionic triode electron guns with directly heated hairpin tungsten cathode are still in wide use in many microscopes for their low requirements on operation conditions, high emitted currents and low service costs. In ISI Brno we started with simulations of thermionic electron guns for electron beam welding machine [1] used also for experiments with electron beam micromachining. Numerical simulations enable us to understand space charge effects and it can help us to improve the geometry of gun electrodes.

For simulations we use the program EOD (Electron Optical Design) [2]. Computation of potential is based on accurate finite element method in large meshes. Space charge distribution is set from tracing test particles and the potential is iteratively improved. From the field intensity near the cathode surface we determine new estimation of the emission current limited by space charge. After each change of emission current the space charge distribution must be recalculated and emission current estimation improved as well as the area of the emitting region on cathode. If the space charge effect is significant, the area of this region depends not only on Wehnelt bias but also on the emitted current.

Figure 1 shows one of the calculated electron guns. Its geometry and beam properties are described in [3]. Electron gun is simulated as rotationally symmetric problem, the cathode is simulated as a cylinder with a spherical cap. Emitted beam is calculated as space charge limited. For accelerating voltage 10 kV and Wehnelt bias -121 V the calculated emission current is 93 μA. A detail of one pencil of rays is in Figure 2.

1. J. Dupák, I. Vlček, M. Zobač, Vacuum **62**, 159-164, 2001.
2. B. Lencová and J. Zlámal, Microsc. Microanal. **13**, Suppl 3 (2007), p. 2
3. R. Lauer in "Advances in Optical and Electron Microscopy Vol. **8**" ed. R. Barer and V. E. Cosslett, (Academic Press Inc, 1982).
4. This work was supported by Grant Agency of AS CR, grant no. IAA100650805 and AV0Z20650511.

M. Luysberg, K. Tillmann, T. Weirich (Eds.): EMC 2008, Vol. 1: Instrumentation and Methods, pp. 557–558, DOI: 10.1007/978-3-540-85156-1_279, © Springer-Verlag Berlin Heidelberg 2008

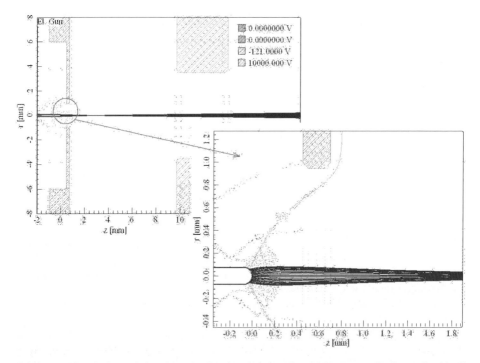

Figure 1. Simulation of the thermionic electron gun in EOD. From the heated cathode approximated by a cylinder with a spherical cap the electron beam is emitted. The beam energy is 10 kV, Wehnelt bias -121 V. Beam was simulated as monochromatic with starting energy close to 0 eV.

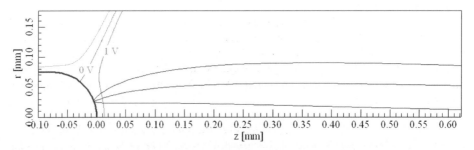

Figure 2. Detail of pencil of rays near the cathode for beam energy 0.24 eV. Equipotential lines -1 V, 0 V and 1 V are shown.

Scintillation SE Detector for Variable Pressure Scanning Electron Microscope

J. Jirak[1,2], P. Cernoch[1], V. Nedela[2], J. Spinka[1]

1. Dept. of Electrotechnology FEEC BUT, Udolni 53, 60200, Czech Republic
2. Institute of Scientific Instruments of ASCR, Kralovopolska 147, 61264, Czech Rep.

jirak@feec.vutbr.cz

Keywords: VPSEM, secondary electrons detector, scintillation detector

Scanning electron microscopes that work with a pressure in the range from several hundred up to several thousands Pa in the specimen chamber (VPSEM) bring the possibility to observe wet specimens, phenomena on phase interfaces and insulators without charging artifacts. Gas ambient in the specimen chamber of these microscopes also enables to amplify signal electrons in the process of impact ionization [1]. In the mentioned process, not only secondary and backscattered electrons are amplified but also products of collisions of primary electrons with gas atoms and molecules. Detection of secondary electrons by a scintillation detector placed directly in the specimen chamber requires the pressure of the order of several Pa at the maximum in the surroundings of the scintillator. At this pressure it is possible to attach a voltage of several kV to a thin conductive layer on the scintillator without complications with discharging processes. In the created electrostatic field, secondary electrons gain energy for efficient scintillations in the scintillator. A scintillation detector of secondary electrons operating at the pressure range up to few thousands Pa in the specimen chamber, that utilizes a room separated from the specimen chamber and evacuated to pressure of several Pa for the scintillator, is described by several authors [2, 3].

In our detector schematically pictured in Figure 1, the room with the scintillator is separated from the specimen chamber by two apertures with holes diameters of several hundreds micrometers. The room between the apertures is pumped by a rotary pump and the scintillator room by a turbomolecular pump. This arrangement allows to reach the pressure below 5 Pa in the scintillator room at the pressure up to 1000 Pa in the specimen chamber. Voltages on the electrodes, E1 and E2, and the apertures, A1 and A2, are in the range from several tens up to several hundreds V. The created electrostatic field allows to pass secondary electrons from the specimen chamber to the scintillator room where they are accelerated to the scintillator by a positive voltage of a value up to 10 kV attached to the thin conductive layer on the scintillator.

In Figure 2a, b, c, d, we demonstrate the wide application scale of this detector. Our next work is oriented to the improvement of the detection stability and efficiency.

1. G. D. Danilatos, Sydney: Academic Press (1990), 103 p. (ISBN 0-12-014678-9).
2. W. Slowko, Vacuum **Vol. 63** (2001), p. 457 – 461.
3. M. Jacka, M. Zadrazil, F. Lopour, Scanning **Vol. 25** (2003), p. 243 – 246.
4. This research is supported by project MSM 0021630516 and Institute of Scientific Instruments of the Academy of Sciences of the Czech Republic.

M. Luysberg, K. Tillmann, T. Weirich (Eds.): EMC 2008, Vol. 1: Instrumentation and Methods, pp. 559–560, DOI: 10.1007/978-3-540-85156-1_280, © Springer-Verlag Berlin Heidelberg 2008

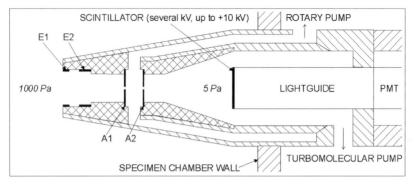

Figure 1. Arrangement of scintillation secondary electron detector; PMT - photomultiplier, E1, E2 - electrodes, A1, A2 - apertures.

Figure 2a, b, c, d. Examples of specimens images of (a) resolution test specimen "tin on carbon" (Agar S1937U), (b) passivated integrated circuit, (c) biological specimen, (d) polyester cloth. Pressure in specimen chamber of 500 Pa, room temperature, probe current of 100 pA, acc. voltage of 20 kV.

The stability of retained austenite in supermartensitic stainless steel (SMSS) examined by means of SEM/EBSD

M. Karlsen[1], J. Hjelen[1], Ø. Grong[1], G. Rørvik[2], R. Chiron[3] and U. Schubert[4]

1. Norwegian University of Science and Technology, Dep. of Materials Science and Engineering, N-7491 Trondheim, Norway
2. StatoilHydro Research Centre, Rotvoll, N-7005 Trondheim, Norway
3. CNRS-PMTM laboratory, 93430 Villetaneuse, France
4. Carl Zeiss NTS GmbH, Carl-Zeiss-Strasse 56, 73447 Oberkochen, Germany

Jarle.Hjelen@material.ntnu.no
Keywords: SEM, EBSD, supermartensitic stainless steel, retained austenite

Pipelines of supermartensitic stainless steels (SMSS) are used for sub-sea oil and gas transportation in the North Sea [1]. The microstructure of SMSS is in the as-received condition characterised by martensite and retained austenite. This austenite forms during inter-critical annealing, providing a metastable phase on subsequent cooling to room temperature. However, the stability of the retained austenite depends on the steel chemical composition and the applied heat treatment [2].

The aim of the present work is to study the thermal and mechanical stability of one particular SMSS using electron backscatter diffraction (EBSD) in a Zeiss Supra 55 VP FESEM equipped with NORDIF CD200 detector and EDAX/TSL EBSD software. Figure 1 presents i) secondary electron (SE) images and ii) EBSD phase maps from two samples that have been inter-critically annealed at 625°C for a) 5 minutes and b) 30 minutes. The SE images disclose topographical details of the microstructures due to austenite precipitation during inter-critical annealing. The two microstructures also appear to be relatively identical. However, the information provided by EBSD shows that the structure that was inter-critically annealed at 5 minutes are fully martensitic while the other one contains about 15 vol.% of retained austenite. SMSS samples in the as-received condition have also been deformed using a specially designed in-situ tensile testing device in combination with EBSD mapping inside the SEM [3]. Figure 2 provides separate EBSD grain maps of the i) martensite and ii) austenite at a) zero strain and b) after 10 % deformation. At zero strain the material contains about 10 vol. % of retained austenite. However, the austenite transforms to martensite during plastic deformation, and the reaction proceeds gradually with increasing plastic strains until approximately 3 vol. % of retained austenite is left in the material at 10 % deformation.

This study shows that SEM/EBSD can be applied to unravel the underlying crystal structures of the austenite and martensite in SMSS, thus providing valuable information about the thermo-mechanical stability of austenite phase.

1. G. Rørvik, S. M. Hesjevik and S. Mollan, "Stainless Steel World Conference & Expo Maastricht", the Netherlands (2005), poster 5089.
2. P.D. Bilmes, M. Scholari and C.L. Llorente, "Mater. Char.", **volume 46** (2001), p. 285.
3. M. Karlsen., J. Hjelen, Ø. Grong, G Rørvik, R. Chiron, U. Shubert and E Nilsen, "Mater. Sci. Tech.", **volume 24** (2008), p. 64.

M. Luysberg, K. Tillmann, T. Weirich (Eds.): EMC 2008, Vol. 1: Instrumentation and Methods, pp. 561–562, DOI: 10.1007/978-3-540-85156-1_281, © Springer-Verlag Berlin Heidelberg 2008

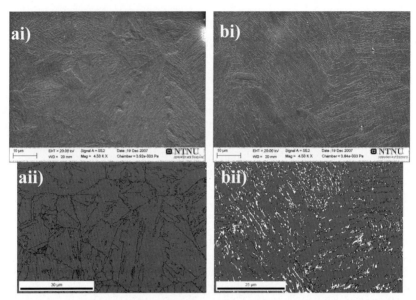

Figure 1. i) SE images and ii) EBSD phase maps (retained austenite in white) of samples being inter-critically annealed at 625°C for a) 5 minutes and b) 30 minutes

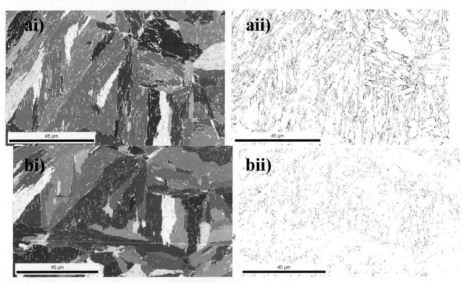

Figure 2. EBSD grain maps of the martensite (left) and retained austenite (right) at a) zero strain and b) after 10 % plastic deformation.

In-situ EBSD studies of hydrogen induced stress cracking (HISC) in pipelines of super-duplex stainless steel

M. Karlsen[1], J. Wåsjø[1], J. Hjelen[1], Ø. Grong[1], G. Rørvik[2], R. Chiron[3] and U. Schubert[4]

1. Norwegian University of Science and Technology, Dep. of Materials Science and Engineering, N-7491 Trondheim, Norway
2. StatoilHydro Research Centre, Rotvoll, N-7005 Trondheim, Norway
3. CNRS-PMTM laboratory, 93430 Villetaneuse, France
4. Carl Zeiss NTS GmbH, Carl-Zeiss-Strasse 56, 73447 Oberkochen, Germany

Jarle.Hjelen@material.ntnu.no
Keywords: SEM, EBSD, in-situ tensile testing, HISC

Hydrogen induced stress cracking (HISC) can cause critical failures in pipeline steels employed for sub-sea transportation of oil and gas [1,2]. The hydrogen, being introduced into the pipelines from the cathodic protection system, can in combination with applied stress/strain make the material susceptible to hydrogen embrittlement [2].

The objective of the present work is to gain knowledge of the HISC fracture mechanism in pipelines made of super-duplex stainless steel (SDSS). Samples were pre-charged with hydrogen and examined during plastic deformation by using the combined technique of in-situ tensile testing and electron backscatter diffraction (EBSD) characterisation in a Zeiss Supra 55 VP FESEM equipped with NORDIF CD200 detector and EDAX/TSL EBSD software. The work was carried out with a specially designed in-situ tensile testing device that can be operated inside the SEM without interfering with objective lens of the microscope [3]. Thus, the progressive microstructure evolution can be explored during plastic deformation.

The microstructure of SDSS is composed of austenite and ferrite grains with a phase balance of approximately 50 vol. % of each phase. The austenite phase is an important microstructual component contributing to the mechanical properties of the steels. However, the austenite grains are also potential traps for hydrogen. Therefore, the amount and condition of the austenite phase may have great impact on the susceptibility to hydrogen embrittlement.

Figure 1a) and 1b) present secondary electron (SE) images of the sample surface after 11 % and 13 % plastic deformation, respectively. The deformation direction is parallel to the horizontal direction in these images. Figure 1a) and 1b) show the presence of two cracks. The first crack is readily seen while the second one is captured at its initial stage of propagating through the surface of the sample.

Figure 2 a) shows an EBSD phase map of the microstructure in Figure 1b), where the austenite and ferrite are given in gray and black, respectively. Figure 2b) presents a gray scaled EBSD Taylor map of the ferrite phase in the examined the region. In this case a low Taylor factor, equivalent to high Schmid factor, is given in dark gray.

The present investigation confirms that microstructure behaves in a brittle manner where the cracks are perpendicular to the tensile direction and the number of cracks

M. Luysberg, K. Tillmann, T. Weirich (Eds.): EMC 2008, Vol. 1: Instrumentation and Methods, pp. 563–564, DOI: 10.1007/978-3-540-85156-1_282, © Springer-Verlag Berlin Heidelberg 2008

increases with increasing deformation. The cracks are mainly located in the ferrite phase and they probably initiate from the deformed neighbouring austenite grains. In most cases the cracks appear in ferrite grains exhibiting relatively high potential for slip activity.

1. V. Olden, C. Thaulow, R. Johnsen and E. Østby, "Scripta Mater.", **volume 57** (2007), p. 615.
2. G. Rørvik, S. M. Hesjevik and S. Mollan, "Effect of microstructure on the HISC susceptibilityof supermartensitic stainless steels", Stainless Steel World Conference & Expo Maastricht, the Netherlands (2005), poster 5089.
3. M. Karlsen., J. Hjelen, Ø. Grong, G Rørvik, R. Chiron, U. Shubert and E Nilsen, "Mater. Sci. Tech.", **volume 24** (2008), p. 64.

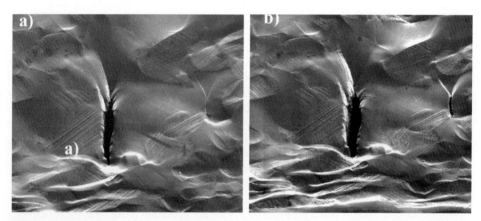

Figure 1. Secondary electron (SE) images of the microstructure following a) 11 % and b) 13 % deformation.

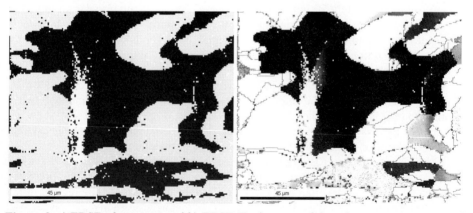

Figure 2. a) EBSD phase map and b) EBSD Taylor map of the microstructure presented in Figure 1 b).

E-beam hardening SEM glue for fixation of small objects in the SEM

S. Kleindiek, A. Rummel and K. Schock

Kleindiek Nanotechnik GmbH, Aspenhaustrasse 25, 72770 Reutlingen, Germany

kleindiek@nanotechnik.com
Keywords: Electron microscopy, manipulation, tensile testing

A new E-beam hardening glue is introduced. It is very viscous and has limited wettability. This allows precise application of small amounts of glue with the tip of a micro- or nanomanipulator. The glue hardens in less than two minutes electron beam exposure at normal working conditions (30 kV, 0,35 nA), see figures 1 and 2.

The tensile strength of the glue was measured by connecting the tips of two tungsten wires with 20 µm diameter and 1 mm length. Applying a sidewards movement of 100 µm of one tip with the help of a micromanipulator (MM3A-EM) the connection was ruptured, see figure 3. Knowing the Young's modulus of tungsten (400 MPa) the corresponding force can be calculated (2 mN) and thus the tensile strength of the connection was derived to be approximately 100 MPa. This is similar to the tensile strength of epoxy.

The new SEM glue opens up new applications like manipulation and secure fixing of small objects, micro tensile experiments etc. Application examples will be presented.

M. Luysberg, K. Tillmann, T. Weirich (Eds.): EMC 2008, Vol. 1: Instrumentation and Methods, pp. 565–566, DOI: 10.1007/978-3-540-85156-1_283, © Springer-Verlag Berlin Heidelberg 2008

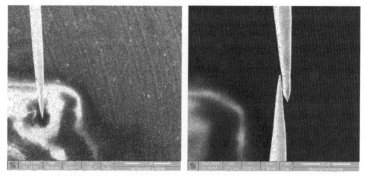

Figure 1. High-resolution micrograph of the glueing process. The glue is picked up from a reservoir with the tip of a micro manipulator and transported to a second tip as a test object. Due to the high viscosity of the glue wettening is easily controllable. Hardening takes place within less than two minutes under normal E-beam conditions (30 kV, 0,35 nA).

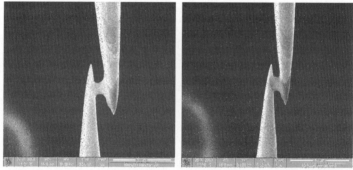

Figure 2. Due to the high viscosity of the glue wettening is easily controllable. Hardening takes place within less than two minutes under normal E-beam conditions (30 kV, 0,35 nA).

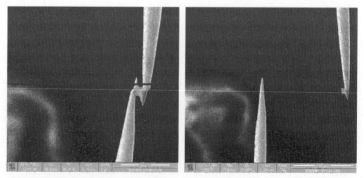

Figure 3. The tensile strength of the glue was measured by rupturing the tips of two tungsten wires with 20 μm diameter and 1 mm length connected by the glue. Applying a sidewards movement of 100 μm the connection was ruptured.

Development of the charging reduction system by electron beam irradiation for scanning electron microscopes

Y. Kono, O. Suzuki and K. Honda

Metrology Inspection Division, JEOL Ltd., Akishima Tokyo, 196-8558, Japan

yukohno@jeol.co.jp
Keywords: SEM, charge

In the observation of insulating materials with a scanning electron microscope (SEM), charging contrast degrades the image quality. Several methods against this surface charging, for example, gas injection, appropriate setting of accererating voltage and etc, have been done so far. All most of them, however, have some restriction for SEM imaging. As a new method, we developed a charging reduction system to reduce the influence of negative charging by irradiating tilted electron beams. This method has not been used with SEM imaging, but same kind of attempts were made in the field of AES [1, 2].

Yields of the secondary electrons and reflected electrons depend on the incident angle and the accelerating voltage of primary electron beams. If the primary electron beams are irradiated in low accelerating voltage and low incident angle, it is possible to increase the radiating electrons more than the irradiating ones. Then positive charges are supplied to the specimen, which can reduce the negative charging on the specimen. This method is so unique that its principle uses differencies of the secondary electron yeilds of the primary electron beam and the tilted electron beam.

Figure 1 shows a schematic diagram of the experiment. There are two columns. One is an observation column and the other is a charging reduction column which has been devoloped for the above purpose. A charging reduction beam is defocused on the specimen to cover the observation region. Its accelerating voltage can be varied from 500V to 1000V and current density can be varied from $0.01A/m^2$ to $40A/m^2$. Observation beams scan the observation region line by line. Figure 2 shows the irradiation timing of the charging reduction beams. There are intervals between one line-scanning and the next one, which we call blanking time. The charging reduction beams are irradiated during this blanking time so as not to deteriorate the SEM image from its SE signals.

A mask with Cr patterns on SiO_2 was used in this experiment. The L&S is 6um, and Cr pattern width is 2um. The Cr pattern is isolated from the ground potential. Observation beam accelerating voltage is 1000V, probe current is 50pA and incident angle is normal to the specimen surface. Observation region is 20um×16um square. Charging reduction beam accelerating voltage was 1000V and incident angle is 50 degree to the normal.

Figure 3 shows an experimental result of the negative charge reduction of the SEM image. The left image was taken without charging reduction beam, and the right one was taken with it of which current density was $1.2A/m^2$. The negative charging contrast

M. Luysberg, K. Tillmann, T. Weirich (Eds.): EMC 2008, Vol. 1: Instrumentation and Methods, pp. 567–568, DOI: 10.1007/978-3-540-85156-1_284, © Springer-Verlag Berlin Heidelberg 2008

568

appeared in the left, which disappeared in the right. This results show that charging contrasts of insulating materials can be reduced dramatically by the irradiation of tilted electron beams.

1. S. Ichimura, H. E. Bauer, H. Seiler and S. Hofmann, SURFACE AND INTERFACE ANALYSIS **14** (1989), p. 250.
2. S. Ichimura, Journal of The Surface Science Society of Japan **24** (2003), p. 207 (in Japanes)

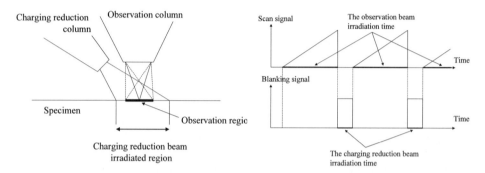

Figure 1. The schematic diagram of the experiment. The charging reduction beams and the observation beams were irradiated one after another.

Figure 2. The schematic diagram of the irradiation timing of the charging reduction beams and observation beams. The charging reduction beams are irradiated during blanking time.

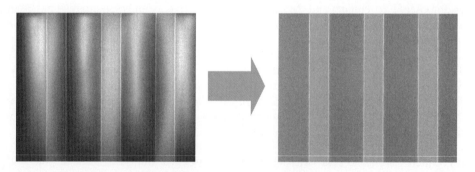

Figure 3. The SEM images of Cr pattern on SiO$_2$. The left image was taken without the charging reduction beam and the right one was taken with it.

Aberrations of the cathode lens combined with a focusing magnetic/immersion-magnetic lens

I. Konvalina, I. Müllerová and M. Hovorka

Institute of Scientific Instruments AS CR, v.v.i., Královopolská 147, 612 64 Brno, Czech Republic

konvalina@isibrno.cz
Keywords: cathode lens, aberrations, focusing lens

The cathode lens (CL) is a frequently used electron optical element in the photo-emission electron microscopy (PEEM), low energy electron microscopy (LEEM), and recently in the scanning electron microscopy (SEM) [1]. Lenc and Müllerová [2,3] studied the basic optical parameters of the CL as such, as well as in combination with a magnetic focusing lens, and derived approximate analytical expressions for the axial aberration coefficients for both cases. For the combination, the sequential arrangement of electrostatic and magnetic fields was considered.

Several authors have claimed the aberration coefficients decreasing for overlapped fields of the lenses under combination [4,5]. In order to compare the aberration coefficients for the sequential fields we employed the EOD software [6] and found a perfect agreement between the analytical and numerical results. Then the EOD software was used for calculation of the aberration coefficients for the overlapped fields.

The simulation setups are shown in Figure 1. The axial magnetic field in the specimen plane reaches $B = 436$ µT with the magnetic focusing lens and $B = 336$ mT with the immersion-magnetic lens (working distance of 3 mm, diameter of the CL electrodes of 12 mm and the aperture bore of 500 µm). The spherical and chromatic aberration coefficients were calculated for the landing energy E_L varying from 10 keV to only 1 eV with the primary beam energy of $E_P = 10$ keV (see Table I and Figure 2).

It is well known and also evident from Table I and Figure 2 that the aberrations are smaller when the specimen is immersed in a strong magnetic field without any electrostatic field. This is somewhat paid by the image interpretation complicated owing to signal trajectories heavily deformed in the magnetic field. However, strong electrostatic field on the sample surface dominates the aberrations and eliminates the advantage of the immersion-magnetic lens.

We can conclude that at lowest electron energies in the CL mode the spatial relation between electrostatic and focusing magnetic fields becomes unimportant and when using overlapped fields we gain only negligible improvement.

1. I. Müllerová and L. Frank, Adv. Imaging Electron Phys. 128 (2003), p. 309.
2. M. Lenc and I. Müllerová, Ultramicroscopy 41 (1992), p. 411.
3. M. Lenc and I. Müllerová, Ultramicroscopy 45 (1992), p. 159.
4. A. Khursheed, Optik 113 (2002), p. 67.
5. M. Mankos, D. Adler, L. Veneklasen and E. Munro, Surface Science 601 (2007), p. 4733.
6. B. Lencová and J. Zlámal, Microscopy and Microanalysis 13 (2007), p. 2.

M. Luysberg, K. Tillmann, T. Weirich (Eds.): EMC 2008, Vol. 1: Instrumentation and Methods, pp. 569–570, DOI: 10.1007/978-3-540-85156-1_285, © Springer-Verlag Berlin Heidelberg 2008

7. Supported by EUREKA grant no. OE08012 and by GAASCR grant no. IAA100650803.

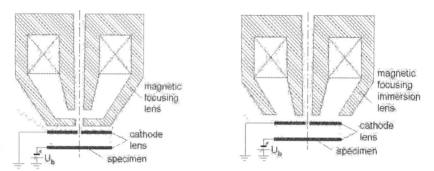

Figure 1. Setups used in simulations: combination of the cathode lens with the magnetic focusing lens (left) and with the magnetic-immersion lens (right).

Table I. The chromatic and spherical aberration coefficients for the systems shown in Figure 1, $E_P = 10$ keV.

	magnetic lens		immersion-magnetic lens	
E_L (eV)	C_S (mm)	C_C (mm)	C_S (mm)	C_C (mm)
10000	21.106	11.343	1.3391	1.7669
5000	13.802	5.6209	0.764	0.91636
1000	2.7879	0.83407	0.22183	0.22066
500	1.2007	0.352	0.12053	0.11843
100	0.13113	0.048274	0.025206	0.026943
50	0.046135	0.021676	0.013013	0.014645
10	0.0051398	0.0044887	0.0036335	0.0040505
5	0.0025427	0.0024935	0.0021919	0.0023693
1	0.00063018	0.0006448	0.00061936	0.00063849

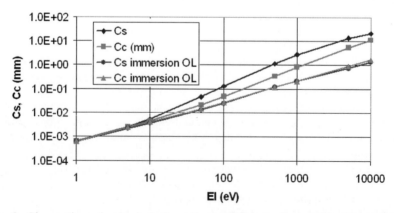

Figure 2. Chromatic and spherical aberration coefficients for combinations of the CL and the magnetic/immersion-magnetic lens as a function of the landing energy E_L.

Identification possibilities of micro/nanoparticles and nanocomposites in forensic practice

M. Kotrly[1], I. Turkova[1] and V. Grunwaldova[2]

1. Institute of Criminalistics Prague (ICP), POBox 62/KUP, Strojnicka 27, 170 89
Praha 7, Czech Republic
2. Zentiva, a.s. Praha, U Kabelovny 130, 102 37 Praha 10, Czech Republic

kup321@mvcr.cz

Keywords: forensic science, SEM, nanoparticles

Field of nanoparticles and nanocomposites is nowadays one of the most dynamic and developing fields both in industrial applications and in research and development. Products based on these particles or the ones forming only their content are becoming quite common. These are namely food-stuff industry, inks, lacquers and pigments, textile application, dental materials, building industries, car industry and the like. For these reasons are these materials increasingly encountered in forensic laboratories, they often present special acceptable marks that can be utilised for identification specification. That is why it is necessary to determine and specify them as precisely as we are able to implement it. This fact stems from reality showing that the majority of material analysis carried out in forensic science deals with determination, characterisation and comparison of practically any materials possibly contacting either persons or objects relating to criminal activities.

Electron microscopy (usually coupled with electron microanalysis EDS/WDS) is one of widely used methods performing material forensic analysis (often used as a screen method), for this reason is also applied to the field of micropacticles and nanocomposites. These are both classic sets with thermal cathode and progressive field emission sources enabling to meet needful discrimination. Based on experiments performed it turns out that current sets with Schottky emission source are capable of examining very effective file of information concerning component morphology showing the size of first tens nm.

For determination of detection limits performing identification of nanoparticles and nanocomposites were carried out several series of experiments containing compo with different content of typical nanoparticles and nanocomposites. By means of SEM, FE-SEM nanoparticles and nanocomposites in a mixture were predicted without problems. Further identification was performed using EDS/WDS, micro XRF and micro XRD analyses.

For an exact verification of nanoparticles morphology are successfully used methods of mathematical morphology implemented into image analysis. These procedures are quite common in light microscopy, for applications to nano field it is necessary to use TEM images or FE-SEM if need be. The image is subsequently thresholded (using system HSI) and morphological parameters of particles are calculated. Furthermore, for subsequent comparison are used statistical methods.

M. Luysberg, K. Tillmann, T. Weirich (Eds.): EMC 2008, Vol. 1: Instrumentation and Methods, pp. 571–572, DOI: 10.1007/978-3-540-85156-1_286, © Springer-Verlag Berlin Heidelberg 2008

Completely separate part presents in forensic practise the need of determination of marking and trap materials. For these purposes is applied automatic analysis SEM/EDS which is used for gun shot residues aiming at determining precise identification of microparticles and nanoparticles occurrence in abrasions and in relicts of marking and traps materials. The system is applicable to particles comprising elements with a higher atomic number (Fe is an assumed limit). Principle of this method lies in automatic detection of particles with higher brightness than is an adjusted threshold and in a subsequent automatic EDS analysis using the system. Fully automated software is functioning including classification according to quantitative element analysis. A minimal size of particles is limited only by actual magnification while having used an electron microscope.

Software Link-ISIS GSR for EDS analysis Link a SEM Vega was utilized for a series of experiments. System was tested for nanoparticles Strontium Lanthanum Manganese Oxide, which were blended into a mixture of silicate contaminators in the amount of below about 0.01%. This mixture was applied to a clothing, a swab (abrasion) was recovered and then sampling on a standard stub with a carbon target ensued. A certain drawback is a considerable time demand – surface 12 mm in diameter can be analysed in dependence on particles numbers and on magnification lasting till dozens of hours.

The latest innovations in forensic laboratories are systems of ion microscopy (FIB – Focused Ion Beam Technology) enabling to carry out material deposition basically at atomic level and "cutting through" of nanocomposites particles. These systems are increasingly appearing also outside high- specialized industrial centres and high tech labs, which are caused both by system miniaturization and by a considerable price decrease. FIB systems often coupled with a classic set SEM, or more precisely FE-SEM, enable in forensic laboratories the examination of inner structure composition of microparticles (gun shot residues and post-blast particles for the determination of their origin and distribution models), a detailed study of micro and nanocomposite materials (more exact material comparison), an exact analysis of crossing strokes when scrutinizing papers, etc.

For a specific field of forensic analyses it is necessary not only to anticipate new analytic procedures but also taking into account specific issues to modify them fundamentally.

1. Acknowledgements: Microanalytical methods at Institute of Criminalistics Prague were supported by grant-aided projects of the Czech Republic Ministry of Interior RN 19961997008, RN 19982000005, RN 20012003007, RN 20052005001, VD20062008B10 a VD20072010B15.

Mass thickness determination of thin specimens using high-resolution scanning electron microscopy

V. Krzyzanek and R. Reichelt

Institute of Medical Physics and Biophysics, University Hospital, University of Münster, Robert-Koch-Str. 31, D-48161 Münster, Germany

krzyzane@uni-muenster.de

Keywords: STEM, SEM, mass thickness determination

Dedicated scanning transmission electron microscopes (STEM) are a very valuable tool allowing for quantitative measurements of thin samples, e.g., for mass determination of macromolecular assemblies [1,2]. Also commercial field-emission high-resolution scanning electron microscopes (SEM) equipped with a very sensitive annular dark-field (ADF) detector [3] and dedicated software [4,5] are most suitable for quantitative studies.

We report on preliminary quantitative investigations made with Epon sections of various thickness at room temperature for which a high-resolution "in-lens" SEM (S-5000; Hitachi Ltd., Japan) equipped with a home-made ADF detector [3] was used.

Figure 1 shows measurements with a ~0.6 µm thick section having a wedge-shaped edge with an angle of 13°. The wedge covers the thickness range of ~0 to ~0.6 µm. Typically, the ADF detector signal increases nonlinearly up to a specific specimen thickness and decreases continuously above this thickness (cf. Figure 1b). For Epon, an electron energy of 30 keV and the given detection geometry of our SEM, the decrease of the ADF signal with increasing thickness, i.e. the contrast reversal, starts at a thickness of ~290 nm. Figure 1c displays the thickness profile calculated from data shown in Figure 1a and Figure 1b. The close agreement of experimental and Monte Carlo simulated data proves the adequacy of the Monte Carlo simulation program for electron scattering [4] up to about 0.5 µm. Figures 1d-e show the atomic force microscope (AFM) topograph and a related height profile which confirm the wedge-shaped edge of a section.

To account quantitatively for beam damage thin Epon sections were employed to estimate the beam induced mass loss (Figure 2). The plot of the residual mass vs. total applied electron dose shows almost linear decrease of the mass in the measured dose range. As expected the comparison of this data and measurements with 80 keV electrons [6] shows at equal low doses a higher mass loss at 30 keV.

Presently, studies with other organic and anorganic materials like silicon are in progress with the aim to pave the way for quantitative studies of various nanostructures.

1. S.A. Müller and A. Engel, Micron **32** (2001), p. 21.
2. J.S. Wall and M.N. Simon, Methods in Molecular Biology **148** (2001), p. 589.
3. V. Krzyzanek, H. Nüsse and R. Reichelt, in "Proc. Microscopy Conference 2005", (Paul Scherrer Institut, Villigen) (2005), p. 49.
4. V. Krzyzanek and R. Reichelt, Microsc. Microanalysis **9** (Suppl. 3) (2003), p. 110.

M. Luysberg, K. Tillmann, T. Weirich (Eds.): EMC 2008, Vol. 1: Instrumentation and Methods, pp. 573–574, DOI: 10.1007/978-3-540-85156-1_287, © Springer-Verlag Berlin Heidelberg 2008

574

5. V. Krzyzanek, S.A. Müller, A. Engel, and R. Reichelt, in "Proc. 16th Internat. Microscopy Congress 2006", vol. 2, eds. H. Ichinose and T. Sasaki (Public. Comm. IMC16, Sapporo) (2006), p. 959.
6. R. Reichelt and A. Engel, J. Microsc. Spectrosc. Electron. **10** (1985), p. 491.
7. We kindly acknowledge the help of Mrs. Ursula Malkus for the skilled preparation of Epon sections and Mrs. Andrea Ricker for the AFM measurements.

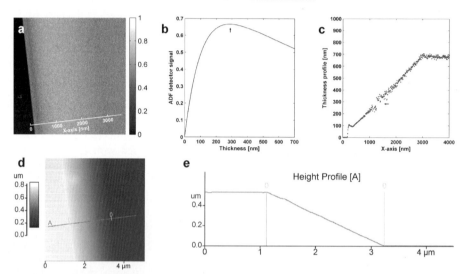

Figure 1. Wedge-shaped Epon section with a wedge angle of ~13°. Calibrated ADF image; its intensity corresponds to the fraction of scattered electrons (a). Graph of fraction of scattered electrons vs. thickness estimated by Monte Carlo simulation program MONCA [4] (b). Average thickness profile (c) calculated from (a) and the data from (b). Arrows in (b,c) indicate the specific thickness. AFM topograph (d) and height profile (e) of the wedge-shaped edge.

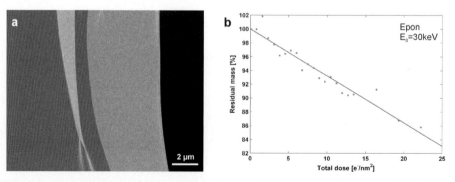

Figure 2. Determination of mass loss for Epon: ADF micrograph of a partially folded ~52 nm thick Epon section (a) and the residual mass vs. the applied total electron dose (b). The residual mass was calculated from preliminary dose series of 21 micrographs using the unfolded section on the left part of the micrograph (a).

Benefits of Low Vacuum SEM for EBSD Applications

K. Kunze[1], St. Buzzi[2], J. Löffler[2], J.-P. Burg[3]

1. Electron Microscopy ETH Zurich (EMEZ), 8093 Zurich, Switzerland
2. Laboratory of Metal Physics and Technology, ETH Zurich, 8093 Zurich, Switzerland
3. Geological Institute, ETH Zurich, 8092 Zurich, Switzerland

karsten.kunze@emez.ethz.ch
Keywords: electron backscatter diffraction, orientation mapping, charge compensation, drift

Electron backscatter diffraction (EBSD) is a SEM-based method for local measurements of crystal orientations on the surface of bulk samples. It has found widespread fields of applications for microtexture analysis, phase discrimination, orientation mapping and others.

Routine applications of EBSD are typically performed at conventional SEM high voltage (10…30kV), long working distance (10…20mm) and rather high beam currents (some nA). On non-conductive materials, charge balance must usually be enabled by conductive coatings. Because of the small information depth of EBSD on steeply tilted sample surfaces, coatings are restricted to thin layers (few nm) of amorphous carbon, and are barely enough to prevent charge built-up on the surface, in particular for high resolution scans.

Under optimum conditions, modern installations at FEG-SEMs allow orientation mappings with (apparent) spatial resolution around 10nm using frame rates in the range of 20fps, where higher rates are reached at cost of resolution [1,2,3]. In many instances, high resolution orientation mappings suffer from drift over the acquisition time of one to several hours, even on well-contacted and well-mounted metallic specimen.

Applications of EBSD may benefit from low vacuum conditions in the SEM chamber for the following reasons [4]:

- Charge compensation on poorly conductive materials
- Drift reduction even for nominally good conductive materials

This study presents an analysis of the effects of low vacuum conditions on EBSD pattern quality, spatial resolution of orientation mappings, scan stability with respect to true dimensions and drift:

a) The sharpness of EBSD patterns generally deteriorates with increasing chamber pressure. The decrease of pattern quality is reduced for shorter beam gas path length (BGPL) using an additional pressure limiting aperture (PLA) attached at the GAD cone, and by usage of water vapour instead of nitrogen (Figure 1).

b) The spatial resolution of orientation mappings is hardly reduced at chamber pressure of 30…60Pa as compared to high vacuum conditions.

c) Orientation maps on nominally conductive samples, which suffered from severe drift problems under high vacuum, approach the true dimensions of the scanned areas when acquired under low vacuum conditions (Figure 2).

M. Luysberg, K. Tillmann, T. Weirich (Eds.): EMC 2008, Vol. 1: Instrumentation and Methods,
pp. 575–576, DOI: 10.1007/978-3-540-85156-1_288, © Springer-Verlag Berlin Heidelberg 2008

Data and examples are obtained and presented for EDAX-TSL Digiview and Hikari EBSD systems installed on a FEI Quanta 200 FEG SEM.

1. F.J. Humphreys, J. Mat. Sci. **36**(2001), 3833–3854
2. S. Zaefferer, Materials Science Forum **495-497**(2003). 3-12.
3. S. Zaefferer, Ultramicroscopy **107**(2007), 254-266.
4. A. Faryna, K. Sztwiertnia, K. Sikorski,, Europ. Ceramic Soc. **26**(2006) 2967-2971.
5. We kindly acknowledge support by Rene de Kloe (Ametek-EDAX Tilburg) and Ellen Baken (FEI Eindhoven).

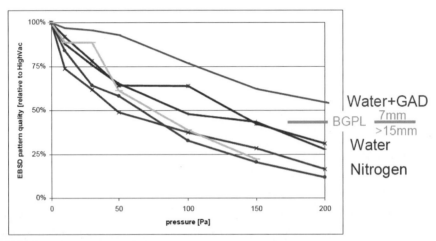

Figure 1. Deterioration of EBSD pattern quality with increasing chamber pressure for standard SEM configuration (Nitrogen and Water) and with reduced BGPL using the GAD cone (Water+GAD). Pairs of lines refer to two different samples (ferritic steel and calcite marble, uncoated).

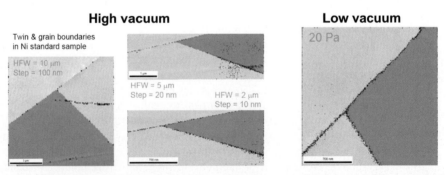

Figure 2. Orientation mappings (colour for orientation, gray level for confidence index) of a grain/twin boundary triple junction. Scan distortion (apparent compression in Y-direction) increases with decreasing scanning step size for acquisition under high vacuum. Distortion is significantly reduced with barely reduced spatial resolution for acquisition under low vacuum (20 Pa).

In-situ combination of SEMPA, STM, and FIB for magnetic imaging and nanoscale structuring

J. Mennig[1], J. Kollamana[1], S. Gliga[1], S. Cherifi[2], F. Matthes[1], D.E. Bürgler[1], C.M. Schneider[1]

1. Institut für Festkörperforschung (IFF-9) and JARA-FIT, Research Center Jülich, Germany
2. CNRS – Institut Néel, Nanoscience department, Grenoble, France

j.mennig@fz-juelich.de

Keywords: SEMPA, STM, magnetization of surfaces

In the field of spintronics and its application in data storage and processing, research on nanomagnetism attains more and more attention. Therefore, we have constructed a versatile UHV instrumentation for the investigation of magnetism and magnetotransport phenomena on the nanometer scale, the *nanospintronics cluster tool.* The system consists of three interconnected vacuum chambers. The first chamber is a preparation unit containing 7 evaporators for thin film deposition, RHEED and LEED for structural analysis, and MOKE, AES and XPS to investigate magnetic and chemical properties of the films. The second chamber houses a low temperature scanning tunneling microscope (STM) that we plan to operate in a spin-resolved mode in order to image the magnetic structure on an atomic scale in a temperature range from 4 to 300 K. The third part of the UHV system combines a scanning electron microscope with polarization analysis (SEMPA) with a focused ion beam (FIB). Thus, it is possible to create objects on the nanometer scale and analyze the magnetic properties of their surfaces *in-situ*. The sample stage of the SEM can be cooled down to 30 K and features electrical contacts for *in-situ* four-point measurements and the investigation of spin-torque driven magnetization dynamics.

As a first test we investigated nanocrystalline Co-disks with diameters from 0.5 to 5 µm and a thickness of 30 nm using our SEMPA [1]. The disks are defined by e-beam lithography and coated with a thin Au-film, which was removed by argon sputtering *in-situ* before the measurements. Due to the low crystalline anisotropy, the domain structure is dominated by the circular shape. The SEMPA images show magnetic domain structures linked to the disk diameter. Small disks with a diameter below 2 µm display only vortices (Figure 1a). Larger disks have more complex domain structures, containing e. g. cross-tie domain walls (Figure 1b). This type of domain structure consists of an alternating series of vortices and anti-vortices, mostly observed in extended thin films. Furthermore, we used our FIB to write nanoscaled patterns in the larger Co disks with the objective of stabilizing complex vortices structures (Figure 2).

In the field of spin-polarized STM, we first aim at mapping the magnetization structure of ultra-thin, 1-4 monolayer (ML) thick Fe films grown on a W(110) single crystal. First STM images (Figure 3) recorded at 78 K visualize the growth mode of Fe films *in-situ* prepared at room temperature. Starting from the bare W surface, Fe

M. Luysberg, K. Tillmann, T. Weirich (Eds.): EMC 2008, Vol. 1: Instrumentation and Methods, pp. 577–578, DOI: 10.1007/978-3-540-85156-1_289, © Springer-Verlag Berlin Heidelberg 2008

adatoms first nucleate and form compact islands. With increasing coverage the islands grow and coalesce to a nearly completely closed first ML. The growth of the following MLs clearly occurs along a preferred growth direction as can be deduced from the elongated island shape (Figure 3). As a next step, we want to utilize Fe coated W tips to image the magnetic domain structure and to compare spin-polarized STM data and SEMPA measurements taken from the same Fe wedge [2].

In summary, we presented first results characterizing the potential of our nanospintronic cluster tool, which combines several microscopies with a variety of preparation and analysis methods.

1. S. Cherifi, J. Mennig, et al., to be submitted (2008).
2. M. Bode, R. Pascal, and R. Wiesendanger, J. Vac. Sci. & Tech. A **15** (1997)

Figure 1. SEMPA images of Co-disks. a) Diameter 1 μm: the left one contains two vortices and the right one a single, centered vortex. b) Diameter 5 μm: the close-up of the right domain wall shows a cross-tie domain wall. The arrows show the direction of the magnetization.

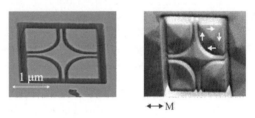

Figure 2. SEM (left) and SEMPA (right) images of a structure, written with FIB in a Co-disk with a diameter of 6 μm. The finest elements have a size of 50 nm, the SEMPA measurements shows a vortex in each segment.

Figure 3. STM images (90 x 90 nm^2) of 0.05 ML Fe (left), 0.72 ML Fe, and 1.05 ML Fe on W(110) recorded at 78 K.

Characterisation of the subgrain structure of the aluminium alloy AA6082 after homogenization and hot forming by EBSD

S. Mitsche[1], P. Sherstnev[2], C. Sommitsch[2,3], T. Ebner[4] and M. Hacksteiner[4]

1. Institute for Electron Microscopy, Graz University of Technology, Steyrergasse 17, A-8010 Graz, Austria
2. Christian Doppler Laboratory for Materials Modelling and Simulation, University of Leoben, Franz-Josef-Strasse 18, A-8700 Leoben, Austria
3. Chair of Metal Forming, University of Leoben, Franz-Josef-Strasse 18, A-8700 Leoben, Austria
4. AMAG rolling GmbH, Postfach 32, A-5282 Ranshofen, Austria

stefan.mitsche@felmi-zfe.at
Keywords: subgrain structure, aluminium alloy, EBSD

The properties of a hot rolled aluminium alloy are strongly dependent on the homogenization and the forming conditions [1,2], because the microstructure of the specimen will be changed by the various processing steps [3]. A powerful tool to investigate the obtained grain – subgrain structure is the backscatter electron diffraction (EBSD) technique [4]. The evolution of the microstructure of the aluminium alloy AA6082 during hot forming is presented in the following.

Specimens of the alloy AA6082 were first homogenized at 570°C for 10h to solute the Mg_2Si-phase and to induce a coagulation of the plate like β-AlFeSi-phase. Rastegaev-cylindrical compression tests of these specimens were performed on a servo-hydraulic thermo-mechanical treatment simulator SERVOTEST. A deformation degree of φ = 0.8 was chosen, because at this deformation a fully stationary subgrain structure was to be expected. The temperatures (500°C and 550°C) and the deformation rates ($1s^{-1}$ and $10s^{-1}$) were selected to have similarly conditions as during hot-rolling with a DUO rolling mill.

Cross sections of the deformed specimens were analysed in a scanning electron microscope (Zeiss DSM 982 Gemini) equipped with an EBSD system from EDAX-TSL (SIT-camera, OIM-software 4.5). An acceleration voltage of 20 kV and a probe current of 3.4 nA were used for all specimens analysing an area of 320 μm x 350 μm with a step size of 1 μm. In Figure 1 the obtained results are displayed as inverse pole figure (IPF) maps, where the white lines represent the high angle boundaries (>15°) and the black lines the subgrain boundaries (between 2° and 15°). Obviously, the size of the subgrains (determined with an imaging analysis program) strongly depends on the temperature and the deformation rate.

Introducing the Zener-Hollomon-parameter Z, which is defined as $Z = \dot{\varphi} \cdot (Q/(R \cdot T))$ ($\dot{\varphi}$..deformation rate, Q..activation energy, R..universal gas constant, T..temperature), a decrease of the subgrains within the elongated high grain boundaries by increasing Z can be observed (see Figure 2). From this figure the line of best fit was determined and

M. Luysberg, K. Tillmann, T. Weirich (Eds.): EMC 2008, Vol. 1: Instrumentation and Methods, pp. 579–580, DOI: 10.1007/978-3-540-85156-1_290, © Springer-Verlag Berlin Heidelberg 2008

580

as a result the size of subgrains in the stationary range δ_{ss} can be described as a function of Z: $\delta_{ss}^{-1} = -0.8274 + 0.0318 \cdot \ln(Z)$

1. J. Hirsch "Virtual Fabrication of Aluminium Products. Microstructural Modelling in Industrial Aluminium Production" WILEY-VCH Verlag GmbH & Co. KGaA, Weinheim, 2006
2. P. Rometsch, S. Wang, A Harriss, P. Gregson, M. Starik, Material Science Forum **396-402** (2002), p.655
3. G. Mromka-Nawotnik, J. Sieniawski, Journal of AMaterial Processing and Technology **162-163** (2005), p. 367
4. F.J. Humphreys, Journal of Material Science **36** (2001), p. 3833

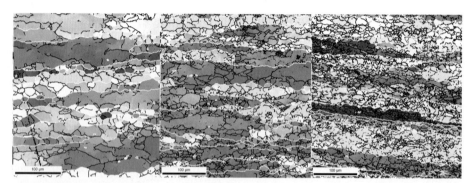

Figure 1. IPF maps of the compressed specimens at T = 550°C and $\dot{\phi}$ = 1s^{-1} (left) respectively $\dot{\phi}$ = 10s^{-1} (centre), as well as T = 500°C and $\dot{\phi}$ = 10s^{-1} (right); White lines mark the high angle boundaries (>15°) and black lines the subgrain boundaries (between 2° and 15°)

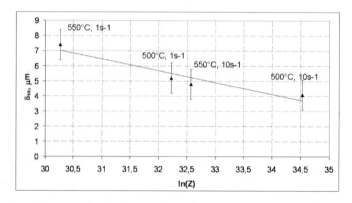

Figure 2. Mean size of the subgrains (experimentally determined) in dependence on the Zener-Hollomon-parameter.

An improved detection system for low energy Scanning Transmission Electron Microscopy

V. Morandi[1], A. Migliori[1], P. Maccagnani[1], M. Ferroni[2] and F. Tamarri[1]

1. CNR-IMM Section of Bologna, via Gobetti 101, 40129 Bologna, ITALY
2. INFM-CNR Sensor Lab and Dept. of Chemistry and Physics for Engineering and
Materials, University of Brescia, via Valotti 9, 25133 Brescia, ITALY

morandi@bo.imm.cnr.it
Keywords: Low Energy STEM, Solid State Detector, Z-Contrast

In the last few years the STEM technique in standard SEMs has become a complementary approach to HAADF-STEM at high energy [1, 4], at least when high resolution is not required, as demonstrated by the availability of STEM attachments for all the commercial SEMs.

Due to the absence of image-forming lenses, in low energy STEM there is no electron optical limitation in collection angles and energy losses. It has been recently demonstrated [1], [2] that it is possible to observe thick specimens (up to 1 μm) or to use very low energies only increasing the range of the collection angles.

Moreover, only with a careful control of the collection angles, and often with the aid of simulations, it is possible to correctly explain the contrast of the images and link their bright and dark features with the compositional variation in the sample [1], [4].

In this abstract we describe an improved STEM detection system, which allows precise variation of the collection angles during the observation, in order to achieve a direct-interpretable Z-contrast image of the specimen features.

Fig. 1 a) reports the layout of the solid-state detector (Patent Pending, n. BO2007A000409), formed by four isolated sectors S1 – S4, each providing a separated signal, which can be recorded separately or combined with one or all the others. The collection angular range varies changing the specimen-detector distance (D), as shown in Fig. 1 b), for two different distances $D = 3.5$ mm, and $D = 8.5$ mm.

Fig. 2 a) shows the HAADF-STEM image obtained at 200 keV with a Tecnai F20 of Au-Pd nanoparticles supported by amorphous carbon. The nanoparticles appear bright as expected for a DF image, independently on the particle size and specimen thickness variation. An image of the same sample obtained with a SEM ZEISS 1525 operating at 20 keV with a commercial STEM detector is reported in Fig. 2 b): the Au-Pd particles appear dark or bright depending on the substrate thickness and on the particle size itself.

Figs. 3 a) – c) report STEM images of a detail of the specimen shown in Fig. 2, obtained at 20 keV, with a SEM ZEISS 1530 using the detector of Fig. 1 a), and collecting the signal for the S1, S2 and S3 sectors separately. It is clearly demonstrated that in this case the image contrast is directly linkable to the specimen features in BF (Fig. 3 a)) as well as in DF images (Fig. 3 b) and c)), showing the capabilities of this detection system.

M. Luysberg, K. Tillmann, T. Weirich (Eds.): EMC 2008, Vol. 1: Instrumentation and Methods,
pp. 581–582, DOI: 10.1007/978-3-540-85156-1_291, © Springer-Verlag Berlin Heidelberg 2008

582

1. V. Morandi and P.G. Merli, Journal of Applied Physics **101** (2007), p. 114917.
2. V. Morandi et al., Applied Physics Letters **90** (2007), p. 163113.
3. C. Probst et al., Micron **38** (2007), p.402.
4. V. Van Ngo et al., Microscopy Today **15** Vol. 2 (2007), p. 12.
5. This work was supported by the EU Project ANNA, contract n. 026134 (RII3).

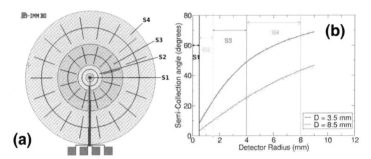

Figure 1. (a) Layout of the low-energy STEM detector; (b) Collection angles for the different sectors S1 – S4 of the detector, for two different detector-specimen distances D = 3.5 mm and D = 8.5 mm

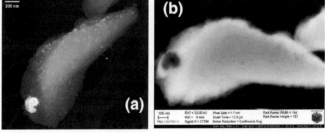

Figure 2. (a) 200 keV HAADF-STEM image of an amorphous carbon particle supporting Au-Pd nanoparticles. (b) 20 keV SEM-STEM image of the same specimen obtained with a commercial STEM detector.

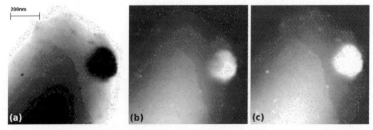

Figure 3. 20 keV SEM-STEM images of the same specimen shown in Fig. 2 obtained with the new STEM detector. (a) Sector S1, collection angles: $0 \div 8°$, the image is of BF type; (b) sector S2, collection angles: $8° \div 23°$, the image is of DF type; (c) sector S3, collection angles: $23° \div 49°$, the image is of DF type.

Thickness determination of thin samples by transmission measurements in a scanning electron microscope

E. Müller

Laboratory for Electron Microscopy and
Center for Functional Nanostructures, University of Karlsruhe, 76128 Karlsruhe, Germany

mueller@lem.uni-karlsruhe.de
Keywords: thickness determination, STEM, slow electrons

Many electron microscopic investigations require the determination of the sample thickness. In this work a method for the determination of the thickness of thin samples by quantitative measurements in the scanning electron microscope (SEM) in transmission mode is proposed. The method is exemplified by the analysis of a GaN lamella prepared by means of focused-ion-beam milling.

Film thickness determinations based on the electron transmission method were reported previously [1] but they require additional arrangements like a Faraday cage and apertures. In this work a Zeiss SEM equipped with a standard scanning transmission electron microscope (STEM) detector is used both for imaging and measuring the transmitted electrons. By comparison with calculated fractions of transmitted electrons, the thickness of the sample can be determined with high lateral resolution. The thickness determined in this way is verified by a line scan in a secondary electron image, tilting the sample in a top-view position as shown in Figure 1a).

The STEM detector operating with connected bright-field and dark-field sectors is positioned below and close to the sample holder. In this way a large amount of transmitted electrons are collected and dynamical scattering effects are averaged out. The fraction of transmitted electrons is given by the ratio of the grey value of the sample and the vacuum region of the image (Figure 1b)).

Assuming a Gaussian distribution of the transmitted electrons with the mean square scattering angle depending on the primary electron energy and the properties of the material [2] the theoretically expected transmission is depicted in Figure 2. The calculated curves are compared with values determined from Monte Carlo simulations [3]. In these considerations corrections were taken into account. One correction is given by the geometry of the detector, concerning the loss of electrons due to the limited capture area. Another correction takes into consideration the detector efficiency at different electron energies. The thickness of the sample determined from these curves at different primary energies is consistent with the result obtained by direct measurement from top view image.

1. H. Niedrig, Optica Acta. **24** (1977), p. 679.
2. V.E. Cosslett and R.N. Thomas, Brit. J. Appl. Phys. **15** (1964), p. 883.
3. N.W.M. Ritchie, Surf. Interface Anal. **37** (2005), p. 1006.

M. Luysberg, K. Tillmann, T. Weirich (Eds.): EMC 2008, Vol. 1: Instrumentation and Methods, pp. 583–584, DOI: 10.1007/978-3-540-85156-1_292, © Springer-Verlag Berlin Heidelberg 2008

4.   This work has been performed within the project Z of the DFG Research Center for Functional Nanostructures (CFN). It has been further supported by a grant from the Ministry of Science, Research and the Arts of Baden-Württemberg (Az: 7713.14-300).

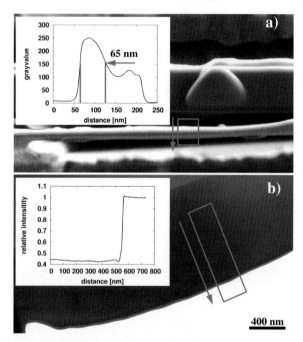

Figure 1. GaN lamella a) top-view SEM imaging with secondary electrons. b) scanning transmission electron imaging in cross section.

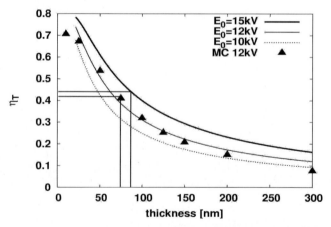

Figure 2. Electron transmission ratio through a thin film of GaN. Continuous lines: calculated values for different primary energies. Triangles: Monte Carlo simulations. Markers show the STEM measurements for the GaN lamella.

Role of the high-angle BSE in SEM imaging

I. Müllerová and L. Frank

Institute of Scientific Instruments ASCR, v.v.i., Královopolská 147, 612 64 Brno, Czech Republic

ilona@isibrno.cz

Keywords: contrast formation, cathode lens, SEM

For acquisition of secondary (SE) as well as backscattered electrons (BSE) in the scanning electron microscope (SEM), single-channel detection systems are traditionally used. We have designed a multi-channel detector collecting BSE emitted to eight intervals of polar angles, and tested it also in the cathode lens mode (CLM) [1]. First experiments showed the method suitable for observation of polycrystalline targets [2].

The detector is positioned just below the objective lens and consists of a grid on the ground potential, situated 6 mm above the specimen, followed by the multi-channel plate (MCP) and the eight-channel collector at the ground potential (see Figure 1). Thickness of the complete detector is 6 mm. The CLM with the sample negatively biased to U_{sp} enables us to fluently control the landing energy E_L of electrons by means of their deceleration from the primary beam energy E_P while preserving the resolution. The collimation of the signal trajectories toward the optical axis can be controlled via the immersion ratio $k=E_P/E_L$.

In order to map the energy/angle dependences of the BSE contrast, experiments were performed with a polycrystalline nickel. Highest grain contrast was obtained in the CLM for the immersion ratio $k=2.5$ with $E_L=4$ keV, and the detector channels enabled us to separate the crystallinic contrast contribution from the topographic one (Figure 3). In the standard BSE imaging without sample bias only a weak grain contrast was observed for the same landing energy of 4 keV (Figure 4) and no significant improvement was achieved at higher energies (Figure 5).

Figure 2 shows the calculated angle/energy ranges for the micrographs presented. These data allow us to identify the high-angle inelastic BSE as dominant carriers of the grain contrast. According the highest contrast in channel B (Figure 3) the BSE down to 1 keV can be termed the optimum signal species. The H channel, acquiring the high-angle elastic BSE, shows an image even more surface sensitive than the side-attached SE detector images in Figures 4 and 5. In the channels A and B in the standard SEM mode the lack of high-angle BSE seems responsible for the reduced grain contrast.

Acquisition of the high-angle BSE, combined with some energy selection, has proven itself an efficient tool for observation of even unsmooth polycrystalline targets.

1. I. Müllerová and L. Frank, Adv. Imaging Electron Phys. **128** (2003), p. 309.
2. I. Müllerová, I. Konvalina and L. Frank, Materials Transactions **48** (2007), p. 940.
3. Specimen was provided by Dr. Jiří Buršík (IPM ASCR Brno). The study is supported by the EUREKA grant no. OE08012 and by the GAASCR grant no. IAA100650803. Technical aid of Mr. Pavel Klein (ISI ASCR Brno) is kindly acknowledged.

M. Luysberg, K. Tillmann, T. Weirich (Eds.): EMC 2008, Vol. 1: Instrumentation and Methods, pp. 585–586, DOI: 10.1007/978-3-540-85156-1_293, © Springer-Verlag Berlin Heidelberg 2008

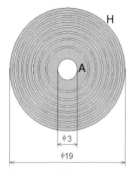

Figure 1. The collector geometry.

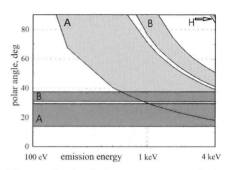

Figure 2. Angle/energy ranges of channels (light-grey fields hold for the CLM).

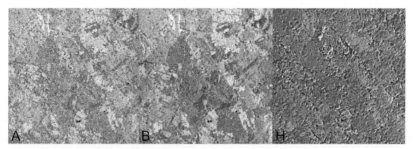

Figure 3. Ni surface imaged in the CLM (E_P=10 keV, E_L= 4 keV, U_{sp}=6 kV).

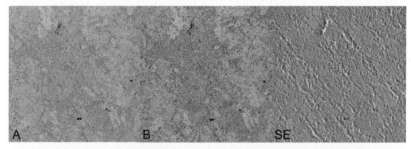

Figure 4. Standard SEM imaging of the same field of view (E_P=E_L=4 keV, U_{sp}=0).

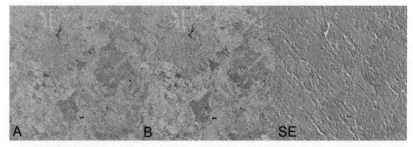

Figure 5. Standard SEM mode again (E_P=E_L=10 keV, U_{sp}=0, field width 150 μm).

Experimental and simulated signal amplification in variable pressure SEM

V. Neděla[1], P. Jánský[1], B. Lencová[1,2], J. Zlámal[2]

1. Institute of Scientific Instruments ASCR v.v.i, Královopolská 147, 61264 Brno, Czech Republic
2. Institute of Physical Engineering, Faculty of Mechanical Engineering, Brno University of Technology, Technická 2, 616 96 Brno, Czech Republic

vilem@isibrno.cz

Keywords: VP-SEM, gas ionisation cascade, EOD, Monte-Carlo

The presence of gases (mostly water vapour with pressure range from 1 Pa to over 2000 Pa) in the specimen chamber of a variable pressure scanning electron microscope (VP-SEM) makes completely different conditions for the detection of signal electrons than for the conventional SEM, and specially designed detectors must be used. In the high pressure conditions, gas ionisation cascade amplifies the signal of secondary electrons (SE), accelerated by the applied field of detection electrode of the ionisation detector [1].

In order to understand the signal generation in the gas environment, the program EOD [2] used for field computation and ray tracing was extended with a plug-in allowing to include collision phenomena (electron gas interactions) using Monte-Carlo method. The first computed results are compared with experimental measurement and with the results obtained from the numerical model published by Meredith at al. [3] and Thiel at al. [4], see Figure 1.

The experimental measurement of the dependence of the signal amplification on the water vapour pressure in the specimen chamber was determined in VP-SEM AQUASEM II [5], equipped with a combined YAG-BSE and ionisation detector of SE, see Figure 2. A small gold sample was placed on a carbon stub and a deep hole in a carbon, situated sufficiently away from the gold, was used for a probe current measurement. Following experimental conditions was set: accelerating voltage 20 kV, probe current 25 pA, gas path length 6 mm and potential applied to the detection electrode of the ionisation detector 300 V.

1. G.D. Danilatos, Foundations of Environmental Scanning Electron Microscopy, Sydney, Academic Press, 1988, p. 249.
2. B. Lencová, J. Zlámal, Microsc. Microanal. **13**, Suppl 3 (2007), p. 2.
3. P. Meredith, at al., Scanning **18** (1996), p. 467.
4. B. Thiel, et al., J. Microsc. **187** (1997), p. 143.
5. V. Neděla, Microsc. Res. Tech. **70** (2007), p. 95
6. This work was supported by the Academy of Sciences of the Czech Republic, grants KJB 200650602 and AVO Z20650511.

M. Luysberg, K. Tillmann, T. Weirich (Eds.): EMC 2008, Vol. 1: Instrumentation and Methods, pp. 587–588, DOI: 10.1007/978-3-540-85156-1_294, © Springer-Verlag Berlin Heidelberg 2008

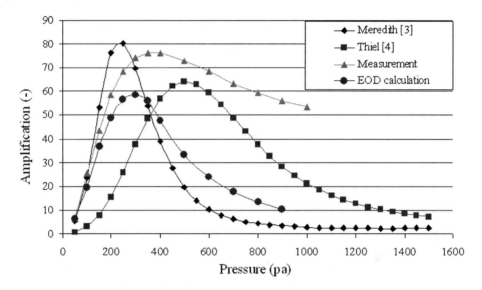

Figure 1. Signal amplification obtained from the numerical models published by Meredith at al. [3] and Thiel at al. [4] compared with new results calculated by program EOD and experimental measurement (accelerating voltage 20kV, probe current 25 pA, gas path length 6 mm, ionisation detector 300 V, water vapour environment).

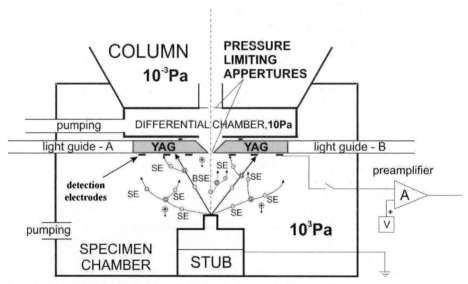

Figure 2. Configuration of the combined YAG-BSE detector and ionisation detector of SE in VP-SEM AQUASEM II.

Study of highly-aggressive samples using the variable pressure SEM

J. Runštuk, V. Neděla

Institute of Scientific Instruments ASCR, Královopolská 147, 61264 Brno, Czech Republic

vilem@isibrno.cz

Keywords: VP-SEM, battery mass, aggressive samples

It is widely known that basic advantage of variable pressure scanning electron microscope (VP-SEM) is the presence of high pressure environment (mostly water vapour with pressure range from 1Pa to over 2000 Pa) in the specimen chamber [1]. These conditions enable the study of non-conductive samples free of charging artefacts as well as wet samples containing different volumes of liquid [2].

Unfortunately, the presence of high pressure environment together with vapours or aerosols released by beam bombardment from aggressive samples (battery material, hydrocarbon containing samples, etc.), can damage specimen chamber equipment like detectors etc., differentially pumped chamber and vacuum system of VP-SEM. These chemicals make problems and create impurities inside the column (along the way of the pumped gas) and cause significant decrease of resolution.

Above mentioned problems may by overcome by the use of a newly designed apparatus for the study of aggressive samples in VP-SEM, see Figure 1. A specimen separating chamber (SSC), where the aggressive sample is placed, is situated inside the specimen chamber of VP-SEM. The SSC, as a main part of this apparatus is separately pumped by a rotary pump with a filtering system. The value of SSC pressure is given by a compromise between the gas flow out to rotary pump and the gas flow into the SSC through a small aperture. This aperture, situated above the sample on the optical axis of the primary beam, must be sufficiently large so as to ensure the necessary gas flow through the SSC and drain off all released aggressive vapours and aerosols to the pumping system. Additionally, the diameter of this aperture must not excessive by limit the field of view. The pressure in the SSC must be high enough to maintain the natural state of observed samples but low enough to guarantee acceptable scattering of primary electrons. The signal of secondary electrons can be detected inside the SSC, the backscattered electrons can be detected although outside the SSC. In this experiment the ionisation detector with electrode system situated inside the SSC was used.

The sample, the paste for cover of lead-acid storage battery electrode (mainly a mixture of small particles of Pb, PbO, $PbSO_4$) saturated by H_2SO_4, was observed in water vapour environment, see Figure 2. Experimental conditions were adjusted to keep liquid phase in the water inside of SSC. It was approximately 2300 Pa in the specimen chamber of the microscope TESLA BS 343. Accelerating voltage was 15 kV, gas path length was approximately 4 mm and ionisation potential applied to the detection electrode of the ionisation detector was 300 V. This old microscope is equipped with

M. Luysberg, K. Tillmann, T. Weirich (Eds.): EMC 2008, Vol. 1: Instrumentation and Methods, pp. 589–590, DOI: 10.1007/978-3-540-85156-1_295, © Springer-Verlag Berlin Heidelberg 2008

590

tungsten cathode and hydration system, which guarantee high humidity environment inside the specimen chamber of VP-SEM.

1. G. D. Danilatos, Foundations of Environmental Scanning Electron Microscopy, Sydney, Academic Press, (1988), p. 249.
2. V. Neděla, Methods for additive hydration allowing observation of fully hydrated state of wet samples in environmental SEM. Microscopy research and technique vol. **70** no. 2 (2007), p. 95.
3. This work was supported by the Academy of Sciences of the Czech Republic, Grant No. KJB 200650602 and by grant No. MSM 0021630516. I would like to thanks to Ing. J. Špinka for his help during experiments and to Doc. M. Calábek for his help to preparation of the sample, from Dept. of Electrotechnology FEEC BUT, Czech Republic.

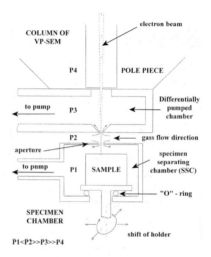

Figure 1. Configuration of the apparatus for the study of aggressive samples.

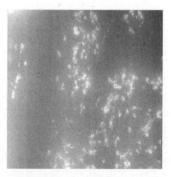

Figure 2. Sample of the paste for cover of lead-acid storage battery electrode (mainly a mixture of small particles of Pb, PbO, PbSO4) saturated by H2SO4. Accelerating voltage 15 kV, gas path length 4 mm, ionisation detector 300 V, water vapour environment, field of view 50 μm.

Characterization of the focusing properties of polycapillary X-ray lenses in the scanning electron microscope

J. Nissen[1], D. Berger[1], B. Kanngießer[2], I. Mantouvalou[2] and T. Wolff[2]

1. Zentraleinrichtung Elektronenmikroskopie , Technische Universität Berlin, Strasse des 17. Juni 135, 10623 Berlin, Germany
2. Institut für Optik und Atomare Physik, Technische Universität Berlin, Hardenbergstr. 36, 10623 Berlin, Germany

joerg.nissen@tu-berlin.de
Keywords: polycapillary X-ray lenses, 3D Micro-XRF

Polycapillary X-ray lenses are composed of several thousands of hollow glass capillaries which transport X-rays by total reflection [1]. They have the ability to focus X-rays with spot diameters of only several tens of μm. One recent application is 3D micro X-ray fluorescence spectroscopy (3D Micro-XRF), in which two X-ray lenses are used in a confocal setup, one in the excitation channel and one in the detection channel. The foci of both lenses overlap and form a probing volume which can be moved through a sample. Knowing the exact characteristics of the lenses, quantitative element analysis in this well defined micro-volume is feasible.

Till now characterization measurements have only been performed in the case of the lens focusing parallel radiation. Assumptions have been made, that in the other case, when isotropic radiation from a spot source is transported to a detector, the intensity distribution in the focal plane is Gaussian, as well [2]. In this paper this assumption is confirmed by measurements using a X-ray spot source generated by an electron beam.

For this purpose a polycapillary lens was installed in front of a Si(Li)-EDX-detector in a scanning electron microscope (Hitachi S-2700), Figure 1. As reference sample a multielement glass standard (NIST 1412) was used.

By adjustment of working and detector distances, the sample is positioned in the optimum focus of the lens. By scanning the electron beam over the sample, quantitative element mappings for different X-ray energies are recorded. Because the sample is homogenous, the mappings are identical to the 2D-intensity profiles of the fluorescence radiation accepted by the X-ray lens. From these profiles the FWHM of the focus at different X-ray energies is derived. The FWHM decreases with increasing X-ray energy because of the decreasing angle of total reflection, Figure 2.

Using this experimental setup, the transmission of the X-ray lens can also be determined. Figure 3 shows the ratio of the X-ray intensity integrated over the mappings with and without lens which is maximized for about 7.5 keV-X-rays. The gain is the corresponding ratio of the X-ray intensities in the middle of the focus spot.

1. IFG Institute for Scientific Instruments www.ifg-adlershof.de
2. W. Malzer, B. Kanngießer, Spectrochim. Acta, Part B 2005, 60, 1334-1341.

M. Luysberg, K. Tillmann, T. Weirich (Eds.): EMC 2008, Vol. 1: Instrumentation and Methods, pp. 591–592, DOI: 10.1007/978-3-540-85156-1_296, © Springer-Verlag Berlin Heidelberg 2008

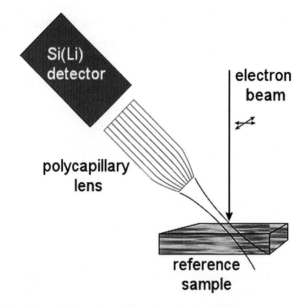

Figure 1. Position of the polycapillary lens in the SEM

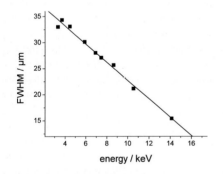

Figure 2. FWHM of the X-ray lens as function of energy

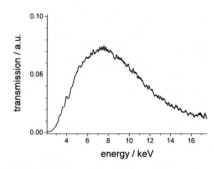

Figure 3. Transmission of the X-ray lens as function of energy

Numerical Simulation of Signal Transfer in Scintillator-Photomultiplier Detector

L. Novák

Institute of Scientific Instruments, v.v.i., AS CR, Královopolská 147, Brno, CZ

lnovak@isibrno.cz
Keywords: scintillation detector, single electron response, numerical simulation

The numerical simulation of the signal transfer in the scintillator-photomultiplier detector based on the statistics of single-electron responses was created.

The experimental data describing the characteristics of the detection channel were acquired by means of a modified detector of the Everhart-Thornley type [1] equipped with P47 or YAP scintillator, PMMA light guide and bi-alkali photocathode. The statistics of the number of photoelectrons emitted from the photocathode of the photomultiplier as well as the pulse height distribution of the photomultiplier output pulses were measured. The experimental data confirmed by the results of earlier studies [2], [3], [4] were consequently used as the input parameters for the simulation of the detection channel. The influence of the preamplifier and the A/D converter on the detected signal was taken into account as well.

The numerical simulation of the signal transfer in the detector is based on randomly generated numbers which are used in the inversion distribution functions describing the processes occurring in the detection channel. These numbers determine the amount of electrons leaving the specimen of the predefined line-scan (Fig. 1A). The number of detected electrons (Fig. 1B) is processed with the use of another set of random numbers. The described method is also used to determine the shape of the signal resulting from individual parts of the detector (Fig. 1C). Convolution with the function representing the time response of the analogue filter forms the output signal of the preamplifier (Fig. 1D and 1E). Different modes of signal processing such as the signal integration or the counting of pulses can be applied to the simulated signal. Digitalization of the signal in the A/D converter is computed according to a chosen rate of digitalization (Fig. 1F). Signal to noise ratio evaluated from the final signal can be used for comparison of different detector settings and different ways of signal processing.

1. T.E. Everhart, R.F.M. Thornley, J. Sci. Instr. **37** (1960), p. 246-248.
2. W. Baumann, L. Reimer, Scanning **4** (1981), p. 141-151.
3. C.W. Oatley, J. Microsc. **139** (1985), p. 153-166.
4. Zs. Kajcsos, W. Meisel, P. Griesbach, P. Gütlich, Ch. Sauer, R. Kurz, K. Hildebrand, R. Albrecht and M.A.C. Ligtenberg, Nucl. Instr. and Meth. in Phys. Res. B **93** (1994), p. 505-515.
5. Supported by: FEI Czech Republic, Ltd.

M. Luysberg, K. Tillmann, T. Weirich (Eds.): EMC 2008, Vol. 1: Instrumentation and Methods, pp. 593–594, DOI: 10.1007/978-3-540-85156-1_297, © Springer-Verlag Berlin Heidelberg 2008

594

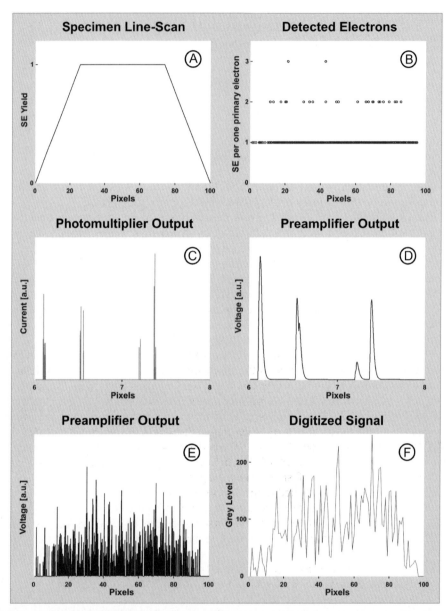

Figure 1. Numerical simulation of the detector signal (primary current 10 pA, dwell time 1µs): A – SE yield of the specimen, B – detected electrons originating from the specimen (40% collection efficiency), C – photomultiplier output signal (single-electron responses are visible in this detailed picture), D – preamplifier output signal (10 MHz bandwidth; detailed picture), E – preamplifier output signal, F – final signal after 8-bit digitization

In-situ Electrical Measurements on Nanostructures in a Scanning Electron Microscope

M. Noyong[1], K. Blech[1], F. Juillerat[2], H. Hofmann[2] and U. Simon[1]

1 RWTH Aachen, Institute of Inorganic Chemistry, Landoltweg 1, 52074 Aachen, Germany.
2 EPFL, Laboratoire de Technologie des Poudres (LTP), Lausanne, Switzerland.

mnoyong@ac.rwth-aachen.de

Keywords: gold nanoparticles, 1D assembly, SEM, nanomanipulation, nanoprobing, nanoprober, SEM-workbench

Scanning Electron Microscopy (SEM) is an essential tool for characterization of small objects down to the nanoscale, i. e. assembled nanoparticles, nanorods or nanostructured surfaces. These materials are important for future applications in the fields of nanooptics or nanoelectronics.

SEM closes the gap between on the one hand the "microworld" with techniques like light microscopy (LM) and on the other hand the "true atomic world" with methods like scanning probe microscopy (SPM) and transmission electron microscopy (TEM). Besides imaging, those methods allow further working directly on samples, which means either manipulation setups in light microscopy or manipulation by using scanning probe tips itself. But with limited resolution in LM or limited amount of probing tips in SPM manipulating and electrical addressing objects on the nanoscale is still a challenge with huge technical effort.

This contribution will show our versatile nanomanipulation setup, which is based on a commercial SEM. With its imaging abilities it allows characterization of sub 10 nm objects. Furthermore, a manipulation setup was added for addressing of nanostructured materials. [1] It has the ability to work with up to four manipulators. Those manipulators can be used independently from each other for manipulating or probing single objects and allows electrical *in-situ* measurements.

The functionality of this setup was shown by contacting 44 nm gold nanoparticles on siliconoxide/silicon wafers. [2] Those nanoparticles were arranged in one dimensional chains by dip coating of a previously structured substrate. [3] By using the nanoprobers different lengths of particle chains were addressed and electrically characterized.

1. M. Noyong, K. Blech, A. Rosenberger, V. Klocke and U. Simon, Meas. Sci. Technol. 18 (2007), p. N84
2. K. Blech, M. Noyong, F. Juillerat, H. Hofmann, U. Simon, J. Nanosci. Nanotech. 8 (2008), p. 461
3. F. Juillerat, H. Solak, P. Bowen and H. Hofmann, Nanotechnology 16 (2005), p. 1311

M. Luysberg, K. Tillmann, T. Weirich (Eds.): EMC 2008, Vol. 1: Instrumentation and Methods, pp. 595–596, DOI: 10.1007/978-3-540-85156-1_298, © Springer-Verlag Berlin Heidelberg 2008

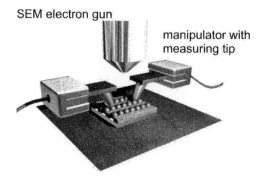

Figure 1. Graphical sketch illustrating the measuring setup of two nanomanipulators in SEM addressing a 1D arrangement of nanoparticles (not drawn to scale).

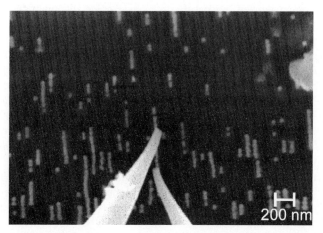

Figure 2. SEM image of gold nanoparticle arrangements in a previously structured substrate. A chain of 12 nanoparticles is addressed by two manipulation tips. [2]

3D Sculptures From SEM Images

R. Pintus[1], S. Podda[2] and M. Vanzi[2,3]

1. 3D Model Lab, Sardegna Ricerche, Loc Piscinamanna, Pula (CA) 09010, Italy
2. Telemicroscopy Lab, Sardegna Ricerche, Loc Piscinamanna, Pula (CA) 09010, Italy
3. University of Cagliari, Piazza D'Armi –DIEE Cagliari 09123 Italy

ruggero.pintus@diee.unica.it
Keywords: Scanning electron microscopy, photometric stereo, 3D reconstruction

In a general framework for exploiting the automation capabilities in modern SEM, two main branches has been investigated: the web-based remote microscopy [1] and the 3D metrology by Photometric Stereo (PS) . Focusing on this second topic, in a previous paper [2] an Automatic Alignment Procedure for a 4-Source Photometric Stereo (PS) technique was presented for metrically reconstructing the third dimension in the Scanning Electron Microscope (SEM). The method, developed on a ESEM Quanta FEI-200 platform [3] equipped by standard ETH and solid state annular BSE detectors, has been completely automated, and made easily implemented on any standard equipped SEM.

The 3D recovery process entails: first the acquisition of four BSE images of the sample, taken from the same viewpoint but under different lighting direction; second the computation of the surface gradient and finally the depth map is recovered by integration of the gradient vector field. For an exhaustive description of the methodology refer to [2].

A possibly marginal but fascinating fallout of the new technique is now the possibility to engrave real 3D sculptures of the microscopic objects imaged at the SEM. Moreover, the technique employed, the electron version of Photometric Stereo, preserves from the original optical domain one of its most peculiar features: the capability to separately store the pure 3D shape and the pure "colour" (the so called *albedo*) of the original object. This means that any 3D colour printer would be enabled to reproduce a SEM object complete of its BSE emission map, that mainly depends on compositional contrast. For our applications the Zcorp mod:Z450 3D printer, working with chalk, glue and colour, has been employed.

In fig. 1 two examples are shown. Fig. 1a and 1b are respectively the original BSE images of a detail of a Italian 1Euro coin (the Leonardo's Vitruvian Man face) and the detail of a damaged solid state GaN LED. Fig 1c and 1d are the corresponding 3D maps while 1e and 1f are the photo of the 3D printed object. The size of the model is 15x15cm. The result is that few micron samples have been magnified and made tangible.

Last, the construction of sculptures maybe helps to appreciate the peculiarities of PS as a 3D technique suitable for SEM: to return a 3D numerical model of a real object, complete of its colours. It is different from stereoscopy, that reconstructs the *perception* of the 3rd dimension under binocular view, and measures the depth upon identifying corresponding objects and making parallax computation. It is also different from

M. Luysberg, K. Tillmann, T. Weirich (Eds.): EMC 2008, Vol. 1: Instrumentation and Methods, pp. 597–598, DOI: 10.1007/978-3-540-85156-1_299, © Springer-Verlag Berlin Heidelberg 2008

598

tomography, that is technically and computationally quite heavy, and where the 3D object is then "painted" by applying some photographic image onto the reconstructed skeleton. The PS approach, enabling 3D printing from standard SEM imaging, may open a new "dimension" in enjoying the microscopic world.

1. F. Mighela, C. Perra: Proceedings of IMTC2006 *p.: 530-535;*
2. Pintus R., Podda S. and Vanzi M.: Accepted for publication in IEEE Transactions on Instrumentation and Measurement.
3. The Quanta 200 User's Operation Manual 3th edition 2002

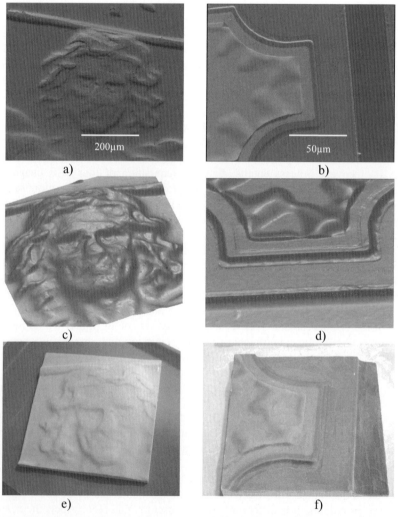

Figure 1. BSE image used for the 3D reconstruction a) detail of a coin; b) demaged solid state GaN LED; c) and d) 3D maps; e) and f) photographs of the 3D printed object

Influence of tilt of sample on axial beam properties

T. Radlička, B. Lencová

Institute of Scientific Instruments of the ASCR, v.v.i., Královopolská 147
612 64 Brno Czech Republic

radlicka@isibrno.cz

Keywords: parasitic aberration, misalignment aberrations

A sample in a vacuum chamber of a scanning electron microscope equipped with a cathode lens should be perpendicular to the axis because the energy of primary electrons is about ten eV. In real situations the tilt of the sample about one or two degrees is possible. The sample is an electrode and its tilt breaks the rotation symmetry of the system. It causes that parasitic field is present [1], which easily influences the properties of a low energy beam.

When the system is not rotationally symmetric, the field cannot be computed as a 2D problem and a general 3D problem must be solved. Fortunately, Sturrock showed [1] that in case of small deviations the method based on perturbation theory can be used and the field can be calculated in 2D. If the perturbation is described by a small parameter, in case of tilted sample by the angle of tilt α, $\delta z = r \cos\varphi \tan\alpha \approx \alpha r \cos\varphi$. The perturbation has the symmetry of a dipole that can be computed in 2D [2]. Assuming only perturbations linear in α, Sturrock's principle gets the perturbed potential in form $\Phi + \alpha r F_1 \cos\varphi$, where F_1 is the solution of the partial differential equation for a dipole [3] with boundary condition $F_1(r, z_s) = E_z(r, z_s)$ on the sample and zero boundary condition on the others electrodes.

The properties of the electron beam are influenced by the parasitic dipole field. Using aberration theory we can find for the axial beam

$$\delta x_i = \alpha D + \alpha C (3x_i'^2 + y_i'^2) + C_S(x_i'^2 + y_i'^2)x_i', \quad \delta y_i = -2\alpha C x_i' y_i' + C_S(x_i'^2 + y_i'^2)y_i' .$$

The misalignment aberration coefficients are given by [2]

$$D = -\frac{1}{2\sqrt{\phi_i}\,h_i'} \int_{z_o}^{z_i} \frac{f_1 h}{\sqrt{\phi}} dz, \qquad C = \frac{1}{\sqrt{\phi_i}\,h_i'^3} \int_{z_o}^{z_i} \frac{f_1 h}{\sqrt{\phi}} \left[\frac{3}{64}\left(\frac{\phi'}{\phi}\right)^2 h^2 + \frac{1}{8}h' + \frac{9}{16}hh'' \right] dz$$

where $h(z)$ is the imaging paraxial trajectory fulfilling in the object $h = 0$, $h' = 1$, ϕ is the potential on the axis, $f_1 = F_1(z, r = 0)$, and C_S is the spherical aberration coefficient in the image.

We have applied this procedure on an objective of a scanning electron microscope equipped with a cathode lens [4], see Figure 1. It consists from a unipotential electrostatic lens and a cathode lens. The electron energy was 10 keV in front of the objective, the energy of primary electrons was 10 eV on the sample. The potential on the central electrode is chosen so that the plane $z = -60$ mm is imaged on the sample. The calculations of the unperturbed field, the gradient of the field on the sample and the

M. Luysberg, K. Tillmann, T. Weirich (Eds.): EMC 2008, Vol. 1: Instrumentation and Methods, pp. 599–600, DOI: 10.1007/978-3-540-85156-1_300, © Springer-Verlag Berlin Heidelberg 2008

image properties of the system including the spherical aberration were done in the program EOD[5]. The calculations of perturbed field are in experimental stage now and they will be implement in EOD soon. The aberration coefficients D and C were calculated in MATLAB. Figure 2 shows the aberration figure in the Gaussian image plane, the beam is shifted by 14.5 μm in direction of axis x for the tilt of 1 degree.

1. P. Sturrock, Philos. Trans. Roy. Soc. London **243** (1951), p. 387
2. E. Munro, J. Vac. Sci. Technol. B **6** (1988) p. 941
3. P. W. Hawkes, E. Kasper, Principles of Electron Optics. Academic Press London (1989)
4. I. Vlček, B Lencová and M. Horáček, Proc. Microsc. Conf. Davos (2005), p. 59
5. B. Lencová and J. Zlámal, Microsc. Microanal. **13**, Suppl 3 (2007), p. 2
6. This work was supported by Grant Agency of AS CR, grant no. IAA100650805 and AS CR, grant no. AV0Z20650511

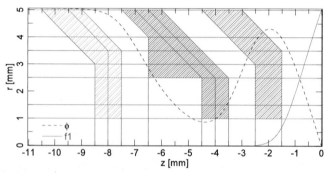

Figure 1. Unipotential objective lens with cathode lens, the maximum of the axial potential is 10 kV and the maximum of f_1 is 6106 V/m.

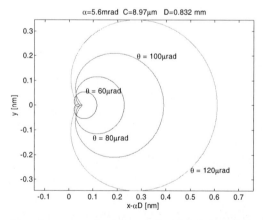

Figure 2. Intersection of rays coming from an axial point source placed in the plane $z = -60$ mm with the image plane, θ is the beam angle in the object.

Experimental determination of the total scattering cross section of water vapour and of the effective beam gas path length in a low vacuum scanning electron microscope.

J. Rattenberger[1], J. Wagner[1], H. Schröttner[1], S. Mitsche[1], M. Schaffer[1], A. Zankel[1]

1. Institute for Electron Microscopy, Graz University of Technology, Steyrergasse 17, A-8010 Graz, Austria

johannes.rattenberger@felmi-zfe.at

Keywords: ESEM, low vacuum scanning electron microscopy, total scattering cross section, skirt effect, beam gas path length

Low Vacuum Electron Microscopy enables the investigation of non-conductive samples without special preparation procedures. The imaging gas inside the specimen chamber is responsible for the contrast formation by gas amplification and the generated positive gas ions suppress charging of the sample. But the gaseous environment inside the chamber is limiting the capability of the microscope by elastic and inelastic collisions of the primary beam electrons (PEs) with the gas molecules. This so called skirt effect degrades the signal to noise ratio by generating gaseous secondary electrons (SEs) as well as SEs from regions far away from the focused probe. Therefore the primary beam loses exponentially electrons to a broadly dispersed skirt along the beam gas path length (BGPL) (1).

The main parameter which describes the scattering of the PEs is the total scattering cross section (σ [m^2]) of the imaging gas. This physical constant depends on the energy of the primary electrons and the type of the imaging gas (e.g. water vapour).

The investigations were performed on a FEI ESEM Quanta 600 under low vacuum conditions (p < 130Pa) using the large field detector (LFD). An external electrometer (Keithley 616) was used for probe current measurements. A new designed faraday cup was used, which shields the cup from scattered PEs, generated gaseous SEs and positive or negative gas ions (Figure 1).

With this breadboard construction, the average number of interaction per electron between the PEs and the gas molecules (m [a.u.]) can be calculated by using Formula 1. (Figure 2). The linear dependency of the average number of collisions per electron on the Working Distance (WD) was used to calculate σ, by determining the slope of the best fit straight line (k [1/m^2]) (Formula 2) (Figure 2). The total scattering cross sections for electron energies between 5 and 30 keV are shown in Figure 3.

The significant electron current loss starts in the region around the pressure limiting aperture (PLA) (2). The different gas density zones between the specimen and the electron column are depending on the microscope type and the chamber pressure (3). By using the equation of the best fit straight line it is possible to calculate the effective BGPL (Formula 3.). The differences between 10mm WD and the effective BGPL for low vacuum pressure conditions (20 keV primary electron energy) are shown in Figure 4. As was expected the effective BGPL increases with increasing chamber pressure.

M. Luysberg, K. Tillmann, T. Weirich (Eds.): EMC 2008, Vol. 1: Instrumentation and Methods, pp. 601–602, DOI: 10.1007/978-3-540-85156-1_301, © Springer-Verlag Berlin Heidelberg 2008

602

1. G. D. Danilatos, Scanning Microscopy, Vol. 4, No. 4 (1990), p. 799
2. G. Danilatos, Mikrochim Acta 114/115, (1994), p. 143
3. R. Gauvin et. al., Scanning Vol. 24, (2002), p. 171

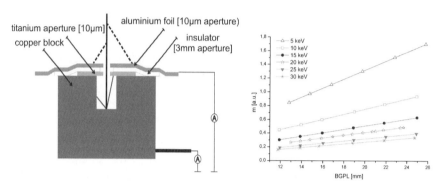

Figure 1. (left) Schematic drawing of the new designed Faraday Cup
Figure 2. (right) m [a.u.] vs. BGPL [mm] (70Pa water vapour)

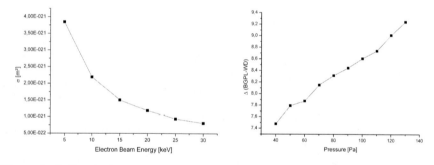

Figure 3. (left) σ [m²] vs. electron beam energy [keV]
Figure 4. (right) difference between BGPL and WD [mm] vs. chamber pressure [Pa] at 10mm WD

$$I = I_0 \cdot e^{-\frac{p}{k_B \cdot T} \cdot \sigma \cdot BGPL} = I_0 \cdot e^{-m} \qquad \text{Formula 1. (3)}$$

I fraction of unscattered beam current [A], I_0 beam current (high vacuum) [A], k_B Boltzmann constant [J/K], p pressure [Pa], T temperature [K]

$$\sigma = \frac{m \cdot k_B \cdot T}{p \cdot BGPL} = \frac{k \cdot k_B \cdot T}{p} \qquad \text{Formula 2.}$$

$$BGPL = \frac{m}{k} \qquad \text{Formula 3.}$$

Response function of the semiconductor detector of backscattered electrons in SEM

E.I. Rau[1], S.A. Ditsman[1], F.A. Luk'yanov[2], R.A. Sennov[2]

1. IPMT RAS, 142432, Chernogolovka, Moscow district, Russia
2. Department of Physics, Moscow State University, 119899, Moscow, Russia

rau@phys.msu.ru
Keywords: scanning electron microscopy, backscattered electron detectors

Practically all modern scanning electron microscopes (SEM) are equipped with semiconductor or scintillator detectors of backscattered electrons (BSE). Therefore it is important to know main characteristics of such detectors. The signal from BSE detector is a product of BSE current and average energy minus net energy reflected from the detector and with account for the net energy loss in the "dead" subsurface layer. The BSE detected signal I_β from, for example, semiconductor silicon detector is usually expressed by the following relation [1]:

$$I_\beta = I_o \eta d\Omega C \left(1 - \eta_{Si} \frac{\overline{E}_{Si}}{\overline{E}}\right) \cdot \frac{\overline{E} - E_{th}}{E_i} \qquad (1)$$

where I_0 is the electron beam current, η and $\overline{E}$ are the coefficient and average energy of BSE, respectively, $d\Omega$ is the BSE collection solid angle of the detector of a given geometry, η_{Si} and $\overline{E}_{Si}$ are the coefficient and energy of electrons backscattered from the silicon detector surface, E_{th} is the minimal threshold energy for the detector below which BSE make no contribution into the signal, E_i is the energy of generating the electron-hole pair in Si (E_i=3.65 eV), C is the detector efficiency equal to the tilt angle tangent of the dependence of the signal I_β on the incoming electrons energy E.

However, experiments show that expression (1) is valid for monoenergetic electron beam only, i.e. at $\overline{E} = E_0$ and at E_{th}= const. In fact, the BSE with various partial energies arrive to the detector, their current being determined by the whole energy spectrum of the electrons backscattered from the sample. Therefore E_{th} cannot be taken as a constant. The value of E_{th} is determined not only by the mass thickness of the protecting film but also by the particular number of BSE with their corresponding average energy within the given interval of the energy spectrum.

It is shown by the analysis that for an "ideal" silicon detector of BSE, in which there are no leakage currents and the efficiency of collecting the carriers at the p-n junction is 100% (C=1 in relation (1)), the value of the detected BSE signal can be expressed by the following semi-empirical relation:

$$I_\beta = I_0 \frac{E}{E_i} \eta d\Omega \cdot \left[0.8 + 0.2\exp\left(\frac{-4d}{R}\right)\right] \cdot \exp\left[-4.6\left(\frac{d}{R}\right)^2\right] \cdot \left(1 - \frac{d}{R}\right)^{0.6} \qquad (2)$$

M. Luysberg, K. Tillmann, T. Weirich (Eds.): EMC 2008, Vol. 1: Instrumentation and Methods, pp. 603–604, DOI: 10.1007/978-3-540-85156-1_302, © Springer-Verlag Berlin Heidelberg 2008

where d is the thickness of the passivated oxide film of the silicon detector (dead layer), R is the penetration depth of the detected electrons in oxide bulk materials. The expression (2) for response function of BSE detector is deduced by taking into account energy losses by incident electrons in the dead layer [2,3] and energy removed by BSE on silicon wafer. For the monoenergetic electron beam in (2) one should take $E = E_0$,

$$R_{SiO_2}[nm] = R_0 = 28.5 \cdot E_0^{1.67} keV \text{ [4], and in the case of detecting BSE, one has in}$$

(2) the following: $R = \overline{R} = 0.462 R_0$, $E = \overline{E} = 1.08 E_0 \cdot (1 - Z^{-0.3})$, where Z is atomic number of the target material. Experiments are in good agreement with the proposed semi-empirical model. Some experimental results concerning response characteristics of Si BSE detector is presented in Fig. 1-2.

1. L. Reimer. Image Formation in Low-Voltage Scanning Electron Microscopy. SPIE Press. Washington: 1993. p. 33
2. H.-J. Fitting. Phys. Stat. Sol. (a). V. 26 (1974) p. 525.
3. H. Seiler. J. Appl. Phys. V. 54 (1983) p. R1.
4. K. Kanaya, S. Okayama. J. Phys. D: Appl. Phys. V. 5 (1972). p. 43.

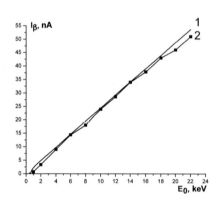

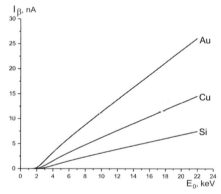

Figure 1. Si detector responsivity from 1-25keV monoenergetic electrons: calculated curve (1) at I_0=0.01 nA and experimental data (curve 2).

Figure 2. Calculated response characteristics of BSE detector for bulk Au, Cu, Si targets.

Main principles of microtomography using backscattered electrons

E.I. Rau

IPMT RAS, 142432, Chernogolovka, Moscow district, Russia

rau@phys.msu.ru

Keywords: scanning electron microscopy, backscattered electrons, microtomography

Up till now the main method for visualizing subsurface microstructures has been to vary the accelerating voltage of the scanning electron microscope (SEM) and to detect the backscattered electrons (BSE) [1]. But when this technique is applied the depth layers cannot be separated distinctly, because the images are overlapped by the background of upper and lower layers of the 3D structure.

The layer-by-layer visualization of the higher quality is provided by BSE-microtomography, when the images are produced with energy filtration of BSE, i.e. the images are formed within definite intervals of BSE energies [2, 3]. Figure 1 shows how electrons are selected according to their energies and depths. The curve A represents schematically BSE energy spectrum dN/dE from a bulk homogeneous sample, the curve B shows the BSE energy spectrum from a layered structure, when the film of the Δx thickness is located at the depth t inside the sample matrix. The dependences of the number of electrons dN scattered at a certain depth inside the dx layer are represented in another plane: dN/dx from the bulk matrix (curve A') and from the inhomogeneous structure (curve B').

Consider the spectrometer energy window ΔE which corresponds to the BSE electrons in the energy interval ΔE backscattered from the layer Δx at the depth t. This electron packet has the index 1 in figure 1. Its position on the energy axis E/E_0 is chosen by the spectrum inhomogeneity over the section B. It can be assumed that this spectrum anomaly corresponds to the interface (heteroboundary) between two layers at the depth t. The number of electrons emerged from this depth is determined by the curve 1'. In the ideal case, if the spectrometer had been tuned at the given energy window, a clear image of the deep layer should have been produced. However, it can be seen in the figure that a certain portion of BSE having the same energy but backscattered from the matrix layer and from a different depth (electron packet 3), generating thus the accompanying background (noise) in the image of the inhomogeneous layer Δx in the homogeneous matrix, which is a drawback of earlier experiments on BSE-tomography. BSE of different energies backscattered from the matrix material and coming up from the same depth (denoted by index 2) do not get into the energy window, i.e. they are not detected and make no contribution into the signal. Naturally, with the standard technique, when the integral number of BSE emerging from all the depths and having the whole spectrum of energies are detected, the discrimination of layers and the contrast of the layer images are much worse, which can be seen from the integrals of the curves A' and B' represented in the plane (dN/dx)–(x/R) in the same figure. In this case the variation of

M. Luysberg, K. Tillmann, T. Weirich (Eds.): EMC 2008, Vol. 1: Instrumentation and Methods, pp. 605–606, DOI: 10.1007/978-3-540-85156-1_303, © Springer-Verlag Berlin Heidelberg 2008

SEM accelerating voltage makes it possible to see deep layers only against the background of all the upper layers. In addition, the curve $E_{BSE} = f(X_{BSE})$ that schematically shows the dependence of BSE energy on the depth x from which BSE emerge is represented in figure 1 in the plane $(E/E_0) - (x/R_0)$.

However, this technique also has some limitations that prevent us from obtaining clear images of a certain subsurface layer of the structure [4]. In order to improve BSE-microtomography method, in the present paper we suggest that the procedure of obtaining clear layer-by-layer images of the subsurface microstructures should be supplemented by the following operations.

It is known that by detecting the electrons scattered and emerging at various definite angles their energy spectrum is deformed (see, e.g. [5]). As the detection angle is increased the spectrum maximum corresponding to the most probable BSE energy is shifted toward the higher energy region. In this case each detection angle corresponds to a particular most probable escape depth of BSE [6]. Thus, high-quality BSE-microtomography requires the following algorithm of operations to be performed: 1. By selection of primary electrons energy E_0 is found and it pre-sets the most probably depth of BSE that is approximately equal to the depth of the microstructure layer of interest. 2. In accordance with this depth, BSE detection angle is determined, and the BSE spectrometer is set at this angle. 3. The choice of the appropriate position of the energy window ΔE in the spectrometer sets the ultimate parameter of the experiment in order to obtain the image of the microinhomogeneity at the given depth under the surface.

If the sample composition is unknown, the three mentioned parameters are found by scanning them in succession during experiments.

1. H. Seiler. Scanning Electron Microscopy (IITRI, Chicago). V.1 (1976) p. 9.
2. E.I. Rau, V.N.E. Robinson. Scanning. V. 18 (1996) p. 556.
3. H. Niedrig, E.I. Rau. Nuclear Instruments and Methods. V. B142 (1998) p. 523.
4. M. Yasuda, Y. Suzuki, H. Kawata, Y. Hirai. Jap. J. Appl. Phys. V. 44 (2005) p. 5515.
5. P. Gerard, J. Balladore, J. Martinez, A. Ouabbou. Scanning. V. 17 (1995) p. 377.
6. M. Yasuda, H. Kawata, K. Murata. J. Appl. Phys. V. 77 (1995) p. 4706.

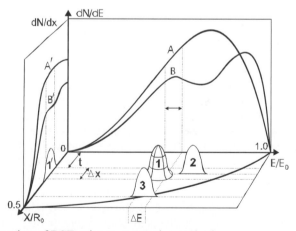

Figure.1. Explanation of BSE-microtomography method.

Considerations of some charging effects on dielectrics by electron beam irradiation

E.I. Rau[1], E.N. Evstaf'eva[2], R.A. Sennov[2] and E. Plies[3]

1. IPMT RAS, 142432, Chernogolovka, Moscow District, Russia
2. Department of Physics, Moscow State University, 119899 Moscow, Russia
3. University of Tuebingen, Auf der Morgenstelle 10, D-72076 Tuebingen, Germany

rau@phys.msu.ru
Keywords: scanning electron microscopy, charging, dielectrics

Some aspects of charging effects on insulator materials are presented. Based on experimental results and the model of two opposite charged layers, an analysis of the time evolution of the electron yield curves $\sigma = f(E_0, t)$, accumulated charges $Q(t)$ and surface potential $V_S(t)$ of insulators during irradiation is given. E_0 is the initial energy of the primary electrons (PE) which are retarded to a lower landing energy by the negative surface potential. The results obtained in our experiments in SEM are consistent with the model for the formation of the double layer of charges under irradiation: one positively charged layer as thick as the depth from which the secondary electrons (SE) escape, and one negatively charged layer whose thickness is determined by the penetration depth of the PE [1,2,3,4].

Determining, comparing and analyzing several characteristics at the same time aid to understand the controversial topic on time constants of charging the dielectric targets and the reasons for discrepancies between experimental and some theoretical estimates [5,6]. It is shown that the real time taken to establish the equilibrium state of charging is two - three orders greater than the value calculated using the theory of the dependence of the SE emission coefficient on the energy and density of the irradiating electron current. Moreover, the time constant of charging the dielectrics cannot be judged from the SE emission current kinetics, since this characteristic is not always in accordance with the time taken both by the surface potential to grow up to the established quasi-static equilibrium state and by accumulation of charges. The extent of difference in behaviour of characteristics under discussion is determined by the target material and the energy of electrons.

The work presents also the observed anomalous effect of splitting the peak of the response of the SE distribution in energy, whose shift determines the value of the high-voltage charging potential of the target. Figure 1 shows two typical energy spectra of polycrystalline Al_2O_3. The measurements were carried out at an PE energy $E_0 = 6$ keV, a probe current $I_0 = 2n$ A and a TV-scan area of 50×50 μm^2 using the annular electro-static toroidal electron spectrometer [4]. The spectra (1) and (2) were recorded 20 s and 40 s respectively after the beginning of the irradiation. These spectra are influenced by local field effects of the charged specimen and the response function of the used spectrometer detector unit (resolution $\Delta E = const \times E$). Nevertheless the SE peak of

M. Luysberg, K. Tillmann, T. Weirich (Eds.): EMC 2008, Vol. 1: Instrumentation and Methods, pp. 607–608, DOI: 10.1007/978-3-540-85156-1_304, © Springer-Verlag Berlin Heidelberg 2008

608

spectrum (1) clearly indicates that the surface potential of the specimen is $V_S \approx -1.9$ kV after 20 s.

The splitting effect of spectrum (2) can be explained by nonuniform distributions of charges and potentials on the irradiated area of the dielectric. When the nominal scan area of 50×50 μm^2 is already negatively charged after some area scans and a certain time the PE are deflected in the local field directly above this charged surface, thereby increasing the irradiated area. But the charging in this fringe area is still weaker after 40 s. Hence the retarding of the PE and the acceleration of the SE are smaller there. Therefore the low-energy peak of spectrum (2) is generated by SE from the fringe area. The non-vertical landing in this area may also contribute to the weaker charging in this region due to the well-known higher SE yield for oblique incidence. The consequences of this charging mechanism are: the lateral nonuniformity in the distribution of the potential and electric field inside the irradiated target volume, the origin of electron-induced conductivity, drift and diffusion of the charges, and finally, the quasi-static equilibrium state of charging the sample takes a longer time to be established.

1. J. Cazaux, J. Appl. Phys. **85** (1999), p. 1137.
2. J. Cazaux, Scanning **26** (2004), p. 181.
3. A. Melchinger and S. Hofmann, J. Appl. Phys. **78** (1995), p. 6224.
4. E. Rau, S. Fakhfakh, M. Andrianov, E. Evstaf'eva, O. Jbara, S. Rondot and D. Mouze, Nucl. Instr. Meth. Phys. Res. **B** (2008), in press.
5. X. Meyza, D. Goeuriot, C. Guerret-Piécourt, D. Tréheux and H.-J. Fitting, J. Appl. Phys. **94** (2003), p. 5384.
6. L.Frank, M.Zadrazil, I.Mullerova. Scanning, 23 (2001). P.36.

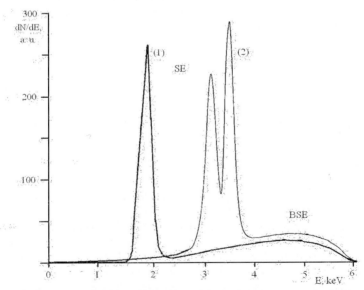

Figure 1. Spectra of secondary electrons (SE) and backscattered electrons (BSE) of polycristalline Al_2O_3. Primary electron energy $E_0 = 6$ keV. For more details see text.

The reduction of pileup effects in spectra collected with silicon drift detectors

T. Elam, R. Anderhalt, A. Sandborg, J. Nicolosi, and D. Redfern

EDAX Inc., 91 McKee Drive, Mahwah, NJ, 07430

Del.redfern@ametek.com

Keywords: EDS, silicon drift detectors, spectrum artefacts

Pileup artifacts appear in energy-dispersive X-ray spectra at high count rates when X-rays arrive at the detector with time separations less than the resolving time of the pulse processor. These artifacts often appear as extra peaks in the spectrum and can mask (or be mistaken for) weak peaks of trace elements. Hardware detection and rejection of multiple x-ray events is most effective at longer processing times. The usage of the silicon drift detector with its high throughput has made it necessary to use faster processing times which have not been as effective at reducing the pileup artifact in the spectrum. Algorithms have been used to reduce the pileup artifacts (sum peaks and elevated background levels) in different fields of analysis [1; 2] and the same approach is used here in EDS.

Recent improvements in high-speed digital discrimination have improved rejection of near-simultaneous events leading to pile-up at very high count rates. This new capability also improves the predictability of pileup rejection, which is essential for accurate modeling and reliable removal of the inevitable events that get past the pileup inspection. We have successfully reduced pileup artifacts by a combination of hardware changes and software correction.

A spectrum collected at a high count rate and high deadtime clearly has more pileup artifacts than a spectrum collected with a lower count rate (Figure 1). Although hardware detection and rejection of pileup events has been optimized it is not possible to prevent pileup artifacts from being displayed in the spectrum. The spectrum can be processed after collection to further remove the effects of pileup as shown in figure 2. A possible trace element (K K-alpha) is obscured by the pileup artifacts and is only visible in the processed spectrum.

Software post-processing can remove the events that cannot be rejected by hardware, provided the correction is based on a realistic model of the hardware response. A combination of hardware optimized for both rejection and predictability, together with a minimum of software correction, is the ideal solution.

1. R. P. Gardner and L. Wielopolski, Nuclear Instruments and Methods, **Vol. 140**, pp289-296, 1977.
2. Q. Bristow, R. G. Harrison, "Theoretical and Experimental Investigations of the Distortion in Radiation Spectra Caused by Pulse Pileup", Nuclear Geophysics, **Vol. 5**, No.1/2, pp. 141-186, 1991.

M. Luysberg, K. Tillmann, T. Weirich (Eds.): EMC 2008, Vol. 1: Instrumentation and Methods, pp. 609–610, DOI: 10.1007/978-3-540-85156-1_305, © Springer-Verlag Berlin Heidelberg 2008

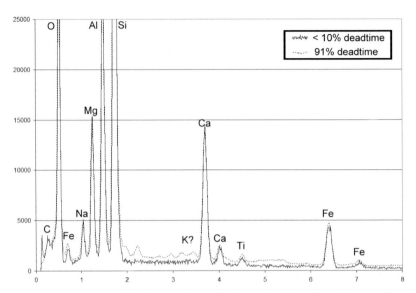

Figure 1. Spectra of a basaltic glass collected at low deadtime (no pileup artifacts) and at a very high deadtime which shows many pileup effects. Unlabeled peaks in the high deadtime spectrum are sum peaks (including, at a minimum: Al+O, Si+O, Al+Al, Al+Si, Si+Si).

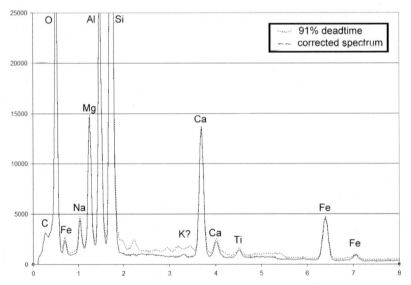

Figure 2. Spectra of the same sample showing the high deadtime spectrum with a corrected spectrum derived from post-processing of the same spectrum. The processed spectrum contains fewer artifacts and also shows a small K peak. The sample is reported to have 0.25 weight % K_2O.

High-temperature oxidation of steel in the ESEM with subsequent scale characterisation by Raman microscopy

A. Reichmann[1], P. Poelt[1], C. Brandl[1], B. Chernev[1] and P. Wilhelm[1]

1. Institute for Electron Microscopy, Graz University of Technology, Steyrergasse 17, A-8010 Graz, Austria

angelika.reichmann@felmi-zfe.at

Keywords: hot corrosion, ESEM, Raman microscopy

Stainless steels used as materials for technical processes at elevated temperatures are protected against high temperature corrosion by the formation of a Cr-rich oxide layer [1]. It is very important that this protective scale is developed very fast at the beginning of the technical process, so that a chemical attack of the steel, which is exposed to different gas atmospheres, is prevented [2].

The ESEM (Environmental Scanning Electron Microscope) equipped with a heating stage enables the direct observation of changes of the morphological structure of the specimen during heating with high magnification and high depth of focus. Moreover, the investigations can be carried out with different gases in the specimen chamber. These gases are used for amplifying the secondary electrons generated from the metal substrate, balancing the negative charge on the surface and they can react with the sample itself [3].

After the in situ investigation of the high temperature oxidation the composition of the oxide scales has to be characterised. Raman Microscopy is an excellent complementary technique to SEM/EDX to identify thin layers of oxide corrosion products on metal surfaces [4].

In our work we investigated the high temperature oxidation of the austenitic steel 353MA, used for heat exchanger tubes. Figure 1 shows the scale formation of the polished steel in the ESEM with a heating rate of 2 K/min till 973K in air at a pressure of 133Pa. At about 623 K the first grains appeared at the polished surface. They grew along lines due to scratches originating from the polishing procedure. At the end of the experiment the whole surface was covered by a dense oxide layer interrupted by the original grain boundaries, which got visible as deep groves. Figure 2 shows a BSE image of the created scale after the high temperature study. The oxidation products were analysed by Raman Microscopy and consist mainly of Cr_2O_3 and spinels. The structures, which formed lines, contain Fe_2O_3, too.

In conclusion we can state that the ESEM is a proper instrument for the investigation of the early stages of scale formation of stainless steels during high temperature oxidation. The identification of different phases could be improved by the development of a BSE detector for high temperature experiments.

1. P.Kofstad, High Temperature Corrosion (Ch.12) Elsevier, London (1988).
2. C. Ostwald and H.J. Grabke, Corrosion Science **46** (2004), 1113-1127

M. Luysberg, K. Tillmann, T. Weirich (Eds.): EMC 2008, Vol. 1: Instrumentation and Methods, pp. 611–612, DOI: 10.1007/978-3-540-85156-1_306, © Springer-Verlag Berlin Heidelberg 2008

612

3.   B.Schmid, N. Aas, Ø. Grong and R. Ødegård, Scanning **23** (2001), 255-266
4.   D.J. Gardiner, C.J. Littleton, K.M. Thomas and K.N. Strafford, Oxidation of Metals **27** Nos.1/2 (1987), 57-72

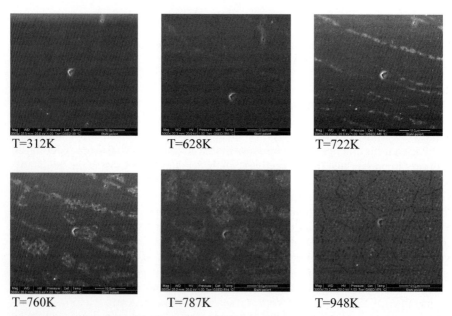

T=312K   T=628K   T=722K

T=760K   T=787K   T=948K

Figure 1. Scale formation during heating in dependence on the temperature

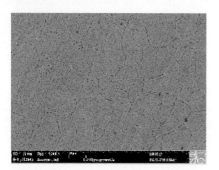

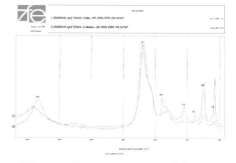

Figure 2. BSE image of the scale   **Figure 3.** Raman spectrum of the scale

Method to determine image sharpness and resolution in Scanning Electron Microscopy images

B.Rieger, G.N.A van Veen

1. FEI Company, Eindhoven, The Netherlands
2. Quantitative Imaging Group, Faculty of Applied Sciences, TU Delft, The Netherlands

bernd.rieger@fei.com

Keywords: SEM, image sharpness, apparent resolution

Image sharpness is very important for the perceived quality of any image. Under certain assumptions sharpness can be related to resolution. For SEM images it lies typically in the nanometer range. If image formation is simplified to a linear system with a point spread function (PSF) we can measure the system blur and relate it to the minimal resolvable distance using a (Rayleigh/Abbe or Sparrow) criterion.

A fully automatic, noise robust image processing algorithm is introduced to find, extract and measure suitable edges from SEM images. The extracted edges of blurred step edges are assumed to have the shape of error functions. A parametric fit to these edges reveals the standard deviation of the underlying Gaussian function, and thus is a measure for the sharpness, i.e. the PSF.

It needs to be stressed that the sharpness in images is measured, not the performance of the machine per se (spot size). However, the image sharpness and related resolution is probably the single most important feature for any microscopy user. Our simplified image formation and analysis model does not take into account the more general problem of resolution definition in images. This would require an object model, a statistical estimation process and most important a full understanding of the sample beam interactions and detector properties. In light microscopy where this is greatly simplified, there has been recent success to tackle this issue in the case of single molecule fluorescence imaging [1,2].

Our fully automated algorithm is based on the extraction of edge profiles and fitting error functions to them. The edges are found in regions of high contrast change. Edge profiles are extracted perpendicular to the found edge line. An error function fit returns the only relevant fitting parameter, the Gaussian σ. A typical image after edge detection and profile extraction is shown in Figure 1. The selected edge profiles must represent a 'nice' edge profile (shown in green in Figure 1), which are a subset of all initially extracted edges (show in red). The algorithm returns the average σ fit to a number of edges (typically 500-1000). It can handle anisotropic pixel sizes in x and y scan directions and can compute the sharpness in different image orientations to detect astigmatism, compare Figure 2.

We tested the consistency and robustness of the algorithm on simulated SEM images [3] with addition of noise, vibrations, defocus, astigmatism and edge effect. For all distortions except the edge effect our method performed very well (within a few percent of the simulations). A higher sharpness is expected in this case as we measure the image

M. Luysberg, K. Tillmann, T. Weirich (Eds.): EMC 2008, Vol. 1: Instrumentation and Methods, pp. 613–614, DOI: 10.1007/978-3-540-85156-1_307, © Springer-Verlag Berlin Heidelberg 2008

sharpness and the edge effect gives rise to a steeper slope. Tests on real SEM images showed also expected values.

1. R.J. Ober, S. Ram and S.E. Ward, Biophysical Journal, **86**(2):1185-1200, 2004.
2. R.J. Ober, Z. Lin and Q. Zou, IEEE Transactions on Signal Processing, **51**(10):2679-2691, 2003.
3. We kindly acknowledge the helpful discussions with Pybe Faber, Eric Bosch, Bart Buijsse and Alexander Henstra from FEI Company, the Netherlands and Andras Vladra and Petr Cizmar from National Institute of Standards, USA for the access to many simulated test images.

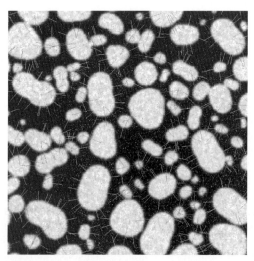

Figure 1. Part of a simulated SEM image. The red lines are rejected edge profiles and green profiles are suitable to measure the edge sharpness.

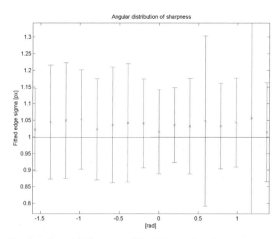

Figure 2. Angular distribution of the edge sharpness, simulated value 1.0.

Instrumentation of an electron microscope for lithography and analysis of devices over a wide dimensional range

G. Rosolen

CSIRO, Information and Communications Technologies Centre,
RadioPhysics Laboratory, Vimiera Road, Marsfield, NSW 2121, Australia

Grahame.Rosolen@csiro.au

Keywords: scanning electron microscope, electron beam lithography

The scanning electron microscope (SEM) is a versatile instrument for imaging structures with dimensions ranging from nanometres to millimetres. The use of scanned electron beams for lithography purposes is also an invaluable tool for fabricating devices. In particular electron beam lithography (EBL) affords a resolution not available with other pattern definition techniques.

A LEO 440 SEM has been modified to function as an electron beam lithography instrument whilst still maintaining the imaging capability within the one instrument. It has been used to both fabricate and image devices ranging in size from the sub-micron up to millimetre dimensions. The instrument comprises a purpose built pattern generator which is integrated with the sample stage and electron optics of the SEM. The electron optics column has been modified with the addition of an electrostatic beam blanking capability. Conventional 2D drafting packages such as AutoCAD are used to define the device patterns and custom software has been written to convert CAD output format into the lithography data files required for patterning.

Direct write electron beam lithography is typically used to fabricate devices with dimensions in the sub-micron range because it offers this fine linewidth capability. Although electron beam lithography has the advantage of high resolution compared with optical lithography it has a low throughput because it is a serial process as the entire pattern must be exposed sequentially. Optical lithography is a parallel exposure technique and offers the advantage of the ability to expose entire patterns at once. However, optical lithography requires the additional fabrication of a mask for defining the patterns to be exposed. This mask is often fabricated using electron beam lithography.

In situations where the optimum topology of the devices to be fabricated is unknown it can be very time consuming and expensive to cycle through the iterative loop of defining the device topology, having optical masks made and then patterning the devices using conventional optical lithography. The masks may also quickly become obsolete as the improvements are made to the device topology. In situations where fine linewidth is also required this process may involve a combination of optical lithography and direct write EBL [1].

The instrument has been successfully used to fabricate devices ranging in size from the sub-micron up to millimetre dimensions, where all of the device geometry is exposed with the electron beam. As a consequence the time taken from a design idea

M. Luysberg, K. Tillmann, T. Weirich (Eds.): EMC 2008, Vol. 1: Instrumentation and Methods,
pp. 615–616, DOI: 10.1007/978-3-540-85156-1_308, © Springer-Verlag Berlin Heidelberg 2008

through to a test device is minimised because the entire device geometry is written by the electron beam. If the design is faulty or not optimal no mask is wasted because the design is coded as a data file in the lithography instrument rather than a physical mask which needs to be fabricated. This approach is best suited to low volume research and development fabrication runs where only a limited number of test devices are required for experimentation purposes.

A further advantage of combining the patterning and visual analysis in the one instrument is the ability to image the fabricated devices with the electron microscope as it affords high resolution, a wide magnification range and wide depth of focus. However, another important advantage of using the same instrument is that the same lithography data files that were used to spatially position the devices on the sample may also be used to locate these devices again for detailed analysis. This saves considerable time spent navigating around a sample trying to locate the devices of interest. It also removes any uncertainty regarding which device is being analysed. The flexibility of the instrument images is demonstrated by the images of two devices fabricated with the instrument as shown in Figure 1.

These devices have widely varying sizes and linewidth requirements. The image in Figure 1a) is of a series of diffraction gratings. The sample area is microscopic and sub-micron resolution is required. In contrast the image in Figure 1b) is one element from a set of 16 Hadamard function masks. Each mask covers an area of 1 mm x 1mm and the pattern areas are defined to an accuracy of 2μm.

1. G.C. Rosolen, "Optically Variable Devices Fabricated by Electron Beam Lithography", SPIE Vol **527**, 2004, pp.318-323.

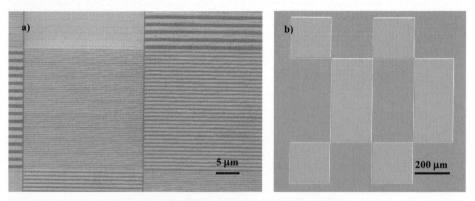

Figure 1.a) Image of a series of diffraction gratings with feature sizes from sub-micron to several microns. b) Image of a Hadamard function mask pattern 1 mm x 1 mm.

Ultra-low energy, high-resolution scanning electron microscopy

L.Y. Roussel[1], D.J. Stokes[1], R.J. Young[2] and I. Gestmann[1]

1. FEI Company, Building AAE, PO Box 80066, 5600 KA Eindhoven, The Netherlands
2. FEI Company, 5350 NE Dawson Creek Drive, Hillsboro, OR 97124, USA

Laurent.Roussel@fei.com

Keywords: high-resolution, low energy, SEM

Traditionally, the use of high primary beam energies in the SEM (up to 30 keV) helps to minimise the diameter of the primary electron beam, and the best 'resolution' of a microscope tends to be specified on this basis.

At these higher energies, transmission techniques such as STEM-in-SEM can be used for high-resolution, high-contrast imaging of thinned specimens. For bulk materials, the rather high penetration range of primary electrons leads to a large interaction volume within the specimen. Given that backscattered electrons are generated from large depths and across the width of the interaction volume, and that these give rise to type-II secondary electrons, a significant loss of spatial resolution can result, especially for soft materials. Furthermore, the electron signal from thin layers or small features can be overwhelmed by the signal from the underlying bulk material, making their visualisation difficult or impossible. Hence it has long been appreciated that, to observe fine surface structures of bulk materials, imaging at low energies is an essential requirement [1, 2]. This has the advantage of enabling electrically insulating materials to be imaged without a conductive coating. Usually, low energy (or low voltage) SEM involves literally controlling the voltage applied to primary electrons leaving the source. But low energy electrons are more susceptible to lens aberrations and scattering in the column, and so there is a loss of resolution associated with this. Nonetheless, the gain in useful information is the key factor, and improvements in electron optics and column design continue to make this less of an issue.

An alternative strategy is to apply a negative bias to the specimen, to decelerate primary electrons on arrival at the specimen surface [3]. Thus the initial primary beam voltage can remain high, with the advantage that the beam remains tightly focused, but the landing energy is significantly reduced, hence reducing the penetration depth of the primary electrons. This is an equivalent situation to using low beam energy, preserving surface sensitivity, but without the associated loss of resolution.

Contrast mechanisms and other imaging aspects differ at low landing energies with and without the application of a stage bias. New technology advances relative to low landing-energy electrons will also be presented. Figure 1 shows the result of decelerating a 4.05 keV primary electron beam to just 50 eV landing energy E_L by applying a bias $Vs = 4.00$ keV to the specimen. Figure 2 shows a highly surface-sensitive, ultra-high resolution image of part of an electronic circuit, the fine features of which were not visible without the use of the beam deceleration technique (primary

M. Luysberg, K. Tillmann, T. Weirich (Eds.): EMC 2008, Vol. 1: Instrumentation and Methods, pp. 617–618, DOI: 10.1007/978-3-540-85156-1_309, © Springer-Verlag Berlin Heidelberg 2008

618

electron landing energy $E_L = 1$ keV). Figs 3 (a) and (b) show the same area of an etched steel specimen, demonstrating different information content using secondary electron and backscattered. Primary beam energy $E_0 = 5$ keV, negative specimen bias $V_s = 2$ kV, to give landing energy $E_L = 3$ keV.

1. E.D. Boyes, Microscopy and Microanalysis, 2000. **6**(4): p. 307-316.
2. J. Goldstein et al, Scanning Electron Microscopy and X-Ray Microanalysis.Third Edition. 2003, New York: Plenum.
3. Mullerova and L. Frank, Scanning, 1993. **15**: p. 193-201.
4. Grateful thanks to colleagues and to Freescale for permission to use Fig. 3.

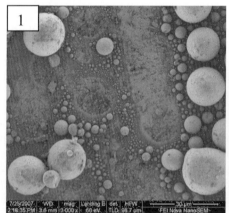

Figure 1. Image of tin balls using primary electron landing energy $E_L = 50$ eV

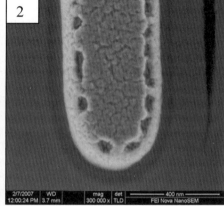

Figure 2. UHR image of a de-passivated integrated circuit imaged, $E_L = 1$ keV, revealing fine surface detail

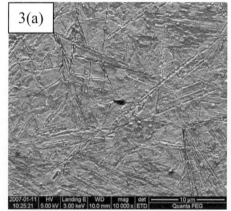

Figure 3. (a) Secondary electron and (b) backscattered electron images of etched steel, primary beam landing energy $E_L = 3$ keV

Non-destructive 3D imaging of the objects internal microstructure by microCT attachment for SEM

A. Sasov

SkyScan, Kartuizersweg 3B, 2550 Kontich, Belgium

sasov@skyscan.be

Keywords: SEM, microtomography, microCT, 3D reconstruction

The standard SEM image displays the surface of the object as a two-dimensional flat picture. An inexpensive microCT attachment for any SEM has been developed to add new imaging modality for visualization and measurement of the true 3D microstructure inside the object without any physical cut or sample preparation [1,2].

Fig.1 (left) shows block-diagram of the microCT attachment for SEM. An objective lens of the SEM (1) focuses electron beam (2) on the surface of a metal target (3). The target produces X-ray radiation (4) which passes through the object (5) and collected by X-ray camera (6). To acquire different angular projections the sample mounted on the axis of precision rotation stage. X-ray magnification can be adjusted by changing distance between the target and the sample using integrated motorized linear stage. All parts of the system except of X-ray camera combined in one microscanner, which can be easily installed in the SEM instead of standard object stage (Fig.1, right). The X-ray camera uses cooled back-illuminated CCD for direct X-ray detection. The microCT attachment doesn't require any connection to SEM electronics and works independently.

During scanning a computer controls the object rotation and acquires a number of X-ray angular shadow projections of the object's internal microstructure. A special program back-projects in the computer memory all acquired images. The combined information corresponds to the complete internal 3D structure of the object obtained non-destructively. It can be displayed as virtual slices in any orientation or as a realistic three-dimensional visual model, which includes not only surface information (as in conventional SEM images), but also all internal object details. Specially developed software allows calculating numerical parameters of internal morphology in 2D and 3D.

Fig.2 shows application of microCT attachment for SEM for glass fibre in epoxy composite material. Standard SEM image in SE mode (Fig.2 left) allows to see only object surface. X-ray transmission image (Fig.2 middle) displays all glass fibres inside the sample. 3D reconstruction based on the number of X-ray transmission images allows creating virtual slices through the sample (Fig.2 right) and building complete 3D model of the internal microarchitecture with possibilities for virtual slicing in any direction. Fig.3 shows microCT reconstruction from the ceramic sample. Note that standard SEM image (Fig.3 left) not allows to get right estimation of porosity, but using number of X-ray views (Fig.3 middle) microCT attachment can reconstruct thin planar virtual slices (Fig.3 right) and calculate true internal morphological parameters.

The developed microCT attachment for SEM opens new possibility for non-destructive 3D imaging in wide range of applications using any conventional SEM [3].

M. Luysberg, K. Tillmann, T. Weirich (Eds.): EMC 2008, Vol. 1: Instrumentation and Methods, pp. 619–620, DOI: 10.1007/978-3-540-85156-1_310, © Springer-Verlag Berlin Heidelberg 2008

620

1. J. Cazaux et al. Journal de Physique IV **03**, C7 (1993) C7-2099-C7-2104
2. A Sasov, Journal of Microscopy, vol.147, pt.2 (1987), pp.169-192.
3. http://www.skyscan.be/products/SEM_microCT.htm

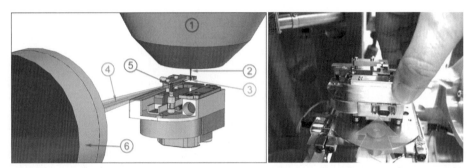

Figure 1. Left - block-diagram of the microCT attachment for SEM (explanations - in text) ; right - installation of microscanner on the object manipulator of JEOL JSM7000F

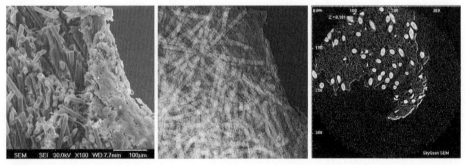

Figure 2. MicroCT in SEM application example. Sample - glass fibres in epoxy matrix composite material. Left - SEM image in SE mode, middle - X-ray image through the sample, right - one of reconstructed virtual slices. Pixel size in X-ray images is 705nm.

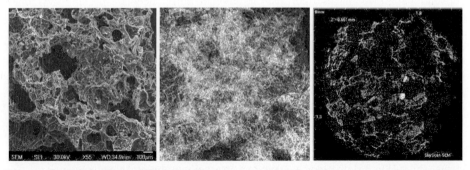

Figure 3. MicroCT in SEM application example. Sample - porous heat-resistant ceramic material. Left - SEM image in SE mode, middle - X-ray image through the sample, right - one of reconstructed virtual slices. Pixel size in X-ray images is 2.8μm.

A novel use of rf-GD sputtering for sample surface preparation for SEM: its impact on surface analysis

K. Shimizu, T. Mitani, P. Chapon

1. University Chemical Laboratory, Keio University, 4-1-1 Hiyoshi, Yokohama , Japan
2. Center for Testing and Analysis, Keio University, 3-14-1 Hiyoshi, Japan
3. Jobin-yvon Horiba, 16-18 rue du Canal, Longjimeau Cedex, France

shimizuk@econ.keio.ac.jp

Keywords: SEM, sample preparation, rf-GD sputtering

Through recent advances in ultra-high resolution FE-SEM with novel electron optics and multi-detection systems [1], where SEs, high angle BSEs and channeling BSEs are detected separately, the world of practical surface analysis may be changed drastically.

In order to realize the enormous potentials of such ultra-high resolution FE-SEMs, however, sample surface preparation is of key importance; preparation of clean and "undamaged" or "nearly damage-free" surfaces is essential. Here, it is demonstrated that the surfaces of required quality can be prepared readily, quickly and routinely by rf-GD sputtering where both conductive and non-conductive surfaces are sputtered very stably with Ar^+ ions of very low energies, <50 eV, and high current density of ~100 mAcm^{-2}. Due to the very low energies of Ar^+ ions, samples are sputtered without "significant formation of altered surface layers". While the high current density of Ar^+ ions ensures sputtering to proceed at very high rates, $1 \sim 10$ μm min^{-1}, making sample treatment time extremely short, normally less than 1 min including sputtering for less than 10 s.

Cross-sectional examination of a flash memory device is taken here as a typical example. Highly flat cross-section of the device was cut using an FEI Quanta 3D FIB and examined using a Carl Zeiss ULTRA 55 FE-SEM in the SE 2 imaging mode at an accelerating voltage of 1.50 kV. Subsequently, the FIB-cut surface was given rf-GD sputtering treatment for 5 s and examined again under otherwise similar conditions. The FIB-cut surface is highly flat and, therefore, the contrast in the SE 2 image is the material contrast only. In the absence of the edge contrast, which is the main contrast in the SE 2 image, the device structure cannot be revealed clearly (Fig.1-a). After rf-GD sputtering, it appears that the image quality has been improved dramatically as shown in Fig.1-b. Sputtering removes a thin damaged layer covering the FIB-cut surface; further it creates sharp steps along the boundaries of different materials due to differences in the sputtering rates. The edge contrast, associated with these newly-created sharp boundaries, allows the device structure, such as gates on silicon, tungsten plugs and a thin TiN layer surrounding the plugs, Al wiring, SiO_2 and silicon nitride layers, to be revealed clearly. Even the boundaries of SiO_2 layers, deposited under slightly different conditions, have been revealed successfully at, for example, the location indicated by the arrow b. A layer indicated by the other, unlabelled arrow is silicon nitride.

1. H.Jaksch; Materials World, October, 1996.

M. Luysberg, K. Tillmann, T. Weirich (Eds.): EMC 2008, Vol. 1: Instrumentation and Methods, pp. 621–622, DOI: 10.1007/978-3-540-85156-1_311, © Springer-Verlag Berlin Heidelberg 2008

622

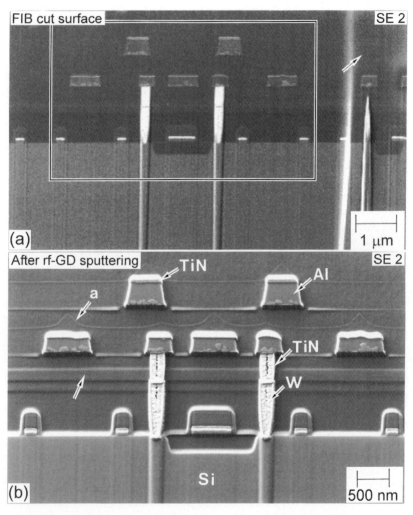

Figure 1. Scanning electron micrographs of cross-section of a flash memory device: (a) FIB-cut surface; (b) after 5 s rf-GD sputtering.

Development of an ultra-fast EBSD detector system

M. Søfferud[1], J. Hjelen[1], M. Karlsen[1], T. Breivik[2], N.C. Krieger Lassen[3], R. Schwarzer[4]

1. Norwegian University of Science and Technology (NTNU), Dept. of Materials Science and Engineering, N-7491 Trondheim, Norway
2. NTNU, Dept. of Geology and Mineral Resources, N-7491 Trondheim, Norway
3. Sofievej 9, DK-2840 Holte, Denmark
4. Kappstrasse 65, D-71083 Herrenberg, Germany

Jarle.Hjelen@material.ntnu.no
Keywords: EBSD, Offline indexing

Today's Electron Backscatter Diffraction (EBSD) systems are based on online pattern acquisition and indexing. The new-developed fast, high-sensitivity CCD camera used in the new EBSD detector has now surpassed the speed of available indexing software. The new EBSD system developed by Jarle Hjelen and co-workers has been installed at the Department of Materials Science and Engineering at NTNU, Norway. To reach very high speeds, Niels Christian Krieger Lassen proposed the idea of streaming the acquired diffraction patterns to a hard drive as a means to overcome the online-indexing bottleneck. This method has shown to have several advantages over current commercially available systems.

Online EBSD systems discard the patterns once they have been indexed, making it impossible to re-index with the exact same raw data if the results for some reason turn out to be unsatisfactory. When performing online EBSD characterization, results depend on an accurate calibration and present phases must be known a priori. With the current offline system, every single diffraction pattern is available for examination and re-indexing with optimized settings. The exact same area can be repeatedly indexed while experimenting with different Hough transform settings, phases, calibrations and pattern processing. Previously, this has not been possible without considering variables such as sample contamination and beam stability.

The new offline system is able to collect patterns at 750 patterns per second (pps), considerably higher than any other available system. This has several important advantages. Valuable microscope time is freed, because time-consuming indexing can be done at any workstation. Furthermore, it is now achievable to perform real-time EBSD-characterization combined with dynamic in-situ thermo-mechanical experiments. Up till now, only sequential sample manipulation and EBSD characterization has been possible, at best. At the Department of Materials Science and Engineering, NTNU, a custom designed sub-stage enables in-situ deformation experiments at elevated or sub-zero temperatures combined with EBSD [1]. Efforts are being made to accommodate the offline system to real-time EBSD characterization. Such a system would revolutionize the way technologically important phenomena such as phase transformations, recrystallization and texture development, are studied.

The hardware used includes a Zeiss Gemini ULTRA 55 FESEM equipped with a NORDIF UF750 EBSD detector and accompanying software. For offline indexing

M. Luysberg, K. Tillmann, T. Weirich (Eds.): EMC 2008, Vol. 1: Instrumentation and Methods, pp. 623–624, DOI: 10.1007/978-3-540-85156-1_312, © Springer-Verlag Berlin Heidelberg 2008

SEMdif Viewer software was used [2]. This software continuously extracts and indexes the stored patterns as if they were streamed directly from the camera.

Since the maximum speed of pattern acquisition (pps) is proportional to the beam current, it is essential to have a SEM with high probe current capability to reach very high scan speeds. The Zeiss ULTRA 55 has a maximum current of 40 nA, sufficient to easily reach 750 pps at low camera gain. At high gain, the same speed can be reached at 16 nA. While high beam currents are known to deteriorate electron image resolution with a thermal W emitter SEM, Humphreys has shown that this has a neglectable effect on spatial EBSD resolution when using a FESEM [3].

Figure 1a shows a crystal orientation map from a sample of recrystallized nickel. The microscope was operated at an accelerating voltage of 20 kV, a beam current of 40 nA, a tilt angle of 70° and a working distance of 22 mm. The acquired diffraction patterns had a resolution of 96x96 pixels. Erroneous points in the map are only found at grain boundaries, giving a high hit rate of approximately 99 %. **Figure 1b** shows a screenshot from the EBSD detector software. Note the acquisition frame rate of 750 pps and the scan time of 24 seconds. Some electronic gain is applied to adjust diffraction pattern intensity. Additionally, singular patterns used for calibration are recorded separately with higher resolution and longer exposure times.

In summary, this new EBSP acquisiton system is capable of collecting crystallographic information in a fast, flexible and reliable way.

1. Karlsen, M., et al., Materials Science and Technology, **24** (2008), p. 64-72.
2. R. A. Schwarzer, Archives of Metallurgy and Materials **53** (2008) in press
3. F. J. Humphreys, J. Mater. Sci. **36** (2001), pp. 3833-3854.

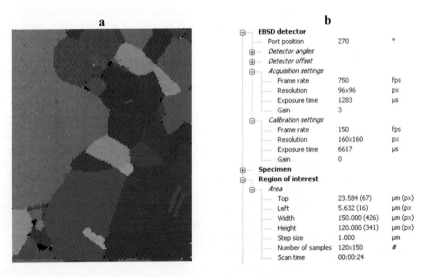

Figure 1. (a) Crystal orientation map of nickel, (b) screenshot from detector software showing the EBSD detector settings and details on scanned area.

Future prospects on EBSD speeds using a 40 nA FESEM

M. Søfferud[1], J. Hjelen[1], M. Karlsen[1], D. Dingley[2], H. Jaksch[3]

1. Norwegian University of Science and Technology, Dept. of Materials Science,
N-7491 Trondheim, Norway
2. EDAX Inc., Mahwah, NJ 07430, U.S.A.
3. Carl Zeiss NTS GmbH, 73447 Oberkochen, Germany

Jarle.Hjelen@material.ntnu.no
Keywords: EBSD, OIM, High speed acquisition, Offline indexing

This paper describes a method of estimating the maximum obtainable mapping speed by the use of a new ultra-fast EBSD detector system [1], both with current equipment and with even faster CCDs that will be available in the future.

The system consists of a Zeiss Gemini ULTRA 55 FESEM equipped with a NORDIF UF750 EBSD detector. For post-processing EDAX TSL OIM 5.2 software has been used. The beam current on this microscope can only be adjusted in coarse steps. Thus, the beam current was set to the maximum value of 40 nA, and to simulate lower beam currents, shorter camera exposure times were used to study the effect on indexing quality. According to Humphreys, varying the beam current from 10 to 40 nA has a minor effect on spatial EBSD resolution when using a FESEM [2]. Several 400x400 μm^2 scans with a step size of 2 μm were carried out on a sample of recrystallized nickel (20 kV, 40 nA, WD = 20mm) with exposure times of 1283, 1000, 700, 500, 350 and 300 μs, corresponding to theoretical speeds of 750, 960, 1300, 1800, 2500 and 3200 patterns per second (pps), respectively. Camera gain was optimized for each exposure time, varying from low at long exposure times to high at short exposure times. At the applied pattern resolution of 96x96 pixels, the maximum frames per second of the CCD camera is 750. The quoted speeds are therefore theoretical (derived from the exposure times), giving an indication of what speeds can be obtained in the future with 40 nA and with even faster CCDs.

Figure 1 gives information on how well the EBSD system performs when using a beam current of 40 nA and short exposure times, alternatively, what results can be expected when using lower currents and a constant speed of 750 pps. According to Figure 1 98 % of the patterns are correctly indexed down to an exposure time of 500 μs at 40 nA (1800 pps theoretical speed), equivalent to 750 pps/1283 μs at 16 nA. Grain maps obtained at 500 and 350 μs (corresponding to 1800 and 2500 pps, resp.) are shown in Figures 1a and 1b, respectively. The threshold value for correctly indexed points was set by discarding values with a Confidence Index (CI) lower than or equal to 0.05, effectively removing unindexed points and random orientation noise.

In recrystallized nickel, it is shown that the NORDIF UF750 offline EBSD system together with the EDAX TSL OIM software delivers 98 % correct indexing at 45 patterns per second per nA.

M. Luysberg, K. Tillmann, T. Weirich (Eds.): EMC 2008, Vol. 1: Instrumentation and Methods,
pp. 625–626, DOI: 10.1007/978-3-540-85156-1_313, © Springer-Verlag Berlin Heidelberg 2008

1. M. Søfferud, J. Hjelen, M. Karlsen, N. C. Krieger Lassen, R. Schwarzer (2008) to be presented at 14th EMC 2008.
2. F. J. Humphreys, J. Mater. Sci. **36** (2001), pp. 3833-3854.

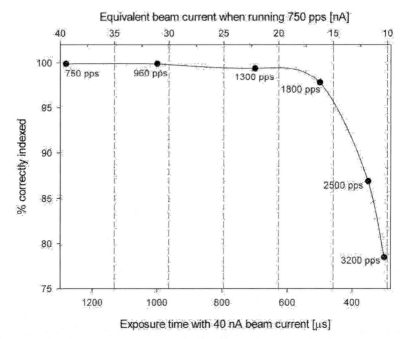

Figure 1. Fraction correctly indexed (threshold value CI>0.05) as a function of exposure time at 40 nA using reference sample of recrystallized nickel. Theoretical speeds in patterns per second are shown next to the data points.

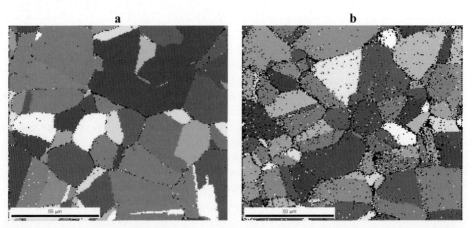

Figure 2. Grain map of nickel at 40 nA and 500 μs (**a**) and 300 μs (**b**) exposure times. Indexing rates are 97.8 % and 86.9 %, respectively. Scale bar is 150 μm.

High pressure imaging in the environmental scanning electron microscope (ESEM)

D.J. Stokes[1], J. Chen[2], W.A.J. Neijssen[1,2], E. Baken[1] and M. Uncovsky[3]

1. FEI Company, Achtseweg Noord 5, 5600 KA Eindhoven, the Netherlands
2. FEI Company, 690 Bibo Road, Pudong, Shanghai 201203, China
3. FEI Company, Podnikatelska 6, 612 00 Brno, The Czech Republic

Debbie.Stokes@fei.com

Keywords: ESEM, room temperature, high pressure

We discuss advances in technology that allow high-quality secondary electron imaging at high pressures in the (E)SEM. The importance of this is the ability to observe hydrated specimens at realistic temperatures whilst maintaining a suitably high relative humidity. This new capability has a great potential for *in situ* imaging of hydrated nano-structured materials and systems at room temperature.

Since becoming popular more than a decade ago, scanning electron microscopes (SEM) with extended pressure modes such as the environmental SEM (ESEM) have continued to evolve. The latest systems offer uncompromised performance over an unprecedented range of sample chamber vacuum conditions. Instruments are now available that provide near-nanometer resolution in all vacuum modes and the ability to operate at pressures as high as 4 kPa (~30 torr).

Electron microscopes utilizing water vapor as the imaging gas have the advantage of enabling moist and liquid specimens to be thermodynamically controlled, as well as allowing dry specimens to be hydrated *in situ*. So far, imaging of hydrated materials has required cooling of the sample to just above freezing to achieving adequate high humidity within the pressure range of a gaseous secondary electron detector. In practice, the maximum gas pressure is physically limited by an increase in primary electron beam scattering, and hence a decrease in image quality, at elevated pressure (i.e. pressures greater than ~600 Pa). Hence high pressure imaging requires the use of high beam currents, which increases the beam diameter, limiting the resolution.

However, with the recently introduced, commercially available, High Pressure detector for the ESEM, it is possible to get high quality secondary electron (SE) imaging at high pressures (in the kilopascal regime), using a relatively low beam current (sub-100 pA). This is achieved by optimizing boundary conditions that govern electron beam scattering in the gas, the energy distribution of electrons in the gas cascade, dielectric breakdown of the gas and detector collection efficiency [1].

Thus high pressure ESEM now enables imaging of surface modification processes occurring at temperatures higher than previously possible, at 100% relative humidity and above, in addition to the ability to observe hydrated materials at room temperature [2]. Figs 1 and 2 help to demonstrate the latter possibility. Furthermore, the detector is capable of continuous operation over an extensive pressure range (e.g. 0.1 to 20 torr),

M. Luysberg, K. Tillmann, T. Weirich (Eds.): EMC 2008, Vol. 1: Instrumentation and Methods, pp. 627–628, DOI: 10.1007/978-3-540-85156-1_314, © Springer-Verlag Berlin Heidelberg 2008

and hence is very versatile for work such as cooling/heating experiments at specific humidities.

Combining the High Pressure ESEM detector with a STEM-in-ESEM system [3] ultimately enables surface-sensitive electron imaging to be combined with transmitted electron imaging of hydrated and liquid-phase samples at room temperature, and therefore holds promise for the high-resolution study of systems such as nano-particles in suspension.

1. M. Toth *et al*, Applied Physics Letters, 2007. **91**(Article No. 053122).
2. D.J Stokes, infocus (Proceedings of the RMS), 2006. **2**(June 2006): p. 64-72.
3. D.J. Stokes and E. Baken, Imaging & Microscopy, 2007. **02/2007**: p. 17-20.

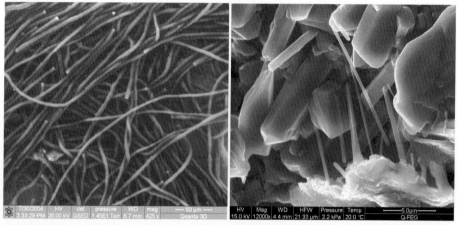

Figure 1. Secondary electron image of cyanobacteria, imaged in water vapor at pressure $p = 1.93$ kPa (14.5 torr), temperature $T = 24°C$

Figure 2. Secondary electron image of gypsum, imaged in water vapor at pressure $p = 2.2$ kPa (16.5 torr), temperature $T = 20°C$

Cathodoluminescence spectrum-imaging in the scanning electron microscope using automated stage control

D.J. Stowe[1], P.J. Thomas[2] and S.A. Galloway[1]

1. Gatan, U.K., 25 Nuffield Way, Abingdon, Oxfordshire, OX14 1RL. U.K.
2. Gatan Research and Development, 5794 West Las Positas Blvd, Pleasanton, CA 94588 USA

dstowe@gatan.com
Keywords: Cathodoluminescence, spectrum-imaging, stage control

Traditionally, cathodoluminescence spectrum-images have been acquired in the scanning electron microscope by using the microscope scan coils to sequentially deflect the electron beam to each pixel to be probed (Figure 1a). This approach has proven to be highly successful in the microanalysis of a wide range of materials. For example, applications ranging from understanding variations in the alloy concentration of compound semiconductors [1], mapping stress fields in semiconductors and ceramics [2 and 3], studying plasmonic effects in novel structures [4], to evaluating trace impurity distributions in geological minerals [5] have been studied effectively.

In the cathodoluminescence spectrum-imaging technique, the most efficient manner of collecting emitted photons is by placing a highly precise collection mirror, which subtends a large solid angle, immediately above the sample with an aperture through which the electron beam may pass to stimulate a specimen. In order to maximise the detection sensitivity for dispersive spectral analysis, the specimen is located at the focal point of the collection mirror in x, y and z, and is coincident with the zoom axis of the microscope at an acceptable working distance. However, in some applications using beam control spectrum-imaging this configuration places unwanted constraints upon an experiment. For example, the inherent variation in the collection efficiency of the system with beam deflection from the optic axis can result in a field of view of a few hundred microns, which in some cases is insufficient. Also, small spectral artefacts associated with the optics alignment have been observed. In a directly optically coupled system using a 300 mm, F# 4.2 monochromator and using a 600 grooves/mm diffraction grating, a systematic spectral shift of ~0.01 nm/μm across the field of view has been observed [3]. In many instances the limited field of view and spectral artefacts are inconsequential. However, where it is necessary to acquire data sets with supreme spectral accuracy, such as high accuracy stress mapping in ceramics, small spectral artefacts add unwanted errors. Furthermore, in specimens where wide area coverage is required, beam deflection is inappropriate.

In order to overcome such issues, we report the development of a stage-controlled spectrum-imaging system. In this case, the microscope stage is moved under computer control to sequentially position each pixel of the spectrum-image at the centre of the field of view (Figure 1b). In this manner, the optimal experimental configuration is maintained at each pixel. Stage-controlled stepping has the advantage of overcoming

M. Luysberg, K. Tillmann, T. Weirich (Eds.): EMC 2008, Vol. 1: Instrumentation and Methods, pp. 629–630, DOI: 10.1007/978-3-540-85156-1_315, © Springer-Verlag Berlin Heidelberg 2008

the collection optics issues that limit the field of view and also minimises spectral artefacts; this, however, is achieved at the expense of acquisition speed. Furthermore, this approach also brings added flexibility and additional benefits to multi-signal spectrum imaging, whereby CL and EDS spectra can be recorded simultaneously from one master platform.

Though stage controlled spectrum-imaging holds some advantages over the beam controlled technique, the maximum spatial resolution is limited by the minimum stage movement and a significantly increased time was required between each pixel's data acquisition. Thus, the stage- and beam-control of spectrum-imaging may be regarded as complimentary techniques with the ideal solution in the low magnification regime being a hybrid approach. Stage controlled spectrum imaging systems are common in electron probe micro-analyzer systems; however, this is the first development of this kind for cathodoluminescence applications in the scanning electron microscope.

1. X.L. Wang, D.G. Zhao, D.S. Jiang, H. Yang, J.W. Liang, U. Jahn and K. Ploog, J. Phys. Condens. Matter **19** (2007) 176005.
2. M. Avella, F. Pommereau, J. Jimenez, J.P. Landesman, B. Liu, A. Rhallabi, Indium Phosphide and Related Materials Conference Proceedings, 2006 International Conference on (2006) 159.
3. R.I. Todd, D.J. Stowe, S.A. Galloway and P.R. Wilshaw *submitted to the J. Eur. Ceram. Soc.*
4. E.J.R Vesseur, R. De Waele, M. Kuttge, A. Polman, Nano Letters **7** (Issue 9) (2007) 2843.
5. E.P. Vicenzi, T. Rose, M. Fries, A. Steele and C. Magee, Microsc. Microanal. **12** (Suppl. 2) (2006) 1518.

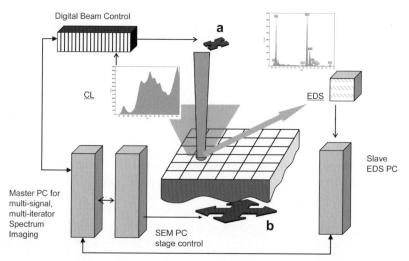

Figure 1. A schematic representation of a multi-signal, spectrum imaging system using a) beam-control and b) stage-control.

Low voltage, high resolution SEM imaging
for mesoporous materials

O. Takagi[1], Shuichi Takeuchi[1], Atsushi Miyaki[1], Hiroyuki Ito[1], Hirofumi Sato[1], Yukari Dan[1], Mine Nakagawa[1], Sho Kataoka[2], Yuki Inagi[2] and Akira Endo[2]

1. Hitachi High-Technologies Corporation, 24-14 Nishi Shimbashi, 1-chome, Minato-ku, Tokyo 105-8717, Japan
2. National Institute of Advanced Industrial Science and Technology (AIST)

takagi-osamu@nst.hitachi-hitec.com
Keywords: mesoporous silica, low voltage SEM, retarding, scan integration

Mesoporous silicas (MPSs), which possess highly ordered structures with a pore size of 2-15 nm, must be widely applied to catalysts, adsorbents, membranes, and sensors. Direct SEM observation of MPSs provides detail information on the external and internal structures, though it consistently faces charge-up problems of insulating silica frameworks. Several skills such as choosing low resistance substrate or replica method succeed to avoid charge-up phenomena [1]. In this contribution, high resolution, direct SEM imaging of MPSs is tried under the condition of low voltages. A cold FEG SEM, which employs the snorkel type objective lens, retarding device [2] and E cross B (ExB) filter [3] for detecting secondary electron (SE) are used for this study.

Figures 1 are images of MPS (SBA-15 [4]) particles with/without a retarding voltage applied. A Retarding Voltage technique uses a voltage applied to the specimen and decelerates the accelerating voltage for achieving high-resolution at low voltages. The retarding image (Figure 1a) reveals highly ordered hexagonal structures of mesopores clearly when compared to a non-retarding image (Figure 1b).

Figures 2 are surface images of MPS (SBA-16 [5]) films on the glass substrate with different scanning conditions. Figures 2a/2b are observed by traditional slow scan and fast-scan image-integration, respectively. The ExB filter minimizes incident beam illumination due to high SE detection efficiency, however, charge-up phenomena still influence the lack of microasperity in Figure 2a and image drifting in Figure 2b. Optimum beam scan settings satisfy both smaller beam illumination time per pixels and good S/N ratio without image drifting (Figure 2c). The surface has two kinds of domains, one consists approximately 10nm uniform pores perpendicular to the surface plane, and the other is a uniform striped pattern parallel to the surface plane. Figure 3 shows a cross-sectional image of SBA-16 film formed in the cylindrical microcapillary tube (200um in diameter) by fracturing. The cross sectional image reveals that the highly ordered structure continues overall of the film without complicated preparation.

1. C-W. Wu, Y. Yamauchi, T. Ohsuna, K. Kuroda, J. Mater. Chem., **16** (2006), p.3091-3098.
2. A. Muto, et al., Proc. Microsc. Microanal. 9, **2** (2003), p.146-147.
3. M. Sato, H. Todokoro, K. Kageyama, Proc. SPIE, **2014** (1993), p.17-23.
4. A. Sayari, B. H. Han, Y.Yang, J. Am. Chem. Soc. **126** (2004), p.14348-14349.
5. S. Kataoka, Akira Endo, Atsuhiro Harada, Takao Ohmori, Mater. Lett., **62** (2008) p.723-726.

M. Luysberg, K. Tillmann, T. Weirich (Eds.): EMC 2008, Vol. 1: Instrumentation and Methods, pp. 631–632, DOI: 10.1007/978-3-540-85156-1_316, © Springer-Verlag Berlin Heidelberg 2008

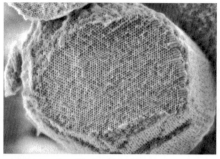 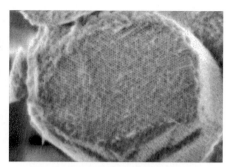

Figure 1a. Retarding image
(Landing voltage:500V)

Figure 1b. Non-retarding Image
(Accerelating voltage:500V)

Figure 1. SBA-15 particles with/without retarding

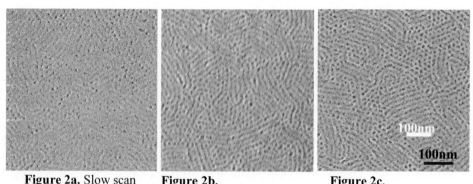

Figure 2a. Slow scan

Figure 2b.
Fast-scan image-integration

Figure 2c.
Optimised beam scan

Figure 2. SBA-16 films on the glass substrate (Landing voltage 800V)

Figure 3. cross-sectional image of SBA-16 film formed in the cylindrical microcapillary tube (Accerelating voltage 800V)

New developments in state of the art silicon drift detectors (SDD) and multiple element SDD

R. Terborg[1], M. Rohde[1]

1. Bruker AXS Microanalysis GmbH, Schwarzschildstr. 12, 12489 Berlin, Germany

ralf.terborg@bruker-axs.de

Keywords: Silicon drift detector, EDS, spectral imaging

Over the past few years, Silicon Drift Detectors (SDD) have become more and more popular in the field of X-ray detection, especially for microanalysis applications, and are about to replace lithium-drifted silicon detectors (Si(Li)). The functional principle and properties have been described elsewhere [1].

Successive improvements have made it possible for state of the art SDD, as represented by the forth generation XFlash 4000 series, to exceed the performance of Si(Li) in almost every respect that matters to analytical EDS. High resolution type SDD produce energy resolutions of 125eV or even better at Mn-Kα. Optimised detector radiation entrance windows [2] provide excellent low energy response with minimal shelf and tail and leads to an energy resolution of ≤ 58eV at F-Kα and ≤ 48eV at C-Kα, Fig. 1. Improved electronics with pulsed charge restoration methodology maintain this energy resolution even at extremely high count rates, allowing all XFlash detectors to achieve full specification at 100 kcps input count rate.

Multiple element SDD structures with four separate detectors integrated on one chip provide even higher count rates without increasing pile-up or dead time. Quad SDD with four separate detectors are capable to deliver up to 1,000,000 cps output count rate. Typical multi element structures consist of four 10mm² and 15mm² SDD, offering effective detector areas of 40 and 60mm², respectively.

Advanced versions of the QUAD XFlash 4040 show energy resolutions of 125eV and 47eV at Mn-Kα and C-Kα respectively, both measured at 400,000 cps. Such a detector is an ideal device for high energy resolution spectrometry as well as ultra high speed mapping applications. A spectral image consisting of 800x600 pixels and a total of 73,000,000 counts obtained within 165s acquisition time shows a clear separation of boron and carbon containing phases, Fig. 2.

The specially designed QUAD XFlash 4060 with four 15 mm² SDD centred around a beam path of a scanning electron beam can be placed between pole piece and sample within a standard SEM in order to cover an extremely large solid angle of about 0.9 sr, Fig. 3.

1. R. Terborg, M. Rohde, Microsc. Microanal. **10** (Suppl.2) (2004), 942
2. A. Niculae et al., Microsc. Microanal. **13** (Suppl.2) (2007), 1430

M. Luysberg, K. Tillmann, T. Weirich (Eds.): EMC 2008, Vol. 1: Instrumentation and Methods, pp. 633–634, DOI: 10.1007/978-3-540-85156-1_317, © Springer-Verlag Berlin Heidelberg 2008

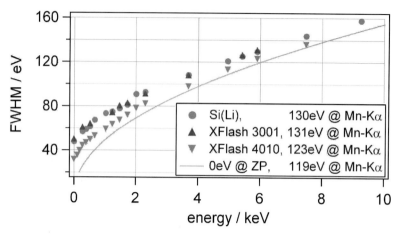

Figure 1. Energy resolution (FWHM) of different SDD in comparison with a Si(Li) and a theoretical detector without electronic noise.

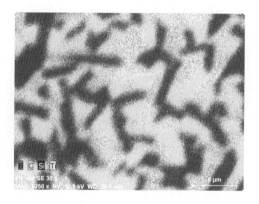

Figure 2. A 800x600 pixels spectral image containing a total of 73,000,000 counts acquired within 165s.

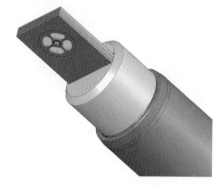

Figure 3. Flat Quad detector (Quad XFlash 4060) which covers a solid angle of 0.9 sr when placed 5mm above the sample.

SEM in forensic science

I. Turkova[1], M. Kotrly[1]

1. Institute of Criminalistics Prague (ICP), POBox 62/KUP, Strojnicka 27, 170 89 Praha 7, Czech Republic

kup321@mvcr.cz

Keywords: forensic science, SEM, microanalysis

Forensic investigation traces of crime scenes are very popular in current TV serial stories. But demonstration of forensic scientist's work is very schematic and authors mix a lot of special fields of examinations together. All equipment has a special effect and it is possible to be used for special determination samples. SEM with EDS/WDS makes is possible to observe topography surface and morphology samples and examination of chemical components. Physical laboratory of Institute of Criminalistics Prague use SEM especially for examination of inorganic samples, rarely for biology material.

Major attention has been given to examination of GSR (gunshot residue) particles. GSR is a term for microscopic particles, predominantly for micron size range and genesis after shooting. Flue gas cloud is dispersed out of weapon and drops off in the vicinity. GSR particle contamination is on shooter hands, his clothing, in the vicinity and sometimes on neighbouring subjects. Particles have special morphology and chemical constitution. Physical aspects of sedimentation of GSR particles and possibilities of secondary contamination have been performed by research project of Institute of Criminalistics Prague (ICP). Identification and analysis particles GSR are major questions forming axis investigation and comparison by SEM.

For the purpose of performing analyses are also submitted traces containing other sorts of thermogenetic particles, such as post- blast residues generating by striking a flint, including cases of breaking into a safe, overcoming of obstacles by using abrasive cutting.

Petrologic and mineral samples constitute a great deal of examination encompassing single crystals where morphology is not utterly clear under optical microscope and for further identification it is essential a large magnification and the determination of chemical compounds. The identification and analysis of single sample fractions of petrologic nature embracing compact samples and already made polished sections is one of other fields in forensic examination.

Examination of morphology fibres is possible by using electron microscopy (scrutinizing longitudinal morphology and deformation, including cross cuts), studying inorganic components in material fibres, and also the examination of morphology trichome materials.

Drivers´ duty to keep the lights switched on while driving is valid through all Europe. In a traffic accident it is frequently unclear whether the driver was keeping lights switched on. Elucidation of discrepancies encountered in witness testimonies and the suspect's testimonies lies in proving the fact that lights were switched on at the

M. Luysberg, K. Tillmann, T. Weirich (Eds.): EMC 2008, Vol. 1: Instrumentation and Methods, pp. 635–636, DOI: 10.1007/978-3-540-85156-1_318, © Springer-Verlag Berlin Heidelberg 2008

moment of the accident. Smelted glass dust from a road accident on a wolfram filament and morphological changes incurred to its filament are used as a proof that lights were switched on at the moment of the its smash.

Another significant area is glass examination encompassing road accidents, burglary, bottle used as a weapon, etc., in these cases the content of microelements is examined and glass particles recovered, e.g. from a suspects clothing.

Completely separate part presents writing tools superposition analysis (crossing strokes). Electron microanalysis as a complementary method in handwriting examination is applicable where toner particles develop a thicker film that is identifiable on paper pulps fibres. In this regard there are usually encountered no problems with toners of laser printers on which polymer melts are generated and these ones being a carrier of black toner for an identifiable layer. For inkjet printers and stamping inks it is possible to identify easily inks containing mineral pigments (by some producers marked as pigment inks). Moreover, ICP in cooperation with other external facilities performs determination of artwork authenticity (paintings, plastics). Analyses of recovered pigments were performed both in the form of polished sections to preserve all layer stratigraphy and in the form of separate single generic samples. Based on the outputs, analyses were one of the key evidence for identification of forgeries and originals.

Among standardized methods are also so-called "unknown samples". These are particularly morphology evaluation, identification and determination of the origin of found microparticles and fragments if need be and their possible comparison with reference samples constituting either different relics of inorganic character or sawdust.

Detection and analyses of nanoparticles, nanocomposites and nanofibres rank among latest trends in forensic analysis.

1. Acknowledgements: Microanalytical methods at Institute of Criminalistics Prague were supported by grant-aided projects of the Czech Republic Ministry of Interior RN 19961997008, RN 19982000005, RN 20012003007, RN 20052005001, VD20062008B10 a VD20072010B15.

Secondary electron imaging due to interface trapped charges for a buried SiO$_2$ microstructure

Hai-Bo Zhang, Wei-Qin Li, Xing Wu and Dan-Wei Wu

Department of Electronic Science and Technology, Xi'an Jiaotong University, Xi'an 710049, People's Republic of China

hbzhang@mail.xjtu.edu.cn

Keywords: interface trapped charge, microstructure, scanning electron microscopy (SEM)

The surface local electric field (LEF) caused by electron beam (e-beam) irradiation can be utilized to form the useful SEM contrast for buried microstructures [1]. However, its formation has been poorly understood. The aim of this work is to clarify the secondary electron (SE) imaging mechanism of a buried SiO$_2$ trench with interface trapped charges, shown in Figure 1, based on the point irradiation model of a 1.9 keV negative charging e-beam. Electron scattering and transport have been simulated with the Monte Carlo and finite difference methods in an axisymmetric system.

Electrons first accumulate near the surface due to the repulsion from trapped charges, then migrate downwards to form the electron beam induced current (EBIC), and finally leave holes in a region outside the electron range (~100 nm) as seen in Figures 2(a) and (b). The positive surface LEF in both z and r directions can thus be generated even in the negative charging case with the surface potential profile shown in Figure 2(c). Figure 3 indicates that some of emitted SEs redistribute to the surface due to the LEF.

Interface trapped charges increase the SE redistribution rate α (I_{RE}/I_{SE}) [1,2] and therefore decrease the SE imaging current I_S, as illustrated in Figure 4. Conversely, without trapped charges, the positive surface LEF is weak and the SE redistribution rate is lower. Therefore, the middle dark contrast of the SEM image in Figure 1(b) can be observed for the trench. In comparison, a longer time of irradiation generates more trapped charges, increasing image contrast in experiments [1,3]. However, the SiO$_2$ thickness or electron mobility may affect image contrast. A much thicker SiO$_2$ layer or extremely low electron mobility will cause the stronger surface LEF and therefore the higher SE redistribution rate. This can reduce the image contrast between the trench with trapped charges and the other region without the underlying charges. Moreover, the irradiation time required for forming interface trapped charges will be prolonged [4].

It is interesting to note that the SE imaging described here will most probably be applicable to detecting interface defects or charges buried in insulators.

1. K. Ura, Denshi Kenbikyo **36** (2001), p. 53 (in Japanese).
2. H.B. Zhang, F.J. Feng and K. Ura. Chin. Phys. Lett. **20** (2003), p. 2011.
3. T. Koike, T. Ikeda, M. Miyoshi, K. Okumura and K. Ura, Jpn. J. Appl. Phys. **41** (2002), p. 915.
4. H.B. Zhang, D.Y. Li and W.Q. Li, Rev. Sci. Instrum. **78** (2007), Art. No. 126105.
5. This work was supported by the National Natural Science Foundation of China under Grant No. 60476018. We would also like to thank Dr. Katsumi Ura, Professor Emeritus of Osaka University, for his constant encouragement and advice.

M. Luysberg, K. Tillmann, T. Weirich (Eds.): EMC 2008, Vol. 1: Instrumentation and Methods, pp. 637–638, DOI: 10.1007/978-3-540-85156-1_319, © Springer-Verlag Berlin Heidelberg 2008

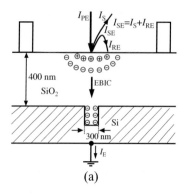
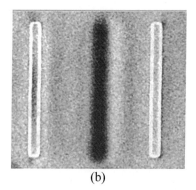

(a)   (b)

Figure 1. (a) A simplified model of the specimen and currents in the process of e-beam point irradiation, and (b) the relevant SEM image [1] in which the middle dark region corresponds to the buried trench with interface trapped charges at side walls.

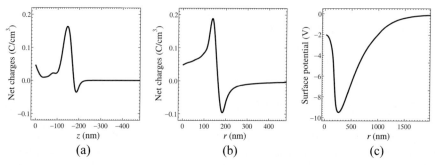

Figure 2. Simulated net charge profiles at (a) $r = 0$ and (b) $z = 0$, and surface potential at (c) $z = 0$. The trapped charge (electron) density was set to be 4×10^{13} cm^{-2}.

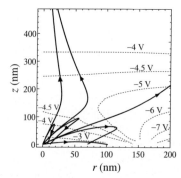
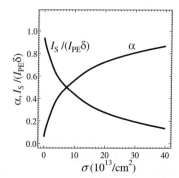

Figure 3. SE trajectories (solid) and spatial equipotential lines (dotted) near the surface.

Figure 4. The SE distribution rate α and relative SE imaging current vs. the trapped charge (electron) density σ.

HRSEM Secondary Electron Doping Contrast: Theory based on Band Bending and Electron Affinity Measurements

I. Zhebova[1], M. Molotskii[1], Z. Barkay[2], G. Meshulam[1], E. Grunbaum[1] and Y. Rosenwaks[1]

1. School of Electrical Engineering- Department of Physical Electronics, Faculty of Engineering, Tel Aviv University, Ramat-Aviv, 69978, Israel
2. Wolfson Applied Materials Research Center, Tel-Aviv University, Ramat-Aviv, 69978, Israel

irinazhe@post.tau.ac.il

Keywords: HRSEM, KPFM, Secondary Electron emission, dopant contrast, electron affinity, band bending

The secondary electron (SE) emission flux in a high-resolution scanning electron microscope (HRSEM) is a powerful tool for delineation of electrically active dopant concentration, built-in potentials and surface electric fields in semiconductor junctions [1,2]. In almost all SE images p-doped regions appear brighter than in n-doped regions (Figure 1).

We present a quantitative theory of SE doping contrast in the p- and n-sides of a pn junction. In our model the contrast $C(pn) = [(B_p \lambda_{esc}^{(p)} / B_n \lambda_{esc}^{(n)}) - 1]$ is a function of the SE escape probability B and their escape length λ_{esc}. B depends on the energy distribution of the SE as well as on the electron affinity χ. An electric field due to positive or negative surface charge shifts the SE energy distribution to higher or lower energies on the p-side or on the n- side, respectively [1]. The surface electric field increases (decreases) the escape length in a p (n)-type semiconductor [3]; hence $\lambda_{esc}^{(p)}$ is higher than $\lambda_{esc}^{(n)}$. The contrast is also affected [2] by changes in the built-in potential U_{bi} as well as the potentials U_{ss} due to band bending on the p and n-sides. The effective electron affinity values on the p and n- side are: $\chi_p = \chi_0 - U_{SS}^{(p)}$, $\chi_n = \chi_0 - U_{bi}^{(n)} - U_{SS}^{(n)}$, where χ_0 is the ideal crystal affinity.

It is found that λ_{esc} depends exponentially on the surface electric field; hence its contribution to the SE contrast is larger by one order of magnitude than the contribution of the effective electron affinity and the energy distribution function.

The calculated contrast (Figure 4) is based on band bending and electron affinities extracted from our Kelvin probe force microscopy (KPFM) measurements (Figures 2 and 3). Figure 2 shows a typical CPD measurement of $p^{++}n$ junction before and following HRSEM measurement, and Figure 3 the resulting band bending scheme of the n-side of the junction. It can be seen from Figure 4 that our model gives a logarithmic

M. Luysberg, K. Tillmann, T. Weirich (Eds.): EMC 2008, Vol. 1: Instrumentation and Methods, pp. 639–640, DOI: 10.1007/978-3-540-85156-1_320, © Springer-Verlag Berlin Heidelberg 2008

weak dependence of the contrast on acceptor concentration as reported by many groups (see for example 2,4).

The effect of carbon contamination, as observed in Figure 2, as well as oxygen contamination will be presented and discussed.

1. P. Kazemian, S. A. M. Mentink, C. Rodenburg, and C. J. Humphreys, J.Appl.Phys. **100** (2006), p. 054901; Ultramicroscopy. **107** (2007), p. 140
2. C.P. Sealy, M. R. Castell and P. R. Wilshaw, J. Electron Microscopy. **49(2)** (2000),
3. p. 3113.
4. E.Schreiber and H.-J.Fitting. J.Electr. Spectr. Relat.Phenom. **124** (2002), p.25
5. M.El-Gomati, F.Zaggout, H.Jayacody, *et al.* Surf.Interface Anal. **37** (2005), p.901.

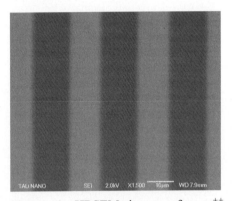

Figure 1. HRSEM image of a $p^{++}n$ sample ($p=10^{19}cm^{-3}$; $n=10^{17}cm^{-3}$)

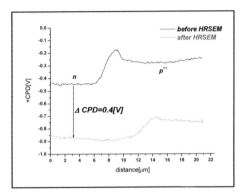

Figure 2. KPFM measured surface potential of $p^{++}n$ Si (100) junction before (solid) and following (dashed) HRSEM measurement.

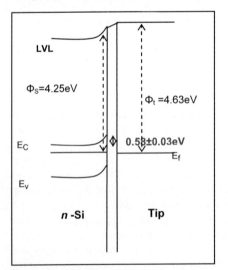

Figure 3. Band structure of n-Si extracted from the measurement in figure 2.

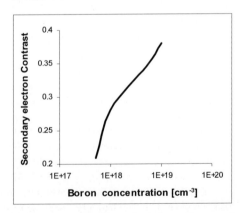

Figure 4. The theoretical dependence of contrast on acceptor concentration in p^{+}-Si.

3D EBSD-based orientation microscopy and 3D materials simulation tools: an ideal combination to study microstructure formation processes

S. Zaefferer

Max-Planck-Institut für Eisenforschung, Max-Planck-Str. 1, 40237 Düsseldorf, Germany

s.zaefferer@mpie.de

Keywords: 3D orientation microscopy, 3D Monte Carlo simulation, nucleation mechanisms

In the last 3 years we have developed a technique for 3-dimensional (3D) orientation microscopy based on serial sectioning using a focused ion beam microscope (FIB) and characterisation of the sections using the well-known electron backscatter diffraction (EBSD) technique [1]. The system runs fully automated and delivers 3D orientation maps with a resolution down to 50 x 50 x 50 nm³ and a maximum observable volume in the order of 50 x 50 x 50 μm³. The technique allows studying a wide range of metals and has proven to be particularly well suited for heavily deformed or fine-grained metals. Figure 1 shows, as an example the microstructure of an electrodeposited nickel-cobalt film (see [2] for further information).

A challenge faced during development of the system was to obtain a 3D spatial resolution which is as good as the lateral 2D resolution of the EBSD technique, i.e. in the order of 50 nm. To this end a number of technical details had to be solved first. However, at the end the resolution goal could only be achieved applying software algorithms for slice alignment. The basic idea behind these algorithms will be shortly lined out in the presentation.

The 3D materials description obtained with the new technique allows new insights into materials: first, the true 3-dimensional morphology of microstructures can be determined which is particularly interesting for non-homogeneous materials where stereological equations do not give a reasonable materials description from 2D sections. Second, grain boundaries can be characterised in full, including the misorientation and the grain boundary inclination. Finally, 3D data may now be used as input for realistic materials modelling.

The latter point is particular important for recrystallisation studies: the nucleation process of recrystallisation is still not well understood and cannot be observed directly by any existing technique. Also the here presented 3D technique does not permit this as it is a destructive technique. Instead, 3D Monte-Carlo (MC) modelling of subgrain growth allows studying the important parameters of the nucleation process. To this end real, measured microstructures are used as input, together with models on the energy and mobility of subgrain boundaries. The simulation is then able to reveal the formation of nuclei and the evolution of the stored energy (i.e. the driving force of recrystallisation) during the nucleation and growth process. These techniques have been used to re-investigate the nucleation process of the cube texture in heavily cold rolled Fe

M. Luysberg, K. Tillmann, T. Weirich (Eds.): EMC 2008, Vol. 1: Instrumentation and Methods, pp. 641–642, DOI: 10.1007/978-3-540-85156-1_321, © Springer-Verlag Berlin Heidelberg 2008

36% Ni, an fcc alloy with high stacking fault energy. Figure 2 shows the measured 3D microstructure on the left side and a series of snapshots of the 3D MC simulation next to it. The first line displays the crystal orientations, in the second maps of the so-called kernel average misorientation are shown, which may be taken as a measure of the stored deformation energy.

1. S. Zaefferer, S. Wright, D. Raabe, Met. Mater. Trans. 39A (2008) 374-389
2. A. Bastos, S. Zaefferer, D. Raabe, J. Microscopy (2008), in print

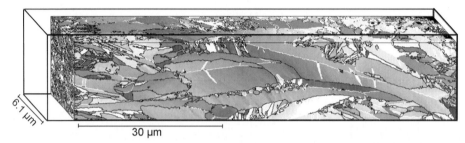

Figure 1. 3D orientation map of a fine-grained, electrodeposited Ni-Co thin film. The colours indicate crystal directions pointing out of the paper plane in an inverse pole figure colour scheme. The measurement consists of 61 slices of 100 nm thickness.

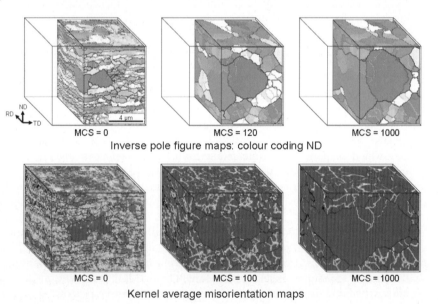

Figure 2. Measured 3D orientation maps (MSC=0) and 2 stages of a 3D Monte-Carlo simulation showing the progress of cube nucleus growth in a heavily cold rolled Fe 36% Ni sample. The inverse pole figure maps show the crystal orientation (red: cube orientation {001}<100>), the kernel average misorientation maps display the stored energy. MCS: Monte-Carlo steps.

Capturing Sub-Nanosecond Quenching in DualBeam FIB/SEM Serial Sectioning.

W.J. MoberlyChan & A.E. Gash

Lawrence Livermore National Laboratory, CMELS, Livermore, CA USA, 94550.

moberlychan2@llnl.gov

Keywords: dualbeam FIB, SEM serial sectioning

Many events that happen (intentional or not) during thin film processing can be captured by cross section TEM. Dynamic possibilities often argue for *in situ* TEM, with two caveats: *in situ* experiments require care, and some reactions observed *in situ* do not capture materials phenomena the same as an *ex situ* manufacturing line. During TEM sample preparation, FIB/SEM serial sections add statistical meaning to a final single TEM image. Dynamic FIB/SEM movies also help interpret reactions. 3D data is often reconstructed to a tomogram of 3D microstructure. However, if events happen differently in the 3rd (z) dimension, this generally is due to local events happening at slightly different times. The "movie form" of a 3D dataset enables temporal differences to be considered in conjunction with spatial differences all at the nanoscale. Dynamic imaging often leads the observer to ask: "What is happening?" Interpretation of movies must abide two kinetic laws: Whatever happens, happens; and whatever happens first, happens first. The FIB/SEM dissection occurs long after any reaction(s) happened, but the movie provides dynamic clues as to what/how/when reactions happen.

Figures 1 and 2 are cross section TEM "before and after" a post-deposition thermal process of Ni/Al multilayers, which have energetic applications [1]. Figures 1b and 2b are SADPs from the multilayer stack. The dominant Ni rings sharpen as grains grow during reaction, however Figure 1 suggests partial reaction occurred during deposition. The reaction is typically self-sustaining [1] and the front propagates faster than a nanosecond per nanometer. Figure 3 presents 8 frames of FIB/SEM serial section, but cannot supplant the dynamic effect of watching the movie. Serial sectioning exhibits the (dark) Al as rougher, and the (bright) Ni as continuous and conforming. A single SEM image does not see as well as TEM, neither resolution nor various contrasts. (SEM contrast is reverse of BF-TEM.) However, the sequential SEM movie establishes the roughness of each Al layer compared to the Ni layers, and determines voids are isolated only in Al layers. *In situ* experiments have the concern that the energy of the imaging electron beam may influence the reaction, and FIB/SEM serial sectioning can have a similar artifact. SEM of a quick-fractured multilayer also observes holes (Figure 4a), and holes form through the topmost layers (Figure 4b) as Al vaporizes. The energy of sputter deposition often causes atomic mixing of a few atom layers as atoms deposit, even before post-deposition interdiffusion processing [2]. As Ni and Al mix, they react to release energy (heat) sufficient to melt and even vaporize Al in vacuum. Molten Al beads at the nanoscale and is trapped by the subsequent Ni layer. If deposition provides enough energy to start the reaction, then why doesn't it self propagate throughout the whole deposition process? In this case the metal substrate acts as a thermal sink. Thus

M. Luysberg, K. Tillmann, T. Weirich (Eds.): EMC 2008, Vol. 1: Instrumentation and Methods, pp. 643–644, DOI: 10.1007/978-3-540-85156-1_322, © Springer-Verlag Berlin Heidelberg 2008

644

the heat of reaction can be quenched as fast as it starts (picoquenching), and the partially reacted Al is frozen in place by the subsequent deposition of a near-perfect Ni layer.

1. Gavens, A.J., T.P. Weihs, et al, J. Appl. Phys. V87 1255-1263 (2000). (www.rntfoil.com)
2. Thanks to: H. Louis, B. Clemens, K. Holloway, & P. Dorsey for discussions on multilayers. This work was performed for the U.S. Dept. of Energy by Lawrence Livermore National Laboratories under contract No. DE-AC52-07NA27344. UCRL-PROC-402235.

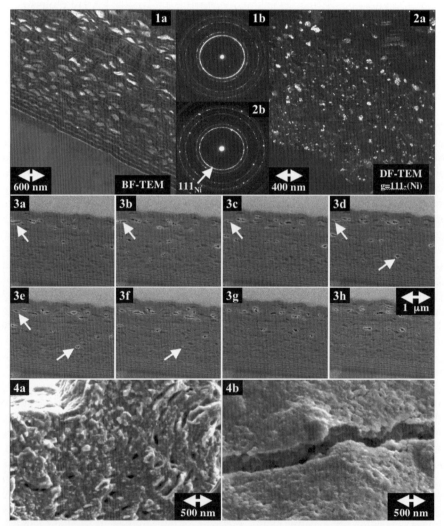

Figure 1. Bright Field TEM and Diffraction of as-deposited Ni-Al multilayer. **Figure 2.** Dark Field TEM and SAD after energetic reaction. Sharper polycrystalline rings of Ni arise from grain growth. **Figure 3.** 8 frames of SEM movie from FIB/SEM serial sectioning. Darker Al layers are rough and have voids. **Figure 4.** SEM of as-deposited multilayer exhibits voids due to Al vaporization; (a) fracture section, and (b) top-down.

Deformation mechanisms in 1D nanostructures revealed by *in situ* tensile testing in an SEM/FIB

D.S. Gianola[1], R. Mönig[1], O. Kraft[1] and C.A. Volkert[1,2]

1. Institute for Materials Research II, Forschungszentrum Karlsruhe, Hermann-von-Helmholtz-Platz-1, 76344 Karlsruhe, Germany
2. Institut für Materialphysik, Universität Göttingen, Friedrich-Hund-Platz 1, 37077 Göttingen, Germany

Dan.Gianola@imf.fzk.de

Keywords: nanowires, mechanical behavior, in situ testing

Plasticity in extremely small volumes is fundamentally different than in large materials; the law of averages gives way to discrete processes that dominate the response. Probing the mechanical response and uncovering the underlying deformation mechanisms of diminishingly small structures in the micro- and nanoscale requires new strategies and approaches that circumvent difficulties associated with handling, gripping, loading, and measuring small specimens. The need for *in situ* experiments that give a one-to-one correlation between mechanical response and deformation morphology is exacerbated by the fact that electron optics are needed to image and manipulate nanostructures. Tensile experiments are the preferred modality at larger scales since they apply a homogeneous stress state and are less sensitive to boundary conditions, easing interpretation. However, results obtained using the currently employed techniques at the nanoscale (e.g. nanoindentation, micro-compression testing) are clouded by these issues.

Here we describe efforts to conduct quantitative *in situ* tensile experiments on 1D nanostructures in a dual-beam scanning electron microscope and focused ion beam (SEM/FIB). Specimen manipulation, transfer, and alignment are performed using an *in situ* manipulator, independently-controlled positioners, and the FIB. Gripping of specimens is achieved using electron-beam assisted Pt deposition. Local strain measurements are obtained using digital image correlation of SEM images taken during testing. Examples showing results for single-crystalline metallic nanowires and nanowhiskers, having diameters between 30 and 300 nm, will be presented in the context of size effects on mechanical behavior and the influence of defects on the accommodation of plasticity in small volumes.

M. Luysberg, K. Tillmann, T. Weirich (Eds.): EMC 2008, Vol. 1: Instrumentation and Methods, pp. 645–646, DOI: 10.1007/978-3-540-85156-1_323, © Springer-Verlag Berlin Heidelberg 2008

Figure 1. Experimental setup for nanowire tensile testing in a dual-beam SEM/FIB instrument. The nanomanipulator for transferring and aligning nanowires is shown.

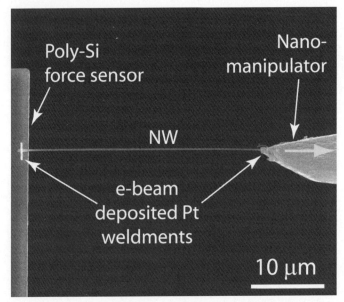

Figure 2. SEM image of a Cu nanowire attached to a MEMS load cell for tensile testing. The Cu nanowire has a diameter of approximately 230 nm.

Focused Ion Beam Tomography of Insulating Biological and Geological Materials

B.M. Humbel[1], D.A.M. de Winter[1], C.T.W.M. Schneijdenberg[1], B.H. Lich[2], M.R. Drury[3] and A.J. Verkleij[1]

1. University Utrecht, EMSA, Cellular Architecture and Dynamics, 5384 CH Utrecht, The Netherlands
2. FEI Company, Achtseweg Noord 5, 5600 KA Eindhoven, The Netherlands
3. Earth Sciences, Faculty of Geosciences, Utrecht University, NL-3584 CD Utrecht, The Netherlands.

B.M.Humbel@uu.nl

Keywords: FIB-SEM, Tomography Forsterite, HUVEC, Aluminium Oxide Polycrystal, Insulator

The analysis of large volumes at high resolution gains more and more importance. Thanks to transmission electron tomography is has become obvious that a two-dimensional representation can lead to incomplete interpretation and understanding of three-dimensional structures. Especially in cell biology new insight has been gained in the connection of intracellular organelles and their biogenesis [1-3].

Another aspect often gets forgotten. Most of the samples are very heterogeneous and only a small part of a large volume is in the focus of research. As shown previously only in a very small area of the pituitary gland of zebra fish embryo prolactin is synthesised [4]. Therefore methods for correlative microscopy are of prime importance.

The combined focused ion beam scanning electron microscope is a new tool, which has the potency to search the area of interest at low magnification. Once found one can zoom in and analyse the structure at high resolution. Due to the ion beam searching is not restricted to two dimensions but slices of the material can be removed until the area to be investigated has surfaced.

Next to unearthening structures, the ion beam can also be used for real tomography by the so-called slice and view process. A thin layer of the surface is ablated with the ion beam and then an image of the newly generated surface is taken with the electron beam. This process has been automated and therefore large volumes can be analysed fast and easily. The volumes can be 10 times larger than by TEM tomography, while accepting compromises in terms of resolution.

To overcome imaging problems like intensity gradients and shadowing effects, we designed a new milling strategy. Here we will show our preliminary results achieved in biomedical and geological research and outline future developments.

1. Trucco, R.S. Polishchuk, O. Martella, A. Di Pentima, A. Fusella, D. Di Giandomenico, E. San Pietro, G.V. Beznoussenko, E.V. Polishchuk, M. Baldassarre, R. Buccione, W.J.C. Geerts, A.J. Koster, K.N.J. Burger, A.A. Mironov and A. Luini, Nat. Cell Biol., **6** (2004), p. 1071-1081.
2. B.J. Marsh, N. Volkmann, J.R. McIntosh and K.E. Howell, Proc. Natl. Acad. Sci. USA, **101** (2004), p. 5565-5570.

M. Luysberg, K. Tillmann, T. Weirich (Eds.): EMC 2008, Vol. 1: Instrumentation and Methods, pp. 647–648, DOI: 10.1007/978-3-540-85156-1_324, © Springer-Verlag Berlin Heidelberg 2008

3. H.J. Geuze, J.L. Murk, A.K. Stroobants, J.M. Griffith, M.J. Kleijmeer, A.J. Koster, A.J. Verkleij, B. Distel and H.F. Tabak, Mol. Biol. Cell, **14** (2003), p. 2900-2907.

4. H. Schwarz and B.M. Humbel, in *Electron Microscopy: Methods and Protocols*, J. Kuo, Editor, (Humana Press Inc, Totowa, NJ) 2007, p. 229-256.

5. We are grateful to NWO Groot, FEI Company and Utrecht University for generous funding to purchase the DualBeam instrument. Further support was given by the European Network of Excellence, FP6, and the Dutch Cyttron project. We also like to thank E. van Donselaar and K. Vocking for preparing the HUVECs and Drs. M. Lebbink making the models.

Redeposition and differential sputtering of La in TEM samples of LaAlO$_3$ / SrTiO$_3$ multilayers prepared by FIB

Eduardo Montoya[1,2], Sara Bals[1], Gustaaf Van Tendeloo[1]

1. EMAT, University of Antwerp, Groenenborgerlaan 171, B-2020 Antwerp, Belgium
2. Instituto Peruano de Energía Nuclear, Av. Canadá 1470, Lima 41, Perú

Sara.Bals@ua.ac.be
Keywords: sample preparation, FIB

We report the results of a double-cross-sectional study of a FIB prepared TEM specimen, of a LaAlO$_3$ / SrTiO$_3$ multilayer. The purpose of the study is to gain information about the thickness profile of the specimen, as well as on the presence of redeposition and other possible preparation artefacts. The TEM sample was prepared by the method of internal lift out. The crude lamella is attached onto a halved TEM grid and thinned at glancing angles, using a FIB beam of 30 kV and 100 pA. Finally it is cleaned with low beam current and energy (29 pA, 5 kV), in an environment practically free of confinement, with minimal expected redeposition.

The procedure for the double-cross-sectional experiment consists of covering both faces of a TEM lamella with sputtered gold (Rubanov & Munroe 2003) and embedding it into a platinum brick. This platinum brick is FIB sliced to prepare cross-sectional *lamellae*, which, after internal lift out and welding to new grids, are thinned and cleaned for TEM examination.

In figure 1 it is shown that the use of FIB allowed obtaining qualitatively good HRTEM images. Figure 2a is a HAADF-STEM image of the region of interest. The arrow indicates the direction of the 5 kV ion beam during the final cleaning step. Figure 2b is a false colour EFTEM map of Ti and La, corresponding to the same region of interest. Note that the La is spread inside the amorphous layers remaining after the 5 kV FIB cleaning. The observed spreading of La suggests a redeposition origin for the amorphous layers. Figure 2c is an EFTEM map of Ga from the same region of interest shown in figure 2. It is accepted that redeposited material is always rich in Ga (Giannuzzi *et al.* 2005). Gallium enrichment can be barely appreciated in the contours of the cross section in figure 3. The relatively high Ga background is expected because the use of FIB milling to prepare the double cross sectional lamella.

Figure 3 shows intensity scan-plots across the multilayer for the marked regions of the HAADF-STEM and the EFTEM maps in figure 2. The scan-plots correspond to the same location of the same region of interest. The LAO layers as well as the STO layers are clearly visible. Even the signals corresponding to layers with thicknesses of one or two unit cells are clearly detected (see black arrows). The EFTEM line profiles show additional signals of La and Ti, labelled by question marks in figure 3a, which were not detected in the HAADF – STEM image. These signals are probably originating from redeposited amorphous material, which exhibiting lower density, leads to a relatively low HAADF-STEM signal. The curves in figure 3b are intensity scan-plots taken along the layers. The HAADF-STEM (blue) curve shows a constant thickness region,

M. Luysberg, K. Tillmann, T. Weirich (Eds.): EMC 2008, Vol. 1: Instrumentation and Methods, pp. 649–650, DOI: 10.1007/978-3-540-85156-1_325, © Springer-Verlag Berlin Heidelberg 2008

650

corresponding to the cross-section of the specimen, delimited by steeply rising walls, corresponding to a steeply change in the Z – contrast caused by the presence of the gold deposited to protect the lamella. The increased La intensity at the edges indicates that the amorphous layers on the surfaces of the sample are enriched with La. Monte Carlo simulations indicate that for every sputtered atom of La, there are also 9.71 atoms of O, 2.05 atoms of Sr, 1.18 atoms of Ti and 1.16 atoms of Al sputtered. Hence, an enrichment of La in the amorphous layer is to be expected.

1. Rubanov S, Munroe PR. Mater. Lett. **57** (2003) p. 2238.
2. L.A. Giannuzzi, B.I. Prenitzer, and B.W. Kempshall in "Introdution to focused ion beams instrumentation, theory, techniques and practice", ed. L. A. Gianuzzi and F.A. Stevie, (Springer) (2005), p. 13.
3. The authors are grateful to: **1)** M. Huijben and G. Rijnders of the MESA+ group at the University of Twente (Nl) for the growth of the multilayers. **2)** IAP programme – Belgian State - Belgian Science Policy. **3)** Fund for Scientific Research – Flanders.

Figure 1. HRTEM image of the LAO/STO multilayer. The thickness of the LAO layer marked by the arrow is 2 unit cells.

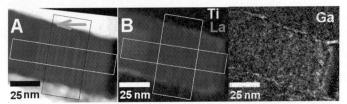

Figure 2. (A): HAADF-STEM image of the ROI. (B): EFTEM map of Ti and La of the same ROI. (C): map of Ga of the same region of interest presented in figure 5.

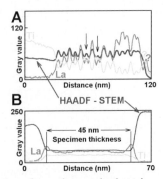

Figure 3. Intensity scan - plots from the marked regions of figure 2. (A): across the multilayer (left-right). (B): along the layers (bottom-up).

Fabrication and characterization of highly reproducible, high resistance nanogaps made by focused ion beam milling

T. Blom[1], K. Welch[2], M. Strømme[2], E. Coronel[1] and K. Leifer[1]

1. Division for Electron Microscopy and Nanoengineering, Department of Engineering Sciences, The Angstrom Laboratory, Uppsala University, Box 534, SE-751 21 Uppsala, Sweden
2. Division for Nanotechnology and Functional Materials, Department of Engineering Sciences, The Angstrom Laboratory, Uppsala University, Box 534, SE-751 21 Uppsala, Sweden

tobias.blom@angstrom.uu.se

Keywords: nanocontact, nanogap fabrication, FIB structuring, high resistance

Conductivity and charge transport mechanisms of single molecules and nanoparticles are topics of intense debate. In order to measure these properties, there is an urgent need for reliable and reproducible nanocontacts, with electrode spacings in the range of a few 10 nm and below. Possible applications derived from the study of the transport properties of single molecules and nanoparticles are for example in the field of molecular electronics.

The nanocontacts in this work are fabricated by using a combination of the structuring techniques; electron beam and photo lithography and focused ion beam (FIB) milling [1]. As the final step in the fabrication process, ~25 nm gaps with extremely high open gaps resistances are created. These high resistances are reached although Ga ions are implanted during the FIB sputter process which could make the gap slightly conducting. The Ga ion implantation is demonstrated in a local TEM analysis of the nanogap structure, see figure 2. In spite of Ga ion implantation during FIB structuring, high quality nanogaps with highest 'empty gap' resistances in the order of 500-1000 TOhm can be achieved. These resistance values are among the highest measured for nanogaps and therefore, the gaps can be used to investigate high resistance molecules. The nanogaps can be connected to external circuits either by using a probe station or by wire bonding the contact pads to a chip holder. The latter enables experiments at cryo temperatures or in in-situ studies inside the focused ion beam microscope.

The molecules or particles can be connected to the nanocontacts by electrophoretic trapping where an AC voltage is applied across the nanogap [2]. The nano-electrode fabrication process yields reproducible nano-electrodes and allows for the production of large numbers of electrodes. This is a requirement for the study of molecular electronics. Our approach consists of covering the nanocontacts with a self assembled monolayer (SAM) in order to get a well defined and reproducible surface. The voids in this SAM layer are filled in by dithiolated conductive molecules which are meant to be

M. Luysberg, K. Tillmann, T. Weirich (Eds.): EMC 2008, Vol. 1: Instrumentation and Methods, pp. 651–652, DOI: 10.1007/978-3-540-85156-1_326, © Springer-Verlag Berlin Heidelberg 2008

characterized by this setup. When a gold nanoparticle is trapped in a gap, it will be bridged by the thiolated molecules sticking out from each electrode surface.

Finally, we present our first results on I-V measurements on conductive molecules in between such nano-electrodes and our approach to trap one single nanoparticle in each nanogap, see figure 2.

1. T. Blom, K. Welch, M. Strömme, E. Coronel and K. Leifer, Nanotechnology **18** (2007), p. 285301.
2. A. Bezryadin, C. Dekker, G. Schmid, Appl. Phys. Lett. **71** (1997), 1273.

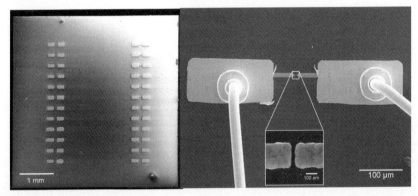

Figure 1. Left: SEM micrograph of a chip containing 24 gold nanocontacts. Right: SEM micrograph of one nanocontact with wire bonded pads. The inset shows a magnification of a 25 nm FIB milled nanogap.

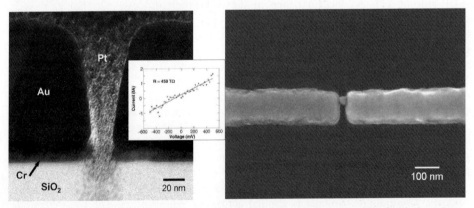

Figure 2. Left: TEM image of the cross section of a 20 nm gap. The platinum comes from the TEM sample preparation technique done in the FIB instrument. The inset shows an I-V measurement of an empty 25 nm gap where the resistance was calculated to ~450 TOhm. Right: SEM micrograph of a single 30 nm gold nanoparticle trapped in a gap.

TEM sample preparation on photoresist

F. Cazzaniga[1], E. Mondonico[2], E. Ricci[1], F. Sammiceli[1], R. Somaschini[1], S. Testai[1], M. Zorz[2]

1. STM6, Via Olivetti 2, 20041 Agrate Brianza(MI),Italy
2. FEI Italia S.r.l., Viale Bianca Maria 21, 20122 Milano, Italy

francesco.cazzaniga@st.com

Keywords: transmission electron microscope (TEM), Focused ion beam (FIB), electronic assisted deposition, Photoresist

In the last years the coming of the 193 nm lithography requested to the characterization technique a continuous improvement. Scanning electron microscopy (SEM), with Atomic force microscopy (AFM), remained the main techniques for the photoresist characterization, but the energy of the electron beam has drastically dropped down. In fact the materials used in the 193 nm lithography are easily modified by the electron beam if the energy is higher then 1 kV.

For the Critical Dimension measurements, performed in-line, the new instruments typically work at energies around 600-800 V while the SEMs, typically used in the laboratories to control the photoresist profile, are able to work at low energies like 500 V or even lower.

In these working conditions the resolution required for the characterization of 32-20 nm technologies is hard to achieve and a lot of artefacts normally occur; furthermore results of cross section analysis are strongly dependent on sample preparation (cleaving) and small structures are not easy to characterize.

Until now the transmission electron microscope (TEM) has not been considered as a tool to pervasively characterize these materials even if, due to their high energy, almost all the electrons pass through the sample and nearly no modification would occur.

This lack of interest is probably due to the difficulties encountered in TEM sample preparation.

In this work we present the possibility to use focused ion beam (FIB)-SEM to prepare samples for TEM even in presence of photoresist with negligible damage.

The fundamental solution proposed to avoid the damage due to FIB cuts consists in protecting the photoresists with an electronic assisted deposition made at low kV (less then 1 kV).

This solution, compared to the shield obtained by sputter deposition of metallic materials [1], guarantees the possibility to prepare TEM sample modifying only the region of interest on the whole wafer.

Different electron beam conditions have been evaluated in order to find the most efficient match of current and energy to deposit the covering layer on photoresist with negligible damage.

The same conditions of the SEM have been maintained during all the TEM sample preparations, to reduce the electron beam damage.

M. Luysberg, K. Tillmann, T. Weirich (Eds.): EMC 2008, Vol. 1: Instrumentation and Methods, pp. 653–654, DOI: 10.1007/978-3-540-85156-1_327, © Springer-Verlag Berlin Heidelberg 2008

654

TEM lamellas have been prepared on samples with photoresist defined with lines or holes for 65-45 nm technology. The comparison between TEM and SEM analyses doesn't show visible variation in the photoresist profile (see Figure1) with, on the other hand, a big improvement in resolution and material definition. (see Figure 2)

1.   J.S.Clarke, M.B.Schmidt, to be published
2.   We kindly acknowledge P.Targa, G.Limonta, F.Roveda, for TEM analyses and T.Crudeli.

Figure 1. SEM and TEM images with the same magnification of photoresist lines from the same wafer

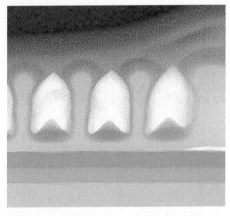

Figure 2. TEM image of holes opened in photoresist for 45 nm technology

Advanced FIB preparation of semiconductor specimens for examination by off-axis electron holography.

D. Cooper[1], R. Truche[1], A.C. Twitchett-Harrison[2], P.A. Midgley[3] and R.E. Dunin Borkowski[4]

1. CEA LETI - Minatec, 17 rue des Martyrs, 38054 Grenoble, Cedex 9, France.
2. Department of Materials, Imperial College London, Exhibition Road, London, SW7 2AZ, UK.
3. University of Cambridge, Department of Materials Science, Pembroke Street, Cambridge, CB3 3QZ, U.K.
4. Rafal Dunin-Borkowski, Center for Electron Nanoscopy, Technical University of Denmark DK-2800 Kongens Lyngby, Denmark

david.cooper@cea.fr

Keywords: off-axis electron holography, FIB, semiconductors, LASER anneal, annealing

Off-axis electron holography is a TEM based technique that can be used to reconstruct phase and amplitude images of a specimen from an interference pattern that is formed by using an electron biprism. As the phase of an electron is very sensitive to changes in potential in a specimen, such as from the presence of dopants, electron holography can be used to fulfil the requirements of the semiconductor industry for a technique that can quantitatively map dopants with nm-scale resolution [1].

Focused ion beam (FIB) milling is a promising technique that can be used to prepare state of the art semiconductor devices for examination by electron holography due to its ease of use and excellent site specificity. However, the FIB introduces many artefacts into the specimens in the form of Ga implantation and damage of the specimen surfaces that are observed in the form of an amorphous layer and electrically 'inactive' layer [2]. The presence of the electrically 'inactive' thickness results in experimental phase measurements across the *p-n* junctions that are much smaller than are predicted by theory.

Focused ion beam milling is known to generate defects deep into the specimen, and it is proposed that these defects lie in mid-band gap energy states in GaAs and Si, giving rise to the thick electrically 'inactive' layers that are observed experimentally. We will show that by annealing the specimens *in situ* in the transmission electron microscope, the concentration of these defects in the specimens is reduced, and the dopants are re-activated thereby increasing the experimental phase change measured across the junction. The electrically inactive thickness is reduced and an improvement in the signal-to-noise ratio is observed in the phase images.

Figure 1. shows phase images of a 240-nm-thick GaAs specimen containing a symmetrical 1×10^{18} cm^{-3} doped *p-n* junction (a) before and (b) after a 500 °C *in situ* anneal. The improvement in the signal-to-noise ratio is clear. Figure 2.(a) shows the phase measured across a 300-nm-thick GaAs specimen before and after annealing showing the increase in the phase measured across the junctions. Figure 2.(b) shows the

M. Luysberg, K. Tillmann, T. Weirich (Eds.): EMC 2008, Vol. 1: Instrumentation and Methods, pp. 655–656, DOI: 10.1007/978-3-540-85156-1_328, © Springer-Verlag Berlin Heidelberg 2008

step in phase measured across a series of junctions as a function of the crystalline specimen thickness measured using convergent beam electron diffraction (CBED) both before and after a 500 °C anneal. By extrapolating the gradients to the x-axis the electrically 'inactive' thickness of the specimen is shown to have been reduced from 80 to 17 nm on each surface after annealing. We will also show that by annealing Si p-n junctions *in situ* at a temperature of only 300 °C, the electrically inactive layer is reduced from 25 to 5 nm [3]. New results from using a LASER to anneal the specimens will also be presented.

We will show that by reducing the operating voltage of the FIB we can reduce the electrical damage to both Si and GaAs specimens. Finally, we will discuss the effects of using ions other than Ga, such as Si, Ar and O to prepare semiconductor specimens for examination using off-axis electron holography.

1. International Technology Roadmap for Semiconductors, 2005 ed. http://public.itrs.net
2. A.C. Twitchett, R.E. Dunin-Borkowski, R.J. Hallifax, R.F. Broom and P.A. Midgley, Phys. Rev. Lett. **88**, 2383021 (2002).
3. D. Cooper, A.C. Twitchett, P.K Somodi, P.A. Midgley, R.E. Dunin-Borkowski, I. Farrer and D.A. Richie, Appl. Phys. Lett. **88**, 063510 (2006).

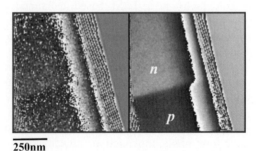

250nm

Figure 1. Shows phase image of a 240-nm-thick GaAs specimen containing a *p-n* junction recorded (a) immediately after ion milling and (b) after a 500 °C *in situ* anneal.

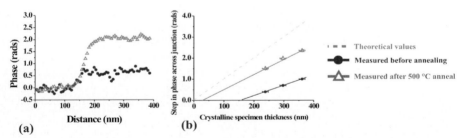

(a)　　(b)

Figure 2. (a) Shows phase profiles acquired across a GaAs specimen containing a *p-n* junction before and after an in situ anneal at 500 °C. (b) Shows the step in phase measured across a series of *p-n* junctions as a function of the specimen thickness measured by CBED.

Comparison of ion- and electron-beam-induced Pt nanodeposits: composition, volume per dose, microstructure, and in-situ resistance

R. Córdoba[1], J.M. De Teresa[2], A. Fernández-Pacheco[1,2], O. Montero[1], P. Strichovanec[1], A. Ibarra[1], and M.R. Ibarra[1,2]

1. Instituto de Nanociencia de Aragón, Universidad de Zaragoza, Zaragoza, 50009, Spain
2. Instituto de Ciencia de los Materiales de Aragón, Universidad de Zaragoza-CSIC, Facultad de Ciencias, Zaragoza, 50009, Spain

rocorcas@unizar.es
Keywords: Dual beam, IBID, EBID, Resistance, TEM

Electron- and ion-beam-induced deposition (EBID and IBID) of metallic materials is one major application of "dual beam" systems [1]. Previous studies on EBID of Pt have shown the decrease of the deposition rate as a function of the beam energy [2,3]. Therefore in the case of IBID of Pt, the deposition rate was found to vary little [2,3]. The composition and microstructure of the Pt deposits have been studied by means of energy-dispersive X-ray (EDX) analysis and transmission electron microscopy (TEM) [2,4,5]. These studies showed that the deposits consist of Pt-rich inclusions in a carbonaceous matrix.

A systematic study of the composition, volume per dose, microstructure, and electrical transport properties of EBID and IBID of Pt were performed in a dual beam equipment (Nova 200 NanoLab) as a function of the beam energy and current [6]. The composition was measured by means of an EDX detector (Oxford Instruments). The volume per dose was calculated after performing cross-sections of the deposited material measuring the deposit thickness. The microstructure was investigated by TEM, (JEOL 1010, 200 kV). Finally, the in-situ deposit resistance was controlled by means of two electrical microprobes (Kleindiek).

The atomic Pt content in EBID decreases from 17 to 11% as a function of the incident electron-beam energy, whereas the volume per dose (EBID) dramatically decreases by a factor four. This can be explained by the decrease in the amount of secondary electrons reaching the sample surface. For IBID, the atomic Pt content also decreases from 27 to 17% as a function of the incident ion-beam energy whereas the Ga content is quite constant (10%), which was related by the sputtering rates. However, the volume per dose for IBID increases (from 0.4 to 0.7 $\mu m^3/nC$), which would be explained by the slight changes in the energy dependence of the secondary electron yield. We may conclude that the ion-induced Pt deposits are several orders of magnitude more conductive than the electron-induced ones (see Figure 1), which is correlated with the percolation of Pt nanocrystal grains embedded in a carbon amorphous matrix (see Figure 2).

M. Luysberg, K. Tillmann, T. Weirich (Eds.): EMC 2008, Vol. 1: Instrumentation and Methods, pp. 657–658, DOI: 10.1007/978-3-540-85156-1_329, © Springer-Verlag Berlin Heidelberg 2008

658

1. L. A. Giannuzzi, and F. A. Stevie in "Introduction to Focused Ion Beams", ed. Springer, (Boston) (2005), p. 357.
2. R. M. Langford, T. X. Wang, and D. Ozkaya, Microelectron. Eng. **84** (2007), p. 784.
3. S. Lipp et al., Microelectronics and Reliability **36** (1996), p. 1779.
4. L. Peñate-Quesada, J. Mitra, and P. Dawson, Nanotechnology **18** (2007), p. 215203.
5. S. Frabboni, G. C. Gazzadi, and A. Spessot, Physica E **37** (2007), p. 265.
6. J. M. De Teresa et al., Nanotechnology, submitted.
7. We kindly acknowledge the help of financial support by Spanish Ministry of Science (through project MAT2005-05565-C02-01 and –02 including FEDER funding), and the Aragon Regional Government and experimental help and discussions with Dr. J. Arbiol (Serveis Cientifico-Tecnics, Universitat de Barcelona), Dr. J. Sesé and L. Serrano.

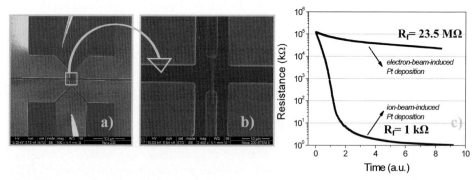

Figure 1. In-situ measurement of the electrical resistance during EBID and IBID deposits on Si_3N_4. a) Image of the microprobes contacting micropatterned Al pads with a gap in the center; b) Image of the grown Pt deposit contacting the Al pads; c) Electrical resistance versus deposition time.

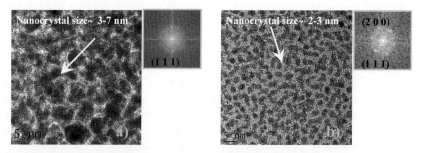

Figure 2. HRTEM images of IBID a) and EBID b), which clearly gives evidence for the different microstructure in each case. Insets are the FFT of platinum nanocrystal domains.

In-line FIB TEM sample preparation induced effects on advanced fully depleted silicon on insulator transistors

V. Delaye, F. Andrieu, F. Aussenac, C. Carabasse

CEA-LETI, MINATEC, 17 rue des Martyrs, 38054 Grenoble Cedex 9 France

vincent.delaye@cea.fr

Keywords: FIB-SEM, dual beam, TEM sample preparation, wafer return, TEM, in-line, CMOS, FDSOI, transistor.

Focused ion beam coupled with scanning electron microscope dualbeam (FIB-SEM) is now widely used for fast *in situ* lift out transmission electron microscopy (TEM) sample preparation in the semiconductor industry. This instrument allows examination of full wafers up to 300 mm and the *in situ* sample lift out chunk method [1] can avoid destructive characterization. A wafer return strategy for a low cost in-line process control or step by step TEM analysis has been thus considered [2, 3].

Recently we presented results relating to the impact of such sample preparation method on electrical performances of contiguous transistors from 32 nm node fully depleted silicon on insulator (FDSOI) wafers during critical front-end of line process steps [4]. The three process steps involved were: after active area patterning (Step 1), during gate stack formation (Step 2) and before raised source drain epitaxial growth (Step3). We proved that there was no measured electrical impact on dies without lift out and on pattern 500 µm far from a FIB crater. Within a same pattern, less than 500 µm far from a crater, the electrical results were depending on the process step (Table 1).

In order to study the FIB sample preparation local induced defectivity, we prepared (Figures 1 and 2) and observed TEM samples with transistor 70 µm far from a FIB crater at each step. A reference sample, with a transistor located a few millimeters far from a FIB crater, has been observed for comparison (Figure 3 to 6).

When sample lift out occurs just after active area patterning (Step 1), the active silicon layer is damaged below and at each side of the gate (Figure 4). This is due to the amorphisation of the 10 nm thin silicon layer during the FIB processing. During the integration process, the total thermal budget induced re-crystallisation of this layer in mono or poly-silicon, depending on the distance from the FIB crater. Electrical results seem to show that further than 180 µm the silicon channel is mono-crystalline and consequently the devices are functional.

At step 2, the contiguous transistors are not functional. We assume that a thin tungsten alloy layer has been locally formed during the FIB-assisted deposition of the protective layer with tungsten hexacarbonyl gas (dark area, Figure 1). That must be the reason why the gate stack etching has been stopped and the transistor un-patterned (Figure 5).

Amorphisation occurred only in the source/drain region at Step 3 because the channel was protected by the gate stack during sample preparation (Figure 6). The transistors are functional but with degraded performances.

M. Luysberg, K. Tillmann, T. Weirich (Eds.): EMC 2008, Vol. 1: Instrumentation and Methods, pp. 659–660, DOI: 10.1007/978-3-540-85156-1_330, © Springer-Verlag Berlin Heidelberg 2008

After proving that there was no measured electrical impact on dies without lift out and on pattern 500 μm far from a FIB crater, we now show the nature of *in situ* FIB sample preparation local impact. Thereby, we are now able to choose the process steps where sample extraction should take place with the lowest impact.

1. T. Tessner, European Semiconductors, 26(9), September 2004
2. N. Bicaïs-Lépiney et al., Proceedings of SPIE vol. 6152, Metrology, inspection, and process control for microlithography XX (2006)
3. D. Mello et al., Nuclear Instruments and Methods in Physics Researsh B 257 (2007) 805-809
4. V. Delaye et al., Microelectron. Eng. (2008), article in press : doi:10.1016/j.mee.2008.01.092

Table 1. Electrical performances of contiguous transistors (*Vt : threshold voltage*)

Process Step number	Description	Electrical performances of other dies Transistors > 500 μm far from a FIB crater	Electrical performances in the same pattern Transistors between 180 μm and 500 μm far from a FIB crater
Step 1	After active area patterning	No measured impact on performances	Work with lower Vt
Step 2	During gate stack formation		Do not work
Step 3	Before raised source drain epitaxial growth		Work with lower Vt

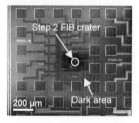

Figure 1. SEM pattern view at Step 2

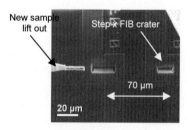

Figure 2. Sample lift out of contiguous transistors using the chunk method

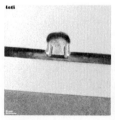

Figure 3. TEM view of Ref. transistor

Figure 4. TEM view of Step 1 transistor

Figure 5. TEM view of Step 2 transistor

Figure 6. TEM view of Step 3 transistor

The development of cryo–FIBSEM techniques for the sectioning and TEM analysis of the cell-biomaterial interface.

H.K. Edwards,[1] M.W. Fay,[2] C.A. Scotchford,[1] D.M. Grant[1] and P.D. Brown[1]

1. School of Mechanical, Materials and Manufacturing Engineering,
2. University of Nottingham Nanotechnology and Nanoscience Centre, both at the University of Nottingham, University Park, Nottingham, NG7 2RD, UK.

michael.fay@nottingham.ac.uk
Keywords: FIBSEM, cryo-FIBSEM, cell, biomaterial

The imaging and chemical mapping of the interface between cells and biomaterials using transmission electron microscopy (TEM) has often proved to be an interesting a challenge, due to the limitations of standard specimen preparation techniques. In the past, ultramicrotomy has mainly been used to section tissue, however harder materials such as the interface between bone and implanted biomaterials have also been sectioned [1]. This has provided an effective and generally reliable methodology for the analysis of materials which are close in material hardness, However, the preparation of ultrathin samples of soft cells attached to hard materials may introduce cutting artefacts into the specimen. Also, ultramicrotomy requires cells to be chemically fixed, dehydrated and embedded in resin prior to cutting, consuming time and resources. Recently, cryo focused ion beam scanning electron microscopy (cryo-FIBSEM) coupled with cryo transfer for TEM has emerged as an alternative pathway to ultramicrotomy for the preparation of such samples.

Cryo-FIBSEM is potentially a powerful tool for the analysis of soft and hydrated biological materials, but as a technique still undergoing development and relatively few instances of its use in the area of the life sciences have been demonstrated. However, notable examples of the applicability of this method include the analysis of samples such as yeast cells [2], lymphoid tumours [2], tobacco plant petals [3] and human epithelia [4]. Cryo-FIBSEM has also been employed for the preparation of TEM specimens such as E. Coli [5]. However, despite the benefits cryo-FIBSEM offers over ultramicrotomy in terms of processing time and artefact formation, TEM sections of hydrated cells attached to biomaterials have not been achieved using this technique to date. This is in part due the practical problems associated with the ion beam deposition of Pt or W under cryo conditions, which leads to the production of ineffective protective coatings, with consequent limitation to the process of site-specific specimen lift-out.

Through the cryo fixation of cells attached to biomaterial foils and the cryo FIB milling of access trenches, these problems may be overcome and ultrathin specimens through the cell-biomaterial bi-layer may be produced in an H-bar style. To develop this methodology, model systems of human mesenchymal stem cells (MSCs) and human osteoblasts (HOBs) on Ti have been examined. MSCs and HOBs were seeded onto Ti foils 0.025-0.25 mm in thickness and cultured for three days. The preparation of FIB

M. Luysberg, K. Tillmann, T. Weirich (Eds.): EMC 2008, Vol. 1: Instrumentation and Methods, pp. 661–662, DOI: 10.1007/978-3-540-85156-1_331, © Springer-Verlag Berlin Heidelberg 2008

cross-sections was carried out using an FEI Quanta 200 3D FIBSEM fitted with a PT2000 Quorum cryo-transfer unit. For cryo-FIBSEM, both unfixed cell/Ti samples and samples fixed in 3 % glutaraldehyde and stained with OsO₄ were washed in phosphate buffered saline solution before being plunged into slush nitrogen and transferred into the cryotransfer unit, which was held at -140 °C. The samples were transferred into the FIBSEM chamber, also held at -140 °C, and after gentle sublimation at -90 °C to remove any ice crystal formation, were sputter coated with Pt. Following deposition of a further W based protective layer, trenches were milled with sequential ion beam currents ranging from 20 nA down to 100 pA (Figure 1). The specimens containing electron transparent lamellae were then transferred to a Jeol 2100f TEM using a Gatan cryo transfer holder and assessed using bright field and scanning TEM and energy dispersive X-ray analysis (Figure 2).

Preliminary results will be presented and the practicalities associated with this methodology will be discussed along with an appraisal of artefact generation.

1. A. Palmquist, T. Jarmer, L. Emanuelsson, R. Brånemark, H. Engqvist and P. Thomsen, Acta Orthadaedica **79** (2008) p.78.
2. J. Heymann, M. Hayles, I. Gestmann, L. Giannuzzi, B. Lich and S. Subramaniam, J. Structural Biology **155** (2006) p.63.
3. M. Hayles, D. Stokes, D. Phifer and K. Findlay, J. Microscopy **226** (2007) p.263.
4. J. McGeoch, J. Microscopy **227** (2007) p.172.
5. M. Marko, C. Hsieh, R. Schalek, J. Frank and C. Mannella, Nature Methods **4** (2007) p.215.
6. This work was supported by the EPSRC under grant EP/E015379/1.

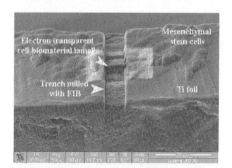

Figure 1. Secondary electron images of cell-biomaterial bi-layer lamellae produced through cryo-FIBSEM milling.

Figure 2. HAADF image of a lamella of a fixed, unstained mesenchymal stem cell attached to Ti, produced through cryo-FIBSEM milling.

Three-slit interference experiments with electrons

S. Frabboni[1,2], G.C. Gazzadi[2] and G. Pozzi[3]

1. Department of Physics, University of Modena and Reggio Emilia and CNR-INFM-S3, Via G. Campi 213/a, 41100 Modena, Italy
2. CNR-INFM-S3, Via G. Campi 213/a, 41100 Modena, Italy
3. Department of Physics, University of Bologna, v. B. Pichat 6/2, 40127 Bologna, Italy

stefano.frabboni@unimore.it

Keywords: fib nanofabrication, interference, diffraction

Multiple-slit interference experiments with electrons have been reported in the pioneering experiments due to Jönsson [1] who was able to produce slits in the micrometer range and to observe them using a dedicated electron optical bench. Today advances in technology make it possible to perform these experiments using commercial instrumentation: a focused ion beam (FIB) machining device allows an easy fabrication of the slits in the nanometer range, and a conventional transmission electron microscope (TEM) can perform the role of the diffraction camera [2].

Aim of this work is the realization of a three slits interference experiment exploiting this combined FIB/TEM approach.

The slits were fabricated by FIB milling using a dual beam apparatus (FEI Strata DB235M) on a commercial silicon-nitride membrane window commonly used for TEM sample preparation [3]. The sample consists of a 3 mm-diameter, 200 µm-thick silicon frame, with a 100 µm x 100 µm square window at the centre, covered with a 500 nm-thick silicon-nitride membrane. The membrane thickness was chosen to minimize electron transmission from regions other than the opened slits. To open the slits, a 10 pA beam, corresponding to a nominal spot-size of 10 nm, was scanned along single pixel lines, 1.5 µm long and spaced by 200 nm, each line for 108 s. The passage through the membrane was monitored by detecting change in brightness of the ion-induced secondary electron emission. From the width value we notice the remarkably high aspect ratio (slit depth/slit width) obtained with the FIB milling technique. The specimen is then inserted in the TEM and observed in both the imaging mode and the low-angle electron diffraction mode with an effective camera length of about 100 m. It must be stressed here that we use a conventional 200-keV microscope, equipped with a standard LaB$_6$ electron source. In order to have enough lateral coherence at the plane of the slits, the condenser aperture and the spot size were selected as small as possible consistent with good signal-to-noise ratio at reasonable exposure time, 120 seconds in the present experiment.

The TEM image given in Figure 1(a) shows that the transmittance of the slits is reasonably uniform over a mean width and length of 30 nm and 1600 nm, respectively, with a spacing, d, of 240 nm. The relative low angle diffraction pattern, Figure 1(b), shows that on both sides of the bright transmitted beam (the 500 nm thick film is unfortunately not completely opaque to 200 keV electrons) the wide diffraction image of the slit is modulated by the three-slit interference pattern. Fringe visibility is

M. Luysberg, K. Tillmann, T. Weirich (Eds.): EMC 2008, Vol. 1: Instrumentation and Methods,
pp. 663–664, DOI: 10.1007/978-3-540-85156-1_332, © Springer-Verlag Berlin Heidelberg 2008

enhanced by slightly defocusing the diffraction lens, Figure 1(c). Here the image of the bright central beam broadens to give an out-of-focus Fresnel image of the large area illuminated by the electron beam, whereas at its centre the interference pattern of the three slits is still visible and with a better contrast than in Figure 1 (b). This can be explained by noting that, as the slits are much smaller that the illuminated area, their diffraction image still falls in the Fraunhofer range. The line scan, five pixels wide, recorded through the interference maxima of Figure 1(c), is shown in Figure 1(d). The equally spaced interference maxima are clearly observed in this intensity plot, which also shows the secondary maxima due to the three-beam interference.

1. C. Jönsson, Z. Phys. 161, 454-474 (1961), English translation : Am. J. Phys. 42, 4-11 (1974).
2. S. Frabboni, G.C. Gazzadi and G. Pozzi, Am. J. Phys. **75** (2007) p.1053
3. SPI Supplies (http://www.2spi.com), product number 4091SN-BA.

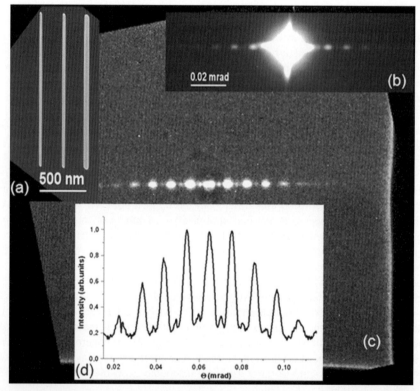

Figure 1. (a) TEM image of the three slits. (b) Low angle diffraction pattern of the three slits. (c) Out-of focus low angle diffraction pattern showing both the image of the silicon nitride membrane(gray background) and the three slits interference fringes. (d) Line scan, five pixels wide, of the interference pattern reported in (c).

Contrast in ion induced secondary electron images

Lucille A. Giannuzzi[1], Mark Utlaut[2], and Lynwood Swanson[1]

1. FEI Company, 5350 NE Dawson Creek Drive, Hillsboro, OR USA 97124
2. Department of Physics, University of Portland, Portland, Oregon USA 97203

lucille.giannuzzi@fei.com
Keywords: FIB, ion contrast, stopping power, secondary electrons

Previous reports have discussed properties of ion induced secondary electron images (ISE) where it has been generally accepted that ISE and electron induced secondary electron images yield complementary material or elemental contrast [1,2]. In addition, it has also been reported that secondary electron (SE) emission in ISE images decreases with increasing atomic number [3,4]. In this report, anomalies to the previous work are presented. In particular, it is observed that the ISE contrast obtained from 30 keV Ga^+ focused ion beam (FIB) imaging does not follow a monotonic variation as a function of atomic number, and in some cases, the contrast increases as atomic number increases.

Al, Ti, Cr, Cu, Mo, and Ag were sputtered coated onto (100) Si in order of increasing atomic number [5]. Each layer was deposited to ~ 100-200 nm in thickness and consisted of nanocrystalline grains. The small grain size was useful to avoid channelling contrast effects in the measurements. The multi-layered sample was cross-sectioned with conventional focused ion beam (FIB) techniques using an FEI Helios NanoLab 600 DualBeam instrument. A cross-section specimen was also prepared for (scanning) transmission electron microscopy ((S)TEM) and the identity of the layers were confirmed using an FEI Tecnai F30 using high angle annular dark field (HAADF) STEM and x-ray energy dispersive spectrometry. ISE images were obtained at a 45° incidence angle using a 9.7 pA beam current of Ga^+ ions operating at 30 keV. Extreme care of the contrast measurements from the ISE images was taken. Specifically, contrast due to ion channelling effects as well as any "creeping crud" artefacts due to Ga^+ intermetallic phase formation during ion scanning was avoided [6,7].

Figure 1 shows the relative contrast measurements obtained from the ISE image [8]. The contrast was normalized to the brightest contrast layer (i.e., Cu) in the image. Note that the lines joining the individual points in figure 1 are for visualization purposes only. Indeed, the ISE contrast does not show a monotonic contrast variation as a function of atomic number and shows an increase in contrast for Cu compared to lower atomic number elements of Al, Si, Ti, and Cr.

1. K. Nikawa, J. Vac. Sci. Technol. B9 (1991), p. 2566.
2. Y. Sakai et al., Appl. Phys. Lett., 73 (1998), p. 611.
3. T. Ishitani and H. Tsuboi, Scanning, 19 (1997) p. 489.
4. T. Ishitani et al., J. Electron Microsc., 51 (2002) 207.
5. Prepared by North Carolina State University Nanofabrication Facility, Raleigh, NC 27695.
6. M.W. Phaneuf et al., Microsc. Microanal. 8 (Supp 2), (2002) p. 52.

M. Luysberg, K. Tillmann, T. Weirich (Eds.): EMC 2008, Vol. 1: Instrumentation and Methods, pp. 665–666, DOI: 10.1007/978-3-540-85156-1_333, © Springer-Verlag Berlin Heidelberg 2008

7. J.R. Michael, Microsc. Microanal., 12 (Supp 2), (2006), p. 1248 CD
8. Lucille A. Giannuzzi et al., to appear in Microsc. Microanal., 14 (Supp 2) 2008.

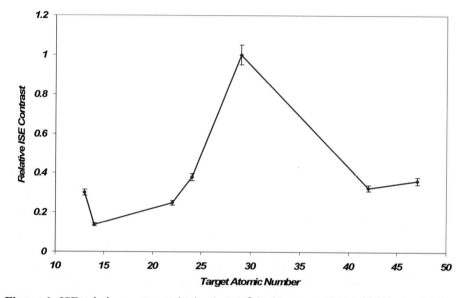

Figure 1. ISE relative contrast obtained at 45° incidence angle of Al, Ti, Cr, Cu, Mo, and Ag sputtered coated onto (100) Si. The individual points are joined by lines only to aid the eye.

Quantitative in situ thickness determination of FIB TEM lamella by using STEM in a SEM

U.Golla-Schindler

1. Institute for Mineralogy, University of Muenster, Corrensstr. 24, 48149 Muenster, Germany

golla@uni-muenster.de

Keywords: SEM, STEM, FIB, thickness determination

One of the practical problems of preparing a FIB lamella for use in a TEM is knowing the thickness attained during the fine polishing process. A quantitative in situ method, which enables the thickness determination with accuracy in the range of some nm would therefore be very helpful. The aims of the following project are first the development of a method which enables the in situ thickness measurement of thin free supported layers and second the design of a combined STEM-UNIT which enables the use of this method for the in situ thickness measurement during the fine polishing process. It is known that the transmission T, of thin films of the mass thickness $x = \rho \cdot t$ (ρ is the density and t the thickness of the thin film) can be described by an exponential law [1]:

$$T(E_0, \alpha, \beta) = \frac{I(E_0, \alpha, \beta)}{I_0} = e^{\frac{-x}{x_K(E_0, \alpha, \beta)}}$$

where $I(E_0, \alpha, \beta)$ and I_0 are the intensities with and without the specimen, respectively, α is the illumination semiangle, β the collection semiangle, E_0 the energy of the incident electrons and $x_K(E_0, \alpha, \beta)$ the contrast thickness. The first test specimen for the proof of concept are thin Al-layers supported on a TEM grid with different mass thickness. The thickness of the thin films was determined during the preparation process and confirmed by using transmission electron microscopy, the LIBRA 200FE and energy filtered TEM. The film thickness t can be determined by a measurement of the relative specimen thickness t/λ_i depending on the mean free path of inelastic scattering λ_i which can be calculated using the equation [2]:

$$\lambda_i \approx \frac{106 \cdot F \cdot \left(\dfrac{E_0}{E_m}\right)}{\ln\left(2 \cdot \beta \cdot \dfrac{E_0}{E_m}\right)}$$

where λ_i is given in nm, β in mrad, E_0 in keV, E_m the mean energy loss in eV and F is a relativistic factor (0.618 for E_0=200 keV).

M. Luysberg, K. Tillmann, T. Weirich (Eds.): EMC 2008, Vol. 1: Instrumentation and Methods, pp. 667–668, DOI: 10.1007/978-3-540-85156-1_334, © Springer-Verlag Berlin Heidelberg 2008

668

We have developed a series of STEM-UNITs, which can be accommodated on every conventional SEM specimen table. In all versions the STEM signal is detected with the normal chamber SE detector [1,3] and therefore delivers TV-rate images that decrease significantly the contamination due to the short dwell times in comparison to the alternative detection using a semiconductor device. The transmission $T = S(E_0, \alpha, \beta)/S_0$ results from the converted SE signals at meshes with the film $S(E_0, \alpha, \beta)$ and at meshes without the film S_0 Figure 1. For the assignment of the sample thickness additionally the contrast thickness has to be known. That can be determined by using a standard specimen, which has the same chemistry and a well-known thickness. For a thin Al film of a thickness of 50 nm we obtained a contrast thickness of 4,3 $\mu g/cm^2$ which was then used to determine a thickness of 22 nm for the second Al film.. This is in a good agreement with the expected film thickness of 20 nm.. This good result for the first test gives confidence that this method will end up with an accuracy in the range of some nm. The next step will now be to combine the STEM-UNIT used for the transmission measurements with a specially-designed micromanipulator that enables the rotation of the cut TEM lamella relative to the incident electron beam. This will enable the thickness to be determined by measuring T during the fine polishing of the TEM lamella [4].

1. U. Golla, B. Schindler and L. Reimer, Journal of Microscopy **173** (1994) p. 219-225.
2. R.F. Egerton Electron Energy-Loss Spectroscopy in the Electron Microscope. Plenum Press, New York (1986), p.305.
3. U. Golla-Schindler, Proceedings of the 13th European Microscopy Congress Antwerpen Vol. I p. 409-410 (De 10211977, EP 1347490).
4. I thank I. Sobchenko (Institut für Materialphysik Prof. Schmitz) for preparing the free supporting films, Kleindiek Nanotechnik GmbH and Carl Zeiss NTS GmbH for their aid.

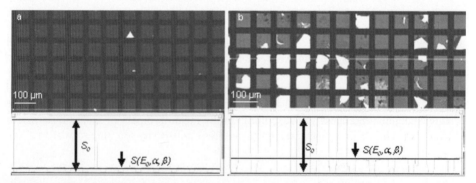

Figure 1. SEM images of free supporting thin Al films recorded in the transmission mode a) sample with known thickness used to assign x_k b) sample where the film thickness was obtained.

Analysis of ion diffusion in multilayer materials by depth profiling in a Crossbeam FIB-SIMS microscope

A. Hospach, A.M. Malik, W. Nisch, and C. Burkhardt

NMI Natural and Medical Sciences Institute at the University of Tuebingen, Germany

a.hospach@fz-juelich.de

Keywords: secondary ion mass spectrometry, SEM, focused ion beam, polyimide, ion diffusion

The potential of secondary ion mass spectrometry combined with scanning electron microscopy (Figure 1a) has been investigated for the characterisation of barrier layers which are used to avoid ion diffusion in polymers. These barrier layers are playing amongst others an important role as encapsulation materials of micro chips for biomedical applications. If a micro chip is implanted in vivo, e.g. a microelectrode array sensor (Figure 1b) [1], it is exposed to aggressive body fluids. To protect the devices from chemical attacks caused by these fluids it is necessary to cover them with dense, flexible and biocompatible materials.

Unfortunately, there is no material which meets all requirements. Hence, there are approaches to combine the advantages of several materials in a multilayer encapsulation and to decrease the pinhole probability of one layer, e.g. HMDSO- (hexamethyldisiloxane) alternating layers or PI-Ti-Au-Ti-PI (polyimide titanium gold) sandwich. SIMS depth profiles of this two film systems are shown in Figure 2b and 3b.

For FIB-SIMS measurements we used a quadrupole mass spectrometer (QMG 420, Pfeiffer Vacuum GmbH, Germany) which was attached directly to a FIB-SEM microscope (Crossbeam Neon® 40, Carl Zeiss SMT AG, Germany). The main advantages of SIMS compared to other analytical techniques are its high depth resolution, its high detection limit and the possibility to detect fragments of molecules. In combination with a focused ion beam microscope a high lateral resolution is added. To get a comparison to other mass spectrometers some benchmark data were determined which are also dependent from the used ion source (in this case Gallium). A depth resolution ($\Delta Z_{84\text{-}16}$) of 20 nm, a mass resolution ($R=m/\Delta m$) of 300 and a lateral resolution of 50 nm was achieved. Masses from 1 (hydrogen) up to 207 (lead) could be detected. Depth profiles on multilayers could be resolved with a lateral resolution of $100\times100\text{nm}^2$, however, bigger areas show better sensitivity for depth profiles. The detection limit of Sodium is 170ppb (in polyimide).

To simulate the body fluids and the involving problem of ion diffusion into the implant a saline solution was used in our experiments. The diffusion of these saline ions (sodium, potassium) into the polymer could be shown by SIMS depth profiles. The encapsulation material was polyimide. Figure 3a/b shows the results of combined SEM imaging of the FIB cross section and the corresponding SIMS depth profile.

1. K. Molina-Luna, M.M. Buitrago, B. Hertler, M. Schubring, F. Haiss, W. Nisch,
2. J.B. Schulz, A.R. Luft, Cortical stimulation mapping using epidurally implanted thin-film microelectrode arrays, Journal of Neuroscience Methods 161 (2007) 118–125.

M. Luysberg, K. Tillmann, T. Weirich (Eds.): EMC 2008, Vol. 1: Instrumentation and Methods, pp. 669–670, DOI: 10.1007/978-3-540-85156-1_335, © Springer-Verlag Berlin Heidelberg 2008

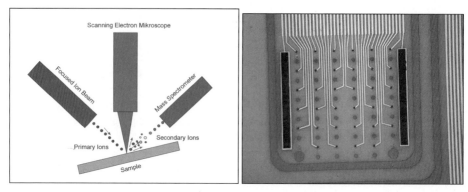

Figure 1a/b. Schematic configuration of a crossbeam FIB/SEM microscope combined with a mass spectrometer (left) and a micrograph of the microelectrode array with polyimide encapsulation (right).

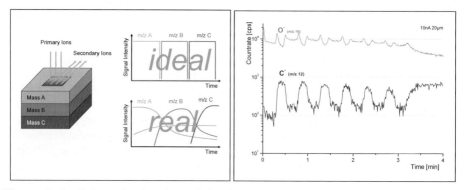

Figure 2a/b. Schematic drawing of the SIMS sputter process with resulting depth profiles (left) and a real depth profile of a HMDSO 14x multilayer with alternating carbon content (right).

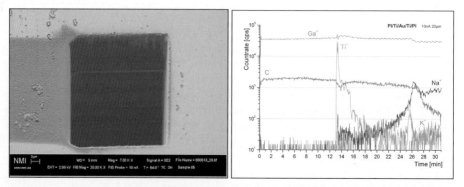

Figure 3a/b. SEM image of a FIB sputter crater in a polyimide sandwich separated by a 300nm thin Ti-Au-Ti layer (left) and the corresponding SIMS depth profile (right).

Growth of In$_2$O$_3$ islands on Y-stabilised ZrO$_2$: a study by FIB and HRTEM

J.L. Hutchison[1], A. Bourlange[2], R. Egdell[2] and A. Schertel[3]

1. Department of Materials, University of Oxford, Parks Road, Oxford OX1 3PH, UK
2. Department of Chemistry, University of Oxford, Mansfield Road, Oxford OX1 3PF, UK
3. Carl Zeiss SMT AG, Carl-Zeiss-Strasse 56, 73447 Oberkochen, Germany

john.hutchison@materials.ox.ac.uk

Keywords: indium oxide, MBE growth, FIB, HRTEM, epitaxy

Indium oxide is an important functional oxide, being used for optically transparent, electrically conducting layers. We have successfully grown high quality In$_2$O$_3$ layers on yttria-stabilised cubic zirconia substrates. Y-stabilised ZrO$_2$ has a cubic (fluorite) structure, with space group Fm3m and lattice parameter a_o = 0.514 nm for the 17% Y doping as used in this study. In$_2$O$_3$ has the bixbyite structure, this being body-centred cubic with space group I213 and a_o = 1.012 nm. There is thus only a very small dimensional mismatch of ~ 1.6% between 2a_o for zirconia and a_o for In$_2$O$_3$. We note that the two structures have essentially similar cation arrays, but in the case of In2O3, ¼ of the anion sites of the common fluorite-type structure are vacant. We have used this close lattice matching as a basis to promote good eptaxial growth. Indium oxide layers were grown on (100) Y-stablised zirconia substrates by plasma-assisted MBE, as described elsewhere [1]. While growth at moderate temperatures, around 650° C produced continuous, well-ordered films [1], growth at higher temperature, typically around 900°C, was found to promote film break-up into a array of micron-sized islands., displaying some remarkable features.

Preliminary investigation by tapping mode atomic force microscopy indicated remarkably uniform azimuthal orientation with respect to the substrate, and a narrow size distribution, as shown in Figure 1. Cross-sectional lamellae suitable for HRTEM were then prepared by focussed-ion-beam milling, using a Zeiss NVision 40 FIB instrument. A secondary-electron SEM image confirmed the excellent alignment of the islands, and in addition showed very clear facetting, as shown in Figure 2. A section was then selected (Figure 3) so as to include several islands, and a lamella was then FIB-cut, thinned and polished to electron transparency before being Pt-welded to a Cu grid for examination at 400 kV in a JEOL 4000EX HREM instrument.

Selected area electron diffraction patterns from an In$_2$O$_3$ island and the underlying substrate confirmed the excellent epitaxy, see Figure 4. High resolution imaging revealed a remarkably sharp interface (Figure 5), with little evidence of cross-diffusion, this being confirmed by EELS microanalysis across the interface. The mismatch appears to be accommodated by means of dislocations.

1. A. Bourlange, D. J. Payne, R. G. Egdell, J. S. Foord, P. P. Edwards, M. O. Jones,A. Schertel, P. J. Dobson, and J. L. Hutchison APPLIED PHYSICS LETTERS **92**, (2008) 092117

672

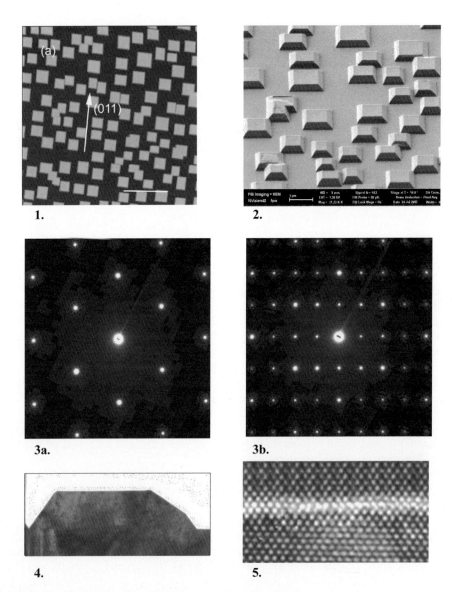

Figure 1. AFM image of In_2O_3 islands on (100) zirconia.
Figure 2. SEM image (SE image) of facetted In_2O_3 islands on zirconia.
Figure 3a. <110> electron diffraction pattern of cubic zirconia; **3b.** <110> diffraction pattern of In_2O_3.
Figure 4. Cross-section of In_2O_3-ZrO_2 interface showing island facetting.
Figure 5. HREM image showing sharp interface.

Carbon nanotubes grown in contact holes for nano electronic applications: how to prepare TEM samples by FIB?

X. Ke[1], S. Bals[1], A. Romo Negreira[2], T. Hantschel[2], H. Bender[2] and G. Van Tendeloo[1]

1. EMAT, University of Antwerp, Groenenborgerlaan 171, B-2020 Antwerp, Belgium
2. IMEC, Kapeldreef 75, B-3001 Leuven, Belgium

Xiaoxing.Ke@ua.ac.be

Keywords: FIB, sample preparation, carbon nanotubes, TEM

Carbon nanotubes (CNT's) grown on silicon-based substrates are of great importance because of their potential use as an interconnect material in next-generation device technologies [1]. Samples in this study consist of layered structures on a silicon substrate, where arrays of so-called "contact holes" are present. CNT's are then grown inside the contact holes after catalyst particles have been deposited in the contacts. In order to investigate the structure of CNT's grown inside the contact holes, transmission electron microscopy (TEM) is used.

Studying the CNT's as well as the catalyst-CNT interface in a cross-section TEM sample requires a dedicated TEM sample preparation technique. An electron-transparent lamella needs to be prepared from a defined zone with minimum damage. Here, focused ion beam (FIB) milling is used for this purpose. Using TEM, we will show that this is a good approach.

For the FIB preparation using a FIB/SEM Dual Beam system, in situ electron beam deposition of a platinum cap layer is applied. This layer is intended to protect the CNT's from the damage that is caused by the usual ion beam deposition of the platinum. The advantage is shown in Figure 1, where a comparison is made for a sample, where platinum was deposited by the ion beam in the FIB without any protection layer (Fig 1a) and with a dual cap layer deposited with the electron and ion beam (Fig 1b,c). After the lamella is extracted by in-situ liftout and mounted onto a TEM grid, it is roughly milled down to a thickness of about 100nm at a voltage of 30kV (Fig 1a,b). However, the ideal sample thickness for high resolution TEM (HRTEM) is approximately 50nm. Since CNT's are easily damaged by high energy milling, final milling is carried out at lower voltage. A voltage of 5kV in combination with small current is applied to the lamella from both sides to obtain a final thickness suitable for HRTEM (Fig 1c).

As-prepared specimens are investigated by advanced TEM techniques, including HRTEM, energy filtered TEM and high angle annular dark field scanning TEM. Figure 2 reveals that CNT's inner shell structure close to catalysts is well preserved inside the contact hole after FIB preparation. These measurements show that the use of the FIB to prepare cross-sectional TEM samples containing CNT's is a very useful approach.

1. F. Kreupl, A.P. Grham, G.S. Duesberg, W. Steinhoegl, M. Liebau, E. Unger and W. Hoenlein, Microelectron. Eng. **64** (2002), p.399-408.

M. Luysberg, K. Tillmann, T. Weirich (Eds.): EMC 2008, Vol. 1: Instrumentation and Methods, pp. 673–674, DOI: 10.1007/978-3-540-85156-1_337, © Springer-Verlag Berlin Heidelberg 2008

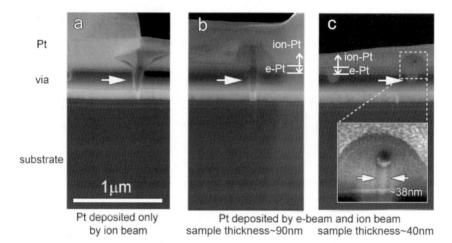

Figure 1. Scanning electron microscopy images of FIB prepared cross-sectional TEM samples. Contact holes containing CNT's are indicated by arrows. The difference between the Pt protection layer deposited only by ion beam and the Pt protection layers deposited by electron beam and ion beam deposition is shown. The enlarged image indicates a CNT.

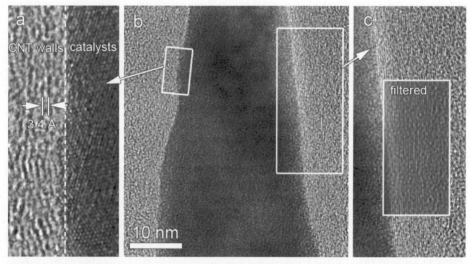

Figure 2. HRTEM images of the contact hole where CNT's are growing around catalysts (a,b). An enlargement of the CNT's left side walls (a) reveals an interlayer distance of 0.34nm. An enlargement of the right side (c) with an inset of a filtered image clearly shows CNT's walls. The CNT's inner shell structure close to the catalyst particle is well preserved.

Advances in 3-dimensional material characterisation using simultaneous EDS and EBSD analysis in a combined FIB-SEM microscope

René de Kloe[1], Hubert Schulz[2] and Felix Reinauer[3]

1. EDAX BV, Ringbaan Noord 103, 5004JC Tilburg, The Netherlands
2. Carl Zeiss NTS GmbH, Carl Zeiss Strasse 56,73447 Oberkochen
3. EDAX Germany, Kreuzberger Ring 6, Wiesbaden, Germany

Rene.de.Kloe@ametek.com
Keywords: 3D, EBSD, EDS, FIB, SEM

Automated Electron Backscatter Diffraction (EBSD) in the Scanning Electron Microscope (SEM) allows the researcher to study the role of crystallographic orientation in the behaviour of polycrystalline materials in great detail. With an estimated information depth of 10-50 nm, high resolution orientation information can be collected in two dimensions (2D) giving information about for example grain properties, textures, and misorientation distributions [1]. When the combined application of EDS and EBSD, "ChI-scan", was introduced in 2003 [2] researchers could more easily investigate multiphase materials and even differentiate phases with identical crystallographic structures.

In order to obtain micro structural and chemical information from the third dimension (3D) in a specimen, EBSD and EDS measurements need to be combined with FIB (Focus Ion Beam) serial sectioning. This allows serial investigations on sections to be produced in-situ. The process automatically repeats milling and EBSD mapping over a desired number of layers. Post processing allows a 3D reconstruction of the microstructure (Figure 1).

During the FIB sectioning, the microstructure and in some cases the even the phases may be modified by interaction with the gallium ions. Integration of the chemical signal from simultaneous EDS and EBSD mapping in these 3D datasets enables successful phase differentiation in multiphase materials and also provides insight in chemical gradients that may be introduced during the analysis (Figure 2).

In this presentation practical aspects of simultaneous EDS and EBSD acquisition in combined SEM-FIB CrossBeam® instruments will be discussed and illustrated with application examples (Figure 3).

1. Schwartz, A. J., M. Kumar, D. P. Field und B. L. Adams (editors): Electron Backscatter Diffraction in Materials Science. Springer, Netherlands, 2000.
2. Nowell, M. M. and S. I. Wright (2004). "Phase differentiation via combined EBSD and XEDS." Journal of Microscopy 213(Pt 3): 296-305F

M. Luysberg, K. Tillmann, T. Weirich (Eds.): EMC 2008, Vol. 1: Instrumentation and Methods, pp. 675–676, DOI: 10.1007/978-3-540-85156-1_338, © Springer-Verlag Berlin Heidelberg 2008

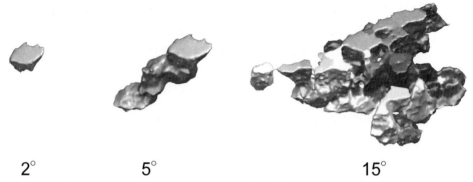

2°　　　　　　5°　　　　　　　　　　　15°

Figure 1. 3D representation of grain clusters with different grain tolerance angles

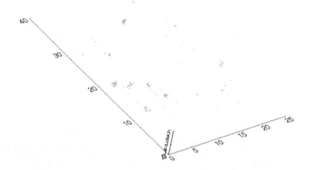

Figure 2. Distribution of Mg in 3D dataset.

Figure 3. Chamber view: CZ CrossBeam® system with EDAX EBSD/EDS analysis system.

Investigation of the effects of the TEM specimen preparation method on the analysis of the dielectric gate stack in GaAs based MOSFET devices.

P. Longo, W. Smith, B. Miller and A.J. Craven

Department of Physics and Astronomy, University of Glasgow, G12 8QQ, UK

p.longo@physics.gla.ac.uk

Keywords: FIB specimen preparation, EELS, STEM, dielectric gate stack, GaAs MOSFET

III-V based MOSFET devices are believed to be one of the candidates to replace Si MOSFET technology. Using processes pioneered by Passlack *et al.* [1], dielectric gate stacks consisting of a template layer of amorphous Ga_2O_3 followed by amorphous GdGaO have been grown on GaAs substrates. Careful deposition of Ga_2O_3 can leave the Fermi Level unpinned. The introduction of Gd is important in order to decrease the leakage current. Electrical properties of such devices are very much dependent upon the quality of the interface and the elemental distribution across the dielectric stack gate region. TEM and, in particular EELS based techniques, are important tools to obtain information on the quality and the composition of such regions in these MOSFET devices.

As the dimensions of the device are reduced, the use of the focused ion beam (FIB) for preparing TEM samples is becoming important. In this paper we want to determine and compare the elemental composition across the GdGaO layer from samples whose TEM specimens were prepared using both conventional cross-sectioning and FIB. The FIB machine used for the experiment is the dual beam type which ensures a better control during the polishing process.

The compositional analysis for both types of TEM samples was performed using EELS spectrum imaging (SI) following the procedure described in [2].

FIB samples were prepared using Ga ion milling with the voltage for the final polishing process as low as 4kV in order to decrease the amount of damage induced during the preparation. "Figure 1" shows a high-resolution TEM image of the GaAs/Ga_2O_3/GdGaO interface region in the FIB prepared sample. The interface looks well preserved, although the GaAs region next to the interface may show some slight damage. "Figure 2" is a HAADF STEM image of the dielectric stack region. The 1nm Ga_2O_3 layer shows up as a result of the better contrast of STEM images.

The EELS SI analysis was performed across the GdGaO layer in two different areas of the FIB prepared sample and the results have been compared to the ones from a sample prepared using conventional cross-sectioning as shown in "Figure 3". The two areas in the FIB sample present a uniform thickness profile ranging from 0.50 down to $0.40t/\lambda$. The area in conventional cross-sectioning prepared sample appears to be much thinner with a thickness profile ranging from 0.27 down to $0.13t/\lambda$.

The datasets from the two different areas in the FIB prepared sample are in good agreement across the GdGaO layer. The agreement between the FIB datasets and the

M. Luysberg, K. Tillmann, T. Weirich (Eds.): EMC 2008, Vol. 1: Instrumentation and Methods, pp. 677–678, DOI: 10.1007/978-3-540-85156-1_339, © Springer-Verlag Berlin Heidelberg 2008

one from the sample prepared by conventional cross-sectioning is also good but only for the first 22nm from the interface with Ga_2O_3. The deviation beyond this may be due to an increasing systematic error with decreasing thickness. For example the ion milled TEM sample could have surface layers with a modified composition. Such layers would have a bigger effect in the thin regions close to the edge of the specimen.

Concluding, the analysis described in this paper shows that FIB samples prepared using voltages down to 4kV, show little if any modification of compositions despite the possibility of Ga implantation. The excellent agreement between the two different areas in the FIB sample shows the high reproducibility and consistency of the analysis.

1. Passlack M, Yu Z, Droopad R, Bowers B, Overgaard C, Abrokwah J, and Kummel AC, J Vacuum Science & Technology B17, 1, 49-52, 1999
2. Longo P, Craven AJ, Scott J, Holland MC and Thayne IG, proceedings of MSM XV, 2007
3. The authors would like to acknowledge EPSRC support under grant EP/F002610

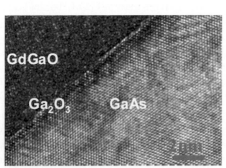

Figure 1. High-resolution TEM image of the interface GaAs/Ga$_2$O$_3$/GdGaO in the sample prepared by FIB.

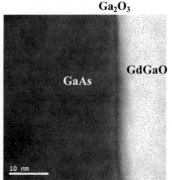

Figure 2. HAADF STEM image of the dielectric gate stack from FIB prepared. The Ga_2O_3 template layer in between the GaAs and the GdGaO.

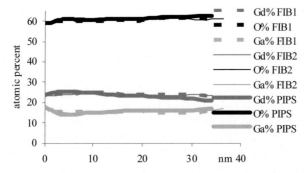

Figure 3. Comparison between the elemental concentration across the GGO layer in two different areas (labelled as FIB1 and FIB2) of the FIB prepared sample and in one area of the cross-sectioned sample (labelled as PIPS).

Manipulation and contacting of individual carbon nanotubes inside a FIB workstation

S.B. Menzel, H. Vinzelberg, T. Gemming

IFW Dresden, P.O.Box 270116, D-01171 Dresden, Germany

s.menzel@ifw-dresden.de

Keywords: nanotubes, micromanipulation, resistance measurement

Carbon Nanotubes (CNTs) have been already applied in several electronic or microelectromechanical devices as well as for microscopic components. However, there are still challenges with respect to (i) the controlled growth of CNTs with well defined properties, (ii) the development of techniques for their precise manipulation or self-alignment, (iii) their filling with other components like metallic clusters, and (iv) the evaluation of the electrical or mechanical properties on individual tubes [1-3].

In present paper the electrical behaviour of individual multiwall CNTs (MWCNTs) were measured between RT and 4.2 K at temperature of liquid-He. Therefore, the MWCNTs were grown on Si(100) substrates using the catalytic CVD process and Fe particles as a catalyst. After inspection of the tubes using the SEM of an FIB workstation (1540XB/ZEISS) some individual tubes were selected for measurements and transferred onto a measurement sample using a 3-axis micromanipulator with rotating tip (MM3 with attached RoTip, Kleindiek Nanotechnik). The arrangement of the micromanipulator inside the FIB machine as well as the approached tip in front of the sample is demonstrated in Fig.1. The manipulation was exclusively done under high vacuum condition ($p<10^{-5}$ mbar) inside the FIB workstation. Contrary to other techniques as, for example, electrophoresis method using liquid solution there is no contamination during manipulation what could influence the electrical measurements. For manipulation the end of the CNT on the opposite side with respect to the substrate was fixed on the tip of the manipulator (Figure 1b), and then the CNT was disconnected from the substrate by ion beam cutting avoiding extensive particle bombardment of the main body of the CNT.

After approaching the CNT was deposited onto the contact fingers (Au/Cr-fingers) of the measurement sample by van-der-Waals interactions and afterwards fixed by platinum deposition preferably using electron beam induced deposition (Figure 2). The CNT was finally disconnected from the tip by FIB cutting and the tip retracted using the linear motor axis.

After preparation the sample was fixed on a carrier and contacted for resistance measurements in the temperature range between RT and 4.2 K using 2-, 3-, or 4-point measurements. For example, results on Ni-filled MWCNTs show that the voltage depends linearly on current at 295 K due to a metallic behaviour of the CNTs. The resistances of such contacted CNTs grown by CVD were in the kΩ range. Furthermore, the resistance increases at decreasing temperature. At low temperatures the current-

M. Luysberg, K. Tillmann, T. Weirich (Eds.): EMC 2008, Vol. 1: Instrumentation and Methods, pp. 679–680, DOI: 10.1007/978-3-540-85156-1_340, © Springer-Verlag Berlin Heidelberg 2008

680

voltage characteristics are non linear and the high contact resistances influenced the resistance measurements.

1. P.M. Ajayan and S. Iijima, Nature **361** (1993), p.333.
2. A.A. Pesetski et al., Appl. Phys. Lett. **88** (2006), p. 113103.
3. J. Orloff et al., High Resolution Focused Ion Beams: FIB and its Application, (Kluwer, NY) (2003).
4. R. Krupke et al., NANO LETTERS **4**[8] (2004), p. 1395.

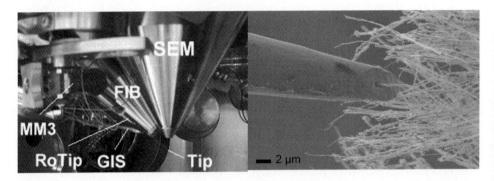

Figure 1. Left: arrangement of the micromanipulator inside the 1540XB workstation; right: approached manipulator tip in front of the CNT sample.

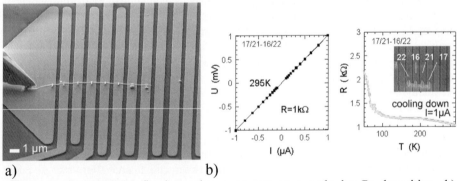

a)   b)

Figure 2. a) MWCNT fixed on the measurement sample by Pt deposition, b) resistance of a Ni-filled MWCNT, left: U(I) curve at 295 K, right: resistance vs. temperature at 1 μA

High volume TEM-sample preparation using a wafer saving in-line preparation tool

U. Muehle[1], S. Jansen[1], R. Schuetten[1], R. Prang[1], R. Schampers[2] and R. Lehmann[2]

1. Qimonda Dresden GmbH & Co OHG, Koenigsbruecker Str. 180, D-01099 Dresden, Germany
2. FEI company, Achtseweg Noord 5, 5600 KA Eindhoven, The Netherlands

uwe.muehle@qimonda.com
Keywords: TEM sample preparation, Lift-Out method, In-line FIB

The semiconductors industry is characterised by a continuous shrinking of device dimensions at a rate of 30% per year in average. As a consequence of this, process monitoring and physical control of new processes and process sequences require an increasing number of TEM-analysis. In addition to this higher volume the TEM-business also needs to deliver results as fast as possible to ensure FAB fluidity and shortest technology development cycles.

Process engineers do appreciate TEM because of its capability to provide detailed structural and -if needed- chemical information but are reluctant to sacrifice a whole 300 mm wafer which contains hundreds of dies. Therefore, a non-destructive TEM sample preparation method is needed where device wafers can be re-injected in the processing flow; this represents a major advantage since a single wafer can be characterized by TEM at each relevant process step and still being fully processed and electrically tested. This enables to characterise very detailed influences of process parameters and distinguish between the effects of unit processes.

To realise this wafer return strategy, TEM samples, containing the region of interest, need to be prepared and extracted from the wafer in the production line. To minimize the impact on the cycle-time of the parent wafer lot that is on hold during this operation, it is essential to keep the overall sample preparation and extraction times as short as possible. Two methods that meet above requirements have been investigated. Both methods use an in-line FEI DA300 DualBeam tool to prepare the TEM samples but differ in the way the samples are extracted: depending on the type of TEM analysis to be performed, the sample is either extracted in-situ in the in-line DualBeam tool or ex-situ in a novel FEI TL150 in-line automated extraction tool.

In the first method a chunk of silicon, containing the region of interest, is prepared and in-situ extracted and transferred to a commercial TEM grid [1]. Final milling can then be performed off-line using a lab DualBeam tool. This way enables to do all critical steps in a lab environment and allows flexible -problem related- variations of the analysis flow. Additional fine milling for special observations, e.g. EELS, is possible.

If the TEM analysis is aimed at just geometrical observations the second, ex-situ extraction method, can be applied. Contrary to the first method, one or multiple TEM samples are now prepared and thinned to final thickness in the 300mm in-line tool. Upon completion of the DA300 process, the wafer and the exact lamella positions list

M. Luysberg, K. Tillmann, T. Weirich (Eds.): EMC 2008, Vol. 1: Instrumentation and Methods, pp. 681–682, DOI: 10.1007/978-3-540-85156-1_341, © Springer-Verlag Berlin Heidelberg 2008

are transferred to the TL150 in-line extraction tool. In this tool, a micromanipulator is automatically driven to the stored lamella position, extracts the lamella and transfers it to a carbon foil grid, where it is ready for TEM inspection. (Fig. 1). The influence on image quality due to overlay with the carbon foil in the TEM-image can be accepted for process monitoring related requests. This method enables to increase speed and efficiency of TEM-sampling significantly due to short operation times at each tool.

Based on several case studies the ability to track the effects of unit processes on semiconductor device structures was demonstrated (fig. 2). This allows reducing the number of wafers, necessary for technology development and fine tuning.

1.   Pokrant, S., Bicaïs-Lepinay, N., Pantel, R., Cheynet, M. ; ICM 16 ; Sapporo 2006

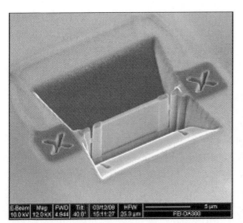

Figure 1. Shape of a thinned sample before extraction (left) and four samples transferred onto the carbon grid (right)

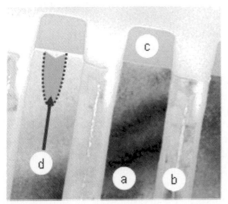
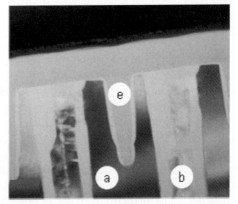

Figure 2. Two TEM images of the same structures taken after different process steps: The silicon substrate (a) is structured with capacitors (b), covered by a protective layer (c), which has to be removed later on in the process. In the left picture the location of transistors, to be processed later, is marked (d), what is done at the right picture (e).

The influence of beam defocus on volume growth rates for electron beam induced platinum deposition

H. Plank[1], M. Dienstleder[1], G. Kothleitner[1], F. Hofer[1]

1. Institute for Electron Microscopy, Graz University of Technology, Graz, Austria

harald.plank@tugraz.at

Keywords: electron beam induced deposition (EBID), platinum deposition, nanofabrication

Electron beam induced deposition (EBID) of conducting (W, Pt, …) or insulating materials (TEOS, …) has attracted considerable attention in recent years as an interesting alternative to ion beam induced deposition (IBID). Especially, soft matter samples or critical applications like polymers or nanoscale devices, respectively, benefit from EBID because of its sputter-free character, the absence of ion implantation, and the reduced thermal stress [1]. One drawback of EBID, however, are the low growth rates (nm^3 / sec) compared to IBID, which originate mainly from the much lower secondary electron yield per incident particle (electrons respectively ions). To push EBID towards its intrinsic limits, extensive experimental and theoretical studies have been carried out, to improve the understanding of the growth processes, which then allow further improvements [2,3]. In this work we demonstrate an enhancement of the volume growth rate for single platinum (Pt) rods by a factor of 2.5 via a systematically introduced defocus during deposition, operating in a mass-transport-limited (MTL) regime (see Figure1). The detailed investigation of the temporal volume growth rate evolution reveals a distinct maximum within the first few seconds, followed by a continuous decay as a result of the interplay between rod height and surface, secondary electron generation zones, and precursor diffusion [4]. The introduction of a defocus delays this maximum and expands higher volume growth rates to the minute scale, which accounts for the increased volumes after same deposition times (see Figure 2). Beside the practical aspect of this enhancement, it is also possible to study very early growth stages, different growth regimes (conical, cylindrical), and their transition on a larger time scale which helps to understand the growth mechanism. By this means, it is shown that the precursor diffusion along almost vertical cylinder walls is strongly involved in the decay of the volume growth rates at increasing rod heights, limiting the efficiency in the MTL regime.

1. R.M. Langford, P.M. Nellen, J. Gierak, Y. Fu, MRS Bulletin **32** (2007), 417.
2. S.J. Randolph, J.D. Fowlkes and P.D. Rack, Crit. Rev. Solid State Mat. Sci. **31** (2006), 55.
3. D. Beaulieu, Y. Ding, Z.L. Wang and W.J. Lackey, J. Vac. Sci. Techn. B **23** (2005), 2151.
4. D.A. Smith, J.D. Fowlkes and P.D. Rack, Nanotechnology **18** (2007), 265308.

M. Luysberg, K. Tillmann, T. Weirich (Eds.): EMC 2008, Vol. 1: Instrumentation and Methods, pp. 683–684, DOI: 10.1007/978-3-540-85156-1_342, © Springer-Verlag Berlin Heidelberg 2008

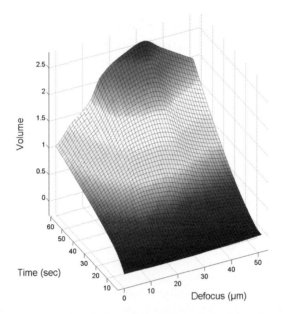

Figure 1. Temporal volume evolution of single Pt rods in dependence on beam defocus.

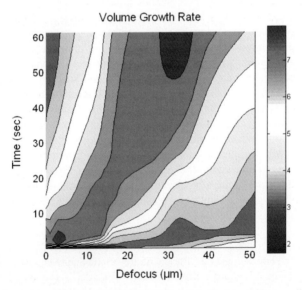

Figure 2. Volume growth rate development in dependence on time and beam defocus.

Reducing of ion beam induced surface damaging using "low voltage" focused ion beam technique for transmission electron microscopy sample preparation

R. Salzer[1], M. Simon[1], A. Graff[1], F. Altmann[1] and L. Pastewka[2] M. Moseler[2]

1. Fraunhofer Institute for Mechanics of Materials, Walter Hülse-Str. 1, 06120 Halle, Germany
2. Fraunhofer Institute for Mechanics of Materials, Wöhlerstrasse 11, 79108 Freiburg, Germany

Roland.Salzer@iwmh.fhg.de
Keywords: FIB preparation, low voltage, MD simulation

The most common technique for target preparation of electron-transparent TEM cross-sections is the focused Ga^+ ion beam (FIB) method [1]. This method enables the production of ultra thin and well shaped samples for analytical and high resolution TEM-investigations. It is known that the irradiation with 30 keV Ga^+ ions not only causes the sputtering of material but it also leave a damaged layer at the irradiated area [1, 2]. For semiconductor materials like silicon the damage level ranges from fully amorphisation near the surface to a defective crystal structure beneath. The surface amorphisation of a lamella disturb the crystallographic information for high resolution TEM. To reduce this amorphisated layer in situ it is feasible to polish the lamellae with low energetic ions [1, 2].

It is of high interest to understand the FIB process in order to predict the amorphisation behaviour, sputter coefficients and their dependency on the crystallographic structure of different materials. But it is very time-consuming to estimate these material properties from experimental investigations because many steps are necessary to obtain one high resolution TEM sample [1]. So the creation of a reliable simulation model of the FIB process is an important scientific issue [3]. In this work the thickness of the amorphisated layer against the acceleration voltage of the ion beam was investigated for silicon single crystals. Classical molecular dynamic simulations based on [4, 5, 6] were used to calculate the FIB process to understand the kinetics of amorphisation. The simulations calculate Ga^+ ions which impinge with different energies (5 keV, 2 keV and 1keV) at 10° to the surface of a silicon crystal (Figure 1 a). Setting the border between amorphisated and undisturbed silicon crystal at 50% of damage level the amorphisation depth becomes ~9 nm, ~4 nm and ~2 nm, respectively (Figure 1 c). For the experimental thickness determination of the amorphisated layer, 5 μm deep trenches were milled into a silicon single crystal with a 30 keV focused Ga^+ beam. The so produced cross-sections were polished with 2 keV, 5 keV, 10keV, 20 keV and 30 keV Ga^+ beams, respectively. Assisted by the electron beam a Platinum bar has been deposited to protect the polished cross-section. For TEM investigation an electron transparent lamella was prepared showing a lateral cut of the protected, polished cross-sections (Figure 2). With TEM investigations we are able to

M. Luysberg, K. Tillmann, T. Weirich (Eds.): EMC 2008, Vol. 1: Instrumentation and Methods, pp. 685–686, DOI: 10.1007/978-3-540-85156-1_343, © Springer-Verlag Berlin Heidelberg 2008

determine the thicknesses of the amorphisated layer for 30 keV, 20 keV, 10 keV , 5 keV and 2 keV to ~ 20 nm, ~ 14.5 nm, ~ 6 nm, ~ 4.6 nm and ~ 1.4 nm, respectively. The assumed ideal case in the simulations is hardly reproducible in the experiment (e.g. the incidence angle). Therefore, the deviation to the simulation data is explainable.

As already used the "low voltage" polishing is a viable method to reduce amorphisation and enhance information yield in TEM investigations. It is shown that the calculations are comparable with the experimental results. The models used to simulate the FIB sputtering process giving a good, qualitative predication and are useful for further investigations.

1. J. Mayer, L. A. Giannuzzi, T. Kamino and J. Michael, MRS **32** (2007) .
2. F. Altmann, et al. Prakt. Metallogr. **43** (2006) 8, 396.
3. L. A. Gianuzzi, B. J. Garrison, J. Vac. Sci. Technol. A **25** (2007) 5, 1417.
4. M. P. Allen and D. J. Tildesley, Oxford University Press (1989)
5. J. Tersoff, Phys. Rev. B **38** (1988), 9902
6. J. F. Ziegler, J.P. Biersack and U. Littmark, in: The Stopping and Range of Ions in Matter, **1**, Pergamon (1985)

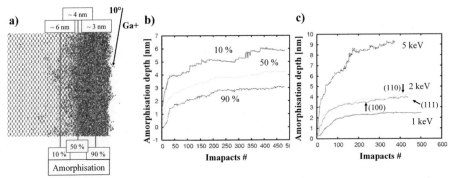

Figure 1. a) The silicon system after ca. 500 impacts with 2 keV Ga$^+$ impinge at 10° to the surface; b) depth for certain damage levels for 2 keV c) amorphisation depth for 1 keV, 2 keV and 5 keV acceleration voltages against the impact number

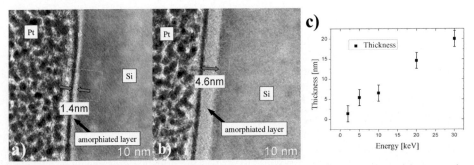

Figure 2. TEM images of the irradiated silicon bulk with the amorphisated layer at the surface and the Platinum protection bar a) polished with 2 keV; b) polished with 5 keV; c) thickness of the amorphisated layer against the energy of the Ga$^+$ ions

DualBeam FIB application of 3D EDXS for superalloy δ-phase characterization

J. Wagner [1], M. Schaffer [1] H. Schroettner [1], S. Mitsche [1], I. Letofsky-Papst [1],
Ch. Stotter [2], and Ch. Sommitsch [2,3]

1. Institute for Electron Microscopy and Fine Structure Research, Graz University of
Technology, Steyrergasse 17, A-8010 Graz, Austria
2. Christian Doppler Laboratory of Materials Modelling and Simulation, University of
Leoben, Franz-Josef-Strasse 18, A-8700 Leoben, Austria
3. Chair of Metal Forming, University of Leoben, Franz-Josef-Strasse 18, A-8700
Leoben, Austria

julian.wagner@felmi-zfe.at
Keywords: Superalloy; Allvac 718 Plus[TM]; Delta-Phase; Microstructure; FIB; 3D-EDXS;

Optimizing the functionality of materials often depends on a precise control of the size, shape, crystal structure and composition of the material being synthesized. But in the age of micro- and nanotechnology it can be a significant challenge to characterize solids in an appropriate way. Therefore many sophisticated analysing methods were established and combined in the past [1, 2]. Automated serial sectioning in combination with physical analytics achieved by using a script-able Dual Beam -Focused Ion Beam system (DB-FIB) equipped with an energy dispersive X-ray spectrometer (EDXS) is only one of the possible characterization opportunities mentioned above [3].

In this work we present an excellent example how powerful EDXS-mapping in three dimensions combined with the advantages of the block lift out technique [4] provides a better understanding of material properties. Today, there are several commonly used alloys like nickel-based superalloys, which consumes up to 30% of the weight in advanced aerospace engines. Some examples of alloys in use are Udimet 720, Waspaloy and Alloy 718, the latter is the most widely used due to its relatively low costs and good formability. Recently, a new nickel-based alloy was developed (Allvac 718 Plus[TM]) by the company ATI Allvac. More details of the extended mechanical properties of this alloy can be found at Bergstrom [5]. Its properties are attributed to the combined effects of chemistry, heat treatment and microstructure. Especially for hot forged gas turbine disks, the influence of δ-phase is important. However, it is well known that the start of δ-phase precipitation strongly depends on the experimental conditions. Thus, an understanding of the phase stability with time and temperature is essential in order to tailor the microstructure of forged turbine blades for high temperature applications.

Among several investigative tools and techniques like electron back scatter diffraction (EBSD), back scatter electron imaging (BSE) and transmission electron microscopy the 3D micro-structural characterization was carried out on the FEI Nanolab Nova200 dualbeam focused ion beam (DB-FIB - FEI company, Eindhoven, The Netherlands) equipped with an energy dispersive Si(Li) X-ray detector (10 mm²) system from EDAX (Mahwah, USA) using the Genesis software version 4.52. Final data

M. Luysberg, K. Tillmann, T. Weirich (Eds.): EMC 2008, Vol. 1: Instrumentation and Methods,
pp. 687–688, DOI: 10.1007/978-3-540-85156-1_344, © Springer-Verlag Berlin Heidelberg 2008

688

visualization was performed using the Amira 3.1 software (Mercury Computer Systems SA). The serial–sectioning thickness was selected to be 500 nm. Due to the lack of an EBSD system in the DB-FIB additional ion beam imaging giving a good channelling contrast was performed and therefore rotating, translating and tilting the sample between the cross-section milling position and the SEM imaging position was required. Figure 1 depicts the 3D models, reconstructed from the experiment. It can be seen in Figure 1b that the plate-like δ-phase appears at both, grain boundaries and twins. δ-phase that appear needle-like have a very small angle to the cross section. Hence, a complete plate-like reconstruction is not possible. The FIB-SEM investigations confirm the plate-like shape of the δ-phase, which agrees well with previous 2D experiments and data from literature.

1. G. Möbus and B. Inkson, *Materialstoday* 10 (2007), 18.
2. P.G. Kotula et al., *Microsc. Microanal.* 12 (2006), 36.
3. M. Schaffer et al., *Ultramicroscopy* 107 (2007), 587.
4. M. Schaffer et al., *Michrochimica Acta and X-ray Microanalysis* online first (2007), DOI 10.1007/s00604-007-0853-5.
5. D.S. Bergstrom et al., The Minerals, Metals & Materials Society, Warrendale (2005) 243.

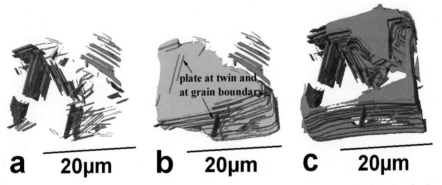

Figure 1. (a) Reconstruction of the δ-phase, (b) twin-reconstruction with the δ-phase and (c) surrounding grain with the δ-phase.

Time-resolved photoemission electron microscopy

Gerd Schönhense

Johannes Gutenberg -Universität, Institut für Physik, Staudingerweg 7,
55099 Mainz, Germany

schoenhense@uni-mainz.de

Keywords: photoelectron microscopy, time resolution, stroboscopic imaging, laser-PEEM

The excellent time structure of Synchrotron radiation and short-pulse lasers has opened the door to a novel way of time-resolved imaging using PEEM. Periodic or repetitive processes can be studied by stroboscopic illumination with the pulsed photon beam. Since the first experiments in 2003, two fields of applications have been established in several groups. One concerns the investigation of fast magnetisation processes like precessional switching, Gigahertz-eigenmodes of ferromagnetic nanostructures or travelling spin waves in thin films. More recently, femtosecond lasers have been used for imaging of localised surface plasmons in nanoparticles and their temporal behaviour in the femtosecond range. In this contribution the state-of-the-art of time-resolved PEEM-imaging is discussed. New contrast mechanisms besides the common X-ray magnetic circular dichroism (XMCD) have been discovered that make the technique independent of Synchrotron radiation. Finally, the potential for dynamic aberration correction will be addressed. The field is reviewed in [1].

In experiments using *Synchrotron radiation* the time resolution is essentially given by the photon pulse width which is typically 50 ps in standard operation down to a few ps in the low-α bunch-compression mode. We succeeded to achieve 15ps total resolution in a "magnetic field pulse pump – photon probe" experiment [2]. The approach has been used by groups at BESSY, the ALS Berkeley and the Swiss Light Source for time-resolved imaging of domain dynamics, precessional modes (centre, wall and vortex modes) in Landau-type flux closure structures in nanoelements. Fig.1 shows the precessional dynamics of an industrial spin-valve element [3]. The local grey level gives access to the time-dependent magnetic response $M(t)$. The magnetisation component in x-direction does not vary homogeneously as is evident from Fig.1 b,c.

If *ultrashort laser pulses* are used for excitation, the time resolution is pushed into the fs range. In a pioneering experiment Schmidt et al. observed lifetime contrast in the 30 fs range in a semiconductor-metal heterostructure [4]. This experiment used "all optical pump-probe". Later, other groups obtained sub-femtosecond time precision in a phase-resolved Mach-Zehnder set-up or observed nano-optical fields [5].

In the pump-probe modes time resolution is achieved by proper synchronisation of pump and probe pulses; for attosecond precision a highly sophisticated handling of ultrashort photon pulses is required. The PEEM itself can be a standard instrument being operated statically. Modification of the microscope opens up further possibilities: *Time-resolved image detection* is facilitated by a 3D (x,y,t)-resolving delayline detector [6]. Local microspectroscopy can be performed by implementing a low-energy drift region into the microscope column, establishing time-of-flight PEEM [1]. Additional

M. Luysberg, K. Tillmann, T. Weirich (Eds.): EMC 2008, Vol. 1: Instrumentation and Methods,
pp. 689–690, DOI: 10.1007/978-3-540-85156-1_345, © Springer-Verlag Berlin Heidelberg 2008

active operation of the lens system paves the way for dynamic aberration correction by circumventing one of the preconditions of Scherzer's theorem.

Time-resolved PEEM bears fascinating *future perspectives*: The XMCD-contrast (Fig.1) requires circularly polarised Synchrotron radiation. However, magnetic contrast can also be achieved in threshold photoemission with UV or visible light, as pioneered by Marx [7] using linearly polarised UV-light. Recently, Nakagawa et al. [8] observed MCD asymmetries of more than 14 % in threshold photoemission using a fs-laser. Thus, very soon studies of magnetisation dynamics can be done in the lab at sub-picosecond time resolution. Alternatively, an imaging spin-filter in the microscope column can yield high magnetic contrast, leading to "time-resolved parallel-imaging SEMPA".

1. G. S., H.J. Elmers, S.A. Nepijko, C. Schneider, Adv. Imaging and El. Phys. **142** (2006) 159
2. A. Krasyuk et al., Phys. Rev. Lett. **95** (2005) 207201
3. F. Wcgclin et al., Phys. Rev. B **76** (2007) 134410
4. O. Schmidt et al., Appl. Phys. B **74** (2002) 223
5. A. Kubo et al., Nano Letters **5** (2005) 1123; M. Aeschlimann et al., Nature **446** (2007) 301
6. A. Oelsner et al., Rev. Sci. Instrum. **72** (2001) 3968
7. G. K. L. Marx et al., Phys. Rev. Lett. **84** (2001) 5888
8. T. Nakagawa et al., Phys. Rev. Lett. **96** (2006) 237402
9. Experiments were funded by BMBF (05KS1 UM1/5, 03N 6500), DFG (SPP 133), and Mat.Wiss. Forschungszentrum Mainz. Thanks are due to H.J. Elmers, C.M. Schneider (FZ Juelich), A. Krasyuk, F. Wegelin and the BESSY staff for good cooperation.

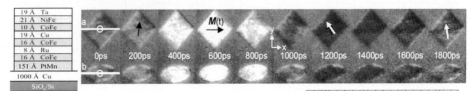

Figure 1. Top: Sequence of stroboscopic XMCD-PEEM images showing the remagnetisation dynamics of two spinvalve elements, cross section top left (sample courtesy NAOMI-Sensitec GmbH, Mainz).

Right: (a) Field pulse at 500 MHz rate (open circles) and calculated magnetic response (MS macrospin model with different damping α; SIM micromagnetic simulation). (b) Measured response at different positions of the square and elliptical platelet as indicated in the inset. (c) Difference in the response of the centres of ellipse and square as compared with the average value. Deviations from the macrospin model are evident (from [3]).

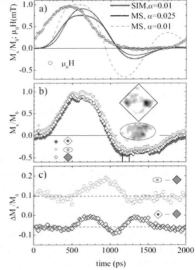

Quantitative 3D imaging of cells at 50 nm resolution using soft x-ray tomography

C. Larabell[1,2], D.Y. Parkinson[2], W. Gu[1], C. Knoechel[2], G. McDermott[1], and M.A. Le Gros[2]

1. Department of Anatomy, University of California at San Francisco, 513 Parnassus, Box 0452, San Francisco CA 94143 USA
2. Physical Biosciences Division, Lawrence Berkeley National Laboratory, 1 Cyclotron Rd., MS 6R2100, Berkeley CA 94720 USA

carolyn.larabell@ucsf.edu

Keywords: x-ray tomography, x-ray microscopy, imaging

Soft X-ray microscopy combines features associated with light and electron microscopy – it is an imaging technique that is both fast and relatively easy to accomplish (like light microscopy) that produces high-resolution, absorption-based images (like electron microscopy). As with light microscopy, one can examine whole, hydrated cells (between 10-15 μm thick), eliminating the need for time-consuming and potentially artifact-inducing embedding and sectioning procedures. Similar to electron microscopy, the image is formed based upon density of cellular components. With soft x-ray microscopy, however, the image is based on the inherent contrast of the organic material in the cell or tissue being examined[1]. Soft x-ray microscopy utilizes photons in the 'water window,' with energies between the K shell absorption edges of carbon (284 eV, λ=4.4 nm) and oxygen (543 eV, λ=2.3 nm). These photons readily penetrate the aqueous environment while encountering significant absorption from carbon- and nitrogen-containing organic material. In this energy range (referred to as the "water window") organic material absorbs approximately an order of magnitude more strongly than water, producing a quantifiable natural contrast and eliminating the need for contrast enhancement procedures to visualize cellular structures[2].

Images are formed using unique optics called zone plates (ZP). An X-ray ZP optic consists of a number of concentric nanostructured metal rings, or zones, formed on a thin X-ray transmissive silicon nitride membrane. The width of the outermost ring determines the spatial resolution of the ZP lens, whereas the thickness of the rings determines the focusing efficiency. In the microscope we utilize, the condenser ZP lens has an overall diameter of 1 cm and an outer zone width of 50 nm. The high-resolution objective ZP lens has a diameter of 63 μm, 618 zones, a focal length of 650 μm at 2.4 nm wavelength, and an outer zone width of 25 nm[2,3]. This produces a numerical aperture of 0.05. The resolution of the instrument is most dependent on the quality and diameter of the objective ZP lens, which is limited by the accuracy of current nanometer fabrication technology. The best ZP available to date is one that generates 15 nm resolution[4].

We are using soft X-ray tomography to examine rapidly frozen whole cells. This technology reveals unique, 3D views of the internal structural organization of cells. By

M. Luysberg, K. Tillmann, T. Weirich (Eds.): EMC 2008, Vol. 1: Instrumentation and Methods, pp. 691–692, DOI: 10.1007/978-3-540-85156-1_346, © Springer-Verlag Berlin Heidelberg 2008

692

imaging in the water window we obtain quantifiable natural contrast of cellular structures without the need for chemical fixatives or contrast enhancement reagents. The time required to collect the data for each cell is less than 3 minutes, making it possible to examine large numbers of cells and to collect statistically significant high-resolution data. In this paper, we show examples of cells imaged using soft x-ray tomography and quantitative analyses of the organelle composition of cells during the cell cycle.

1   M.A. Le Gros, G. McDermott, and C.A. Larabell, Current Opinion in Structural Biology **15** (2005) 593.
2   D. Attwood, Soft Soft X-ray and Extreme Ultraviolet Radiation Principles and Applications, Cambridge University Press (1999).
3   E.H. Anderson, D. Olynick, B. Harteneck, E. Veklerov, G. Denbeaux, W. Chao, A. Lucero, L. Johnson, and D. Attwood, J.Vac Sci Tech B. **18** (2000) 2970.
4   W. Chao, B.D. Harteneck, J.A. Liddle, E.H. Anderson, and D.T. Attwood, Nature **435** (2005) 1210.

Figure 1. Soft X-ray tomography of the yeast, *Schizosaccharo-myces pombe*. Yeast were rapidly frozen and 90 images were collected at 2-degree intervals using x-rays at 517 eV (2.4 nm). Structures are observed based on their inherent contrast; no chemical contrasting agents were used. Images were aligned and tomographic reconstruction accomplished using the software program, IMOD.
Left: Single yeast cell; Right: Yeast just prior to dividing into two cells.

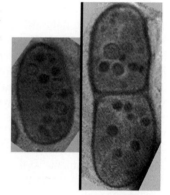

Figure 2. Soft X-ray tomography of the yeast, *Schizosaccharo-myces pombe*. Organelles have been color-coded according to the average x-ray absorption coefficients inside the organelles. The nuclei are colored orange.

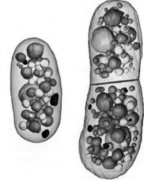

STXM-NEXAFS of individual titanate-based nanoribbon

C. Bittencourt[1], A. Felten[2], X. Gillon[2], J.-J. Pireaux[2], E. Najafi[3], A.P. Hitchcock[3],
X. Ke[4], G. Van Tendeloo[4], C.P. Ewels[5], P. Umek[6] and D. Arcon[6]

1. LCIA, University of Mons-Hainaut, Av. Nicolas Copernic 1, Mons 7000, Belgium
2. LISE, University of Namur, 61 rue de Bruxelles, 5000, Namur, Belgium
3. McMaster University, Hamilton, Ontario L8S 4 M1, Canada
4. EMAT, University of Antwerp, Groenenborgerlaan 171, 2020, Antwerp, Belgium
5. IMN-CNRS, Nantes, France
6. Institute Jozef Stefan, Jamova 39, Ljubdljana 1000, Slovenia

carla.bittencourt@umh.ac.be
Keywords: STXM, TiOx, XPS

In recent years, many low dimensional TiO_2-materials, such as nanowires, nanotubes, and nanorods have been successively synthesized. These novel nanostructures bring to reality the possibility to fine tune chemical reactivity as the system structure and the occupation of the outermost energy levels can be tuned by changing preparation parameters.

In this work, electronic properties of individual TiO_x-pristine and Co doped nanoribbons (NR) prepared by hydrothermal treatment of anatase TiO_2 were studied using STXM-NEXAFS at the Advanced Light Source (ALS). NEXAFS is ideally suited to study TiO_2-based materials because both the O K-edge and Ti L-edge features are very sensitive to the local bonding environment, providing diagnostic information about the crystal structures and oxidation states of various forms of titanium oxides and suboxides. STXM-NEXAFS combines microscopy with spectroscopy allowing the study of the electronic structure of individual nanostuctures with spatial resolution better than 30 nm [1]. In addition, the directional electric field vector ($\bar{E}$) of the x-rays can be used as a "search tool" for the direction of chemical bonds of the atom selected by its absorption edge [2].

Figure 1 illustrate the oxygen $1s$ (K-edge) and titanium $2p$ (L-edge) absorption edge recorded in two experimental geometry: $\bar{E}$ parallel and perpendicular to the principal axis of the nanostructure. The Ti 2p spectra are similar to the reported for the TiO_2 anatase phase [3], with the main discrepancy in the single structure at 460 eV that appears split in the TiO_2. This structure results from transitions to the final state $(2p_{3/2})^{-1} d(3e_g)^1 p^6$, the e_g states are formed by d_z^2 and $d_{x^2-y^2}$ orbitals, which are directed towards ligand anions and are sensitive to deviations from Ti O_h symmetry. Consequently, the absence of splitting of e_g states into d_z^2 and $d_{x^2-y^2}$ suggest that for TiO_x-NR Ti occupies sites with high O_h symmetry in contrast to sites with distorted O_h symmetry in TiO_2-anatase.

Figure 1b shows the normal incidence NEXAFS spectra measured with $\bar{E}$ parallel and perpendicular to the principal axis of the nanostructure. No evidence for anisotropic distribution of Ti sites can be observed. Conversely, the O $1s$ spectra (figure 1a) suggest anisotropic distribution of O sites. The O $1s$ transitions identified as t_{2g} and e_g in the

M. Luysberg, K. Tillmann, T. Weirich (Eds.): EMC 2008, Vol. 1: Instrumentation and Methods,
pp. 693–694, DOI: 10.1007/978-3-540-85156-1_347, © Springer-Verlag Berlin Heidelberg 2008

694

spectra result from transitions to final states, $3d(2t_{2g})^1(1s)^{-1}p^6$ and $3d(3e_g)^1(1s)^{-1}p^6$. The energy separation between t_{2g} and e_g (crystal field splitting) is 2.5 eV, in close agreement with the value measured in the Ti $2p$ spectrum.

The influence of Co doping in the NR electronic structure will be discussed.

1. Felten et al. Nano Lett., *7*, (2007), 2435
2. J. Stohr in NEXAFS Spectroscopy Springer Series in Surface Science 25
3. R. Ruus et al. Solid State Comm., 104 (1997), 199
4. This work is financially supported by Nano2Hybrids (EC-STREP-033311), the PAI 6/08, NSERC (Canada), and the Canada Research Chair program (APH)

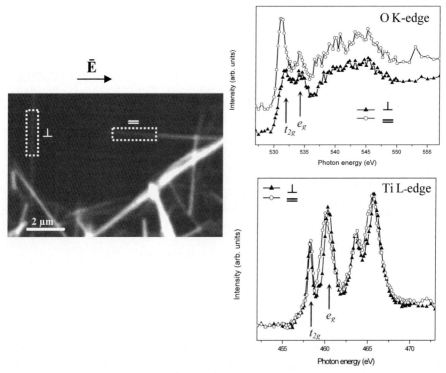

Figure 1. Nexafs spectra recorded with the Ē-vector paralel and perpendicular to the TiO$_x$-NR longitudinal axis (a) O K-edge (b) Ti L-edge.

The fine structure of bioreactor liver tissue seen through the eyes of X-ray micro-computed tomography

C. Fernandes[1], D. Dwarte[1], K. Nagatsuma[2], M. Saito[2], T. Matsuura[2,3], F. Braet[1]

1. Australian Key Centre for Microscopy & Microanalysis, The University of Sydney, NSW 2006, Australia
2. Division of Gastroenterology and Hepatology, Department of Internal Medicine, The Jikei University School of Medicine, Tokyo, Japan
3. Department of Laboratory Medicine, The Jikei University School of Medicine, Tokyo, Japan

f.braet@usyd.edu.au

Keywords: bioreactor, correlative microscopy, liver organoid, tomography, vasculature

X-ray micro-computed tomography (m-CT) is increasingly becoming a standard method to investigate soft biological material in 3-D. Numerous recent studies have been published outlining its use in investigating the fine structure of different organs such as heart, kidney and lung tissue [1]. In previous studies, we successfully imaged intact liver tissue and its associated vasculature via X-ray m-CT [2,3]. The urgency to facilitate patients with severe life treating liver conditions like cirrhosis and hepatitis has resulted in researchers investigating methods to increase recovery rate and delay transplantation by developing artificial livers grown in bioreactors [4]. In line, several studies indicate that the choice of cell culture substrate is central in generating fully functional bioreactor liver tissue. Therefore, we aimed to image and model liver tissue grown on different scaffolds in a bioreactor via X-ray m-CT. Classical SEM and TEM imaging were performed to accumulate cross-correlative structural evidence [5].

Firstly, bioreactors filled with apatite fibres (Fig. 1A), hydroxy apatite beads or cellulose beads were analyzed in detail using the combined X-ray m-CT (Fig. 1B), SEM and TEM approach. 3-D modelling and image analysis revealed that the apatite fibres are the most optimal bioreactor scaffold because of the unique homogeneous structural organization of the fibres (Fig. 1C) and their large surface area available for cell culture. Interestingly, porosity measurements indicated that the total void space available for cell tissue growth is about 41%, 72% and 88% for bioreactors filled with apatite fibres, hydroxy apatite beads and cellulose beads, respectively. Based on the outcomes of the first part of this study we next analyzed liver organoid grown in apatite fibres-packed bioreactors (Fig. 1D). X-ray m-CT analysis of liver organoid revealed the islets of tissue closely aligned around the fibrous scaffold (Fig. 1E). Correlative SEM and TEM studies confirmed the above structural observations (Fig. 1F).

In conclusion, the ability to correlate information in large tissue volumes from the X-ray micro CT models with EM data on the same sample facilitated our knowledge about the structural-functional relationships of bioreactor liver tissue. This approach is an invaluable tool to bridge the gap between bench science and the future development of fully optimized liver organoid bioreactors for therapeutic purposes.

M. Luysberg, K. Tillmann, T. Weirich (Eds.): EMC 2008, Vol. 1: Instrumentation and Methods, pp. 695–696, DOI: 10.1007/978-3-540-85156-1_348, © Springer-Verlag Berlin Heidelberg 2008

696

1. E.L. Ritman. Annu Rev Biomed Eng **6** (2004), p. 185-208.
2. S. Ananda, V. Marsden, K. Vekemans, E. Korkmaz, N. Tsafnat, L. Soon, A. Jones, F. Braet. J Electron Microsc **55** (2006), p. 151-155;
3. F. Braet, K. Nagatsuma, M. Saito, L. Soon, E. Wisse, T. Matsuura. World J Gastroentero **13** (2007), p. 821-825.
4. M. Saito, T. Matsuura, T. Masaki, H. Maehashi, K. Shimizu, Y. Hataba, T. Iwahori, T. Suzuki, F. Braet. World J Gastroentero **12** (2006), p. 1881-1888.
5. F. Braet, E. Wisse, P. Bomans, P. Frederik, W. Geerts, A. Koster, L. Soon, S. Ringer. Microsc Res Techniq **70** (2007), p. 230-242.
6. We kindly acknowlegde the help of Ms. E. Korkmaz & Dr. A Jones of the AKCMM for expert assistance.

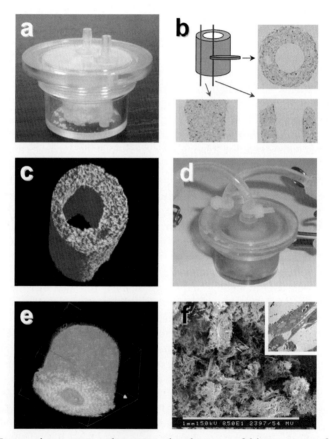

Figure 1. X-ray micro-computed tomography data set of bioreactor scaffold (a-c) & liver tissue (d-f). (**a**) Bioreactor filled with apatite AFS2000 fibers. (**b**) X-ray m-CT serial section images at different orientations. (**c**) 3-D model of the apatite fiber scaffold used for liver cell culture substrate. (**d**) Liver organoid grown on apatite fibre scaffold. (**e-f**) X-ray m-CT (e) and EM observations (f) of liver organoid grown in a bioreactor.

Comparing the Si(Li)-detector and the silicon drift detector (SDD) using EDX in SEM

U. Gernert

Zentraleinrichtung Elektronenmikroskopie (ZELMI), Technische Universität Berlin, Strasse des 17. Juni 135, 10623 Berlin, Germany

ulrich.gernert@tu-berlin.de

Keywords: EDX, SDD, X-ray detector

Recent advances in electron optics and detection systems have enabled the scanning electron microscope (SEM) for usage of the low acceleration voltages. The resulting advantages of better spatial resolution and surface sensitivity are also true for energy dispersive X-ray spectroscopy (EDX). However, the EDX-detectors can still be improved for the use at low energies.

In this context, the comparison of conventional Si(Li)-detector and silicon drift detector (SDD) is described in this paper focusing on the following detector properties: The mechanical and electrical construction, energy resolution, sensitivity at low and high energies, spectrum artefacts and count rates. The experiments were carried out with Bruker XFlash 4010 SDD and with Si(Li)-detectors from eumeX and e2v, all equipped with AP3.3 Moxtek windows.

The field effect transistor (FET) of the used SDD is integrated on the Si-chip. This leads to a small detector capacitance and makes the system insensitive to microphony. The result is an excellent energy resolution (Figure 1). On the contrary, the total electrical capacitance of the Si(Li), whose FET is wired to the crystal, is significantly higher. Furthermore, to get best energy resolution and low noise, the Si(Li) has to be cooled with liquid nitrogen using a cryostat. Both make it sensitive to mechanical vibration. The SDD works at a temperature of –20°C, hence it is not as sensitive to window contamination or icing of the Si-chip. However, long-term experiences do not exist.

In the low energy range (e.g. using the beryllium Kα-line), both types of detectors offer comparable energy resolution and sensitivity, since the influence of the window and the pulse shaping of the signal processor are more important than the properties of the detector. Above X-ray energies of 10 keV, however, the efficiency of the SDD decreases, due to the smaller chip-thickness of 450 μm compared to the Si(Li) crystal thickness of 2 to 4 mm (Figure 2).

The probability of the escape effect is known for both detectors and therefore correctable. The pile-up effect, however, is only partially eliminable. Especially the pile-up peaks of light elements like carbon interfere with X-ray-lines of other elements. For example, the C pile-up peak overlaps with the O Kα-line. The small detector capacitance of the SDD leads to fifty-fold faster shaping time compared to the Si(Li), reducing the pile-up effect considerably (Figure 3).

The best energy resolution of the SDD is obtained for pulse processing with a time constant of 1μs (for Si(Li)s typically 64 μs), and there is no deterioration with count rates up to 60000 cps. In order to produce fast X-ray maps, an input count rate of even

M. Luysberg, K. Tillmann, T. Weirich (Eds.): EMC 2008, Vol. 1: Instrumentation and Methods, pp. 697–698, DOI: 10.1007/978-3-540-85156-1_349, © Springer-Verlag Berlin Heidelberg 2008

698

500000 cps has been tested: Background and pile-up effects increase, but the resolution is still satisfying. Additionally, the SDD is about twenty times faster than the Si(Li) with comparable spectrum quality (Figure 4).

The SDD is not only a nitrogen free EDX-detector, but has reached an energy resolution comparable to that of a Si(Li), even for much higher count rates. The only current drawback is the lower efficiency at high energies.

1.   I kindly acknowledge the support by **Bruker** AXS Microanalysis. Special thanks are due to Tobias Salge.

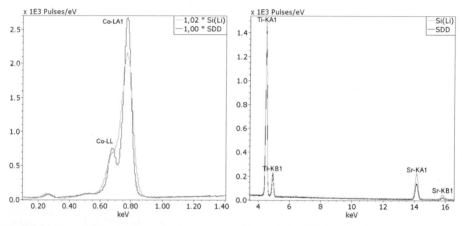

Figure 1. Co-L-lines are separated using a SDD with a resolution of 58 eV @ 677 eV.

Figure 2. Loss of efficiency of the SDD compared to the Si(Li) at higher energy.

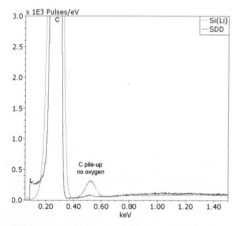

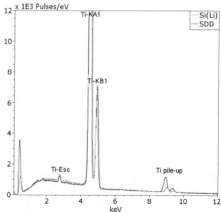

Figure 3. Different intensities of carbon pile-up peak due to shorter shaping time of the SDD. Both spectra have same input count rate.

Figure 4. Input count rate of 500 kcps for the SDD with pile-up effect and peak shift. 40 kcps for the Si(Li).

Enhancing contrast of Al traces on Si substrates using low-voltage SEM-hosted XRM

B.C. Gundrum[1], J.A. Hunt[1]

1. Gatan Research & Development, 5794 W. Las Positas, Pleasanton, CA, USA

bgundrum@gatan.com
Keywords: x-ray microscopy, tomography, semiconductor failure analysis

The use of x-ray microscopy (XRM) to perform semiconductor failure analysis is a well developed application. Here a new imaging technique is presented that extends the capabilities of a SEM-hosted XRM to address die level inspection of Al traces on Si substrates. This is accomplished by imaging with x-ray energies between Al and Si K-absorption edges using a pseudo-monochromatic source.

The SEM-XRM is a commercially available x-ray microscopy attachment for a scanning electron microscope. This system utilizes the electron beam from the SEM to produce a point source of x-rays from a metal target. The sample is placed between the point source and a CCD camera to produce images in projection geometry. The resolution of the system is ultimately limited by the x-ray source size produced by the electron beam and the contrast in the image is the result of both absorption and phase contrast mechanisms. In the standard imaging configuration the system is most sensitive to x-rays around 3-10 keV, and completely insensitive to energies below 2 keV. The absorption contrast produced at the standard imaging energies for a 0.25 µm thick Cu trace is approximately 1% and essentially zero for Al at this x-ray energy due to the long attenuation lengths.

Moving to lower energy x-rays to increase x-ray stopping power of Al is insufficient in itself to produce quality images, as Al and Si have essentially the same attenuation lengths from 2 to 10 keV (Figure 1a). Assuming 0.25 µm thick Al traces and a substrate that has been back thinned to 20 microns, the contrast between the Al traces and the Si substrate is only 2.5% at 4 keV where the absorption length is almost an order of magnitude shorter than at 8 keV. Thus it is difficult and dose intensive to observe a thin layer of Al on a thick layer of Si. However, by imaging with 1.70 keV x-rays, which is in between the absorption edges of Al and Si, the absorption length of Al is almost an order of magnitude shorter than that of Si (Figure 1a). At this x-ray energy the absorption contrast produced by a 0.25 µm thick Al trace is 19%, thus making it clearly visible. To produce a pseudo-monochromatic source suitable for this application an accelerating voltage of 10 kV is used with a Ta target. In Fig. 1b the quantum efficiency curve for the CCD camera and the incident and measured spectrum from the Ta target are plotted as a function of x-ray energy. The sample was thinned to ~20 µm so that the Si transmits ~20% of the x-ray intensity. Figure 2 shows two images of a die having 0.65 µm wide Al traces and W plugs imaged using the standard imaging technique (Figure 2a) and the low keV imaging technique (Figure 2b) where the Al interconnect traces are clearly visible.

M. Luysberg, K. Tillmann, T. Weirich (Eds.): EMC 2008, Vol. 1: Instrumentation and Methods, pp. 699–700, DOI: 10.1007/978-3-540-85156-1_350, © Springer-Verlag Berlin Heidelberg 2008

1. The authors would like to acknowledge Mark Zurbuchen of the Aerospace Corporation for valuable assistance with providing the sample and sample preparation.

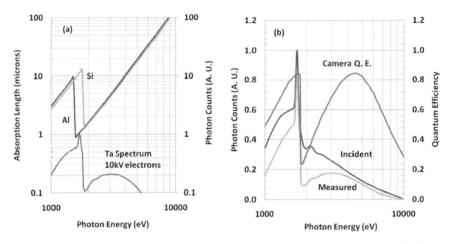

Figure 1. (a) Using 10keV electrons incident on a Ta target a pseudo-monochromatic x-ray source is created with the peak between the absorption edges of Si and Al, thus enhancing the contrast of Al on Si. **(b)** The quantum efficiency of the camera used assists in producing a pseudo monochromatic source due to the absorption edge in the detector.

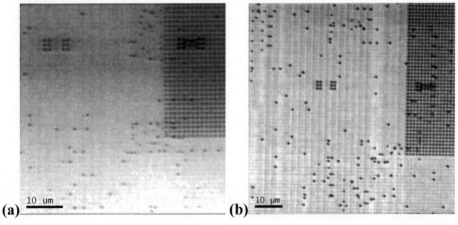

Figure 2. Images of a die from The Aerospace Corporation with W plugs and 0.65 μm wide Al traces. **(a)** Image acquired using the standard imaging, 30 keV electrons incident on a Ta target, technique in the SEM-XRM, and **(b)** same sample with image acquired using 10 keV electrons incident on a Ta target.

An optical demonstration of ptychographical imaging of a single defect in a model crystal

A. Hurst, F. Zhang, and J.M. Rodenburg

Department of Electronic & Electrical Engineering, University of Sheffield, Sir Frederick Mappin Building, Mappin Street, Sheffield S1 3JD, UK

a.hurst@shef.ac.uk
Keywords: diffractive imaging, ptychography, phase retrieval

Ptychographical iterative phase retrieval has been proven to be a promising diffractive imaging technique, especially for use with short wavelength radiation[1-3]. In ptychography, a set of diffraction data are collected from different but overlapping illumination regions. A novel technique called the ptychographical iterative engine (PIE) is used for phase retrieval which has several advantages over conventional Fienup-based methods, including its freedom from conjugate ambiguity, unrestricted field of view, rapid convergence, and the non-requirement of a support or the isolation of the object.

Diffraction from crystalline specimens is dominated by isolated peaks due to the periodic features. This presents a problem when local information, such as a dislocation, is of particular interest. Ptychography allows this problem to be overcome. We report here a simulation result from a model crystal with a single dislocation type defect, and a successful light optical reconstruction wherein a specimen with the same local defect is used.

A diffraction pattern generated from one illumination position on the model crystal is shown in Figure 1, with the resulting reconstruction. To simulate the behaviour of a real detector, the diffraction data were quantized to 16 bits, and noise added. With Poisson noise added at 50% signal amplitude, the reconstruction is still successful, and the 'dislocation' structure can be well resolved.

Our visible light optical experiment, illustrated in Figure 2, uses as a specimen the model of a crystal with a single dislocation, printed onto a photolithographic mask such as is used in modern CMOS microfabrication. The specimen is mounted on a two-axis translation stage so that the incident illumination may be scanned across its surface in a square grid of overlapping positions, while at each position a Fraunhofer diffraction pattern is collected using a 10-bit uncooled CCD. The illumination spot is formed using an aperture in a spread beam of He-Ne (λ=633nm) laser light. The resulting diffraction patterns are processed using the PIE method, with the results shown in Figure 3.

1. J.M. Rodenburg, H.M.L. Faulkner, Appl. Phys. Lett. **85** (2005) 4795-4797
2. J.M. Rodenburg, A.C. Hurst, A.G. Cullis, Ultramicroscopy **107** (2007) 227-231
3. J.M. Rodenburg, A.C. Hurst, A.G. Cullis, B.R. Dobson, F.Pfeiffer, O.Bunk, C.David, K. Jefimovs, I. Johnson, Phys. Rev. Lett. **98** (2007) 034801
4. We gratefully acknowledge the help of Gavin Williams in preparing the optical specimen.

M. Luysberg, K. Tillmann, T. Weirich (Eds.): EMC 2008, Vol. 1: Instrumentation and Methods, pp. 701–702, DOI: 10.1007/978-3-540-85156-1_351, © Springer-Verlag Berlin Heidelberg 2008

702

5. The authors wish to gratefully acknowledge the support of EPSRC for funding this work which was part of the Basic Technology Grant (EP/E034055/1) - Ultimate Microscopy: wavelength limited resolution without high quality lenses.

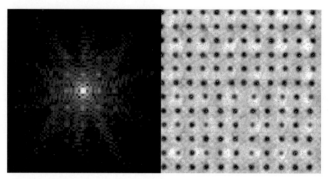

Figure 1. Simulated diffraction pattern (left, log scale) and reconstructed amplitude (right) of a model crystal defect.

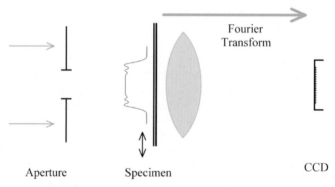

Figure 2. Schematic diagram of the experimental set up. A profile of the soft-edged illumination function is shown.

Figure 3. The reconstructed amplitude (left) and phase (right) of the optical specimen. The reconstructed local defect is ringed.

HRTEM and STXM, a combined study
of an individual focused-ion-beam patterned CNT

A. Felten[1], X. Ke[2], X. Gillon[1], J.-J. Pireaux[1], E. Najafi[3], A.P. Hitchcock[3],
C. Bittencourt[4] and G. Van Tendeloo[2]

1. LISE, University of Namur, 61 rue de Bruxelles, 5000, Namur, Belgium
2. EMAT, University of Antwerp, Groenenborgerlaan 171, 2020, Antwerp, Belgium
3. McMaster University, Hamilton, Ontario L8S 4 M1, Canada
4. LCIA, University of Mons-Hainaut, Av. Nicolas Copernic 1, Mons 7000, Belgium

carla.bittencourt@umh.ac.be

Keywords: Carbon Nanotubes, FIB, HRTEM

Carbon nanotubes (CNTs) are either metallic or semiconducting and are unique in terms of their chemical and thermal stability, strength and elasticity. The combination of all these properties is expected to reshape the development of functional devices. However, the relatively low reactivity of the CNT surface presents a challenge for their integration. One approach to tackle this drawback is to tailor the interface reactivity by creating functional CNT surfaces. In this context, a strategy for enhancement of CNT application is the activation of their surface by grafting functional groups. Several strategies relying on random active site creation have been developed to functionalize CNTs [1]. The lack of precise control over the functionalized location renders such methods unsuitable for applications where different sections of the very same CNT need different reactivity. For example, some sections of the CNT can be used to build up electrical contacts and others for anchoring nanostructures or assembling molecules. Recently, site-selective functionalization of CNTs at a nanoscale spatial resolution was shown to be possible. By using focused-ion-beam (FIB) irradiation high-chemical-reactivity segments can be created along the CNT axis via spatially resolved defect generation, followed by mild chemical treatments [2].

Irradiating CNTs with ions generates defects (vacancy clusters and unsaturated bonds) altering the CNT shape and structure and influencing their electronic properties. In this work, we combine HRTEM and scanning transmission x-ray microscopy (STXM) to evaluate the electronic, structural and chemical changes induced by irradiating a CNT at different ion doses. Different segments of the very same carbon nanotube, supported on a labelled grid commonly used for TEM analysis, were irradiated in a FEI Nova 200 FIB/SEM Dual Beam system equipped with a FIB of 5-30 kV Ga^+, and a SEM. The use of labelled grids allows the identification of the very same CNTs by the different techniques. HRTEM was carried out in JOEL 3000F at EMAT, University of Antwerp, while STXM was carried out at beamline 5.3.2 at the Advanced Light Source, Berkeley.

STXM combines near-edge absorption fine structure spectroscopy with energy resolution of 0.15 eV to microscopy with spatial resolution better than 30 nm allowing studies of isolated nanotubes. Figure 1 compares NEXAFS spectra recorded on different

M. Luysberg, K. Tillmann, T. Weirich (Eds.): EMC 2008, Vol. 1: Instrumentation and Methods, pp. 703–704, DOI: 10.1007/978-3-540-85156-1_352, © Springer-Verlag Berlin Heidelberg 2008

704

segments of the same CNT. The spectrum recorded on the non-irradiated segment (I) shows strong polarization dependence; the intensity of the C1s → π* transition is found to be the highest when the E-vector is perpendicular to the tube axis, while it almost completely vanishes when the E-vector is parallel to the tube axis. The polarization dependence decreases for increasing irradiation time (II, III) probably due to increase in the density of defects (reduction in the number or directional bonds). In fact, in low quality CNTs with high density of sp^2 defects the polarization effect can be minimal [3].

In addition to the gradual broadening of the C1s → π* peak for increasing irradiation time, the K-edge NEXAFS spectra shows a gradual increase in the intensity of the oxygen-related resonance peaks ranging from 287 to 289 eV (Figure 1b) indicating that ion-bombardment induced defects increase the chemical reactivity of the ion-irradiated CNT segment and enable selective functionalization. Moreover, the gradual increase testifies that the density of defects in a CNT-section, i.e., the density of functionalization can be tailored by fine tuning irradiation parameters.

CNT structural modification and the creation of defects due to the irradiation, as probed by STXM and HRTEM, will be discussed.

1. S. Ciraci, S. Dag, T. Yildirim, O. Gülseren and R. T. Senger, J. Phys: Condens. Matter 16 (2004), p. R901

2. M. S. Raghuveer, A. Kumar, M. J. Frederick, G. P. Louie, P. G. Ganesan and G. Ramanath, Adv. Mater. 18 (2006), p. 547

3. E. Najafi, D. Hernández-Cruz, M. Obst, A. Hitchcock, B. Douhard, JJ Pireaux and A. Felten submitted to small

4. This work is financially supported by Nano2Hybrids (EC-STREP-033311), the PAI 6/08, NSERC (Canada), and the Canada Research Chair program (APH).

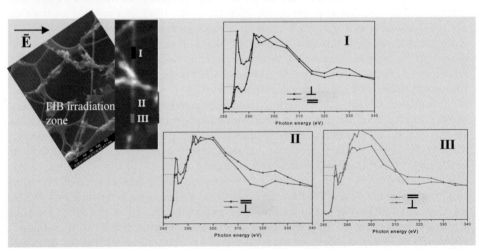

Figure 1: STXM images (left) in the region around the irradiated CNT (see SEM image). Nexafs spectra recorded with the nanotube paralell and perpendicular to the E-vector. Region I: non-irradiated segment, region II: outer region of the irradiation zone, region III: center of the irradiated zone

Compact micro-CT/micro-XRF system for non-destructive 3D analysis of internal chemical composition.

A. Sasov, X. Liu and D. Rushmer

SkyScan, Kartuizersweg 3B, 2550 Kontich, Belgium

sasov@skyscan.be

Keywords: X-ray, microtomography, micro-CT, X-ray fluorescence, micro-XRF

We developed a compact laboratory scanner, which combines X-ray microtomography (micro-CT) with X-ray microfluorescence (micro-XRF). This dual-modality scanner opens possibility for nondestructive three-dimensional volumetric analysis of internal local chemical composition, enhanced by morphological information provided by built-in micro-CT system.

Unlike known microXRF methods based on collimated beam and detector [1,2], our micro-XRF scanner uses full field (two-dimensional) acquisition system with 512x512 pixels energy sensitive detector operated in photon counting mode. It allows detecting two-dimensional photon energy maps in the range of 3...20keV. The detector based on the new type of pinhole optics with square aperture and pyramidal opening on both sides. Such shape allows sensitivity increasing by 25% in the central part of the image and up to 70% in the corners. Block-diagram of the system shown in Fig.1. Two X-ray sources (1) symmetrically illuminate an object (2). An X-ray camera for micro-XRF mode (3) placed in between of excitation sources. An X-ray camera for transmission imaging (4) uses radiation from one x-ray source. The object mounted on a rotation stage (5) for collecting necessary angular views for tomographical reconstruction.

Operator can select up to 8 sets of energy windows, which will be collected independently and simultaneously. By object rotation the scanner acquires all necessary angular two-dimensional views in transmission and fluorescence modes for following 3D reconstruction. The system acquires data in such a way that micro-CT scans and micro-XRF scans match each other exactly in position, magnification and spatial orientation. This makes image registration much easier and more accurate. Micro-CT data is reconstructed with modified Feldkamp algorithm (filtered back-projection) [3]. All micro-XRF datasets are reconstructed by maximum likelihood iterative algorithm [4]. Micro-CT results can be used for absorption correction in micro-XRF datasets. After reconstruction 3D morphological information and 3D chemical composition exactly match each other. For visualization of results chemical information shown in colors overlays micromorphology shown in grayscale.

Good performance of the system has been demonstrated using phantom measurements and real objects. One of the phantom samples is shown in Fig.2, corresponding results - in Fig.3. Top line of the images shows one of the angular views, bottom line contains central reconstructed cross sections. X-ray shadow image (left) and partial micro-XRF images for Fe, Ti, Cu and V are processed independently and finally combined in color-coded maps of internal chemical composition of the object.

M. Luysberg, K. Tillmann, T. Weirich (Eds.): EMC 2008, Vol. 1: Instrumentation and Methods, pp. 705–706, DOI: 10.1007/978-3-540-85156-1_353, © Springer-Verlag Berlin Heidelberg 2008

1. K. Janssens et al. X-Ray Spectrom. **29** (2000), 73-91.
2. A.Hokura et al. Proc. 8th Int Conf. X-Ray Microscopy, pp323-325.
3. L.A. Feldkamp, L.C. Davis and J.W. Kress. J.Opt.Soc.Am.A, **1**, 6 (1984), pp.612-619
4. P.J. La Riviere et al., Optical Engineering, **45**, 7, (2006), pp.07005_1-10.

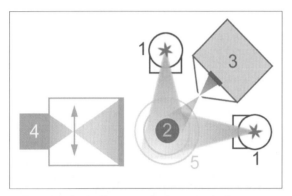

Figure 1. Block-diagram of micro-CT / micro-XRF system (explanations - in text).

Figure 2. Phantom sample.

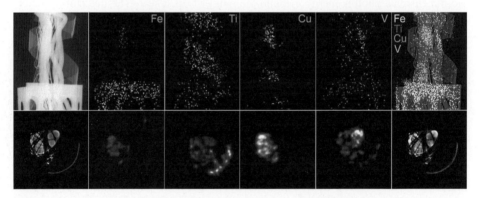

Figure 3. Micro-CT / micro-XRF results from the object shown in Figure 2:
top line - one of angular projections, bottom line - central reconstructed cross section.
left to right: X-ray shadow image (micro-CT), micro-XRF images in Fe, Ti, Cu and V
characteristic lines, combined micro-CT / micro-XRF colour-coded image.

NanoCT: Visualising of Internal 3D-Structures with Submicrometer Resolution

F. Sieker[1], O. Brunke[2]

1. phoenix|x-ray Systems + Services GmbH, Niels-Bohr-Str. 7, 31515 Wunstorf, Germany
2. phoenix|x-ray Systems + Services GmbH, Niels-Bohr-Str. 7, 31515 Wunstorf, Germany

fsieker@phoenix-xray.com
Keywords: nanoCT, high-resolution Computed Tomography, 3D micro-analysis

Current high-resolution X-ray microscopy offers an informative view on the internal microstructure of small devices, allowing the observer to obtain a number of conclusions on the status and internal properties of the sample. However, due to the nature of the resulting two-dimensional images, this technique will give ambiguous information if the three-dimensional structure of the micro assembly is complex. During the past several years, computed tomography has progressed to higher picture resolution and quicker reconstruction of the 3D-volume. More recently it has allowed for a three-dimensional look into the inside of materials with submicron resolution.

phoenix|x-ray's new nanotom is a very compact laboratory system that can analyse samples up to 120 mm in diameter and weighing 1 kg with unique voxel-resolutions down to <500 nm (0.5 microns). It is the first 180 kV nanofocus computed tomography system in the world that is tailored specifically to the highest-resolution applications in the fields of materials science, micro mechanics, electronics, geology and biology.

Computed Tomography at such exceptionally high spatial resolution requires careful consideration of all factors for system design which might influence the final resolution. These special needs demand for a special construction of manipulation, detector and X-ray tube. For example, the nanotom uses a 180 kV high power nanofocus tube that can easily penetrate metal samples. This makes it particularly suitable for nanoCT-examination of sensors, complex mechatronic samples, micro electronic components and material samples of every type, see "Figure 1" and "Figure 2. Other examples include, but are not limited to synthetic materials, ceramics, composite materials, mineral and – of course – even low absorbing organic samples.

The resultant CT volume data set is visualised by slices perpendicular to the three dimensions or compiled in a three-dimensional view, which can be displayed in various ways. It also allows slicing and sectional views in any direction of the volume. As this technique requires only a mouse click, CT will substitute destructive mechanical slicing and cutting in many applications.

The initial CT results obtained with the nanotom demonstrate that it is now possible to analyze the three-dimensional microstructure of materials with submicrometer resolution. Any internal detail that corresponds to a contrast in material, density or porosity can be visualised and data such as distances or the pore volume can be

M. Luysberg, K. Tillmann, T. Weirich (Eds.): EMC 2008, Vol. 1: Instrumentation and Methods, pp. 707–708, DOI: 10.1007/978-3-540-85156-1_354, © Springer-Verlag Berlin Heidelberg 2008

measured. It is possible to image metal phases differently in alloys or to analyse the texture of fibres in composite materials. Another possibility is to simply black out the unwanted material and to analyse the normally hidden pore network of minerals or foam materials alone.

nanoCT widely expands the spectrum of detectable micro-structures. The nanotom opens a new dimension of 3D-microanalysis and will replace more destructive methods – saving cost and time per sample inspected.

Figure 1. nanoCT of fibreglass. Clearly visible are single fibres as well as dark pores in the resin. For better visualisation of the tissue, the resin was blinded out in the left part of the image.

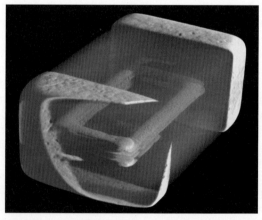

Figure 2. 3D-visualisation of a SMD ceramic chip inductor showing the layers of the internal coil. The sample size is about 2 mm.

Dynamics of nanostructures on surfaces revealed by high-resolution, fast-scanning STM

Flemming Besenbacher

Interdisciplinary Nanoscience Center (iNANO), University of Aarhus,
DK-8000 Aarhus C, Denmark

fbe@inano.dk

Keywords: STM, oxide surfaces, DNA nanostructures

In the era of nanoscience and nanotechnology, numerous examples illustrating the saying "small is different" exist, i.e. nanoclusters exist. Scanning tunneling microscopy (STM) has proven to be a fascinating and powerful technique for revealing the atomic-scale realm of matter,

In this talk I will show how the unique aspect of our Aarhus STM to record time-resolved, high-resolution STM images, visualized in the form of STM movies can be used to obtain important new insight into the dynamics of surface processes and nanostructures [1-3]. First, I will discuss MoS_2 nanoclusters, which are of interest in inorganic nanotubes, in nanoelectronics, and as the active nanoparticles in heterogeneous catalyst. By using STM to systematically map and classify the atomic-scale structure of triangular MoS_2 nanoclusters as a function of size, we observed a distinct size dependence for the cluster morphology and electronic structure driven by the tendency to optimize the sulphur excess present at the cluster edges. [4].

Secondly, I will discuss the reactivity of Au nanoclusters on $TiO_2(110)$ surfaces in three different oxidation states (i) reduced having bridging oxygen vacancies, (ii) hydrated having bridging hydroxyl groups, and (iii) oxidized having oxygen adatoms. From an interplay of STM and DFT results we identify very surprisingly two different gold – $TiO_2(110)$ adhesion mechanisms for the reduced and oxidized supports, and that the adhesion of the Au clusters is strongest on the oxidized support [5].

Finally, the self-assembly of nucleic acid (NA) base molecules on solid surfaces has been investigated since NA base molecules and DNA molecules are particularly interesting as promising building blocks for the bottom-up fabrication of functional supramolecular nanostructures on surfaces. I will discuss the fact that guanine molecules form the so-called G-quartet structure on Au(111) that is stabilized by cooperative hydrogen bonds [6]. Interestingly, cytosine molecules only form disordered structures by quenching the sample to low temperatures, which can be described as the formation of a 2D organic glass on Au(111) [7].

M. Luysberg, K. Tillmann, T. Weirich (Eds.): EMC 2008, Vol. 1: Instrumentation and Methods, pp. 709–710, DOI: 10.1007/978-3-540-85156-1_355, © Springer-Verlag Berlin Heidelberg 2008

1. F. Besenbacher, Reports on Progress in Physics 59 (1996) 1737
2. R. Schaub *et al.*, Science 299, 378 (2003); Science 303 (2004) 511
3. S. Wendt *et al.*, Physical Review Letters 96 (2006) 066107
4. J.V. Lauritsen *et al.,* Nature Nanotechnology 2 (2007) no. 1, 53
5. D. Matthey *et al*, Science 315 (2007) 1692
6. R. Otero *et al.*, Angew. Chem. Int. Ed. 44 (2005) 2270-2275
7. R. Otero *et al.*, Science 319 (2008) 312-315

Spin mapping on the atomic scale

Roland Wiesendanger

Institute of Applied Physics and Interdisciplinary Nanoscience Center Hamburg,
University of Hamburg, D-20355 Hamburg, Germany

wiesendanger@physnet.uni-hamburg.de, www.nanoscience.de

Keywords: Spin-polarized scanning tunnelling microscoppy, magnetic exchange force
microscopy, nanomagnetism

A fundamental understanding of magnetic and spin-dependent phenomena requires the determination of spin structures and spin excitations down to the atomic scale. The direct visualization of atomic-scale spin structures [1-4] has first been accomplished for magnetic metals by combining the atomic resolution capability of Scanning Tunnelling Microscopy (STM) [5, 6] with spin sensitivity, based on vacuum tunnelling of spin-polarized electrons [7]. The resulting technique, Spin-Polarized Scanning Tunnelling Microscopy (SP-STM), nowadays provides unprecedented insight into collinear and non-collinear spin structures at surfaces of magnetic nanostructures and has already led to the discovery of new types of magnetic order at the nanoscale [8]. More recently, the detection of spin-dependent exchange and correlation forces has allowed a first direct real-space observation of spin structures at surfaces of antiferromagnetic insulators [9]. This new type of scanning probe microscopy, called Magnetic Exchange Force Microscopy (MExFM), provides a powerful new tool to investigate different types of spin-spin interactions based on direct-, super-, or RKKY-type exchange down to the atomic level. By combining MExFM with high-precision measurements of damping forces [9] localized or confined spin excitations in magnetic systems of reduced dimensions now become experimentally accessible.

While SP-STM is nowadays well established for revealing atomic spin configurations at surfaces, its application is limited to electrically conducting samples such as magnetic metal films or magnetic semiconductors. In order to map atomic spin structures at surfaces of insulators and to open up the exciting possibility of studying spin ordering effects with atomic resolution while going through a metal-insulator transition, we have developed MExFM. This technique is based on the detection of short-range spin-dependent exchange and correlation forces at very small tip-sample separations (a few Angstroms), in contrast to MFM where the magnetic dipole forces are probed with a ferromagnetic probe tip at a typical tip-to-surface distance of 10-20 nm [10, 11].

An important starting point for achieving the challenging goal of atomic-resolution spin mapping on surfaces of insulators has been the development of non-contact atomic force microscopy (NC-AFM) with true atomic resolution [12]. Nowadays, NC-AFM allows atomically resolved studies of any material system [13], even in the case of curved surface topographies [14]. MExFM combines the capabilities of NC-AFM and atomic-scale spin resolution by making use of an atomically sharp probe tip with a very well defined spin state at its apex. Based on the knowledge gained during the

M. Luysberg, K. Tillmann, T. Weirich (Eds.): EMC 2008, Vol. 1: Instrumentation and Methods,
pp. 711–712, DOI: 10.1007/978-3-540-85156-1_356, © Springer-Verlag Berlin Heidelberg 2008

development of SP-STM in preparing such tips we have recently succeeded in resolving the surface spin structure of the antiferromagnetic insulator NiO(001) [9]: By approaching an out-of-plane magnetized Fe-coated tip very close to the surface atoms, the spin-dependent exchange interaction between the rather localized Ni d-states of the sample and the Fe d-states of the tip leads to a different force or force gradient above Ni atoms with a different orientation of their magnetic moments. As a result, a superperiodicity corresponding to an antiferromagnetically ordered state of the NiO(001) surface is observed in the MExFM image. The apparent height difference between the magnetically non-equivalent Ni-sites only amounts to 1.5 pm, corresponding to the different magnitude of the spin-dependent quantum-mechanical forces felt by the tip above the different Ni atoms. To resolve such tiny signals, the AFM instrument has to be operated at low temperatures and in ultrahigh vacuum.

More recently, MExFM has been applied to the antiferromagnetically ordered ground state of a single atomic layer of Fe on a W(001) substrate [15] for which a direct comparison with SP-STM results [3, 4] could be made. Significant differences in the distance-dependence of the MExFM contrast have been observed between the NiO(001) and Fe/W(001) surfaces.

1. R. Wiesendanger, I. V. Shvets, D. Bürgler, G. Tarrach, H.-J. Güntherodt, J. M. D. Coey, and S. Gräser, Science **255**, 583 (1992); R. Wiesendanger, I. V. Shvets, D. Bürgler, G. Tarrach, G. Güntherodt, H.-J. Güntherodt, and J. M. D. Coey, Europhys. Lett. **19**, 141 (1992)

2. S. Heinze, M. Bode, O. Pietzsch, A. Kubetzka, X. Nie, S. Blügel, and R. Wiesendanger, Science **288**, 1805 (2000)

3. A. Kubetzka, P. Ferriani, M. Bode, S. Heinze, G. Bihlmayer, K. von Bergmann, O. Pietzsch, S. Blügel, and R. Wiesendanger, Phys. Rev. Lett. **94**, 087204 (2005)

4. M. Bode, E. Y. Vedmedenko, K. von Bergmann, A. Kubetzka, P. Ferriani, S. Heinze, and R. Wiesendanger, Nature Materials **5**, 477 (2006)

5. G. Binnig and H. Rohrer, Rev. Mod. Phys. **59**, 615 (1987)

6. R. Wiesendanger, "Scanning Probe Microscopy and Spectroscopy: Methods and Applications", Cambridge University Press, Cambridge 1994

7. R. Wiesendanger, H.-J. Güntherodt, G. Güntherordt, R. J. Gambino, and R. Ruf, Phys. Rev. Lett. **65**, 247 (1990)

8. K. von Bergmann, S. Heinze, M. Bode, E. Y. Vedmedenko, G. Bihlmayer, S. Blügel, and R. Wiesendanger, Phys. Rev. Lett. **96**, 167203 (2006)

9. U. Kaiser, A. Schwarz, and R. Wiesendanger, Nature **446**, 522 (2007)

10. Y. Martin and K. Wickramsinghe, Appl. Phys. Lett. **50**, 1455 (1987); J. J. Saenz, N. Garcia, P. Grütter, E. Meyer, H. Heinzelmann, R. Wiesendanger, L. Rosenthaler, H. R. Hidber, and H.-J. Güntherodt, J. Appl. Phys. **62**, 4293 (1987)

11. A. Schwarz, M. Liebmann, U. Kaiser, R. Wiesendanger, T. W. Noh, and D. W. Kim, Phys. Rev. Lett. **92**, 077206 (2004)

12. F. J. Giessibl, Science **267**, 68 (1995)

13. S. Morita, R. Wiesendanger, and E. Meyer (eds.), *Non-contact Atomic Force Microscopy*, Springer (2002)

14. M. Ashino, A. Schwarz, T. Behnke, and R. Wiesendanger, Phys. Rev. Lett. **93**, 136101 (2004)

15. R. Schmidt, C. Lazo, H. Hölscher, U. H. Pi, V. Cacius, A. Schwarz, R. Wiesendanger, and S. Heinze, submitted.

Researching the structure of the surface of undoped ZnO thin films by means of Atomic Force Microscopy

N. Muñoz Aguirre[1], P. Tamayo Meza[1] and L. Martínez Pérez[2]

1. Instituto Politécnico Nacional, Sección de Estudios de Posgrado e Investigación, Escuela Superior de Ingeniería Mecánica y Eléctrica-UA. Av. Granjas, N°682, Colonia Santa Catarina. Del. Azcapotzalco, CP.02550, México, DF. México.
2. Unidad Profesional Interdisciplinaria en Ingeniería y Tecnologías Avanzadas del Instituto Politécnico Nacional, Av. IPN No. 2580, Col. Barrio La Laguna Ticomán, C.P. 07340, México D.F. México.

nmag804@avantel.net

Keywords: Atomic Force Microscopy, contrast phase images, ZnO thin films

Surface characteristics of the amorphous phase composition of ZnO thin films grown by the water-mist-assisted spray pyrolysis method [1] are identified using Atomic Force Microscopy. The AFM phase contrast mode images show (See Figure 1) no homogeneity of the surface structure. White zones in the phase contrast mode images could be correspond to such amorphous phase composition of ZnO thin films. More aspects related with the no homogeneity of the surface structure are also discussed in this work.

1. L. Martínez Pérez, M. Aguilar-Frutis, O. Zelaya-Angel, and N. Muñoz Aguirre,"Improved electrical, optical and structural properties of undoped ZnO thin films grown by water-mist-assisted spray pyrolysis", Physica Status Solidi a, **203, No. 10,** 2411-2417 (2006).
2. We kindly acknowledge to the Scanning Probe Microscopy Laboratory of the SEPI-ESIME-AZCAPOTZALCO from the INSTITUTO POLITECNICO NACIONAL for using its facilities.
3. This work was partially supported by the proyect SIP-20082271.

M. Luysberg, K. Tillmann, T. Weirich (Eds.): EMC 2008, Vol. 1: Instrumentation and Methods, pp. 713–714, DOI: 10.1007/978-3-540-85156-1_357, © Springer-Verlag Berlin Heidelberg 2008

Topography images   **Contrast phase images**

300 °C

400 °C

500 °C

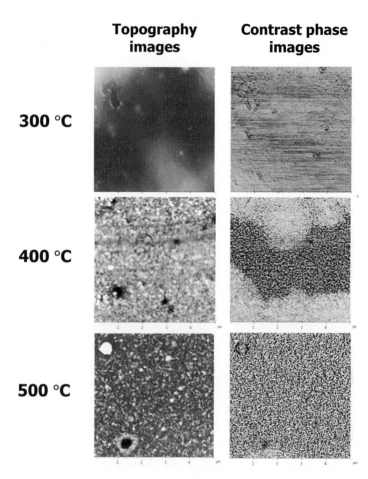

Figure 1. Topographic vs. contrast phase mode AFM surface images of undoped ZnO thin films as a function of its synthesis temperature. In the second row (400 ° C) white zones could be related with amorphous phase of the grains composition. The sizes of the images are $5 \times 5 \times 0.3$ μm.

Improving the structural characterization of supported on glass gold nanoparticles using Atomic Force Microscopy on vacuum conditions

N. Muñoz Aguirre[1], J.E. Rivera López[1], L. Martínez Pérez[2] and P. Tamayo Meza[1]

1. Instituto Politécnico Nacional, Sección de Estudios de Posgrado e Investigación, Escuela Superior de Ingeniería Mecánica y Eléctrica-UA. Av. Granjas, N°682, Colonia Santa Catarina. Del. Azcapotzalco, CP.02550, México, DF. México.
2. Unidad Profesional Interdisciplinaria en Ingeniería y Tecnologías Avanzadas del Instituto Politécnico Nacional, Av. IPN No. 2580, Col. Barrio La Laguna Ticomán, C.P. 07340, México D.F. México.

nmag804@avantel.net

Keywords: Atomic Force Microscopy, tapping mode, Gold nanoparticles

Surface characteristics of gold nanoparticles [1] at ambient conditions and in vacuum of the order of 10^{-4} Torr are compared using Atomic Force Microscopy topographic images in tapping mode. Figure 1 shows the improvement of the right image got at lower pressure than left one. The got image at lower pressure (right) shows clearly the like spheroid shape characteristics of the particles as predicted by the SPR experiments and theory [1]. Profiles and RMS roughness of the particles are also measured.

1. N. Muñoz Aguirre, E. Lopez Sandoval, A. Passian, C. Vazquez López, L. Martínez Pérez and T. L. Ferrell. *"The use of the surface plasmons resonanse sensor in the study of the influence of allotropic cells on water"*, Sensors & Actuators B, **99**, 149-155 (2004).
2. We kindly acknowledge to the SPM Laboratory of the SEPI-ESIME-AZCAPOTZALCO from the INSTITUTO POLITECNICO NACIONAL for using its facilities.
3. This work was partially supported by the proyect SIP-20082271.

M. Luysberg, K. Tillmann, T. Weirich (Eds.): EMC 2008, Vol. 1: Instrumentation and Methods, pp. 715–716, DOI: 10.1007/978-3-540-85156-1_358, © Springer-Verlag Berlin Heidelberg 2008

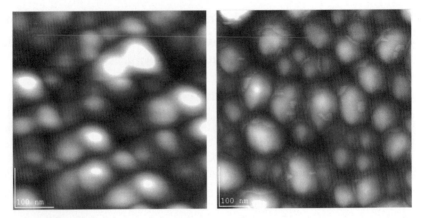

Figure 1. Topographic tapping mode AFM surface images of gold nanoparticles at ambient conditions (left) and pressure of 10^{-4} Torr (right). The size of the images is 500 × 500 × 10 nm.

CO and O₂ chemisorption on Pd₇₀Au₃₀(110) : evolution of the surface studied by in situ STM and complementary surface analysis techniques at elevated pressures

M.A. Languille[1,□], F.J. Cadete Santos Aires[1], B.S. Mun[2,#], Y. Jugnet[1],
M.C. Saint-Lager[3], H. Bluhm[2], O. Robach[4], D.E. Starr[2], C. Rioche[2], P. Dolle[3],
S. Garaudée[3], P.N. Ross[5], J.C. Bertolini[1]

1. IRCELYON. 2 Av. Albert Einstein. 69626 – Villeurbanne cedex. France.
2. ALS/LBNL. One Cyclotron Road, MS 2R0100. Berkeley, CA 94720-8226. USA.
3. Institut Néel, 25, Av. des Martyrs – BP166. 38042 – Grenoble cedex 9. France.
4. CEA-LPCM. 17, R. des Martyrs. 38054 Grenoble cedex 9. France.
5. MSD/LBNL. One Cyclotron Road, MS 2R0100. Berkeley, CA 94720-8226. USA.
□ *Present address:* UCCS. 59655 Villeneuve d'Ascq cedex. France.
Present address: Dept. Appl. Phys. Hanyang Univ., Ansan, Kyeonggi 426-791, Korea.

francisco.aires@ircelyon.univ-lyon1.fr
Keywords: in situ STM, in situ surface analytical techniques, Pd₇₀Au₃₀(110) surface

Elevated pressure *in situ* STM studies were crucial to show that CO adsorption on gold surfaces strongly modifies their surface structures [1-3]. However oxygen dissociation does not occur spontaneously on such surfaces preventing efficient CO oxidation. The addition of an oxygen-dissociative metal (such as Pd) to gold may overcome this difficulty. In this work we present results concerning the behaviour of Pd₇₀Au₃₀(110) under elevated pressures of CO and O₂ and compare them to the results obtained on Au(110) under similar conditions.

The STM and PM-IRRAS in situ experiments were developed at IRCELYON based on a modified MicroLH STM (Omicron) and on a NEXUS spectrometer (Thermo Nicolet), respectively. The XPS experiment was performed at ALS beamlines 9.3.2 & 11.0.2 (Berkeley, USA)) and the SXRD was performed at ESRF beamline BM32 (Grenoble, France) on a reactor developed at the Institut Néel. Under UHV, the outmost surface layer of the Pd₇₀Au₃₀(110) (from Surface Prep Lab) is strongly enriched (above 85% of Au as shown by LEISS).

Pd₇₀Au₃₀(110) under UHV conditions exhibits a (1x1) un-reconstructed surface (Fig. 1a,b). Under CO pressure the surface roughens and a "rice grain" morphology is observed with typical domain sizes around 4 nm and 0.05 nm corrugation (Fig. 1c) that prevails up to 500Torr. PM-IRRAS of CO adsorption on Pd₇₀Au₃₀(110) (at different CO pressures: 0.02Torr to 100Torr) shows three vibration bands : 2115cm⁻¹ (CO top on Au), 2090cm⁻¹ (CO top on Pd) and 1980-1990cm⁻¹ (bridged CO). Complementary studies by XPS show the building up of high energy shoulders on the Pd3d₃/₂ peak with increasing CO pressure. These shoulders can be related to chemical and/or structural effects induced by CO chemisorption. Under O₂ pressure (<1 Torr) XPS reveals a strong oxidation of Pd. At higher pressures (500 Torr) a structure close to a p(2x2) building on a roughened surface is observed by STM (Fig. 1d). SXRD showed that the surface (1x1)

M. Luysberg, K. Tillmann, T. Weirich (Eds.): EMC 2008, Vol. 1: Instrumentation and Methods,
pp. 717–718, DOI: 10.1007/978-3-540-85156-1_359, © Springer-Verlag Berlin Heidelberg 2008

structure disappears under CO pressure. Addition of O_2 (pressure 500Torr) induces the formation of a bulk-like Pd oxide at the surface as shown by the increase of the oxide peak (k=1.47) at T=473K.

The use of complementary in situ techniques enabled us to study the evolution of $Pd_{70}Au_{30}(110)$ surface characteristics upon adsorption of CO and O_2 at elevated pressures, unavailable otherwise. STM shows a strong modification of the surface structure whereas XPS reveals a strong influence of CO and O_2 on the electronic properties of Pd, with oxidation at high pressure/temperature in the latter case. SXRD yields consistent results with both.

1. Y. Jugnet, F.J. Cadete Santos Aires, L. Piccolo, C. Deranlot, J.C. Bertolini, Surface Science **521** (2002), L639.
2. L. Piccolo, D. Loffreda, F.J. Cadete Santos Aires, C. Deranlot, Y. Jugnet, P. Sautet, J.C. Bertolini, Surface Science **566-568** (2004), 995.
3. F.J. Cadete Santos Aires, C. Deranlot, M.A. Languille, L. Piccolo, Y. Jugnet, A. Piednoir, J.C. Bertolini, in CD-ROM : Extended abstracts, questions & answers of the 13th International Congress in Catalysis, (IFP Ed.), 2004.

Figure 1. STM images of the $Pd_{70}Au_{30}(110)$ surface under : (a,b) UHV conditions (20x20nm² and 4.7x3nm² respectively); (c) 1Torr of CO (20x20 nm²); (d) 500Torr of O_2 (10x10nm2).

Height measurements on soft samples: applied force, molecules deformation and phase shift

C. Albonetti[1], N.F. Martínez[2], A. Straub[1], F. Biscarini[1], R. Pérez[3] and R. García[2]

1. ISMN-CNR, Via P. Gobetti 101, 40129 Bologna, Italy
2. IMM-CSIC, Isaac Newton 8, 28760 Tres Cantos, Madrid, Spain
3. FTMC-UAM, Ciudad Universitaria de Cantoblanco, 28049, Madrid, Spain

c.albonetti@bo.ismn.cnr.it
Keywords: atomic force microscopy, soft samples, attractive and repulsive regime

In Atomic Force Microscopy (AFM) the topographic height is related to the tip-surface interaction and, consequently, to the applied force [1]. This phenomenon is expected to be relevant in soft samples, such as organic thin films, where the applied force can deform molecules or, in extreme cases, destroy the film.

To study this effect we used sub-monolayer films of sexithiophene (6T), grown on native Si/SiO_x substrates via high vacuum sublimation [2]. We have observed that the morphology of the films, which is composed by well-defined islands in the sub-monolayer regime, is dependent on the resistivity ρ of the silicon substrate. Low ρ means sparse and dendritic islands, whereas high ρ favors the formation of round-shaped islands with a smaller surface area (Figure 1). The type of doping, n or p, seems to be irrelevant for the island growth.

To investigate the relationship between applied force, molecules deformation and phase shift on these islands, we have measured their height by tapping mode technique. Exploiting cantilevers with different compliance, ranging from soft (low resonant frequency, low elastic constant) to stiff (high resonant frequency, high elastic constant), we applied different force conditions. In addiction, the phase shift provides information on the tip-surface interaction regimes: attractive (high oscillation amplitude, low applied force) or repulsive (low oscillation amplitude, high applied force) [3].

In the case of low ρ ($\leq 0.0015\Omega\cdot cm$), the height of the islands (Figure 2a) is measured $\approx 2.3nm$ (attractive regime, force applied: 17nN). This value is in agreement with the molecule length and the expected orthogonal arrangement of the 6T molecules on the Si/SiO_x surface [4]. Moreover, cumulative height graphs in attractive and repulsive regimes are comparable, with height differences $\approx$ 1-2 Å. Although this difference is close to the instrument sensitivity, its absence in the case of the peak associated to the Si/SiO_x substrate shows that this is a real effect associated with the reduced mechanical strength of the organic layer. Both the small height difference and the large dispersion of the measured heights (as shown by the full half-maximum-width of the peaks FWHM = 3Å) are consistent with theoretical simulations. Increasing the resistivity of the substrate, the islands become smaller, rounded and higher, with an average height up to $\approx 3.7nm$ (Figure 2b-attractive regime, force applied: 17nN) for the highest value of resistivity ($\rho=8$-$12\Omega\cdot cm$). These measurements suggest a second layer grows on top on the islands, with the molecule tilted with respect to the surface normal.

M. Luysberg, K. Tillmann, T. Weirich (Eds.): EMC 2008, Vol. 1: Instrumentation and Methods, pp. 719–720, DOI: 10.1007/978-3-540-85156-1_360, © Springer-Verlag Berlin Heidelberg 2008

In this case, the average island heights measured in the attractive and repulsive regimes (force applied ranging from 300pN to 17nN) are markedly different, in particular in the repulsive regime (from ≈3.7nm to ≈3.2nm, a variation of 10%).

To conclude, we have proved that, for forces up to 20nN, the molecules of the first layer (orthogonal to the surface) are stiff in both regimes while, the molecules of the second layer (tilted respect to the orthogonal direction), are more compressible, leading to the 10% height reduction found in the repulsive regime.

1   R. García and R. Pérez, Surf. Sci. Rep. **47 (6-8)** (2002), p. 197.
2   J.-F. Moulin, , F. Dinelli, M. Massi, C. Albonetti, R. Kshirsagar and F. Biscarini , Nuclear Inst. and Methods in Physics Research B **246 (1)** (2006), p. 122.
3   R. García and A. San Paulo, Phys. Rev. B **60** (1999), p. 4961.
4   N.F. Martínez, W. Kaminski, C.J. Gómez, C. Albonetti, F. Biscarini, R. Perez and R. García "Molecular energy-dissipating processes in oligothiophene monolayers determined by phase-imaging force microscopy" submitted to Advance Materials (2007).

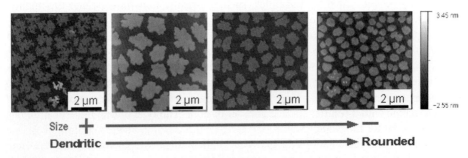

Figure 1 AFM images of the 6T islands for different substrate's resistivity (from left to right: ρ≤0.0015, 1-10, 2-20 and 8-12Ω·cm respectively).

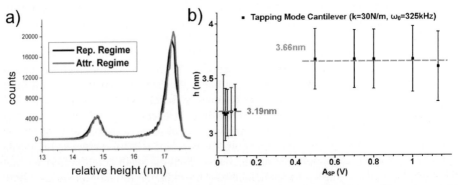

Figure 2 (a) Cumulative height graph in Attractive (A) and Repulsive (R) regimes; (b) Height difference in A and R regimes, measured with stiff cantilever.

Effect of temperature on phase transition of cardiolipin liquid-crystalline aggregates studied by AFM

A. Alessandrini[1,2] and U. Muscatello[1]

1. CNR-INFM-S3 NanoStructures and BioSystems at Surfaces, 41100 Modena, Italy
2. Physics Department, University of Modena and Reggio Emilia, 41100, Modena, Italy

andrea.alessandrini@unimore.it
Keywords: AFM, cardiolipin, inverted hexagonal phase

Cardiolipin is a negatively charged phospholipid that plays an important functional role in bacterial and mitochondrial membranes [1]. Because of the presence of four acyl chains and of a large negative charge, cardiolipin is ready to form either lamellar (L_α) or inverted hexagonal (H_{II}) phases according to several physical parameters. In the H_{II} phase the lipid molecules are arranged in cylindrical structures where the hydrophilic polar groups surround an inner aqueous core and the hydrocarbon tails fill the interstitial regions of the lattice, thus stabilizing the overall structure by the hydrophobic interaction. The cardiolipin phase behavior is relevant in a variety of cellular events [2].

The dependence of the lamellar-to-hexagonal phase transition on pH, the ionization state of phosphatidyl head groups, the salt concentration, the presence or the absence of specific cations has been shown in a number of studies [3]; also temperature variations have an influence on the polymorphic phase behaviour. In this research, the polyphasic behaviour of cardiolipin has been studied as a function of temperature in lipid aggregates dehydrated on a solid support. AFM allows high spatial resolution imaging and the possibility of identifying the formation of local nanoscale defects in the crystalline structures. The nanoscopic observations were conveniently complemented with the data provided by ATR-FTIR that reveals changes in absorption bands related to phase transitions.

AFM examination reveals that, upon dehydration, a transition form a lamellar phase to an inverted hexagonal phase occurs for cardiolipin molecules at all temperatures investigated from 5°C to 39°C. The data provided by ATR-FTIR are in good agreement with AFM observations. The cylinder repeat spacing decreases by about 0.1 nm by varying the preparation temperature from 25°C to 39°C. The value of repeat spacing at 5°C is significantly lower than those observed for temperatures above 25°C. Imaging of local defects in the hexagonal inverted phase can give insights on the details of the nucleation and growth of this phase from the bilayer phase. Besides affecting the repeat spacing of cylinders, the increase of temperature above 25°C leads to the formation of domains ("Figure 1a"). The progressive increase in temperature leads to a progressive decrease in the average area of domains with a parallel increase in their number. The increase in temperature is also associated with a progressive increase in the curvature of cylinder axes that may result in the formation of focal domains or rotational dislocations ("Figure 1b").

M. Luysberg, K. Tillmann, T. Weirich (Eds.): EMC 2008, Vol. 1: Instrumentation and Methods, pp. 721–722, DOI: 10.1007/978-3-540-85156-1_361, © Springer-Verlag Berlin Heidelberg 2008

1. Schlame, M., Rua, D., Greenberg, M.L. *Prog. Lipid Res.* 2000, **39**, 257-288.
2. Siegel, D.P., Epand, R.M. *Biophys. J.* 1997, **73**, 3089-3111.
3. Rand, R.P., Sengupta, S. *Bioch. Biophys. Acta* 1972, **255**, 484-492.

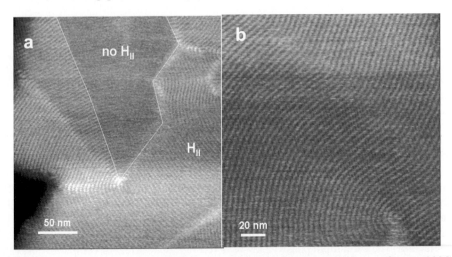

Figure 1. a) AFM image of cardiolipin crystals dehydrated on a mica surface at 32°C. Domains with different orientations and an area where the hexagonal phase is not present are identified; b) AFM image of cardiolipin crystals dehydrated on mica surface at 39°C. Disclination defects are clearly seen (arrow).

Investigating the influence of dynamic scattering on ptychographical iterative techniques

Cheng Liu, T. Walther and J.M. Rodenburg

Department of Electronic and Electric Engineering, University of Sheffield, UK

Cheng.liu@sheffield.ac.uk

Keywords: Phase retrieval, Electron microscopy, Dynamic scattering, Ptychography

The PIE (ptychographical iterative engine) algorithm for solving the diffraction phase problem in electron or X-ray microscopy was proposed several years ago [1]. This approach combines the ideas of traditional ptychography [2] and iterative phase retrieval to produce a powerful new technique to reconstruct the scattered wave front in situations where the illumination radiation can be shifted laterally relative to the specimen. Using diffraction patterns measured at different positions of the incident radiation beam, the complex transmission function of the specimen can be recovered and thus the structure information of the specimen studied becomes available. The feasibility of this algorithm has been verified with both light and X-ray for thin specimens [3]. However, in this algorithm, a multiplicative assumption has been used for the exit and the incident fields, i.e. $\varphi(r, R) = O(r)P(r - R)$, where $O(r)$ is the transmission function of the specimen, which physically represents the exit wave of the object illuminated by a plan wave. $P(r)$ is the field distribution of the illumination. $O(r)$ or $P(r)$ can be moved relative to one another by various distances R. In electron diffraction this assumption is generally satisfied when the specimen is sufficiently thin, or the energy of the illuminating electron is high enough [3]. However, with the increase of the thickness of the specimen, the influence of dynamical scattering of the electron becomes serious and this finally results in the breakdown of the assumption. To investigate the influence of the dynamic scattering on the PIE technique and in order to determine its imaging depth, numerical simulation is much easier to carry out than practical experiments. Software based on the multislice theory has been specially developed in MATLAB and used to investigate the influence of the dynamic scattering on the PIE algorithm. The distribution of the electrons behind a Si <100> sample is first calculated, and then the intensity of its Fourier transform is used for the reconstruction by treating it as the diffraction pattern obtained from a set of 6×6 illumination spots as used in practical reconstruction experiments. The simulation results for 200kV elections show that when the thickness of the silicon crystal is 100 Å, we can still get satisfying reconstruction. Figures 1,2 and 3are the reconstructions for thicknesses of 54.3, 108.6, and 217.2 Å respectively, where images (a) of each figure are the amplitude and phase distributions of the complex exit wave of planar illumination, and images (b) the PIE results. It has been found that the reconstruction produces reliable results up to thickness of 14 nm for Si at 200kV, which corresponds to half the extinction distance of g_{220} along the <100> direction. At this thickness double diffraction of the {220} beams dominates, which renders the kinematic approximation invalid. At higher thicknesses, however, the

M. Luysberg, K. Tillmann, T. Weirich (Eds.): EMC 2008, Vol. 1: Instrumentation and Methods, pp. 723–724, DOI: 10.1007/978-3-540-85156-1_362, © Springer-Verlag Berlin Heidelberg 2008

algorithm seems to recover again somewhat, producing noisy lattice fringes in both amplitude and phase images. The origin of this will also be investigated.

1. H.M.L. Faulkner, J.M. Rodenburg, Ultramicroscopy **103**(2005) 153.
2. J.R. Fienup, Appl. Opt. **21** (1982) 2758.
3. H.M.L. Faulkner and J.M. Rodenburg, Physical Review Letters **93**(2004) 023903.
4. Acknowledgement: The authors are grateful for financial support from EPSRC (Grant No. EP/E034055/1).

(a)   (b)

Figure 1. The exit field of the [100] silicon crystal at 54.3 Å thickness (a), and the PIE reconstructed result (b).

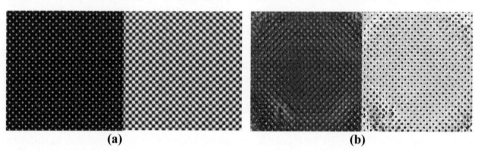

(a)   (b)

Figure 2. The exit field of the [100] silicon crystal at 108.6 Å thickness (a), and the PIE reconstructed result (b).

(a)   (b)

Figure 3. The exit field of the [100] silicon crystal at 217.2 Å thickness (a), and the PIE reconstructed result (b).

Determination of the lateral Resolution of a Cantilever based Solid Immersion Lens Near Field Microscope

T. Merz, K. Rebner, R.W. Kessler

Institut of Applied Research, Process Analysis and Technology, Alteburgstrasse 150, 72762 Reutlingen, Germany

Tobias.merz@reutlingen-university.de
Keywords: Near field imaging, SNOM, SIL

Motivation

The Rayleigh criterion limits the optical characterisation of features smaller than approximately $\lambda/2$. The combination of SPM and near field optical microscopy allows the characterization of the morphology and chemistry of surfaces and cell structures on a nanometer scale. To achieve high lateral resolution, we exploit the localised electromagnetic field of a Solid Immersion Lens (SIL) with a half sphere to get a lateral resolution better than $\lambda/2$ [1].

The lateral resolution of an imaging system can be obtained in different ways. The definition is based on the smallest distance which can be resolved with confidence. In the case of a point, the intensity profile of this point (PSF, point spread function) can be used to estimate the resolution by the full width at half maximum (FWHM). The same applies to lines (LSF, line spread function) at the distance between 12% and 88%. The edge spread function (ESF) determines the resolution of larger structures. To assess the contrast performance of the system, the modulation transfer function (MTF) gives more information about the imaging system.

To apply this methodology to a near field microscope, further aspects have to be considered, such as lack of well defined structures in a range from 10 nm until 100 nm without topology together with a strong contrast for the near field imaging.

In order to overcome these limitations we have used a multilayer stack of GaAs, $Al_{0.65}Ga_{0.35}As$ and $In_{0.33}Ga_{0.67}As$ (provided by BAM), a Fischer Pattern of polymer beads (Kentax) and a specified aluminium oxide membrane (provided by MPI Halle).

The paper will show the design and construction principles of this SIL near field microscope spectrometer and several ways to determine the lateral resolution in the near field.

System Setup

The SIL-SNOM consists of a cantilever holder with the integrated SIL (see Figure 1) which is mounted on top of the piezo scanner[2]. This unit is placed directly into the principle light path of a micro spectral photometer (UMSP Zeiss). An objective with a long working distance and with a high numerical aperture is used to illuminate the SIL. Depending on the focal plane in the SIL, different near field contrasts can be obtained. At illumination under the condition of total internal reflectance (TIR), the near field is then localized around the SIL cone. If a dielectric medium approaches to this field, the near field will be attenuated. The loss of photons can then be detected in the far field.

M. Luysberg, K. Tillmann, T. Weirich (Eds.): EMC 2008, Vol. 1: Instrumentation and Methods, pp. 725–726, DOI: 10.1007/978-3-540-85156-1_363, © Springer-Verlag Berlin Heidelberg 2008

726

When the sample is illuminated by the refracted light of the SIL, TIR does not occur. The SIL will convert the localized near field by scattering or photon tunnel phenomena into a propagating field. This field can now be detected in the far field. The samples are imaged with polychromatic light in reflectance by the SIL-SNOM technique [3].

Results

The resolution limit can be calculated with the BAM sample in three different ways. Using ESF the calculated resolution is 48 nm, for LSP it is 33 nm and for MTF 34 nm. Figure 2 shows a detailed section of the pattern and its intensity cross-section. The LSF of the aluminium membrane pattern can be determined measuring the cross-section of a single tube. The FWHM of the LSF is 45 nm. The paper will show the influence of changes in the focal plane in the SIL, surface roughness and changes in refractive indices on the response functions, factors which also affect the lateral resolution of the system

The advantage of the aperture-less SIL system is the high lateral resolution together with its high transmission efficiency which results in a better S/N ratio and improved contrast.

1. S. M. Mansfield and G. S. Kino, Applied Physics Letters 57, (1990), 2615
2. L. P. Ghislain and V. B. Elings, Applied Physics Letters 72, (1998), 2779
3. T. Merz and R.W. Kessler; SPIE Proceedings, 6631, (2007), 66310V-1
4. We kindly acknowledge the financial support of the German Federal Ministry of Education and Research

Figure 1. Cantilever design with the solid immersion lens

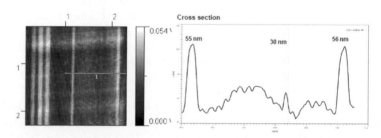

Figure 2. Detail of the BAM-L002 stripe pattern and the LSP of the SIL-SNOM signal

LT-STM manipulation and spectroscopy of single copper and cobalt atoms

E. Zupanič[1], R. Žitko[1], H.J.P. van Midden[1], A. Prodan[1], I. Muševič[1,2]

1. Department of Solid State Physics, Jožef Stefan Institute, Jamova 39, 1000 Ljubljana, Slovenia
2. Faculty of Mathematics and Physics, University of Ljubljana, Jadranska 19, 1000 Ljubljana, Slovenia

erik.zupanic@ijs.si

Keywords: STM, STS, manipulation, spectroscopy.

UHV LT-STM enables not only atomic-scale characterization of solid surfaces but also manipulation of adsorbed atoms and molecules with atomic precision [1]. New artificial nanostructures can be formed in an atom-by-atom process by precisely controlling the quantum-mechanical interactions between the STM tip and the adsorbed species. Using the tip as an analytical tool, novel quantum phenomena can be probed and electronic properties of such nanostructures can be studied. Such experiments require cryogenic temperatures (below 20K), atomically clean and properly ordered surfaces, as well as extreme mechanical and electrical stability of the tip-sample tunneling junction of the STM.

The experiments presented were performed with a Besocke type UHV LT-STM operated at 9K. The Cu(111) and (112) single crystal surfaces were cleaned in-situ by repeated sputter-anneal cycles. An etched polycrystalline W wire was used as the STM tip. Single Cu adatoms were extracted from the sample surface by controlled tip-sample contact under high tunneling bias voltages [2]. Submonolayer amounts of Co atoms were deposited in-situ onto clean Cu surfaces from a Knudsen source or by direct heating of a cobalt wire. Individual adatoms were manipulated into desired nanostructures (Figure 1). The final tip condition was optimized by a controlled tip-forming until the differential conductance spectra (dI/dV) on clean Cu(111) show typical features, such as the onset of the surface state band. Local spectroscopic measurements were performed on individual Cu and Co atoms as well as on artificial nanostructures, by means of the lock-in technique [3].

We have observed that differential conductance spectra, which are proportional to LDOS, are fully reproducible and give an insight into the local electronic properties of artificially formed nanostructures as a function of their form and composition (Figure 2). STS experiments were combined with theoretical simulations of electronic LDOS.

1. S. W. Hla, J. Vac. Sci. Tech. B, **23** (2005), p. 1351.
2. J. Lagoute, X. Liu and S. Foelsch, Phys. Rev. Lett., **95** (2005), p. 136801.
3. S. Foelsch et. Al, Phys. Rev. Lett., **92** (2004), p. 056803.

M. Luysberg, K. Tillmann, T. Weirich (Eds.): EMC 2008, Vol. 1: Instrumentation and Methods, pp. 727–728, DOI: 10.1007/978-3-540-85156-1_364, © Springer-Verlag Berlin Heidelberg 2008

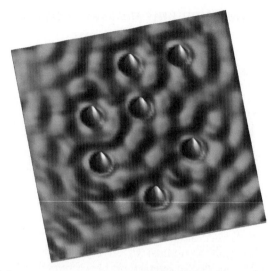

Figure 1. Artificial nanostructure, composed of individual Cu atoms, displaced along the Cu(111) single crystal surface with the STM tip. 14.5 x 14.5 nm, T=9.5 K.

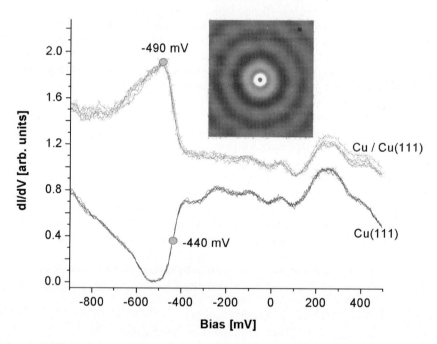

Figure 2. Differential conductance spectra dI/dV, measured on a clean Cu(111) surface (bottom) and on a selected Cu adatom on Cu(111) surface (top). The circles in the inset mark the respective tip positions during the measurements.

3D atomic-scale chemical analysis of engineering alloys

A. Cerezo, E.A. Marquis, D.W. Saxey, C. Williams, M. Zandbergen, G.D.W. Smith

Department of Materials, University of Oxford, Parks Road, Oxford, OX1 3PH, UK

alfred.cerezo@materials.ox.ac.uk

Keywords: atom probe, precipitation, grain boundary segregation, aluminium alloys, steels

The 3-dimensional atom probe (3DAP) permits the 3D reconstruction of atomic-scale chemical variations within conductive materials, often termed atom probe tomography (APT). Position-sensitive time-of-flight mass spectrometry is used to locate (to sub-nanometre resolution) and chemically identify single atoms removed from the surface of a needle-shaped specimen (end radius 50 - 100 nm) [1].

APT has proved to be extremely powerful for studying the ultra-fine microstructures of advanced engineering alloys [2]. The advent of the local electrode atom probe (LEAP™, Imago Scientific Instruments) has greatly increased the analysis speed and volume, from 20 nm × 20 nm × 100 nm in 1 day, to 100 nm × 100 nm × 500 nm in a few hours [3]. There is now an overlap between the larger field-of-view of the LEAP and that of high-resolution analytical TEM. Mass resolution in initial LEAP designs was lower than that available from reflectron-based 3DAP instruments [4], but the latest generation of LEAP instruments incorporate a patented reflectron design, to give both large field-of-view and high mass resolution [5]. Figure 1 compares mass spectra with and without the reflectron; the higher mass resolution of the reflectron instrument allows full separation of elements, giving increased sensitivity and accuracy.

This paper will demonstrate the capability of APT for the detailed study of engineering alloys, using examples from some of the current work being carried out at Oxford, including the microstructure of oxide-dispersion strengthened EUROFER steel, a candidate material for structural materials in fusion power plant. Figure 2 gives an example of the type of information available from the technique: this sequence of images obtained from a Al-Mg-Si (6xxx series) alloy aged at 180 °C for different times shows the precipitation of the β'' phase that causes hardening. Focussed ion-beam (FIB) techniques, have made site-selective specimen fabrication more routine, enhancing the study of lower-density microstructural features, such as grain boundaries. Figure 3 shows a 3DAP analysis of a grain boundary in a Pd-modified stainless steel.

1. M.K. Miller, A. Cerezo, M.G. Hetherington and G.D.W. Smith, "Atom probe field–ion microscopy", (Oxford University Press, Oxford) (1996).
2. M.K. Miller, "Atom probe tomography", (Plenum Press, New York) (2000).
3. T.F. Kelly, T.T. Gribb, J.D. Olson, R.L. Martens, J.D. Shepard, S.A. Wiener, T.C. Kunicki, R.M. Ulfig, D.R. Lenz, E.M. Strennen, E. Oltman, J.H. Bunton and D.R. Strait, Microsc. Microanal. **10** (2004), p. 373.
4. A. Cerezo, T.J. Godfrey, S.J. Sijbrandij, P.J. Warren, & G.D.W. Smith, Rev. Sci. Instrum. **69** (1998), p. 49.
5. P. Panayi. International patent application WO2006/120428, published 16th November 2006.

M. Luysberg, K. Tillmann, T. Weirich (Eds.): EMC 2008, Vol. 1: Instrumentation and Methods, pp. 729–730, DOI: 10.1007/978-3-540-85156-1_365, © Springer-Verlag Berlin Heidelberg 2008

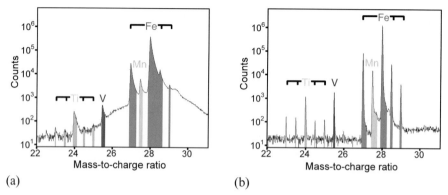

(a) (b)

Figure 1. Comparison between the mass spectra obtained in the LEAP™ analysis of the same steel (a) without and (b) with a reflectron energy-compensating lens.

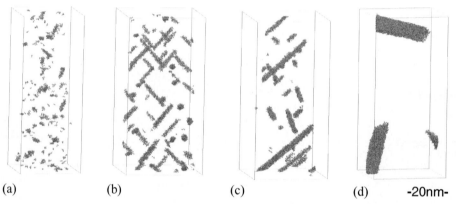

(a) (b) (c) (d) -20nm-

Figure 2. 3D atom maps of Mg (green) and Si (blue) in an Al - 0.5 %Mg - 1.0 %Si alloy aged at 180°C for (a) 10 minutes, (b) 30 minutes, (c) 18 hours and (c) 400 hours. Only solutes within clusters or precipitates are shown.

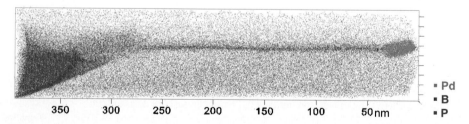

Figure 3. 3D atom map from a Pd-modified stainless steel, after sensitisation at 650 °C for 24 hours, showing grain boundary segregation of P and the formation of second-phase particles on the boundary.

New Applications for Atom-Probe Tomography in Metals, Semiconductors and Ceramics

Thomas F. Kelly[1], David J. Larson[1], Roger L. Alvis[1], Peter H. Clifton[1],
Stephan S.A. Gerstl[1], Rob M. Ulfig[1], Daniel Lawrence[1], David P. Olson[1],
David A. Reinhard[1], and Krystyna Stiller[2]

1. Imago Scientific Instruments Corp., 5500 Nobel Drive, Madison, WI USA 53711
2. Dept. of Physics, Chalmers University of Technology - SE-412 96 Göteborg, Sweden

tkelly@imago.com

Keywords: atom-probe tomography, microanalysis, three-dimensional imaging

As atom-probe tomography has expanded in accessibility and capability in the past decade, many new applications have developed for a spectrum of materials types: advanced metals, silicon-based semiconductor structures, compound semiconductors, and even ceramics. Examples of recent analyses will be shown for materials in each of these categories.

These advances in applications have been made possible by major developments in instrumentation, specimen preparation, and methods correlation with TEM and SIMS. These topics will also be discussed to illustrate their beneficial impact on applications development.

An example of a metal application is shown in Figure 1. Here an a_2-Ti_3Al phase is in the process of dissolving into the γ-TiAl phase. This figure illustrates the use of APT for both compositional and structural imaging.

1   J. Angenete, K. Stiller and V. Langer, Oxidation of Metals 60 (2003) 47.
2   M. J. Galtrey, R. A. Oliver, M. J. Kappers, C. McAleese, D. Zhu, and C. J. Humphreys, P. H. Clifton, D. Larson, and A. Cerezo, Appl. Phys. Lett. 92, 041904 (2008).

M. Luysberg, K. Tillmann, T. Weirich (Eds.): EMC 2008, Vol. 1: Instrumentation and Methods,
pp. 731–732, DOI: 10.1007/978-3-540-85156-1_366, © Springer-Verlag Berlin Heidelberg 2008

Figure 1. Two-phase TiAl alloy which shows a$_2$-Ti$_3$Al (mostly red) phase being dissolved into the γ-TiAl phase (mostly blue). (Ti atoms shown red, Al atoms shown blue). Laser pulsing of atom probes makes it possible to analyze semiconductor and dielectric materials. Figure 2 shows an analysis from an oxidized NiAl diffusion coating [1]. The concentration (excluding H, which accounted for 0.74%) was calculated to be 37.4±0.03% Al – 62.6.0±0.03% O, which is slightly O enriched.

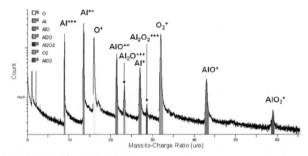

Figure 2. Mass spectrum from alumina.

A C-plane GaN-based quantum well structure [2] is shown in Figure 3. These data demonstrate non-uniform distributions of In in the quantum wells which is suspected to be the cause of less-than-expected quantum efficiency for light conversion.

Figure 4 is an APT image of the SiGe source/drain region of a strain-enhanced-mobility MOS transistor. The oxide spacer is depicted by an O isoconcentration surface. B out-diffusion from the SiGe region alters the channel length of the device.

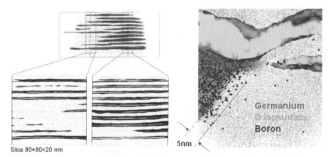

Figure 3. (Left above) Top is a side view of an image with isoconcentration surfaces set at 2% In. Bottom are slices through this image as shown from two locations.

Figure 4. (Right above) APT image of the source/drain region of a strain-enhanced transistor.

Pulsed laser atom probe tomography analysis of advanced semiconductor nanostructures

M. Müller[1], A. Cerezo[1], G.D.W. Smith[1], L. Chang[2]

1. Department of Materials, University of Oxford, Parks Road, OX1 3PH, UK
2. Department of Material Science and Engineering, National Chiao Tung University, 1001 Ta Hsueh Road, Hsinchu, Taiwan, 300, ROC

michael.mueller@materials.ox.ac.uk
Keywords: atom probe tomography, quantum dots, semiconductor superlattice

Atom probe tomography (APT) provides atomic-scale, three-dimensional chemical and microstructural data. APT has widely been used to study metallic materials, engineering alloys and surface–related phenomenon. The use of pulsed laser assisted field evaporation [1,2] expands the range of applications towards less conductive materials, such as semiconductor materials and nanostructures. This paper presents latest investigations of two different types of advanced semiconductor heterostructure materials for photonics applications.

Firstly, we investigated a system of buried In(Ga)As quantum dots overgrown with GaAs produced using molecular beam epitaxy. We were able to image staggered quantum dots (Figure 1) and to evaluate a significant number of single dots within the substantial atom probe datasets obtained. As described in other studies e.g. [3,4] a significant incorporation of gallium was confirmed in both the dots and in the wetting layers. On average, the quantum dot composition was found to be around $In_{0.22\pm0.01}Ga_{0.78\pm0.01}As$. It was observed that the quantum dots exhibit an indium-rich core that can be laterally shifted to the edge of the dot. As represented in atom probe data, the quantum dots appear lens-shaped with an elliptic base possessing an average axis ratio of around 1:1.5 and an average dot height of 3.5 ± 0.3 nm. Detailed investigations of cross-sectional indium profiles give evidence for a truncated pyramid shape of the dots as suggested by other studies [5].

We have also obtained preliminary atom probe results from a GaSb/InAs superlattice structure used in IR detectors, as shown in Figure 1. Due to the characteristic evaporation behaviour of this material, an atom density correction procedure was required to compensate for evaporation artefacts. Concerning the mutual intermixing (i.e. the incorporation of both gallium and antimony in the InAs layers and indium and arsenic in the in the GaSb layers), it appears that gallium and antimony are incorporated to a greater degree in the InAs wells than vice versa. Moreover, different interfaces and transitions zones between the diverse layers could be identified showing fairly sharp interfaces when InAs is grown onto GaSb, but in contrast more diffuse interfaces when GaSb is grown on InAs.

The results show that laser-pulsed APT can provide unique insight into the nano-scale chemistry of up-to-date semiconductor materials. Hence, APT can be considered as an analytic technique complementary to e.g. TEM and STM. By combining these

M. Luysberg, K. Tillmann, T. Weirich (Eds.): EMC 2008, Vol. 1: Instrumentation and Methods, pp. 733–734, DOI: 10.1007/978-3-540-85156-1_367, © Springer-Verlag Berlin Heidelberg 2008

734

different techniques a more holistic and comprehensive approach of analyzing semiconductor materials can be achieved.

1. G. L. Kellogg and T. T. Tsong, J. Appl. Phys. **51** (1980), p. 1184.
2. T. T. Tsong, S. B. McLane and T. J. Kinkus, Rev. Sci. Instrum. V **53** (1982), p. 1442.
3. J. Tersoff, Phys. Rev. Lett. **81** (1998), p. 3183.
4. Ch. Heyn, A. Schramm, T. Kipp and W. Hansen, J. Cryst. Growth 301-302 (2007), p. 692.
5. L. G. Wang, P. Kratzer, N. Moll and M. Scheffler, Phys. Rev. B Condens. Matter Mater. Phys. **62** (2000), p. 1897.

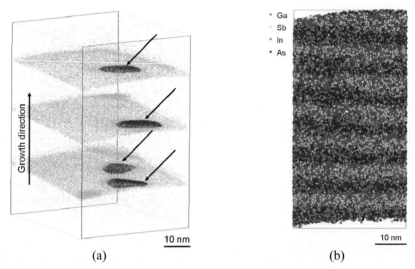

(a)   (b)

Figure 1. 3D atom maps from (a) the In(Ga)As quantum dot material showing the indium atoms (green) of the wetting layers and the quantum dots (arrowed). Image (b) shows an atom probe atom map of the GaSb/InAs superlattice structure.

Low Energy Electron Microscopy: A 10 Year Outlook

Rudolf M. Tromp

IBM T.J. Watson Research Center, 1101 Kitchawan Road, P.O. Box 218, Yorktown Heights, NY 10598, USA

rtromp@us.ibm.com
Keywords: LEEM, PEEM, microscopy

Low Energy Electron Microscopy (LEEM), and its close cousin Photo Electron Emission Microscopy (PEEM) have evolved from curiosities in the hands of physicists, to powerful tools for dynamic materials analysis. Invented in 1962, and first successfully realized in 1985, there are about 30 combined LEEM/PEEM instruments in the world today, in addition to about an equal number of PEEM-only instruments. Most of these instruments follow a design that is now almost 20 years old. The most advanced instruments are located at synchrotron radiation facilities, but they are relatively few in number, and not readily available to the general user. In the meantime, the field of electron microscopy is undergoing revolutionary advances fueled by dual breakthroughs in electron energy filtering and aberration correction. Similarly, the field of quantum optics is developing ever more powerful light sources in the Vacuum Ultra Violet (VUV) and soft X-ray ranges that do not depend on massive investments in national infrastructure (i.e. synchrotrons), but deliver their photons in a standard laboratory. These advances from two very different fields will combine to set the stage for the development of LEEM and PEEM over the next decade. Aberration correction will improve LEEM resolution from 5-10 nm today, to 1.5-2 nm in the near future, enough to resolve individual unit cells in the famous Si(111)-(7x7) surface, or –probably more important- sufficient to resolve the structure of nanoscale features such as magnetic domain walls. In PEEM the spatial resolution will improve from ~ 20 nm today to 4-5 nm, while at the same time improving transmission by a factor ~10. Energy filtering, today mostly used by synchrotron-based instruments, will be ubiquitous, powerful, simple, and relatively inexpensive. For PEEM there will be a broad choice of light sources, not just the ever-present Hg discharge lamp of today. We will obtain not just the atomic structure of surfaces using LEED-IV information with nm spatial resolution, nor just the spatial arrangement of structural domains, quantum well structures, or other multi-phase arrangements, as well as their evolution in time; but we will also learn the spatially resolved electronic structure from nano-ARUPS, Angle Resolved Ultraviolet Photoelectron Spectroscopy at the nanoscale, as well as the chemical makeup of the surface with nano-XPS, X-ray Photoelectron Spectroscopy at the nanoscale, all in one compact instrument, in a standard laboratory environment. Some of the ingredients of this new paradigm are in place today, others need more work. This talk will give an overview of where we are, and where we need more work and more investment.

M. Luysberg, K. Tillmann, T. Weirich (Eds.): EMC 2008, Vol. 1: Instrumentation and Methods, pp. 735–736, DOI: 10.1007/978-3-540-85156-1_368, © Springer-Verlag Berlin Heidelberg 2008

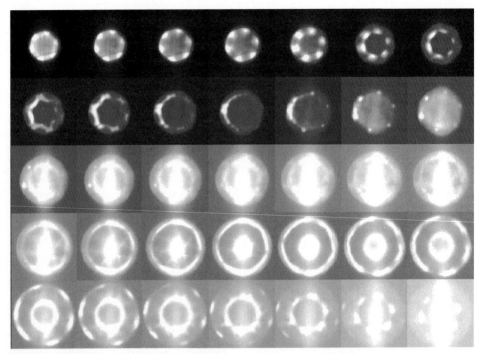

Figure 1. HeI (top two rows) and HeII (bottom three rows) VUV photoelectron angular distributions measured with increasing kinetic energy, using an energy filtered LEEM/PEEM instrument with novel electron energy filter design. The sample consisted of ~ 1.5 monolayer of graphene on SiC(0001). The data presented here show the presence of the graphene bandstructure, with well-developed σ and π bands.

The sample was fabricated *in-situ* in the LEEM/PEEM system, and characterized by both Low Energy Electron Diffraction and LEEM imaging prior to photoemission analysis. This is the first example of lab-based bandstructure analysis in an energy-filtered LEEM/PEEM instrument to date.

Imaging of Surface Plasmon Waves in Nonlinear Photoemission Microscopy

Frank-J. Meyer zu Heringdorf[1], N.M. Buckanie[1], L.I. Chelaru[1,2], N. Raß[1]

1. Universität Duisburg-Essen, Lotharstrasse 1, 47057 Duisburg, Germany
2. present address: Institute of Solid State Research, 52425 Jülich, Germany

meyerzh@uni-essen.de
Keywords: Photoemission Microscopy, Surface Plasmon Polaritons, Pump Probe Microscopy

Photoemission electron microscopy (PEEM) is excellently suited for studying electronic excitations at surfaces by using fs laser pulses. When the photon energy of the used fs laser pulses (E~3.1eV) is too low for threshold photoemission, photoemission must proceed via two photon photoemission (2PPE) through a virtual or real intermediate state. In our earlier work [1] we demonstrated that localized plasmon resonances (LSPs) in small Ag particles act as intermediate states for 2PPE and lead to enhanced photoemission. Here we focus on larger Ag islands, in which propagating surface plasmon polariton waves (SPPs) dominate the 2PPE signal [2]. Figure 1 shows an example for the resulting SPP contrast. Panel (a) shows a bright Ag island on a ($\sqrt{3}$ x $\sqrt{3}$)-Ag reconstructed Si(111) background in regular threshold photoemission. In 2PPE PEEM, shown in Fig. 1 (b), the contrast is completely changed. Panel (b) was recorded using p-polarized fs laser pulses in a grazing incidence geometry, i.e. the k-vector of the light, k_{light}, forms an angle of 74° with the surface normal. The incidence direction of the light k_{light}, and the in-plane component of the electric field vector are indicated at the bottom left of Fig. 1 (b).

Several features in panel (b) of Fig. 1 are noteworthy. First, the island is surrounded by diffraction fringes (feature "A"): due to the grazing incidence geometry, the diffracted light impinges on the surface and modulates the photoemission yield of the Ag/Si background. The most prominent feature in Fig 1 (b), however, is the 2PPE yield modulation on top of the island. This modulation is caused by the superposition of three electric fields at the surface. Whenever a fraction of the in-plane component of the electric field of the laser pulse is perpendicular to the edge of a Ag island, a SPP wave is excited that propagates away from the islands' edge. In the triangular geometry of Fig. 1, two of such waves are excited, labeled as SPP_1 and SPP_2. The wavelength of the SPPs (λ_{SPP} ~ 360nm for excitation with λ=400nm laser pulses [2]) does not explain the observed modulation of the 2PPE yield with a periodicity of ~2.5 μm. To explain the experimental result it is necessary to consider the electric field of the fs laser pulse as well: While the pulse does start the two SPP waves SPP_1 and SPP_2, it also propagates across the island, just like the SPP waves, due to the aforementioned grazing incidence geometry. The SPPs and the in-plane component of the laser pulse thus behave rather similar, but they propagate at different speeds (the SPP waves propagate much more slowly than the laser light). At each point in time and space, the total electric field E_{total} is then simply the superposition of the three electric fields E_{SPP1}, E_{SPP2} and E_{light}, and the 2PPE

M. Luysberg, K. Tillmann, T. Weirich (Eds.): EMC 2008, Vol. 1: Instrumentation and Methods, pp. 737–738, DOI: 10.1007/978-3-540-85156-1_369, © Springer-Verlag Berlin Heidelberg 2008

yield Y_{2PPE} is proportional to $E_{total}{}^4$ due to the nature of the 2PPE process. To further complicate the situation, it is important to consider that the laser pulse passes the Ag island in less than 50fs, i.e. the PEEM detector is incapable of resolving the propagation of the laser pulse and SPP waves in time and integrates over the whole propagation process. The 2PPE PEEM signal is then a time-averaged beat pattern which gives access to fundamental properties of the propagating SPP wave, like the SPP k-vector and the SPP propagation length. The intensity modulation in the center of the island is caused by superposition of the two SPP waves and illustrates that also SPP interaction and interference are accessible to 2PPE PEEM.

Due to the dispersion relation of SPPs, the excitation of a SPP by light requires a k-vector that can be found at the edge of the Ag island. Vice versa, if a propagating SPP arrives at the edge of a Ag island, it can be converted back into light and modulate the electric field in the surrounding of the island. In Fig. 1 (b), the corresponding field modulation behind the island is the feature denoted "B". The intensity modulation is clearly correlated with the beat pattern that is visible on top of the island.

In conclusion, while the detailed understanding of some of the described features requires computer simulations to reproduce the beat patterns, 2PPE PEEM is a strikingly simple method to study field enhancement effects, SPP wave propagation and the coupling of light into and out of nanostructures.

1. L. Chelaru, M. Horn- von Hoegen, D. Thien and F. Meyer zu Heringdorf, Phys. Rev. B **73** (2006), p. 115416
2. L. Chelaru and F. Meyer zu Heringdorf, Surf. Sci. **601** (2007), p. 4541
3. F. Meyer zu Heringdorf, L. Chelaru, S. Möllenbeck, D. Thien and M. Horn von Hoegen, Surf. Sci. **601** (2007), p. 4700

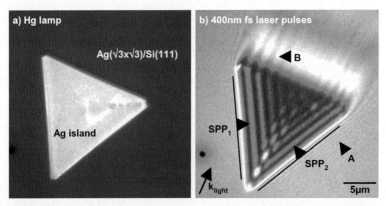

Figure 1. Threshold photoemission and two photon photoemission microscopy images of a Ag island on a Si(111) surface. (a) threshold photoemission (b) illumination with fs laser pulses. The Ag island is surrounded by a pronounced diffraction pattern (feature "A"). The brightness modulation on top of the island is a beat pattern, formed between two SPP waves (SPP₁ and SPP₂) with the illuminating laser pulse. Behind the island, the electric field is modulated (feature "B"). Panel (b) has been plotted using a logarithmic lookup table.

High resolution surface analysis of metallic and biological specimens by NanoSIMS

C.R.M. Grovenor

Department of Materials, University of Oxford, Parks Road, Oxford OX1 3PH, UK

chris.grovenor@materials.ox.ac.uk
Keywords: NanoSIMS, surface analysis, high sensitivity

In the past decade, significant progress has been made in the development of Secondary Ion Mass Spectrometry (SIMS) instrumentation with very high spatial resolution. One of these developments has been the Cameca NanoSIMS50, which offers high sensitivity SIMS analysis with lateral resolution well below 100nm. Instruments of this kind allow us to carry out key analytical experiments that would have been considered impossible only a few years ago, and these studies are starting to provide some startling new data on scientifically and commercially important materials.

This presentation will include some of the recent data that we have obtained in Oxford using our NanoSIMS to study a diverse set of important materials problems. In each case, I will discuss the key scientific problem we are studying, the question of reliable sample preparation, the advantages and limitations of NanoSIMS analysis for that particular combination of sample and problem, and the novel data that can be obtained. The kind of data that can be obtained can be seen in the specific examples given in Figures 1 and 2.

The Oxford NanoSIMS group have studied cracking phenomena in the stainless steels used in critical applications in nuclear reactors [1]. Figure 1 gives a typical example of the kind of high resolution analysis of the chemistry around crack tips that can easily be achieved on metallographically polished bulk samples. The duplex nature of the oxide in the crack is very clear in the FeO and CrO molecular ion images, as is the penetration of the 50 – 100 nm oxide 'fingers' along deformation bands into the left hand grain. Trace elements like S and B can also be mapped with similar resolution.

A second example of the kind of analysis we are undertaking on our NanoSIMS is given in Figure 2. Here we are mapping the uptake of BrdU in cancerous cells by direct imaging of the $^{79}Br^-$ ion image. On the right hand side is an example of the optical fluorescent image from a similar cell. This is the technique used conventionally to map the distribution of BrdU, but the spatial resolution is much better in the NanoSIMS image. This kind of experiment is being developed to map the uptake of new drugs that are not themselves fluorescent and cannot easily be tagged with fluorescent markers.

1. S. Lozano-Perez, M.R. Kilburn, T. Yamada, T. Terachi, C.A. English and C.R.M. Grovenor, Journal of Nuclear Materials **374** (2008) p.61
2. The author acknowledges the contribution of INSS (Japan) for providing the sample in Figure 1, and Sergio Lozano-Perez, Martin Christlieb and Khim-Heng Lau for the images.

M. Luysberg, K. Tillmann, T. Weirich (Eds.): EMC 2008, Vol. 1: Instrumentation and Methods, pp. 739–740, DOI: 10.1007/978-3-540-85156-1_370, © Springer-Verlag Berlin Heidelberg 2008

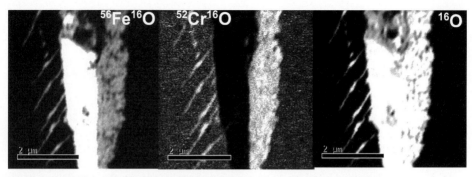

Figure 1. High-resolution NanoSIMS image of a crack in a 304 stainless steel stress corrosion cracked in simulated pressurised water reactor water at 320°C.

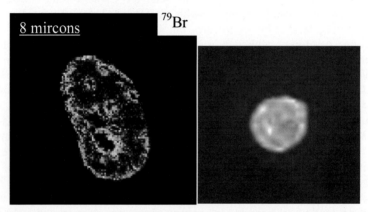

Figure 2. Bromine ion image from the nucleus of a single HeLa cell exposed to BrdU, compared to on the right hand side an optical fluorescent image of BrdU distribution in a similar cell.

Elemental distribution profiles across Cu(In,Ga)Se$_2$ solar-cell absorbers acquired by various techniques

D. Abou-Ras[1], C.A. Kaufmann[1], A. Schöpke[1], A. Eicke[2], M. Döbeli[3], B. Gade[4], T. Nunney[5]

1. Hahn-Meitner-Institut Berlin, Glienicker Strasse 100, 14109 Berlin, Germany
2. ZSW Stuttgart, Industriestr. 6, 70565 Stuttgart, Germany
3. Paul Scherrer Institute, c/o ETH-Hönggerberg, 8093 Zürich, Switzerland
4. Thermo Fisher Scientific, Im Steingrund 4-6, 63303 Dreieich, Germany
5. Thermo Fisher Scientific, Imberhorne Lane, East Grinstead, West Sussex, UK

daniel.abou-ras@hmi.de

Keywords: Cu(In,Ga)Se$_2$, thin-film solar cells, RBS, SIMS, SNMS, XPS, Auger, EDX

Thin-film solar cells with Cu(In,Ga)Se$_2$ absorbers offer a low-cost perspective at high photovoltaic performance. These solar cells generally consist of a ZnO/CdS/Cu(In,Ga)Se$_2$/Mo thin-film stack on a glass substrate. The Cu(In,Ga)Se$_2$ absorber layers do not exhibit homogeneous compositions but gradients of the Ga and the In concentrations across the layer thickness. The analysis of these gradients is important in order to gain information on the optoelectronic properties of the Cu(In,Ga)Se$_2$ thin films, since CuInSe$_2$ has a different band-gap energy (1.0 eV) to CuGaSe$_2$ (1.7 eV).

For research and development, various techniques are generally applied in order to determine the chemical composition of the Cu(In,Ga)Se$_2$ absorber layers. However, these techniques have not yet been compared with each other in terms of their capabilities to determine qualitatively or quantitatively the elemental distributions across the absorber layers. In the present work, results are shown obtained by energy-dispersive X-ray spectrometry (EDX) in a scanning (SEM-EDX) and in a transmission electron microscope (TEM-EDX), X-ray photoelectron (XPS), secondary ion-mass (SIMS) and sputtered neutral mass (SNMS), Auger electron (AES) and also Rutherford backscattering spectrometry (RBS), all performed on nominally identical Cu(In,Ga)Se$_2$ absorber layers from the same production run, deposited on Mo-coated glass substrates.

The RBS data was simulated by use of the RUMP software, assuming a stack of six Cu(In,Ga)Se$_2$ layers with different compositions and thicknesses. The XPS, AES, SIMS, and SNMS measurements required etching by means of an Ar-ion beam. The SNMS, SIMS, AES and EDX results were quantified by use of the Cu, In, Ga, and Se concentrations obtained by X-ray fluorescence analyses. The calculated concentrations of Cu and Se from SIMS depth profiles appear to be affected close to the sample surface and to the Cu(In,Ga)Se$_2$/Mo interface by the known matrix effects.

All of these methods reproduce quantitatively similar Ga and In elemental distribution profiles across the Cu(In,Ga)Se$_2$ thin film, as shown in Figure 1. The (in part) substantial differences between the quantification results can be attributed to different sensitivity factors used in the evaluation of the data.

M. Luysberg, K. Tillmann, T. Weirich (Eds.): EMC 2008, Vol. 1: Instrumentation and Methods, pp. 741–742, DOI: 10.1007/978-3-540-85156-1_371, © Springer-Verlag Berlin Heidelberg 2008

1. This work was supported in part by the EFRE project ANTOME. The authors are grateful to R. Caballero, K. Sakurai, N. Blau, B. Bunn, C. Kelch, M. Kirsch, P. Körber, and T. Münchenberg for solar-cell processing.

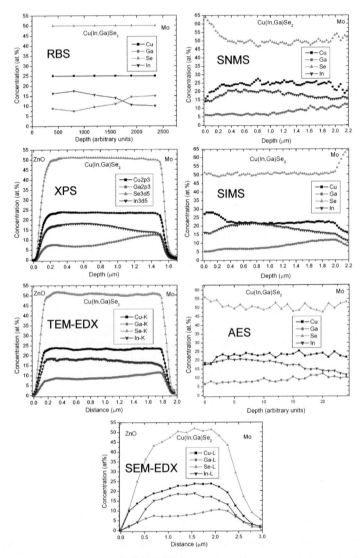

Figure 1. Elemental distribution profiles of Cu, Ga, Se, and In signals across Cu(In,Ga)Se$_2$ thin films on Mo-coated glass substrates, as obtained by RBS, SNMS, XPS, SIMS, TEM-EDX, AES and SEM-EDX.

High resolution Kelvin force microscopy

Matthias A. Fenner, John Alexander and Sergei Magonov

Agilent Technologies Inc., 4330 W. Chandler Blvd., Chandler, AZ 85226, USA

matthias_fenner@agilent.com

Keywords: Atomic Force Microscopy, Kelvin Force Microscopy, Materials Science

Kelvin Force Microscopy (KFM) is a very powerful tool for mapping of surface charges, surface potentials, and doping profiles. [1] This technique is implemented in amplitude modulation and frequency modulation Atomic Force Microscopy (AFM) modes. [2] In many applications surface electric properties are measured with two-pass technique in which a "spill over" of topographic response to the probe motion is reduced by lifting the probe over a sample surface during detection of electric signals. Such approach has also severe limitations in sensitivity and lateral resolution due to a remote probe position during the lift scan. A separation of topographic and electrostatic responses is also possible by operating topography and electric-response servo loops at different frequencies that enable single-pass KFM with simultaneous studies of sample topography and surface potential. The latter is measured by voltage applied to the probe that nullifies its electrostatic interaction with a sample.

We consider several practical implementations for the electric-response servo loop. In search for one that provides best spatial resolution [3] and highest sensitivity. Different inputs and frequencies were applied in this search. We have also employed AFM probes with different cantilever geometries (Figure 1, left and right) and tip dimensions. The results of this study will be presented and illustrated by KFM images of different materials: semiconductor structures with different level of doping, polymer composites, graphite, Au (111) and fluoroalkanes. Doped electric passes (Figure 2, right), negatively charged self-assemblies of fluoroalkane layer on graphite (Figure 3, right) and contamination patches, which are grown on the graphite surface in air (Figure 4, right) are visualized in surface potential images. It was possible to achieve lateral resolution better than 5 nm and sensitivity of few tens of milliVolts.

1. M. Nonnenmacher, M.P. O'Boyle, H.K. Wickramasinghe, Appl. Phys. Lett. **58**, 1991, 58, p. 2921
2. U. Zerweck, C. Loppacher, T. Otto, S. Grafström, L.M. Eng, Phys Rev B **71**, 2005, p. 125424
3. J. Colchero, A. Gil, A.M. Baro, Phys. Rev. B. 64, 2001, p. 245403

M. Luysberg, K. Tillmann, T. Weirich (Eds.): EMC 2008, Vol. 1: Instrumentation and Methods, pp. 743–744, DOI: 10.1007/978-3-540-85156-1_372, © Springer-Verlag Berlin Heidelberg 2008

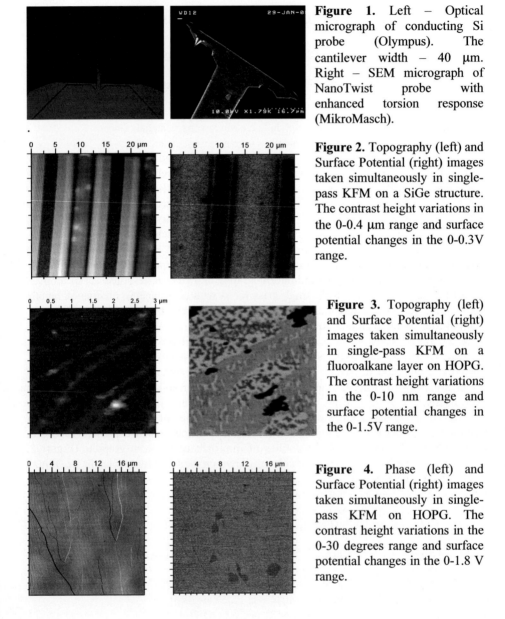

Figure 1. Left – Optical micrograph of conducting Si probe (Olympus). The cantilever width – 40 μm. Right – SEM micrograph of NanoTwist probe with enhanced torsion response (MikroMasch).

Figure 2. Topography (left) and Surface Potential (right) images taken simultaneously in single-pass KFM on a SiGe structure. The contrast height variations in the 0-0.4 μm range and surface potential changes in the 0-0.3V range.

Figure 3. Topography (left) and Surface Potential (right) images taken simultaneously in single-pass KFM on a fluoroalkane layer on HOPG. The contrast height variations in the 0-10 nm range and surface potential changes in the 0-1.5V range.

Figure 4. Phase (left) and Surface Potential (right) images taken simultaneously in single-pass KFM on HOPG. The contrast height variations in the 0-30 degrees range and surface potential changes in the 0-1.8 V range.

High resolution in interferometric microscopy

Marc Jobin and Raphael Foschia

Ecole d'Ingénieurs de Genève, University of Applied Science, Geneva,
SWITZERLAND

marc.jobin@hesge.ch

Keywords: interference microscopy, scanning force microscopy, atomic step

Interferometric optical microscopes (IOM) are very powerful 3D metrology tools which use integrated interferometers inside optical objectives. In Phase Shift Mode (PSM) [1], they can reach subnanometer vertical resolution but the lateral resolution, as any far-field optical system, is limited by diffraction to typically 0.5 um. They have a widespread use in microfabrication industries (microelectronics and MEMS).

We have made our own interferometric microscope (Fig. 1) in order to *1)* integrate it to other instruments such as an Atomic Force Microscope or a Nanoindentor [2] and *2)* develop new modes and reconstruction algorithms to address specific tasks [3].

As for any microscopy technique, the resolution issue is of prime importance. To increase the lateral resolution, we have used high luminance blue LED [4] and more sophisticated techniques like "4-pi", structured illumination, etc... have been proposed as well. These improvements are however not competitive with the lateral resolution one can achieve with mechanical profilers (AFM or stylus). We have then integrated our IOM to a commercial AFM. An example of use is shown in Fig. 2. We have investigated several mathematical filters to compare the roughness obtained by both instruments.

The vertical resolution specification of IOM is often given in terms of *rms* roughness (Rq) on a supposedly infinitely flat mirror. Recently, we have shown [5] that the vertical resolution can be brought to true atomic level provided a suitable averaging frame time is chosen. As an example, atomic steps of graphite (HOPG) are shown in Fig. 3. The smallest steps correspond to single atomic steps. We have developed a simple model which shows the ability of IOM to image atomic steps in PSM as a function of the illumination wavelength and the CCD quantification (n-bits).

1. See for example : D. Malacara,M. Servin, Z. Malacara, Interferogram Analysis for Optical Testing, Optical Engineering Series, Taylor & Francis, 2005, ISBN978-1-57444-682-1
2. M. Jobin, Ph. Passeraub and R. Foschia, Proc. SPIE 6188, 6188OT (2006)
3. M. Jobin and R. Foschia, Microsc. Microanal, 12, 1774 (2006)
4. M. Jobin and R. Foschia, Proc. SPIE 6672, 667205 (2007)
5. M. Jobin, R. Foschia, Measurement (2008), doi:10.1016/j.measurement.2007.12.006

M. Luysberg, K. Tillmann, T. Weirich (Eds.): EMC 2008, Vol. 1: Instrumentation and Methods, pp. 745–746, DOI: 10.1007/978-3-540-85156-1_373, © Springer-Verlag Berlin Heidelberg 2008

Figure 1. Setup of the interferometric microscope in its stand-alone version, where the sample sits on the piezo translator. When integrated into the AFM, the piezo translator holds the interferometric objective.

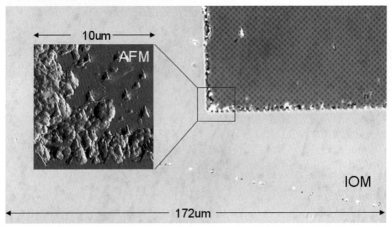

Figure 2. Lateral resolution : example of integrated AFM/IOM imaging. The sample is silicon on which (top-right) an array of sharp silicon tips has been microfabricated.

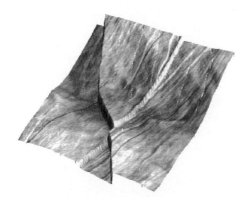

Figure 3. Vertical resolution: atomic step imaging of HOPG.

Effects of annealing on the microstructural evolution of copper films using texture analysis

A. Moskvinova[1], S. Schulze[1], M. Hietschold[1], I. Schubert[2], R. Ecke[2], S.E. Schulz[2]

1. Institute of Physics, Solid Surfaces Analysis Group,
Chemnitz University of Technology, D-09126, Chemnitz, Germany
2. Center for Microtechnologies, Chemnitz University of Technology,
D-09107, Chemnitz, Germany

anastasia.moskvinova@etit.tu-chemnitz.de
Keywords: Copper electroplating; post deposition annealing; SEM; EBSD; XRD.

Successful copper investigation is a key for the newest technologies. Today copper becomes a more and more widespread material in interconnect technology. Its main advantages are a lower resistivity, high conductivity and high purity. Thin copper films were grown by electrodeposition on copper seed layers which were grown by MOCVD and PVD on different barrier layers such as MOCVD titanium-nitride and sputtered tantalum-nitride coated silicon wafers. The bulk copper films were then subjected to i) self-annealing at room temperature, ii) vacuum annealing and iii) N_2 annealing at various temperatures periods of time.

Using EBSD, the average grain size was measured to be from 0,12 to 0,45µm and dependent of method of anneal for all films. Microstructure after deposition seed layer is not stable and the film recrystallizes at room temperature within several days after deposition. During this recrystallization, new texture components can develop via several mechanisms: normal growth of small randomly oriented grains at the expense of (111) grains, abnormal growth of grains with a high orientation such as (100) and twining of (111) grains. Fig. 1 shows XRD and EBSD measurements which indicated that CVD TiN/PVD Cu has a strong orientation (111). After N_2 annealing all samples have a low sheet resistance, large grain size and preferred orientation in the areas (200) and (311). Increasing temperature leads to decreased sheet resistance and increased grain size (Fig. 2) and randomly oriented grains.

M. Luysberg, K. Tillmann, T. Weirich (Eds.): EMC 2008, Vol. 1: Instrumentation and Methods, pp. 747–748, DOI: 10.1007/978-3-540-85156-1_374, © Springer-Verlag Berlin Heidelberg 2008

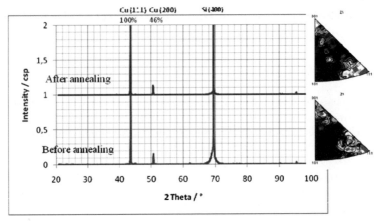

Figure 1. XRD-Spectrum and EBSD inverse pole figure CVD TiN/PVD Cu before and after vacuum annealing

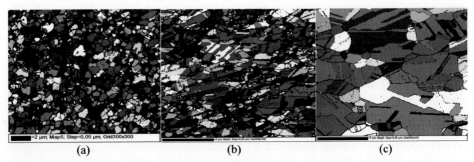

Figure 2. SEM micrographs of sputtered TaN/PVD Cu a) before annealing, average of grain size 0,19μm; b) after vacuum annealing 150°C, average of grain size 0,20μm; c) after N_2 annealing 400°C, average of grain size 0,45μm

Characterisation of Ga-distribution on a silicon wafer after inline FIB-preparation using inline ToFSIMS

U. Muehle[1], R. Gaertner[1], J. Steinhoff[2], W. Zahn[3]

1. Qimonda Dresden GmbH & Co, Physical Failure Analysis, Koenigsbruecker Str. 180, D-01099 Dresden, Germany
2. ASML Netherlands B.V., De Run 6501, 5504 DR Veldhoven
3. Westsaechsiche Hochschule Zwickau, Inst. F. Oberflächentechnologien und Mikrosysteme, Dr.-Friedrichs-Ring 2A, D-08056 Zwickau, Germany

uwe.muehle@qimonda.com
Keywords: Focus Ion Beam, ToFSIMS, Ga-contamination

The continuous miniaturisation of microelectronic devices requires an increasing amount of TEM-based process control [1]. Especially during the development of new generations of memory products the unit processes with all their parameters has to be aligned to each other very carefully. Geometrical and physical control of this complexity generates a large number of TEM-analysis, which have to be performed very fast in order to shorten learning cycles and the feedback loop to production. Recent developments of the manufacturer of preparation equipment allow to extract a TEM-sample from a wafer in the production line without scrapping it [2, 3]. This sequence is based on the application of a dual beam inline FIB-tool operating on a 300mm-wafer between two unit process steps of the production. To enable this it is necessary that the preparation tool in the production line fulfils all requirements of a clean room tool in terms of particle generation, wafer logistic etc..

It was shown, that the gallium ions of the focussed ion beam penetrate the material in direct and lateral direction depending on acceleration voltage [4, 5]. Due to the action of Ga in the Si the electrical properties in the environment of the extracted sample the influenced region has to be known. Furthermore it has to be guaranteed, that Ga diffusion during following thermal process steps keeps restricted. The ToF-SIMS method can be involved in the cleanroom too for several purposes [6] and allows to determine remaining Ga after FIB-preparation [4].

For the samples preparation a DualBeam FIB of the type DA300 by FEI was used. The gallium analysis was performed using a ToFSIMS of the type "TOF SIMS 300R" by Iontof, enabled to work on whole wafers including navigation on base of KLA-files. The Ga-contamination was analysed on the surface on planar Si-surfaces, on wafers with a 600nm thick SiO_2 layer on the top, simulating the situation after deposition of a thick oxide and on real product wafers, where the production process was interrupted. After analysis of the Ga-contamination depending on the distance of sample extraction up to 10mm, the wafer were subjected to a thermal treatment using parameters of real production, followed by an additional ToFSIMS analysis. This should simulate the influence of further production steps on primary inserted gallium. The results were

M. Luysberg, K. Tillmann, T. Weirich (Eds.): EMC 2008, Vol. 1: Instrumentation and Methods, pp. 749–750, DOI: 10.1007/978-3-540-85156-1_375, © Springer-Verlag Berlin Heidelberg 2008

amended by depth profiles before and after the thermal treatment in a defined distance from the location of sample extraction. [7]

A very surprising result was the detection of detection of about $1.8*10^{10}$ Ga-atoms per cm^2 on the wafer surface at a location, which was situated below the liquid metal ion source (LMIS) after the activation of the Ga-source, but without any application using the Ga-beam on the wafer. A probable explanation is an exposure of the wafer with Ga-atoms from the LMIS due to imperfect suppression in the LMIS in blanket- state that leads to a basic contamination.

Figure 1 as an example for the data shows a comparison of the Ga-contamination on the product wafer depending on the distance from the location of sample extraction directly after FIB-work and after an additional anneal. The reduction of Ga during the anneal is the result of diffusion into the depth, where the complex structures with their broad variety of phase boundaries give rise to high averaged diffusion constants.

The results show that at minimum the direct neighbours of the die, which is used for analysis, are influenced by Ga. This must be taken into account at any electrical measurements after finalization of the production. The main application of inline sample extraction method is the field of research and development, where the sampled wafer will not be lost and no part of it leaves the factory.

1. Vallett, D.P.; IEEE Transact. **1** (2002); p. 117-121
2. http://www.fei.com/uploadedFiles/Documents/Content/2006_06_DefectAnalyzer_pb.pdf
3. Pokrant, S., Bicaïs-Lepinay, N., Pantel, R., Cheynet, M. ; ICM 16 ; Sapporo 2006
4. Muehle, U., Steinhoff, J., Hillmann, L.; mc 2007; Saarbruecken 2007
5. Kato, N.; ICM 16; Sapporo 2006
6. Grehl, T., Möllers, R., Niehus, E., Rading, D.; SIMS XVI, proceedings; Kanazawa 2007
7. Gärtner, R.; Diploma Thesis; Westsächsische Hochschule Zwickau; 2007

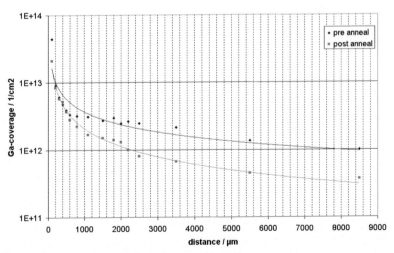

Figure 1. Comparison of Ga-contamination at different distances from the location of samples extraction after FIB-work and after a followed anneal.

First results in thin film analysis based on a new EDS software to determine composition and/or thickness of thin layers on substrates

K. Sempf[1], M. Herrmann[1] and F. Bauer[2]

1. Fraunhofer-Institut für Keramische Technologien und Systeme, Winterbergstr. 28, D-01277 Dresden
2. Oxford Instruments GmbH, Otto-von-Guericke-Ring 10, D-65205 Wiesbaden

kerstin.sempf@ikts.fraunhofer.de
Keywords: Thin film, multiple layers, EDS analysis, SEM

In nanotechnology it is required to get information about thin layers and their chemical composition by very small lateral resolution. Thin film means a microscopically thin layer of material that is deposited onto a metal, ceramic or semiconductor substrate. Thin films having a thickness of less than 1 micron can be conductive or dielectric (non-conductive). Thin films can be used, for example, as top metal layer on a chip and multiple coating on magnetic disks. By means of FESSEM especially thin layers and structures in nanoscale can be observed and the elemental composition or thickness of each layers determined. It is quite difficult to realize small dimensions (e.g. 5-100 nm) and low thicknesses (1-100 nm) with a chemical composition that is mainly based on light elements. The experiments should show whether the EDS/ThinFimID method is suited for element identification and failure analysis.

Thus, it was the motivation to develop and evaluate a new analytical tool for thin film analysis in SEM. The ThinFilmID program is an additional option for Oxford's INCA energy dispersive analysis system. At IKTS the references for validation were produced in thin film technique, and the measurements were performed with FESEM.

A special solvability tool makes it possible to simulate analysis and estimated spectra for different conditions. Additionally, it allows to measure thickness and chemical composition. An optimum in microscope condition setup will be calculated by software.

Different samples were produced and validated. Multi-layer systems in a range between 2 and 200nm were measured und verified by additional methods. The ThinFilmID program is compatible with all SEMs using INCAEnergy and in the end it's a non destructive method.

First measurements were done on well known thin film systems with silicon wafer as substrate. For comparison, thickness of layers was measured by different methods:
- Ellipsometry
- EDX-analysis with ThinFilmID (TFID)
- Cross section cuttings produced and measured with dual beam SEM (FIB)

M. Luysberg, K. Tillmann, T. Weirich (Eds.): EMC 2008, Vol. 1: Instrumentation and Methods, pp. 751–752, DOI: 10.1007/978-3-540-85156-1_376, © Springer-Verlag Berlin Heidelberg 2008

Conclusion: Results from all three different methods are very close together. Mean thickness values over all methods are e.g. 9.753±0.502 for ZrO_2 and 10,087±0,278 for TiO_2.

List of layer systems (spin coating process) on Si-wafer :

Substrate	Layer / layer systems	Chemistry / thickness	Thickness measurement
Si_wafer	Single layer Si/**TiO_2**	TiO_2 (10nm)	Ellipsometry / TFID
Si-Wafer	Single layer Si/**ZrO_2**	ZrO_2 (10nm)	Ellipsometry / TFID
Si-Wafer	Multi layer Si/ **TiO_2/ ZrO_2**	TiO_2 (10nm)/ ZrO_2 (10nm)	Ellipsometry / TFID / FIB
Si-Wafer	Multi layer Si/ **TiO_2/ ZrO_2**	TiO_2 (20nm)/ ZrO_2 (20nm)	Ellipsometry / TFID / FIB
Si-Wafer	Multi-layer **Si/Si_3N_4/ TiO_2**	Si_3N_4 (100nm)/ TiO_2 (10nm)	Ellipsometry / TFID

Example:
1. Ellipsometry – substrate Si / layer 1 : ZrO_2 10.2nm / layer 2 : TiO_2 10.2nm
2. Direct measurement on cross section in STEM-mode with FIB:

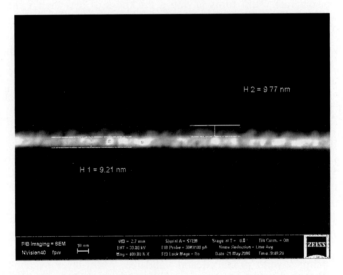

3. EDX analysis with ThinFilmID (TFID):

Sample Description:						
Layer	Element	Weight%	Atomic%	Density(g/cm3)	Thickness(nm)	Sigma(nm)
- TiO_2	Ti	59.95	33.33	4.23	10.29	0.79
	O	40.05	66.67			
- ZrO_2	Zr O_2	74.03	33.33	5.68	9.85	0.84
	O	25.97	66.67			
Substrate	Si	100	100			

Calibration of RHEED patterns for the appraisal of titania surface crystallography

T. Tao, R. Walton, H.K. Edwards, M.W. Fay[1], D.M. Grant and P.D. Brown

1. School of Mechanical, Materials and Manufacturing Engineering,
2. The Nottingham Nanotechnology and Nanoscience Centre,
University of Nottingham, University Park, Nottingham, NG7 2RD, UK

emxtt@nottingham.ac.uk
Keywords: TiO_2, RHEED

Titanium and its alloys are commonly used as biomaterials due to their various beneficial surface properties, especially the formation of chemically stable passivating oxide films which can provide for excellent corrosion resistance and reduce metal ion release. Titanium oxide is a ceramic layer consisting typically of TiO_2 which can exhibit different polymorphs for different annealing conditions [1].

A range of titanium samples annealed at different temperatures and for different time periods have been appraised for the dynamic evolution of the titanium oxide near surface structure. Information on the near surface crystal structure of the TiO_2/Ti surfaces was acquired using reflection high energy electron diffraction (RHEED). Additionally, the complementary technique of secondary electron imaging (SEI) using an FEI, XL30 SEM has been used to give further insight into the development of those titanium oxide surfaces.

Illustration of the results of the combined RHEED & SEI characterisation of the sample set is shown in Figure 1. As the annealing temperature increases, the crystal structure of titanium oxide changes from amorphous, through polycrystalline anatase, to a mixture of anatase and rutile, and finally to polycrystalline rutile (Figure 1a). As the annealing time period increases, larger grain sizes and slightly preferred orientation of the grains become apparent (Figure 1b).

Cross-sectional TEM is required to investigate the depth profile of the oxides formed on the Ti samples. For a Ti sample annealed at 900 ºC for 45 mins, cross-sectional TEM (XTEM) analysis was performed to identify the sub-oxides below the surface. Figure 2 illustrates bright field TEM images from the cross sectioned sample with associated selected area electron diffraction patterns. Figure 2 (a) indicates a total oxide thickness of $\sim$ 7 μm in this instance and varying levels of electron transparency and porosity in the coating. Based on the analysis of the diffraction patterns, Figure 2 (b) shows an area encompassing (i) an upper coating, correlated almost exclusively with rutile; (ii) a lower coating, showing good correlation with rutile and a little evidence of Ti_2O; and (iii) the substrate, being single crystal β-Ti.

1. E. Lautenschlater & P. Monaghan. Titanium and titanium alloys as dental materials. Biomaterials, 1993. **27**: p. 245-253
2. Acknowlegement: this work was supported by the EPSRC under grant EP/E015379/1.

M. Luysberg, K. Tillmann, T. Weirich (Eds.): EMC 2008, Vol. 1: Instrumentation and Methods, pp. 753–754, DOI: 10.1007/978-3-540-85156-1_377, © Springer-Verlag Berlin Heidelberg 2008

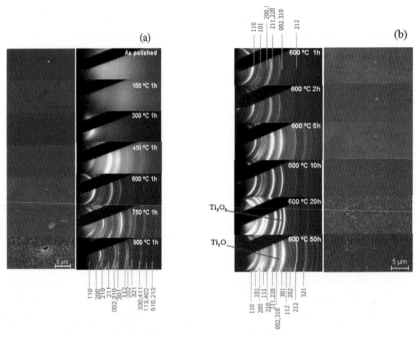

Figure 1. Combined SEI & RHEED patterns from polished titanium samples annealed for (a) different annealing temperatures and (b) different annealing time periods (hkl indices relate to TiO₂ (rutile) and diffraction rings attributable to other sub-oxide phases are shown)

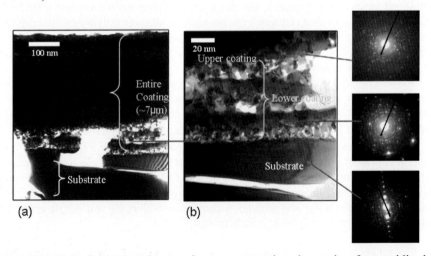

Figure 2. Bright field TEM images from cross sectioned sample of an oxidised Ti substrate, annealed at 900 °C for 45 mins. (a) Entire coating, and (b) Enlarged region of the oxide / Ti substrate interface

Surface orientation dependent termination and work-function of in situ annealed strontium titanate

N. Barrett[1], L.F. Zagonel[1], A. Bailly[2], O. Renault[2], J. Leroy[1],
J.C. Cezar[3], N. Brookes[3], Shao-Ju Shih[4], D. Cockayne[4]

1. CEA DSM/DRECAM/SPCSI, CEA Saclay, 91191 Gif sur Yvette, France
2. CEA LETI Minatec, 17 rue des Martyrs, 38054 Grenoble cedex 9, France
3. ESRF, 6 rue Jules Horowitz, BP 220, 38043 Grenoble cedex 9, France
4. Department of Materials, Oxford University, Parks Road, Oxford OX1 3PH UK

nick.barrett@cea.fr

Keywords: Strontium Titanate, surface termination, X-PEEM.

The combination of high brightness photon source and aberration corrected energy filtering has allowed new progress in the field of electron emission microscopy for nanoscience and nanotechnology. We first present the principles of spectromicroscopy, and in particular the use of an electrostatic PEEM column together with an aberration corrected double hemispherical energy analyzer. The accurate focus tracking of the instrument allows imaging at the secondary electron threshold and across specific core levels at chosen kinetic energies. The high lateral and energy resolutions give an imaging capability with full chemical state sensitivity. [1,2]

The principles will be illustrated by a study of the surface chemistry of Nb doped $SrTiO_3$ ceramic. We have imaged the polycrystalline surface at the Sr $3d_{5/2}$, Ti $2p_{3/2}$ and O $1s$ levels. As seen in Figure 1, a contrast clearly determines different grain surface terminations and is due to the low photoelectron escape depth provided by using tuned synchrotron radiation (ID08 beam line at the ERSF). Each grain in the instrument's field of view is considered as a single crystal sample submitted to the same preparation procedure and therefore a broad range of surface orientation is examined. The degree of $SrTiO_3$ surface polarity is expected to be a function of the cations present in the surface. These results are correlated with the spatially resolved work function contrast in order to quantify the grain termination and orientation dependence of the basic electronic properties of the ceramics. A method for analysing these data is proposed based on the stereographic projection plot of the core level intensity as function of the grain orientation, which was determined by electron backscatter diffraction (EBSD). The plots show a good correlation between both: close grain orientations lead to similar core level signal intensities.

1. M. Escher et al., J. Phys.: Condens. Matter **17** (2005) S1329.
2. O. Renault et al., Surf. Sci. 601 (2007) 4727.

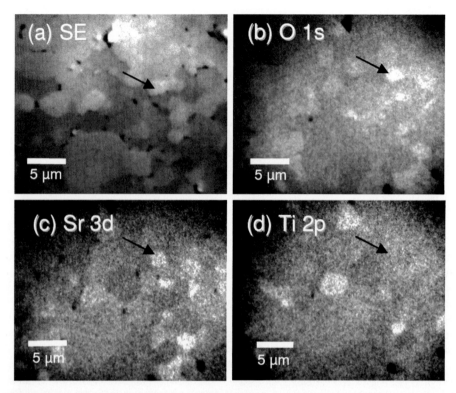

Figure 1. (a) Secondary electrons at a kinetic energy of 4.3 eV. Core level images with respective binding energies: (b) O 1s, 530.5 eV; (c) Sr 3d, 266.5 eV; (d) Ti 2p, 206.0 eV. The arrow indicates a grain in the [100] direction.

Structure determination of zeolites by electron crystallography

Junliang Sun[1,2*], Daliang Zhang[1,2], Zhanbing He[1], Sven Hovmöller[1], Xiaodong Zou[1,2*], Fabian Gramm[3,4], Christian Baerlocher[3] and Lynne B. McCusker[3]

1. Structural Chemistry, Stockholm University, SE-106 91 Stockholm, Sweden
2. Berzelii Centre EXSELENT on Porous Materials, Stockholm University, SE-106 91 Stockholm, Sweden
3. Laboratory of Crystallography, ETH Zürich, CH-8093 Zürich, Switzerland
4. Electron Microscopy ETH Zurich - EMEZ, CH-8093 Zurich, Switzerland

junliangs@struc.su.se

Keywords: IM-5, zeolite Beta, SAED, HRTEM, Electron crystallography

Many zeolite structures have remained unsolved for a long time because of their structural complexity, the size of the crystallites or the presence of defects or impurities. By combining electron microscopy and X-ray powder diffraction data, some of them have been solved (e.g. the high-silica zeolite IM-5, which was first reported in 1998 [1] and recently solved using a charge-flipping structure solution algorithm [2]), while for other zeolites, good X-ray powder diffraction data are hard to obtain (e.g. polymorph A or B of zeolite Beta). Here we demonstrate a complete structure determination of IM-5 (one of the most complicated zeolited) and polymorph B of zeolite Beta [3] using electron crystallography alone. This shows the power and advantage of structure determination by electron microscopy compared with the X-ray diffraction techniques. The method is general and can be applied to both zeolites and other materials, where the crystals are too small or the structure too complicated to be solved from X-ray powder diffraction data alone [4]. It is particularly useful for structures containing defects.

For IM-5, the space group was determined to be $Cmcm$ by combining SAED patterns and HRTEM images. The unit cell and the Bravais lattice type were obtained from a tilt series of SAED patterns using the program Trice [5] ($a = 14.3$ Å, $b = 57.4$ Å, $c = 20.1$ Å with C-centering). Its Laue class was determined to be mmm as the SAED patterns are related by mmm symmetry. Thus, IM-5 has a C-centered orthorhombic structure. Based on the extinction rules of the SAED patterns, there are only three possible space groups: $Cmc2_1$, $C2cm$ and $Cmcm$. The three space groups can be distinguished from the projection symmetries of the HRTEM images. The projection symmetries were determined from CRISP [6] to be close to pmg, pmm and cmm for the [100], [010] and [001] projections, respectively, and this is consistent with the space group $Cmcm$.

To determine the atomic positions, amplitudes and phases of 144 independent reflections were obtained from the HRTEM images taken along the [100], [010] and [001] directions. A 3D potential map was calculated from these reflections by inverse Fourier transformation (Fig. 1). [7] All 24 unique Si positions could be determined from the peaks in the 3D potential map. Oxygen atoms were added between each Si-Si pair and the geometry was optimized using the distance least-squares program DLS-76 [8].

M. Luysberg, K. Tillmann, T. Weirich (Eds.): EMC 2008, Vol. 1: Instrumentation and Methods, pp. 757–758, DOI: 10.1007/978-3-540-85156-1_379, © Springer-Verlag Berlin Heidelberg 2008

758

This final structure model deviates on average by 0.16Å for Si and 0.31Å for O from that refined using X-ray powder diffraction data.

The structure of polymorph B of zeolite Beta can be determined in a similar way. The unit cell and Laue class were obtained from a tilt series of SAED patterns. The space group was uniquely determined to be $C2/c$ from the systematic absences and the symmetry of HRTEM images along the [1-10] direction. The 3D potential map could be reconstructed from one single HRTEM image by applying the symmetry $C2/c$, and all Si atoms were founded from this 3D map (Fig. 2).

We have shown that the unit cells and space groups of two zeolites could be determined from SAED and HRTEM data. The final Si positions were obtained with reasonable accuracy (0.1-0.2 Å) by 3D reconstruction of HRTEM images followed by a distance least-squares refinement. The technique demonstrated here is general and can be applied to not only zeolites, but also to other complicated structures.

1   Benazzi, E., Guth, J.L., Rouleau, L., *PCT WO* **1998**, 98/17581.
2   Baerlocher, Ch., Gramm, F., Massüger, L., McCusker, L.B., He, Z.B., Hovmöller, S., Zou, X.D. *Science* **2007**, *315*, 1113.
3   Corma, A., Moliner, M., Cantín, A., Díaz-Cabañas, M.J., Jordá, J.L., Zhang, D.L., Sun, J.L., Jansson, K., Hovmöller, S., Zou, X.D. *Chem. Mater.* **2008**, in press.
4   Wagner, P., Terasaki, O., Ritsch, S., Nery, J.G., Zones, S.I., Davis, M.E., Hiraga, K.J. *Phys. Chem. B* **1999**, *103*, 8245.
5   Zou, X.D., Hovmöller A., Hovmöller S. *Ultramicroscopy* **2004**, *98*, 187.
6   Hovmöller, S. *Ultramicroscopy* **1992**, *41*, 121.
7   Oleynikov, P. http://www.analitex.com/.
8   Baerlocher, Ch., Hepp, A., Meier, W.M. *DLS-76*, ETH Zurich, Switzerland, 1976.

Figure 1. A 3D potential map reconstructed from 144 reflections using the program eMap [6]. The Si net is superimposed. Green is outside and blue inside the walls.

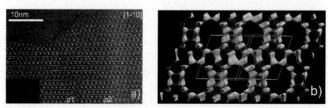

Figure 2. (a) HRTEM image of zeolite beta polymorph B taken along the [1-10] direction. Insets are (from left to right) Fourier transform, and averaged images with $p1$ and $p2$ symmetry imposed, respectively. (b) A 3D potential map reconstructed from the HRTEM image in (a). All 9 unique Si atoms were obtained from this 3D map (after Ref. 3).

3D electron diffraction of protein crystals: data collection, cell determination and indexing

D.G. Georgieva[1], L. Jiang[1], H.W. Zandbergen[2], S. Nicolopoulos[3] and J.P. Abrahams[1]

1. Department of Biophysical and Structural Chemistry, Leiden University, P.O. Box 9502 2300 RA Leiden, The Netherlands
2. Kavli Institute of Nanoscience Delft, Delf University, Lorentzweg 1, 2628 CJ Delft, The Netherlands.
3. Nanomegas SPRL, Blvd Edmond Machtens no.79, Bruxelles, Belgium

abrahams@chem.leidenuniv.nl
Keywords: protein, crystallography, precession electron diffraction

Relative to the number of elastic scattering events, X-rays are three orders of magnitude more damaging to organic molecules than electrons, so for structure determination of 3D nano-crystals of proteins and other damage-prone molecules, electron diffraction should be an attractive alternative. For 2D protein crystals, electron diffraction is the only option. For 3D protein crystals, limitations in collecting diffraction data of sufficient quality and dynamical scattering have so far frustrated the promise electrons hold in this respect. Here we report that these problems could, to a certain extent, be overcome by combining modern flash-freezing techniques, low dose diffraction techniques (microdiffraction) and precession of the electron beam. Our procedures, specifically aimed at gathering high-resolution, 3D reciprocal space data, allowed electron diffraction up to 2.1 Å resolution of 3D protein nano-crystals (lysozyme, see figure 1) and even beyond 1 Å resolution for nano-crystals of complex organic pharmaceuticals. The resulting diffraction patterns of non-oriented crystals could be indexed.

Although currently X-ray diffraction is most certainly the first choice for structural studies of beam sensitive macro-crystals, we show electron diffraction to have potential and may in future solve problems associated with X-ray crystallography:

- Protein crystals of nanometre size which are too small for single crystal X-ray diffraction experiments, can be studied with electron diffraction techniques

- Sample preparation requirements for electron microscopy studies are significantly relaxed in comparison with X-ray single and powder diffraction. In the case of protein crystals the development of plunge freezing techniques allows vitrification of nano-crystals without additional cryo-protectants, while for preservation of macro-crystals in X-ray crystallography, the use of cryo-protectants is essential. Furthermore, the optimal cryo-protecting conditions need to be identified for each case individually. The requirements of crystal powder samples for homogeneity and crystal size consistency in the case of synchrotron powder diffraction are irrelevant for electron diffraction studies

- Diffraction beyond to 1Å can be observed in single pharmaceutical nanocrystals obtained from powders using electron diffraction. Moreover, compared to synchrotron powder diffraction which only yields one-dimensional diffraction information of bulk

M. Luysberg, K. Tillmann, T. Weirich (Eds.): EMC 2008, Vol. 1: Instrumentation and Methods, pp. 759–760, DOI: 10.1007/978-3-540-85156-1_380, © Springer-Verlag Berlin Heidelberg 2008

powders, electron diffraction studies provide two-dimensional diffraction information of individual nano-crystals. Based on the electron diffraction patterns it should be possible to identify different crystal phases in pharmaceutical samples and to detect impurities even if they are in very small quantities since individual crystals can be studied.

- We show that unit cell parameters can be deduced from the data using a brute force search. We consider it likely that subsequent steps in crystallographic structure determination (indexing, integrating, scaling, merging and phasing) are also possible using electron diffractograms.

- We show our methods to be compatible with electron precession. We consider this relevant, since for subsequent steps in structure determination, precession electron diffraction will be beneficial. It reduces the dynamical diffraction, resulting in more quasi-kinematical data, which should facilitate the data analysis when the protein or pharmaceutical crystals are thicker than 100 nm. Furthermore, precession allows the collection of fully recorded reflections, further facilitating the merging of intensity data obtained from different diffraction patterns.

The high beam sensitivity of protein and pharmaceutical crystals does not allow full 3D data collection from a single nano-crystal and diffraction patterns sets from many different crystals are needed in order to build up a complete dataset and solve the structure. The merging of data of randomly oriented single-shot diffractograms is not trivial and requires the development of novel routines, while existing programmes may need to be adapted for further integration and processing. Such procedures are currently under development.

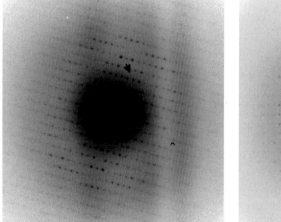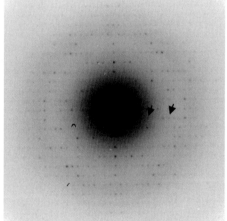

Figure 1. Electron diffraction patterns of vitrified 3D lysozyme nano-crystals acquired from different crystallographic zones. The systematic absences shown with arrows indicate that the patterns do not suffer from strong systematic dynamic effects.

Self-assembly of cholesterol-based nonionic surfactants in water. Unusual micellar structure and transitions

Ludmila Abezgauz[1], Irina Portnaya[1], Dganit Danino[1]

Department of Biotechnology and Food Engineering, Technion – Israel Institute of Technology, 32000 Haifa, Israel

ludaa@techunix.technion.ac.il
Keywords: nonionic surfactants, discs structures, cryo-TEM

Polyoxyethylene cholesteryl ethers (ChEO$_m$) are a class of nonionic surfactants with an "inverse" structure: rigid hydrophobic tail and long hydrophilic head groups. These surfactants are nontoxic, biodegradable and used in cosmetic and pharmaceutical industries. Unexpectedly for nonionic surfactants, several reports indicated that the viscosity of low concentration solutions of ChEO$_m$ increase several orders of magnitude upon raise of temperature or upon mixing with polyoxyethylene dodecyl ether (C$_{12}$EO$_m$) surfactants [1, 2]. The micellar assembly behavior of these systems has not yet adequately studied.

Using cryogenic-transmission electron microscopy (cryo-TEM) and isothermic titration calorimetry (ITC) we studied the self-assembly of ChEO$_m$ in aqueous solutions, and primarily examined the effect of temperature and additives on the structural behavior of ChEO$_{10}$.

We found that the basic assemblies of ChEO$_{10}$ in aqueous solution at 20°C are uniform disk-like aggregates, 16 nm in diameter. Disk elements are rarely found in binary surfactant-water systems, and their formation is explained by the unique nature of this surfactant: the balance between rigid cholesteryl tail and flexible polyoxyethylene head-structure permit a flat geometry. Upon increase of temperature or concentration these disks grow into long connected ribbons (Figure 1). The latter is the result of water extracting from the hydrated layer of the polyoxyethylene group, therefore reducing the head-group area and promoting micellar growth. The interaction of ChEO$_{10}$ with additives strongly depends on the additives nature. The structural transitions we found upon addition of the polyoxyethylene cholesteryl ether surfactant ChEO$_3$ (a vesicle-forming amphiphile) are similar to those detected upon temperature and concentration increase, but enhanced. The disks transform into ribbons upon low quantity addition of C$_{12}$EO$_3$ (0.1-0.2 wt %) and further develop into a saturated network of ribbons in the presence of only 0.32 wt% of C$_{12}$EO$_3$. On the other hand, addition of ionic surfactants leads to a continuous shrinkage of the structures and formation of spherical micelles. In the case of inorganic salts (which affect only the hydrophilic part) we found growth of disk into ribbons, independent on the location of the salt in the lyotropic series. Affecting the hydrophobic part by addition of cholesterol to ChEO$_{10}$ molecules, cause the disk-like aggregates to transform into vesicles (Figure 2).

M. Luysberg, K. Tillmann, T. Weirich (Eds.): EMC 2008, Vol. 1: Instrumentation and Methods, pp. 761–762, DOI: 10.1007/978-3-540-85156-1_381, © Springer-Verlag Berlin Heidelberg 2008

762

1   Acharya, D.P., Kunieda, H., Journal of Physical Chemistry B, 2003. 107: 10168-10175.
2   Sato, T., Hossain, M.K., Acharya, D.P., Glatter, O., Chiba, A., Kunieda, H., Journal of Physical Chemistry B, 2004. 108: 12927-12939.

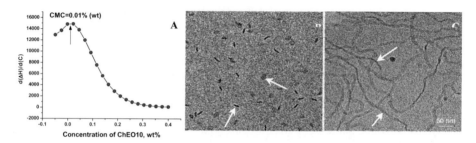

Figure 1. (A) The CMC were determined from ITC measurements at 20°C. (B and C) Cryo-TEM images of the structures of 0.5wt% ChEO$_{10}$ at increasing temperatures. (B) The disc-like aggregates that exist at 20°C gradually grow into flat, long ribbons at 60°C (C). Arrows in both images show the several orientations of the flat structures in the vitrified ice.

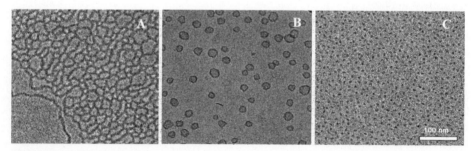

Figure 2. (A) Formation of a saturated network of ribbons in the presence of 0.32 wt% C$_{12}$EO$_3$. (B) solubilization of cholesterol (up 0.07 wt %) leads to 2-dimensional growth, and formation of vesicles. (C) Addition of CTAB (cetyltrimethyl ammonium bromide) shrinks of the disk structures, until eventually, with 3.0 wt% CTAB spherical micelles form.

Quantitative study of anode microstructure related to SOFC stack degradation

A. Faes[1,2], A. Hessler-Wyser[1], D. Presvytes[3], A. Brisse[4], C.G. Vayenas[3]
and J. Van herle[2]

1. Interdisciplinary Centre for Electron Microscopy (CIME), Ecole Polytechnique Fédérale de Lausanne (EPFL), CH-1015 Lausanne, Switzerland
2. Industrial Energy Systems Laboratory (LENI), EPFL, CH-1015 Lausanne, Switzerland
3. Department of Chemical Engineering, Laboratory of Chemical and Electrochemical Processes, University of Patras, GR-26504 Patras, Greece
4. EDF – EIfER, DE-76131 Karlsruhe, Germany

antonin.faes@epfl.ch

Keywords: Solid oxide fuel cells, anode degradation, image quantification analysis, TPB

As the performances of Solid Oxide Fuel Cells (SOFC) get attractive, long term degradation becomes the main issue for this technology. Therefore it is essential to localise the origin of degradation and to understand its processes in order to find solutions and improve SOFC durability.

The electrode microstructure ageing, in particular nickel grain coarsening at the anode side, is known to be a major process to cause performance loss [1]. The increase in nickel particle size will diminish the Triple Phase Boundary (TPB), where fuel oxidation takes place, and decrease the anode electronic conductivity. These two effects degrade the electrochemical performance of the fuel electrode.

Degradation is defined as the decrease of potential at constant current density with time in %/1000h or mV/1000h. This study is based on HTceramix® anode supported cells tested in stack conditions from 100 to more than 1000 hours.

The anode microstructure has been characterized by Scanning Electron Microscopy (SEM). As the back scattered electron yield coefficients of nickel and yttria stabilized zirconia (YSZ) are very close, the contrast of the different phases (Ni, YSZ and pores) is low. Various techniques are used to enhance the contrast [2, 3]. A new technique is presented here using impregnation and SEM observation based on secondary electron yield coefficients to separate the phases.

Image treatment and analysis is done with an in-house Mathematica® code. Image treatment (Figure 1) follows four steps: 1. inhomogeneous background correction, 2. double thresholding, 3. cleaning of the binary images and 4. reconstruction of a three-phase image. Image analysis gives information about phase proportion, particle size, particle size distribution, contiguity and finally a new procedure is developed to compute TPB density.

A model to describe the coarsening of the nickel particles is also developed. The model assumes an exponential growth of the nickel particles. Using a particle population balance, it estimates the growth of the nickel particles and the concomitant drop in the TPB length. This model is in very good agreement with experimental data,

M. Luysberg, K. Tillmann, T. Weirich (Eds.): EMC 2008, Vol. 1: Instrumentation and Methods, pp. 763–764, DOI: 10.1007/978-3-540-85156-1_382, © Springer-Verlag Berlin Heidelberg 2008

especially for relatively low fuel cell operation times (up to 100-200 hours). This model can be used in the estimation of operational parameters of the anode electrode such as the degradation rate using fundamental parameters of the cermet anode like the anode overpotential and the work of adhesion of the nickel particles on the YSZ substrate. This model gives the portion of stack degradation that corresponds to anode performance decrease due to particle sintering.

Finally this study gives the possibility to isolate the degradation coming from the anode sintering and compare to the full SOFC stack degradation.

1. D. Simwonis, F. Tietz, and D. Stoever, Solid State Ionics **132** (2000) p. 241.
2. K.R. Lee, S.H. Choi, J. Kim, H.W. Lee, and J.H. Lee, Journal of Power Sources **140** (2005) p. 226.
3. C. Monachon, A. Hessler-Wyser, A. Faes, J. Van herle, and E. Tagliaferri, A quick method for characterizing Nickel-Yttria stabilized Zirconia cermet microstructure by scanning electron microscopy, Journal of the American Ceramic Society, Submitted.

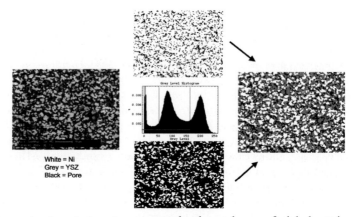

Figure 1. Image treatment steps to separate the three phases of nickel, yttria stabilized zirconia and porosity.

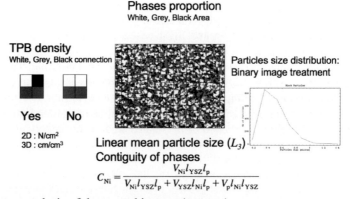

Figure 2. Image analysis of the treated images (see text).

New considerations for exit wavefunction restoration under aberration corrected conditions.

S.J. Haigh[1], L-Y. Chang[1], H. Sawada[2], N.P. Young[1] and A.I. Kirkland[1]

1. University of Oxford, Department of Materials, Parks Road, Oxford, OX1 3PH, U.K.
2. JEOL Ltd., 1-2 Musashino 3-chome, Akishima, Tokyo 196, Japan

sarah.haigh@materials.ox.ac.uk
Keywords: aberration correction, TEM, higher order

Transmission electron microscopes (TEM) fitted with electron optical elements capable of compensating for the coherent aberrations up to 3rd order spherical aberration are becoming increasingly widely available. These instruments facilitate use of a wider range of imaging conditions where the compensatable aberrations are additional variables. For example, a negative spherical aberration and over focus imaging has been found to give increased contrast because the linear and non-linear imaging terms add constructively [1].

However, the majority of the work on optimal imaging conditions considers only the radially symmetric aberration coefficients (defocus, third order spherical aberration and in some cases fifth order spherical aberration) whereas for practical operation it is frequently the higher order non-radially symmetric parasitic aberrations that limit the interpretable resolution of an aberration corrected microscope. A critical element of the corrector alignment is the accurate measurement of all the higher order aberration coefficients, typically up to fifth order spherical aberration. Although these high order aberrations cannot be removed during imaging, an accurate knowledge of their values allows their compensation during *a posteriori* restoration of the exit wavefunction.

This high order compensation is performed by applying a phase plate to the conventionally restored exit wavefunction in Fourier space. The high order compensation phase plate shown in figure 1 can be used to remove the higher order aberration coefficients. The values of these coefficients were measured experimentally using the CEOS software for the JEOL aberration corrected 2200MCO (S)TEM.

Higher order aberration compensation has been performed on the exit wavefunction restoration shown in figure 2. The specimen is β-Si_3N_4 and the restoration was performed using a linear Wiener filter approach [2] from a focal series of 20 aberration corrected images acquired on the 2200MCO with a focal step of 5nm. The restored exit wavefunction clearly shows the Si-N column separation at a distance of 0.098nm but at this resolution the need for higher order aberration compensation is not observable. However, high order compensation is necessary where the resolution limit ismuch better than 0.1nm. Thus, higher order aberration compensation is an important step where the resolution of the restoration is improved to 0.071nm by including images acquired under tilted illumination conditions.

M. Luysberg, K. Tillmann, T. Weirich (Eds.): EMC 2008, Vol. 1: Instrumentation and Methods, pp. 765–766, DOI: 10.1007/978-3-540-85156-1_383, © Springer-Verlag Berlin Heidelberg 2008

Direct electron optical aberration correction can also reveal the presence of off-axial aberrations and these can also be removed from the exit wavefunction by applying an appropriate real space phase plate during restoration.

1. C.L. Jia, M. Lentzen and K. Urban, Science **299** (2003), p. 870.
2. A.I. Kirkland and J.L. Hutchison in "Science of microscopy", ed. P.W. Hawkes and J.C.H. Spence, (Springer, New York) (2007), p. 696.
3. We kindly acknowledge financial support from EPSRC.

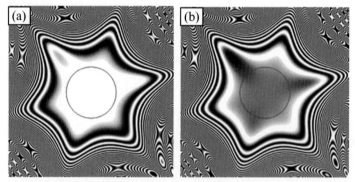

Figure 1. (a) Real and (b) imaginary parts of the phase plate used to compensate for the effects of higher order aberration coefficients as measured experimentally for the 2200MCO by the CEOS software. The circle is included to indicate a resolution of 0.1nm.

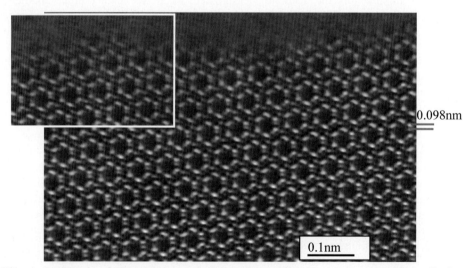

0.098nm

0.1nm

Figure 2. Phase of the exit wavefunction for a β-Si_3N_4 specimen where all the aberrations up to 5th order spherical aberration have been compensated during the restoration procedure. The inset (top left) shows the restoration before higher order correction.

High quality electron diffraction data by precession

Sven Hovmöller[1], Daliang Zhang[1,2], Junliang Sun[1,2], Xiaodong Zou[1,2*]
and Peter Oleynikov[1]

1. Structural Chemistry, Stockholm University, SE-106 91 Stockholm, Sweden
2. Berzelii Centre EXSELENT on Porous Materials, Stockholm University, SE-106 91 Stockholm, Sweden

svenh@struc.su.se

Keywords: Precession, electron diffraction, quantification

Electron precession [1,2] has been proposed as a method for collecting high-quality, near kinematical electron diffraction data. Here we present a quantitative investigation of data quality of $K_2O \cdot 7Nb_2O_5$ (projection symmetry *4mm*, $a = 27.5$Å), using electron precession, compared to standard selected area electron diffraction (SAED) data.

The diffraction data quality should be sufficient for a) identifying the compound studied, b) solving an unknown crystal structure, c) finding not just the heavier atoms, but all atoms in the structure and finally d) refining accurate atomic positions.

For a) identifying the compound studied, it is usually sufficient to obtain accurate unit cell dimensions - no need for quantitative intensities. This is equally well done with SAED as with precession, but in both cases may have to involve a rather precise calibration of the microscope, including non-uniform magnification [3].

By b) solving an unknown crystal structure, we normally mean to find not necessarily all, but at least sufficiently many of the heaviest atoms, to within about 0.2 Ångström of their correct positions. For a metal oxide, the structure is solved if the metal atoms are found. High-resolution transmission electron microscopy (HRTEM), combined with crystallographic image processing, is a powerful method for solving unknown crystal structures. This has been shown in many cases on compounds ranging from alloys [4] and metal oxides to zeolites [5,6] and even proteins. However, it may be difficult to obtain sufficiently good HRTEM images and thus it is desirable to be able to solve crystal structures directly from only ED data.

In order to c) find all atoms in a structure, it is necessary to have a large number of reflections of high quality. This is accomplished if the data goes to high resolution;, and if the data is nearly kinematical (or alternatively if very advanced corrections can be made for multiple diffraction effects [7].) The SAED pattern in Fig 1a goes to 1.10 Å resolution (h-max = 25), while the 1.1° precession pattern in Fig. 1b goes to 0.76 Å resolution (h-max = 36). The total number of unique reflections that are observed above the background is doubled for precession compared to SAED.

Precession electron diffraction patterns were obtained with the SpinningStar from NanoMEGAS, Brussels, mounted on a JEOL 2000CX TEM, equipped with a 16-bit CCD-camera. Electron diffraction patterns were quantified with ELD from Calidris, Sollentuna. Internal R-value (R_{merge}) for symmetry-related reflections was about 5%.

The structure was solved using Sir97 and refined using SHELX.

M. Luysberg, K. Tillmann, T. Weirich (Eds.): EMC 2008, Vol. 1: Instrumentation and Methods, pp. 767–768, DOI: 10.1007/978-3-540-85156-1_384, © Springer-Verlag Berlin Heidelberg 2008

All metal atoms were easily found in the very clear projected potential map obtained with precession data using Sir97, see Fig. 2. Sir97 could also solve the structure from the SAED data.

The metal atom positions were refined against the precession data. The R-value was much lower (19%) using precession than SAED data (28%). The atomic co-ordinates for the niobium atoms were very close to those obtained by X-ray diffraction for the isomorphous compound $Tl_2O \cdot 7Nb_2O_5$ [9]; on average within 0.04 Ångström.

We are currently trying to localize the oxygen atoms.

1. R.J. Vincent, P.A. Midgley, Ultramicroscopy **53** (1994) 271-282.
2. Oleynikov,P. Hovmöller,S. and Zou,X.D. Precession electron diffraction: observed and calculated intensities. Ultramicroscopy **107** (2007), 523-533
3. Capitani,GC, Oleynikov,P., Hovmöller,S., Mellini, M. **106** (2006) 66-74
4. Zou,X.D, Mo,Z.M., Hovmöller,S., Li, X.Z. and Kuo,K. Acta Cryst. **A59** (2003) 526-539.
5. F. Gramm, Ch. Baerlocher, L.B. McCusker, S.J. Warrender, P.A. Wright, B. Han, S.B. Hong, Z. Liu, T. Ohsuna and O. Terasaki, *Nature* **444** (2006) 79-81
6. Baerlocher,C., Gramm,F., Massüger,L., McCusker,L.B., He,Z.B., Hovmöller,S. and Zou,X.D.. Science **315** (2007) 1113-16
7. J. Jansen, D. Tang, H. W. Zandbergen and H. Schenk *Acta Cryst.* **A54** (1998) 91-101

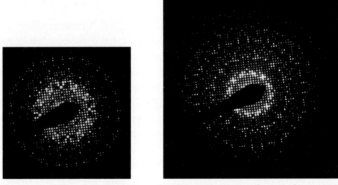

Figure 1. Electron diffraction patterns of $K_2O \cdot 7Nb_2O_5$ using (left) SAED and (right) 1.1° precession, at the same scale. The *4mm* symmetry is evident in both patterns, but the data goes to much higher resolution with precession.

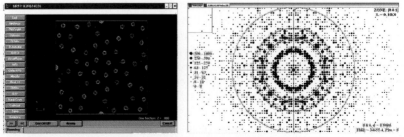

Figure 2. (left) The Nb atoms solved by Sir97. The kinematical electron diffraction pattern (right), calculated by eMap from AnaliTEX.

Optimal noise filters for high-resolution electron microscopy of non-ideal crystals

K. Ishizuka[1], P.H.C. Eilers[2] and T. Kogure[3]

1. HREM Research Inc, 14-48 Matsukazedai, Higashimatsuyama, 355-0055 Japan
2. Utrecht University, 3508 TC Utrecht, The Netherlands
3. University of Tokyo, 7-3-1 Hongo, Bunkyo-ku, Tokyo, 113-0033 Japan

ishizuka@hremresearch.com

Keywords: noise filter, Wiener filter, non-ideal crystal, nano particle

Noise filtering such as a Wiener filter or a background subtraction filter discussed by Kilaas [1] will work for an ideal case where a relatively large single crystal is present in an image. However, these filters do not work for non-ideal crystals, such as cylindrical crystals and nano particles. We propose here two techniques that make these filters applicable even for such materials.

Wiener filter or a background subtraction filter is a filter that works in Fourier space [1]. We may write an observed signal F_o in Fourier transform as a sum of a true signal F_c due to crystal parts and a background F_b due to non-crystal parts: $F_o = F_c + F_b$. Here, the signal F_c from crystals will yields Bragg peaks, while the background F_b will distribute rather uniformly. The Wiener filter seeks a solution that minimizes the summed square difference between the true signal F_c and its estimate $\hat{F}_c \equiv M_W F_o$ resulting

$$ \hat{F}_c = M_W F_o = \frac{|F_c|^2}{|F_o|^2} F_o \approx \frac{|F_o|^2 - |\hat{F}_b|^2}{|F_o|^2} F_o, $$

where $\hat{F}_b$ is an estimate of the background. Here, we set the mask M_w to zero, when an estimated background is stronger than an observed Fourier transform: $|F_o| - |\hat{F}_b| \le 0$. This will substantially suppress speckle noise.

Previously, the background has been estimated as an azimuthal average of the Fourier transform of the whole image by assuming that the background varies slowly [1]. However, the filter with such a radial background does not work for non-ideal crystals, such as randomly oriented nanoparticles or cylindrical crystals. Figure 1 shows such a case, where an image was taken from a cylindrical clay mineral, crysotile. A Wiener filtered image using a radial background is reproduced in Fig. 1c. Although this is seemingly good, its difference from the original image clearly demonstrates an insufficient extraction of the structure (Fig. 1d). This is because the signal from such materials appears at equidistant points from the origin in Fourier transform as shown in Fig. 1b. We can resolve the problems by using a smoothed two-dimensional background estimated based on P-spline fitting [2], and dividing an image into small areas that is processed separately [3]. Fig. 1e shows a Wiener filtered image using a set of two-

M. Luysberg, K. Tillmann, T. Weirich (Eds.): EMC 2008, Vol. 1: Instrumentation and Methods, pp. 769–770, DOI: 10.1007/978-3-540-85156-1_385, © Springer-Verlag Berlin Heidelberg 2008

dimensional backgrounds estimated over each 64x64 pixels. The difference from the original image (Fig. 1f) does not show any structural features. This filter is also useful to extract nano particles embedded in an amorphous substrate.

1. R. Kilaas, *J. Microscopy* 190 (1997) 45-51.
2. P.H.C. Eilers et al, *Computational Statistics and Data Analysis* 50 (2006) 61-76.
3. K. Ishizuka, P.H.C. Eilers and K. Kogure, M&M2007, 902-903CD.

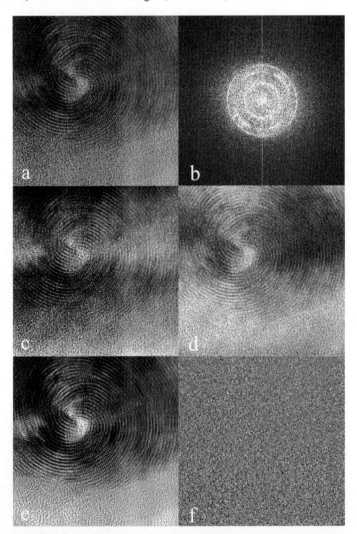

Figure 1. Optimal Wiener filter. (a) and (b): original image of crysotile and its Fourier transform. (c) and (d): conventional Wiener filter with radial background and its difference from the original image. (e) and (f): new optimal Wiener filter with local 2D background and its difference from the original image.

Noise considerations in the application of the transport of intensity equation for phase recovery

S. McVitie[1]

1. Department of Physics and Astronomy, University of Glasgow, Glasgow, G12 8QQ, United Kingdom

s.mcvitie@physics.gla.ac.uk
Keywords: transport of intensity equation, noise, phase imaging

The transport of intensity equation (TIE) [1] can be used for phase recovery in the transmission electron microscope (TEM) and has been applied successfully in a number of areas e.g. magnetic imaging [2]. However in some cases it has been noted that the technique is very sensitive to low spatial frequency noise which can be much larger than the phase variation to be measured [3,4]. In this paper possible sources of noise are investigated.

The solution of the TIE may be written in the case where there is no amplitude contrast as: $\phi = 2\pi\nabla_\perp^{-2}[I(\mathbf{r}_\perp,\Delta f) - I(\mathbf{r}_\perp,-\Delta f)]/(\lambda\Delta f I(\mathbf{r}_\perp,0))$. Therefore three images (I) taken normally at focus and either side of focus $\pm\Delta f$ can be used to calculate the phase. λ is the electron wavelength in the TEM, $\nabla_\perp^{-2}$ represents the inverse Laplacian operator.

To illustrate the contributions from noise, consider the difference of two images of free space taken 100μm either side of focus as shown in Figure 1(a) with the intensity variation shown in Figure 1(b). The signal variation observed has a typical profile representing noise from the source and detector. Applying the TIE algorithm results in a phase image as shown in Figure 2 with associated linetrace. It is clear that the phase image is dominated by low spatial frequency variations. However this is an image of a vacuum and should be perfectly flat. The observed frequency variation highlights the fact that the TIE solution is effectively a low frequency filter. The phase image in Figure 2(a) has a deviation of 0.16 rad and maximum and minimum values of 0.44 and –0.32 respectively. The phase image given in such cases represents the low frequency components of the random noise and differs for each reconstruction.

The effect can be further quantified by considering a single spatial frequency component in the difference image. In 1-D assume this has the form $I_k = A_k\sin(2\pi kx)$ with spatial frequency k and amplitude A_k. The associated phase component can be written as $\phi_k = (A_k/k^2)\sin(2\pi kx)/(4\pi\lambda\Delta f)$. The important points here are that each spatial frequency component of the phase is weighted inversely by k^2 and Δf. As the spatial frequencies present in digital images are a multiple of the lowest frequency it is apparent the weighting factor becomes very significant in suppressing higher spatial frequencies. Thus a white noise type of signal variation as seen in Figure 1 is heavily filtered by the TIE method resulting in a reconstruction which is dominated by low frequency variations.

Further general points regarding this contribution from noise can be linked to magnification/pixel spacing and defocus. In the case of the former for the same noise

M. Luysberg, K. Tillmann, T. Weirich (Eds.): EMC 2008, Vol. 1: Instrumentation and Methods,
pp. 771–772, DOI: 10.1007/978-3-540-85156-1_386, © Springer-Verlag Berlin Heidelberg 2008

spectrum a smaller pixel spacing will reduce the phase amplitude i.e. higher magnification is better. In terms of defocus the recovered phase amplitude reduces with defocus so that a small defocus will result in a larger phase variation noise for the same noise distribution. Examples from sample phase reconstructions will be given.

1.   D. Paganin and K. A. Nugent, Phys. Rev Lett. 80 (1998), p2586.
2.   S. Bajt et al, Ultramicroscopy **83** (2000), p67.
3.   M. Beleggia et al, Ultramicroscopy **102** (2004), p37.
4.   K. Ishizuka and B. Allman, J. Electron Microscopy **54** (2005), p191.

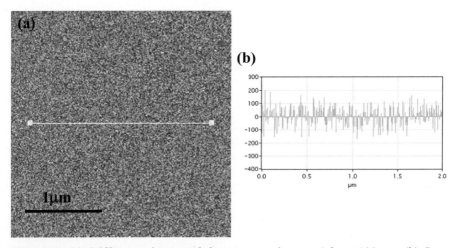

Figure 1. (a) Difference image of free space taken at $\Delta f = \pm100\mu$m. (b) Intensity variation along line indicated in (a).

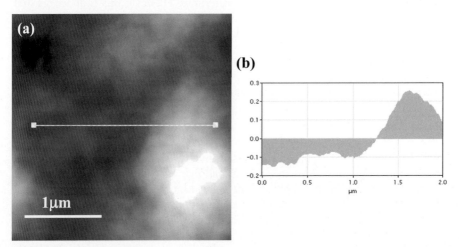

Figure 2. (a) TIE phase reconstruction image from free space at 100μm defocus. (b) Intensity variation along line indicated in (a) with vertical scale in radians.

elmiX – An Electron Microscopy Software Collection for Data Analysis and Education

A. Reinholdt and T.E. Weirich

Gemeinschaftslabor für Elektronenmikroskopie, RWTH Aachen, Ahornstr. 55, D-52074 Aachen, Germany

reinholdt@gfe.rwth-aachen.de

Keywords: software, education, GNU/Linux live-CD

One major drawback in teaching processing and analysis of electron microscopy data is the limited accessibility of software for students and lecturers outside an electron microscopy laboratory. Commonly used software is generally commercial, the available demo versions of these programs lack often for important program features. Suitable alternatives of public domain or freeware programs are difficult to find or often do not exist. Another disadvantage is the limitation that prominent EM software is usually bound to a certain operating system. In order to bridge this gap we have compiled the GNU/Linux live-CD *elmiX* based on the Debian sid distribution [1, 2, 3] for PCs with Intel compatible CPUs (Pentium II and later) and at least 256 MB RAM. All programs are available for immediate use with manuals and internet links to the web pages of the respective authors after rebooting the PC from the CD. The current version of *elmiX* (March 2008) contains the following electron microscopy software:

CASINO (D. Drouin) [4]
Crystal (M. Otten) [5]
CTF Explorer (M.V. Sidorov) [6]
Digital Micrograph (Gatan) [7]
EELS Model (J. Verbeeck) [8]
Electron Direct Methods (L. Marks & R. Kilaas) [9]
ImageJ (W. Rasband) [10]
Java Electron Crystallography Package (X.Z. Li) [11]
JEMS Student Edition (P. Stadelmann) [12]
Monte Carlo Simulations of electron-solid interactions (D. Joy) [13]
Powder Cell (W. Kraus, G. Nolze) [14]
Process Diffraction (J. Labar) [15]
Space Group Explorer (Calidris) [16]
VESTA (K. Momma, F. Izumi) [17]

For use in teaching and for exercises the *elmiX* software collection comes with a number of GNU/Linux office programs (including PDF writing capability and printer support), a browser for access to various web resources, a free and open source multimedia and flash player and the unit converter tool NumericalChameleon [18]. So far the live-CD contains only a few tutorials, scientific papers and sample files distributed together with the programs. For future releases of the CD it is planned to include material (lectures, tutorials, images, etc.) provided by the user community of

M. Luysberg, K. Tillmann, T. Weirich (Eds.): EMC 2008, Vol. 1: Instrumentation and Methods, pp. 773–774, DOI: 10.1007/978-3-540-85156-1_387, © Springer-Verlag Berlin Heidelberg 2008

774

elmiX. The always latest version of the live-CD can be obtained as ISO-image from the project web-site at *http//:www.elmix.org*

1. R. Stallman and GNU team (1984). GNU project, http://www.gnu.org
2. L. Torvalds and Kernel team (1991). http://www.kernel.org
3. I. Murdock and Debian team (1993). http://www.debian.org
4. http://www.gel.usherbrooke.ca/casino/index.html
5. M. Otten, private communication
6. http://clik.to/ctfexplorer
7. http://www.gatan.com/
8. http://webh01.ua.ac.be/eelsmod/eelsmodel.htm
9. http://www.numis.northwestern.edu/edm/
10. http://rsb.info.nih.gov/ij/
11. http://www.unl.edu/CMRAcfem/XZLI/programs.htm
12. http://cimewww.epfl.ch/people/stadelmann/jemsSE/jemsSEv3_2710u2008.htm
13. http://web.utk.edu/~srcutk/htm/simulati.htm
14. http://www.bam.de/de/service/publikationen/powder_cell_a.htm
15. http://www.mfa.kfki.hu/~labar/ProcDif.htm
16. http://www.calidris-em.com/archive.htm
17. http://www.geocities.jp/kmo_mma/crystal/en/vesta.html
18. http://www.jonelo.de/java/nc/

Speed considerations when performing particle analysis and chemical classification by SEM/EDS

S. Scheller[1]

1. Bruker AXS Microanalysis GmbH, Schwarzschildstr 12, 12489 Berlin, Germany

samuel.scheller@bruker-axs.de

Keywords: Particle Analysis, Feature Detection, Chemical Classification

Particle analysis by SEM/EDS is an integration combining SEM image acquisition, processing and evaluation with the EDS analysis and chemical classification of the feature of interest. The ever growing hunger for information and representative data in the information age has also pushed SEM based particle analysis for more data in less time for various reasons. Some applications use this method for quality assurance or grading, others purely to make better profits by adding another control mechanism in production. To keep up with production and increasing recognition of the method has put pressure on the speed of this type of analysis.

To reduce the time of particle analysis is a complex problem that presents manufacturers of particle analysis solutions with problems that constantly push against the limitations of current technology. A new and faster technology merely shifts the bottleneck to a different component in the system. A typical example is the introduction of the analytical SDD detector, which allows much higher count-rates at equal or better energy resolution [1], and is up to 10 times faster than conventional Si(Li) detectors [2]. This improvement will invariably move the particle analysis bottleneck to the pulse processor, quantification software, image acquisition or even stage movement.

The place the bottleneck moves to however is dependant on the application. Take an example particle sample which can be analyzed using relatively high probe current (e.g. > 5 nA) and would be packed densely enough to fit 100 or more particles of interest per frame. A typical high count EDS spectrum acquisition and quantification would take the bulk of the time of the measurement as can be seen in Fig. 1. In an example with a high number of particles per field any improvement in count rate would have a direct and linear effect on time to result. Hence a high throughput EDS detector such as Bruker's 4th generation XFlash 4000 series detector, would prove as an ideal solution for an increase in particle analysis throughput (see Fig. 2). Furthermore the resolution of these detectors is $\leq 125eV$ at Mn Kα up to more than 100,000 counts per second; hence a better statistic and resolution in less time with no trade-off.

In some cases this increase is still not sufficient and working with lower spectrum counts might be considered. If quantification accuracy for particles of interest should stay high, a *two-tier* classification approach could be used. This is especially useful if not all particles in a frame are of interest. A low count spectrum would exclude any false positives, and the particles of interest would be evaluated with a second high count spectra.

M. Luysberg, K. Tillmann, T. Weirich (Eds.): EMC 2008, Vol. 1: Instrumentation and Methods, pp. 775–776, DOI: 10.1007/978-3-540-85156-1_388, © Springer-Verlag Berlin Heidelberg 2008

If the samples have only a few particles per sample, the amount of time spent collecting spectra and evaluating them is insignificant. Most of the time would be spent driving the stage and collecting images. It is in such cases where stage movements and image acquisition are the bottlenecks and spectrum acquisition plays a much smaller role. Empty fields need to be discarded as quickly as possible, and stage movements must be fast and/or few. The trade-off between pixel size and number (or magnification and resolution) versus the number of stage movements becomes important. Furthermore the speed of acquisition of each frame and even the acquisition of each pixel starts to play an ever increasing role. This provides another technological possibility for speed improvement using different backscattered electron (BSE) detectors. There are two BSE detector technologies in use today, both of which have strong and weak points and must therefore be carefully chosen to meet the needs of the application.

It is important to look at particle analysis as a solution based on a system, where any change to one part of the system will affect the performance of the other. For the best particle analysis solution, it is necessary to look at the proposed application and customize the system to suit its purpose. It is the only way to get the best performance out of a particle analysis solution.

1. R. Terborg, M. Rohde, Microsc. Microanal. 12 (Supp 2) (2006) 1408 CD.
2. N.W.M. Ritchie and D.E. Newbury, Microsc. Microanal. 12 (Supp 2)(2006) 860 CD.

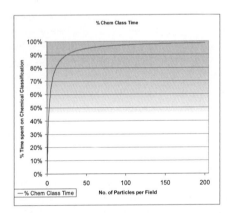

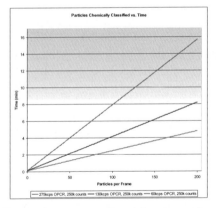

Figure 1. The samples particle density impacts heavily on analysis time. Only barren samples or a particle density of less than 2 particles per field are affected by image acquisition and processing.

Figure 2. Number of particles versus time, doubling output count rate (OPCR) implies an analysis time reduction to about half (plus image processing constant)

Morphological characterization of particles with very broad size distributions using program MDIST

M. Slouf [1,2], M. Lapcikova[1], H. Vlkova[1], E. Pavlova[1], J. Hromadkova[1]

1. Institute of Macromolecular Chemistry AS CR, Heyrovskeho namesti 2, 16206 Praha 6, Czech Republic
2. Member of Consortium for Research of Nanostructured and Crosslinked Polymeric Materials (CRNCPM)

slouf@imc.cas.cz

Keywords: electron microscopy, image analysis, polyethylene and polystyrene

Morphological characterization of particles, i.e. the analysis of particle sizes and/or shapes, is quite a frequent problem of microscopic analyses. Most samples contain particles of similar sizes and shapes. If the analyzed particles have dimensions within one order of magnitude, then their size and shape distributions can be obtained by means of a *standard image analysis* using standard commercial software. In these simple cases, the image analysis procedure can be divided into three steps: (i) separation of particles and background, (ii) calculation of particle morphological parameters and (iii) calculation of size/shape distributions.

However, a number of real systems contain particles, whose dimensions differ by more than 2-3 orders of magnitude. In these systems the smallest and the biggest particles cannot be observed simultaneously in one micrograph. At high magnifications, the smallest particles are readily observable, but the biggest particles have sizes comparable with the real width of image and, consequently, they frequently touch the image boundaries and must be excluded. At low magnifications the biggest particles are easily analyzed, but the smallest particles become almost invisible and, as a result, their morphological analysis is imprecise or even impossible.

Correct morphological analysis of particles with very broad size distribution requires *two* sets of micrographs. From low-magnification micrographs we obtain morphological parameters of bigger particles, while from high-magnification micrographs we obtain parameters of smaller particles. The first two steps of image analysis, i.e. (i) separation of particles from background and (ii) calculation of particle morphological parameters, are the same as in the case of a standard image analysis. The first and the second step are carried out with standard software and the only difference consists in that they are performed twice, both for low- and high-magnification micrographs. In the third step of image analysis, the two separated data sets from previous steps are combined to calculate correct size/shape distributions. The calculation is complicated by the fact that small and big particles should be determined from the same area of the specimen so that the distribution was correct. However, this is rarely fulfilled as the low-magnification micrographs show larger areas of the specimen than high-magnification ones. Therefore both data sets have to be assigned different statistical weights depending on the number of images and their magnifications.

M. Luysberg, K. Tillmann, T. Weirich (Eds.): EMC 2008, Vol. 1: Instrumentation and Methods, pp. 777–778, DOI: 10.1007/978-3-540-85156-1_389, © Springer-Verlag Berlin Heidelberg 2008

778

The statistical weights of each micrograph and final size/shape distributions can be calculated either manually or by means of a small program, which was developed in author's laboratory and called MDIST. Manual calculation of size distributions, for example in a spreadsheet program like Excel, is time-consuming and error-prone. In contrast, program MDIST is fast and simple. It is a set of command-line scripts, which can be adjusted for a particular application. MDIST outputs both statistical parameters and graphs showing size/shape distributions (Fig. 1). MDIST results will be shown and compared with standard image analyses using two real systems of synthetic polymers, which contained particles with broad size distributions: (a) in vivo polyethylene wear particles with sizes 0.05 to 10 μm isolated on polycarbonate membranes and (b) HIPS polymer containing polybutadiene inclusions with sizes from 0.1 to 12 μm.

1. Financial support through grants MSMT 2B06096 and MPO FT-TA3/110 is gratefully acknowledged.

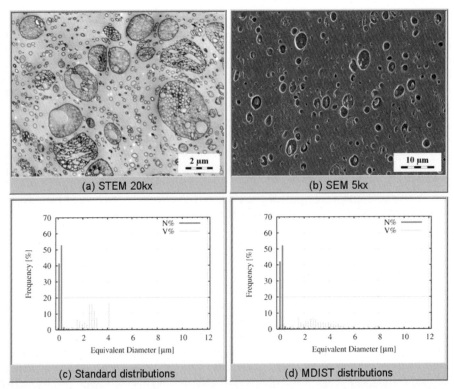

Figure 1. Sample output of program MDIST. (a) High-magnification STEM micrograph of HIPS polymer, (b) low-magnification SEM micrograph of the same polymer, (c) number (N%) and volume (V%) size distributions of polybutadiene inclusions, calculated only from STEM micrographs by a standard image analysis, (d) the same distributions calculated from both STEM and SEM micrographs by MDIST.

Multiple protein structures in one shot:
maximum-likelihood image classification in 3D-EM

S.H.W. Scheres[1] and J.M. Carazo[1]

1. Centro Nacional de Biotecnología, CSIC. Calle Darwin 3, 28049, Madrid, Spain.

scheres@cnb.csic.es

Keywords: macromolecular machines, structural heterogeneity, XMIPP

Many tasks in the living cell are performed by macromolecular assemblies encompassing various binding interactions and accompanied by conformational changes. The 3D-EM approach holds the promise of being able to visualize these "molecular machines" in their various functional states. In the single-particle reconstruction approach, individual molecules are visualized with an electron microscope, and the resulting projections, often numbering tens of thousands, are combined in a three-dimensional density map. As the technique imposes no restrictions on the range of existing conformations in the sample, information about various functional states is often available in a single experiment. However, the combination of images in a three-dimensional reconstruction requires that they represent projections of identical three-dimensional objects and that their relative orientations be known. As the problems of conformational and orientational classification are strongly intertwined, the coexistence of different conformations or ligand binding states in the sample has seriously limited the applicability of the 3D-EM approach [1].

We will describe a classification approach based on maximum likelihood principles for handling structurally heterogeneous data [2]. From a theoretical point of view, maximum-likelihood estimators are particularly well suited for the problem at hand because, as the amount of experimental data increases, these methods yield results with smaller variances than alternative estimators [3]. Furthermore, previous applications of the maximum-likelihood approach to related problems in the 3D-EM field have demonstrated its particular robustness to high levels of noise [4-7]. The most important difference of the maximum-likelihood approach compared to conventional (maximum cross-correlation) approaches is the underlying statistical data model. Whereas maximum cross-correlation approaches implicitly assume noiseless data, our maximum likelihood approach employs an explicit model for the experimental noise, which is assumed to be white, zero-mean and additive Gaussian.

We will show that our approach permits the use of initial reference maps that are devoid of any prior information (or bias) about the structural variability in the sample. Such bias-free seeds are obtained by performing a single iteration of likelihood optimization for K randomly drawn subsets of the data, using a low resolution representation of the average structure as single reference. Therefore, classification based on likelihood optimization becomes unsupervised because it depends only on a preliminary reconstruction of the data and a rough estimate for the number of existing classes, which determines the choice of the value of K. Taken together with the

M. Luysberg, K. Tillmann, T. Weirich (Eds.): EMC 2008, Vol. 1: Instrumentation and Methods, pp. 779–780, DOI: 10.1007/978-3-540-85156-1_390, © Springer-Verlag Berlin Heidelberg 2008

intrinsically combined treatment of orientational and conformational assignment, this suggests that, in contrast to existing approaches, classification by likelihood optimization may be applicable to a broad range of structurally heterogeneous data sets. We will demonstrate the effectiveness of this approach for two macromolecular assemblies with different types of conformational variability: the Escherichia coli 70S ribosome and Simian virus 40 (SV40) large T-antigen.

Finally, we will discuss an alternative data model for the image formation process, which allows modelling non-white (or coloured) noise, as well as the effects of the microscope's point spread function [8]. Using this approach, we were able to visualize, for the first time, 3D density for a double-stranded DNA probe protruding from a dodecameric complex of Simian Virus 40 large-T antigen.

1. E.V. Orlova, & H.R. Saibil, Curr. Opin. Struct. Biol. **14** (2004), p. 584-590.
2. S.H.W. Scheres et al. Nat Methods **4** (2007), p. 27-29.
3. J.A. Rice, Mathematical Statistics and Data Analysis, Duxbury Press, Belmost (1995).
4. F.J. Sigworth, J. Struct. Biol. **122** (1998), p. 328-339.
5. Z. Yin, et al. J. Struct. Biol. **144** (2003), p. 24-50.
6. S.H.W. Scheres, et al. J. Mol. Biol. **348** (2005), p. 139-149.
7. S.H.W. Scheres, et al. Bioinformatics **21** (Suppl. 2) (200), p. ii243-ii244
8. S.H.W. Scheres et al. Structure **15** (2007), p. 1167-1177.
9. We kindly acknowledge Drs. H. Gao, J. Frank, M. Valle, R. Núñez-Ramírez, Y. Gómez-Llorente and C. San Martín for providing experimental data, and Drs. G.T Herman and P.P.B. Eggermont for help with the mathematics. We thank the Barcelona Supercomputing Center (Centro Nacional de Supercomputación) for providing computer resources. This work was funded by the EU (FP6-502828; UE-512092), the US NIH (HL740472), the Spanish CICyT (BFU2004-00217), MEC (CSD2006-00023 and BIO2007-67150-C03-03), FIS (04/0683) and CAM (S-GEN-0166-2006).

Compensation and evaluation of errors of 3D reconstructions from confocal microscopic images

M. Čapek[1,2], P. Brůža[1,2], L. Kocandová[1,2], J. Janáček[1], L. Kubínová[1], and R. Vagnerová[3]

1. Institute of Physiology, Academy of Sciences of the Czech Republic, v.v.i.,
Vídeňská 1083, 142 20 Prague 4 – Krč, Czech Republic
2. Faculty of Biomedical Engineering, Czech Technical University in Prague,
nám. Sítná 3105, 272 01 Kladno 2, Czech Republic
3. Institute of Histology and Embryology, First Faculty of Medicine,
Charles University in Prague, Albertov 4, 128 00 Praha 2, Czech Republic

capek@biomed.cas.cz

Keywords: error assessment, volume reconstruction, confocal microscopy

Volume (3D) reconstruction enables us to visualize, and subsequently analyse biological specimens that are greater than field of view and/or thicker than maximal depth of scanning of a used optical instrument. We use a laser scanning confocal microscope capable to acquire a 3D digital representation of the specimen [1].

Our approach to 3D reconstruction consists of the following steps (Fig. 1): A) Specimen preparation and embedding into a paraffin block. B) Cutting the specimen into physical slices by a microtome and simultaneous recording pictures from the cutting plane of the paraffin block by a USB light microscope (a portable Dino-Lite USB digital microscope with resolution of 1280×1024 pixels). This step is performed so as to have image information about the original size and shape of the specimen prior its cutting. C) Acquisition of overlapping fields of view (spatial tiles) from all physical slices by a confocal microscope. D) Horizontal merging (mosaicking) of the spatial tiles into a sub-volume representing a physical slice. E) Vertical merging of sub-volumes of successive physical slices into volume representation of the whole specimen, employing information from USB light microscope images. Vertical merging is based on B-spline elastic registration of images of successive physical slices. The reason for applying the elastic registration lies in the possibly extensive deformations of large specimens caused by their cutting and manipulation during preparation. F) Image enhancement of optical sections. Finally, the resulting volume is visualized with the help of the hardware accelerated volume rendering.

The elastic algorithm in Step E) solves efficiently problems regarding mutual deformations of successive physical slices, however, it introduces a difficulty: The alignment is performed step by step using a mosaic image of one physical slice as a reference (fixed), and the mosaic image of the second physical slice as registered (floating). As a consequence, objects/structures in the floating image are transformed according to objects/structures in the fixed image. Thus, when the floating image is registered with respect to the fixed one that is highly deformed, this deformation is propagated by elastic registration through the rest of the volume. We solve this problem by recording the shape of investigated objects by using a USB light microscope prior to

M. Luysberg, K. Tillmann, T. Weirich (Eds.): EMC 2008, Vol. 1: Instrumentation and Methods,
pp. 781–782, DOI: 10.1007/978-3-540-85156-1_391, © Springer-Verlag Berlin Heidelberg 2008

782

specimen cutting. The shape of objects is then restored after elastic registration using information in the captured pictures.

Comparison of images from the confocal microscope and from the USB light microscope helps to quantify the extent of deformations, using, for example, a stereological point counting method [1] (a Point Grid module in the Ellipse SW, www.vidito.com) for measuring areas of specimen slices.

We show results using a 17-day-old embryo of a laboratory Norway rat here. Ten confocal images of physical slices of the embryo were aligned with corresponding images of these slices captured by the USB light microscope by elastic registration. We measured areas of embryo slices from confocal data both before and after alignment and also from USB light microscope data. The measurements revealed partial deformations of cut physical slices (decreasing / increasing of areas measured from original confocal data with respect to USB light microscope data up to 92.7% / 105.3%), but an extensive one as well (decreasing the area of the embryo slice in confocal data to 55% only). However, even this dramatic deformation was successfully compensated by elastic alignment. Thus, our method for compensation of deformations using accessorial data from the USB light microscope turns up to be efficient and providing us with realistic visualization of biological structures under study.

1. A. Diaspro, editor: Confocal and Two-Photon Microscopy, Foundation, Applications, and Advances, New York: Wiley-Liss, Inc., 2002.
2. We kindly acknowledge the support by MŠMT ČR grants No. MSM6840770012 and No. LC06063, by grants of the Academy of Sciences of the Czech Republic No. A100110502, No. A500200510 and No. AV0Z 50110509, and by GAČR grant No. 102/08/0691.

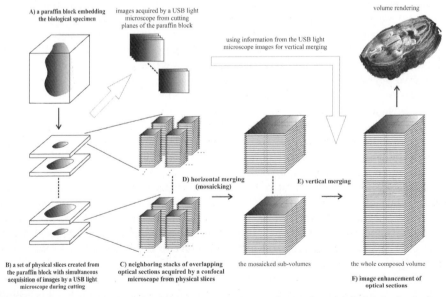

Figure 1. An overview of 3D volume reconstruction of a thick biological tissue specimen using a confocal laser scanning microscope and a USB light microscope.

High content image-based cytometry as a tool for nuclear fingerprinting

W.H. De Vos[1], B. Dieriks[1], G. Joss[2], and P. Van Oostveldt[1]

1. Department of Molecular Biotechnology, University of Ghent, Coupure links 653, 9000 Ghent, Belgium
2. Department of Biological Sciences, Macquarie University, North Ryde, NSW 2109, Sydney, Australia

winnok.devos@ugent.be
Keywords: cytometry, high content screening, segmentation

Cytomics aims at understanding the functional relationships between cellular phenotypes (cytome) and metabolic pathways (proteome) that result from a combination of genetically defined mechanisms (genome) and environmental conditions [1,2]. Although flow-cytometry is able to measure the optical properties of single cells at a rate of >1000 cells per minute it has a limited capability of mapping individual events. To accurately quantify (sub-) cellular characteristics within a natural context there is a fast-growing need for image-based cytometry. Images, obtained with fluorescence microscopy, provide the exact information on signal intensity, location and distribution of specific molecules within intact cell systems (tissue or monolayers) and allow for investigating cellular properties in relation to the cell-ecological context [3].

Previously, we have developed a cytometric approach for scoring DNA lesion endpoints in confocal images of murine fibroblasts [3]. We now present a generalized approach for multivariate phenotypic profiling of individual nuclei using automated fluorescence mosaic microscopy and optimized digital image processing tools. An indefinite number of fields, z-slices and channels can be analyzed; the only prerequisite is the presence of a nuclear counterstain, which is used for the generation of masks. To anticipate for erroneous segmentation of clustered nuclei in dense cell cultures we implemented an iterative conditional segmentation (ics) algorithm that uses both morphological and intensity information from the image (Figure 1). The method makes use of a priori knowledge about the size and shape of nuclei in stringent feedback selection of correctly segmented nuclei. Depending on the degree of clustering, segmentation performance varies between 95% and 100%. Complete analysis of nuclei and subnuclear features for a region of 25 images of 1000x1000 pixels, 3 z-slices and 3 channels only takes ~ 3 minutes or ~ 0.7sec/nucleus. Our method is insensitive to scaling, illumination heterogeneity and variability or non-uniformity of staining.

We have successfully applied our system in cell cycle analysis, scoring of transfection efficiency and assessment of (localized) DNA damage in response to genotoxic stress and ionizing radiation.

M. Luysberg, K. Tillmann, T. Weirich (Eds.): EMC 2008, Vol. 1: Instrumentation and Methods, pp. 783–784, DOI: 10.1007/978-3-540-85156-1_392, © Springer-Verlag Berlin Heidelberg 2008

784

1. T. Bernas, G. Gregori, E. Asem and J. Robinson, Molecular & Cellular Proteomics **5** (2006), p. 2.
2. G. Herrera, L. Diaz, A. Martinez-Romero, A. Gomes, E. Villamon, R. Callaghan and J. O'connor, Toxicology In Vitro **21** (2007), p. 176
3. G. Valet, J. Leary and A. Tarnok, Cytometry A **59** (2004), p. 167.
4. P. Baert, G. Meesen, S. De Schynkel, A. Poffijn and P. Van Oostveldt, Micron **36** (2005), p. 321.

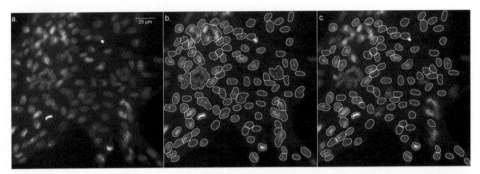

Figure 1. Improved segmentation of dense nuclear images using the ics-procedure. a. Original image of DAPI counterstained human fibroblast nuclei. b. Outlines of general segmentation result superimposed onto the original image. c. Outlines of iterative conditional segmentation superimposed onto the original image. Incorrectly segmented nuclei are marked (*).

Measurement of surface area of biological structures, based on 3D microscopic image data

L. Kubínová[1] and J. Janáček[1]

1. Department of Biomathematics, Institute of Physiology ASCR, v.v.i., Vídeňská 1083, 14220 Prague, Czech Republic

kubinova@biomed.cas.cz

Keywords: surface area, stereology, confocal microscopy, electron tomography

Surface area represents an important geometrical parameter of biological structures observed at different microscopic levels, e.g. surface area of tissues and cells involved in gas exchange investigated by light microscopy, or surface area of membranes involved in membrane transport studied at the ultramicroscopic level. Proper measurement of surface area is a difficult task as its unbiased estimation from practically two-dimensional (2D) thin histological or ultrathin sections requires randomized direction of cutting such physical sections, which is often technically demanding, inefficient and sometimes impossible. This problem can be solved by applying methods for surface area measurement based on evaluation of three-dimensional (3D) image data which can be acquired by some of the contemporary microscopic techniques, such as confocal microscopy or transmission electron tomography. Furthermore, surface area measurement is particularly sensitive to resolution and noise present in digitized microscopic images which requires careful tuning of methods applied.

We developed software implementation of a number of methods for surface area estimation based on evaluation of 3D image data, namely interactive stereological methods and automatic or semi-automatic digital methods. All methods were implemented in special plug-in modules of *Ellipse* SW environment (ViDiTo, Košice, Slovakia).

The stereological methods were based on using computer-generated, properly randomized virtual linear test probes in the form of spatial grids of "fakir" straight lines [1] or cycloids [2], and interactive marking of intersections of these test probes and surface area of microscopic structures under study. We studied variances of the relevant surface area estimators using several arrangements of spatial grids of lines. We showed that cycloid probes yield lower variance due to the grid orientation than fakir probes while the residual variance of the estimator (due to the grid translation) using fakir probes can be decreased in practice more easily than that of the cycloid-based estimator requiring to increase the height of cycloids which can be limited by the physical dimensions of the image volume.

Further, we implemented the automatic digital methods, based on Crofton formula [3] or surface triangulation [4] which were applied to automatically or semi-automatically segmented images. We checked the sensitivity of these methods to image

M. Luysberg, K. Tillmann, T. Weirich (Eds.): EMC 2008, Vol. 1: Instrumentation and Methods, pp. 785–786, DOI: 10.1007/978-3-540-85156-1_393, © Springer-Verlag Berlin Heidelberg 2008

pre-processing, which is usually necessary for successful segmentation of structures under study.

The methods for surface area measurement were demonstrated on 3D images acquired by confocal microscopy or electron tomography techniques. They were tested on geometrical models and different types of biological surfaces, namely tobacco cell wall, internal surface of conifer needles and walls of rat skeletal muscle fibres.

We compared the above stereological and image analysis methods for surface area estimation from the point of view of their applicability, efficiency and precision. We conclude that there is no absolutely universal method which would be optimal for all types of structure. Automatic methods are faster than interactive stereological methods but require automatic segmentation of analyzed objects from the images or at least a high contrast between the object and the background. Further, automatic digital methods for surface area estimation, especially triangulation method, are sensitive to processing of microscopic image data, e.g. degree of smoothing which is used for reduction of noise in acquired images. Therefore, they require careful testing and adjusting to the given type of microscopic structure.

1. L. Kubínová and J. Janáček, J. Microsc. **191** (1998), p. 201.
2. A.M. Gokhale, R.A. Evans, J.L. Mackes and P.R. Mouton, J. Microsc. **216** (2004), p. 25.
3. F. Meyer, J. Microsc. **165** (1992), p. 5.
4. L. Kubínová, J. Janáček and I. Krekule in "Confocal and Two-photon Microscopy", ed. A. Diaspro, (Wiley-Liss, New York) (2002), p. 299.
5. The presented study was supported by the Academy of Sciences of the Czech Republic (grants A600110507, A100110502, A500200510 and AV0Z50110509) and Ministry of Education, Youth and Sports of the Czech Republic (research program LC06063).

The imaging function for tilted samples: simulation, image analysis and correction strategies

V. Mariani, A. Schenk, A. Engel and A. Philippsen

Maurice E. Müller Institute for Structural Biology,
University of Basel, Klingelbergstr. 70, 4056 Basel, Switzerland

valerio.mariani@unibas.ch

Keywords: tilt, imaging function, CTF correction, electron tomography, electron crystallography

Several electron microscopy techniques involve tilting of the observed sample. Under this conditions, the defocus varies continuously in the direction perpendicular to the tilt axis. The Point Spread Function (PSF) is not space-invariant, and consequently the imaging function cannot be described by a convolution.

A description of the imaging process in tilted geometry for weak-phase objects is introduced: the Tilted Contrast Imaging Function (TCIF) [1]:

$$Q(\mathbf{p}) = i\left\{ e^{-iW_0(\mathbf{p})}\, \Phi\left(\mathbf{p} - \tfrac{1}{2}\mathbf{d}p^2\lambda\,Tan(\alpha)\right) - e^{iW_0(\mathbf{p})}\, \Phi\left(\mathbf{p} + \tfrac{1}{2}\mathbf{d}p^2\lambda\,Tan(\alpha)\right) \right\}$$

Information coming from two different points of the Fourier transform of the sample function (Φ) is collapsed into a single point in the Fourier transform of the observed image (Q). The separation of the two points increases with spatial resolution. For a tilt angle (α) of zero degrees, the TCIF collapses to the classical Contrast Transfer Function (CTF).

Three simulation procedures to apply the TCIF transformation are presented, each representing a different compromise between computational speed and accuracy. These simulations are then compared to experimental images of tilted samples. The premature disappearance of Thon rings in images collected at high tilt angles is shown to be an implicit effect of the TCIF and not due to a drop of the signal-to-noise ratio of the data: this phenomenon is accurately described without introducing any quenching envelope function. The splitting of diffraction peaks in power spectra of tilted images of 2D crystals is also demonstrated to be attributable to the TCIF.

The imaging function for tilted samples can only be described as a convolution in an approximate way at low spatial frequencies. Correcting the effect of the imaging process in tilted images using a deconvolution-based approach introduces phase errors that increase as the approximation loses accuracy and become relevant at high spatial frequencies. A theoretical inversion of the TCIF is proposed but shown to be computationally intractable at this stage. A number of available correction strategies for Electron Crystallography and Tomography are then discussed and their accuracy and performance are evaluated. These methods include CTF correction in stripes parallel to the tilt axis for which the defocus is almost constant, and other heuristic correction schemes.

M. Luysberg, K. Tillmann, T. Weirich (Eds.): EMC 2008, Vol. 1: Instrumentation and Methods, pp. 787–788, DOI: 10.1007/978-3-540-85156-1_394, © Springer-Verlag Berlin Heidelberg 2008

788

1. A. Philippsen, H. A. Engel and A. Engel, Ultramicroscopy **107(2-3)** (2007), p. 202-212.

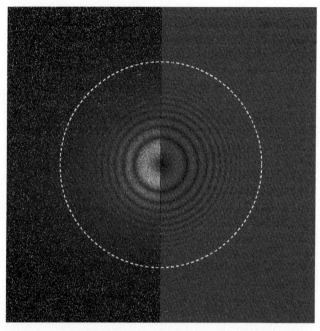

Figure 1. Comparison of an experimental (left) and a simulated (right) tilted image of carbon film collected at a tilt angle of 60 degrees with a defocus at the centre of the image of 5.65 μm. The disappearance of the Thon rings in the experimental image is correctly predicted in the simulation making use of the TCIF only, without any quenching envelope function, and shown to be an effect of the tilted imaging function and not of a drop in the signal-to-noise ratio of the data. The dotted line corresponds to a spatial frequency of 10 Å^{-1}.

4D-Microscopy

Alexander A. Mironov, Massimo Micaroni, Galina V. Beznoussenko

Department of Cell Biology and Oncology, Consorzio Mario Negri Sud,
Via Nazionale 8, 66030, S. Maria Imbaro (Chieti),
Italy

mironov@negrisud.it
Keywords: 4D imaging, correlative microscopy, FRET, electron microscopic tomography

The routine light microscopy provides only a 2D imaging. Using optical sectioning and consecutive 3D-reconstruction one gets 3D imaging. A 3D imaging in time (3D video microscopy) gives a 4D-imaging. Finally, combination of a 4D-imaging of two fluorochromes (usually CFP and YFP) determines a 5D-imaging. After fixation and staining of the same cell, addition of a red fluorochrome (the correlative light-light microscopy) will give a >5D-imaging. After production of 5D-movies one should fix cells at a define stage of the life cycle of an organelle just observed and then analyze the same organelle under a light or electron microscope using an additional immunolabeling, including FRET, and more precise 3D-reconstruction with an electron microscopic tomography of the very same organelle.

We are presenting the way to combine double-color video microscopy with an additional immunofluorescent, including FRET, and then immunoelectron microscopic labeling, and subsequent tomographic 3D-reconstruction of the very same organelle just observed with a time-lapse double color analysis. Cells were transfected with two fusion proteins (one should be tagged with CFP, another construct has to be tagged with YFP), for example, with Sec23-CFP (one of the subunits of COPII) and VSVG-YFP (a routine cargo protein) on a coordinated cover slip, the cell of interest (with a define localization on the coordinated grid) was examined with the 5D-video microscopy with the help of a digitalized microscope. At defined moment, the cell was fixed by addition of 4% formaldehyde +0.05% glutaraldehyde on 0.15 M HEPES (pH 7.2) into the chamber for 5 min. Fixation stopped movement of all organelles inside the cell. During fixation the optical sectioning of the cell was performed on the laser scanning confocal microscope for both fluorochromes. Next, bleaching of YFP and the next measurement of CFP signal examined the fixed cell with FRET. The cell was, then, permeabilized with 0.1% saponin and additionally labeled with monoclonal antibodies against ßCOP and GM130 which then were subsequently detected with anti-IgG Fab fragments conjugated with Cy3 and Cy5 and Z-staking was performed of the same organelle for the subsequent 3D-reconstruction. Finally, cells were stained with the polyclonal antibodies against the luminal portion of VSVG, which were detected by immunoperoxidase amplification of nano-gold-gold enhancement procedure. Then samples were embedded into Epon and the cells of interest were sectioned tangentially to obtained 200-nm sections suitable for tomography. Using labeling for VSVG and the exact positioning of the organelle inside the cell we have found the transport carrier just leaving the ER exit site and made 3D-

M. Luysberg, K. Tillmann, T. Weirich (Eds.): EMC 2008, Vol. 1: Instrumentation and Methods,
pp. 789–790, DOI: 10.1007/978-3-540-85156-1_395, © Springer-Verlag Berlin Heidelberg 2008

reconstruction with the help of EM tomography. The method gives visualization of 4 different proteins in 3D after their video analysis.

1. We kindly acknowledge the help of Teleton (Italy) in the funding of the research.

Strategies for high content imaging screening and analysis of primary neurons

S. Munck[1] and W. Annaert[2]

1. Light Microscopy and Imaging Network LiMoNe VIB,Department of Molecular and Developmental Genetics,K.U.Leuven, Campus Gasthuisberg, Herestraat 49, box 602, 3000 Leuven,Belgium
2. Membrane Trafficking Lab VIB,Department of Molecular and Developmental Genetics,K.U.Leuven, Campus Gasthuisberg, Herestraat 49, box 602, 3000 Leuven, Belgium

Sebastian.munck@med.kuleuven.be

Keywords: screening, HCS, Neuron, quantitative image analysis

Approaching biological questions by means of quantitative analysis is of increasing interest. Methods of rapid analysis and diagnostics have been developed a lot in the past. One example is screening methods used by pharmaceutical companies. To date several devices are on the market for high content screening task on an imaging basis. Practically that means it acquires pictures of for example cells or cellular organelles and uses image analysis tools to analyze these sets of pictures. Normally multiwell plates are used for screening, to test several parameters in parallel. Until recently publications on the topic of high content screening are rather rare and lack significant methodological information. This is probably due to the fact that this field was dominated by pharmaceutical companies and therefore knowledge was classified. Only few examples peak out from suppliers highlighting the potential of the technique (for review on that topic see [1]). As currently the Life Science field and especially neuroscience is moving more in that direction it is time to catch up. Very recently descriptions of quantitative methods for analysis of in vitro neurite outgrowth have been published [2,3]. However mainly neuronal cell lines have been used and studies on primary neurons are scarce (for example [4]). Cell models are rather convenient but are not displaying all the features of real neurons. Apart from that we have encountered that the neurons react much more sensitive to treatments and drugs than do cell lines. Therefore we see it inevitable to work directly with primary neurons. Here we would like to describe strategies to use these cells in imaging based cellular screens. These include protocols for stained cell lines over living neurons derived from transgenic mice and how to overcome problems like feeder layers. Moreover we describe approaches to analyze later stages of neurons and extract quantifiable data from highly differentiated neurons that have already formed dense networks.

M. Luysberg, K. Tillmann, T. Weirich (Eds.): EMC 2008, Vol. 1: Instrumentation and Methods, pp. 791–792, DOI: 10.1007/978-3-540-85156-1_396, © Springer-Verlag Berlin Heidelberg 2008

1. Oliver Rausch. High content cellular screening. Current Opinion in Chemical Biology, **10** (2006), 316–320.
2. Vibor Laketa, Jeremy C. Simpson, Stephanie Bechtel, Stefan Wiemann and Rainer Pepperkok. High-Content Microscopy Identifies New Neurite Outgrowth Regulators. Molecular Biology of the Cell **18** (2007), 242–252.
3. Mitchell PJ, Hanson JC, Quets-Nguyen AT, Bergeron M, Smith RC. A quantitative method for analysis of in vitro neurite outgrowth. J Neurosci Methods **30** (2007) 350-62.
4. Min Hua, Mark E. Schurdakb, Pamela S. Puttfarckena, Rachid El Kouhena,
5. Murali Gopalakrishnana, Jinhe LiHigh content screen microscopy analysis of Aβ1–42-induced neurite outgrowth reduction in rat primary cortical neurons: Neuroprotective effects of α7 neuronal nicotinic acetylcholine receptor ligands Brain research **1151** (2007), 227–235.

Modelling and analysis of clustering and colocalization patterns in ultrastructural immunogold labelling of cell compartments based on 3-D image data

A. Vyhnal[1]

1. Department of Biomathematics, Institute of Physiology, ASCR, v.v.i., Prague, Czech Republic

avyhnal@biomed.cas.cz

Keywords: cellular compartments, random point fields, electron tomography.

In biological investigations requiring statistical analysis of immunogold clustering, inhibition, colocalization or mutual inhibition patterns the concept of random point field theory is used. We use a generalization, to the 3-D case of electron tomography data, of the results obtained in [1] in which the clustering or colocalization of gold particles was detected using various characteristics of the distribution of distances between them. Pair correlation and cross-correlation functions [2, 3, 4, 5] are used for the exploratory analysis. Second-order reduced K (or Ripley's K) functions [6, 7] are used for testing the statistical significance of observed events. In general it is necessary to use simultaneously several quantities of different types to analyse and characterize a given point structure. In particular, this means that basing conclusions upon mean values or distance functions, or second-order quantities (such as the pair correlation function) only, is not sufficient. None of these quantities alone can, in general, determine uniquely the distribution of a point structure, unless it arises from a Poisson point field. Therefore, a combined use of intensities, a distance distribution and a second-order quantity is necessary [8]. Confidence intervals of function values are estimated by Monte Carlo simulations of the Poisson field for independent particles [1].

Random point fields are used when the heterogeneity is distributed discretely in the space, as is the case of the discrete distribution of gold particles in cell components. In an informal formulation, to each Borel set in a Euclidean space a random variable that characterizes the random number of points contained in this set is assigned. All fields are supposed to be simple, i.e., there are no multiple points within infinitesimal spatial domains. In biological applications a random point field is usually also supposed to be homogeneous, i.e., the mean value of the number of points is proportional to the Lebesgue measure of the respective set; the constant of proportionality is called the intensity. Next, the second-order measure for two different Borel sets is defined as the mean value of the product of the corresponding random variables. If we write this product as a summation, over all unordered pairs of points, of the product of indicator functions and if we restrict ourselves only to differing points we obtain the factorial moment measure. The density of this measure is the second-order product density. In our application we suppose all fields to be isotropic, i.e., this density depends only on the distance; we can define the pair-correlation function as the second-order product density divided by the square of the intensity. This function has the intriguing

M. Luysberg, K. Tillmann, T. Weirich (Eds.): EMC 2008, Vol. 1: Instrumentation and Methods, pp. 793–794, DOI: 10.1007/978-3-540-85156-1_397, © Springer-Verlag Berlin Heidelberg 2008

properties, which are used in the assessment of the statistical properties of a point pattern appearing in our 3-D electron tomography data of immunogold labelling, as follows. Large values of the pair-correlation function at a certain distance indicate that the corresponding interpoint distances appear frequently in the pattern and vice versa; a maximum of this function correspond to the distance that occurs most frequently. If there are poles at infinitesimally small distances, it indicates the presence of a cluster in the spatial pattern of gold particles. If this function has the form of a step-like function it indicates that this pattern has a hard-core distance, i.e., the interpoint distances smaller than a certain bound are forbidden. In this case the K-function of Ripley at a positive distance is defined as the mean number of points within this distance from a typical point of the studied spatial point pattern, excluding this point itself. The relation to the pair-correlation function, i.e., the fact that in 3-D it is proportional to the derivative of the K-function divided by the square of the distance, is used.

We deal with the spatial point pattern that appears in the electron tomography data of an immunogold labelling experiment; we first test the null hypothesis, i.e., the complete space randomness is present. By the language of the random point field theory it means that the point pattern forms the homogeneous Poisson random field. More complicated random fields are then used for the assessment of the statistical properties of the studied spatial pattern of gold particles. These include the Poisson inhomogeneous fields, Poisson hard-core fields, Neymann-Scott fields and the Markov and Gibbs point fields. As an estimator for the pair-correlation function and other characteristics we use the formulae of [3] which are generalized to the 3-D case

1. A. Philimonenko, J. Janacek P. Hozak, Journal of Structural Biology 132 (2000), 201–210
2. D. Stoyan et al., "Stochastic Geometry and its Applications, 2nd Edition", (Wiley, New York, 1996).
3. D. Stoyan et al., "Fractals, Random Shapes and Point Fields", (Wiley, New York, 1994).
4. K. Sobczyk et al., "Stochastic Modeling of Microstructures", (Birkhäuser, Berlin, 2001);
5. N. Cressie, "Statistics for Spatial Data", (Wiley, New York, 1993)
6. B. Ripley, "Spatial Statistics", (Wiley, New York, 1980)
7. P. Diggle, "Statistical Analysis of Spatial Point Patterns", (Academic Press, London, 1983)
8. D. König, S. Caravajal-Gonzalez, A. Downs, J. Vassy, J. Rigaut, Journal of Microscopy 161 (1990), 405-433

9. The presented study was supported by the Grant Agency of the Czech Republic (grant 204/05/H023), Academy of Sciences of the Czech Republic (grant AV0Z50110509), and Ministry of Education, Youth and Sports of the Czech Republic (research program LC06063).

Modern Methods of TEM Specimen Preparation in Material Science

H.J. Penkalla

Forschungszentrum Jülich, Institute for Energy Research, 52425 Jülich, Germany

h.j.penkalla@fz-juelich.de

Keywords: electro polishing, sandwich preparation, FIB target preparation

The aim of electron microscopy is to generate good micrographs, which enables accurately studies of the structure of a specimen. Therefore, it cannot be over emphasised that good specimen preparation is the key to produce good electron micrographs. Consequently, it is essential to perfect a suitable specimen preparation technique before embarking upon a detailed micro structural investigation.

The TEM specimen preparation of bulk material or surface layers is mainly a thinning process, which can be classified in mechanical thinning (sawing, grinding or polishing), chemical thinning (electro polishing) and ion milling (carry off by Ar or Ga).

The mostly used method to prepare metallic bulk materials is the electro polishing of thin slices to electron transparent foils. From the initial test piece a slice of about 1x1 cm and about 0.5 to 1 mm thickness is cut by diamond disc saw or wire saw. The aim of the subsequent grinding a polishing process is to achieve a thin foil of a homogeneous thickness of <100 μm. The grinding must be done on both sides of the sawed slice and is subdivided in steps using SiC grinding paper with different granulations, starting with 240 and finished with 1200 grid. In order to insure parallelism of the slice the use of lapping and polishing fixtures is recommended. In order to avoid surface artefacts, as grinding force only the weight of the grinding fixture is applied. The final polished thin foil is subsequently punched to discs of 3mm diameter.

The grinded and punched discs are now electro polished using a single-stage jet polisher. The disc with a maximum thickness of 0.1 mm is placed between two immersed jets. A pump system sends a flow of electrolyte through the nozzles against the specimen. An electric circuit is established with a DC-power supply, with a cathode placed in the electrolyte and with the specimen connected as anode, and the material will be removed by electrolytic excavitation from the specimen. This results in the forming of a small hole in the centre of the specimen with an electron transparent surrounding.

The process of electro polishing is controlled by different parameters such as: Selection of electrolyte: anode voltage and anode current density and fluid-dynamic parameters. The electrolyte solution consists of two components, the oxidising agent and a solvent for oxidising products. A list of other receipts for other materials can be found e.g. in [1]. A help for the choice of the optimum voltage is the voltage – current curve. In many cases the curve shows a plateau where the current density is more or less independent on the voltage and where the electro polishing occurs. The fluid dynamic

M. Luysberg, K. Tillmann, T. Weirich (Eds.): EMC 2008, Vol. 1: Instrumentation and Methods, pp. 795–796, DOI: 10.1007/978-3-540-85156-1_398, © Springer-Verlag Berlin Heidelberg 2008

parameters are flow rate, nozzle diameter and electrolyte viscosity, which must be optimised for the actual application. A good description is given in [2].

For the investigation of prepared cross sections in TEM the sandwich preparation method becomes more and more important, which requires a pre-preparation of a composite before the thinning process. From the interesting surface layer two thin slices (about 0.1 to 0.2 mm) are cut and then mounted layer by layer, using epoxy glue for fixing. This sandwich is arranged in two ceramic half cylinders and enveloped with a metallic (or ceramic) tube of 3 mm outer diameter. Half cylinders and tube stabilise the composite of the sandwich.

In a further step the composite is cut to thin slices, which are thinned similar as described for the electro polishing process. Before the last thinning by ion milling the specimen must be dimpled up to a residual thickness of < 5mm in the centre of the disc. The final thinning occurs with a low angle ion milling machine. The thinning and the transparency of the specimen can be observed by a special microscope. A description of all steps of the sandwich method and a video demonstration is given in [3,4].

In last years the application of focused ion beam (FIB) machines for the TEM specimen preparation becomes more and more important and allows the preparation of materials which cannot be handled by conventional methods. A further important aspect is the possibility of target preparation, which is demonstrated here with a preparation of a crack tip and its surrounding in a metallic sample. The complete process and a video demonstration can be found in [4].

In a FIB a Ga ion beam is used to sputter the material. The ion beam is controlled similar the electron beam in a SEM and the resolution of about 5 to 10 nm allows a very precisely cutting with high beam current or imaging with low beam current by excitation of secondary electrons.

There are two types of FIB, single beam and dual beam machines. The last type allows the observation of the milling process by an electron beam similar as SEM. Two methods of preparation techniques by FIB are used: the window method and the "lift-out" method. With the lift-out method a small lamella is cut directly from the surface of a supplied test piece. For target preparation (e.g. precipitations or grain boundaries) pre-preparation such as metallographic cross section is required.

The details of the thinning process are shown in [4,5] . The last step of the thinning process by FIB is the lift-out of the lamella. Two methods are used today: the ex-situ-lift-out under an external optical microscope and the in-situ-lift-out directly under observation in the FIB chamber. For transportation lamella is fixed on a W-needle and then on a special TEM grid.

FIB is a versatile and easy instrument for the TEM specimen preparation. Its presence in laboratories is a new one, but will expand in the near future.

1. K.C. Thompson-Russel, J.W. Edington, Pract. Electron Microscopy in Material Science, Monograph 5, MacMillan, Philips Techn. Lib, Eindhoven, 1977
2. B.J. Kestel, MRS Symp. Proc. Vol 199, Boston, 1990, p 356
3. H.J. Klaar, C.A. Huang, Prakt. Metallogr. 6, Vol 31, Carl Hanser Verlag, München, 1994
4. H.J. Penkalla, Internet under http://www.fz-juelich.de/ief/ief-2/datapool/VideoTEM/
5. R. Alani, R.J. Mitro, P.R. Swann, MRS Symp. Proc. Vol 480, Boston, 1997, p .187

Ultramicrotomy in biology and materials science: an overview

H. Gnaegi[1], D. Studer[2], E. Bos[3], P. Peters[3], J. Pierson[3]

1. Diatome Ltd, P.O. Box 1164, CH-2501 Biel, Switzerland
2. Institute of Anatomy, Baltzerstrasse 2, CH-3000 Bern, Switzerland
3. National Cancer Institute, Plesmanlaan 121, 1066 CX Amsterdam, The Netherlands

helmut.gnaegi@diatome.ch
Keywords: ultramicrotomy, diamond knife, sample preparation

A breakthrough in ultrathin sectioning was the introduction of diamond knives [1]. Glass knives are suitable for the sectioning of acrylate or epoxy resin embedded biological samples. However, diamond knives allow the sectioning of a great variety of samples. No matter whether these samples are soft and hard (of biological or of materials origin), or whether they are sectioned at room or at low temperatures. Diamond knives are used for wet and dry sectioning (the latter eg is mandatory for secondary ion mass spectroscopy SIMS [2]). Furthermore diamond knives are long lasting and very sharp-egded. The radius curvature of high quality diamond cutting edges is less than 5nm [3].

During manufacturing diamond knife characteristics (knife angle, polishing quality, surface properties, etc) are adapted to the needs of a particular application. The requirements are manifold: Sections as thin as 15nm may be needed for industrial materials such as metals, crystals, semiconductors, etc, while sections as thick as 5µm are required for optical microscopy [4]. Serial sections are often used in optical microscopy for 3D reconstruction [5].

Polymer samples are microtomed at room temperature or at cryo temperatures, according to their glass transition temperature. Epoxy and acrylic resins are used for embedding biological samples. These samples are sectioned at room temperature. The sections may undergo compression (compression is the shortening of the section compared to the sample size in the cutting direction).

Less compression and hence a better structure preservation is achieved by reducing the wedge angle from 45° to 35° or even to 25° [6]. However, reducing the wedge angle leads to a more sensitive cutting edge. An alternative is to oscillate the knife. This approach allows compression reduction and improved structure preservation [7, 8]. Applications for the oscillating knife are rather rigid polymers and biological samples in epoxy or acrylate resins.

Ultramicrotomy of hard materials samples is an alternative to the established preparation techniques (such as polishing, FIB, ion thinning etc). Hard material samples are embedded in a rigid epoxy resin [9]. Metallic coatings on brittle substrates may be processed without embedding [10]. If a sample is resin embedded, sections of a thickness of 35nm may be achieved, non embedded samples may be sectioned as thin as 15nm.

M. Luysberg, K. Tillmann, T. Weirich (Eds.): EMC 2008, Vol. 1: Instrumentation and Methods, pp. 797–798, DOI: 10.1007/978-3-540-85156-1_399, © Springer-Verlag Berlin Heidelberg 2008

For the sectioning of metals very small sample blocks (width <50µm) are trimmed. Feed is set at 15-30nm. One has to take into account that a number of metals attack diamond chemically. In this case the cutting edge worns out rapidly.

In biology sucrose infiltrated and frozen hydrated samples are processed by cryo-ultramicrotomy. The sucrose infiltrated samples are sectioned at about 150K, the sections are thawn and thereafter immunolabelled [11]. For the pick-up of the sections a metal loop with a droplet of sucrose/methyl-cellulose solution is used.

Cryo-ultramicrotomy of frozen hydrated samples at 120K gains increasing interest in the biologists community. The sample preparation by high pressure freezing, cryo-ultramicrotomy, cryo-transfer and cryo-TEM preserves living matter close to the living state [12]. Ultrathin cryosections serve for high resolution imaging and 3D reconstruction [13, 14]. Typical artefacts in cryosectioning of frozen hydrated samples are: compression, chatter, knife marks and crevasses [15]. Optimising sectioning and the environment conditions of the microtome reduce these artefacts (eg. 20% relative humidity). Substantial reduction of ice crystal contamination on knife and sections is achieved by a newly developed glovebox surrounding the microtome: the cryosphere. Section pick-up, transfer of the sections onto the grid and the pressing for flattening is difficult. Micromanipulators for the handling of the section ribbons and the grids make cryosectioning easier and less depending on the operators skill. An antistatic device (ionizer) eliminates electrostatic charging in the cryochamber and improves the gliding of cryosections on the knife surface [16]. A new ionizer/charger type is used to electrostatically fix cryosections on the carbon film of the grid. Flat carbon films are mandatory if cryosections are generated for tomography.

High quality diamond knives are not just used for generating sections, but also for surfacing all kinds of biological and material specimens for AFM investigation 17, 18]. Extremely smooth surfaces with a roughness of a few nm are obtained. For achieving the best possible surface quality the sections are cut as thin as possible.

1. H. Fernández-Morán, Experimental Cell Research 5/1 (1953), p. 255-256.
2. J.L. Guerquin-Kern, T.D. Wu, C. Quintana and A. Croisy, BBA 1724/3 (2005), p. 228-238.
3. K. Lickfeld, Journal of Ultrastructural Research 93 (1985), p. 101-115.
4. O.L. Reymond, Basic Applied Histochemistry 30 (1968), p. 487-494.
5. M.J.F. Blumer, P. Gahleitner, T. Narzt, C. Handl and B. Ruthensteiner, Journal of Neuroscience Methods 120 (2002), pl 11-16.
6. J J.C. Jésior, Scanning Microscopy Supplement 3 (1989), p. 147-153.
7. D. Studer and H. Gnaegi, Journal of Microscopy 197 (2000), p. 94-100.
8. J.S. Vastenhout and J.D. Harris, Microscopy Today (2006), p. 20-21.
9. P. Swab and R.E. Klinger, Materials Res. Soc. Symp. Proceedings (1988), p. 229-234.
10. P. Schubert-Bischof and T. Krist, Microscopy and Microanalysis Proceedings (1997), p. 359.
11. K.T. Tokuyasu, Journal of Cell Biology 57 (1973), p. 551-565.
12. Al-Amoudi A. et al. (2004). EMBO J. Vol. 23, p. 3583-3588.
13. C.E. Hsieh, A. Leith, C.A. Mannella, J. Frank and M. Marko, J. Struct. Biol. (2006), p. 1-13.
14. Al-Amoudi, D.C. Diaz, M. Betts and A.S. Frangakis, Nature 450 (2007), p. 832-837.
15. Al-Amoudi, D. Studer and J. Dubochet, Journal of Structural Biology 150 (2005) p. 109-121.
16. M. Michel, H. Gnaegi and M. Müller, Journal of Microscopy 166 (1992), p. 43-56.
17. P.H. Vallotton et al., Journal of Biomaterials Science, Polymer Edition 6 (1994), p. 609-620.
18. N. Matsko and M. Müller, Journal of Structural Biology 146 (2004), p. 334-343.

Preparation of Biological Samples for Electron Microscopy

H. Schwarz[1] and B.M. Humbel[2]

1. Max-Planck-Institut für Entwicklungsbiologie, Spemannstr. 35, 72076 Tübingen, Germany
2. Electron Microscopy & Structure Analysis, Dept. Molecular Cell Biology, Utrecht University, Padualaan 8, 3584 CH Utrecht, The Netherlands

heinz.schwarz@tuebingen.mpg.de

Keywords: Chemical fixation, Cryofixation, Dehydration

Biological structures of living organisms are highly dynamic, they are always in an aqueous environment and the components are mainly made of elements of low atomic number: H, C, O, N, P, S. Last but not least most of the organism are rather large. The physical properties of electrons define the preparation methods of biological samples. Electrons have a short mean free pathway of about 1 cm in air and even shorter in denser structures, therefore the electron microscope needs to be evacuated. Thus an ideal sample for electron microscopy is tiny, does not move, is rigid, contains no liquids and has a high contrast. In addition for transmission electron microscopy the sample must be thin. The opposing properties and demands of electrons versus biology led to cumbersome path on biological sample preparation. In the beginning viruses and bacteria were studied [1, 2]. They were adsorbed on support film and initially directly imaged with the electron beam. The poor contrast was improved by heavy metal shadowing. For the visualization of molecules and virus particles in the 50ies the spreading technique introduced by Kleinschmidt [3] as well as the negative staining technique pioneered by Brenner & Horne [4] were cornerstones.

Already in 1945 Porter et al. [5] cultivated isolated chicken cells on the support film and fixed them in vapour of osmium tetroxide. For the first time remarkable ultrastructural details of a cell were visualised: cytoskeleton, protrusions, mitochondria and cell-cell contacts between the fibroblast and nerve cells. The next important step was the introduction of embedding of chemically fixed tissue in electron beam resistant resins [6, 7], the ultramicrotomy [8], positive staining with heavy metal salts such as uranyl acetate and lead nitrate or lead citrate [9, 10] and the introduction of glutaraldehyde as a primary fixative by Sabatini [11]. These were the prerequisites for the explosion of biological knowledge in the 50ties and 60ties of the last millennium. Up to now most of the text book illustrations originate from this era.

With the wealth of data doubts about the reliability grew. Many publications turned-up about changes of morphology, loss of lipids and destruction of proteins. As early as 1960, Fernandez-Moran [12] suggested cryo-electron microscopy as the solution. He introduced cryo-fixation, imaging of frozen-hydrated samples and freeze-substitution combined with low-temperature embedding. Many years had to come before his vision became true: A real challenge was the application of cryomethods in biological specimen preparation to proof the quality of the up to then established classical methods

M. Luysberg, K. Tillmann, T. Weirich (Eds.): EMC 2008, Vol. 1: Instrumentation and Methods, pp. 799–800, DOI: 10.1007/978-3-540-85156-1_400, © Springer-Verlag Berlin Heidelberg 2008

of chemical fixation, room temperature dehydration and resin embedding followed by heat polymerisation.

Here, major breakthroughs were the validation of freeze-substitution [13] and the introduction of freeze-fracturing [14, 15], high pressure freezing [16, 17], imaging of biological structures in amorphous ice [18], cryo-fixation and freeze-substitution followed by low-temperature embedding in methacrylates [19, 20] and reliable cryosectioning of native cryofixed biological material, analysed by cryo-electron microscopy [21].

Moreover, cryosectioning of chemically fixed material infiltrated in high concentration of sucrose introduced by Tokuyasu [22] allowed the widespread application of immunolocalisation of epitopes within biological specimen on thawed cryosections parallel to the on-section labelling of resin embedded material.

Finally, the inspection of frozen-hydrated cryosections in a cryo-TEM (CEMOVIS) allowed demonstrating the biological structure in a life-like state. Thus, the benefits and drawbacks of the traditional methods of aldehyde fixation, room temperature dehydration and resin embedding could be evaluated.

1. G.A. Kausche, E. Pfankuch and H. Ruska, Naturwiss. **27** (1939), p. 292.
2. M. von Ardenne "Elektronen-Übermikroskopie", (Springer Verlag, Berlin) (1940)
3. A. Kleinschmidt and R. K. Zahn, Zeitschr. Naturforsch. **14b** (1959), 770.
4. S. Brenner and R. W. Horne, Biochim. Biophys. Acta **34** (1959), 103.
5. K. R. Porter, A. Claude and E. F. Fullam, J. Exp. Med. **41** (1945), 233.
6. A. M. Glauert and R. H. Glauert, J. Biophys. Biochem. Cytol. **4** (1958), 191.
7. J. H. Luft, J. Biophys. Biochem. Cytol. **9** (1961), 409.
8. K. R. Porter and J. Blum, Anat. Rec. **117** (1953), 685.
9. E. S. Reynolds, J. Cell Biol. **17** (1963), 208.
10. J. H. Venable and R. Coggeshall, J. Cell Biol. **25** (1965), 407.
11. D. D. Sabatini, K. Bensch and R. Barnett, J. Cell Biol. **17** (1963), 19.
12. H. Fernandez-Moran, Ann. N. Y. Acad. Sci. **85** (1960), 689.
13. A. Van Harreveld and J. Crowell, Anat. Rec. **149** (1964), 381.
14. R. L. Steere, J. Biophys. Biochem. Cytol. **3** (1957), 45.
15. H. Moor, K. Mühlethaler, H. Waldner et al., J. Biophys. Biochem. Cytol. **10** (1961), 1
16. U. Riehle, ETH Diss. Nr. 4271 Zürich (1968)
17. M. Müller and H. Moor, Science of Biological Specimen Preparation 1983, 131 (SEM Inc., AMF O'Hare) (1984)
18. J. Dubochet. J. Lepault, R. Freeman, J. Berriman and J. Homo, J. Microsc. **128** (1982), 219.
19. E. Carlemalm, R. M. Garavito and W. Villiger, J. Microsc. **126** (1982), 123.
20. B. Humbel, T. Marti and M. Müller, Beitr. Elektronenmikroskop. Direktabb. Oberfl. **16** (1983), 585.
21. A. Al-Amoudi, L. P. O. Norlen and J. Dubochet, J. Struct. Biol. **148** (2004), 131.
22. K. T. Tokuyasu, J. Cell Biol. **57** (1973), 551.

Web sample preparation guide
for transmission electron microscopy (TEM)

Jeanne Ayache[1], Luc Beaunier[2], Jacqueline Boumendil[3], Gabrielle Ehret[4]
and Danièle Laub[5]

1. Jeanne Ayache, CNRS-UMR 8126, LM2C-Institut Gustave Roussy
94805 Villejuif Cedex, France
2. Luc Beaunier, CNRS-UPR 15, LISE, Université P. et M. Curie,75252 Paris
3. Previously Université Claude Bernard-Lyon1, Villeurbanne, France
4. Previously Institut Physique et Chimie des Matériaux de Strasbourg, France
5. Danièle Laub, EPFL-CIME, Station 12, 1015 Lausanne, CH

ayache@igr.fr
Keywords: sample preparation, TEM, interactive guide, website, Physics and biology

Sample preparation is of central importance for the characterization of materials by transmission electron microscopy (TEM). As a guide to researchers seeking practical and up-to-date information on TEM sample preparation, we have created the Internet website **http://temsamprep.in2p3.fr/** "Figure 1".

This website is dedicated to TEM sample preparation in Materials Science, Earth Science and Biology. Designed in French and now translated into English, this website is accessible free of charge.

The website is an interactive didactic guide to suggest the most suitable TEM sample preparation technique. It is based on materials properties, types of TEM observation and analysis "Figure 2".

At present, the guide covers 36 sample preparation techniques with essential information about the basic principles, materials, procedure, artefacts, advantages, drawbacks, as well as the specific equipment required for each technique. It has been designed to include new techniques as they are being developed.

The website also contains a database with pictures of different types of materials and of artefacts produced by sample preparation techniques. This database has been designed to be interactive and will be supplemented by new pictures of various materials from the users community.

This guide is completed by two books series, available in French for now and soon in English [1, 2]. Volume 1 is dedicated to methodology of the material approach and covers theoretical and practical aspects of sample preparation for TEM.

Volume 2 is dedicated to technical hints. 36 different pre-preparation, preparation and post-preparation protocols are developed.

1,2. J. Ayache, L. Beaunier, J. Boumendil, G. Ehret and D. Laub in "Guide de Préparation des échantillons pour la microscopie électronique en transmission" **Volume 1** Méthodologie, **Volume 2** Techniques, ed. *Les Publications de l'Université de Saint Etienne* (2007) France.

M. Luysberg, K. Tillmann, T. Weirich (Eds.): EMC 2008, Vol. 1: Instrumentation and Methods, pp. 801–802, DOI: 10.1007/978-3-540-85156-1_401, © Springer-Verlag Berlin Heidelberg 2008

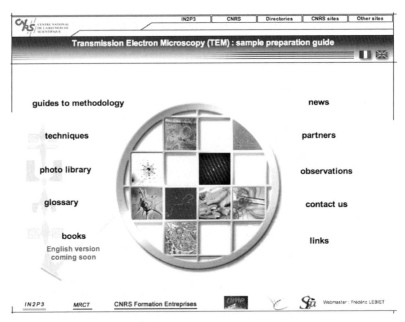

Figure 1. Image of the website gate.

Figure 2. Partial image of the didactic guide. The 5 techniques containing only the sign √ correspond to the possible techniques as regards the in coming material criteria chosen by the user, referring to the material and/or the TEM characterization. The cross indicates an impossibility of using this particular technique for these specific criteria among the different choices of properties. They are listed on when they are activated.

Novel carbon nanosheets as support for ultrahigh resolution structural analysis of nanoparticles

André Beyer[1], Christoph Nottbohm[1], Alla Sologubenko[2], Inga Ennen[1], Andreas Hütten[1], Harald Rösner[3], Joachim Mayer[2], and Armin Gölzhäuser[1]

1. Fakultät für Physik, Universität Bielefeld, Postfach 10 01 31, 33501 Bielefeld, Germany
2. Gemeinschaftslabor für Elektronenmikroskopie, RWTH Aachen, Ahornstrasse 55, 52074 Aachen and Ernst Ruska-Centre for Microscopy and Spectroscopy with Electrons, Research Centre Jülich 52425 Jülich, Germany
3. Institut für Nanotechnologie, Forschungszentrum Karlsruhe, Postfach 3640, 76021 Karlsruhe, Germany

beyer@physik.uni-bielefeld.de

Keywords: self assembled monolayer, ultrathin carbon film, transmission electron microscopy, scanning transmission electron microscopy, nanoparticles

The resolution in transmission electron microscopy (TEM) has reached values as low as 0.08 nm. However, these values are not accessible for very small objects in the size range of a few nanometers or lower as they have to be placed on some support, which contributes to the overall electron scattering signal, thereby blurring the contrast. Here, we report on the use of nanosheets made from cross-linked aromatic self-assembled monolayers (SAM) as TEM sample supports.

The fabrication process is schematically shown in Fig. 1. First, we prepared a SAM of 4´-[(3-Trimethoxysilyl)propoxy]-[1,1´-biphenyl]-4-carbonitril (CBPS) on silicon nitride. We cross-linked these SAMs with electrons, dissolved the silicon nitride substrate in hydrofluoric acid (HF) and deposited the monolayer onto a TEM grid. The single 1.6 nm thick nanosheet can cover the grid and is free-standing within the micron-sized openings.

It is difficult to detect the 1-2 nm thick nanosheet with a scanning electron microscope. In high magnification micrographs (Fig. 2a), the monolayer membrane is completely featureless and the frame appears white due to high contrast. Some monolayer membranes will rupture during the fabrication process. One example is shown in Fig. 2c, where the ruptured part of the membrane is folded into one corner. In these cases the monolayer can clearly be seen. Improved imaging of the monolayer is possible by employing a He ion microscope.

Despite its thinness, the sheet is stable under the impact of the electron and He ion beam. TEM micrographs taken from nanoclusters onto these nanosheets show highly increased contrast in comparison to images taken from amorphous carbon supports (Fig. 3). In scanning transmission electron microscopy with nanosheet support, a size analysis of sub-nanometer Au clusters was performed and single Au atoms were resolved.

M. Luysberg, K. Tillmann, T. Weirich (Eds.): EMC 2008, Vol. 1: Instrumentation and Methods, pp. 803–804, DOI: 10.1007/978-3-540-85156-1_402, © Springer-Verlag Berlin Heidelberg 2008

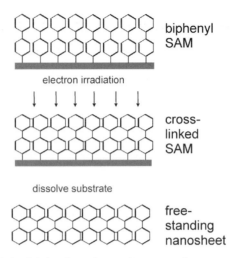

Figure 1. Schematic of the fabrication of monolayer membranes.

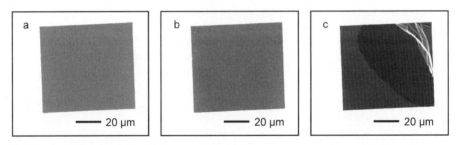

Figure 2. SEM images of nanosheet suspended over rectangular openings in a silicon sample: (a) a clean membrane, (b) with some etch residues, (c) a ruptured membrane

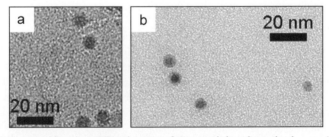

Figure 3. High magnification TEM image of Co particles deposited on a 15 nm carbon film (a) and on a nanosheet (b) under otherwise identical conditions.

A new automated plunger for cryopreparation of proteins in defined - even oxygen free - atmospheres

F. Depoix, U. Meissner and J. Markl

Institute of Zoology, Johannes Gutenberg-University, 55099 Mainz, Germany

depoix@uni-mainz.de

Keywords: Cryoelectronmicroscopy, Specimen preparation oxgenfree, Plunge freezing oxgenfree

We study the structure and function of hemocyanins. They are giant extracellular oxygen carriers in the hemolymph of many molluscs and arthropods. Since some of these blue, copper-containing proteins show the highest cooperativity in nature (h = 10), one of our goals is to understand the chemomechanical interaction between the different substructures during allosteric oxygen binding.

X-ray crystal structures of hemocyanin subunits are available. Moreover, several 10 Å cryo-EM structures of the whole molecules have been recently published [1-3]. They allow the fitting of crystal structures to obtain overall molecular models, which can lead to understand the inner molecular signal transfer.

It is now crucial to obtain such 3D reconstructed quaternary structures in different functional states, notably the fully oxygenated and the fully deoxygenated form.

This can either be done by an in-silico post processing of cryo-specimen with mixed populations of both states via single particle analysis, or by preparation of homogeneous populations of defined states prior to shock-freezing. Ideally both methods can be combined.

The present study follows the second strategy, using a new, fully automated plunger for cryofixation in atmospheres of defined oxygen content (25% O_2/75% N_2 for oxy- and 100% N_2 for deoxy-hemocyanin). In contrast to existing, commercially available systems it has the following advantages:

- Possibility to work with preincubated, oxygen-free specimen under oxygen-free conditions during the whole process of plungefreezing.
- Small volume of the plunging chamber to quickly establish stable atmospheres.
- Control of all parameters relevant for cryofixation (O_2, humidity, temperature, blotting time...)

With the new cryoplunger we already produced series of excellent grids. With several hemocyanin types prepared both in 25% and 0% oxygen, we observed significant structural differences. A preliminary example (tarantula hemocyanin) is shown in Figure 1-3. Other data will be shown at the poster session of the congress by A. Moeller and C. Gatsogiannis.

Even if this device has been designed for the study of respiratory proteins, it could also be very useful for preparations under clean room conditions or in case of oxygen-sensitive samples.

M. Luysberg, K. Tillmann, T. Weirich (Eds.): EMC 2008, Vol. 1: Instrumentation and Methods, pp. 805–806, DOI: 10.1007/978-3-540-85156-1_403, © Springer-Verlag Berlin Heidelberg 2008

1. Meissner et al. (2007), Micron **38 (7)**, 754-765;
2. Martin et al. (2007), J. Mol. Biol. **366**, 1332-1350
3. Gatsogiannis et al. (2007), J. Mol. Biol. **374 (2)**, 465-486

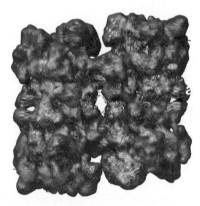

Interface a↔a

Figure 1.
Tarantula (*Eurypelma californicum*) hemocyanin.
This hemocyanin is composed of four hexamers which are connected via several molecular interfaces. During oxygenation, the hexamers are dislocated with respect to each other, leading to changes at the interfaces.

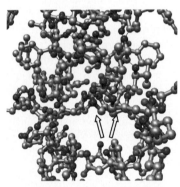
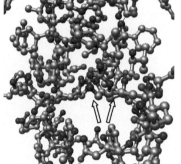

Figure 2.
Interface a↔a in the presumed oxy-state of the hemocyanin.
Several histidins that might be involved in allosteric interaction [2] form a dense cluster.

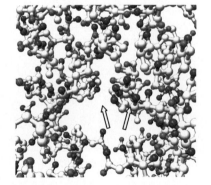

Figure 3.
Interface a↔a in the presumed deoxy-state of the hemocyanin.
The histidins move apart.

A novel method for precipitates preparation using extraction replicas combined with focused ion beam techniques

M. Dienstleder[1,2], H. Plank[1,2], G. Kothleitner[1,2], F. Hofer[1,2]

1. Centre for Electron Microscopy, Graz, Austria
2. Institute for Electron Microscopy, Graz University of Technology, Graz, Austria

martina.dienstleder@felmi-zfe.at

Keywords: focused ion beam, extraction replicas, plane view preparation

The nanoscale analysis of small precipitates embedded in a matrix via electron energy loss spectroscopy (EELS) or energy-dispersive x-ray spectroscopy (EDXS) in transmission electron microscopes (TEM) can be hampered because of matrix signals, contributing to the data of interest. Conventional preparation techniques such as Ar^+ ion milling or focused ion beam (FIB) preparation are unable to remove the matrix, asking for different ways of sample preparation. Extraction replicas represent an alternative, where pre-etched sample surfaces are covered with a conducting film (replica) like carbon, silicon, alumina oxide, polymer films or others [1-4]. These samples are then etched from the backside, extracting the precipitates from the matrix, while they are still fixed on the replica. However, such samples can show some drawbacks: *i*) replica related problems during preparation (e.g. mechanical stress) can prevent a successful extraction; *ii*) the chemistry of the replica is often overlapping with the precipitant's composition and complicates reliable analytical investigations; *iii*) the precipitates are often too thick and non-electron-transparent for TEM / EELS investigations, which requires further thinning steps (e.g. by FIB). In this work we demonstrate the application of two different replica materials, which are cast from solution and cured via UV and / or temperature, in a concept combining FIB steps. The first approach is a temporary transfer-substrate with a cross-linking polymer that can withstand subsequent etching. After precipitate extraction, an analytically non-interfering metal layer is sputtered, and the polymer is then removed completely, leading to a carbon-free replica. The second approach entails a permanent substitution of the substrate by a material that has a defined chemical nature as well as excellent mechanical properties. FIB technology is then used for either approach to prepare plane-view lamellas, followed by a final thinning step to meet the requirements for analytical TEM. The suitability of both replica materials for TEM sample preparation was characterized with respect to optimal preparation conditions. This novel method will be demonstrated for the analysis of chromium-carbide precipitates in a steel and tested via TEM and EELS and EDXS.

1. G. Petzow "Metallographisches Keramographisches Plastographisches ÄTZEN"
2. L. Reimer „Elektronenmikroskopische Untersuchungs- und Präparationsmethoden (1967)
3. C.P. Scott, D. Chaleix, P. Barges, V. Rebischung, Scripta Materialia **47** (2002), 845
4. K.M. Pickwick and R.H. Packwood, Metallography **9**, (1976), 245

M. Luysberg, K. Tillmann, T. Weirich (Eds.): EMC 2008, Vol. 1: Instrumentation and Methods, pp. 807–808, DOI: 10.1007/978-3-540-85156-1_404, © Springer-Verlag Berlin Heidelberg 2008

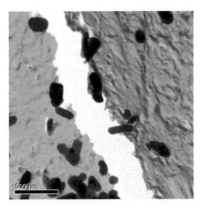

Figure 1. TEM BF image of a classical extraction replica of precipitates in a carbon film.

Figure 2a-c. SEM images, FIB plane view preparation

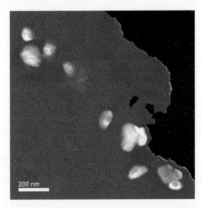

Figure 3. Cr L_{23} jump ratio image of extracted and FIB milled precipitates in a 12% Cr-steel.

An appraisal of FIBSEM and ultramicrotomy for the TEM analysis of the cell-biomaterial interface

H.K. Edwards,[1] M.W. Fay,[2] S.I. Anderson,[3]
C.A. Scotchford,[1] D.M. Grant[1] and P.D. Brown[1]

1. School of Mechanical, Materials and Manufacturing Engineering,
2. University of Nottingham Nanotechnology and Nanoscience Centre,
3. University of Nottingham Advanced Microscopy Unit, School of Biomedical Sciences, Queen's Medical Centre, all at the University of Nottingham, NG7 2RD, UK.

michael.fay@nottingham.ac.uk

Keywords: FIBSEM, cryo-FIBSEM, ultramicrotomy, cell, biomaterial

Transmission electron microscopy (TEM) analysis of the interface between biological cells and biomaterials in cross-section is a challenging process. Although ultramicrotomy has developed as a powerful tool for the sectioning of tissues and even the biomaterial-bone interface [1], the sectioning of samples containing a hardness mismatch, such as soft cells attached to a biomaterial, proves more difficult to achieve without incorporating cutting artefacts into the cells. Hence, biomaterial substrates are often removed from biological samples before microtomy, leaving only the surface cells to be imaged [2]. The preparation of samples for the chemical analysis of cell-biomaterial interface presents an interesting materials science problem.

An alternative to ultramicrotomy for the preparation of TEM specimens is focused ion beam scanning electron microscopy (FIBSEM), which has long been employed it the analysis of semiconductor and engineering materials, but has recently seen more applications in the area of the life sciences [3]. The FIBSEM lift-out technique allows a site-specific cross-sectional specimen to be removed from a bulk sample and attached to a TEM grid for further analysis. This method may be applied to the cell-biomaterial system for chemically fixed samples embedded in a thin layer of resin. An advantage of FIBSEM is that the sectioning of delicate or composite samples may be achieved without the sample compression associated with ultramicrotomy. However, it may result in its own set of artefacts such as Ga^+ ion implantation, surface amorphisation and curtaining effects, whereby a section finish, particularly of a soft material, is affected by differential ion beam milling through the sample. In this context, the applicability of FIBSEM to the sectioning of biological cells may be extended through the use of a cryo-transfer system, which allows chemically un-fixed and hydrated biological specimens to be rapidly frozen in slush liquid nitrogen, transferred into the FIBSEM chamber and subsequently imaged and milled at low temperature. This technique may also be used, in principle, for the preparation of cross-sectional TEM specimens, provided a cryo-transfer system is available for the transport of samples from the FIBSEM to the TEM. Cryo-FIBSEM avoids the problems associated with chemical fixation and dehydration as sources of artefact, but has its own set of potential artefact problems such as frost accumulation and accentuated curtaining effects.

M. Luysberg, K. Tillmann, T. Weirich (Eds.): EMC 2008, Vol. 1: Instrumentation and Methods, pp. 809–810, DOI: 10.1007/978-3-540-85156-1_405, © Springer-Verlag Berlin Heidelberg 2008

810

In this context, an appraisal of ultramicrotomy, FIBSEM and cryo-FIBSEM for the preparation of cross-sectional specimens of cell-biomaterial bi-layers is presented. To assess the viability of these techniques for the chemical analysis of the interface between attached cells and biomaterials, and their associated artefacts, model systems of mesenchymal stem cells (MSCs) and human osteoblasts (HOBs) on Ti have been employed.

Ti foils 0.025 - 0.25 mm in thickness were used as substrates for MSC and HOB attachment. Cell-Ti samples for FIBSEM lift-out and ultramicrotomy were fixed in 3 % glutaraldehyde, washed in phosphate buffered saline solution (PBS), post-fixed with 1 % OsO_4 and potassium ferrocyanide, dehydrated and embedded in epoxy based resin. Ultramicrotomy was carried out using a Reichert Jung Ultracut Ultramicrotome, whilst the preparation of FIB cross-sections was carried out using an FEI Quanta 200 3D FIBSEM fitted with a Quorum PT2000 cryo-transfer unit and Omniprobe micromanipulator. For cryo-FIBSEM, both fixed and OsO_4 stained and unfixed cell-Ti samples were washed in PBS before being plunged into slush nitrogen and transferred into the evacuated cryotransfer unit held at -140 °C. The samples were transferred into the FIBSEM chamber, which was also held at -140 °C and after gentle *in-situ* sublimation at -90 °C to remove any ice crystal formation, electron transparent regions were milled into the specimens. For FIBSEM and cryo-FIBSEM, the samples were sputter coated with Pt and Au and the FIBSEM gas injector system was used to apply protective W coatings. Electron transparent specimens were inspected using a JEOL 2100f TEM.

1. A. Palmquist, T. Jarmer, L. Emanuelsson, R. Brånemark, H. Engqvist and P. Thomsen, Acta Orthadaedica **79** (2008) p.78.
2. S.I. Anderson, Ph.D. thesis: The effect of silicon, silica and silicates on the osteoblast in vitro, University of Nottingham, (2001)
3. J. Heymann, M. Hayles, I. Gestmann, L. Giannuzzi, B. Lich and S. Subramaniam, J. Structural Biology **155** (2006) p.63.
4. This work was supported by the EPSRC under grant EP/E015379/1.

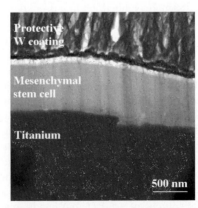

Figure 1. Bright field TEM image of an electron transparent lamella prepared by cryo FIBSEM of a fixed mesenchymal stem cell attached to Ti.

Observation of the structure of aqueous polymers with cryo-SEM

A. Fujino[1], M. Yamashita[1] and N. Satoh[1]

1. Research Laboratories, KOSÉ Corporation, 1-18-4 Azusawa, Itabashi-ku, Tokyo 174-0051, Japan

a-fujino@kose.co.jp

Keywords: aqueous polymer, cryo-SEM, cryo-ultramicrotome

Aqueous polymers are included in cosmetics such as milky lotions and creams to make them smooth to the touch while they are put on one's face, to increase their viscosity and to keep them stable. Properties and qualities of cosmetics depend on the structure of aqueous polymers. For development and warranty of cosmetics, it is important to reveal the structure of aqueous polymers in water when cosmetics are produced.

The structure of polymers can be generally analyzed via using nuclear magnetic resonance (NMR), light scattering, microscopes and so on. Especially, the cryo-scanning electron microscope (cryo-SEM) method enables to observe the polymer as three-dimensional structure without troublesome handling [1-3]. Also it is effective to observe the materials including water. We have already reported on the observations of o/w emulsions by cryo-SEM [4]. However the results showed the difficulties of ice formings to observe aqueous polymer solutions that contained water more than 90%. In this study, we report about the several examinations that the structure of aqueous polymer in water was ovserved.

Aqueous polymer solutions consisted of 2wt% Acrylates / C10-30 alkyl acrylate crosspolymer were used. Samples were frozen rapidly by the metal contact method with gold-coated copper plates cooled in liquid nitrogen. Then parts of samples were cut up with a cryo-ultramicrotome (Leica EM, FC6) to make fresh planes, and then observed in a cryo-SEM (JSM-6700F/Alto2500) at accelerating voltage 1kV. Also another parts of samples were cut using a cryo-ultramicrotome to prepare the cross-section of the sample.

Fig.1 showed a cryo-SEM image of polymer solutions that was observed vertically against the plane cut by a cryo-ultramicrotome without any influence of ice. Aqueous polymers in water revealed fine network structures as like intertwined strings.

Fig.2 demonstrated a cryo-SEM image of cross-section of frozen polymer solutions, which revealed the structures of aqueous polymers from the metal contacted part toward the inside.

1. M. E. Fray, A. Pilaszkiewicz, W. Swieszkowski and K. J. Kurzydlowski, European Polymer Journal **43** (2007) 2035-2040
2. J. P. M. van Duynhoven, I. Broekmann, A. Sein, G. M. P. van Kempen, G. J. W. Goudappel and W. S. Veeman, Journal of Colloid and Interface Science **285** (2005) 703-710.

M. Luysberg, K. Tillmann, T. Weirich (Eds.): EMC 2008, Vol. 1: Instrumentation and Methods, pp. 811–812, DOI: 10.1007/978-3-540-85156-1_406, © Springer-Verlag Berlin Heidelberg 2008

812

3. D. J. Stokes, J. Y. Mugnier and C. J. Clarke, Journal of Microscopy **213** (2004) 198-204
4. M. Yamashita, K. Kameyama, R. Kobayashi, A. Asahina, S. Aita, and K. Ogura, Journal Electron Microscopy **45** (1996) 461-462.

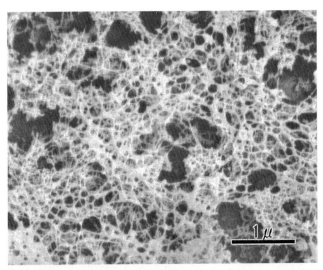

Figure 1. SEM image of aqueous polymer frozen via metal contact method with gold coated copper plates cooled in liquid nitrogen.

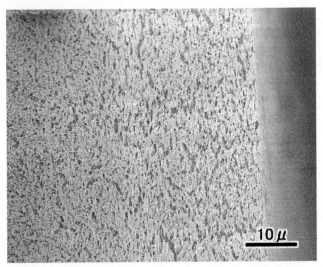

Figure 2. SEM image for the cross section of aqueous polymer via metal contact method with gold-coated copper plates cooled in liquid nitrogen.

Lipid nanotube encapsulating method in low voltage scanning transmission electron microscopy

H. Furusho[1], Y. Mishima[2], N. Kameta[3], M. Yamane[4], M. Masuda[5], M. Asakawa[5], I. Yamashita[4,6], A. Takaoka[1], and T. Shimizu[3,5]

1. Research Center for Ultra-High Voltage Electron Microscopy, Osaka University, 7-1 Mihogaoka Ibaraki, Osaka, Japan
2. CBM, CNRS, Rue Charles Sadron F45071, Orléans, France
3. SORST, JST, Tsukuba Central 5, 1-1-1 Higashi, Tsukuba, Ibaraki, Japan
4. CREST, JST, 4-1-8 Honcho, Kawaguchi, Saitama, Japan
5. NARC, AIST, Tsukuba Central 5, 1-1-1 Higashi, Tsukuba, Ibaraki, Japan
6. GSMS, NAIST, 8916-5 Takayama, Ikoma, Nara, Japan

furusho@uhvem.osaka-u.ac.jp
Keywords: lipid nanotube, sample fixation, low voltage STEM, ferritin

Lipid nanotube (LNT) [1] encapsulating method is an advanced sample fixation technique by encapsulating nanomaterials in homogeneous hollow cylinders of LNTs (shown in "Figure 1"). In characterizing nanomaterials with transmission electron microscopy (TEM), this method provides some advantages over usual method, in which samples are fixed on a supporting thin film.

The beam loss of tilted membrane critically influences the quality of energy filtering (EF) TEM and TEM computed tomography (CT). Using LNT encapsulating method, total sample thickness keeps uniform and thin when the sample is tilted or rotated around the LNT axis. This property of LNT addresses the problem of the supporting film whose thickness is relatively increased by tilting.

LNT encapsulating method has other merits; size controllability, high stability, good reproducibility, and so on. Using the glycolipid nanotubes (Glc-Cn-COOH, the structure model is shown in "Figure 2"), some applications of 300 kV TEM analysis of ferritin iron cores were already reported in refs. 2-3.

In this work, high-contrast images of ferritin and LNTs were successfully recorded by applying LNT encapsulating method to low voltage scanning transmission electron microscopy (LV-STEM) [4]. Using a cryo transfer holder as cooling sample (with liquid nitrogen, sample temperature was about 100K), thermal damage of low energy beam (0.5-30 kV) was decreased enough to take high magnification LV-STEM images without collapsing samples. Examples of LV-SEM/STEM high-contrast images are shown in "Figure 3".

On the other hand, we succeeded in recording a series of tilted LV-STEM images without a cryo transfer holder. It indicates that LNTs have enough durability for LV-STEM-CT analysis even at room temperature.

Based on these results and other LV-STEM image properties, we will discuss the possibility of using LNT encapsulating method.

M. Luysberg, K. Tillmann, T. Weirich (Eds.): EMC 2008, Vol. 1: Instrumentation and Methods, pp. 813–814, DOI: 10.1007/978-3-540-85156-1_407, © Springer-Verlag Berlin Heidelberg 2008

814

1. vT. Shimizu et al.,Chem. Rev. **105**(2005) 1401.
2. H. Furusho et al., Proc. IMC16, Sapporo (2006) 683.
3. H. Furusho et al., Jpn. J. Appl. Phys., **47**(2008) 394.
4. Takaoka et al., J. Electron Microsc., **55**(2006) 157.

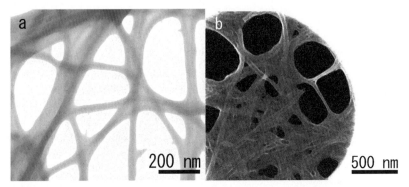

Figure 1. Low magnification EM images of LNTs lying over the microgrid holes. Nanomaterials are encapsulated in LNTs. (a) 300 kV EF-TEM image. (b)30 kV SEM image.

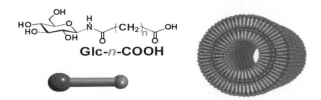

Figure 2. The structure model of LNT (glycolipid nanotube). The cylinder size, inner and outer surface properties are controllable by changing the molecular structure of the lipid molecules.

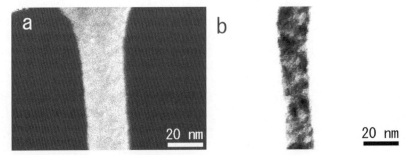

Figure 3. LV-STEM images of ferritin encapsulated in LNTs. (a) Cryo-SEM mode. (b) Cryo-STEM mode.

Microscopy observation of food biopolymers and related sample preparation methods

C. Gaillard

Microscopy core facility laboratory, Research Unit Biopolymers Interactions Assemblies (BIA), French National Institute for Agronomic Research – 44 000 Nantes, France

cedric.gaillard@nantes.inra.fr

Keywords: biopolymers, atomic force microscopy, transmission electron microscopy

Microscopy techniques represent useful tools for the investigation of food and non-food products from agricultural sources, and the understanding of the constituents organisation on the larger scale range, from the nanometre to the millimetre. As the biopolymers (polysaccharides, proteins, lipid chains) are the mean constituents of food products, the ability to characterise their assembly mechanisms during their biosynthesis or during the development of formulated systems is crucial for any research activities from agricultural raw materials.

Biopolymers samples are soft, deformable materials and so they are very sensitive to the technique of imaging used for their characterisation. Specific observation conditions and sample preparation methods have to be adapted to take into account both the nature of the polymer (sensitive chemical bonding or internal water) and the probe effect of the microscopy technique (solid tip, electron beam…) [1]. Moreover, using a particular microscopy technique alone is often not able to give unambiguously the true characteristics of a biopolymer (morphology, size distribution, conformation, assemblies). Coupling between different microscopy techniques can improve a more realistic characterisation of biopolymers and of their assemblies but requires sample preparation ways to be developed specifically on the same biopolymer sample for each microscopy technique [2,4].

Here, we aim to present some examples of characterisation of food biopolymers using different microscopy techniques (AFM, conventional TEM, cryo-TEM, SEM), focusing on the relation that exists between the apparent biopolymer morphology and the related sample preparation method and microscopy technique, see "Figure 1".

1. C. Gaillard, G. Fuchs, C.J.G. Plummer and P.A. Stadelmann, Micron **38** (2007), p. 522.
2. J.P. Douliez, L. Navailles, F. Nallet and C. Gaillard, ChemPhysChem **9** (2008), p.74.
3. C. Gaillard, B. Novales, J. François and J.P. Douliez, Chemistry of Materials **20** (2008), p. 1206.
4. A. Barakat, C. Gaillard, D. Lairez, L. Saulnier, B. Chabbert and B. Cathala, Biomacromolecules **9** (2008), p. 487.

M. Luysberg, K. Tillmann, T. Weirich (Eds.): EMC 2008, Vol. 1: Instrumentation and Methods, pp. 815–816, DOI: 10.1007/978-3-540-85156-1_408, © Springer-Verlag Berlin Heidelberg 2008

816

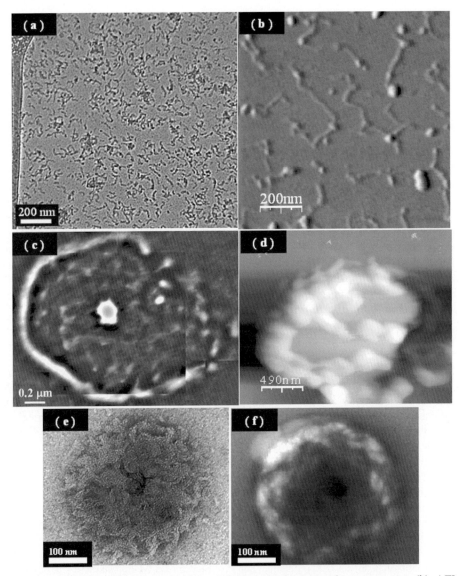

Figure 1. (a) Cryo-TEM image of fractal-shaped whey protein aggregates; (b) AFM height image of sugar beet pectins; (c) Energy-filterd image of an hybrid silicon/lipid particle and (d) the corresponding AFM height image; (e) Negative stained TEM image of a lipidic vesicular assembly and (f) the corresponding cryo-TEM image.

Preparation of SiC/SiC thin foils for TEM observations by wedge polishing method

M. Gec[1], T. Toplišek[1], V. Šrot[2], G. Dražić[1], S. Kobe[1], P.A. van Aken[2], M. Čeh[1]

1. Nanostructured Materials, Jožef Stefan Institute, Ljubljana, Slovenia
2. Stuttgart Center for Electron Microscopy, Max Planck Institute for Metals Research, Stuttgart, Germany

medeja.gec@ijs.si

Keywords: wedge polishing, transmission electron microscopy, SiC/SiC

Composite materials based on continuous crystalline SiC fibers embedded in submicron SiC matrix exhibit excellent mechanical properties, which make this composite suitable material for the first wall of future fusion reactor [1]. The interface between the SiC fibers and the submicron SiC matrix plays an important role in preventing failure of such composite material under mechanical and thermal load. In order to improve the contact between the fibers and the matrix the fibers in our study were coated with a thin layer of diamond like carbon (DLC) using physical vapor deposition. Our preliminary studies showed that after thermal treatment at 1300°C for 3h in argon atmosphere, a reaction layer formed between the thin layer and the matrix [2]. However, the transmission electron microscopy (TEM) specimens' preparation by conventional ion-milling resulted in more or less complete erosion of the matrix. Also, the fibers were not sufficiently thinned, which made detailed TEM investigation of the fiber/matrix interface not possible. In view of this the aim of our work was to use wedge polishing method for the preparation of the SiC/SiC thin foils in order to improve the sample quality for detailed TEM analysis. The comparison with the ion-milled thinned SiC/SiC thin foils is also presented.

The SiC/SiC thin foils for TEM observations were prepared in two ways: by conventional ion-milling and by wedge polishing method. For ion-milling the specimens were thinned at 4 keV and 10° incident angle in a Bal-Tec RES 010 ion-miller. For mechanical preparation of the SiC/SiC thin foils a wedge automatic polishing device Allied MultiPrep was used [3, 4]. The SiC/SiC specimens embedded into epoxy resin were mounted in such a way to obtain cross-sections of the fibers (Fig. 1a). The mechanical thinning was performed on a diamond lapping films (Buehler, Ultra-Prep) at a very small wedge angle of 1°. During the mechanicals thinning the thickness of the samples was monitored and controlled using interference fringes on optical microscope (Fig. 1b). The final polishing was performed on both sides of the sample with 0.02 μm colloidal silica on a felt-covered platen in order to achieve the very thin and clean wedge. So prepared specimens were removed from the glass support and mounted on a Ta grid for TEM observation. A high-angle annular dark field scanning transmission electron microscopy (HAADF-STEM) images were acquired using the Zeiss SESAM transmission electron microscope (TEM) operated at an

M. Luysberg, K. Tillmann, T. Weirich (Eds.): EMC 2008, Vol. 1: Instrumentation and Methods, pp. 817–818, DOI: 10.1007/978-3-540-85156-1_409, © Springer-Verlag Berlin Heidelberg 2008

818

accelerating voltage of 200 kV. The microscope is equipped with a monochromator and a high-transmissivity in-column MANDOLINE energy filter.

An ion-milled specimen is shown in figure 2a. One can easily observe that the SiC matrix was preferentially etched away during the ion-milling of the sample and that the thickness of embedded SiC fibers is varying within each fiber. Contrary, in wedge prepared sample the matrix surrounding the fibers is retained thus producing numerous high quality fiber/matrix interface regions for TEM investigations. It was concluded that the wedge polishing method for preparation of the SiC/SiC thin foils for TEM investigations is far superior as compared to the conventional ion-milling. The method seems very suitable for any multiphase specimens with large hardness differences. Further analytical electron microscopy investigations (electron energy-loss spectroscopy and energy-dispersive X-ray spectroscopy) are in progress and will be discussed.

1. W. Zhang, T. Hinoki, Y. Katoh, A. Kohyama, T. Noda, T. Muroga and J. Yu, Journal of Nuclear Materials 1577-1581 (1998), p. 258-263.
2. T. Toplišek, G. Drazić, S. Novak and S. Kobe, Scanning 30 (2008), 35.
3. M. Gec, V. Šrot, J.H. Jeon, P.A. van Aken and M. Čeh, 7MCM Proc. (2007), 251-252.
4. P.M. Voyles, J.L. Grazul and D.A. Muller, Ultramicroscopy (2003), 251-273.
5. The authors acknowledge financial support from the European Union under the Framework 6 program under a contract for an Integrated Infrastructure Initiative. Reference 026019 ESTEEM. We thank Ute Salzberger (MPI, Stuttgart) for helpful discussions the wedge polishing method.

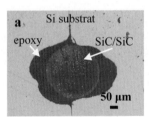
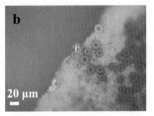

Figure 1. (a) Optical micrograph of tripod polished specimen prepared as a cross-section. (b) Interference fringes at the edge of the specimen.

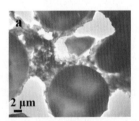
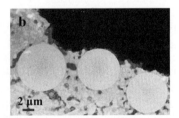

Figure 2. (a) TEM image of ion-milled SiC/SiC. (b) HAADF-STEM image of SiC/SiC fibers prepared by tripod polishing.

Serial-section Polishing Tomography

J.A. Hunt[1], P. Prasad[1], E. Raz[1]

1. Gatan Research & Development, 5794 W. Las Positas, Pleasanton, CA, USA
2. Gatan FA Products Division, 5794 W. Las Positas, Pleasanton, CA, USA

jhunt@gatan.com
Keywords: serial-section, polishing, tomography, optical microscopy

Tomographic reconstruction of a wide range of materials can be accomplished via mechanical polishing serial-sectioning – alternately polishing away material and imaging the remaining polished surface with optical (or electron) microscopy. Reconstruction of the acquired data is straightforward provided alignment issues and polishing artefacts are controlled. The resulting three-dimensional datasets can be of large areas of many mm^2 with sub-micron resolution.

Ultimate reconstruction resolution is limited primarily by: (1) resolution of the microscope (~0.5 μm for the light microscope used here); (2) image pixel size; and (3) ability to align the image sections (typically sub-pixel). Several imaging artefacts must be controlled to produce accurate image reconstructions, including: (1) polishing scratches; (2) polish debris remaining on surface or in voids during imaging; (3) microscopy contrast that confuses reconstruction such as specular contrast from surface roughness).

The Gatan Centar Frontier, a computer-controlled polishing system designed for failure analysis specimen preparation including semiconductor delayering, was straightforward to modify for this application. The system features a robotic arm that can precisely control the polishing force and orientation of the sample onto a mechanical polishing wheel. Polishing rate and progress can be monitored through closed-loop positional feedback on the robotic arm and via light microscope observation. The robotic arm can rotate the polished sample surface from the polishing wheel to the optical microscope and back with a repeatability of about a micron and essentially no image field rotation. Translation of the sample enables field stitching to extend the field-of-view to the centimetre level.

Tomography acquisition involves indicating imaging resolution (XY) and field-of-view (FOV), polishing depth per slice and number of sections (Z), and selection of polishing media optimized for speed versus surface quality. The PC board example in Figure 1 has an XY pixel size of 3 μm and FOV of 1.4 mm x 3 mm, and 150 sections of 1.68 μm each. Polishing used a 0.5 μm diamond paper and required ~30 s per slice, in addition to a dead time of ~20 s per slice for sample rotation and cleaning, imaging and stitching 4 FOV, and bringing the sample back to the polishing table.

Tomography reconstruction involves identifying and correcting image artefacts and aligning the images in XY. For the example shown in Figure 2, the as-acquired data required no additional image alignment, but some polishing scratches were removed using the third dimension to identify and correct intensity outliers.

M. Luysberg, K. Tillmann, T. Weirich (Eds.): EMC 2008, Vol. 1: Instrumentation and Methods, pp. 819–820, DOI: 10.1007/978-3-540-85156-1_410, © Springer-Verlag Berlin Heidelberg 2008

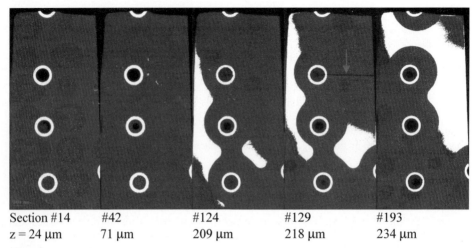

Section #14	#42	#124	#129	#193
z = 24 μm	71 μm	209 μm	218 μm	234 μm

Figure 1. Selected sections illustrating sectioning progress of a 1.4mm X 3.0mm printed circuit board. Arrow shows polishing artifact (scratch).

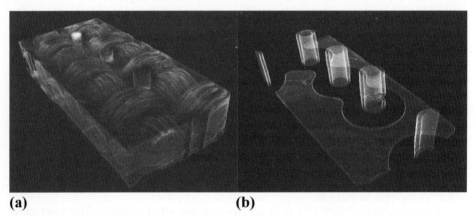

(a)   **(b)**

Figure 2. Reconstructed volume from data in figure 1 (first 150 slices shown). **(a)** Intensities of fiberglass weave displayed. Individual 7-10 mm fiber clearly shown in all orientations. **(b)** Intensities of metal vias displayed.

Visualization of detergent resistant membrane rafts in human colorectal cancer cells with correlative confocal and transmission electron microscopy

K. Jahn[1], E.P. Kable[1] and F. Braet[1]

1. Australian Key Centre for Microscopy and Microanalysis, The University of Sydney, NSW 2006, Australia

k.jahn@usyd.edu.au
Keywords: DRM, correlative microscopy, actin cytoskeleton

Detergent resistant membrane domains (DRM) can for the first time be visualised in the context of their native environment (i.e. attached to a cell) using correlative fluorescence and electron microscopy (CFEM). DRMs found on whole mounted colorectal cancer cells have a size range that is consistent with that widely accepted for lipid rafts (50-100nm). However, micron sized domains have also been observed. Furthermore, CFEM provides a tool to study DRMs on living cells yet still allows them to be prepared for high-resolution transmission electron microscopy (TEM) and electron tomographic analysis.

Membrane rafts have been shown to play a pivotal role in regulating key cell biological processes, such as signal transduction, cellular transport and cell survival [1]. Cholesterol is the main component of membrane rafts [2] which are also known to be highly transient structures [3]. A possible subfraction of membrane rafts are DRMs which additionally are resistant to extraction with Triton-X-100 at 4°C and are commonly studied using sucrose gradient centrifugation [4]. However, the direct visualization of membrane rafts in their native environment with either fluorescence or electron microscopy has so far been proven technically rather cumbersome [5].

To study DRMs on whole mounted cells we cultured them directly on TEM grids and removed the soluble membrane fraction by Triton X-100 extraction. Subcellular structures of interest (i.e. the actin cytoskeleton and G_{M1}, a lipid raft marker ganglioside) can subsequently be labeled with fluorescent probes or antibodies. The distribution of both, actin and rafts can then be visualized with confocal microscopy. The subsequent preparation for TEM includes glutaraldehyde fixation, tannic acid and uranyl acetate treatment, dehydration in graded series of ethanol and HMDS drying [6]. Cells imaged with confocal can then be located in the TEM, allowing high resolution imaging of nanometer sized DRMs.

DRMs could be preserved both on the apical and basolateral side of cells, allowing the direct observation of single actin filaments that tethered the DRMs to the intricate cytoskeleton. Two DRM groups can be distinguished with regard to the domain diameter in colorectal cancer cells (Caco-2). There are firstly circular domains of only 50 – 100nm in diameter and secondly larger and rather irregular shaped domains of micrometers in diameter. Treatment of Caco-2 with the cholesterol depleting drug M-b-CD results in a change in domain size and distribution as shown in Figure 1, suggesting

M. Luysberg, K. Tillmann, T. Weirich (Eds.): EMC 2008, Vol. 1: Instrumentation and Methods, pp. 821–822, DOI: 10.1007/978-3-540-85156-1_411, © Springer-Verlag Berlin Heidelberg 2008

822

a high cholesterol content of the observed domains. Furthermore, CFEM revealed the co-localization of lipid raft marker GM1 with DRM domains, suggesting that DRMs exist in the cell instead of being induced by detergent treatment.

The developed, fast and straightforward sample preparation protocol for CFEM allows an excellent overall preservation of actin structures and membrane domains in whole mounted cells. It also facilitates the direct visualization of DRMs and possibly lipid rafts in their native environment (i.e. attached to a cell). Future electron tomography studies will allow us to dissect the 3D relationship between actin filaments and lipid rafts down to 2-3 nanometer resolution.

1. V. Michel and M. Bakovic, *Biol Cell* **99** (2007) 129.
2. K. Simons and E. Ikonen, *Nature* **387** (1997) 569.
3. M. Edidin, Annu Rev Biophys Biomol Struct **32** (2003) 257.
4. D. A. Brown and J. K. Rose, *Cell* **68** (1992) 533.
5. B. C. Lagerholm, G. E. Weinreb, K. Jacobson et al., *Annu Rev Phys Chem* **56** (2005) 309.
6. K. A. Jahn, D. Barton, and F. Braet, Méndez-Vilas A, Labajos-Broncano L, eds. Current Issues on Multidisciplinary Microscopy Research and Education **III** (2007)

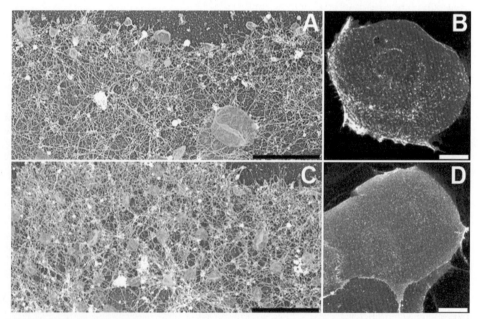

Figure 1: DRMs tethered to the actin cytoskeleton of whole mounted control (A) and M-b-CD treated (C) Caco-2 cells were visualized with TEM. The confocal images B and D show the lipid raft / GM1 distribution in control (B) and M-b-CD treated (D) Caco-2 cells. A, C imaged with the Phillips CM12 at 120kV, scale bar 1µm. Images B, D imaged with the Nicon C1 confocal microscope, 60x water objective. scale bar 20µm.

Contribution of low tension ion-milling to heterostructural semiconductors preparation

M. Korytov, O. Tottereau, J.M. Chauveau et P. Vennéguès

CRHEA-CNRS, Rue Bernard Grégory, Sophia Antipolis, 06560 Valbonne, France

Maxim.Korytov@crhea.cnrs.fr
Keywords: low tension ion-milling, heterostructural semiconductors

Specimens of superior quality are required for quantitative Transmission Electron Microscopy. The classical preparation of heterostructural semiconductors includes stages of mechanical polishing followed by ion-milling up to the transparency of the sample. The last stage is critical, as incorrectly carried out ion-milling brings to the appearance of the different types of preparation defects. Ionic radiation can create crystalline defects which will superpose over the intrinsic defects of the studied materials. The elevation of temperature induced by ion-milling can also cause transformations of phases or metastable phase amorphisation. The most often consequence of the ion-milling is the formation of an amorphous layer on the surface of the sample which can vastly perturb the quantitative analysis. We shall show afterwards that low tension ion-milling allows avoiding some of these defects.

ZnO and ZnO based compound is the object of numerous recent researches since these materials have various potential applications as in the field of the optoelectronic as in the domain of the spintronic. However studies of ZnO material by Transmission Electron Microscopy are quite limited because of the sample preparation difficulties. ZnO was found to be very sensitive ion-milling. On figure 1a one can see a plan view TEM image of a ZnO epitaxial film grown on sapphire. The sample was prepared by ion-milling at 2 kV, which brought to the formation of crystalline defects. It's possible to eliminate these defects by using a chemical attack of the thinned sample (figure 1b). But for a cross-section sample, the small thickness of the studied layers and the different chemical natures of layers do not allow to use chemical attacks. By using a final ion-milling step at very low tension (up to 100V), one can eliminate most part of defects (figure 1c). Low tension ion-milling as well allows avoiding amorphous layer formation on the sample surface.

We should also report the advantages of the diamond powder deposition for cross-section preparation of samples including surface areas of interest. Generally ionic beam attacks the specimen uniformly, which leads to formation of round hole with only few transparent zones close to the surface. Moreover change of local thickness drastically influence quantitative analyzes.

Solid diamond particles, preliminary deposited on the sample surface, locally protect the sample that leads to oblong columns formation (figure 2). Those columns have almost constant thickness which allows analysing quantitatively extensive structures.

1. We kindly acknowledge GATAN society for successful collaboration.

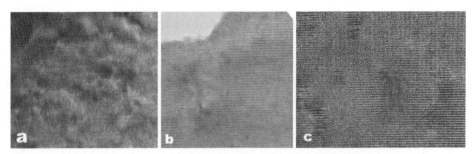

Figure 1. ZnO plan view: a) after ion-milling at 2 kV, b) after ion-milling at 2 kV following by HNO_3 chemical attack, c) after ion-milling step at 100V

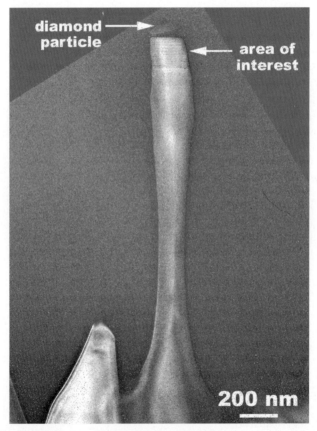

Figure 2. Oblong column obtained by diamond deposition

Metallographic characterization of MgH₂-Mg system

M. Vittori Antisari[1], A. Montone[1], A. Aurora[1], M.R. Mancini[1],
D. Mirabile Gattia[1], L. Pilloni[1]

1. Department, ENEA – C.R. Casaccia, Via Anguillarese 306, 00123 Rome, Italy

montone@casaccia.enea.it

Keywords: magnesium hydride, microscopy, desorption

Magnesium represents an important candidate for solid state hydrogen storage even if the hydrogen desorption temperature is too high for most applications.

The study of the MgH₂-Mg phase transformation in powder samples is so of primary importance and the support of metallographic information can be really significant.

We have developed a method for studying this phase transformation by cross sectional samples SEM observation of partially transformed material. This method is based on the peculiar features of this system where the MgH₂ phase is insulating while Mg is a metallic conducting phase. This difference can induce a contrast between the two phases owing to the different secondary emission yield. The contrast is particularly high at low voltage where the emissivity of the insulating phase is limited to the unity by the surface charging effect [1].

The sample has to be embedded in order to allow particle sectioning and surface polishing. In order to exploit this contrast in practical observation the embedding medium should be able to disperse to the ground the charge injected by the electron beam in the metallic phase in order to keep the surface neutrality of this phase. In this configuration the insulating MgH₂ will charge under the electron beam and, at low voltage, where the emissivity is larger than unity, the steady state secondary emission is unity and preserves the surface neutrality, as it can be seen from the schematic drawing in Figure 1 [1]. On the contrary the metallic phase is free of providing an emission larger than unity owing to the ability of dispersing to the ground the excess charge.

Experimental work has been carried out on MgH₂ powders processed by ball milling in order to introduce and disperse 10 wt% of Fe acting as reaction accelerator and partially desorbed. The average fraction of metallic Mg deriving from hydrogen desorption is of the order of 10wt%. The powders have been embedded in an Al matrix with a procedure already developed for the preparation of powder samples for TEM observations [2].

Images reported in Figure 2a and 2b represent clearly the contrast situation. In Figure 2a a BSE image taken at 20 kV is reported. The presence of three phases with different average atomic number is clearly evidenced even if the contrast between Mg and MgH₂ is quite feeble. An opposite situation is observed in Figure 2b taken at 1 kV with the SE signal. The contrast between the two Mg based phases is quite strong and allows to study the details of the microstructure, while only the larger Fe particles are clearly observed, probably owing to reduced backscattering yield at low voltage [3]. Summarizing, the complete information is provided by the comparison of the two images since the localization of the three phases is so clearly displayed. It is quite

M. Luysberg, K. Tillmann, T. Weirich (Eds.): EMC 2008, Vol. 1: Instrumentation and Methods, pp. 825–826, DOI: 10.1007/978-3-540-85156-1_413, © Springer-Verlag Berlin Heidelberg 2008

evident that the phase transformation begins at the Fe particles and that the Mg-MgH$_2$ phase boundary moves outwards from these nucleation centres.

1. J. Cazaux, Scanning **26** (2004), p. 181-203.
2. A. Montone, M. Vittori Antisari, Micron **34** (2003), p. 79–83
3. L. Frank, R. Stekly, M. Zadrazil, M.M. El-Gomati, I. Mullerova, Mikrochimica Acta **132** (2000), p. 179.

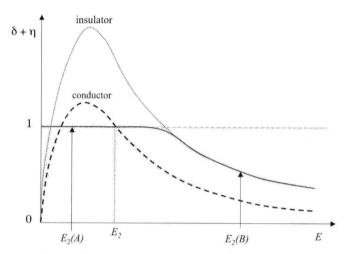

Figure 1. Schematic representation of SE yield of a conducting (dash line) and insulating (dot line) phase vs primary electron energy. The charge accumulation at the surface of the insulating phase influences the SE yield, so that, at steady state, the observed emission of an insulator follows the full line.

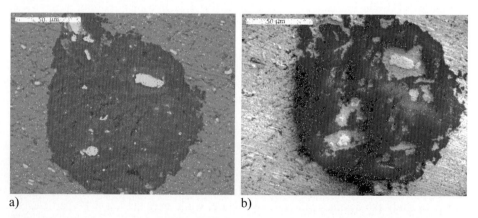

a)   b)

Figure 2. SEM image of a particle of MgH$_2$ with 10wt% Fe catalyst milled for 10 hours and partially desorbed taken with BSE detector at 20 kV (a). SE image of the same particle obtained at 1 kV showing clearly the Mg, the MgH$_2$ and Fe phases (b).

LSM tomography of 2-cell mouse embryo

M.A. Pogorelova[1,2], V.A. Golichenkov[1] and V.N. Pogorelova[2]

1. Biological Faculty, Moscow State University, Leninskie gory, Moscow, 117234, Russia

2. Institute of Theoretical and Experimental Biophysics, RAS, Pushchino, Moscow province, 142290, Russia

pogm2007@rambler.ru

Keywords: mouse early embryo, LSM, 3-DR

Changes in the cell volume play an important role in regulation of key cellular functions, including metabolism [1], protein synthesis [1], gene expression [2] and cell death [3]. The early embryo membrane is highly permeable for water, that makes the cell very sensitive to osmotic shock [4]. Qualitative relationships between osmotic characteristics of the incubation medium and the volume of the mammalian early embryo have been shown experimentally. Nevertheless, it is difficult to find the quantitative proportion between these characteristics, because it is hard to measure the volume of objects with such small sizes. This problem may be solved by three-dimensional reconstruction (3-DR).

The keeping of the intact volume (shape) of the embryo compartments was based on freeze-drying technique [5-7]. Several variants of osmotic shock were modeled by changes of NaCl contents in the incubation medium. After cryofixation in liquid propane and subsequent low-temperature dehydration, the embryo was immersed in the Epon medium. A Z-stack of optical slices at a step of 1 μm between the layers was obtained in a confocal microscope (Zeiss 510, Germany). Figure 1 demonstrates the examples of optical section obtained through the diametral plane of 2-cell mouse embryo in the mode of laser scanning microscope (LSM). 3-DR was performed in the 3ds max medium. Figure 2 illustrates the comparative results of the reconstruction of the embryo after hyperosmotic shock.

Our data indicates that a long-term hypoosmotic shock result in the embryo volume recovery. A hyperosmotic medium induced irreversible shrinkage of embryonic cells. Anisotonic conditions induced both changes in the volume parameters and qualitative transformation of the shape. We observed the formation of outgrowths in the medium with a high NaCl content (Figure 1C, Figure 2B). These morphological structures are also called "blebs" [8]. Incubation in Dulbecco's standard medium initiated a gradual decrease in the blastomere volume. The effect was less pronounced than in the hypertonic medium.

The obtained quantitative data correspond to the qualitative effects observed in experiments in vitro. Note, that the developed technology of LSM tomography of mouse early embryo allows us to measure the cell volume corresponding to the like life state.

M. Luysberg, K. Tillmann, T. Weirich (Eds.): EMC 2008, Vol. 1: Instrumentation and Methods, pp. 827–828, DOI: 10.1007/978-3-540-85156-1_414, © Springer-Verlag Berlin Heidelberg 2008

828

1. K. Anbari and R.M.Schultz, Mol. Reprod. **35** (1993), pp. 24–28.
2. A. Benzeev, BioEssays **13** (1991), pp. 207-212.   ,
3. Cohen G.M., Sun X.M., Snowden R.T., Dinsdale D. and Skilleter D.N., Biochem. J. **286** (1992), pp. 331–334.
4. A.D.C. Macknight, Ren. Physiol. Biochem. **11** (1988), pp. 114–141.
5. A.G. Pogorelov, B.L. Allachverdov, , I.V. Burovina, G.G. Mazay and V.N. Pogorelova, J. Microscopy **12** (1991), pp. 24–38.
6. A.G. Pogorelov, Katkov I.I. and V.N. Pogorelova, CryoLetters **28** (2007), pp. 403-408.
7. A.G. Pogorelov, V.N. Pogorelova and Katkov I.I., J. Microscopy (2008), in press.
8. M.M. Perry and M.H.L. Snow, Dev. Biol. **45** (1975), pp. 372–377.

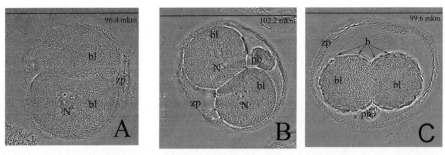

Figure 1. Optical section obtained through the diametral plane of 2-cell mouse embryo after a 15 min incubation in: (A) hypotonic, (B) isotonic and (C) hypertonic conditions; b- blebs, bl- blastomere, pb-polar body, zp- zona pellucida, N- nucleus.

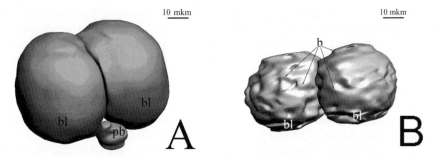

Figure 2. Two-cell mouse embryos images obtained with 3-D reconstruction. (A) Control embryo cryofixed immediately after extraction from the oviduct; (B) 30 min incubation in hypertonic conditions; b- blebs, bl- blastomer, pb-polar body.

Microwave-assisted sample preparation for life science

J.A. Schroeder

Central EM-Lab, Department of Pathology, Regensburg University Hospital,
F-J-Strauss Allee 11, 93053 Regensburg, Germany

josef.schroeder@klinik.uni-regensburg.de

Keywords: microwave-assisted, rapid embedding, fixation

Microwave (MW) irradiation has been successfully applied in organic chemistry providing stunning reaction accelerations, higher yields and product purities, and many clinical laboratory methods, especially in histopathology diagnosis for turnaround time reduction of processed samples [1]. The effects of MW result from material-wave interactions and show a dichotomous nature: thermal effects in polar substances attributed to "dielectric heating", and non-thermal "specific MW effects", which are still a controversial issue [2].

The favourable effects of MW can be used both for improving the accessibility and diffusion of different reagents during fixation and further resin embedding of "difficult specimens" like insects, plants, bacterial spores, parasitic organisms, bone enclosed soft tissues, and time reduction of all sample processing steps (especially resin polymerization), but also for immunolabelling and antigen retrieval methods on epoxy resins.

We report the results of MW-assisted rapid tissue sample processing collected with a semi-automatic (Milestone/Sorisole, Italy, Figure 1) and the latest, fully-automatic (Leica/Vienna, Austria, Figure 2) MW-tissue processor for routine diagnostic use. This technology cuts the usual 3-5 days turnaround time down to approx. 3–6 hours, enabling a "same-day" EM-diagnosis in urgent clinical settings or emerging infectious agents (e.g. tumours, SARS, B. anthracis) [3]. Examples of MW-assisted processed tissues showing excellent preservation of ultrastructure will be presented (Figure 3).

In the Milestone Rapid Electron Microscope MW device (REM) the vial, containing baskets with the samples immersed in the processing solution, is placed in a special carrier which locates the vial in a defined position in the MW cavity. A non-contact infrared temperature sensor measures the current solution temperature in the vial, which is the critical parameter to monitor the magnetron wattage power output (max. 700W). This is controlled via a feedback loop during the continuous MW irradiation of the sample [4]. The slope of the temperature rise/stabilization and the time for each processing step can be defined on a dedicated touch screen monitor. Each solution change to the next process step must be done manually by the user.

This change is carried out automatically by a robotic reagent system in the Leica automatic MW tissue processor (AMW) which is a great benefit for saving laboratory manpower and time [5]. The mono-mode MW chamber provides homogeneous MW distribution at the sample location without hot and cold spots. Thus water loads are not required and virtually 100% of the MW-radiation energy (restricted to 30W) is absorbed by the processing fluids and the specimens. Additionally, a dedicated pulse mode is available to maximize the benefits of the MW-assisted processing.

M. Luysberg, K. Tillmann, T. Weirich (Eds.): EMC 2008, Vol. 1: Instrumentation and Methods, pp. 829–830, DOI: 10.1007/978-3-540-85156-1_415, © Springer-Verlag Berlin Heidelberg 2008

830

1. L. Kok and M. Boon, "Microwaves for the Art of Microscopy" (Coulomb, Leiden) (2003).
2. A. de la Hoz et al., J Microwave Power Electromagn Energy **41** (2007), p. 44-64.
3. J.A. Schroeder et al., Micron **37** (2006), p. 577-90.
4. F. Visinoni et al., J Histotechnol **21** (1998), p. 119-24.
5. Website Leica: www.em-preparation.com
6. I kindly acknowledge the help of B. Voll and H. Siegmund (EM-Lab Regensburg).

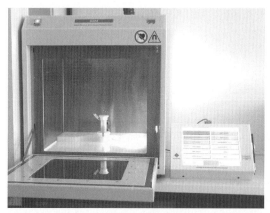

Figure 1. Rapid Electron Microscope microwave device (Milestone REM)

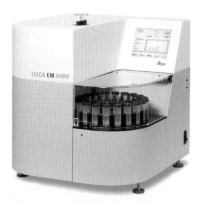

Figure 2. Automatic microwave tissue processor (Leica AMW)

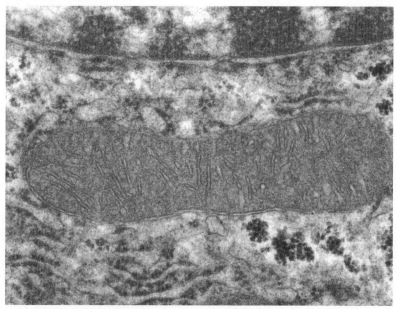

Figure 3. Liver cell detail. Microwave embedding, Leica AMW. Total sample processing time: 5 hours. Note the excellently preserved ultrastructure of the mitochondrium, nucleus, RER, and glycogen particles. Orig. magnification: 20,000x.

Comparison TEM specimen preparation of perovskite thin films by conventional Ar ion milling and tripod polishing

E. Eberg[1], A.T.J. van Helvoort[2], B.G. Soleim[2], A.F. Monsen[2], L.C. Wennberg[2],
T. Tybell[1] and R. Holmestad[2]

1. Department of Electronics & Telecommunication, Norwegian University of Science
& Technology (NTNU), 7491, Trondheim, Norway
2. Department of Physics, Norwegian University of Science & Technology (NTNU),
7491, Trondheim, Norway

a.helvoort@phys.ntnu.no
Keywords: TEM specimen preparation, thin films, perovskites,

Perovskites are a class of materials showing a number of interesting properties such as ferroelectricity, ferromagnetism, high-T_c superconductivity, gigantic magneto-resistance or combinations of these [1]. These properties can be exploited in thin film devices in which the perovskite films can be grown with high quality and precision using a range of deposition techniques. In order to implement these materials into thin film devices, it is necessary to understand size- and interface effects [2]. TEM is an important tool for achieving this understanding.

A challenge is to obtain a representative thin specimen for a TEM study. Argon ion milling as the final step in TEM specimen preparation can lead to artefacts in the TEM samples, for example loss of crystallinity at the substrate/thin film interface, phase transformations and formation of a damaged surface layer on the specimen. These artefacts are especially deteriorating when advanced microscopy techniques are required such as high resolution electron microscopy (HREM) and high resolution annular dark field scanning TEM (HR-ADF STEM). These techniques require ultrathin and clean crystalline samples. Tripod polishing is an alternative specimen preparation procedure. This pure mechanical specimen preparation technique has successfully been applied to for example Si and GaAs [3].

Here the effects of conventional Ar ion milling and tripod polishing specifically to perovskite thin films have been investigated and compared. The focus of the study was to obtain high quality TEM specimens in both plane view and cross section for a range of different perovskite thin films ($PbTiO_3$, $SrRuO_3$ and $LaFeO_3$). All these films were grown on 001-oriented $SrTiO_3$ substrates by off-axis RF magnetron sputtering.

Both conventional Ar ion milling and tripod polishing can result in large thin areas required for a TEM study. However, the quality of tripod polished specimens was superior to that of specimens made by ion milling. The differences between the two techniques and specific specimen preparation artefacts are material dependent. For example the $SrTiO_3$ substrates are less sensitive to ion milling than thin films of $SrRuO_3$ (Figure 1) or $LaFeO_3$ (Figure 2), in which a phase transformation occurs during the ion milling process. The results from this study contribute to optimise the TEM specimen preparation routine and improve advanced TEM studies of perovskite thin films.

M. Luysberg, K. Tillmann, T. Weirich (Eds.): EMC 2008, Vol. 1: Instrumentation and Methods, pp. 831–832, DOI: 10.1007/978-3-540-85156-1_416, © Springer-Verlag Berlin Heidelberg 2008

832

1. C.H. Ahn, J.-M. Triscone and J. Mannhart, Nature **424** (2003), p. 1015.
2. N. Nakagawa, H.Y. Hwang and D.A. Muller, Nature Materials **5** (2006), p. 204.
3. See for example: H. Okuno, M. Takeguchi, K. Mitsuishi, X.J. Guo and K. Furuya, Journal of Electron Microscopy **57** (2008), p. 1.
4. This work was supported by the Research Council of Norway (NFR) and by the NFR NANOMAT Nationally Co-ordinated Project in Oxides for Future Information and Communication Technology, 158518/431. TT acknowledges specifically support from NFR via contract number 162874/v00.

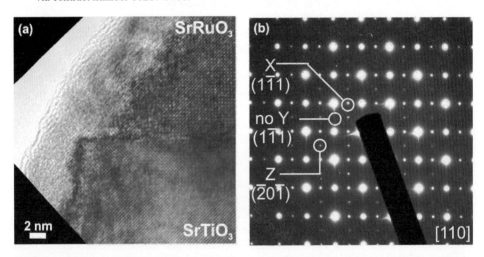

Figure 1. 100 nm thick SrRuO$_3$ thin film on SrTiO$_3$ substrate. (a) HREM of cross sectional specimen made by cooled Ar ion milling depicting surface damage layer variation between thin film and substrate. (b) Selected area electron diffraction pattern of tripod polished specimen in plane view showing only two orientation and not three as found in ion milled specimens (not shown).

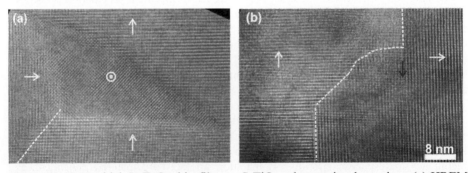

Figure 2. 40 nm thick LaFeO$_3$ thin film on SrTiO$_3$ substrate in plane view. (a) HREM of specimen made by tripod and short Ar ion milling depicting 3 different crystal orientations in the thin film (b) HREM of tripod polished specimen showing only two orientations (White arrows indicate c-direction, black arrow an antiphase boundary and dashed lines grain boundaries).

In-situ temperature measurements on TEM-specimen during ion-milling

M. Wengbauer, J. Gründmayer, J. Zweck

University of Regensburg, Institute of Experimental and Applied Physics,
93040 Regensburg, Germany

martin.wengbauer@physik.uni-regensburg.de

Keywords: ion milling, temperature rise, specimen preparation, TEM

In order to be able to investigate a specimen in a Transmission Electron Microscope (TEM) it has to be electron transparent. Electron transparency is achieved for common materials such as semiconductors or metals when the specimen's thickness is below 100 nm, for HRTEM purposes more in the range below 20 nm. A commonly used method for final thinning of TEM-specimens is ion-milling. However, this kind of preparation always bears the risk of modifying the object to be analysed due to the high energy ion bombardment, which ranges from several 100 eV up to 10 keV. This is very problematic for heat-sensitive materials such as GaMnAs, a ferromagnetic semiconductor, which changes its Curie temperature when its temperature is brought above approximately 150 – 200 °C. The intention of the experiments presented below is to give a good estimation of the temperature rise during the ion-milling process.

The temperature is measured by a thin thermocouple attached to one side of the specimen. This thermocouple consists of two thin wires (Chromel as positive, Alumel as negative electrode) of 0.07 mm diameter each, which are welded on one end. The welded end is fixed directly in the centre of the specimen with a two-component glue.

In consequence of using a thermocouple there are different problems to bear in mind and, as far as possible, to solve: e.g. it is not possible to rotate the specimen during the ion-milling process, which would be done under real conditions. The particles used for milling are noble gas ions and so charges are directed onto the sample and the thermocouple, which affects the measurements. Last but not least the thermocouple and the glue change the heat capacity and the heat transfer rate of the whole system. Some of these problems are system immanent and cannot be solved completely, but can be minimized. For example, in order to minimize the effect of the charges on the thermocouple, one can simply use the ion guns from the opposite specimen side only.

Figure 1 and 2 show the first measurements. They were done on a silicon carbide specimen (a material which is often used for heteroepitactic growth of GaN, for example), ground down to 100µm, followed by a dimpler treatment to end up with a dimple on one side of about 30µm depth, leaving approximately 70 µm of material (see Figure 3 for experimental setup). On the opposite side the thermocouple is installed. Figure 1 shows the temperature characteristics on the sample under the following conditions: acceleration voltage 2.7 keV, 10° etching angle, and both guns from the same side, opposite to the thermocouple. The temperature rises in about 45 seconds from room temperature to 63 °C and then remains nearly constant. After the ion guns

M. Luysberg, K. Tillmann, T. Weirich (Eds.): EMC 2008, Vol. 1: Instrumentation and Methods,
pp. 833–834, DOI: 10.1007/978-3-540-85156-1_417, © Springer-Verlag Berlin Heidelberg 2008

834

are turned off, an exponential decrease of the temperature can be noticed, as expected. Figure 2 shows results from the same sample but with an acceleration voltage of 5 keV. The main difference to figure 1 is now the much larger temperature rise up to 180 °C. Another difference is that the temperature does not reach a constant value, even when measured for a longer time. Instead, it keeps rising slightly but constantly up to 189 °C during the > 200 s exposure period.

As a first result of the measurements, it is shown that the acceleration voltage has a strong influence on the temperature, as was expected. Further experiments have also shown that the etching angle and the thermal contact between sample and sample holder do affect the temperature rise. The temperature increase is expected to be even more severe in the perforated region which is suitable for TEM investigations, since there the heat conduction is limited towards one side of the specimen only (due to the perforation). Additionally, the mass of the remaining specimen is very small in the transparent region, which can – for a constant heat capacity – lead to even higher temperature.

As a conclusion it can be said that temperature rise during ion milling is obviously present and the temperatures which can be reached are not negligible, especially for heat-sensitive materials. For those, special effort is necessary to minimize temperature rise during preparation.

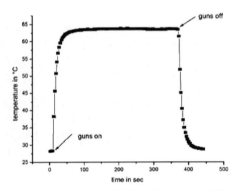

Figure 1. 100µm thin silicon carbide sample at 2.7 keV acceleration voltage, 10° etching angle, both ion guns from same side, opposite to the thermocouple: maximum temperature 63 °C.

Figure 2. The same sample as in fig. 1 under similar conditions but at 5.0 keV acceleration voltage: maximum temperature 189 °C.

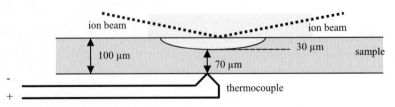

Figure 3. Experimental setup.

Author Index

Subject Index

Printing: Krips bv, Meppel, The Netherlands
Binding: Stürtz, Würzburg, Germany